The Real Numbers and Real Analysis

Ethan D. Bloch

The Real Numbers and Real Analysis

Ethan D. Bloch
Mathematics Department
Bard College
Annandale-on-Hudson, NY 12504
USA
bloch@bard.edu

ISBN 978-1-4899-9834-7 ISBN 978-0-387-72177-4 (eBook)
DOI 10.1007/978-0-387-72177-4
Springer New York Dordrecht Heidelberg London

Mathematics Subject Classification (2010): 26-01

Printed on acid-free paper

Springer is part of Springer Science+Business Media (www.springer.com)

Dedicated to my two wonderful children Gil Nehemya and Ada Haviva,
for whom my love has no upper bound

Contents

Preface

The Origin of This Book

This text grew out of two types of real analysis courses taught by the author at Bard College, one for undergraduate mathematics majors, and the other for students in the mathematics section of Bard's Masters of Arts in Teaching (M.A.T.) Program. Bard's undergraduate real analysis course is a standard introductory course at the junior–senior level, but the M.A.T. real analysis course, as explained below, is somewhat less standard. The author was therefore unable to find an existing real analysis textbook that exactly met the needs of the students in the M.A.T. course, and so this text was written to fill the gap. To make this text more broadly useful, however, it has been written in a way that makes it sufficiently flexible to meet the needs of a standard undergraduate real analysis course as well, though with a few distinguishing features.

One of the principles on which Bard's M.A.T. Program was founded is that secondary school teachers need, in addition to sufficient training in pedagogy, a substantial background in their subject areas. In the Bard M.A.T. Program in Mathematics, not only are all students required to have completed the equivalent of a B.A. in mathematics to enroll in the program, but they are required to take four mathematics courses in the M.A.T. Program, one of which is in real analysis. The M.A.T. mathematics courses are different from standard first-year mathematics graduate courses, in that rather than directing the students toward more advanced mathematical topics, the emphasis is on giving the students an advanced look at the material taught in secondary school mathematics courses. For example, it is important for prospective teachers of calculus to have a good understanding of the properties of the real numbers (including decimal expansion), and a detailed look at logarithmic, exponential and trigonometric functions, none of which is usually treated in detail in standard undergraduate real analysis courses. Of course, a prospective teacher of calculus must also have a good grasp of limits, differentiation and integration, as found in any real analysis course. By contrast, it is not as important for prospective secondary teachers to spend valuable course time on some standard introductory real analysis topics such as sequences and series of functions. Hence, the focus of a real analysis course for M.A.T. students is somewhat different from a standard undergraduate real analysis course.

This text contains all the material needed for both a standard introductory course in real analysis and for variants of such a course aimed at prospective teachers. It is the hope of this author that for each intended audience, this text will offer a clear, accessible and interesting exposition of this beautiful material.

Audience

This text is aimed at three target audiences:

1. Mathematics majors taking a standard introductory real analysis course;
2. Prospective secondary school mathematics teachers taking an introductory real analysis course;
3. Prospective secondary school mathematics teachers taking a second real analysis course.

For undergraduate mathematics majors taking an introductory real analysis course, this text covers all the standard topics that are typically treated in an introductory single-variable real analysis book. The order of the material is slightly different than usual (with sequences being treated after derivatives and integrals), and as a result a few of the proofs are different, but all the standard topics are present, as well as a few extras.

For prospective secondary school mathematics teachers taking an introductory real analysis course, this text has, in addition to the standard topics one would encounter in any undergraduate real analysis course, a thorough treatment of the properties of the real numbers, and an equally thorough treatment of logarithmic, exponential and trigonometric functions. Additionally, the book contains some historical information that a mathematics teacher could use to enliven a calculus course.

For prospective secondary school mathematics teachers taking a second real analysis course (for example, M.A.T. students in mathematics who have already had an undergraduate real analysis course), this text has, in addition to a review of the basic topics of real analysis (limits, derivatives, integrals, sequences), a development of the real numbers starting with the Peano Postulates, a detailed discussion of the decimal expansion of real numbers via least upper bounds, a thorough treatment of logarithmic, exponential and trigonometric functions, and additional topics not usually found in introductory real analysis texts (for example, a discussion of π in terms of the circumference and area of circles, and a proof of the equivalence of various theorems such as the Extreme Value Theorem and the Bolzano–Weierstrass Theorem with the Least Upper Bound Property). It is the belief of this author that for those M.A.T. students who have already had an undergraduate course in real analysis, the proper training for prospective teachers is not to offer a course in more advanced topics in analysis (for example, Lebesgue measure and integration, or metric spaces), but rather to discuss in more detail those aspects of single-variable real analysis that are most directly related to the topics that teachers encounter in secondary schools.

Pedagogical Concerns

Regardless of any particular choices in the selection and order of the material in this text, at heart this text is a detailed and rigorous introduction to real analysis designed for students who have not previously studied the subject.

Some of the pedagogical concerns of this text are as follows.

Slow and Steady

Though it is fun to rush straight to the most exciting results in a subject, and it is tempting to skip over the details of some proofs (either because they seem too routine or because they seem too long), in the author's experience the best way for students to learn the basics of a technical subject such as real analysis is to work through all the details of the subject slowly and steadily. Students need to be challenged, but in an introductory text such as this one it is best to leave the challenges to the exercises, not the proofs of theorems. Most proofs in this text are written out in full detail, and when details are omitted, that is stated explicitly. When previous results in the text are used in a proof, those results are always referenced. Most other real analysis texts of this length cover more material than we do; our aim is not to fit as much material as possible into the book, but to provide sufficient material for a one-semester course (with a few options for the instructor), and to cover that material as thoroughly as possible.

Careful Writing

Every effort has been made to provide clearly and carefully written definitions, theorems and proofs throughout the book. As seen in the author's book *Proofs and Fundamentals: A First Course in Abstract Mathematics* (Birkhäuser, Boston, 2000), the author views the careful writing of proofs as an important part of both teaching and learning rigorous mathematics, and he has attempted to adhere to the advice he gave about writing in that book.

Minimal Technicalities

Though real analysis is technical by nature, this text attempts to keep technical concepts to a minimum. For example, we omit discussion of limit inferior and limit superior, because we can accomplish everything we need without it.

When technicalities are kept to a minimum, the result is that some particularly slick proofs are not available, which is unaesthetic to the experienced mathematician. For the sake of student learning, however, it it better to use a minimum number of technicalities repeatedly than to have the shortest or cleverest proof of each theorem. For example, there are some theorems involving continuity, differentiation and integration (such as the Extreme Value Theorem and the Intermediate Value Theorem) that can be proved very efficiently by using sequences, but which we prove using only the basic properties of the real numbers (and in particular the Least Upper Bound Property).

Another example of keeping technicalities to a minimum is our choice of Dedekind cuts rather than Cauchy sequences for constructing the real numbers from the rational numbers; the method of Dedekind cuts is slightly longer, but it avoids both sequences and equivalence classes. (Of course, there is a discussion of Cauchy sequences in this text, but it is in its natural place in the chapter on sequences, which is well after the chapter on the construction of the real numbers.)

Features

There are many undergraduate books in real analysis, but the author has not found any with the exact same choice of material and pedagogical concerns as this text.

Some of the distinguishing features of this text are as follows.

Thorough Treatment of the Real Numbers

At the heart of real analysis are the properties of the real numbers. Whereas most introductory real analysis texts move as quickly as possible to the core topics of calculus (such as limits, derivatives and integrals) by giving relatively brief treatments of the axioms for the real numbers and the consequences of those axioms, this text emphasizes the importance of the properties of the real numbers as the basis of real analysis. Hence, the real numbers and their properties are developed in more detail than is found in most other introductory real analysis texts. The goal of the text is for students to have a thorough understanding of the fundamentals of real analysis, not to cover as much ground as possible.

Multiple Entryways

A particularly distinctive feature of this text is that it offers three ways to enter into the study of the real numbers.

Entry 1, which yields the most complete treatment of the real numbers, begins with the Peano Postulates for the natural numbers, and then leads to the construction of the integers, the rational numbers and the real numbers, proving the main properties of each set of numbers along the way.

Entry 2, which is more efficient than Entry 1 but more detailed than Entry 3, skips over the axiomatic treatment of the natural numbers, and begins instead with an axiomatic treatment of the integers. It is first shown that inside the integers sits a copy of the natural numbers, and after that the rational numbers and the real numbers are constructed, and their main properties proved.

Entry 3, which is the most efficient approach to the real numbers, starts with an axiomatic treatment of the real numbers. It is shown that inside the real numbers sit the natural numbers, the integers and the rational numbers. This approach is the one taken in most contemporary introductions to real analysis, though we give a bit more details about the natural numbers, integers and rational numbers than is common.

The existence of the three entryways into the real numbers allows for great flexibility in the use of this text. For a first real analysis course, whether for mathematics majors or prospective secondary school mathematics teachers, Entry 3 should be used; for a second real analysis course for prospective secondary school mathematics teachers, or as supplementary reading for a standard introductory real analysis course, Entry 1 or Entry 2 should be used. No matter which entry is used, all students end up knowing the same properties of the real numbers, and hence are equally prepared for the subsequent material.

Follows Order of Material in Calculus Courses

Undergraduate real analysis courses are often organized according to the goal of preparing the students for more advanced mathematics courses. Such a design, however, does not necessarily lead to the best pedagogical approach. Whereas many of the more advanced aspects of real analysis are quite abstract, the motivation for introductory real analysis is the need for a rigorous foundation for calculus. Given that the students in an introductory real analysis course have already had courses in calculus, and given that pedagogically it is best to relate new material to that which is already familiar, this text presents the basic material in real analysis in an order that is closer to that encountered in calculus courses than is found in most real analysis books.

Sequences Later in the Text

In standard calculus courses, sequences usually receive very minimal treatment, and are discussed only as much as they are needed as partial sums of series. In real analysis, by contrast, sequences are a very important tool, and they are treated in great detail in most real analysis texts. Moreover, in most such texts sequences are located right after the preliminary treatment of the real numbers, and prior to the discussion of limits, derivatives and integrals, both because the definition of the convergence of sequences is viewed as slightly easier to learn than the definition of the convergence of functions, and because some of the major theorems about sequences (such as the Bolzano–Weierstrass Theorem) are used in the proofs of some important theorems about continuous functions, derivatives and integrals (such as the Extreme Value Theorem).

In this text, by contrast, sequences are treated after the chapters on limits, derivatives and integrals, similarly to the order of material in a calculus course. Whereas sequences are used in many real analysis books in the proofs of some of the important theorems concerning functions, it turns out that all such theorems can be proved without the use of sequences, where instead of using the Bolzano–Weierstrass Theorem and similar results, a direct appeal is made to the Least Upper Bound Property, or to direct consequences of that property. As such, it is possible to treat continuous functions, derivatives and integrals without the added technicality of sequences, and in the order familiar from calculus courses. Moreover, the use of sequences in proofs where they are not needed, while sometimes making for short and clever proofs, may at times obscure the essential ideas of the theorem being proved.

Of course, wherever they are placed, sequences are a very important topic in real analysis, and they are given a thorough treatment in this text, with all the usual theorems proved.

Integration via Riemann Sums

There are two standard ways of defining the Riemann Integral that are found in introductory real analysis texts: via Riemann sums, and via upper and lower integrals. The latter approach is used by many (if not most) current introductory real analysis books, and it gives a fast route to proving the important theorems about integrals. However, given that the treatment of integrals in calculus courses is via Riemann sums, this text also uses that approach in its definition of integrals, so that students can understand the rigorous treatment of integrals in terms of what they had previously seen in calculus courses.

Equivalence of Various Theorems with the Least Upper Bound Property

Every student in a real analysis course learns that the Least Upper Bound Property is at the heart of what the real numbers consist of, and it is the basis for the proofs of many of the main theorems of real analysis, such as the Extreme Value Theorem and the Bolzano–Weierstrass theorem. Many of these theorems, for example the two just mentioned, are in fact logically equivalent to the Least Upper Bound Property, and in this text we present a proof of this logical equivalence, which is not commonly found in real analysis books.

Thorough Discussion of Transcendental Functions

Logarithmic, exponential and trigonometric functions are familiar to students from precalculus and calculus courses. Whereas most introductory real analysis books either ignore these functions or give them a cursory treatment, in this text these functions are defined rigorously, and their basic properties are proved in detail. Of particular note is our treatment of the sine and cosine functions; these functions are trickier to define rigorously than logarithms and exponentials, but nonetheless deserve a thorough exposition.

Discussion of Area and Arc Length

The main motivation for the development of the definite integral is to compute areas of certain regions in the plane. However, the very important fact that the definite integral of a non-negative function yields the area under the graph of the function, while regularly asserted, is rarely proved in introductory real analysis books (indeed, the concept of area is rarely defined rigorously), which leads not only to a gap in rigor but also to an incomplete understanding of the concept of area. In this text, we give a thorough discussion of area and arc length, starting with geometric definitions of these concepts, and then proofs that in appropriate cases, they can be computed via definite integrals.

More about π

Students are familiar with the number π from a very young age, where it is discussed in the context of the circumference and area of circles. The number π is also introduced into the study of trigonometric functions in precalculus and calculus courses. In real analysis, if the trigonometric functions are to be studied, then the number π cannot be avoided. In this text, a particularly detailed treatment of π is given, in order to clarify the relation between the geometric approach to this number (via the circumference of circles) and analytic approach to it (via the definition of the trigonometric functions using integrals).

Reflections for Every Section

The heart of mathematics is the details, and students in a real analysis course quite naturally get very caught up in the ε's and δ's. However, it is useful at times to step back from the details and ask broader questions, such as: why are things done as they are; why are some aspects of the material straightforward and other aspects not; whether all the hypotheses of the theorems are really needed; and whether there might be an easier way to define or prove things. In real analysis, moreover, it is also helpful to compare the way things are done in that course with the way they were done (or not done) in calculus courses that the student took previously. Hence, every section of this text concludes with some very brief remarks that look back upon the material in the section, often in the context of what the student has seen prior to real analysis. The main purpose of these remarks is not, however, simply to convey the author's thoughts about the material, but is rather to encourage the reader to engage in her own similar reflections upon the material discussed in this text, and upon other mathematical ideas encountered subsequently.

Historical Remarks for Every Chapter

The material in this book is presented in the logical order of development that is now standard for real analysis, but which is quite different from the way the subject developed historically. Though it would be very inefficient to learn the details of real analysis in the order in which it occurred historically, because it took mathematicians a rather circuitous route to reach the understanding we have today, it is nonetheless beneficial for mathematicians to know something about how important topics such as calculus arose. Such historical context is especially valuable for prospective teachers, though it can benefit all students of real analysis, not in understanding the details of rigorous definitions and proofs, but in seeing the bigger picture. Hence, each chapter concludes with a historical discussion of the material in the chapter.

The author is not a historian, and he hopes that the historical material provided is both useful and informative. For a thorough and engaging treatment of the history of mathematics in general, the reader is referred to [Kat98]. Because of the availability of the wonderful website [OR], which has extensive biographical information on every mathematician about whom the reader has heard (and many others as well), the historical material in this text does not include biographical information (other than dates of birth and death) about the mathematicians who developed real analysis.

Errors

In spite of the author's best effort, there will inevitably be some errors in this text. If the reader finds any such errors, it would be very helpful if she would send them to the author at `bloch@bard.edu`. An updated list of errors is available at `http://math.bard.edu/bloch/rnra_errata.pdf`.

Acknowledgments

As with many texts in mathematics, this book developed out of lecture notes, which were first used in the fall of 2006 at Bard College. The first draft of this text made partial use of class notes taken by Bard student Matt Brophy in Math 361 in the fall of 2000.

I would like to thank James Belk, John Cullinan, Cliona Golden, Mark Halsey, Sam Hsiao, Greg Landweber, Benjamin Stevens, Rebecca Thomas, George Vaughan and Japheth Wood for their valuable assistance and extremely helpful comments on various drafts. Bard students Lionel Barrow, Arlene Campbell, Adam Chodoff, Jennifer D'Arcy, Robert Feinsinger, Alexandros Fragkopoulos, Wui-Ming Gan, Beth Goldberg, Matthew Goodell, Elias Halloran, Liz Jimenez-Martinez, Durrell Jones, Morgon Kanter, Kimberly Larie, Ryan McCann, Paul McLaughlin, Mona Merling, Andrija Perunicic, Tomasz Przytycki, Viriya Ratanasangpunth, Ben Selfridge, Derek Tingley, Zhexiu Tu, Alexandru Vladoi, Jordan Volz, Changwei Zhou and YuGai Zhu have found many errors in various drafts, and provided useful suggestions for improvements.

It is impossible to acknowledge every source for every idea, theorem or exercise in this text. Most of the theorems, proofs and exercises are either standard or are variants of standard results; some of these the author first encountered as a student, others were learned from a variety of sources. The following are texts that the author consulted regularly, and which notably influenced the writing of this book. Real analysis texts: [Gor02], [Lay01], [Pow94], [Ros68], [Sto01], [TBB01], [Tre03], [Wad00]; calculus text: [Spi67]; history of mathematics texts: [Bar69], [Boy49], [Boy91], [Coo05], [Edw79], [Fer08], [Jah03], [Kat98], [OR].

Some parts of Chapter 1 in this text are a revised version of material in [Blo00], which has been removed from [Blo10], the second edition of that book; this material is important for the construction of the real numbers, a topic that is at the heart of real analysis, but was somewhat inappropriately included in the more elementary, and earlier written, textbook [Blo00] due to the author's perhaps excessive enthusiasm for this material. Additionally, although it is assumed that the reader taking a course in real analysis is familiar with the core material from a book on proofs, sets and functions such as [Blo10], there is nonetheless some overlap in the treatment of induction and recursion between that book and this text, the material being important for both.

My appreciation goes to Elizabeth Loew, Senior Editor of Mathematics at Springer, for providing me with very helpful support. I would like to thank the unnamed reviewers for their many useful suggestions for improving the book. Thanks also go to Nathan Brothers and the copyediting and production staff at Springer for their terrific work on the book; to Martin Stock and Brian Treadway for help with LaTeX; and to Pedro Quaresma for assistance with his very nice LaTeX commutative diagrams package DCpic, with which the commutative diagram in this text was composed. Some of the illustrations in this text were drawn using Mathematica.

I would very much like to thank the Einstein Institute of Mathematics at the Hebrew University of Jerusalem, and especially Professor Emanuel Farjoun, for hosting me during a sabbatical when parts of this book were written.

Lastly, I would like to thank my amazing wife Nancy and my wonderful children Gil and Ada for their support during my work on this book.

Ethan Bloch
Annandale-on-Hudson, NY
August 2010

To the Student

The Material

Calculus has a number of important aspects: intuition, computation, application and rigor. Standard introductory calculus courses in American universities and colleges treat the first three of these well, but gloss over the fourth. At base, a first course in real analysis, such as a course that uses this text, is an exposition of the rigorous ideas that make calculus work.

Of course, if all that a course in real analysis did was to verify that everything done in calculus is correct, it would hardly be worth the effort, because most of us are willing to take on faith that the people who developed calculus got it right. In fact, in the course of giving rigorous definitions and proofs of the main concepts of calculus, a real analysis course introduces the student to many fascinating and powerful new concepts and techniques of proof. These ideas, for example rigorous definitions of limits and continuity, turn out to be useful in both the further study of real analysis, as well as in fields such as topology and complex analysis. As such, a first course in real analysis is both a completion of the study of calculus commenced in introductory calculus courses, and also an entrance into further advanced study in a number of branches of mathematics.

The history of calculus is virtually the opposite of how we present it in a modern real analysis course. Calculus started with the ideas of derivatives and integrals, as well as their applications; the details at the time were not entirely rigorous by modern standards. As new phenomena was discovered, the need for a more rigorous treatment of derivatives and integrals was understood, and that led to the formulation of the definition of limits, and the use of limits as the basis for all other aspects of calculus. Finally, as the properties of limits were explored, it was realized that completely rigorous proofs regarding limits could only be obtained if we had a rigorous treatment of the real numbers. Today, we learn real analysis in its logical, as opposed to historical, order, which means starting with a detailed look at the real numbers, then limits and continuity, and then derivatives and integrals, followed by additional topics such as sequences and series.

This text, which is designed for a one-semester course in real analysis, has more material than can typically be covered in one semester, in order to accommodate different choices of emphasis by the instructor (or by yourself, if you are reading this book on your own). An outline of the text is as follows.

Chapters 1 and 2: Construction of the Real Numbers and Properties of the Real Numbers

There are two standard ways to discuss the real numbers rigorously: either start with an axiomatic treatment of the natural numbers or the integers, and then construct the real numbers, or start with an axiomatic treatment of the real numbers. The former approach is treated in Chapter 1, and the latter in Section 2.2. The properties of the real numbers are explored in the rest of Chapter 2.

Chapters 3–5: Limits and Continuity, Differentiation and Integration

These three chapters contain a rigorous treatment of the core material from calculus.

Chapters 6 and 7: Limits to Infinity and Transcendental Functions

These two chapter contain optional material, the first covering some topics usually found in a second-semester calculus course, for example l'Hôpital's Rule and improper integrals, and the second providing a rigorous treatment of logarithmic, exponential and trigonometric functions.

Chapters 8 and 9: Sequences and Series

Sequences and series, often discussed in a second-semester calculus course, take on a more important role in real analysis, and are core topics in a first course in the subject.

Chapter 10: Sequences and Series of Functions

This chapter is a fitting close to the book, because it involves many of the ideas that were discussed in earlier chapters. The highlights of this chapter include a treatment of Taylor series, and an example of a function that is continuous everywhere but differentiable nowhere.

Prerequisites

To be ready for an introductory real analysis course, a student must have taken the standard calculus courses, must have some experience writing rigorous mathematical proofs, and must be familiar with the basic properties of sets, functions and relations (as found, for example, in [Blo10]). Because many students find the material in real analysis a bit harder to learn than the material in other standard junior–senior level proofs-based mathematics courses such as abstract algebra, it is often recommended, though not necessarily required, that students have taken another junior–senior level proofs-based course prior to studying real analysis, for the sake of having more experience with rigorous mathematical proofs.

Notation

The notation used in this book is, as much as possible, standard. We assume that the reader is familiar with basic notation involving sets and functions, for example unions and intersections of sets, and we will not review all such notation here. However, because not all mathematical notation is entirely standardized (a tribute to the decentralized ethos of the mathematical community), we list here a few items of notation that we will use, but which might not have been encountered previously by the reader. Of course, we will define a lot of new notation throughout this book.

$\mathbb{N}$	natural numbers
$\mathbb{Z}$	integers
$\mathbb{Q}$	rational numbers
$\mathbb{R}$	real numbers
$\mathbb{N} \cup \{0\}$	non-negative integers
$\mathbb{Z} - \{0\}$	non-zero integers
$(0, \infty)$	positive real numbers
$[0, \infty)$	non-negative real numbers
$\mathbb{R} - \{0\}$	non-zero real numbers
$A \subseteq B$	subset, not necessarily proper
$A \subsetneq B$	proper subset
$A - B$	set difference
1_A	identity map on the set A
$f\|_A$	restriction of the function f to the set A
f^{-1}	inverse function of a bijective function f

Exercises

Mathematics is learned by actively doing it, not by passively reading about it. Certainly, listening to lectures and reading the textbook is an important part of any mathematics course, but the real learning occurs when the student does exercises, which provide an opportunity for the student to make use of the concepts discussed in the course, and to see whether she understands these concepts.

The exercises in this text have been arranged in order so that in the course of working on an exercise, a student may use any previous theorem or exercise (whether or not she did it), but not any subsequent result (unless stated otherwise). If an exercise makes use of a previous exercise, that previous exercise is sometimes noted at the end of the exercise where it is used, in case the reader has not done the previous exercise; in most cases it is not mentioned if previous theorems, lemmas and the like proved in the text are used because it is assumed that the reader has read the text (though in a few particularly tricky exercises a relevant result from the text is mentioned as a hint).

Many of the exercises are used in the text, and are labeled as such. There are two reasons why so many of the exercises are used in the text: to streamline some of the lengthier proofs by leaving manageable parts to the students, and to provide exercises that are actually useful, as opposed to exercises that exist simply for the sake of having exercises (though there are plenty of those too, to provide sufficient practice for the student).

In some of the exercises the reader is asked to prove statements that have a number of similar cases, for example statements involving functions that are increasing or decreasing. For such exercises it is acceptable to do one case in detail, and to say that the other cases are similar and that the details are omitted, as long as you are sure that the details really are similar.

Finally, we note that there is a large variation in the level of difficulty of the exercises, ranging from some that are very straightforward (being slight variants of proofs in the text), to some that are quite challenging, and with many in between. No attempt has been made to rate the difficulty of the exercises, such a rating being necessarily subjective.

Writing Proofs

Doing exercises is an important aspect of learning mathematics; writing the exercises carefully makes them even more effective. Advanced mathematics is not always easy, and everyone makes honest mathematical errors in the process of learning such material. There is no reason, however, to have avoidable mistakes due to carelessness and poor writing if sufficient care is taken.

One of the most common, if not the most common, source of errors when students first encounter real analysis involves problems with quantifiers. A number of very important definitions in real analysis, for example the definitions of limits of functions, continuity, uniform continuity, and limits of sequences, involve two quantifiers in a given order. The need to prove statements that involve multiple quantifiers is what makes real analysis a bit harder for many undergraduates to grasp upon first encounter than linear algebra and abstract algebra, which also require proofs, but which do not have such complications with quantifiers. The best way to be careful with quantifiers is to write proofs very carefully and precisely. In particular, when proving a statement with quantifiers, it is crucial to deal with the quantifiers in the exact order in which they are given in the statement that is being proved.

Another very common error for beginners in real analysis, which also does not occur as much in subjects such as linear algebra and abstract algebra, is the need to distinguish between one's scratch work and the actual proof. A proof must always start with what we know, and deduce that which we want to prove. It is a common logical error to try to prove something by assuming the thing we are supposed to prove, and then working backwards until we arrive at something that we already know (for example that something equals itself). Such "backwards proofs" are, for reasons this author does not understand, extremely common in high school mathematics, though there the arguments are often sufficiently simple, and reversible, so that no harm is done other than conveying the incorrect impression that "backwards proofs" are actual proofs, which they are not. By contrast, in real analysis it is crucial not to confuse "backwards proofs" with real proofs. For example, as we will see very clearly in Example 3.2.3, the ε–δ proofs discussed in Section 3.2 often require first some scratch work that is "backwards," and then a rather different-looking proof that is "forwards." Hence, it is very important in the proofs that you write for the exercises in this book that you distinguish between how you think of a proof, which can be any combination of "backwards," "forwards" and anything else, and how you write the final draft of the proof, which must be very precise in going from what we assume to what we want to prove.

A few additional points about writing mathematical proofs are the following:

- Strategize the outline of a proof before working out the details; the outline of a proof is determined by what is being proved, not by what is hypothesized.
- Use definitions precisely as stated.
- Do not omit steps in proofs; when in doubt, prove it.
- Justify each step in a proof, citing the appropriate results from the text as needed.
- If a step in a proof is skipped, for example because it is very similar to a previous step, state explicitly that that is the case.
- Use correct grammar, including full sentences and proper punctuation.
- Use "=" signs properly.
- Proofs should stand on their own; check your proofs by reading them as if they were written by someone else.

See [Blo10, Section 2.6], [Gil87], [Hig98], [KLR89] and [SHSD73] for further discussion of writing mathematics.

The bottom line is to write your proofs very carefully, because doing so will help you learn the material in this book.

hand is the danger that conveying the incorrect impression that "backwards proofs" are actual proofs, which they are not. By contrast, proof analysis [illegible] is [illegible] confuse "backwards proofs" with real proofs. For [illegible] as we will see [illegible] clearly [illegible] 1.3 [illegible] the sort of proofs discussed in Section 2.2 often require [illegible] some scratch work that is "backwards," but then write a different-looking proof that [illegible]. Hence it is very important to present it [illegible] you write for the exercises in this book that you distinguish between [illegible] and a [illegible] proof, which should only [illegible] "backwards" [illegible], and nothing else, but have your proof [illegible] final draft of the proof, which must be very precise [illegible] what was assumed [illegible] what we wanted to prove.

A few additional points [illegible] writing mathematics [illegible] the following:

- [illegible] the reader [illegible] relevant [illegible] of the author [illegible] [illegible] mentioned [illegible] proved [illegible]
- Use definitions [illegible] as stated.
- [illegible] what [illegible] prove it.
- [illegible] appropriate results from the [illegible] proved.
- If [illegible] skipped [illegible] similar to a previous step [illegible] identify that it is the case.
- Use correct grammar [illegible] all sentences [illegible] proper punctuation.
- [illegible] pronouns.
- Proofs should stand on their own [illegible] check your proofs [illegible] difficult [illegible] if they [illegible] written by someone else.

See [illegible] [KLR89] and [SHSD] [illegible] writing mathematics.

[illegible] your proofs very [illegible] become [illegible] you [illegible] the material in this book.

To the Instructor

This text, which is designed for a one-semester course in single-variable real analysis, has more material than can typically be covered in one semester, in order to provide flexibility for the instructor. Moreover, this text has been designed to accommodate three different types of real analysis courses, each for a different target audience. Suggested course outlines for these three audiences are given below, though of course each instructor should be guided by her own choices more than by the author's suggestions.

Standard Introduction to Real Analysis

This course is a traditional first course in real analysis for mathematics majors, and for other students (for example physics majors) who want to be prepared for advanced work in mathematics. This course covers all the typical single-variable topics, albeit with a slightly more thorough treatment of the properties of the real numbers than usual, and with sequences placed after differentiation and integration.

Chapter 2, Properties of the Real Numbers: 2.1, 2.2, 2.3, 2.4, 2.5, 2.6.
Chapter 3, Limits and Continuity: 3.1, 3.2, 3.3, 3.4, 3.5.
Chapter 4, Differentiation: 4.1, 4.2, 4.3, 4.4, 4.5.
Chapter 5, Integration: 5.1, 5.2, 5.3, 5.4, 5.5, 5.6, 5.7.
Chapter 8, Sequences: 8.1, 8.2, 8.3, 8.4.
Chapter 9, Series: 9.1, 9.2, 9.3, 9.4, 9.5.
Chapter 10, Sequences and Series of Functions: 10.1, 10.2, 10.3, 10.4, 10.5.

Introduction to Real Analysis for Prospective Secondary School Teachers

This course is for prospective secondary school teachers who have not previously studied real analysis, for example undergraduate mathematics education majors or M.A.T. students in mathematics who did not study real analysis as undergraduates, and who seek a good understanding of the material from calculus that they will be teaching. This course focuses on core topics such as properties of the real numbers (important for any mathematics teacher separately from its role in calculus), limits, derivatives and integrals, and it also has extra topics of direct concern to future teachers of calculus such as exponential and logarithmic functions (at the expense of part of the treatment of series).

Chapter 2, Properties of the Real Numbers: 2.1, 2.2, 2.3, 2.4, 2.5, 2.6.
Chapter 3, Limits and Continuity: 3.1, 3.2, 3.3, 3.4, 3.5.
Chapter 4, Differentiation: 4.1, 4.2, 4.3, 4.4, 4.5, 4.6.
Chapter 5, Integration: 5.1, 5.2, 5.3, 5.4, 5.5, 5.6, 5.7, 5.9.
Chapter 7, Transcendental Functions: 7.1, 7.2.
Chapter 8, Sequences: 8.1, 8.2, 8.3, 8.4.
Chapter 9, Series: 9.1, 9.2, 9.3, 9.4, 9.5.

Second Course in Real Analysis for Prospective Secondary School Teachers

This course is for prospective secondary school teachers, for example M.A.T. students in mathematics who already had an introductory course in real analysis as undergraduates, who seek a more thorough understanding of the material from calculus that they will be teaching. In addition to a review of core topics such as limits, derivatives and integrals, this course includes a detailed look at topics of direct concern to future teachers of calculus but not usually treated fully in many introductory real analysis courses, for example the construction of the real numbers, transcendental functions, improper integrals, area, arc length and π.

Chapter 1, Construction of the Real Numbers: 1.1, 1.2, 1.3 (or 1.4 instead of the previous two sections), 1.5, 1.6, 1.7.
Chapter 2, Properties of the Real Numbers: 2.1, 2.3, 2.5, 2.6, 2.7, 2.8.
Chapter 3, Limits and Continuity: Review as needed.
Chapter 4, Differentiation: Review as needed, 4.6.
Chapter 5, Integration: Review as needed, 5.8, 5.9.
Chapter 6, Limits to Infinity: 6.1, 6.2, 6.3, 6.4.
Chapter 7, Transcendental Functions: 7.1, 7.2, 7.3, 7.4.
Chapter 8, Sequences: Review as needed, 8.3, 8.4.
Chapter 9, Series: Review as needed, 9.4, 9.5.
Chapter 10, Sequences and Series of Functions: 10.1, 10.2, 10.3, 10.4, 10.5.

1

Construction of the Real Numbers

1.1 Introduction

Real analysis—which in its most basic form is the rigorous study of the ideas in calculus—takes place in the context of the real numbers, because the real numbers have the properties needed to allow things such as derivatives and integrals to work as we would like them to. A rigorous study of derivatives and integrals requires a rigorous treatment of the fundamental properties of the real numbers, and that is the topic of this chapter and the next.

Inside the set of real numbers (which intuitively form the complete "number line") there are three familiar sets of numbers: the natural numbers (intuitively $1,2,3,\ldots$), the integers (intuitively $\ldots,-2,-1,0,1,2,\ldots$) and the rational numbers (the fractions). We use the standard symbols $\mathbb{N}$, $\mathbb{Z}$, $\mathbb{Q}$ and $\mathbb{R}$ to denote the natural numbers, the integers, the rational numbers and the real numbers, respectively. These sets are subsets of one another in the order $\mathbb{N} \subseteq \mathbb{Z} \subseteq \mathbb{Q} \subseteq \mathbb{R}$, where each set is a proper subset of the next.

This text offers three ways to enter into the study of the real numbers.

Entry 1, which starts in Section 1.2 in the present chapter, and which offers the most complete treatment, begins with axioms for the natural numbers, and then leads to constructions of the integers, the rational numbers and the real numbers, proving the main properties of each set of numbers along the way.

Entry 2, which starts in Section 1.4 in the present chapter, skips over the axiomatic treatment of the natural numbers, and begins instead with an axiomatic treatment of the integers. It is shown that inside the integers sits a copy of the natural numbers. After that, the rational numbers and the real numbers are constructed, and their main properties proved. This approach is a bit shorter and simpler than that of Entry 1, though still more detailed than that of Entry 3.

Entry 3, which starts in Section 2.2 in the next chapter, commences with an axiomatic treatment of the real numbers. This approach, which is the one taken in most introductory real analysis books, is the most efficient route to the core topics of real analysis, but it gives the least insight into the number systems. In Section 2.4

it is shown that inside the real numbers sit the natural numbers, the integers and the rational numbers, with all their expected properties.

All three entries lead to proofs of the same properties of the real numbers, and as such all three are reasonable starting points in the study of real analysis. The reader should now proceed to the entry of her choice, after first reading the following definition, which is needed to define concepts such as addition, multiplication and negation of real numbers.

Definition 1.1.1. Let S be a set. A **binary operation** on S is a function $S \times S \to S$. A **unary operation** on S is a function $S \to S$. △

Let S be a set, and let $*: S \times S \to S$ be a binary operation. If $x, y \in S$, the correct way to write the result of doing the operation $*$ to the pair (x, y) would be $*((x, y))$. However, because such notation is both quite cumbersome to write, and does not resemble the way we write familiar binary operations such as addition of numbers, we will write $x * y$ instead of $*((x, y))$. Similarly, if $\neg: S \to S$ is a unary operation, and if $x \in S$, the correct way to write the result of doing the operation $\neg$ to x would be $\neg(x)$, but we write the more familiar $\neg x$.

Additionally, let $T \subseteq S$ be a subset. We say T is "closed" under the binary operation $*$ if $x * y \in T$ for all $x, y \in T$. A similar definition holds for a subset being closed under a unary operation. We note that this use of the term "closed" will be employed only informally, and only occasionally, in contrast to the very important use of this same word in Definition 2.3.6, where closed intervals are defined. (A much more general, and very important, use of the word "closed" can be found in any introductory topology text, for example [Mun00].)

1.2 Entry 1: Axioms for the Natural Numbers

The simplest, and most fundamental, set of numbers is the set of natural numbers, that is, the numbers $1, 2, 3, 4, \ldots$. In this section we will give an axiomatic treatment of these numbers, and in subsequent sections we will construct all the other familiar sets of numbers (integers, rational numbers, real numbers) in terms of the natural numbers.

A good axiomatic system is one that assumes as little as possible, and from which as much as possible can be proved. To make an efficient axiomatization for the natural numbers, we need to strip these numbers down to their bare essentials. Intuitively, we know various things about the natural numbers, such as the existence and basic properties of the binary operations addition and multiplication, and the relation less than. How few of these notions can we take as axioms, from which we can deduce everything else that we need to know about the natural numbers? It turns out that very little is needed for an axiomatization of the natural numbers—neither addition nor multiplication, nor the relation less than (they will all be constructed from our axioms).

The standard axiomatization of the natural numbers, known as the Peano Postulates, is based upon the notion of proof by induction. We assume that the reader is familiar with proof by induction, at least informally. We will review the practical use of such proofs in Section 2.5; at present we need proof by induction for theoretical purposes.

In its most bare-bones form, the natural numbers will consist of a set (denoted $\mathbb{N}$), a distinguished element (denoted 1) and a unary operation on the set (denoted $s\colon \mathbb{N} \to \mathbb{N}$). Intuitively, the function s takes each natural number to its successor, which we would normally think of as being the result of adding 1 to each natural number, though that cannot be formally stated quite yet, because we do not have the notion of addition in our axioms for the natural numbers. The Peano Postulates require that three entities $\mathbb{N}$, 1 and s satisfy a few simple properties. One of these properties, listed as Part (c) of Axiom 1.2.1 below, is just the formal statement that proof by induction works.

It is rather surprising, upon first encounter, that we can get away with assuming so little about the natural numbers. For example, we make no axiomatic assumption about addition and multiplication; these operations will be constructed using the Peano Postulates. As the reader will see from the details of our development of the various number systems, the Peano Postulates are incredibly powerful.

Axiom 1.2.1 (Peano Postulates). *There exists a set* $\mathbb{N}$ *with an element* $1 \in \mathbb{N}$ *and a function* $s\colon \mathbb{N} \to \mathbb{N}$ *that satisfy the following three properties.*

a. There is no $n \in \mathbb{N}$ *such that* $s(n) = 1$.
b. The function s *is injective.*
c. Let $G \subseteq \mathbb{N}$ *be a set. Suppose that* $1 \in G$, *and that if* $g \in G$ *then* $s(g) \in G$. *Then* $G = \mathbb{N}$.

Observe that it does not say in the Peano Postulates (Axiom 1.2.1) that the set $\mathbb{N}$ is unique, though in fact that turns out to be true; see Exercise 1.2.8 for details. We can therefore make the following definition.

Definition 1.2.2. The set of **natural numbers**, denoted $\mathbb{N}$, is the set the existence of which is given in the Peano Postulates. △

Part (a) of the Peano Postulates says, intuitively, that 1 is the "first" number in $\mathbb{N}$. Parts (a) and (b) together are needed to ensure that $(\mathbb{N}, 1, s)$ is infinite. To see why, let $M = \{1, p\}$, and let $s\colon M \to M$ be defined by $s(1) = p$ and $s(p) = p$. It is straightforward to see that Parts (a) and (c) of the postulates hold for this M, 1 and s, even though M is not what we would intuitively want to call the set of natural numbers; of course, this function s does not satisfy Part (b) of the Peano Postulates. Using the same set M but with $s(1) = p$ and $s(p) = 1$, shows that a finite set may satisfy Parts (b) and (c) of the postulates, but not Part (a). Hence, to ensure that a set satisfying the Peano Postulates truly models the natural numbers, we need both Parts (a) and (b) of the postulates (or something like them). That we need something like Part (c) of the postulates seems reasonable, because we will need to use proof by induction in a number of our proofs about the natural numbers.

We cannot *prove* that $\mathbb{N}$ is precisely what our intuition tells us it should be, because we cannot prove things about our intuition. The best we can do, and we will indeed do this, is to prove that $\mathbb{N}$ satisfies all the basic properties we can think of for the natural numbers. Formally, we simply define the natural numbers to be the set $\mathbb{N}$ given in the Peano Postulates.

Our first result about the natural numbers is the following simple lemma, which certainly fits in with our intuitive sense that for every natural number other than 1, there is another natural number that precedes it. The proof of the following lemma is a typical use of the Peano Postulates.

Lemma 1.2.3. *Let $a \in \mathbb{N}$. Suppose that $a \neq 1$. Then there is a unique $b \in \mathbb{N}$ such that $a = s(b)$.*

Proof. We start with uniqueness. Suppose that there are $n, m \in \mathbb{N}$ such that $a = s(n)$ and $a = s(m)$. Then $s(n) = s(m)$. By Part (b) of the Peano Postulates we know that s is injective, and therefore $n = m$.

To prove existence, let

$$G = \{1\} \cup \{c \in \mathbb{N} \mid \text{there is some } b \in \mathbb{N} \text{ such that } s(b) = c\}.$$

We will use Part (c) of the Peano Postulates to prove that $G = \mathbb{N}$, which will immediately imply the existence part of this lemma. It is clear that $G \subseteq \mathbb{N}$ and that $1 \in G$. Now let $n \in G$. We need to show that $s(n) \in G$. Let $p = s(n)$. To show that $p \in G$, we will show that $p \in \{c \in \mathbb{N} \mid \text{there is some } b \in \mathbb{N} \text{ such that } s(b) = c\}$. Let $b = n$. Then $s(b) = p$ by the definition of p. It follows that $p \in G$, and therefore $s(n) \in G$. Hence $G = \mathbb{N}$. □

We now want to define the binary operations addition and multiplication for the natural numbers, using only the Peano Postulates, and results derived from these postulates (see Section 1.1 for the definition of binary operations). However, before we can define these binary operations, which are given in Theorem 1.2.5 and Theorem 1.2.6 below, we need to prove the following theorem, which provides the main tool in our definitions of addition and multiplication. This theorem, called Definition by Recursion, allows us to define a function with domain $\mathbb{N}$ by defining the function at 1, and then defining it at $n+1$ in terms of the definition of the function at n. (See Section 2.5 for further discussion of the practical use of Definition by Recursion, as well as some examples.) It is important to recognize that recursion, while intimately related to induction, is not the same as induction (though it is sometimes mistakenly thought to be); the essential difference is that induction is used to prove statements about things that are already defined, whereas recursion is used to define things.

The proof of the following theorem is our most difficult proof involving the natural numbers, and to avoid interrupting the flow of the present section, this proof is given in Section 2.5 (where the theorem is restated as Theorem 2.5.5). However, because the proof of this theorem relies upon nothing other than the Peano Postulates (Axiom 1.2.1), and because we will need this theorem very soon to define addition and multiplication on the natural numbers, it is important that the reader not skip over the statement of this theorem; the reader who wishes to read the proof now can safely skip ahead and read the proof of Theorem 2.5.5, and then return to this point.

Theorem 1.2.4 (Definition by Recursion). *Let H be a set, let $e \in H$ and let $k\colon H \to H$ be a function. Then there is a unique function $f\colon \mathbb{N} \to H$ such that $f(1) = e$, and that $f \circ s = k \circ f$.*

The equation $f \circ s = k \circ f$ in the statement of Theorem 1.2.4 means that $f(s(n)) = k(f(n))$ for all $n \in \mathbb{N}$. If $s(n)$ were to be interpreted as $n+1$ (as indeed it will be in Theorem 1.2.5), then $f(s(n)) = k(f(n))$ would mean that $f(n+1) = k(f(n))$, which looks more like the familiar form of definition by recursion. Additionally, the equation $f \circ s = k \circ f$ can be expressed by saying that the following diagram is "commutative," which means that going either way around the square yields the same result. We will not be making use of commutative diagrams in this text, but in some parts of mathematics they are very useful.

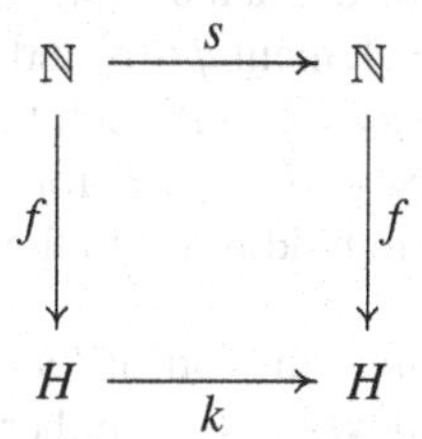

Now that we have Definition by Recursion available to us, we are ready to define addition on $\mathbb{N}$, as given in the following theorem. Though it might not be evident at first why we choose the two properties of addition listed in this theorem rather than other, more common, properties of addition, with hindsight they turn out to work well, allowing for nice proofs of other properties of addition.

Theorem 1.2.5. *There is a unique binary operation $+\colon \mathbb{N} \times \mathbb{N} \to \mathbb{N}$ that satisfies the following two properties for all $n, m \in \mathbb{N}$.*

a. $n + 1 = s(n)$.
b. $n + s(m) = s(n + m)$.

Proof. To prove uniqueness, suppose that there are two binary operations $+$ and $\oplus$ on $\mathbb{N}$ that satisfy the two properties of the theorem. Let

$$G = \{x \in \mathbb{N} \mid n + x = n \oplus x \text{ for all } n \in \mathbb{N}\}.$$

We will prove that $G = \mathbb{N}$, which will imply that $+$ and $\oplus$ are the same binary operation, which is what we need to show for uniqueness. It is clear that $G \subseteq \mathbb{N}$. By Part (a) applied to each of $+$ and $\oplus$ we see that $n + 1 = s(n) = n \oplus 1$ for all $n \in \mathbb{N}$, and hence $1 \in G$. Now let $q \in G$. Let $n \in \mathbb{N}$. Then $n + q = n \oplus q$ by hypothesis on q. It then follows from Part (b) that $n + s(q) = s(n + q) = s(n \oplus q) = n \oplus s(q)$. Hence $s(q) \in G$. We now use Part (c) of the Peano Postulates to conclude that $G = \mathbb{N}$.

For existence, we start by observing that for $p \in \mathbb{N}$, we can apply Theorem 1.2.4 to the set $\mathbb{N}$, the element $s(p) \in \mathbb{N}$ and the function $s\colon \mathbb{N} \to \mathbb{N}$, to deduce that there is a unique function $f_p\colon \mathbb{N} \to \mathbb{N}$ such that $f_p(1) = s(p)$ and $f_p \circ s = s \circ f_p$. Let $+\colon \mathbb{N} \times \mathbb{N} \to \mathbb{N}$ be defined by $c + d = f_c(d)$ for all $(c, d) \in \mathbb{N} \times \mathbb{N}$. Let $n, m \in \mathbb{N}$. Then $n + 1 = f_n(1) = s(n)$, which is Part (a), and $n + s(m) = f_n(s(m)) = (f_n \circ s)(m) = (s \circ f_n)(m) = s(f_n(m)) = s(n + m)$, which is Part (b). □

Part (a) of Theorem 1.2.5 says that the function s works exactly as we had initially thought of it intuitively. From now on, we will often write $n+1$ in places where we would have written $s(n)$.

Using addition, we now turn to the definition of multiplication.

Theorem 1.2.6. *There is a unique binary operation* $\cdot\colon \mathbb{N}\times\mathbb{N}\to\mathbb{N}$ *that satisfies the following two properties for all* $n,m\in\mathbb{N}$.

a. $n\cdot 1 = n$.
b. $n\cdot s(m) = (n\cdot m)+n$.

Proof. We leave uniqueness to the reader in Exercise 1.2.1.

Let $q\in\mathbb{N}$. Let $h_q\colon \mathbb{N}\to\mathbb{N}$ be defined by $h_q(m) = m+q$ for all $m\in\mathbb{N}$. Applying Theorem 1.2.4 to the set $\mathbb{N}$, the element $q\in\mathbb{N}$ and the function $h_q\colon\mathbb{N}\to\mathbb{N}$, implies that there is a unique function $g_q\colon\mathbb{N}\to\mathbb{N}$ such that $g_q(1) = q$ and $g_q\circ s = h_q\circ g_q$. Let $\cdot\colon\mathbb{N}\times\mathbb{N}\to\mathbb{N}$ be defined by $c\cdot d = g_c(d)$ for all $(c,d)\in\mathbb{N}\times\mathbb{N}$. The proof that the two properties of the theorem hold is left to the reader in Exercise 1.2.1. □

We will, as usual, write "*nm*" instead of "$n\cdot m$," except in cases of potential ambiguity (for example, we will write "$1\cdot 1$" rather than "11"), or where the $\cdot$ makes the expression easier to read.

The following theorem gives some of the very familiar properties of addition and multiplication of natural numbers. The main technique of proof for these properties is Part (c) of the Peano Postulates. The different parts of the theorem have been arranged so that to prove each, it is permissible to use everything stated previously, but not subsequently. This same strategy of using only previously stated results will hold throughout this text, except when otherwise noted.

Theorem 1.2.7. *Let* $a,b,c\in\mathbb{N}$.

1. *If* $a+c=b+c$, *then* $a=b$ *(Cancellation Law for Addition).*
2. $(a+b)+c = a+(b+c)$ *(Associative Law for Addition).*
3. $1+a = s(a) = a+1$.
4. $a+b=b+a$ *(Commutative Law for Addition).*
5. $a+b\neq 1$.
6. $a+b\neq a$.
7. $a\cdot 1 = a = 1\cdot a$ *(Identity Law for Multiplication).*
8. $(a+b)c = ac+bc$ *(Distributive Law).*
9. $ab=ba$ *(Commutative Law for Multiplication).*
10. $c(a+b) = ca+cb$ *(Distributive Law).*
11. $(ab)c = a(bc)$ *(Associative Law for Multiplication).*
12. *If* $ac=bc$ *then* $a=b$ *(Cancellation Law for Multiplication).*
13. $ab=1$ *if and only if* $a=1=b$.

Proof. We will prove Parts (1), (5), (6) and (12), leaving the rest to the reader in Exercise 1.2.2.

(1) Let

$$G = \{z \in \mathbb{N} \mid \text{if } x, y \in \mathbb{N} \text{ and } x + z = y + z, \text{ then } x = y\}.$$

We will show that $G = \mathbb{N}$, which will imply the desired result. Clearly $G \subseteq \mathbb{N}$. To show that $1 \in G$, let $j, k \in \mathbb{N}$ and suppose that $j + 1 = k + 1$. Then $s(j) = s(k)$ by Theorem 1.2.5 (a), and so $j = k$ by the injectivity of s (Part (b) of the Peano Postulates). Hence $1 \in G$. Now let $r \in G$. Further, let $j, k \in \mathbb{N}$, and suppose that $j + s(r) = k + s(r)$. By Theorem 1.2.5 (b) we deduce that $s(j + r) = s(k + r)$. Hence $j + r = k + r$ by the injectivity of s. Because $r \in G$ we deduce that $j = k$. Therefore $j + s(r) = k + s(r)$ implies $j = k$, and it follows that $s(r) \in G$. We deduce that $G = \mathbb{N}$ by Part (c) of the Peano Postulates.

(5) Suppose that $a + b = 1$; we will derive a contradiction. There are two cases. First, suppose that $b = 1$. Then $1 = a + b = a + 1 = s(a)$, which is a contradiction to Part (a) of the Peano Postulates. Now suppose that $b \neq 1$. By Lemma 1.2.3 there is some $x \in \mathbb{N}$ such that $s(x) = b$. By Theorem 1.2.5 (b) we then have $1 = a + b = a + s(x) = s(a + x)$, again a contradiction to Part (a) of the Peano Postulates.

(6) Let

$$H = \{z \in \mathbb{N} \mid \text{if } y \in \mathbb{N} \text{ then } z + y \neq z\}.$$

We will show that $H = \mathbb{N}$. Clearly $H \subseteq \mathbb{N}$. By Part (5) of this theorem we know that $1 + k \neq 1$ for all $k \in \mathbb{N}$, and it follows that $1 \in H$. Now let $r \in H$. Suppose further that there is some $k \in \mathbb{N}$ such that $s(r) + k = s(r)$. By Part (4) of this theorem we see that $k + s(r) = s(r)$, and then by Theorem 1.2.5 (b) we deduce that $s(k + r) = s(r)$. Because s is injective (Part (b) of the Peano Postulates), it follows that $k + r = r$. Using Part (4) of this theorem again we deduce that $r + k = r$, which is a contradiction to the fact that $r \in H$. Hence there is no $k \in \mathbb{N}$ such that $s(r) + k = s(r)$, and we deduce that $s(r) \in H$. Hence $H = \mathbb{N}$.

(12) Let

$$H = \{x \in \mathbb{N} \mid \text{if } y, z \in \mathbb{N} \text{ and } xz = yz, \text{ then } x = y\}.$$

Let $j, k \in \mathbb{N}$, and suppose that $1 \cdot k = jk$. Suppose further that $j \neq 1$. By Lemma 1.2.3 there is some $t \in \mathbb{N}$ such that $j = s(t)$. By Parts (4), (7) and (9) of this theorem and Theorem 1.2.6 (b), we see that $k = 1 \cdot k = jk = kj = ks(t) = kt + k = k + kt$, which is a contradiction to Part (6). Therefore $j = 1$, and hence $1 \in H$.

Now let $r \in H$. Let $j, k \in \mathbb{N}$, and suppose that $s(r)k = jk$. As before, by previous parts of this theorem and Theorem 1.2.6 (b), we deduce that $kr + k = jk$. If $j = 1$ then by previous parts of this theorem we see that $k + kr = k$, which is a contradiction to Part (6) of this theorem. Hence $j \neq 1$. By Lemma 1.2.3 there is some $p \in \mathbb{N}$ such that $j = s(p)$. Therefore $kr + k = s(p)k$, and using the same ideas as before we see that $kr + k = kp + k$. By previous parts of this theorem it follows that $rk = pk$. Because $r \in H$, it follows that $r = p$. Therefore $s(r) = s(p) = j$. Hence $s(r) \in H$, and we deduce that $H = \mathbb{N}$. □

Observe that Theorem 1.2.7 (3) states that the function s is just what we intuitively thought it would be. From now on we will write $a+1$ instead of $s(a)$ for $a \in \mathbb{N}$.

Addition and multiplication are the two most important binary operations on the natural numbers. The most important relations on the natural numbers are less than and less than or equal to, to which we now turn.

Definition 1.2.8. The relation $<$ on $\mathbb{N}$ is defined by $a < b$ if and only if there is some $p \in \mathbb{N}$ such that $a + p = b$, for all $a, b \in \mathbb{N}$. The relation $\leq$ on $\mathbb{N}$ is defined by $a \leq b$ if and only if $a < b$ or $a = b$, for all $a, b \in \mathbb{N}$. △

We will write $a > b$ to mean the same thing as $b < a$, and similarly for $a \geq b$.

As expected, if $a, b \in \mathbb{N}$ and $a < b$, then the element $p \in \mathbb{N}$ such that $a + p = b$ is unique; the proof of this fact is left to the reader in Exercise 1.2.3. The following theorem gives some of the basic properties of $<$ and $\leq$.

Theorem 1.2.9. *Let $a, b, c, d \in \mathbb{N}$.*

1. *$a \leq a$, and $a \not< a$, and $a < a+1$.*
2. *$1 \leq a$.*
3. *If $a < b$ and $b < c$, then $a < c$; if $a \leq b$ and $b < c$, then $a < c$; if $a < b$ and $b \leq c$, then $a < c$; if $a \leq b$ and $b \leq c$, then $a \leq c$.*
4. *$a < b$ if and only if $a + c < b + c$.*
5. *$a < b$ if and only if $ac < bc$.*
6. *Precisely one of $a < b$ or $a = b$ or $a > b$ holds (Trichotomy Law).*
7. *$a \leq b$ or $b \leq a$.*
8. *If $a \leq b$ and $b \leq a$, then $a = b$.*
9. *It cannot be that $b < a < b + 1$.*
10. *$a \leq b$ if and only if $a < b + 1$.*
11. *$a < b$ if and only if $a + 1 \leq b$.*

Proof. We will prove Parts (2), (6), (7), (8), (9) and (10), leaving the rest to the reader in Exercise 1.2.4.

(2) There are two cases. First, suppose that $a = 1$. Then certainly $1 \leq a$. Second, suppose that $a \neq 1$. By Lemma 1.2.3 there is some $p \in \mathbb{N}$ such that $a = p + 1$. Hence $a = 1 + p$, which means that $1 < a$, and therefore $1 \leq a$.

(6) We first show that no two of $a < b$ and $a = b$ and $a > b$ can hold simultaneously. Suppose that $a < b$ and $a = b$. It then follows from Part (3) of this theorem that $a < a$, which is a contradiction to Part (1) of this theorem. A similar argument shows that it is not the case that $a > b$ and $a = b$. Now suppose that $a < b$ and $b < a$. By Part (3) of this theorem we deduce that $a < a$, again a contradiction.

We now show that at least one of $a < b$ or $a = b$ or $a > b$ holds. Let

$$G = \{x \in \mathbb{N} \mid \text{if } y \in \mathbb{N}, \text{ then } x < y \text{ or } x = y \text{ or } x > y\}.$$

We will show that $G = \mathbb{N}$, and the desired result will follow. We start by showing that $1 \in G$. Let $j \in \mathbb{N}$. By Part (2) of this theorem we know that $1 \leq j$. It follows that $1 = j$ or $1 < j$. Hence $1 \in G$. Now suppose that $k \in G$; we will show that $k + 1 \in G$.

Let $j \in \mathbb{N}$. By hypothesis on k we know that $k < j$ or $k = j$ or $k > j$. First suppose that $k < j$. Then there is some $p \in \mathbb{N}$ such that $k + p = j$. If $p = 1$, then $k + 1 = j$; if $p \neq 1$, then by Lemma 1.2.3 and Theorem 1.2.5 (a) there is some $r \in \mathbb{N}$ such that $p = r + 1$, which implies that $k + (r + 1) = j$, which by Theorem 1.2.7 (2) (4) implies $(k + 1) + r = j$, and from that we deduce that $k + 1 < j$. Next suppose that $k = j$. Then by Part (1) of this theorem it follows that $k + 1 > k = j$. Finally, suppose that $k > j$. Once again we know that $k + 1 > k$, and by Part (3) of this theorem it follows that $k + 1 > j$. Putting all three cases together, we see that one of $k + 1 < j$ or $k + 1 = j$ or $k + 1 > j$ always holds. Hence $k + 1 \in G$, and by Part (c) of the Peano Postulates we conclude that $G = \mathbb{N}$.

(7) & (8) These follow directly from Part (6).

(9) Suppose that $b < a < b + 1$. Then there are $g, h \in \mathbb{N}$ such that $b + g = a$ and that $a + h = b + 1$. Then $(b + g) + h = b + 1$. By Theorem 1.2.7 (2) (4) we see that $(g + h) + b = 1 + b$, and by Theorem 1.2.7 (1) we then conclude that $g + h = 1$. This last statement contradicts Theorem 1.2.7 (5).

(10) First suppose that $a \leq b$. Suppose further that $a \geq b + 1$. Then by Part (3) of this theorem we deduce that $b + 1 \leq b$, which is a contradiction to Parts (1) and (6) of this theorem. Second, suppose that $a < b + 1$. Suppose further that $a > b$. It follows that $b < a < b + 1$, which is a contradiction to Part (9) of this theorem. □

Part (9) of Theorem 1.2.9 says that the natural numbers are "discrete," a feature not shared by the rational numbers or the real numbers. The integers are also discrete in the sense of Theorem 1.2.9 (9), though there is a prominent difference between the natural numbers and the integers, which is, intuitively, that the latter "goes to infinity" in two directions, whereas the former does so in only one direction. This last property of the natural numbers, combined with the notion of discreteness, form the intuitive basis for the following theorem, which says that every non-empty subset of $\mathbb{N}$ has a smallest element (though not necessarily a largest one).

Theorem 1.2.10 (Well-Ordering Principle). *Let $G \subseteq \mathbb{N}$ be a non-empty set. Then there is some $m \in G$ such that $m \leq g$ for all $g \in G$.*

Proof. Suppose that there is no $m \in G$ such that $m \leq g$ for all $g \in G$. We will derive a contradiction. Let

$$H = \{a \in \mathbb{N} \mid \text{if } n \in \mathbb{N} \text{ and } n \leq a, \text{ then } n \notin G\}.$$

It follows from the definition of H that $H \cap G = \emptyset$. We will show that $H = \mathbb{N}$, and it will then follow that G is empty, the desired contradiction.

Suppose that $1 \notin H$. Then there is some $q \in \mathbb{N}$ such that $q \leq 1$ and $q \in G$. By Theorem 1.2.9 (2) (8) we see that $q = 1$. Hence $1 \in G$. It then follows from Theorem 1.2.9 (2) that G has an element, namely, the number $m = 1$, such that $m \leq g$ for all $g \in G$, which is a contradiction to our hypothesis that no such element exists. Therefore $1 \in H$.

Now suppose that $a \in H$. Suppose further that $a+1 \notin H$. Then there is some $p \in \mathbb{N}$ such that $p \leq a+1$ and $p \in G$. If it were the case that $p \leq a$, then we would have a contradiction to the fact that $a \in H$. Hence, by Theorem 1.2.9 (6) we deduce that $a < p$. Therefore $a < p \leq a+1$, and it follows from Part (9) of the same theorem that $p = a+1$. Hence $a+1 \in G$. Now let $x \in G$. Suppose that $x < a+1$. Then $x \leq a$ by Theorem 1.2.9 (10). Because $a \in H$ it follows that $x \notin G$, which is a contradiction. Hence $a+1 \leq x$ by Theorem 1.2.9 (6). We now have a contradiction to the fact that no element such as $a+1$ exists in G. It follows that $a+1 \in H$, and hence that $H = \mathbb{N}$. □

Reflections

We have taken the Peano Postulates for the natural numbers as axiomatic, but in fact it is possible to derive the existence of a system satisfying the Peano Postulates from the Zermelo–Fraenkel axioms for set theory; see [Vau95, Chapters 2–3] or [Mor87, Chapter 5] for details. Of course, one has to start somewhere axiomatically, and from the perspective of real analysis it makes sense to start with the Peano Postulates, which describe a simple and familiar set of numbers, rather than with the much less intuitive Zermelo–Fraenkel axioms for set theory, which would also take us rather far afield from calculus.

The material in this section looks a bit easier than it really is, because we omitted the proof of Definition by Recursion (Theorem 1.2.4), which is trickier and lengthier than any other proof in the section. The proof of Definition by Recursion is to be found in Section 2.5, where it is seen by all readers of the text, no matter which entry they used. However, although the proof of Definition by Recursion was omitted from the present section for organizational purposes, and also to allow for a smoother development of the natural numbers, it is important to stress the great importance of Definition by Recursion—it is needed to define addition and multiplication for the natural numbers; the reader who wants to see all the details of a rigorous treatment of the natural numbers right now should jump ahead and read the proof of Definition by Recursion before proceeding to the next section.

Exercises

Exercise 1.2.1. **[Used in Theorem 1.2.6.]** Fill in the missing details in the proof of Theorem 1.2.6.

Exercise 1.2.2. **[Used in Theorem 1.2.7.]** Prove Theorem 1.2.7 (2) (3) (4) (7) (8) (9) (10) (11) (13).

Exercise 1.2.3. **[Used in Section 1.2.]** Let $a, b \in \mathbb{N}$. Suppose that $a < b$. Prove that there is a unique $p \in \mathbb{N}$ such that $a + p = b$.

Exercise 1.2.4. **[Used in Theorem 1.2.9.]** Prove Theorem 1.2.9 (1) (3) (4) (5) (11).

Exercise 1.2.5. **[Used in Exercise 1.3.3.]** Let $a, b \in \mathbb{N}$. Prove that if $a + a = b + b$, then $a = b$.

Exercise 1.2.6. Let $b \in \mathbb{N}$. Prove that

$$\{n \in \mathbb{N} \mid 1 \leq n \leq b\} \cup \{n \in \mathbb{N} \mid b+1 \leq n\} = \mathbb{N}$$
$$\{n \in \mathbb{N} \mid 1 \leq n \leq b\} \cap \{n \in \mathbb{N} \mid b+1 \leq n\} = \emptyset.$$

Exercise 1.2.7. Let $A \subseteq \mathbb{N}$ be a set. The set A is **closed** if $a \in A$ implies $a+1 \in A$. Suppose that A is closed.

(1) Prove that if $a \in A$ and $n \in \mathbb{N}$, then $a+n \in A$.
(2) Prove that if $a \in A$, then $\{x \in \mathbb{N} \mid x \geq a\} \subseteq A$.

Exercise 1.2.8. [Used in Section 1.2.] Suppose that the set $\mathbb{N}$ together with the element $1 \in \mathbb{N}$ and the function $s\colon \mathbb{N} \to \mathbb{N}$, and that the set $\mathbb{N}'$ together with the element $1' \in \mathbb{N}'$ and the function $s'\colon \mathbb{N}' \to \mathbb{N}'$, both satisfy the Peano Postulates. Prove that there is a bijective function $f\colon \mathbb{N} \to \mathbb{N}'$ such that $f(1) = 1'$ and $f \circ s = s' \circ f$. The existence of such a bijective function proves that the natural numbers are essentially unique.

The existence of the function f follows immediately from the existence part of Theorem 1.2.4; the trickier aspect of this exercise is to prove that f is bijective. To do that, find an inverse for f by using the existence part of Theorem 1.2.4 again, and then prove that the function you found is an inverse of f by using the uniqueness part of Theorem 1.2.4.

1.3 Constructing the Integers

The natural numbers have many nice properties, but there are some things obviously missing from them. There is nothing in the natural numbers that plays the role of the number zero (Theorem 1.2.7 (6) rules out any such number), and there is nothing that plays the role of negative numbers (the definition of less than for the natural numbers means that if any two natural numbers are added, the result is a number greater than each of the original two numbers). We could simply try to adjoin zero and the negative numbers to the set of natural numbers by brute force, but doing so would leave us unsure that we are on safe ground, unless we assume additional axiomatic properties, which we would prefer not to do. Instead, we will *construct* a new set of numbers, the integers, out of the set of natural numbers, and we will show that this new set contains a copy of the natural numbers together with zero and the negatives of the natural numbers, and that this new set of numbers obeys all the expected properties.

The construction and proofs in this section are quite different from those seen in Section 1.2. For the sake of brevity, we will use the results of Section 1.2 without always giving explicit references to theorems and lemmas whenever we use standard properties of the natural numbers (for example, the fact that $a+b = b+a$ for all $a, b \in \mathbb{N}$). A crucial tool needed in the construction of the integers, which we did not use in Section 1.2, is the concept of equivalence relations and equivalence classes.

The intuitive idea in our construction is that we can think of an integer as given by an expression of the form "$a-b$," where $a, b \in \mathbb{N}$. Because we do not have the operation subtraction on $\mathbb{N}$, we replace "$a-b$" with the pair (a, b). It could happen, however, that "$a-b$" equals "$c-d$" for some $a,b,c,d \in \mathbb{N}$, where $a \neq c$ and $b \neq d$. Then both pairs (a,b) and (c,d) ought to represent the same integer. To take care of this problem we define the following relation on $\mathbb{N} \times \mathbb{N}$.

Definition 1.3.1. The relation $\sim$ on $\mathbb{N} \times \mathbb{N}$ is defined by $(a,b) \sim (c,d)$ if and only if $a+d=b+c$, for all $(a,b),(c,d) \in \mathbb{N} \times \mathbb{N}$. △

Lemma 1.3.2. *The relation $\sim$ is an equivalence relation on $\mathbb{N} \times \mathbb{N}$.*

Proof. We will prove reflexivity and symmetry, leaving transitivity to the reader in Exercise 1.3.2. Let $(a,b),(c,d) \in \mathbb{N} \times \mathbb{N}$. We note that $a+b=b+a$, and hence $(a,b) \sim (a,b)$. Therefore $\sim$ is reflexive. Now suppose that $(a,b) \sim (c,d)$. Then $a+d=b+c$. Hence $c+b=d+a$, and therefore $(c,d) \sim (a,b)$. It follows that $\sim$ is symmetric. □

We are now ready for the definition of the set of integers, together with addition, multiplication, negative and the relations less than, and less than or equal to. We will use the standard symbols $+$, $\cdot$, $-$, $<$ and $\leq$ in the definition, though we need to be careful to note that these symbols formally mean different things when used with the integers from when they are used with the natural numbers. (Moreover, it is not possible to "add" a natural number and an integer using addition as we have defined it; this problem will be resolved when we see that a copy of the natural numbers sits inside the set of integers.) Recall the definition of binary and unary operations given in Section 1.1.

Definition 1.3.3. The set of **integers**, denoted $\mathbb{Z}$, is the set of equivalence classes of $\mathbb{N} \times \mathbb{N}$ with respect to the equivalence relation $\sim$.

The elements $\hat{0}, \hat{1} \in \mathbb{Z}$ are defined by $\hat{0} = [(1,1)]$ and $\hat{1} = [(1+1,1)]$. The binary operations $+$ and $\cdot$ on $\mathbb{Z}$ are defined by

$$[(a,b)] + [(c,d)] = [(a+c,b+d)]$$
$$[(a,b)] \cdot [(c,d)] = [(ac+bd,ad+bc)]$$

for all $[(a,b)],[(c,d)] \in \mathbb{Z}$. The unary operation $-$ on $\mathbb{Z}$ is defined by $-[(a,b)] = [(b,a)]$ for all $[(a,b)] \in \mathbb{Z}$. The relation $<$ on $\mathbb{Z}$ is defined by $[(a,b)] < [(c,d)]$ if and only if $a+d < b+c$, for all $[(a,b)],[(c,d)] \in \mathbb{Z}$. The relation $\leq$ on $\mathbb{Z}$ is defined by $[(a,b)] \leq [(c,d)]$ if and only if $[(a,b)] < [(c,d)]$ or $[(a,b)] = [(c,d)]$, for all $[(a,b)],[(c,d)] \in \mathbb{Z}$. △

As is often the case with definitions involving equivalence classes, we need to check whether the binary operations $+$ and $\cdot$, the unary operation $-$, and the relation $<$ are well-defined. For example, we defined $[(a,b)] + [(c,d)]$ to be $[(a+c,b+d)]$, where $[(a,b)],[(c,d)] \in \mathbb{Z}$, but for this definition to make sense, we need to verify that if $[(a,b)] = [(x,y)]$ and $[(c,d)] = [(z,w)]$, then $[(a+c,b+d)] = [(x+z,y+w)]$. In other words, we need to verify that the sum of the equivalence classes $[(a,b)]$ and $[(c,d)]$ depends only upon the equivalence classes themselves, and not upon the particular elements that are used to represent the equivalence classes. As seen in the following lemma, everything works out well. (We do not need to deal with the relation $\leq$ in the lemma, because that is defined in terms of $<$.)

Lemma 1.3.4. *The binary operations* $+$ *and* $\cdot$, *the unary operation* $-$, *and the relation* $<$, *all on* $\mathbb{Z}$, *are well-defined.*

Proof. We will show that $+$ and $<$ are well-defined; the other parts of the lemma are left to the reader in Exercise 1.3.3.

Let $(a,b),(c,d),(x,y),(z,w) \in \mathbb{N}\times\mathbb{N}$. Suppose that $[(a,b)] = [(x,y)]$ and $[(c,d)] = [(z,w)]$.

By hypothesis we know that $(a,b) \sim (x,y)$ and $(c,d) \sim (z,w)$. Hence $a+y = b+x$ and $c+w = d+z$. By adding these two equations and doing some rearranging we obtain $(a+c)+(y+w) = (b+d)+(x+z)$, and we deduce that $[(a+c,b+d)] = [(x+z,y+w)]$. Therefore $+$ is well-defined.

Now suppose that $[(a,b)] < [(b,d)]$. Therefore $a+d < b+c$. Adding $b+x = a+y$ and $c+w = d+z$ to this inequality, we obtain $a+d+b+x+c+w < b+c+a+y+d+z$. Canceling yields $x+w < y+z$, and it follows that $[(x,y)] < [(y,w)]$. This process can be done backwards, and hence $[(x,y)] < [(y,w)]$ implies $[(a,b)] < [(b,d)]$. Therefore $[(a,b)] < [(b,d)]$ if and only if $[(x,y)] < [(y,w)]$, which means that $<$ is well-defined. □

We now turn to the basic algebraic properties of the integers. The idea of the proof of the following theorem is to rephrase things in terms of natural numbers, and then use the appropriate facts we have already proved about the natural numbers. As is usual, we will write "xy" instead of "$x\cdot y$," except in cases of potential ambiguity, or for ease of reading. We will write $x > y$ to mean the same thing as $y < x$, and similarly for $x \geq y$.

Theorem 1.3.5. *Let* $x,y,z \in \mathbb{Z}$.

1. $(x+y)+z = x+(y+z)$ *(Associative Law for Addition).*
2. $x+y = y+x$ *(Commutative Law for Addition).*
3. $x+\hat{0} = x$ *(Identity Law for Addition).*
4. $x+(-x) = \hat{0}$ *(Inverses Law for Addition).*
5. $(xy)z = x(yz)$ *(Associative Law for Multiplication).*
6. $xy = yx$ *(Commutative Law for Multiplication).*
7. $x\cdot\hat{1} = x$ *(Identity Law for Multiplication).*
8. $x(y+z) = xy+xz$ *(Distributive Law).*
9. *If* $xy = \hat{0}$, *then* $x = \hat{0}$ *or* $y = \hat{0}$ *(No Zero Divisors Law).*
10. *Precisely one of* $x < y$ *or* $x = y$ *or* $x > y$ *holds (Trichotomy Law).*
11. *If* $x < y$ *and* $y < z$, *then* $x < z$ *(Transitive Law).*
12. *If* $x < y$ *then* $x+z < y+z$ *(Addition Law for Order).*
13. *If* $x < y$ *and* $z > \hat{0}$, *then* $xz < yz$ *(Multiplication Law for Order).*
14. $\hat{0} \neq \hat{1}$ *(Non-Triviality).*

Proof. We will prove Parts (2), (9) and (12), leaving the rest to the reader in Exercise 1.3.5. Suppose that $x = [(a,b)]$, that $y = [(c,d)]$ and that $z = [(e,f)]$, for some $a,b,c,d,e,f \in \mathbb{N}$.

(2) Using the definition of addition of integers, we see that $x+y = [(a,b)] + [(c,d)] = [(a+c,b+d)] = [(c+a,d+b)] = [(c,d)] + [(a,b)] = y+x$, where the middle equality holds by Theorem 1.2.7 (4).

(9) Suppose that $xy = \hat{0}$ and that $x \neq \hat{0}$. We will deduce that $y = \hat{0}$. Using the definition of multiplication of integers, we see that $xy = [(ac+bd, ad+bc)]$. It then follows from Exercise 1.3.4 (1) that $a \neq b$ and $ac+bd = ad+bc$. By Theorem 1.2.9 (6) we know that either $a < b$ or $a > b$. First, suppose that $a < b$. Then there is some $g \in \mathbb{N}$ such that $a+g = b$. Hence $ac + (a+g)d = ad + (a+g)c$. By rearranging and canceling we deduce that $d = c$. Exercise 1.3.4 (1) then implies that $y = \hat{0}$. The case where $a > b$ is similar to the previous case, and we omit the details.

(12) Suppose that $x < y$. Then $[(a,b)] < [(c,d)]$, and therefore $a+d < b+c$. Hence $(a+d)+(e+f) < (b+c)+(e+f)$, and therefore $(a+e)+(d+f) < (b+f)+(c+e)$. It follows that $[(a+e,b+f)] < [(c+e,d+f)]$, and hence $[(a,b)]+[(e,f)] < [(c,d)]+[(e,f)]$, which means that $x+z < y+z$. □

We now have two sets of numbers, namely, the natural numbers (as defined in Section 1.2) and the integers (as constructed in the present section). As we defined these two sets, they are entirely disjoint. Of course, intuitively we think of the natural numbers as sitting inside the integers, and we now show that although formally these two sets of numbers are distinct, a copy of the natural numbers sits inside the integers, and this copy is just as good as the original. To find this copy of the natural numbers, we start with the following very simple definition.

Definition 1.3.6. Let $x \in \mathbb{Z}$. The number x is **positive** if $x > \hat{0}$, and the number x is **negative** if $x < \hat{0}$. △

As we now make precise in the following theorem, the set of positive integers can be viewed as a copy of the natural numbers, a fact that certainly fits with our intuition. More formally, we will show that there is a bijective function between the natural numbers and the set of positive integers that preserves the number 1, the binary operations addition and multiplication, and the relation less than.

Theorem 1.3.7. *Let $i\colon \mathbb{N} \to \mathbb{Z}$ be defined by $i(n) = [(n+1,1)]$ for all $n \in \mathbb{N}$.*

1. *The function $i\colon \mathbb{N} \to \mathbb{Z}$ is injective.*
2. *$i(\mathbb{N}) = \{x \in \mathbb{Z} \mid x > \hat{0}\}$.*
3. *$i(1) = \hat{1}$.*
4. *Let $a,b \in \mathbb{N}$. Then*
   a. *$i(a+b) = i(a)+i(b)$;*
   b. *$i(ab) = i(a)i(b)$;*
   c. *$a < b$ if and only if $i(a) < i(b)$.*

Proof. We prove Parts (2) and (4a), leaving the rest to the reader in Exercise 1.3.6.

(2) Let $y \in i(\mathbb{N})$. Then $y = [(p+1,1)]$ for some $p \in \mathbb{N}$. By Part (a) of the Peano Postulates we know that $p \neq 1$. It then follows from Exercise 1.3.4 (3) that $[(p+1),1] > \hat{0}$, and we deduce that $y \in \{x \in \mathbb{Z} \mid x > \hat{0}\}$. Hence $i(\mathbb{N}) \subseteq \{x \in \mathbb{Z} \mid x > \hat{0}\}$.

Now let $z \in \{x \in \mathbb{Z} \mid x > \hat{0}\}$. Again using Exercise 1.3.4 (3) we deduce that $z = [(q,1)]$ for some $q \in \mathbb{N}$ such that $q \neq 1$. By Lemma 1.2.3 we know that $q = c+1$ for some $c \in \mathbb{N}$, and hence $z = [(c+1,1)]$, which in turn implies that $z \in i(\mathbb{N})$. We deduce that $i(\mathbb{N}) \supseteq \{x \in \mathbb{Z} \mid x > \hat{0}\}$.

(4a) Using the definitions of addition for $\mathbb{Z}$, the relation $\sim$ and the function i, together with Theorem 1.2.7, we see that

$$\begin{aligned} i(a)+i(b) &= [(a+1,1)]+[(b+1,1)] = [((a+1)+(b+1),1+1)] \\ &= [(((a+b)+1)+1,1+1)] = [((a+b)+1,1)] = i(a+b). \end{aligned}$$

□

Because Theorem 1.3.7 implies that from the point of view of addition, multiplication and the relation less than, we can identify the natural numbers with the positive integers, we can therefore dispense with the set of natural numbers as a separate entity (except when we need it in proofs), because we have a copy of the natural numbers inside the integers that works just as well. Everything that was proved about the natural numbers in Section 1.2 that involve only addition, multiplication and the relation less than (and therefore also the relation $\leq$, which is derived from $<$), still holds true when we think of the natural numbers as the set of positive integers. For example, the reader is asked in Exercise 1.3.10 to give a detailed proof, using Theorem 1.3.7, that the Well-Ordering Principle (Theorem 1.2.10), which was stated for $\mathbb{N}$ in Section 1.2, still holds when we think of $\mathbb{N}$ as the set of positive integers.

From this point on, we will use the symbol $\mathbb{N}$ to denote the set of positive integers, and we will dispense with the notation $\hat{0}$ and $\hat{1}$, and simply write 0 and 1 instead. If we let $-\mathbb{N}$ denote the set of negative integers, then it follows from the Trichotomy Law (Theorem 1.3.5 (10)) that $\mathbb{Z} = -\mathbb{N} \cup \{0\} \cup \mathbb{N}$, and that these three sets are pairwise disjoint.

We conclude this section with a few more properties of the integers that we will need later on. In contrast to the properties in Theorem 1.3.5, the proofs of which required going back to the definition of the integers as equivalence classes, the properties in the following lemma do not require the use of the definition of the integers, but rather are proved directly from the properties in Theorem 1.3.5. As such, Theorem 1.3.5 contains more fundamental properties of the integers. Indeed, in Section 1.4 (which the reader of the current section should skip), it is seen that what we have stated in the present section as Theorem 1.3.5 is taken as part of the axioms for the integers, if one starts with such axioms rather than the Peano Postulates. As is usual, we will write "$-xy$" when we mean "$-(xy)$."

Lemma 1.3.8. *Let $x,y,z \in \mathbb{Z}$.*

1. *If $x+z = y+z$, then $x = y$ (Cancellation Law for Addition).*
2. $-(-x) = x$.
3. $-(x+y) = (-x)+(-y)$.
4. $x \cdot 0 = 0$.
5. *If $z \neq 0$ and if $xz = yz$, then $x = y$ (Cancellation Law for Multiplication).*
6. $(-x)y = -xy = x(-y)$.

7. $xy = 1$ *if and only if* $x = 1 = y$ *or* $x = -1 = y$.
8. $x > 0$ *if and only if* $-x < 0$, *and* $x < 0$ *if and only if* $-x > 0$.
9. $0 < 1$.
10. *If* $x \leq y$ *and* $y \leq x$, *then* $x = y$.
11. *If* $x > 0$ *and* $y > 0$, *then* $xy > 0$. *If* $x > 0$ *and* $y < 0$, *then* $xy < 0$.

Proof. We will prove Parts (2), (6), (8) and (9), leaving the rest to the reader in Exercise 1.3.11.

(2) By the Inverses Law for Addition, we know that $(-x) + [-(-x)] = 0$. It follow that $x + \{(-x) + [-(-x)]\} = x + 0$. By the Associative and Identity Laws for Addition it follows that $\{x + (-x)\} + [-(-x)] = x$, and hence by the Inverses Law for Addition we see that $0 + [-(-x)] = x$. We now use the Commutative and Identity Laws for Addition to deduce that $-(-x) = x$.

(6) By the Inverses Law for Addition, we know that $x + (-x) = 0$. We then use the Distributive Law to deduce that $yx + y(-x) = y[x + (-x)] = y \cdot 0$. It follows from Part (4) of this lemma that $yx + y(-x) = 0$. By the Commutative Law for Multiplication we obtain $xy + (-x)y = 0$. By the Inverses Law for Addition we know that $xy + (-xy) = 0$. Hence $xy + (-x)y = xy + (-xy)$, and using the Commutative Law for Addition we see that $(-x)y + xy = (-xy) + xy$. Using Part (1) of this lemma we deduce that $(-x)y = -xy$. A similar argument shows that $x(-y) = -xy$.

(8) Suppose that $0 < x$. Then by the Addition Law for Order we see that $0 + (-x) < x + (-x)$. By the Commutative, Identity and Inverses Laws for Addition we deduce that $-x < 0$. A similar argument shows that $-x < 0$ implies $0 < x$, and also shows that $x < 0$ if and only if $-x > 0$.

(9) By Non-Triviality we know that $0 \neq 1$. It therefore follows from the Trichotomy Law that either $0 < 1$ or $0 > 1$, but not both. Suppose that $0 > 1$. By Part (8) of this lemma we deduce that $0 < -1$. It follows from the Multiplication Law for Order that $0 \cdot (-1) < (-1)(-1)$. The Commutative Law for Multiplication together with Part (4) of this lemma imply that $0 < (-1)(-1)$. By Part (6) of this lemma together with the Identity Law for Multiplication we see that $(-1)(-1) = -[(-1) \cdot 1] = -(-1)$, and by Part (2) of this lemma we deduce that $(-1)(-1) = 1$. We conclude that $0 < 1$, which is a contradiction to our assumption that $0 > 1$. Therefore $0 > 1$ is impossible, and hence $0 < 1$. □

Whereas the proof of Theorem 1.3.8 makes use of only the properties of the integers given in Theorem 1.3.5, it turns out that not all properties of the integers can be deduced from that theorem. For the following result, which shows that the integers are intuitively "discrete," we need to make use of not only Theorem 1.3.5 (and Theorem 1.3.8, which is derived from Theorem 1.3.5), but also Theorem 1.2.9 (9), which by Theorem 1.3.7 holds for the set $\mathbb{N}$ when viewed as a subset of $\mathbb{Z}$. The following theorem cannot be deduced from Theorem 1.3.5 alone because the set of real numbers $\mathbb{R}$ satisfies all the properties in Theorem 1.3.5, and yet the following theorem would not be true if $\mathbb{Z}$ were replaced with $\mathbb{R}$.

Theorem 1.3.9. *Let $x \in \mathbb{Z}$. Then there is no $y \in \mathbb{Z}$ such that $x < y < x+1$.*

Proof. Suppose that there is some $y \in \mathbb{Z}$ such that $x < y < x+1$. By the Addition Law for Order we see that $x+[(-x)+1] < y+[(-x)+1] < [x+1]+[(-x)+1]$, and then by repeated use of the Associative, Commutative, Identity and Inverses Laws for Addition we deduce that $1 < y+[(-x)+1] < 1+1$. By Theorem 1.3.8 (9) we know that $0 < 1$, and it then follows from the Transitive Law that $0 < y+[(-x)+1]$. Hence 1 and $y+[(-x)+1]$ are both in $\mathbb{N}$, and we therefore have a contradiction to Theorem 1.2.9 (9). □

Finally, we note that because $1 \in \mathbb{Z}$, we can now use all the familiar integers such as the number 2, which is defined to be $1+1$; the number 3, which is defined to be $2+1$; the number 4, which is defined to be $3+1$; and so on.

The reader who has read Sections 1.2 and 1.3 should now skip Section 1.4, where an axiomatic approach to the integers is taken as an alternative to what we saw in the present section and the previous one, and should proceed straight to Section 1.5 for the construction of the rational numbers.

Reflections

At first glance this section might appear to be much ado about nothing. We defined the natural numbers in Section 1.2, and the reader might wonder why we could not simply define the integers by letting the negative numbers be the negations of the natural numbers, and letting zero be an additional number that works as expected with respect to addition. That approach, while in principle workable, turns out to be not really all that simple. First, there is no operation of negation defined for the natural number in Section 1.2, so we cannot use such an operation until we define it, and defining the negation of the natural numbers is tantamount to already having the negative numbers defined. It would be possible to proceed formally by defining a set, called the set of negative natural numbers, by simply have one such number corresponding to each element of the set of natural numbers. However, doing so leads to the problem of defining addition on the set of integers, that is, the set containing the natural numbers, the formally defined negative natural numbers and a formally defined zero. Such a definition would require various cases, depending upon which type of numbers are being added, with the main problem occurring when adding a natural number and a negative natural number. To define the sum of such numbers, it would be necessary to find the difference between the given natural number and the natural number that corresponds to the given negative number, which could be done via Exercise 1.2.3. It would then be necessary to define multiplication on the set of integers, which again requires various cases, and also to prove Theorem 1.3.5 using these definitions of addition and multiplication, at which point things become more or less as complicated as the method used in the text. Hence, we might as well stick with the method in the text, because it does not involve a lot of cases, because after some familiarity it is seen to be quite natural and because it is analogous to the construction of the rational numbers from the integers used in Section 1.5.

Exercises

Exercise 1.3.1. Let $\approx$ be the relation on $\mathbb{N} \times \mathbb{N}$ defined by $(a,b) \approx (c,d)$ if and only if $a^2d = c^2b$, for all $(a,b),(c,d) \in \mathbb{N} \times \mathbb{N}$, where n^2 is an abbreviation for $n \cdot n$.

(1) Prove that $\approx$ is an equivalence relation.
(2) List or describe all the elements in $[(2,3)]$.

Exercise 1.3.2. [Used in Lemma 1.3.2.] Complete the proof of Lemma 1.3.2. That is, prove that the relation $\sim$ is transitive.

Exercise 1.3.3. [Used in Lemma 1.3.4.] Complete the proof of Lemma 1.3.4. That is, prove that $\cdot$ and $-$ for $\mathbb{Z}$ are well-defined. The proof for $\cdot$ is a bit more complicated than might be expected. [Use Exercise 1.2.5.]

Exercise 1.3.4. [Used in Theorem 1.3.5 and Theorem 1.3.7.] Let $a,b \in \mathbb{N}$.

(1) Prove that $[(a,b)] = \hat{0}$ if and only if $a = b$.
(2) Prove that $[(a,b)] = \hat{1}$ if and only if $a = b+1$.
(3) Prove that $[(a,b)] = [(n,1)]$ for some $n \in \mathbb{N}$ such that $n \neq 1$ if and only if $a > b$ if and only if $[(a,b)] > \hat{0}$.
(4) Prove that $[(a,b)] = [(1,m)]$ for some $m \in \mathbb{N}$ such that $m \neq 1$ if and only if $a < b$ if and only if $[(a,b)] < \hat{0}$.

Exercise 1.3.5. [Used in Theorem 1.3.5.] Prove Theorem 1.3.5 (1) (3) (4) (5) (6) (7) (8) (10) (11) (13) (14).

Exercise 1.3.6. [Used in Theorem 1.3.7.] Prove Theorem 1.3.7 (1) (3) (4b) (4c).

Exercise 1.3.7. Let $x,y,z \in \mathbb{Z}$.

(1) Prove that $x < y$ if and only if $-x > -y$.
(2) Prove that if $z < 0$, then $x < y$ if and only if $xz > yz$.

Exercise 1.3.8. [Used in Exercise 1.5.9.] Let $x \in \mathbb{Z}$. Prove that if $x > 0$ then $x \geq 1$. Prove that if $x < 0$ then $x \leq -1$

Exercise 1.3.9.

(1) Prove that $1 < 2$.
(2) Let $x \in \mathbb{Z}$. Prove that $2x \neq 1$.

Exercise 1.3.10. [Used in Section 1.3.] Prove that the Well-Ordering Principle (Theorem 1.2.10), which was stated for $\mathbb{N}$ in Section 1.2, still holds when we think of $\mathbb{N}$ as the set of positive integers. That is, let $G \subseteq \{x \in \mathbb{Z} \mid x > 0\}$ be a non-empty set. Prove that there is some $m \in G$ such that $m \leq g$ for all $g \in G$. Use Theorem 1.3.7.

Exercise 1.3.11. [Used in Lemma 1.3.8.] Prove Theorem 1.3.8 (1) (3) (4) (5) (7) (10) (11).

1.4 Entry 2: Axioms for the Integers

There are two standard approaches to defining the integers: starting with the Peano Postulates for the natural numbers and then constructing the integers from the natural numbers, or taking the integers axiomatically as an ordered integral domain with an additional condition. The former approach, which we took in Sections 1.2 and 1.3, is the longer of the two, but is more revealing of the inner workings of the natural numbers. The latter approach, which we take in the present section, is somewhat shorter, and therefore provides a slightly quicker route to constructing the real numbers, though it still involves the constructing the rational numbers and the real numbers, as in the other approach. Both approaches ultimately lead to the same facts about the natural numbers and the integers, though they work in reverse order. The Peano Postulates for the natural numbers, which are taken axiomatically in Section 1.2, are a theorem in the present section, and the axioms for the integers given in the present section are theorems in Sections 1.2 and 1.3. In other words, whether we start with the natural numbers and construct the integers, or whether we start with the integers and find the natural numbers inside them, we obtain the same sets of numbers. Hence, if you have read Sections 1.2 and 1.3, you should skip the present section. Our treatment of the integers in the present section, which is less commonly used than starting with the Peano Postulates but is nonetheless a nice alternative, follows [Dea66, Chapter 3].

What is it that characterizes the integers? Certainly, the integers have two binary operations, namely, addition and multiplication, as well as the unary operation negation, and the relation less than, and these operations and relation satisfy a number of standard properties (for example, the fact that $x+y=y+x$ for all integers x and y). However, these aspects alone do not characterize the integers, because we will see that the set of rational numbers and the set of real numbers have these same operations and relation, which satisfy these same properties. What distinguishes the integers from the rational numbers and the real numbers is the "discreteness" of the integers, which intuitively means that for each integer, there is a unique integer "right below it," namely, the integer minus 1, and a unique integer "right above it," namely, the integer plus 1. Phrased another way, if a is an integer, then there is no integer between a and $a+1$. We will prove in Theorem 1.4.6 that the integers are discrete in this sense. It turns out, however, that this approach to discreteness does not characterize the integers among ordered integral domains, as seen by Exercise 1.4.7. Hence, we will need to use a different characterization of discreteness in the axioms for the integers, as discussed below.

In order to state our axioms for the integers, we start with the following definition involving some algebraic properties of addition, multiplication and less than. If the reader has studied abstract algebra, she has probably encountered the concept of an integral domain, though perhaps not an ordered integral domain, which is simply an integral domain as standardly defined in abstract algebra together with some additional properties for the order relation, and the interaction of this relation with addition and multiplication. We do not assume that the reader is already familiar with integral domains, and we will give the complete definition here.

In the following definition, we use the notion of a binary operation and a unary operation on a set, as defined in Section 1.1. As is usual, we will write "xy" instead of "$x \cdot y$," except in cases of potential ambiguity (for example, we will write "$1 \cdot 1$" rather than "11"), or where the $\cdot$ makes the expression easier to read. We will write $x > y$ to mean the same thing as $y < x$.

Definition 1.4.1. An **ordered integral domain** is a set R with elements $0, 1 \in R$, binary operations $+$ and $\cdot$, a unary operation $-$ and a relation $<$, which satisfy the following properties. Let $x, y, z \in R$.

a. $(x+y)+z = x+(y+z)$ (Associative Law for Addition).
b. $x+y = y+x$ (Commutative Law for Addition).
c. $x+0 = x$ (Identity Law for Addition).
d. $x+(-x) = 0$ (Inverses Law for Addition).
e. $(xy)z = x(yz)$ (Associative Law for Multiplication).
f. $xy = yx$ (Commutative Law for Multiplication).
g. $x \cdot 1 = x$ (Identity Law for Multiplication).
h. $x(y+z) = xy+xz$ (Distributive Law).
i. If $xy = 0$, then $x = 0$ or $y = 0$ (No Zero Divisors Law).
j. Precisely one of $x < y$ or $x = y$ or $x > y$ holds (Trichotomy Law).
k. If $x < y$ and $y < z$, then $x < z$ (Transitive Law).
l. If $x < y$ then $x+z < y+z$ (Addition Law for Order).
m. If $x < y$ and $z > 0$, then $xz < yz$ (Multiplication Law for Order).
n. $0 \neq 1$ (Non-Triviality). △

The Non-Triviality axiom might seem, well, trivial, but it is very much needed, because otherwise we could have an ordered integral domain consisting of a single number 0, which is not what we would want the set of integers to be.

As previously stated, the properties of an ordered integral domain do not alone characterize the integers. It will be seen from Definition 2.2.1 and Lemma 2.3.2 (15) that the real numbers are an ordered integral domain as well. In order to distinguish the integers from all other ordered integral domains, we will need one additional axiom, to which we now turn, starting with the following definition.

Definition 1.4.2. Let R be an ordered integral domain, and let $A \subseteq R$ be a set.

1. The relation $\leq$ on R is defined by $a \leq b$ if and only if $a < b$ or $a = b$, for all $a, b \in R$.
2. The set A has a **least element** if there is some $a \in A$ such that $a \leq x$ for all $x \in A$. △

If we think of the integers intuitively, we see that some subsets of the integers have least elements (for example, all finite subsets of the integers have least elements), whereas other subsets of the integers do not have least elements (for example, the set of negative integers). In fact, it is precisely the existence of negative integers that prevents all subsets of the integers from having least elements; intuitively, because of the "discreteness" of the integers, every subset of the positive integers has a least element. The real numbers, by contrast, are also an ordered integral domain, but not every subset of the positive real number has a least element (for example, the set of all positive real numbers does not have a least element). It turns out that this fact about subsets of the positive integers is a strong version of "discreteness," and it actually characterizes the integers.

Definition 1.4.3. Let R be an ordered integral domain. The ordered integral domain R satisfies the **Well-Ordering Principle** if every non-empty subset of

$$\{x \in R \mid x > 0\}$$

has a least element. △

We are now ready for our axiomatic characterization of the integers.

Axiom 1.4.4 (Axiom for the Integers). *There exists an ordered integral domain* $\mathbb{Z}$ *that satisfies the Well-Ordering Principle.*

The Axiom for the Integers (Axiom 1.4.4) does not say that the integers are unique, though in fact that turns out to be true; the proof of this fact is given in the first two steps of the proof of Theorem 2.7.1, and discussion of what we mean by uniqueness in this context is found at the start of Section 2.7.

We now turn to some very useful, and very familiar, properties of the integers.

Lemma 1.4.5. *Let* $x, y, z \in \mathbb{Z}$.

1. *If* $x+z=y+z$, *then* $x=y$ *(Cancellation Law for Addition).*
2. $-(-x)=x$.
3. $-(x+y)=(-x)+(-y)$.
4. $x \cdot 0=0$.
5. *If* $z \neq 0$ *and if* $xz=yz$, *then* $x=y$ *(Cancellation Law for Multiplication).*
6. $(-x)y=-xy=x(-y)$.
7. $xy=1$ *if and only if* $x=1=y$ *or* $x=-1=y$.
8. $x>0$ *if and only if* $-x<0$, *and* $x<0$ *if and only if* $-x>0$.
9. $0<1$.
10. *If* $x \leq y$ *and* $y \leq x$, *then* $x=y$.
11. *If* $x>0$ *and* $y>0$, *then* $xy>0$. *If* $x>0$ *and* $y<0$, *then* $xy<0$.

Proof. We will prove Parts (2), (6), (8) and (9), leaving the rest to the reader in Exercise 1.4.1.

(2) By the Inverses Law for Addition we know that $(-x)+[-(-x)]=0$. It follows that $x+\{(-x)+[-(-x)]\}=x+0$. By the Associative and Identity Laws for Addition we deduce that $\{x+(-x)\}+[-(-x)]=x$, and hence by the Inverses Law for Addition we see that $0+[-(-x)]=x$. We now use the Commutative and Identity Laws for Addition to conclude that $-(-x)=x$.

(6) By the Inverses Law for Addition we know that $x+(-x)=0$. We then use the Distributive Law to deduce that $yx+y(-x)=y[x+(-x)]=y \cdot 0$. It follows from Part (4) of this lemma that $yx+y(-x)=0$, and by the Commutative Law for Multiplication we obtain $xy+(-x)y=0$. By the Inverses Law for Addition we know that $xy + (-xy) = 0$. Hence $xy + (-x)y = xy + (-xy)$, and using the Commutative Law for Addition we see that $(-x)y + xy = (-xy) + xy$. Using Part (1) of this lemma we deduce that $(-x)y = -xy$. A similar argument shows that $x(-y) = -xy$.

(8) Suppose that $x > 0$. Then by the Addition Law for Order we see that $x + (-x) > 0 + (-x)$. By the Commutative, Identity and Inverses Laws for Addition we deduce that $0 > -x$, which is the same as $-x < 0$. A similar argument shows that $-x < 0$ implies $x > 0$, and also shows that $x < 0$ if and only if $-x > 0$.

(9) By Non-Triviality we know that $0 \neq 1$. It therefore follows from the Trichotomy Law that either $0 < 1$ or $0 > 1$, but not both. Suppose that $0 > 1$. By Part (8) of this lemma we deduce that $0 < -1$. It follows from the Multiplication Law for Order that $0 \cdot (-1) < (-1)(-1)$, and by the Commutative Law for Multiplication and Part (4) of this lemma we deduce that $0 < (-1)(-1)$. By Part (6) of this lemma, the Identity Law for Multiplication and Part (2) of this lemma we see that $(-1)(-1) = -[(-1) \cdot 1] = -(-1) = 1$. We conclude that $0 < 1$, which is a contradiction to our assumption that $0 > 1$. Therefore $0 > 1$ is impossible, and hence it must be the case that $0 < 1$. □

We note that the proof of Lemma 1.4.5 makes use of only the properties of the integers given in Definition 1.4.1, and it does not make full use of all the properties of the integers. By contrast, the following theorem, which shows that the integers are "discrete" in the intuitive sense, makes use of the full power of the Well-Ordering Principle.

Theorem 1.4.6. *Let $x \in \mathbb{Z}$. Then there is no $y \in \mathbb{Z}$ such that $x < y < x + 1$.*

Proof. First, suppose that $x = 0$. Suppose further that there is some $y \in \mathbb{Z}$ such that $0 < y < 1$. Let

$$S = \{z \in \mathbb{Z} \mid 0 < z < 1\}.$$

We know that $S \neq \emptyset$, because $y \in S$. From the definition of $\mathbb{N}$, we observe that $S \subseteq \mathbb{N}$. It follows from the Well-Ordering Principle that S has a least element. That is, there is some $p \in S$ such that $p \leq x$ for all $x \in S$. By the definition of S we know that $0 < p < 1$. Using the Multiplication Law for Order we see that $0 \cdot p < p \cdot p < 1 \cdot p$, and hence by the Commutative and Identity Laws for Multiplication and Lemma 1.4.5 (4) it follows that $0 < p \cdot p < p$. By the Transitive Law we deduce that $0 < p \cdot p < 1$. It follows that $p \cdot p \in S$, and $p \cdot p < p$, which is a contradiction to the fact that p is the least element of S. We deduce that there is no $y \in \mathbb{Z}$ such that $0 < y < 1$.

We now consider the general case, with arbitrary x. Suppose that there is some $w \in \mathbb{Z}$ such that $x < w < x + 1$. By the Addition Law for Order we deduce that $x + (-x) < w + (-x) < (x + 1) + (-x)$, and then using the Associative, Inverses and Identity Laws for Addition we see that $0 < w + (-x) < 1$. We therefore have a contradiction to the previous paragraph, and hence no such w exists. □

Our final task in this section is to show that the natural numbers sit inside the integers, and behave appropriately. We start with the following definition.

Definition 1.4.7.

1. Let $x \in \mathbb{Z}$. The number x is **positive** if $x > 0$, and the number x is **negative** if $x < 0$.
2. The set of **natural numbers**, denoted $\mathbb{N}$, is defined by

$$\mathbb{N} = \{x \in \mathbb{Z} \mid x > 0\}. \qquad \triangle$$

If we let $-\mathbb{N}$ denote the set of negative integers, then it follows from the Trichotomy Law (Theorem 1.3.5 (10)) that $\mathbb{Z} = -\mathbb{N} \cup \{0\} \cup \mathbb{N}$, and that these three sets are pairwise disjoint.

We want to verify that the set $\mathbb{N}$ as defined in Definition 1.4.7 (2) indeed behaves the way we expect the natural numbers to behave. We will do that by proving that the set $\mathbb{N}$ as we have now defined it satisfies the "Peano Postulates," which were taken as the axioms for the natural numbers in Section 1.2. By proving that the Peano Postulates follow from the axioms for the integers stated in Axiom 1.4.4 we come full circle in our study of the natural numbers and the integers, because in Sections 1.2 and 1.3 it was shown that the axioms for the integers follow from the Peano Postulates (see Theorem 1.2.10 and Theorem 1.3.5).

To be able to state the following theorem, we note that by Exercise 1.4.2 the map s given in the theorem is well-defined.

Theorem 1.4.8 (Peano Postulates). *Let $s\colon \mathbb{N} \to \mathbb{N}$ be defined by $s(n) = n+1$ for all $n \in \mathbb{N}$.*

a. There is no $n \in \mathbb{N}$ such that $s(n) = 1$.
b. The function s is injective.
c. Let $G \subseteq \mathbb{N}$ be a set. Suppose that $1 \in G$, and that if $g \in G$ then $s(g) \in G$. Then $G = \mathbb{N}$.

Proof.

(a) Suppose to the contrary that there were some $x \in \mathbb{N}$ such that $s(x) = 1$. Then $x+1 = 1$, and therefore $x+1 = 0+1$ by the Commutative and Identity Laws for Addition. It follows from Lemma 1.4.5 (1) that $x = 0$. However, we know by Definition 1.4.7 that $x > 0$, which is a contradiction to the Trichotomy Law.

(b) Suppose that $s(n) = s(m)$ for some $n, m \in \mathbb{N}$. Then $n+1 = m+1$, and hence $n = m$ by Lemma 1.4.5 (1). Therefore s is injective.

(c) Suppose that $G \neq \mathbb{N}$. Then $\mathbb{N} - G$ is non-empty. By the Well-Ordering Principle the set $\mathbb{N} - G$ has a least element, which means that there is some $m \in \mathbb{N} - G$ such that $m \leq x$ for all $x \in \mathbb{N} - G$. Observe that $m \neq 1$, because $1 \in G$. By the Trichotomy Law, we know that $m < 1$ or $m > 1$. Suppose that $m < 1$. Because $m \in \mathbb{N}$ we know that $m > 0$. It follows that $0 < m < 1$. Using the Commutative and Identity Laws for Addition we see that $0 < m < 0+1$, which is a contradiction to Theorem 1.4.6. Therefore it could not have been the case that $m < 1$. Hence $m > 1$. Applying Theorem 1.4.6 again, we know that it cannot be the case that $1 < m < 1+1$, and therefore by the Trichotomy Law it must be the case that $m \geq 1+1$. It follows from the Inverses,

Identity and Associative Laws for Addition that $m+(-1) \geq 1$. By Lemma 1.4.5 (9) we know that $0 < 1$, and it then follows from the Transitive Law that $m+(-1) > 0$. Hence $m+(-1) \in \mathbb{N}$.

By Lemma 1.4.5 (9) (8) we see that $-1 < 0$. Using the Addition Law for Order and the Identity Law for Addition we deduce that $m+(-1) < m$. Because m is the least element of $\mathbb{N} - G$, then $m+(-1) \notin \mathbb{N} - G$, and hence $m+(-1) \in G$. By the Associative, Identity and Inverses Laws for Addition, and by hypothesis on G, we deduce that $m = [m+(-1)]+1 = s(m+(-1)) \in G$, which is a contradiction to the fact that $m \in \mathbb{N} - G$. We conclude that $G = \mathbb{N}$. □

Part (c) of the Peano Postulates is a formal statement that proof by induction works. We will discuss proof by induction further in Section 2.5.

Finally, we note that because $1 \in \mathbb{Z}$, we can now use all the familiar integers such as the number 2, which is defined to be $1+1$; the number 3, which is defined to be $2+1$; the number 4, which is defined to be $3+1$; and so on.

Reflections

The reader who has commenced reading this text with the present section need not read either Section 1.2, which has axioms for the natural numbers, or Section 2.2, which has axioms for the real numbers. However, even though it is not necessary for the reader of the present section to read those two other sections in order to have a completely rigorous treatment of the real numbers, it might be of interest to the reader to compare the types of axioms in each of those two other sections with the axioms in the present section. As the reader will see from such a comparison, the axioms in the present section share some of the features of the axioms of each of the two other sections.

What the axioms for the integers in the present section and the axioms for the natural numbers in Section 1.2 have in common is that they are entirely algebraic in nature, and can be found in some texts on algebra. By contrast, the axioms for the real numbers in Section 2.2 include the Least Upper Bound Property, which is at the heart of real analysis, and is rarely discussed in a purely algebraic context. What the axioms for the integers in the present section and the axioms for the real numbers in Section 2.2 have in common is their parallel structure, namely, both start with the axioms for a general algebraic structure with two binary operations and an order relation (ordered integral domains and ordered fields, respectively), and both add one additional axiom to distinguish the particular set of numbers in question (the Well-Ordering Principle and the Least Upper Bound Property, respectively). By contrast, the axioms for the natural numbers in Section 1.2 do not at all resemble the axioms in the two other sections, in that for the natural numbers neither a binary operation nor an order relation is postulated, and the assumptions made seem much more minimal than the two other sets of axioms.

From the above comparison, one could conclude that the axioms in the present section are either the best of both worlds, being purely algebraic and yet parallel to other useful axioms systems, or the worst of both, being neither as minimal as one alternative nor as powerful as the other. It is left to the reader to decide which assessment, if either, is the correct one.

Exercises

Exercise 1.4.1. **[Used in Lemma 1.4.5.]** Prove Lemma 1.4.5 (1) (3) (4) (5) (7) (10) (11).

Exercise 1.4.2. **[Used in Section 1.4.]** Let $n \in \mathbb{N}$. Prove that $n+1 \in \mathbb{N}$.

Exercise 1.4.3. **[Used in Exercise 1.4.8.]** Let $x, y \in \mathbb{Z}$. Prove that $x \leq y$ if and only if $-x \geq -y$.

Exercise 1.4.4. **[Used in Exercise 1.4.6, Exercise 1.4.8 and Exercise 1.5.9.]** Prove that $\mathbb{N} = \{x \in \mathbb{Z} \mid x \geq 1\}$.

Exercise 1.4.5. Let $a, b \in \mathbb{Z}$. Prove that if $a < b$, then $a+1 \leq b$.

Exercise 1.4.6. **[Used in Theorem 2.5.4.]** Let $n \in \mathbb{N}$. Suppose that $n \neq 1$. Prove that there is some $b \in \mathbb{N}$ such that $b+1 = n$. [Use Exercise 1.4.4.]

Exercise 1.4.7. **[Used in Section 1.4.]** Let $\mathbb{Z}[x]$ denote the set of polynomials with integer coefficients and variable x. This set has binary operations $+$ and $\cdot$ as usual for polynomials. The relation $<$, called the **dictionary order** on $\mathbb{Z}[x]$, is defined by $f < g$ if and only if either the degree of f is less than the degree of g, or if the degrees of f and g are equal and if $f \neq g$ and if the highest degree coefficient which differs for f and g is smaller for f, for all $f, g \in \mathbb{Z}[x]$. Let $0, 1 \in \mathbb{Z}[x]$ be the polynomials that are constantly 0 and 1, respectively.

(1) Prove that $\mathbb{Z}[x]$, with $+$, $\cdot$, $<$, 0 and 1 as defined above, is an ordered integral domain.
(2) Let $f \in \mathbb{Z}[x]$. Prove that there is no $g \in \mathbb{Z}[x]$ such that $f < g < g+1$.
(3) Prove that $\mathbb{Z}[x]$ does not satisfy the Well-Ordering Principle.

Exercise 1.4.8. **[Used in Exercise 1.4.10.]** Let $a \in \mathbb{Z}$.

(1) Let $G \subseteq \{x \in \mathbb{Z} \mid x \geq a\}$ be a set. Suppose that $a \in G$, and that if $g \in G$ then $g+1 \in G$. Prove that $G = \{x \in \mathbb{Z} \mid x \geq a\}$. [Use Exercise 1.4.4.]
(2) Let $H \subseteq \{x \in \mathbb{Z} \mid x \leq a\}$ be a set. Suppose that $a \in H$, and that if $h \in H$ then $h + (-1) \in H$. Prove that $H = \{x \in \mathbb{Z} \mid x \leq a\}$. [Use Exercise 1.4.3.]

Exercise 1.4.9. **[Used in Exercise 1.4.10.]** The two standard approaches to defining the integers are the approaches we took in Section 1.3 (where we started with the Peano Postulates for the natural numbers, and constructed the integers from the natural numbers) and in Section 1.4 (where we took the integers axiomatically as an ordered integral domain that satisfies the Well-Ordering Principle). The purpose of this exercise is to present an additional approach to axiomatizing the integers, which is from [Mar61].

The idea for this alternative axiom for the integers is to modify the Peano Postulates so that there is no "first" natural number, but instead every integer has both a successor and a predecessor; proof by induction is done by "going in both directions" at once. We will not give all the details of the equivalence of this approach with the approaches we have taken, because doing so would be as lengthy as Section 1.2 (and of a similar nature), but we will sketch part of the ideas. In this exercise we will start with the alternative axiom, and go as far as showing that the integers as given by this alternative axiom contain a set, called the natural numbers, that satisfy the Peano Postulates. In Exercise 1.4.10 it will be seen that the integers, as defined in the present section, satisfy the alternative axiom.

The alternative axiom for the integers is that there exists a set $\mathbb{Z}$ and a function $s\colon \mathbb{Z} \to \mathbb{Z}$ that satisfy the following five properties.

a. $\mathbb{Z} \neq \emptyset$.
b. The function s is injective.
c. For each $a \in \mathbb{Z}$, there is some $b \in \mathbb{Z}$ such that $a = s(b)$.
d. Let $G \subseteq \mathbb{Z}$ be a set. Suppose that $G \neq \emptyset$, and that $g \in G$ if and only if $s(g) \in G$. Then $G = \mathbb{Z}$.
e. There is some $Q \subseteq \mathbb{Z}$ such that $Q \neq \emptyset$ and $Q \neq \mathbb{Z}$, and that if $q \in Q$ then $s(q) \in Q$.

The following items help clarify the nature of this alternative axiom.

(1) Find an example to show that Property (e) of these axioms cannot be dropped. That is, find an example of a set W and a function that $s\colon W \to W$ that satisfy all the axioms except Property (e).
(2) Prove that s is bijective.
(3) Prove that Property (d) can be replaced with the following condition: Let $G \subseteq \mathbb{Z}$ be a set. Suppose that $G \neq \emptyset$, and that if $g \in G$ then $s(g) \in G$ and $s^{-1}(g) \in G$. Then $G = \mathbb{Z}$. (This reformulation of Property (d) is more clearly seen to be "induction going in both directions.")
(4) Let $Q \subseteq \mathbb{Z}$ be a set satisfying Property (e). Prove that there is some $r \in \mathbb{Z}$ such that $r \notin Q$ and $s(r) \in Q$.
(5) Let Q and r be as in Part (4) of this exercise. Rename the element r as "0." Let $1 = s(0)$. A subset $A \subseteq \mathbb{Z}$ is **natural** if it has the properties that $0 \notin A$, that $1 \in A$, and that if $a \in A$ then $s(a) \in A$. Observe that there exist natural subsets of $\mathbb{Z}$, for example the set Q. We then define $\mathbb{N} \subseteq \mathbb{Z}$ to be the intersection of all natural subsets of $\mathbb{Z}$. First, prove that $\mathbb{N}$ is a natural subset of $\mathbb{Z}$, and that it is a subset of every other natural subset of $\mathbb{Z}$. Second, prove that $\mathbb{N}$ as so defined satisfies the Peano Postulates (as stated in Theorem 1.4.8, though with the function s given by the axioms in this exercise, and not as given in Theorem 1.4.8).

Exercise 1.4.10. This exercise makes use of Exercise 1.4.9. Prove that $\mathbb{Z}$, as defined in the present section, satisfies the alternative axiom given in Exercise 1.4.9.

[Use Exercise 1.4.8.]

1.5 Constructing the Rational Numbers

Although the integers have nicer properties than the natural numbers (for example, we can take negatives of integers), the integers are still not entirely satisfactory from an algebraic viewpoint, because we cannot divide one integer by another and still expect to obtain an integer. To remedy this problem, we will now construct the rational numbers (which intuitively are the fractions) from the integers to allow for division, or, equivalently, to allow for the existence of multiplicative inverses of integers other than zero. We will, as expected, find a copy of the set of integers inside the set of rational numbers.

For the sake of brevity, we will sometimes use the standard properties of the integers (for example the Commutative Law for Addition) that we have previously encountered in Section 1.3 or 1.4 without always giving explicit references to the relevant theorems and lemmas. The constructions and proofs in this section are very much analogous to the construction of the integers from the natural numbers in Section 1.3. (For the reader who has not read Section 1.3, we note that we do not make any use here of that section, and we mention it only by way of analogy for the reader who has read Section 1.3). In particular, we will make crucial use of equivalence relations and equivalence classes in the construction of the rational numbers from the integers.

We want to think of rational numbers as expressions of the form "$\frac{a}{b}$," where $a, b \in \mathbb{Z}$, and where $b \neq 0$. However, because we do not have the operation division on $\mathbb{Z}$, we will replace "$\frac{a}{b}$" in our construction of the rational numbers with the pair (a, b). It certainly happens that "$\frac{a}{b}$" equals "$\frac{c}{d}$" for some $a, b, c, d \in \mathbb{Z}$, where $a \neq c$ and $b \neq d$, and $b \neq 0$ and $d \neq 0$, and in that case we ought to have the pairs (a, b) and (c, d) represent the same rational numbers. To take care of this problem we define the following relation on pairs of integers.

Definition 1.5.1. Let $\mathbb{Z}^* = \mathbb{Z} - \{0\}$. The relation $\asymp$ on $\mathbb{Z} \times \mathbb{Z}^*$ is defined by $(x, y) \asymp (z, w)$ if and only if $xw = yz$, for all $(x, y), (z, w) \in \mathbb{Z} \times \mathbb{Z}^*$. △

Lemma 1.5.2. *The relation $\asymp$ is an equivalence relation.*

Proof. We will prove transitivity, leaving reflexivity and symmetry to the reader in Exercise 1.5.1. Let $(x, y), (z, w), (u, v) \in \mathbb{Z} \times \mathbb{Z}^*$. Suppose that $(x, y) \asymp (z, w)$ and $(z, w) \asymp (u, v)$. Then $xw = yz$ and $zv = wu$. It follows that $(xw)v = (yz)v$ and $y(zv) = y(wu)$, which implies that $(xv)w = (yz)v$ and $(yz)v = (yu)w$, and hence $(xv)w = (yu)w$. We know that $w \neq 0$, and therefore we deduce that $xv = yu$. It follows that $(x, y) \asymp (u, v)$. Therefore $\asymp$ is transitive. □

The set of rational numbers, together with addition, multiplication, negation, multiplicative inverse and the relations less than and less than or equal to, are given in the following definition. We use the standard symbols $+$, $\cdot$, $-$, $<$ and $\leq$ in the definition, though we need to be careful to note that these symbols formally mean different things when used with the rational numbers and when used with the integers. (Moreover, it is not possible to "add" an integer and a rational number using addition as we have defined it; this problem will be resolved when we see that a copy of the integers sits inside the set of rational numbers.) Recall the definition of binary and unary operations in Section 1.1.

Definition 1.5.3. The set of **rational numbers**, denoted $\mathbb{Q}$, is the set of equivalence classes of $\mathbb{Z} \times \mathbb{Z}^*$ with respect to the equivalence relation $\asymp$.

The elements $\bar{0}, \bar{1} \in \mathbb{Q}$ are defined by $\bar{0} = [(0,1)]$ and $\bar{1} = [(1,1)]$. Let $\mathbb{Q}^* = \mathbb{Q} - \{\bar{0}\}$. The binary operations $+$ and $\cdot$ on $\mathbb{Q}$ are defined by

$$[(x,y)] + [(z,w)] = [(xw + yz, yw)]$$
$$[(x,y)] \cdot [(z,w)] = [(xz, yw)]$$

for all $[(x,y)], [(z,w)] \in \mathbb{Q}$. The unary operation $-$ on $\mathbb{Q}$ is defined by $-[(x,y)] = [(-x,y)]$ for all $[(x,y)] \in \mathbb{Q}$. The unary operation $^{-1}$ on $\mathbb{Q}^*$ is defined by $[(x,y)]^{-1} = [(y,x)]$ for all $[(x,y)] \in \mathbb{Q}^*$. The relation $<$ on $\mathbb{Q}$ is defined by $[(x,y)] < [(z,w)]$ if and only if either $xw < yz$ when $y > 0$ and $w > 0$ or when $y < 0$ and $w < 0$, or $xw > yz$ when $y > 0$ and $w < 0$ or when $y < 0$ and $w > 0$, for all $[(x,y)], [(z,w)] \in \mathbb{Q}$. The relation $\leq$ on $\mathbb{Q}$ is defined by $[(x,y)] \leq [(z,w)]$ if and only if $[(x,y)] < [(z,w)]$ or $[(x,y)] = [(z,w)]$, for all $[(x,y)], [(z,w)] \in \mathbb{Q}$. △

We now need to check whether the binary operations $+$ and $\cdot$, the unary operations $-$ and $^{-1}$ and the relation $<$ are well-defined. For example, we defined $[(x,y)] \cdot [(z,w)]$ to be $[(xz, yw)]$, but for this definition to make sense, we need to verify that if $[(x,y)] = [(p,q)]$ and $[(z,w)] = [(s,t)]$, then $[(xz,yw)] = [(ps,qt)]$. That is, we need to verify that the product of the equivalence classes $[(x,y)]$ and $[(z,w)]$ depends only upon the equivalence classes themselves, and not upon the particular elements that are used to represent the equivalence classes. As seen in the following lemma, everything works out well. (We do not deal with the relation $\leq$ in the lemma, because that is defined in terms of $<$.)

Lemma 1.5.4. *The binary operations $+$ and $\cdot$, the unary operations $-$ and $^{-1}$, and the relation $<$, all on $\mathbb{Q}$, are well-defined.*

Proof. We will show that $\cdot$ and $-$ are well-defined, leaving the rest to the reader in Exercise 1.5.2.

Let $(x,y), (z,w), (a,b), (c,d) \in \mathbb{Z} \times \mathbb{Z}^*$. Suppose that $[(x,y)] = [(a,b)]$ and that $[(z,w)] = [(c,d)]$.

By hypothesis we know that $(x,y) \asymp (a,b)$ and $(z,w) \asymp (c,d)$. Hence $xb = ya$ and $zd = wc$. By multiplying these two equations and doing some rearranging we obtain $(xz)(bd) = (yw)(ac)$, and this implies that $[(xz,yw)] = [(ac,bd)]$. Therefore $\cdot$ is well-defined. Also, from $xb = ya$ we deduce that $(-x)b = y(-a)$, and hence $[(-x,y)] = [(-a,b)]$. Therefore $-$ is well-defined. □

The following theorem states the most fundamental algebraic properties of the rational numbers. The idea of the proof of the following theorem is to rephrase things in terms of integers, and then use the appropriate facts we have already seen about the integers. As is usual, we will write "rs" instead of "$r \cdot s$," except in cases of potential ambiguity, or for ease of reading. We will write $r > s$ to mean the same thing as $s < r$, and similarly for $r \geq s$.

Theorem 1.5.5. *Let $r,s,t \in \mathbb{Q}$.*

1. $(r+s)+t = r+(s+t)$ *(Associative Law for Addition).*
2. $r+s = s+r$ *(Commutative Law for Addition).*
3. $r+\bar{0} = r$ *(Identity Law for Addition).*
4. $r+(-r) = \bar{0}$ *(Inverses Law for Addition).*
5. $(rs)t = r(st)$ *(Associative Law for Multiplication).*
6. $rs = sr$ *(Commutative Law for Multiplication).*
7. $r \cdot \bar{1} = r$ *(Identity Law for Multiplication).*
8. *If $r \neq \bar{0}$, then $r \cdot r^{-1} = \bar{1}$ (Inverses Law for Multiplication).*
9. $r(s+t) = rs+rt$ *(Distributive Law).*
10. *Precisely one of $r < s$ or $r = s$ or $r > s$ holds (Trichotomy Law).*
11. *If $r < s$ and $s < t$, then $r < t$ (Transitive Law).*
12. *If $r < s$ then $r+t < s+t$ (Addition Law for Order).*
13. *If $r < s$ and $t > \bar{0}$, then $rt < st$ (Multiplication Law for Order).*
14. $\bar{0} \neq \bar{1}$ *(Non-Triviality).*

Proof. We will prove Parts (4), (7), (10) and (13), leaving the rest to the reader in Exercise 1.5.4. Suppose that $r = [(x,y)]$, that $s = [(z,w)]$ and that $t = [(u,v)]$ for some $x,z,u \in \mathbb{Z}$ and $y,w,v \in \mathbb{Z}^*$. Throughout this proof we will make use of properties of the integers that we saw in Section 1.3 or 1.4.

(4) By the definition of negation and addition of rational numbers we see that $r + (-r) = [(x,y)] + [(-x,y)] = [(xy+(-x)y, yy)] = [(0,yy)] = \bar{0}$, where the last equality holds by Exercise 1.5.3 (1).

(7) Using the definition of $\bar{1}$ and multiplication of rational numbers we see that $r \cdot \bar{1} = [(x,y)] \cdot [(1,1)] = [(x \cdot 1, y \cdot 1)] = [(x,y)] = r$.

(10) We need to show that precisely one of $r < s$, or $r = s$ or $r > s$ holds. There are two cases.

First, suppose that $y > 0$ and $w > 0$, or that $y < 0$ and $w < 0$. Then the Trichotomy Law for the integers states that precisely one of $yz < xw$ or $yz = xw$ or $yz > xw$ holds. If $yz < xw$ then $r > s$ by the definition of $<$ for rational numbers. If $yz = xw$, then $r = s$ by the definition of the relation $\asymp$. If $yz > xw$ then $r < s$ by the definition of $<$ for rational numbers. The second case, where $y > 0$ and $w < 0$, or where $y < 0$ and $w > 0$, is similar to the first case, and we omit the details.

(13) Suppose that $r < s$ and $t > \bar{0}$.

There are now four cases. First, suppose that $y > 0$ and $w > 0$, or that $y < 0$ and $w < 0$; and suppose that $v > 0$. By the definition of $<$ for rational numbers, the inequality $t > \bar{0}$ implies $v \cdot 0 < u \cdot 1$, which implies $0 < u$. Hence $uv > 0$. Again using the definition of $<$ for rational numbers, the inequality $r < s$ implies $yz > xw$. Therefore $(yz)(uv) > (xw)(uv)$, and hence $(zu)(yv) > (xu)(wv)$. Because $v > 0$, then $y > 0$ and $w > 0$ imply $yv > 0$ and $wv > 0$, and $y < 0$ and $w < 0$ imply $yv < 0$ and $wv < 0$. We then deduce from the definition of $<$ for rational numbers that $[(xu,yv)] < [(zu,wv)]$. However, by the definition of multiplication of rational numbers we know that $rt = [(x,y)] \cdot [(u,v)] = [(xu,yv)]$ and $st = [(z,w)] \cdot [(u,v)] = [(zu,wv)]$, and that completes the first case.

The other three cases, which depend upon whether each of y, w and v is positive or negative, are similar to the first case, and we omit the details. □

The properties of the rational numbers listed in Theorem 1.5.5 are satisfied by other number systems as well (for example the real numbers), and are sufficiently important to merit special terminology. Any set with two binary operations that satisfies Parts (1)–(9) is called a "field," and any field that also has a relation that satisfies Parts (10)–(14) of these theorems is called an "ordered field." Fields are studied extensively in abstract algebra.

Although many of the properties of the rational numbers are similar to properties of the integers, we note that there is no analog for the rational numbers of Theorem 1.3.9 or Theorem 1.4.6. Indeed, it is seen in Exercise 1.5.7 that between any two rational numbers there is another rational number.

Because of the way we constructed the rational numbers from the integers, these two sets are entirely disjoint. However, we can find a copy of the set of integers inside the set of rational numbers by identifying each integer x with the rational number $[(x,1)]$, where we think informally of $[(x,1)]$ as representing the fraction $\frac{x}{1}$. We will see in the following theorem that this identification preserves the numbers 0 and 1, the binary operations addition and multiplication and the relation less than.

Theorem 1.5.6. *Let* $i\colon \mathbb{Z} \to \mathbb{Q}$ *be defined by* $i(x) = [(x,1)]$ *for all* $x \in \mathbb{Z}$.

1. *The function* $i\colon \mathbb{Z} \to \mathbb{Q}$ *is injective.*
2. $i(0) = \bar{0}$ *and* $i(1) = \bar{1}$.
3. *Let* $x, y \in \mathbb{Z}$. *Then*
   a. $i(x+y) = i(x) + i(y)$;
   b. $i(-x) = -i(x)$;
   c. $i(xy) = i(x)i(y)$;
   d. $x < y$ *if and only if* $i(x) < i(y)$.
4. *For each* $r \in \mathbb{Q}$ *there are* $x, y \in \mathbb{Z}$ *such that* $y \neq 0$ *and* $r = i(x)(i(y))^{-1}$.

Proof. We prove Part (4), leaving the rest to the reader in Exercise 1.5.5.

(4) Let $r \in \mathbb{Q}$. Then $r = [(x,y)]$ for some $(x,y) \in \mathbb{Z} \times \mathbb{Z}^*$. We then have $i(x) = [(x,1)]$ and $i(y) = [(y,1)]$. Because $y \neq 0$, we see that $[(y,1)] \neq \bar{0}$ by Exercise 1.5.3 (1). Hence $i(y) \in \mathbb{Q}^*$. It then follows from Definition 1.5.3 that $i(x)(i(y))^{-1} = [(x,1)] \cdot [(y,1)]^{-1} = [(x,1)] \cdot [(1,y)] = [(x \cdot 1, 1 \cdot y)] = [(x,y)] = r$. □

Even though technically the set of integers and the set of rational numbers are entirely disjoint, we see from Theorem 1.5.6 that from the point of view of addition, multiplication, negation and the relation less than, the integers can be identified with a subset of the rational numbers. We can therefore dispense with the set of integers as a separate entity (except when we need it in proofs), because we have a copy of the integers inside the rational numbers that works just as well as the original. We will therefore also dispense with the notation $\bar{0}$ and $\bar{1}$, and simply write 0 and 1 instead.

Everything that was proved about the natural numbers and the integers in previous sections, and which involve only addition, multiplication and the relation less than, still hold true when we think of the natural numbers and the integers as part of the set of rational numbers.

We can draw another useful conclusion from Theorem 1.5.6 as well, but first we need the following definition.

Definition 1.5.7. The binary operation $-$ on $\mathbb{Q}$ is defined by $r - s = r + (-s)$ for all $r, s \in \mathbb{Q}$. The binary operation $\div$ on $\mathbb{Q}^*$ is defined by $r \div s = rs^{-1}$ for all $r, s \in \mathbb{Q}^*$; we also let $0 \div s = 0 \cdot s^{-1} = 0$ for all $s \in \mathbb{Q}^*$. The number $r \div s$ is also denoted $\frac{r}{s}$. △

Observe that if $b \in \mathbb{Q}$ then $0 - b = -b$, and if $b \neq 0$ then $\frac{1}{b} = b^{-1}$.

By thinking of the integers as sitting inside the rational numbers, the function i in Theorem 1.5.6 then becomes the function that takes each integer to itself. We can then combine Definition 1.5.7 with Theorem 1.5.6 (4) to see that we now see that for each $r \in \mathbb{Q}$ there are $a, b \in \mathbb{Z}$ such that $b \neq 0$ and $r = \frac{a}{b}$. We can think of each integer $n \in \mathbb{Z}$ as being identified with the fraction $\frac{n}{1}$. Hence, we have now come full circle in our discussion of the rational numbers, and we have recovered our original intuitive notion of rational numbers as fraction. Moreover, by the proof of Theorem 1.5.6 (4) we see that the rational number $\frac{a}{b}$ is the same as the rational number $[(a,b)]$. We can then reformulate Definition 1.5.3 in the following lemma, which we state without proof, because it is just a matter of translating from one notation into another.

Lemma 1.5.8. *Let $a, c \in \mathbb{Z}$ and $b, d \in \mathbb{Z}^*$.*

1. $\frac{a}{b} = \frac{c}{d}$ *if and only if* $ad = bc$.
2. $\frac{a}{b} + \frac{c}{d} = \frac{ad+bc}{bd}$.
3. $-\frac{a}{b} = \frac{-a}{b}$.
4. $\frac{a}{b} \cdot \frac{c}{d} = \frac{ac}{bd}$.
5. *If* $a \neq 0$, *then* $\left(\frac{a}{b}\right)^{-1} = \frac{b}{a}$.
6. *If* $b > 0$ *and* $d > 0$, *or if* $b < 0$ *and* $d < 0$, *then* $\frac{a}{b} < \frac{c}{d}$ *if and only if* $ad < bc$; *if* $b > 0$ *and* $d < 0$, *or if* $b < 0$ *and* $d > 0$, *then* $\frac{a}{b} < \frac{c}{d}$ *if and only if* $ad > bc$.

We can now comfortably use fractions exactly as we learned to use them in elementary school, though with the knowledge that there is a completely rigorous foundation for all that we learned.

Reflections

The rational numbers do not always get the respect they deserve in real analysis courses. In the present chapter, where the real numbers are constructed via a process that starts with an axiomatic treatment of either the natural numbers or the integers, the rational numbers are sometimes viewed as merely a stepping stone on the way to the construction of the real numbers. In Chapter 2, where we start with an axiomatic treatment of the real numbers, the rational numbers are sometimes viewed as merely a subset of the real numbers that does not behave as nicely as the set of all real numbers.

In fact, the rational numbers are extremely important from a variety of perspectives. From the point of view of applications, the rational numbers are very useful for many computations in the real world; indeed, in spite of the availability of modern calculators and computers, it would be a mistake to assume that fractions are no longer an important topic for school mathematics. From a theoretical point of view, the rational numbers play a very important role in abstract algebra and number theory. Even for real analysis the rational numbers are important from a pedagogical point of view. Much of what we do in real analysis makes crucial use of the Least Upper Bound Property of the real numbers, a property not satisfied by the rational numbers. In order to appreciate the role of this property in real analysis, it will be useful for the reader to ask herself whether or not various aspects of real analysis, to be encountered subsequently in this text, also work for the rational numbers.

Exercises

Exercise 1.5.1. [Used in Lemma 1.5.2.] Complete the proof of Lemma 1.5.2. That is, prove that the relation $\asymp$ is reflexive and symmetric.

Exercise 1.5.2. [Used in Lemma 1.5.4.] Complete the proof of Lemma 1.5.4. That is, prove that the binary operation $+$, the unary operation^{-1} and the relation $<$, all on $\mathbb{Q}$, are well-defined.

Exercise 1.5.3. [Used in Theorem 1.5.5 and Theorem 1.5.6.] Let $x \in \mathbb{Z}$ and $y \in \mathbb{Z}^*$.

(1) Prove that $[(x,y)] = \bar{0}$ if and only if $x = 0$.
(2) Prove that $[(x,y)] = \bar{1}$ if and only if $x = y$.
(3) Prove that $\bar{0} < [(x,y)]$ if and only if $0 < xy$.

Exercise 1.5.4. [Used in Theorem 1.5.5.] Prove Theorem 1.5.5 (1) (2) (3) (5) (6) (8) (9) (11) (12) (14).

Exercise 1.5.5. [Used in Theorem 1.5.6.] Prove Theorem 1.5.6 (1) (2) (3).

Exercise 1.5.6. [Used in Exercise 1.6.2, Theorem 1.7.6 and Exercise 1.7.3.] Let $r, s, p, q \in \mathbb{Q}$.

(1) Prove that $-1 < 0 < 1$.
(2) Prove that if $r < s$ then $-s < -r$.
(3) Prove that $r \cdot 0 = 0$.
(4) Prove that if $r > 0$ and $s > 0$, then $r + s > 0$ and $rs > 0$.
(5) Prove that if $r > 0$, then $\frac{1}{r} > 0$.
(6) Prove that if $0 < r < s$, then $\frac{1}{s} < \frac{1}{r}$.
(7) Prove that if $0 < r < p$ and $0 < s < q$, then $rs < pq$.

Exercise 1.5.7. [Used in Lemma 1.6.2, Lemma 1.6.8 and Exercise 1.6.2.]

(1) Prove that $1 < 2$.
(2) Let $s, t \in \mathbb{Q}$. Suppose that $s < t$. Prove that $\frac{s+t}{2} \in \mathbb{Q}$, and that $s < \frac{s+t}{2} < t$.

Exercise 1.5.8. Let $r \in \mathbb{Q}$. Suppose that $r > 0$.

(1) Prove that if $r = \frac{a}{b}$ for some $a, b \in \mathbb{Z}$ such that $b \neq 0$, then either $a > 0$ and $b > 0$, or $a < 0$ and $b < 0$.

(2) Prove that $r = \frac{m}{n}$ for some $m, n \in \mathbb{Z}$ such that $m > 0$ and $n > 0$.

Exercise 1.5.9. [Used in Lemma 1.6.9 and Exercise 1.6.2.] Let $r, s \in \mathbb{Q}$.

(1) Suppose that $r > 0$ and $s > 0$. Prove that there is some $n \in \mathbb{N}$ such that $s < nr$.
[Use Exercise 1.5.6, Exercise 1.5.8, and either Exercise 1.3.8 or Exercise 1.4.4.]

(2) Suppose that $r > 0$. Prove that there is some $m \in \mathbb{N}$ such that $\frac{1}{m} < r$.

(3) For each $x \in \mathbb{Q}$, let x^2 denote $x \cdot x$.
Suppose that $r > 0$ and $s > 0$. Prove that if $r^2 < p$, then there is some $k \in \mathbb{N}$ such that $\left(r + \frac{1}{k}\right)^2 < p$.

1.6 Dedekind Cuts

The rational numbers work very well from the point of view of addition and multiplication, because of the existence of negatives and multiplicative inverses (of non-zero rational numbers), but they are still not satisfactory for doing real analysis.

The most obvious flaw of the rational numbers is that it is not possible to solve—in the set of rational numbers—some polynomial equations with rational coefficients, for example $x^2 - 2 = 0$; we will see a proof of this fact in Theorem 2.6.11. By contrast, as we will see in Theorem 2.6.9, the equation $x^2 - 2 = 0$ does have a solution in the set of real numbers. Hence, the real numbers, to be constructed in Section 1.7 after preliminaries in the present section, are an improvement upon the rational numbers from the point of view of solving polynomial equations. It should be noted, however, that although the real numbers are better than the rational numbers for solving polynomial equations, the real numbers are also not entirely satisfactory in this regard, because there are some polynomial equations with rational coefficients, for example $x^2 + 2 = 0$, that have no solution in the real numbers. (This equation, and indeed all polynomial equations with rational coefficients, have solutions in the complex numbers. A very important theorem, called the Fundamental Theorem of Algebra, states that any polynomial with complex coefficients (which includes all rational and real coefficients) has a root in the complex numbers. Most introductory complex analysis texts include a proof of the Fundamental Theorem of Algebra; see [BC09, Section 53]. We will not discuss complex numbers in this text. For a proof of the Fundamental Theorem of Algebra using topology rather than complex analysis, see [Mun00, Section 56].)

As important as it is to be able to solve polynomial equations, however, that is not the difference between the rational numbers and the real numbers that makes the real numbers an appropriate place to do analysis and the rational numbers inappropriate for analysis. Notice, for example, that the rational numbers are missing not only numbers such as $\sqrt{2}$, which is the root of a polynomial with rational coefficients, but also very important numbers such as π and e, which are not the roots of any polynomial with rational coefficients, and are called "transcendental numbers." See [Ste04, Sections 24.2 and 24.3] for proofs that π and e are transcendental, and see [Jac85, Section 4.12] for a more general result about transcendental numbers that includes both π and e.

The essential difference between the rational numbers and the real numbers is that the real numbers are "complete" and the rational numbers are not. By complete we mean intuitively that there are no "gaps" between sets of numbers that ought not to have gaps between them. In the rational numbers, by contrast, there is, intuitively, a gap where each of the numbers such as $\sqrt{2}$, π and e ought to be, but are not. It is these sorts of gaps that make the rational numbers unfit for real analysis, because some important theorems in real analysis, for example the Extreme Value Theorem and the Intermediate Value Theorem (both proved in Section 3.5), do not work when there are gaps where there ought to be numbers. Technically, the idea of completeness for the real numbers can be formulated in two ways, one involving least upper bounds, and the other involving Cauchy sequences. The Least Upper Bound Property of the real numbers, which we will use directly and indirectly throughout this text, is first discussed in Section 1.7, and is discussed in greater detail in Section 2.6. Cauchy sequences are discussed in Section 8.3, though to ascertain their full importance as a tool for determining completeness the reader will have to wait until she encounters topics such as metric spaces that are beyond the scope of this book.

Before we can discuss the Least Upper Bound Property of the real numbers, we need to have the set of real numbers, and our present task is to construct the real numbers from the rational numbers. This construction is somewhat more complicated than the construction of the integers from the natural numbers (in Section 1.3) or the construction of the rational numbers from the integers (in Section 1.5), and we will need the present section just for preliminaries, to be followed by the actual definition of the real numbers in Section 1.7.

There are two standardly used approaches for constructing the real numbers from the rational numbers. One of the methods also involves Cauchy sequences; more precisely, it uses equivalence classes of Cauchy sequences of rational numbers. The other method, which is the approach we take, uses Dedekind cuts, to be defined shortly; this method does not involve equivalence classes. The Cauchy sequence method is quicker, but it involves the extra burden of learning about Cauchy sequences at this point in the development of the material; the proofs in the Dedekind cut method are lengthier, but have the advantage of involving nothing beyond what we have seen so far about the rational numbers. The original treatment of Dedekind cuts is in [Ded63]; our discussion of Dedekind cuts, using a slight variant of the original approach, follows [Rud53], [Men73] and [Bur67]. See [Sto79, Sections 3.5 and 3.6] or [Str00, Chapter 2] for the Cauchy sequence construction of the real numbers from the rational numbers.

The intuitive idea behind Dedekind cuts is simple, even though their formal definition might appear somewhat technical at first. We want to use the rational numbers to construct the real numbers, and the key observation is that every real number can be characterized by the set of rational numbers that are greater than it. For example, the real number 1 is characterized by the set $\{x \in \mathbb{Q} \mid x > 1\}$. Of course, the number 1 is a rational number, and so the set $\{x \in \mathbb{Q} \mid x > 1\}$ is defined in the realm of rational numbers, which is all we currently have at our disposal. On the other hand, the set $\{x \in \mathbb{Q} \mid x > \sqrt{2}\}$ is not describable using only the rational numbers, because $\sqrt{2}$ is not a rational number. Instead, we define Dedekind cuts to be subsets of $\mathbb{Q}$ that behave as sets of the form $\{x \in \mathbb{Q} \mid x > r\}$ for real numbers r ought to behave, while using a definition that is strictly in terms of the rational numbers. After proving some properties of Dedekind cuts in the present section, we will, in Section 1.7, define the set of real number to be the collection of all Dedekind cuts of $\mathbb{Q}$.

In order to focus on the key ideas in our discussion of Dedekind cuts, and in order to keep the proofs from being any longer than necessary, we will not cite references to the standard (and very familiar) properties of the rational numbers proved in Section 1.5.

Definition 1.6.1. Let $A \subseteq \mathbb{Q}$ be a set. The set A is a **Dedekind cut** if the following three properties hold.

a. $A \neq \emptyset$ and $A \neq \mathbb{Q}$.
b. Let $x \in A$. If $y \in \mathbb{Q}$ and $y \geq x$, then $y \in A$.
c. Let $x \in A$. Then there is some $y \in A$ such that $y < x$. △

The definition of Dedekind cuts is rather abstract, and before proving things about them, we need to verify that they actually exist. As seen in the following lemma, Dedekind cuts are in fact plentiful.

Lemma 1.6.2. *Let $r \in \mathbb{Q}$. Then the set*

$$\{x \in \mathbb{Q} \mid x > r\}$$

is a Dedekind cut.

Proof. Let $D = \{x \in \mathbb{Q} \mid x > r\}$. We will show that D satisfies the three parts of the definition of Dedekind cuts.

(a) We know that $r - 1, r + 1 \in \mathbb{Q}$, and that $r - 1 < r < r + 1$. Hence $r - 1 \notin D$ and $r + 1 \in D$, and therefore $D \neq \emptyset$ and $D \neq \mathbb{Q}$.

(b) Let $m \in D$, and let $y \in \mathbb{Q}$. Suppose that $y > m$. Because $m > r$, it follows that $y > r$. Hence $y \in D$.

(c) Let $n \in D$. Then $n > r$. It follows from Exercise 1.5.7 (2) that $\frac{n+r}{2} \in \mathbb{Q}$, and $r < \frac{n+r}{2} < n$. Hence $\frac{n+r}{2} \in D$. □

Because of Lemma 1.6.2 we know that Dedekind cuts exist, and there are at least as many of them as there are rational numbers. The natural question to ask next is whether all Dedekind cuts are of the form given in Lemma 1.6.2. Of course, if all Dedekind cuts had that form, then it would have been silly to have defined Dedekind cuts in the abstract way that we did, and so the reader might well guess, as will indeed be seen in the following example, that not all Dedekind cuts have the form given in Lemma 1.6.2.

Example 1.6.3. To find a Dedekind cut that is not of the form given in Lemma 1.6.2, the simplest idea is to look at the set of all rational numbers greater than a real number that is not rational; the problem is how to describe such a set without making use of anything other than the rational numbers. In the case of the number $r = \sqrt{2}$, there turns out to be a simple solution to this problem, as we will now see. We note first, however, that we have not yet formally defined what "$\sqrt{2}$" means, nor proved that there is such a real number, though we will do so in Theorem 2.6.9 and Definition 2.6.10. We have also not yet proved that "$\sqrt{2}$" is not rational, a fact with which the reader is, at least informally, familiar; we will see a proof of this fact in Theorem 2.6.11. More precisely, it will be seen in that example that there is no rational number x such that $x^2 = 2$, and this last statement makes use only of rational numbers, so it is suited to our purpose at present. Nothing in our subsequent treatment of "$\sqrt{2}$" in Section 2.6 makes use of the current example, so it will not be circular reasoning for us to make use of these subsequently proved facts here.

Let

$$T = \{x \in \mathbb{Q} \mid x > 0 \text{ and } x^2 > 2\}. \tag{1.6.1}$$

It is seen by Exercise 1.6.2 (1) that T is a Dedekind cut, and by Part (2) of that exercise it is seen that if T has the form $\{x \in \mathbb{Q} \mid x > r\}$ for some $r \in \mathbb{Q}$, then $r^2 = 2$. By Theorem 2.6.11 we know that there is no rational number x such that $x^2 = 2$, and it follows that T is a Dedekind cut that is not of the form given in Lemma 1.6.2. ◊

Example 1.6.3 explains the need for the following definition.

Definition 1.6.4. Let $r \in \mathbb{Q}$. The **rational cut** at r, denoted D_r, is the Dedekind cut

$$D_r = \{x \in \mathbb{Q} \mid x > r\}.$$

An **irrational cut** is a Dedekind cut that is not a rational cut at any rational number. △

Before using Dedekind cuts to form the set of real numbers in Section 1.7, we will take the remainder of the present section to prove some technically useful properties of Dedekind cuts, starting with the following simple lemma that will be used frequently.

Lemma 1.6.5. *Let $A \subseteq \mathbb{Q}$ be a Dedekind cut.*

1. $\mathbb{Q} - A = \{x \in \mathbb{Q} \mid x < a \text{ for all } a \in A\}$.
2. *Let $x \in \mathbb{Q} - A$. If $y \in \mathbb{Q}$ and $y \le x$, then $y \in \mathbb{Q} - A$.*

Proof.

(1) Let $x \in \mathbb{Q} - A$. Let $a \in A$. We know that $x < a$ or $x = a$ or $x > a$. If $x = a$, then x would be in A by the definition of a, which is a contradiction. If $x > a$, then x would be in A by Part (b) of the definition of Dedekind cuts, which is a contradiction. Hence $x < a$. We conclude that $\mathbb{Q} - A \subseteq \{x \in \mathbb{Q} \mid x < a \text{ for all } a \in A\}$.

Now let $y \in \{x \in \mathbb{Q} \mid x < a \text{ for all } a \in A\}$. If $y \in A$, we would then have $y < y$, which is a contradiction. Hence $y \in \mathbb{Q} - A$. We deduce that $\mathbb{Q} - A \supseteq \{x \in \mathbb{Q} \mid x < a \text{ for all } a \in A\}$. Therefore $\mathbb{Q} - A = \{x \in \mathbb{Q} \mid x < a \text{ for all } a \in A\}$.

(2) This part follows easily from Part (1); the details are left to the reader. □

Our next two lemmas about Dedekind cuts are straightforward.

Lemma 1.6.6. *Let $A, B \subseteq \mathbb{Q}$ be Dedekind cuts. Then precisely one of $A \subsetneqq B$ or $A = B$ or $B \subsetneqq A$ holds.*

Proof. If $A = B$ there is nothing to prove, so assume that $A \neq B$. There are now two cases. First, suppose that there is some $a \in A$ such that $a \in \mathbb{Q} - B$. Then by Lemma 1.6.5 (1) we know that $a < b$ for all $b \in B$. By Part (b) of the definition of Dedekind cuts it follows that $b \in A$ for all $b \in B$. Hence $B \subseteq A$. Because we are assuming that $B \neq A$, then $B \subsetneqq A$. The second case is that there is some $d \in B$ such that $d \in \mathbb{Q} - A$, and a similar argument shows that $A \subsetneqq B$; we omit the details. □

Lemma 1.6.7. *Let A be a non-empty family of subsets of $\mathbb{Q}$. Suppose that X is a Dedekind cut for all $X \in A$. If $\bigcup_{X \in A} X \neq \mathbb{Q}$, then $\bigcup_{X \in A} X$ is a Dedekind cut.*

Proof. Let $B = \bigcup_{X \in A} X$. Assume that $B \neq \mathbb{Q}$. We will show that B satisfies the three parts of the definition of Dedekind cuts.

(a) We know that $X \neq \emptyset$ for all $X \in A$, so $B \neq \emptyset$. By hypothesis we know that $B \neq \mathbb{Q}$.

(b) Let $b \in B$, and let $y \in \mathbb{Q}$. Suppose that $y \geq b$. We know that $b \in X$ for some $X \in A$. By Part (b) of the definition of Dedekind cuts applied to X, we see that $y \in X$. Hence $y \in B$.

(c) Let $c \in B$. Then $c \in D$ for some $D \in A$. By Part (c) of the definition of Dedekind cuts there is some $z \in D$ such that $z < c$. Hence $y \in B$. □

The following lemma about Dedekind cuts is somewhat technical, and has a slightly tedious proof, but it will be needed to define addition, multiplication, negation and multiplicative inverse for the real numbers in Section 1.7.

Lemma 1.6.8. *Let $A, B \subseteq \mathbb{Q}$ be Dedekind cuts.*

1. *The set*
$$\{r \in \mathbb{Q} \mid r = a + b \text{ for some } a \in A \text{ and } b \in B\}$$
is a Dedekind cut.
2. *The set*
$$\{r \in \mathbb{Q} \mid -r < c \text{ for some } c \in \mathbb{Q} - A\}$$
is a Dedekind cut.
3. *Suppose that $0 \in \mathbb{Q} - A$ and $0 \in \mathbb{Q} - B$. The set*
$$\{r \in \mathbb{Q} \mid r = ab \text{ for some } a \in A \text{ and } b \in B\}$$
is a Dedekind cut.
4. *Suppose that there is some $q \in \mathbb{Q} - A$ such that $q > 0$. The set*
$$\{r \in \mathbb{Q} \mid r > 0 \text{ and } \frac{1}{r} < c \text{ for some } c \in \mathbb{Q} - A\}$$
is a Dedekind cut.

Proof. We will prove Parts (1), (2) and (4), leaving the remaining part to the reader in Exercise 1.6.3.

(1) Let

$$M = \{r \in \mathbb{Q} \mid r = a + b \text{ for some } a \in A \text{ and } b \in B\}.$$

We will show that M satisfies the three parts of the definition of Dedekind cuts.

(a) We know that $A \neq \emptyset$ and $A \neq \mathbb{Q}$, and $B \neq \emptyset$ and $B \neq \mathbb{Q}$. Let $x \in A$, let $p \in \mathbb{Q} - A$, let $y \in B$ and let $q \in \mathbb{Q} - B$. Then $x + y \in M$, so $M \neq \emptyset$. We know by Lemma 1.6.5 (1) that $p < a$ for all $a \in A$ and $q < b$ for all $b \in B$. It follows that $p + q < a + b$ for all $a \in A$ and $b \in B$. Hence $p + q \in \mathbb{Q} - M$, and so $M \neq \mathbb{Q}$.

(b) Let $c \in M$, and let $y \in \mathbb{Q}$. Suppose that $y \geq c$. We know that $c = a + b$ for some $a \in A$ and $b \in B$. Then $y = [a + (y - c)] + b$. Because $y \geq c$, then $a + (y - c) \geq a$, and hence by Part (b) of the definition of Dedekind cuts we see that $a + (y - c) \in A$. It follows that $y \in M$.

(c) Let $d \in M$. We know that $d = s + t$ for some $s \in A$ and $t \in B$. Applying Part (c) of the definition of Dedekind cuts to A, we see that there is some $g \in A$ such that $g < s$. Then $g + t \in M$, and $g + t < s + t = d$.

(2) Let

$$N = \{r \in \mathbb{Q} \mid -r < c \text{ for some } c \in \mathbb{Q} - A\}.$$

We will show that N satisfies the three parts of the definition of Dedekind cuts.

(a) We know that $A \neq \emptyset$ and $A \neq \mathbb{Q}$. Let $b \in \mathbb{Q} - A$. By Lemma 1.6.5 (2) we deduce that $b - 1 \in \mathbb{Q} - A$. Then $-(b - 1) \in N$. Hence $N \neq \emptyset$. Next, let $a \in A$. Then $-(-a) \notin \mathbb{Q} - A$, and therefore by Lemma 1.6.5 (2) we know that $-(-a) \not< g$ for all $g \in \mathbb{Q} - A$. Hence $-a \in \mathbb{Q} - N$, and so $N \neq \mathbb{Q}$.

(b) Let $d \in N$, and let $y \in \mathbb{Q}$. Suppose that $y \geq d$. It follows that $-d \geq -y$. By the definition of N, we know that $-d < c$ for some $c \in \mathbb{Q} - A$. Then $-y < c$, and therefore $y \in N$.

(c) Let $e \in N$. Then $-e < c$ for some $c \in \mathbb{Q} - A$. Hence $e > -c$. Let $s = \frac{e + (-c)}{2}$. It follows from Exercise 1.5.7 (2) that $s \in \mathbb{Q}$, and $-c < s < e$. Hence $-s < c$. It follows that $s \in N$.

(4) Because $q \in \mathbb{Q} - A$ and $q > 0$, it follows from Lemma 1.6.5 (1) that $0 < q < a$ for all $a \in A$. Let

$$R = \{r \in \mathbb{Q} \mid r > 0 \text{ and } \frac{1}{r} < c \text{ for some } c \in \mathbb{Q} - A\}.$$

We will show that R satisfies the three parts of the definition of Dedekind cuts.

(a) Clearly $0 \notin R$, so $R \neq \mathbb{Q}$. By Exercise 1.5.7 (2) we see that $0 < \frac{q}{2} < q$. Then $\frac{2}{q} = \left(\frac{q}{2}\right)^{-1} \in R$, and hence $R \neq \emptyset$.

(b) Let $w \in R$, and let $y \in \mathbb{Q}$. Suppose that $y \geq w$. We know $w > 0$, and hence $y > 0$. It follows that $\frac{1}{y} \leq \frac{1}{w}$. By the definition of R, we know that $\frac{1}{w} < c$ for some $c \in \mathbb{Q} - A$. Then $\frac{1}{y} < c$, and hence $y \in R$.

(c) Let $m \in R$. Therefore $m > 0$, and $\frac{1}{m} \in \mathbb{Q} - A$, and $\frac{1}{m} < c$ for some $c \in \mathbb{Q} - A$. Hence $0 < \frac{1}{c} < m$. Let $t = \frac{m+\frac{1}{c}}{2}$. It follows from Exercise 1.5.7 (2) that $t \in \mathbb{Q}$ and $0 < \frac{1}{c} < t < m$. Therefore $0 < \frac{1}{t} < c$, and hence $t \in R$. □

Our final lemma of this section might appear somewhat unmotivated, but it will be used in the proof of Theorem 1.7.6, which states that the real numbers satisfy some basic algebraic properties. The proof of the following lemma is not lengthy, but it requires a powerful tool, namely, the Well-Ordering Principle (Theorem 1.2.10 or Axiom 1.4.4).

Lemma 1.6.9. *Let $A \subseteq \mathbb{Q}$ be a Dedekind cut. Let $y \in \mathbb{Q}$.*

1. *Suppose that $y > 0$. Then there are $u \in A$ and $v \in \mathbb{Q} - A$ such that $y = u - v$, and $v < e$ for some $e \in \mathbb{Q} - A$.*
2. *Suppose that $y > 1$, and that there is some $q \in \mathbb{Q} - A$ such that $q > 0$. Then there are $r \in A$ and $s \in \mathbb{Q} - A$ such that $s > 0$, and $y > \frac{r}{s}$, and $s < g$ for some $g \in \mathbb{Q} - A$.*

Proof.

(1) We follow [Rud53]. Because $A \neq \emptyset$ and $A \neq \mathbb{Q}$ there are $w \in A$ and $z \in \mathbb{Q} - A$. By Lemma 1.6.5 (1) we know that $z < w$. Because $w - z > 0$ and $y > 0$, it follows from Exercise 1.5.9 (1) that there is some $n \in \mathbb{N}$ such that $w - z < ny$. Hence $w + n(-y) < z$. By Lemma 1.6.5 (2) we deduce that $w + n(-y) \in \mathbb{Q} - A$.

Let
$$G = \{p \in \mathbb{N} \mid w + p(-y) \in \mathbb{Q} - A\}.$$
We know that $n \in G$, and so $G \neq \emptyset$. By the Well-Ordering Principle (Theorem 1.2.10 or Axiom 1.4.4), we know that there is some $m \in G$ such that $m \leq g$ for all $g \in G$. It follows that $w + m(-y) \in \mathbb{Q} - A$ and that $w + (m-1)(-y) \in A$.

There are now two cases. First, suppose that there is some $e \in \mathbb{Q} - A$ such that $w + m(-y) < e$. We then let $v = w + m(-y)$ and $u = w + (m-1)(-y)$. Clearly $y = u - v$, and $v < e$. Second, suppose that $w + m(-y) \geq q$ for all $q \in \mathbb{Q} - A$. Because $w + \left(m - \frac{1}{2}\right)(-y) \geq w + m(-y)$ then $w + (m - \frac{1}{2})(-y) \in A$. Because $w + m(-y) \in \mathbb{Q} - A$, then Lemma 1.6.5 (2) implies that $w + (m + \frac{1}{2})(-y) \in \mathbb{Q} - A$. We let $u = w + (m - \frac{1}{2})(-y)$, and $v = w + \left(m + \frac{1}{2}\right)(-y)$ and $e = w + m(-y)$. Then $y = u - v$, and $v < e$, where $e \in \mathbb{Q} - A$.

(2) We follow [Men73] and [Bur67] in part. Because $y > 1$ and $q > 0$, then $(y-1)\frac{q}{2} > 0$. By Part (1) of this lemma there are $u \in A$ and $v \in \mathbb{Q} - A$ such that $(y-1)\frac{q}{2} = u - v$ and $v < e$ for some $e \in \mathbb{Q} - A$.

There are now two cases. First, suppose that $v > \frac{q}{2}$. Because $y - 1 > 0$, it follows that $(y-1)v > (y-1)\frac{q}{2}$ and hence $(y-1)v > u - v$. Therefore $yv > u$. Because $v > \frac{q}{2}$

we know that $v > 0$, and hence we conclude that $y > \frac{u}{v}$. We then let $r = u$ and $s = v$, and evidently $s > 0$, and $y > \frac{r}{s}$ and $s < e$.

Second, suppose that $v \leq \frac{q}{2}$. We let $r = u + (\frac{3q}{4} - v)$ and $s = \frac{3q}{4}$. Because $q > 0$ then $s > 0$. Because $\frac{3q}{4} - v > 0$, we see that $r > u$, and hence $r \in A$ by Part (b) of the definition of Dedekind cuts. Also, because $s < q$ and $q \in \mathbb{Q} - A$, then $s \in \mathbb{Q} - A$ by Lemma 1.6.5 (2). Evidently $r - s = u - v = (y - 1)\frac{q}{2}$. Because $q > 0$, we deduce that $(y-1)\frac{3q}{4} > r - s$, and hence $(y-1)s > r - s$, which yields $ys > r$. Because $s > 0$, we conclude that $y > \frac{r}{s}$. □

Reflections

Proving a theorem is fun when either the theorem itself is interesting, or when the proof of the theorem is clever or insightful. Tedious proofs of dry technical results that are of interest only for proving something else yet to come are not particularly appealing, and, unfortunately, that is the nature of much of the present section, which consists of the preliminaries needed for the actual construction of the real numbers seen in the next section. Hence, the reader could not be faulted for skipping some of the details of some of the proofs in the present section upon first reading. However, for the sake of having a complete treatment of the real numbers, detailed proofs have been included, and are available for whoever wants to read them. Treatments of Dedekind cuts in some other books might appear to be more brief than our treatment, but that is only because we have put in details that are often omitted.

Exercises

Exercise 1.6.1. **[Used in Exercise 1.6.5.]** Let $A, B \subseteq \mathbb{Q}$ be Dedekind cuts. Suppose that $A \subsetneqq B$. Prove that $B - A$ has more than one element. If you are familiar with the cardinality of sets, prove that $B - A$ is countably infinite.

Exercise 1.6.2. **[Used in Example 1.6.3.]** Let T be the set defined in Equation 1.6.1.

(1) Prove that T is a Dedekind cut.
(2) Prove that if $T = D_r$ for some $r \in \mathbb{Q}$, then $r^2 = 2$.

[Use Exercise 1.5.6, Exercise 1.5.7 and Exercise 1.5.9 (3).]

Exercise 1.6.3. **[Used in Lemma 1.6.8.]** Prove Lemma 1.6.8 (3).

Exercise 1.6.4. **[Used in Lemma 1.7.4.]** Let $A \subseteq \mathbb{Q}$ be a Dedekind cut, and let $r \in \mathbb{Q}$.

(1) Prove that $A \subsetneqq D_r$ if and only if there is some $q \in \mathbb{Q} - A$ such that $r < q$.
(2) Prove that $A \subseteq D_r$ if and only if $r \in \mathbb{Q} - A$ if and only if $r < a$ for all $a \in A$.

Exercise 1.6.5. What we call a Dedekind cut is often called an "upper cut," to differentiate it from the analogous "lower cut." Both types of cuts are equally valid, and are mirror images of each other, though upper cuts are slightly simpler to use because the product of positive numbers is positive, whereas the product of negative numbers is not negative.

(1) Write a precise definition of lower cuts, modeled upon Definition 1.6.1.
(2) Let $A \subseteq \mathbb{Q}$ be a Dedekind cut. Find an example to show that $\mathbb{Q} - A$ is not necessarily a lower cut.
(3) Let $A \subseteq \mathbb{Q}$ be a Dedekind cut. Prove that if $\mathbb{Q} - A$ is not a lower cut, then Part (a) of the definition of lower cuts does not hold. Deduce that there is some $m \in \mathbb{Q} - A$ such that $x \leq m$ for all $x \in \mathbb{Q} - A$.
(4) Let $A \subseteq \mathbb{Q}$ be a Dedekind cut. Suppose that $\mathbb{Q} - A$ is not a lower cut. Prove that there is a unique element $k \in \mathbb{Q} - A$ such that $\mathbb{Q} - (A \cup \{k\})$ is a lower cut.
(5) Let $A \subseteq \mathbb{Q}$ be a Dedekind cut. Suppose that $\mathbb{Q} - A$ is not a lower cut. Let k be as in Part (4) of this lemma. Prove that $k \leq x$ for all $x \in A \cup \{k\}$.
(6) Let $\mathcal{D}^u$ denote the set of all Dedekind cuts of $\mathbb{Q}$, and let $\mathcal{D}^l$ denote the set of all lower cuts of $\mathbb{Q}$. Prove that there is a bijective function $\phi\colon \mathcal{D}^u \to \mathcal{D}^l$ such that $A \subseteq B$ implies $\phi(A) \supseteq \phi(B)$ for all $A, B \in \mathcal{D}^u$. Because lower cuts are completely analogous to Dedekind cuts, you may assume that the analog of everything that has been previously proved about Dedekind cuts and lower cuts holds with the roles of Dedekind cuts and lower cuts reversed.

[Use Exercise 1.6.1.]

Exercise 1.6.6. [Used in Exercise 1.7.8.] In Definition 1.6.1, Dedekind cuts were defined as subsets of the set $\mathbb{Q}$. However, an examination of this definition reveals that it does not make use of the full features of $\mathbb{Q}$, but only the order relation $<$ on $\mathbb{Q}$. Hence, it is possible to define Dedekind cuts on sets equipped with only order relations, but not necessarily with binary operations such as addition and multiplication.

Let S be a non-empty set, and let $<$ be a relation on S. The relation $<$ is an **order relation** if it satisfies the Trichotomy Law and the Transitive Law, as stated, for example, in Theorem 1.5.5 (10) (11); the set S is an **ordered set** if $<$ is an order relation. For example, the natural numbers, the integers and the rational numbers are all ordered sets. Dedekind cuts can be defined for any ordered set exactly as in Definition 1.6.1.

(1) Given an example of an ordered set for which the analog of Lemma 1.6.2 does not hold.
(2) Find criteria on an ordered set that would guarantee that the analog of Lemma 1.6.2 holds. The criteria must be defined strictly in terms of the order relation.
(3) Verify that the analog of Lemma 1.6.7 holds for arbitrary ordered sets.

1.7 Constructing the Real Numbers

Having done the hard work regarding Dedekind cuts in Section 1.6, we are now ready to use them to define the set of real numbers. We want the set of real numbers to contain a copy of the set of rational numbers, and it should also have numbers such as $\sqrt{2}$, where the set of rational numbers has "gaps." If we consider the fact that each rational number r corresponds to the rational cut D_r, and that there are also irrational cuts such as the one defined in Equation 1.6.1 (which appears as if it wants to be the set of all rational numbers greater than a number that is missing from the rational numbers), then we might guess that each Dedekind cut corresponds to a real number, and each real number corresponds to a Dedekind cut. Such a guess would be correct. Given that we have not yet seen a definition of the real numbers, we cannot prove such an intuitive correspondence between the real numbers and Dedekind cuts of rational numbers, but we can take it as inspiration for the following definition.

Definition 1.7.1. The set of **real numbers**, denoted $\mathbb{R}$, is defined by

$$\mathbb{R} = \{A \subseteq \mathbb{Q} \mid A \text{ is a Dedekind cut}\}. \qquad \triangle$$

We note that in both the definition of the integers in terms of the natural numbers (in Definition 1.3.3), and the definition of the rational numbers in terms of the integers (in Definition 1.5.3), the constructions used equivalence classes. In Definition 1.7.1, by contrast, we have the advantage of not having to use an equivalence relation.

We now turn to the definitions of addition, multiplication, negation, multiplicative inverse, less than and less than or equal to for the real numbers. We start with the last two of these, which are the simplest, and which are needed in the definitions of multiplication and multiplicative inverse.

Given that Dedekind cuts are sets of rational numbers, we define the relation less than on the real numbers in terms of the relation "subset" on sets of rational numbers.

Definition 1.7.2. The relation $<$ on $\mathbb{R}$ is defined by $A < B$ if and only if $A \supsetneqq B$, for all $A, B \in \mathbb{R}$. The relation $\leq$ on $\mathbb{R}$ is defined by $A \leq B$ if and only if $A \supseteq B$, for all $A, B \in \mathbb{R}$. $\triangle$

The following definition of addition and negation for real numbers makes sense because of Lemma 1.6.8 (1) (2).

Definition 1.7.3. The binary operation $+$ on $\mathbb{R}$ is defined by

$$A + B = \{r \in \mathbb{Q} \mid r = a + b \text{ for some } a \in A \text{ and } b \in B\}$$

for all $A, B \in \mathbb{R}$. The unary operation $-$ on $\mathbb{R}$ is defined by

$$-A = \{r \in \mathbb{Q} \mid -r < c \text{ for some } c \in \mathbb{Q} - A\}$$

for all $A \in \mathbb{R}$. $\triangle$

The definition of multiplication and multiplicative inverse for real numbers is a bit more complicated than the definition of addition and negation, because we will need various cases. We start with the following lemma.

Lemma 1.7.4. *Let $A \in \mathbb{R}$, and let $r \in \mathbb{Q}$.*

1. *$A > D_r$ if and only if there is some $q \in \mathbb{Q} - A$ such that $q > r$.*
2. *$A \geq D_r$ if and only if $r \in \mathbb{Q} - A$ if and only if $a > r$ for all $a \in A$.*
3. *If $A < D_0$ then $-A \geq D_0$.*

Proof.

(1) & (2) These are just restatements of Exercise 1.6.4.

(3) Suppose that $A < D_0$. Then $A \supsetneqq D_0$. Because $D_0 = \{x \in \mathbb{Q} \mid x > 0\}$, it follows that there is some $q \in A$ such that $q \le 0$. By Part (b) of the definition of Dedekind cuts, we deduce that $0 \in A$. Hence $0 \notin \mathbb{Q} - A$, and therefore $-0 \notin \mathbb{Q} - A$, which implies that $0 \notin -A$, and hence $0 \in \mathbb{Q} - (-A)$. By Part (2) of this exercise we deduce that $-A \ge D_0$. □

The following definition of multiplication and multiplicative inverse makes sense because of Lemma 1.6.8 (3) (4) and Lemma 1.7.4.

Definition 1.7.5. The binary operation $\cdot$ on $\mathbb{R}$ is defined by

$$A \cdot B = \begin{cases} \{r \in \mathbb{Q} \mid r = ab \text{ for some } a \in A \text{ and } b \in B\}, & \text{if } A \ge D_0 \text{ and } B \ge D_0 \\ -[(-A) \cdot B], & \text{if } A < D_0 \text{ and } B \ge D_0 \\ -[A \cdot (-B)], & \text{if } A \ge D_0 \text{ and } B < D_0 \\ (-A) \cdot (-B), & \text{if } A < D_0 \text{ and } B < D_0. \end{cases}$$

The unary operation $^{-1}$ on $\mathbb{R} - \{D_0\}$ is defined by

$$A^{-1} = \begin{cases} \{r \in \mathbb{Q} \mid r > 0 \text{ and } \frac{1}{r} < c \text{ for some } c \in \mathbb{Q} - A\}, & \text{if } A > D_0 \\ -(-A)^{-1}, & \text{if } A < D_0. \end{cases}$$

△

Having now defined the basic operations and relations on the real numbers, we are ready to prove the most fundamental algebraic properties of these numbers. The proof of the following theorem is lengthy, and tedious in parts, and the reader would not be faulted for skipping a few of the details upon first reading. As is usual, we will write "AB" instead of "$A \cdot B$," except in cases of potential ambiguity, or for ease of reading. We will write $A > B$ to mean the same thing as $B < A$.

Theorem 1.7.6. *Let $A, B, C \in \mathbb{R}$.*

1. $(A+B)+C = A+(B+C)$ *(Associative Law for Addition).*
2. $A+B = B+A$ *(Commutative Law for Addition).*
3. $A+D_0 = A$ *(Identity Law for Addition).*
4. $A+(-A) = D_0$ *(Inverses Law for Addition).*
5. $(AB)C = A(BC)$ *(Associative Law for Multiplication).*
6. $AB = BA$ *(Commutative Law for Multiplication).*
7. $A \cdot D_1 = A$ *(Identity Law for Multiplication).*
8. *If $A \ne D_0$, then $AA^{-1} = D_1$ (Inverses Law for Multiplication).*
9. $A(B+C) = AB+AC$ *(Distributive Law).*
10. *Precisely one of $A < B$ or $A = B$ or $A > B$ holds (Trichotomy Law).*
11. *If $A < B$ and $B < C$, then $A < C$ (Transitive Law).*
12. *If $A < B$ then $A+C < B+C$ (Addition Law for Order).*
13. *If $A < B$ and $C > D_0$, then $AC < BC$ (Multiplication Law for Order).*

14. $D_0 < D_1$ *(Non-Triviality).*

Proof. We will prove the parts of this theorem not in the order in which they are stated (which is a standard way of writing the theorem), but rather in an order chosen so that the proof of each part relies only upon the previously proved parts. As was the case in Section 1.6, to avoid making the proof any longer than necessary, we will not cite references to the standard properties of the rational numbers proved in Section 1.5.

(14) Left to the reader in Exercise 1.7.3.

(1) Using the definition of addition for $\mathbb{R}$ we see that

$$\begin{aligned}(A+B)+C &= \{r \in \mathbb{Q} \mid r = (a+b)+c \text{ for some } a \in A \text{ and } b \in B \text{ and } c \in C\} \\ &= \{r \in \mathbb{Q} \mid r = a+(b+c) \text{ for some } a \in A \text{ and } b \in B \text{ and } c \in C\} \\ &= A+(B+C).\end{aligned}$$

(2) The proof of this part of the theorem is very similar to the proof of Part (1), and we omit the details.

(3) Using the definition of addition for $\mathbb{R}$ we see that

$$A + D_0 = \{r \in \mathbb{Q} \mid r = a+b \text{ for some } a \in A \text{ and } b \in D_0\}.$$

Let $a \in A$. Then by Part (c) of the definition of Dedekind cuts, there is some $c \in A$ such that $c < a$. Then $a - c > 0$, and hence $a - c \in D_0$. Therefore $a = c + (a - c) \in A + D_0$. It follows that $A \subseteq A + D_0$.

Let $d \in A + D_0$. Then $d = s + t$ for some $s \in A$ and $t \in D_0$. By the definition of D_0, we know that $t > 0$. Hence $d > s$. It follows from Part (b) of the definition of Dedekind cuts that $d \in A$. It follows that $A + D_0 \subseteq A$, and hence $A + D_0 = A$.

(4) Using the definition of addition and negation for $\mathbb{R}$ we see that

$$A + (-A) = \{r \in \mathbb{Q} \mid r = a+b \text{ for some } a \in A \text{ and } b \in -A\}.$$

Let $x \in A + (-A)$. Then $x = s + t$ for some $s \in A$ and $t \in -A$. It follows from the definition of $-A$ that $-t < c$ for some $c \in \mathbb{Q} - A$. By Lemma 1.6.5 (2) we know that $-t \in \mathbb{Q} - A$, and hence by Lemma 1.6.5 (1) we see that $-t < s$. Hence $s + t > 0$, and therefore $x > 0$, which implies that $x \in D_0$. We deduce that $A + (-A) \subseteq D_0$.

Now let $y \in D_0$. Then $y > 0$. It follows from Lemma 1.6.9 (1) that there are $u \in A$ and $v \in \mathbb{Q} - A$ such that $y = u - v$, and $v < e$ for some $e \in \mathbb{Q} - A$. It follows that $y = u + (-v)$, and $-(-v) < e$. Therefore $-v \in -A$, and hence $y \in A + (-A)$. It follows that $D_0 \subseteq A + (-A)$, and therefore $A + (-A) = D_0$.

(10) This part of the theorem is just a restatement of Lemma 1.6.6.

(11) This part of the theorem is just a restatement of a standard fact about subsets of sets; see [Blo10, Lemma 3.2.4].

(12) Suppose that $A < B$. Hence $A \supsetneqq B$. Let $x \in B+C$. Then $x = u+v$ for some $u \in B$ and $v \in C$. Then $u \in A$, so $x \in A+C$. It follows that $A+C \supseteq B+C$. There is some $p \in A-B$. Then by Lemma 1.6.5 (1) we know that $p < b$ for all $b \in B$. Let $c \in C$. Then $p+c < b+c$ for all $b \in B$. It follows from Lemma 1.6.5 (1) that $p+c \in \mathbb{Q}-(B+C)$. Because $p+c \in A+C$, we deduce that $A+C \supsetneqq B+C$.

(5) Left to the reader in Exercise 1.7.5.

(6) By Part (10) of this theorem, we know that either $A \geq D_0$ or $A < D_0$, and similarly for B. There are now four cases.

First, suppose that $A \geq D_0$ and $B \geq D_0$. Then

$$\begin{aligned} AB &= \{r \in \mathbb{Q} \mid r = ab \text{ for some } a \in A \text{ and } b \in B\} \\ &= \{r \in \mathbb{Q} \mid r = ba \text{ for some } a \in A \text{ and } b \in B\} = BA. \end{aligned}$$

Second, suppose that $A \geq D_0$ and $B < D_0$. By Exercise 1.7.4 (1) we know that $-B > D_0$. The definition of multiplication of Dedekind cuts, together with the case we have already proved, imply that $AB = -[A(-B)] = -[(-B)A] = BA$.

The other two cases are where $A < D_0$ and $B \geq D_0$, and where $A < D_0$ and $B < D_0$, are very similar to the case just proved, and we omit the details.

(7) Left to the reader in Exercise 1.7.5.

(8) Suppose that $A \neq D_0$. By Part (10) of this theorem we know that either $A > D_0$ or $A < D_0$. First, suppose that $A > D_0$. By Exercise 1.7.2 (2) we know that $A^{-1} > D_0$. It then follows from the definition of multiplication of Dedekind cuts that

$$AA^{-1} = \{r \in \mathbb{Q} \mid r = ab \text{ for some } a \in A \text{ and } b \in A^{-1}\}.$$

Let $x \in AA^{-1}$. Then $x = uv$ for some $u \in A$ and $v \in A^{-1}$. By Lemma 1.7.4 (2) we know that $u > 0$, and by the definition of A^{-1} when $A > D_0$ we know that $v > 0$, and that $\frac{1}{v} < h$ for some $h \in \mathbb{Q}-A$. By Lemma 1.6.5 (2) we see that $\frac{1}{v} \in \mathbb{Q}-A$, and hence by Lemma 1.6.5 (1) we know that $\frac{1}{v} < u$. Hence $1 < uv$, and therefore $x > 1$, which means that $x \in D_1$. It follows that $AA^{-1} \subseteq D_1$.

Now let $y \in D_1$. Then $y > 1$. Because $A > D_0$, we know by Lemma 1.7.4 (1) that there is some $q \in \mathbb{Q}-A$ such that $q > 0$. It then follows from Lemma 1.6.9 (2) that there are $r \in A$ and $s \in \mathbb{Q}-A$ such that $s > 0$, and $y > \frac{r}{s}$, and $s < k$ for some $k \in \mathbb{Q}-A$. We know that $\frac{1}{s} > 0$, and that $\frac{1}{\frac{1}{s}} < k$, and we deduce that $\frac{1}{s} \in A^{-1}$. Hence $\frac{r}{s} = r \cdot \frac{1}{s} \in AA^{-1}$. Because $y > \frac{r}{s}$, it follows from Part (b) of the definition of Dedekind cuts that $y \in AA^{-1}$. Hence $D_1 \subseteq AA^{-1}$, and therefore $AA^{-1} = D_1$.

Next, suppose that $A < D_0$. Then by the definition of multiplicative inverse we see that $A^{-1} = -(-A)^{-1}$. Hence $-A^{-1} = -[-(-A)^{-1}]$, and it follows from Exercise 1.7.4 (2) that $-A^{-1} = (-A)^{-1}$. By Exercise 1.7.4 (1) we know that $-A > D_0$, and by an argument similar to one used in the case where $A > D_0$, we see that $(-A)^{-1} > D_0$. Therefore $-A^{-1} > D_0$, and hence by Exercise 1.7.4 (1) again we deduce that $A^{-1} = -(-A)^{-1} < D_0$. We now know by the definition of multiplication Dedekind cuts, combined with the previous case, that $AA^{-1} = (-A)(-A^{-1}) = (-A)(-A)^{-1} = D_1$.

(9) We follow [Men73]. There are eight cases.

First, suppose that $A \geq D_0$ and $B \geq D_0$ and $C \geq D_0$. By Lemma 1.7.4 (2), we know that every element of each of the sets A, B and C is greater than 0. It follows that every element in $B+C$ is greater than 0, and hence by Lemma 1.7.4 (2) we know that $B+C \geq D_0$.

Using the definition of AB, when $A \geq D_0$ and $B \geq D_0$, we see that

$$AB = \{r \in \mathbb{Q} \mid r = ab \text{ for some } a \in A \text{ and } b \in B\}.$$

It follows that every element of AB is greater than 0. A similar argument shows that every element of each of the sets AC, $A(B+C)$ and $AB+AC$ is greater than 0.

Let $x \in A(B+C)$. As just noted, we know $x > 0$. Also, we know by definition that $x = aq$ for some $a \in A$ and $q \in B+C$, and that $q = b+c$ for some $b \in B$ and $c \in C$. Hence $x = a(b+c) = ab+ac$. Because $ab \in AB$ and $ac \in AC$, we deduce that $x \in AB+AC$. Hence $A(B+C) \subseteq AB+AC$.

Now let $y \in AB+AC$. Then $y = u+v$ for some $u \in AB$ and $v \in AC$. Hence $u = a_1 b$ and $v = a_2 c$ for some $a_1, a_2 \in A$, and $b \in B$ and $c \in C$. If $a_1 = a_2$, then $y = a_1(b+c)$, and so $y \in A(B+C)$. Now suppose that $a_1 \neq a_2$. Without loss of generality, assume that $a_1 > a_2$. Observe that $\frac{ba_1}{a_2} > b$, and hence by Part (b) of the definition of Dedekind cuts we see that $\frac{ba_1}{a_2} \in B$. Because $y = a_2\left[\frac{ba_1}{a_2} + c\right]$, we deduce that $y \in A(B+C)$. Therefore $AB+AC \subseteq A(B+C)$, and hence $A(B+C) = AB+AC$.

The remaining cases all make use of the case that we have already proved, together with Exercise 1.7.4; for the sake of brevity we will not cite that exercise when we use it.

For our second case, suppose that $A \geq D_0$ and $B \geq D_0$ and $C < D_0$. Then $-C > D_0$. There are two subcases, depending upon whether $B+C \geq D_0$ or $B+C < D_0$. First, suppose that $B+C \geq D_0$. By the definition of multiplication of Dedekind cuts, we know that $AC = -[A(-C)]$, and hence $-(AC) = A(-C)$. By the definition of multiplication of Dedekind cuts, the case we have already proved, and Parts (3) and (4) of this theorem, we see that $AB+AC = A[(-C)+(B+C)]+AC = A(-C)+A(B+C)+AC = -(AC)+A(B+C)+AC = A(B+C)$. Second, suppose that $B+C < D_0$. Then $-(B+C) > D_0$. Reasoning similar to the above shows that $AB+AC = AB+\{-[A(-C)]\} = AB+\{-[A[B+[-(B+C)]]]\} = AB+\{-[AB+A[-(B+C)]]\} = AB+\{-[AB]\}+\{-[A[-(B+C)]]\} = -[A[-(B+C)]] = A(B+C)$.

Third, suppose that $A \geq D_0$ and $B < 0$ and $C \geq D_0$. This case is just like the previous case, and we omit the details.

Fourth, suppose that $A \geq D_0$ and $B < D_0$ and $C < D_0$. Then $B+C < D_0$, and $-B > D_0$ and $-C > D_0$ and $-(B+C) > D_0$. Then, using the definition of multiplication of Dedekind cuts, we see that $A(B+C) = -[A[-(B+C)]] = -\{A[(-B)+(-C)]\} = -\{[A(-B)]+[A(-C)]\} = \{-[A(-B)]\}+\{-[A(-C)]\} = AB+AC$.

There are four other cases, which are similar to the cases we have already seen, and which are left to the reader in Exercise 1.7.6.

(13) Suppose that $A < B$ and $C > D_0$. By Parts (4) and (12) of this theorem we see that $D_0 = A+(-A) < B+(-A)$. Hence by Exercise 1.7.2 (1) we see that

$[B+(-A)]C > D_0$. It follows from Parts (12) and (3) of this theorem that $AC + [B+(-A)]C > AC + D_0 = AC$. By Parts (6) and (9) of this theorem we deduce that $\{A+[B+(-A)]\}C > AC$, which by Exercise 1.7.4 (5) implies that $BC > AC$. □

The observant reader will have noticed that the properties of the real numbers listed in Theorem 1.7.6 are identical to the properties of the rational numbers listed in Theorem 1.5.5. Hence, as mentioned in Section 1.5, the set of real numbers is called an "ordered field."

Because the rational numbers and the real numbers share the same algebraic properties listed in Theorem 1.5.5 and Theorem 1.7.6, these properties alone do not suffice to distinguish these two sets of numbers. We now turn to a property of the real numbers that is not satisfied by the rational numbers. This property, which is called the Least Upper Bound Property, and is stated in Theorem 1.7.9 below, in fact characterizes the real numbers when taken together with the algebraic properties stated in Theorem 1.7.6. That is, not only do the real numbers satisfy Theorem 1.7.6 and Theorem 1.7.9, but essentially nothing else satisfies these two theorems, as will be proved in Section 2.7.

In order to state the Least Upper Bound Property, we need the following definition.

Definition 1.7.7. Let $A \subseteq \mathbb{R}$ be a set.

1. The set A is **bounded above** if there is some $M \in \mathbb{R}$ such that $X \leq M$ for all $X \in A$. The number M is called an **upper bound** of A.
2. The set A is **bounded below** if there is some $P \in \mathbb{R}$ such that $X \geq P$ for all $X \in A$. The number P is called a **lower bound** of A.
3. The set A is **bounded** if it is bounded above and bounded below.
4. Let $M \in \mathbb{R}$. The number M is a **least upper bound** (also called a **supremum**) of A if M is an upper bound of A, and if $M \leq T$ for all upper bounds T of A.
5. Let $P \in \mathbb{R}$. The number P is a **greatest lower bound** (also called an **infimum**) of A if P is a lower bound of A, and if $P \geq V$ for all lower bounds V of A. △

As the reader is asked to verify in Exercise 2.3.11, a subset $A \subseteq \mathbb{R}$ is bounded if and only if there is some $M \in \mathbb{R}$ such that $|X| \leq M$ for all $X \in A$; it is always possible to choose M so that $M > 0$. The proof of this fact has to wait until Section 2.3 because that is where we define absolute value.

Further discussion of upper bounds and lower bounds, and least upper bounds and greatest lower bounds, including examples, will be given in Section 2.6.

Before stating and proving the Least Upper Bound Property, we first state and prove the "mirror image" of this property, namely, the Greatest Lower Bound Property. These two properties are completely equivalent, in that each one implies the other. The Least Upper Bound Property is more commonly used, but the definition of Dedekind cuts makes it easier to prove the Greatest Lower Bound Property first.

Theorem 1.7.8 (Greatest Lower Bound Property). *Let $A \subseteq \mathbb{R}$ be a set. If A is non-empty and bounded below, then A has a greatest lower bound.*

Proof. Suppose that A is non-empty and bounded below. Let $L = \bigcup_{X \in A} X$, where the union makes sense because we can think of the elements of $\mathbb{R}$ as Dedekind cuts, which are subsets of $\mathbb{Q}$. We will show that $L = \text{glb} A$. Let $B \in \mathbb{R}$ be a lower bound of A. Because $B \leq X$ for all $X \in A$, then $B \supseteq X$ for all $X \in A$. It follows that $L = \bigcup_{X \in A} X \subseteq B$. Because $B \neq \mathbb{Q}$ by Part (a) of the definition of Dedekind cuts, we see that $L \neq \mathbb{Q}$. We then use Lemma 1.6.7 to deduce that L is a Dedekind cut.

Clearly $X \subseteq L$ for all $X \in A$, and hence $L \leq X$ for all $X \in A$. It follows that L is a lower bound of A. Now let $C \in \mathbb{R}$ be a lower bound of A. Then $C \leq X$ for all $X \in A$, and hence $C \supseteq X$ for all $X \in A$. By a standard property of unions of sets (see [Blo10, Theorem 3.4.5]), we deduce that $C \supseteq \bigcup_{X \in A} X = L$. Hence $C \leq L$, and we conclude that L is the greatest lower bound of A. □

We now see that the Least Upper Bound Property is a straightforward consequence of the Greatest Lower Bound Property (Theorem 1.7.8). Observe that the proof of the following theorem uses nothing other than Definition 1.7.7 and the statement of the Greatest Lower Bound Property; in particular, the proof does not make any mention of Dedekind cuts.

Theorem 1.7.9 (Least Upper Bound Property). *Let $A \subseteq \mathbb{R}$ be a set. If A is non-empty and bounded above, then A has a least upper bound.*

Proof. Suppose that A is non-empty and bounded above. Let

$$U = \{X \in \mathbb{R} \mid X \text{ is an upper bound of } A\}.$$

Then $U \neq \emptyset$, because A is bounded above. Let $B \in A$. Then $B \leq X$ for all $X \in U$, and hence B is a lower bound of U. Because U is non-empty and bounded below, we can apply the Greatest Lower Bound Property (Theorem 1.7.8) to U to deduce that U has a greatest lower bound, say L.

If $C \in A$, then $C \leq X$ for all $X \in U$. Therefore every element of A is a lower bound of U. Because L is the greatest lower bound of U, it follows that $C \leq L$ for all $C \in A$. Hence L is an upper bound of A. Because L is a lower bound of U, then $L \leq X$ for every $X \in U$. We deduce that L is a least upper bound of A. □

We now have a situation very analogous to when we constructed the integers from the natural numbers (in Section 1.3), and the rational numbers from the integers (in Section 1.5), in that we have two sets of numbers, namely, the set of real numbers and the set of rational numbers, which, while informally one set is viewed as containing the other, formally these two sets as constructed are entirely disjoint. As expected, however, we can find a copy of the set of rational numbers inside the set of real numbers by identifying each rational number r with the real number D_r. We will see in the following theorem that this identification preserves the numbers 0 and 1, the binary operations addition and multiplication and the relation less than.

Theorem 1.7.10. *Let $i\colon \mathbb{Q} \to \mathbb{R}$ be defined by $i(r) = D_r$ for all $r \in \mathbb{R}$.*

1. *The function $i\colon \mathbb{Q} \to \mathbb{R}$ is injective.*
2. *$i(0) = D_0$ and $i(1) = D_1$.*
3. *Let $r, s \in \mathbb{Q}$. Then*
   a. *$i(r + s) = i(r) + i(s)$;*

b. $i(-r) = -i(r)$;
c. $i(rs) = i(r)i(s)$;
d. *if* $r \neq 0$ *then* $i(r^{-1}) = [i(r)]^{-1}$;
e. $r < s$ *if and only if* $i(r) < i(s)$.

Proof. Left to the reader in Exercise 1.7.7. □

Again analogously to our previous constructions, we see from Theorem 1.7.10 that even though technically the set of rational numbers and the set of real numbers are entirely disjoint, in fact from the point of view of addition, multiplication, negation, multiplicative inverse and the relation less than, the rational numbers can be identified with a subset of the real numbers. We therefore dispense with the set of rational numbers as a separate entity (except when we need it in proofs), because we have a copy of the rational numbers inside the real numbers that works just as well as the original. We will therefore also dispense with the notation D_0 and D_1, and simply write 0 and 1 instead.

Moreover, not only will we dispense with the Dedekind cut notation from now on, we will not need to use the concept of Dedekind cuts at all after this section. All further properties of the real numbers will now be derived from the statements of Theorem 1.7.6 and Theorem 1.7.9, without any reference to the fact that we proved these two theorems using Dedekind cuts. Our construction of the real numbers proves that there exists a set with the properties that we would expect of the real numbers, but ultimately, it is the properties of the real numbers, not how they are constructed, that matter.

Reflections

As mentioned in Exercise 1.6.5, what we call a Dedekind cut is often called an "upper cut," to differentiate it from the analogous "lower cut." Both types of cuts, which are mirror images of each other, can be used to construct the real numbers. We have chosen to work with upper cuts because they are technically slightly easier to use, due to the fact that the product of positive numbers is positive, whereas the product of negative numbers is not negative. On the other hand, there is an advantage to using lower cuts, which is that they allow for a direct proof of the Least Upper Bound Property, as opposed to upper cuts, which lead naturally to the Greatest Lower Bound Property, and only from there do we arrive at the Least Upper Bound Property. Of course, the Greatest Lower Bound Property is just as good as the Least Upper Bound Property, and it is possible to use the former instead of the latter, but we follow the standard approach used today and focus on the Least Upper Bound Property as the fundamental property.

Our construction of the real numbers follows the traditional route of starting with the natural numbers, constructing the integers from the natural numbers, constructing the rational numbers from the integers and finally constructing the real numbers from the rational numbers. It turns out that there is a more efficient construction of the real numbers directly from the integers; see [A'C], [Str] or [DKO$^+$] for details. We have chosen to stay with the traditional route, however, because the steps used, while less efficient, seem more natural, and because the rational numbers are important in their own right, and deserve to have their properties discussed.

We have now come to the end of our construction of the number systems. It is possible to construct the complex numbers from the real numbers, but we will not need the complex numbers in a book on real analysis. Interestingly, although the complex numbers have many fascinating and important properties, and complex analysis is a subject well worth studying, the construction of the complex numbers from the real numbers is quite simple in comparison to the constructions we have seen in this chapter.

Exercises

Exercise 1.7.1. [Used in Exercise 1.7.7.] Let $r \in \mathbb{Q}$.

(1) Prove that $D_{-r} = -D_r$, using only Definition 1.6.4 and Definition 1.7.3.
(2) Prove that $D_{r^{-1}} = [D_r]^{-1}$, using only Definition 1.7.5 and Definition 1.7.3.

Exercise 1.7.2. [Used in Theorem 1.7.6.] Let $A, B \in \mathbb{R}$. Suppose that $A > D_0$ and $B > D_0$. For this exercise, you may use only results prior to Theorem 1.7.6.

(1) Prove that $AB > D_0$.
(2) Prove that $A^{-1} > D_0$.

Exercise 1.7.3. [Used in Theorem 1.7.6.] Prove Theorem 1.7.6 (14). For this exercise, you may use only results prior to Theorem 1.7.6. [Use Exercise 1.5.6 (1).]

Exercise 1.7.4. [Used in Theorem 1.7.6, Exercise 1.7.5 and Exercise 1.7.6.] For this exercise, use only the properties of real numbers stated in Theorem 1.7.6 (1) (2) (3) (4) (10) (11) (12) (14); it is not necessary to use the definition of real numbers as Dedekind cuts. Let $A, B \in \mathbb{R}$.

(1) Prove that $A > D_0$ if and only if $-A < D_0$, and that $A < D_0$ if and only if $-A > D_0$.
(2) Prove that $-(-A) = A$.
(3) Prove that $-(A+B) = (-A) + (-B)$.
(4) Prove that if $A > D_0$ and $B > D_0$, then $A+B > D_0$, and that if $A < D_0$ and $B < D_0$, then $A+B < D_0$
(5) Prove that $A = (-B) + (A+B) = B + [A + (-B)]$ and $-A = B + [-(B+A)]$.

Exercise 1.7.5. [Used in Theorem 1.7.6.] Prove Theorem 1.7.6 (5) (7). For this exercise, you may use only Parts (1), (2), (3), (4), (10), (11), (12) and (14) of the theorem, and anything prior to the theorem. [Use Exercise 1.7.4.]

Exercise 1.7.6. [Used in Theorem 1.7.6.] Prove the remaining four cases in the proof of Theorem 1.7.6 (9). [Use Exercise 1.7.4.]

Exercise 1.7.7. [Used in Theorem 1.7.10.] Prove Theorem 1.7.10. [Use Exercise 1.7.1.]

Exercise 1.7.8. This exercise makes use of Exercise 1.6.6. Let S be a non-empty ordered set. The **Dedekind set** of S, denoted S^D, is defined by

$$S^D = \{A \subseteq S \mid A \text{ is a Dedekind cut}\}.$$

For example, we know by definition that $\mathbb{Q}^D = \mathbb{R}$. The order relation $<$ on S^D is defined analogously to Definition 1.7.2.

(1) Find an example of an ordered set T for which $T^D = \emptyset$. It is sufficient to state informally the reason why your example works.
(2) Find an example of an ordered set U for which U^D has exactly one element. It is sufficient to state informally the reason why your example works.
(3) Verify that S^D satisfies the Least Upper Bound Property.
(4) What can you say about $\mathbb{R}^D$? It is sufficient to answer this question informally.

1.8 Historical Remarks

When the use of numbers in the ancient world is discussed, it is common to consider the different methods by which numbers were written in different cultures, the most familiar of such methods being Roman numerals. From the perspective of real analysis, however, our interest in the history of numbers is concerned very little with the way in which numbers were written in various periods of history, and instead with what the concept of "number," however written, was understood to mean. In real analysis we make extensive use of the properties of the real numbers, and hence we need to know what these numbers are, but we make no use of how numbers are written. We will discuss the decimal place-value system in detail in Section 2.8, and briefly here, because this notation was historically important in the development of calculus because it facilitated numerical calculation. Nonetheless, from the perspective of real analysis, the ability to write numbers in the decimal place-value system is a nice application of the axioms for the real numbers, but it is not a tool used in proofs. (The decimal place-value system is famously used in Cantor's diagonal argument that shows that the set of real numbers is uncountable, but we will not use that fact. Moreover, as a nice application of sequences, we will provide a different proof, also due to Cantor, of the uncountability of the real numbers that does not involve decimals; see Theorem 8.4.8.)

The real numbers as we now understand them, and as we use them in real analysis, consist of a variety of types of numbers, including natural numbers, zero, negative integers, rational numbers that are not integers, irrational numbers that are algebraic (that is, irrational numbers that are the roots of polynomials with rational coefficients) and irrational numbers that are transcendental (that is, numbers that are not algebraic). We will not discuss the complex numbers in these historical comments, because we do not use them in this text.

Ancient World

The first numbers to be used were the natural numbers, that is, the positive whole numbers, which were used to count objects. Such numbers arose in many ancient cultures, as did methods for adding and multiplying these numbers; the particular algorithms for addition and multiplication used by various cultures are not of interest to us here. The need for fractions also arose in many ancient cultures. The rest of the numbers, however, were slower to be recognized.

In the ancient Middle East and Europe, the word "number" meant positive integers and their ratios. The number 1 was not always viewed as a number; Euclid (c. 325–c. 265 BCE), for example, held that view.

Negative numbers and zero were understood as numbers at some point in ancient China and India; it seems that negative numbers were used before zero was used. In ancient China calculations were done with rods and the abacus, and zero is not needed in that case, because it suffices to have the absence of a rod or bead. Positive and negative numbers, on the other hand, were represented by rods of different colors. Rules for addition and subtraction of signed numbers were known in ancient China in the 2nd century; rules for multiplication and division of signed numbers were available only in 1303 in *Suanshu Chimeng*.

Ancient texts in a variety of cultures had approximate rational values for numbers such as $\sqrt{2}$ and π that we now know are irrational. However, whereas it was known that such approximations were not the exact value of the number, there did not appear to be an awareness in ancient China, Mesopotamia, Egypt or India that it was not possible to achieve an exact rational value; that is, these cultures did not recognize the concept of irrational numbers.

The idea of incommensurable pairs of lengths of line segments was discovered in ancient Greece; ratios of such lengths represent what we call irrational numbers. The fact that $\sqrt{2}$ is irrational, expressed geometrically in terms of the incommensurability of the diagonal and sides of a square, is often attributed to Pythagoras of Samos (c. 569–c. 475 BCE) or his followers, and was presumably known by the time of Theodorus of Cyrene (465–398 BCE), who understood that 3, 5, 6, 7, 8, 10, 11, 12, 13, 14, 15 and 17 do not have rational square roots, a fact referred to by Plato (427–347 BCE). The irrationality of such numbers is proved in Book X of Euclid's *Elements*, which is attributed (at least in part) to Theaetetus of Athens (c. 417–c. 369 BCE), who was a student of Theodorus.

Even though the ancient Greeks were aware of incommensurable lengths of line segments, this discovery did not lead to the adoption of irrational numbers as numbers, perhaps due to the ancient Greek separation of arithmetic from geometry. Aristotle (384–322 BCE), in Book VI of the *Physics*, emphasized the distinction between "number," which was discrete and which had an indivisible unit, and "magnitude," which was "continuous" and which had no indivisibles (hence, for example, line segments were not made up of points). Magnitudes could be lengths of line segments, areas of regions of the plane, time, and other physically meaningful objects. Aristotle used the distinction between number and magnitude as part of his attempt to refute Zeno's paradoxes. Aristotle was very influential in medieval Europe, and the distinction between number and magnitude is one of the Aristotelian ideas that needed to be overcome as part of the development of calculus, and more generally modern science.

In Book VII of Euclid's *Elements* there is a theory of ratios of numbers, which corresponds to what we call fractions. In Book V of the *Elements* there is a theory of proportions, that is, ratios of magnitudes. The two magnitudes in a ratio must be quantities of the same type, though ratios of different types of quantities can be compared. This theory includes definitions of order, addition and multiplication for ratios of magnitudes, and the Archimedean Principle is invoked. The theory of proportions is very important for a number of theorems in the *Elements*, for example it is needed to state the fact that the areas of two circles have the same proportion as the squares of their diameters. In some ways this theory prefigures the idea of Dedekind cuts formulated by Dedekind in 1858 (as discussed below). The theory of proportions in the *Elements* is attributed to Eudoxus of Cnidus (408–355 BCE), and the Archimedean Principle should perhaps be named after Eudoxus rather than Archimedes (287–212 BCE), though this principle is also found in Book I of *On the Sphere and Cylinder* by Archimedes.

Medieval Period

The mathematicians in ancient India, who were more interested in algebra and computing than the ancient Greeks, did not have the philosophical restrictions of the latter in regard to numbers. This freedom to develop numbers eventually helped the development of calculus. Negative numbers and zero were conceptualized, and recognized as numbers, in ancient India. Brahmagupta (598–670), in the *Brahmasphutasiddhanta* of 628, considered zero to be the number obtained by subtracting a number from itself, and gave rules for addition, subtraction, multiplication and division of signed numbers and zero, though he did not understand division by zero. In the *Lilavati* of 1150, Bhaskara II (1114-1185), also known as Bhaskaracharya, built upon the work of Brahmagupta, but recognized the problem of dividing by zero.

The rise of algebra and analytic geometry helped promote the recognition of the real numbers as numbers. The algebra of Abu Ja'far Muhammad ibn Musa Al-Khwarizmi (c. 780–c. 850), in *Al-kitab al-muhtasar fi hisab al-jabr wal-muqabala* of around 825, allowed different types of numbers, rational and irrational, to be treated in a more unified way than before. The Latin rendering of the title of this work led to the word "algebra."

The blurring of the distinction between different types of numbers was further enhanced by Abu Kamil Shuja ibn Aslam (c. 850–930), whose work on algebra influenced Leonardo of Pisa (1170–1250), also known as Fibonacci, who spread Arabic algebra in Europe via his book *Liber abaci* of 1202. Abu Mansur ibn Tahir Al-Baghdadi (c. 980–1037) more explicitly broke down the ancient Greek distinction between number and magnitude by forming a correspondence between numbers and lengths of line segments via multiples of a fixed line segment, where irrational numbers correspond to those line segments that are not rational multiples of the fixed line segment. He also demonstrated the density of the irrational numbers (which is proved in Theorem 2.6.13 (2)).

Nicole Oresme (1323–1382) made two contributions to the development of numbers. First, though he did not invent invent analytic geometry as we use it today, in that he did not associate curves with equations in general, he made progress toward analytic geometry around 1350, when he related the study of variation with representation by coordinates. Second, he inquired about the meaning of raising a number to an irrational power, a question which is not trivial to answer, and which was resolved only with 19th-century rigor.

In parallel with the spread of algebra from the Arabic world to Europe was the spread of the decimal place-value system. Place-value systems for representing numbers (though not written the way we do) were invented separately in ancient Mesopotamia, India, China and Mesoamerica. The Maya system used a symbol for zero as a place-holder, and was base 20. The system in Mesopotamia was base 60, though initially it did not have zero. India and China used base 10 (called "decimal"). The Hindu-Arabic system that we use for writing numbers today, that is, the decimal place-value system, has three aspects: place-value, base 10 and the Hindu-Arabic symbols for the numbers 0–9. The first of these three is by far the most important.

The decimal place-value system for writing whole numbers, including the use of zero (sometimes denoted by a dot and sometimes by our modern symbol), originated in India apparently in the 7th century, and was fully developed by the 8th century. This system might have been inspired by Chinese counting boards which reached India via trade; the written system, in any case, spread from India to China and the Arab world.

The first Arabic work that used the Indian system for writing whole numbers is *Kitab al-jam wal-tafriq bi hisab al Hind* of Al-Khwarizmi, which gave algorithms for addition, subtraction, multiplication, division and more. The Latin translation of this work helped spread the decimal place-value system, and the Latin rendering of Al-Khwarizmi's name eventually became the word "algorithm."

A step forward in the development of the decimal-place value system was due to Abu'l Hasan Ahmad ibn Ibrahim Al-Uqlidisi (920–980), in *Kitab al-fusul fi al-hisab al-Hindi* of 952, who gave algorithms for use with pen and paper, as opposed to those of Al-Khwarizmi which were for dust board, and, more importantly, who introduced decimal fractions. This idea was further developed by Ibn Yahya al-Maghribi Al-Samawal (1130–1180) in 1172, who understood that some numbers cannot be expressed as finite decimal fractions, and can only be approximated by them. Al-Samawal also had a preliminary understanding of proof by induction. Ghiyath al-Din Jamshid Mas'ud al-Kashi (c. 1380–1429) in the early 15th century, had a thorough understanding of decimal fractions.

Gerbert of Aurillac (c. 946–1003), who later became Pope Sylvester II, was one of the first (if not the first) person to introduce the decimal place-value system for writing whole numbers (though not the number zero, and not decimal fractions) to the West. Gerbert learned this material from Arabic teachers while residing in Spain, which was partly under Arabic control at the time. This system was also promoted by Leonardo of Pisa in his widely circulated *Liber abaci*.

The first real appearance of proof by induction, in specific examples though not stated as a general principle, was apparently in the study of combinatorics by Levi ben Gerson (1288–1344) in *Maasei Hoshev* of 1321.

Renaissance

Whereas ancient Greek and medieval European mathematicians had not regarded irrational ratios as numbers, the Indian and Arab mathematicians did not distinguish between rational and irrational numbers. By the 16th century, when Hindu-Arabic algebra was widely adopted in Europe, mathematicians there recognized irrational numbers as numbers, though irrational numbers were still considered distinct from other numbers, having earlier been called numerus surdus by Leonardo of Pisa (surdus is the Latin root for the word absurd, though it originally meant deaf). Negative numbers, which were not used in Europe as late as the start of the 14th century, were accepted in Europe by the 16th century, though with the stigma of being numeri falsi or numeri ficti; in subsequent centuries negative numbers were accepted as numbers without stigma.

The decimal place-value system of writing whole numbers was spread in Europe by, among others, Robert Recorde (1510-1558) in *The Grounde of Artes* of 1543, and Adam Ries (1492–1559) in *Rechenung nach der lenge, auff den Linihen vnd Feder* of 1550, which had the advantage of being printed, that method having been recently invented. (Children in Germany still say "nach Adam Riese," which means "according to Adam Ries," when doing arithmetic.) This system was slow to be accepted, and one of the people who was responsible for the popularity of this system was Simon Stevin (1548–1620), via his widely read and translated work *De Thiende* of 1585. Not only did Stevin promote the use of the decimal place-value system for whole numbers, as did his European predecessors, but he introduced the use of decimal fractions to the West (though he used only finite decimals, so that only certain rational numbers could be represented exactly); decimal fractions were previously known in the Arab world, but Stevin might have been unaware of that. What finally led to the widespread use of decimal fractions was Napier's use of that notation in his work on logarithms. Stevin's notation, which was not exactly the way we write numbers today, was brought into modern form in the English translation of 1616 of Napier's work on logarithms.

In addition to helping popularize the decimal place-value system, Stevin helped expand the notion of what a "number" consists of in *L'Arithmetique* of 1585. Though the ancient Greeks did not consider 1 to be a number, Stevin said that it was, and after that the idea of 1 as a number gained widespread acceptance. However, Stevin viewed 0 as the place where the natural numbers start, but not as a number itself. More importantly, in contrast to the earlier view that the word "number" referred only to fractions, but not irrational numbers, Stevin said that "every root is a number," and did not call roots of numbers by a distinct name such as "absurd" or "surd." Stevin was not the first person to blur the ancient Greek distinction between "number" and "magnitude," but he may have been the first person to have stated explicitly that there is no such distinction.

Franciscus Maurolycus (1494–1575), presumably unaware of the work of Levi ben Gerson, used proof by induction in 1575. Blaise Pascal (1623–1662), apparently aware of the work of Maurolycus, used proof induction in his discussion of what we call Pascal's triangle in 1665, and gave an explanation of this method of proof.

A major step forward in the development of algebra is due to François Viète (1540–1603). In his book *In artem analyticum isagoge* of 1591 he developed an approach to the study of equations that focused on general cases rather than specific examples, and he promoted the use of symbols for variables and constants, which we take for granted today, but which was an innovation at the time. Viète used vowels for variables and consonants for constants; today we follow Descartes and use letters at the end of the alphabet for variables and letters at the beginning of the alphabet for constants. Viète's emphasis on the use of symbols and writing general formulas was eventually incorporated in the development of calculus—the strength of calculus is precisely that a few simple general formulas allow for the solution of many specific problems (some of which were solved by ad hoc methods prior to the invention of calculus).

Seventeenth Century

The invention of analytic geometry independently, and simultaneously in 1637, by Pierre de Fermat (1601–1665) and René Descartes (1596–1650) helped promote the recognition of the real numbers as numbers. The work of the former, though unpublished and less influential, was perhaps more modern in starting with equations and associating curves to them. The work of the latter, published in the appendix *La Géométrie* of the philosophical work *Discours de la méthode pour bien conduire sa raison et chercher la vérité dans les sciences*, always started with geometrically defined curves, and then associated equations to them.

In Euclid's *Elements*, not only is the length of a line segment not represented by a number, but also, whereas numbers (meaning rational numbers) can be multiplied, lengths of line segments cannot be multiplied (the "product" of two line segments is a rectangle, which is a distinct type of geometric object). Descartes, by contrast, defined the product of the lengths a and b of two line segments by the property that the ratio of ab to b is the same as the ratio of a to a fixed length I (which represents the number 1). That is, for Descartes the lengths of line segments can be multiplied just as numbers are multiplied. The distinction between numbers and lengths of line segments had vanished for Descartes; the set of lengths of line segments corresponded to the set of real numbers, a correspondence that was needed to form a correspondence between equations and curves, which is at the heart of analytic geometry. Observe that Descartes' approach to the real numbers as corresponding to the points on the line was not related to the decimal expansion of such numbers.

On the one hand, by the 17th century numbers were viewed abstractly—no longer the number of things in a collection of objects—and all types of number were accepted as genuine numbers. On the other hand, the real numbers were still associated with geometric ideas such as lengths of line segments. This geometric association, combined with the perceived "continuity" of the real number line, allowed mathematicians to have an intuitive idea of limits of sequences of numbers. However, this geometric link led to a reliance on intuition that precluded the need for a more rigorous approach to numbers.

Gottfried von Leibniz (1646–1716) was the first person to distinguish between algebraic numbers (that is, roots of polynomials with rational coefficients, for example $\sqrt{2}$) and transcendental numbers (that is, numbers that are not algebraic, for example π). In particular, Leibniz suggested that π might be transcendental, a fact that was proved only two centuries later.

Nineteenth Century

Carl Friedrich Gauss (1777–1855) used the idea of least upper bounds informally, which was a step forward in the development of the real numbers, but he did not provide a construction of the real numbers; he had the older notion of the real numbers as varying continuously, and took that as the basic intuitive idea of real analysis.

The first person to have attempted to construct the real numbers from the rational numbers might have been Bernard Bolzano (1781–1848), who in the first half of the 19th century defined the real numbers in terms of sequences of rational numbers, though he did not have the correct details. Bolzano made use of the Least Upper Bound Property, which he had proved assuming that Cauchy sequences are convergent, though his attempted proof that Cauchy sequences are convergent was incorrect (he formulated his "proof" prior to his attempt to construct the real numbers, so it was doomed to fail). In spite of its flaws, Bolzano's approach to the real numbers was very insightful for its time, though he did not publish these ideas, and they did not influence subsequent developments.

The attempt by Augustin Louis Cauchy (1789–1857) to put calculus on a firm foundation was a major advance in the development of real analysis, but his view of the real numbers was within the standard understanding of his time, and was not as insightful as Bolzano's approach. In the textbook *Cours d'analyse a l'École Royal Polytechnique* of 1821, Cauchy used the fact that irrational numbers are limits of sequences of rational numbers, but from Cauchy's perspective that was not a definition of irrational numbers but rather an observation about such numbers, which were simply assumed to exist. Cauchy implicitly assumed that Cauchy sequences (as we call them) are convergent, which would be intuitively true for someone who took the then prevalent intuitive view of the real numbers as "continuous," but which from our perspective requires proof, and such a proof requires either a construction of the real numbers or an axiomatization of them.

William Rowan Hamilton (1805–1865), in an effort to clarify the meaning of negative and imaginary numbers in the 1830s, gave a definition of negative numbers using a construction similar to the construction of the integers from the natural numbers found in Section 1.3. He then constructed the rational numbers from the integers, and attempted, though unsuccessfully, to construct the real numbers from the rational numbers. Using the real numbers, however constructed, he provided the modern construction of the complex numbers from the real numbers.

The first proof of the existence of transcendental numbers was due to Joseph Liouville (1809–1882) in 1844; he proved that a specific number (concocted for the purpose, though not otherwise interesting) was transcendental. Charles Hermite (1822–1901) proved that the number e was transcendental in 1873, and Ferdinand von Lindemann (1852–1939) proved that π was transcendental in 1882.

In spite of Bolzano's and Hamilton's earlier work, it was only in the second half of the 19th century that a broader effort was made to put the real numbers on an arithmetic, as opposed to geometric, basis; that is, to have the real numbers be based only upon the rational numbers (which in turn are based upon the integers). More generally, prior to the 19th century numbers were viewed as an example of the notion of "quantity," which was tied to the view of numbers and other mathematical concepts as having meaning in the real world, or at most as abstractions of things having such meaning. By the end of the 19th century that view had changed, and the modern approach to numbers was obtained. This transition involved both the development of the notion of numbers per se, and also the development of set theory as the foundation of mathematics (and in particular as a framework for the axiomatization of the real numbers).

The development of a rigorous treatment of the real numbers was the result of the desire to provide a firm foundation for calculus; without a good understanding of the real numbers, it was not possible to give complete proofs of some important results. For example, Bolzano's and Cauchy's proofs of the Intermediate Value Theorem implicitly used the Monotone Convergence Theorem, though that theorem had not been proved at the time; this theorem was also used implicitly by Cauchy and Riemann in proving that certain types of functions were integrable. To prove the Monotone Convergence Theorem it is necessary to use the fundamental properties of the real numbers.

Possibly the first rigorous construction of the real numbers from the rational numbers was due to Richard Dedekind (1831–1916), who worked out his construction in lectures in 1858, to provide a foundation for a course in real analysis that did not rely upon geometric proofs. Dedekind did not publish these ideas until 1872, in *Stetigkeit und irrationale Zahlen*, when he saw that Heine and Cantor were about to publish their versions of the construction of the real numbers. Dedekind's method, based upon what we now call "cuts," harked back to Eudoxus' approach to proportions as seen in Book V of Euclid's *Elements*. Dedekind's cuts were not exactly the same as what we call "Dedekind cuts" today; Bertrand Russell (1872–1970) used cuts the way we do now.

Another early construction of the real numbers from the rational numbers was due to Karl Weierstrass (1815–1897), who seems to have first presented his ideas on the real numbers in lectures in 1863; he did not publish these ideas. Similarly to Dedekind, Weierstrass presented his ideas in order to provide a foundation for proofs in real analysis, removing all geometric reasoning from analysis, and basing it upon numbers alone. Weierstrass essentially said that an irrational number is by definition an "aggregate" of rational numbers that intuitively converge to a number.

Charles Méray (1835–1911), first in 1869 and in more detail in 1872, defined convergence of sequences of rational numbers by using the Cauchy condition, and then essentially defined irrational numbers as Cauchy sequences that did not converge to rational limits; however, Méray was not entirely rigorous. Georg Cantor (1845–1918), apparently independently of Méray, had the idea of using Cauchy sequences of rational numbers to define the real numbers. This idea was taken up in 1872 by Cantor's colleague at Halle, Eduard Heine (1821–1881), who provided a rigorous treatment of this approach via equivalence classes. Interestingly, Heine pointed out that if his construction is done starting with the real numbers, no additional numbers are obtained.

The rigorous constructions of the real numbers from the rational numbers did not engender immediate universal support. For example, Leopold Kronecker (1823–1891), who opposed Cantor, believed only in numbers that could be finitely constructed from the natural numbers. When Lindeman proved that π is transcendental in 1882, Kronecker said "Of what use is your beautiful investigation of π? Why study such problems when irrational numbers do not exist?"

Even after the real numbers were constructed from the rational numbers, some questions about the real numbers remained. First, if the real numbers are based upon the rational numbers, what are the rational numbers based upon? The rational numbers, which are intuitively much simpler than the real numbers, can be constructed straightforwardly from the integers, which can in turn be constructed straightforwardly from the natural numbers. The question then arose as to the foundation of the natural numbers. The first person to characterize the natural numbers by axioms was apparently Dedekind, in *Was sind und was sollen die Zahlen?* of 1888, though again Dedekind formulated these ideas earlier. Dedekind's approach to the natural numbers, similarly to the more familiar Peano Postulates, started with the number 1 and with an injective but not surjective successor function, though instead of induction Dedekind used the fact that the set of natural numbers is the smallest set with the previous two properties (this approach is similar to the definition of the natural numbers in Section 2.4). Gottlob Frege (1848–1925) published his approach to the natural numbers in the book *Die Grundlagen der Arithmetik* of 1884, though it did not have full details, and was not widely read. Giuseppe Peano (1858–1932) published his widely used postulates in 1889 (these postulates are used in Section 1.2).

A more fundamental issue concerning the real numbers was raised by Frege, who questioned the approach of Weierstrass, Heine, Cantor, Dedekind, et al., asking how they knew that there were no logical contradictions that might arise from our assumptions about this set of numbers. Frege wanted to resolve the matter by basing numbers on logic (he used the notion of cardinal numbers, which in turn was based upon Cantor's idea of two sets having the same cardinality). However, Russell's paradox in set theory got in the way of Frege's program being completed, because that paradox showed that even set theory, as understood at the time, had logical contradictions.

Rather than constructing the real numbers from the rational numbers, which ultimately derive from the axioms for the natural numbers, David Hilbert (1862–1943) took the approach of defining the real numbers by their own set of axioms in *Grundlagen der Geometrie* of 1899 (where Hilbert was primarily concerned with providing a set of axioms for Euclidean geometry, to make complete what was missing from Euclid's *Elements*). Hilbert's original approach to the real numbers had axioms that define an ordered field, together with the Archimedean Principle. Those axioms do not, in fact, characterize the real numbers (the rational numbers also satisfy these properties), and in *Über den Zahlbegriff* of 1900, as well as in later editions of *Grundlagen der Geometrie*, Hilbert added the Axiom of Completeness, which says that the system cannot be made larger while maintaining all the other axioms; adding this axiom does lead to a characterization of the real numbers. Today we replace both the Archimedean Principle and the Axiom of Completeness with the Least Upper Bound Property, which implies both of the other properties. Hilbert did not prove that there was a system that satisfied his axioms, nor did he resolve Frege's concern.

Nonetheless, Hilbert's approach to the real numbers is the one used today in most introductory treatments of real analysis. In this text we take the approaches of Peano and Dedekind in Chapter 1, and the approach of Hilbert in Chapter 2.

2

Properties of the Real Numbers

2.1 Introduction

In order to prove the theorems of real analysis we need to make use of the properties of the real numbers, and hence a rigorous treatment of real analysis requires a rigorous foundation for the real numbers. As discussed in Section 1.1, we offer in this text three ways to enter into the study of the real numbers. The first two ways were given in Chapter 1, namely, Entry 1 in Section 1.2, which begins with an axiomatic treatment of the natural numbers, and Entry 2 in Section 1.4, which begins with an axiomatic treatment of the integers. Starting with either Entry 1 or Entry 2, the culmination of Chapter 1 is the construction of the real numbers, together with the proofs of the core properties of these numbers, stated in Theorem 1.7.6 and Theorem 1.7.9.

In the present chapter we take a more direct route to the main properties of the real numbers. Rather than starting with the natural numbers or the integers and constructing the real numbers, we have Entry 3 in Section 2.2, which begins with an axiomatic treatment of the real numbers. What is taken as axiomatic in Entry 3 is nothing other than what was proved in Theorem 1.7.6 and Theorem 1.7.9; and what was taken as axiomatic in Entry 1 and Entry 2 is now proved in Section 2.4, where it is shown that inside the real numbers sit the natural numbers, the integers and the rational numbers, with the expected properties. The approach taken in the present section, which is the one taken in most introductory real analysis texts, leads as quickly as possible to the core topics in real analysis, but it requires a larger set of hypotheses to be taken as axiomatic, and it provides less insight into the number systems than the approaches in Chapter 1.

The reader who starts the study of the real numbers with Entry 3 in Section 2.2 can safely have skipped Chapter 1, with the exception of Definition 1.1.1 in Section 1.1, which the reader should now read. The reader who has read Chapter 1 (starting in either Section 1.2 or 1.4) should skip Section 2.2 and go straight to Section 2.3.

2.2 Entry 3: Axioms for the Real Numbers

When we think of “all the numbers” that we normally encounter, such as 1, $\frac{7}{8}$, $\sqrt{2}$ and π, and when we think of these numbers as forming the “number line,” what we are really thinking of is the set of real numbers. (The reader might be familiar with the complex numbers, which are extremely useful in various situations, but they are not part of the real numbers—quite the opposite, they contain the real numbers—and we will not make use of complex numbers in this text.)

The set of real numbers is the mathematical universe in which the study of real analysis occurs. The set of rational numbers, by contrast, is not sufficient for our purposes, in the sense that various important theorems in real analysis, for example the Extreme Value Theorem and the Intermediate Value Theorem (both proved in Section 3.5), are not true in the context of the rational numbers. When comparing the set of real numbers with the set of rational numbers, the most immediate difference, informally, is that the real numbers contain various numbers, such as $\sqrt{2}$, that are not found in the rational numbers. However, the advantage of the set of real numbers over the set of rational numbers is not simply that the former has some useful numbers missing from the latter (which, for example, allow certain polynomial equations to be solved), but rather it is that the real numbers taken as a whole do not have “gaps.” Of course, the term “gap” is not a rigorous concept, and the real numbers have so many useful properties that it is not at all obvious which properties to select as a minimal set of properties that characterize the real numbers as distinct from other sets of numbers, and in particular which property (or properties) of the real numbers captures the notion of not having “gaps.”

Some of the needed properties of the real numbers are algebraic in nature, for example properties about addition and multiplication, though it turns out that such properties do not suffice to distinguish the real numbers from some other sets of numbers, for example the set of rational numbers. Hence we will need both algebraic axioms as well as something additional. We start with the former.

The algebraic properties we need involve not only addition and multiplication, but also the relation less than. These algebraic properties are combined in the notion of an ordered field, as defined below. The reader who has studied abstract algebra has probably encountered the concept of a field, though perhaps not an ordered field, which is simply a field as standardly defined in abstract algebra together with some additional axioms about the order relation, and the interaction of this relation with addition and multiplication. We do not assume that the reader is already familiar with fields, and we will give the complete definition here; we will prove all the algebraic properties of fields that we need for our purposes in Section 2.3.

In the following set of axioms, we use the notion of a binary operation and a unary operation on a set, as defined in Section 1.1. As is usual, we will write “xy” instead of “$x \cdot y$,” except in cases of potential ambiguity (for example, we will write “$1 \cdot 1$” rather than “11”), or where the $\cdot$ makes the expression easier to read. We will write $x > y$ to mean the same thing as $y < x$.

Definition 2.2.1. An **ordered field** is a set F with elements $0, 1 \in F$, binary operations $+$ and $\cdot$, a unary operation $-$, a relation $<$, and a unary operation $^{-1}$ on $F - \{0\}$, which satisfy the following properties. Let $x, y, z \in F$.

a. $(x+y)+z = x+(y+z)$ (Associative Law for Addition).
b. $x+y = y+x$ (Commutative Law for Addition).
c. $x+0 = x$ (Identity Law for Addition).
d. $x+(-x) = 0$ (Inverses Law for Addition).
e. $(xy)z = x(yz)$ (Associative Law for Multiplication).
f. $xy = yx$ (Commutative Law for Multiplication).
g. $x \cdot 1 = x$ (Identity Law for Multiplication).
h. If $x \neq 0$, then $xx^{-1} = 1$ (Inverses Law for Multiplication).
i. $x(y+z) = xy+xz$ (Distributive Law).
j. Precisely one of $x < y$ or $x = y$ or $x > y$ holds (Trichotomy Law).
k. If $x < y$ and $y < z$, then $x < z$ (Transitive Law).
l. If $x < y$ then $x+z < y+z$ (Addition Law for Order).
m. If $x < y$ and $z > 0$, then $xz < yz$ (Multiplication Law for Order).
n. $0 \neq 1$ (Non-Triviality). △

The Non-Triviality axiom might seem, well, trivial, but it is very much needed, because otherwise we could have an ordered field consisting of a single number 0, which is not what we would want the set of real numbers to be.

The properties of an ordered field do not alone characterize the real numbers, because the rational numbers are also an ordered field (as was proved in Theorem 1.5.5, for the reader who commenced the study of the real numbers in Chapter 1, and as can be deduced from Axiom 2.2.4 and Corollary 2.4.14 for the reader who commences in the present section). In order to distinguish the real numbers from all other ordered fields, we will need one additional axiom, to which we now turn. This axiom uses the concepts of upper bounds and least upper bounds; while we are at it, we will also define the related concepts of lower bounds and greatest lower bounds. To avoid interrupting the flow of our treatment of the real numbers, we state the following definition without further discussion now, but we will have a full discussion of it in Section 2.6.

Definition 2.2.2. Let F be an ordered field, and let $A \subseteq F$ be a set.

1. The set A is **bounded above** if there is some $M \in F$ such that $x \leq M$ for all $x \in A$. The number M is called an **upper bound** of A.
2. The set A is **bounded below** if there is some $P \in F$ such that $x \geq P$ for all $x \in A$. The number P is called a **lower bound** of A.
3. The set A is **bounded** if it is bounded above and bounded below.
4. Let $M \in F$. The number M is a **least upper bound** (also called a **supremum**) of A if M is an upper bound of A, and if $M \leq T$ for all upper bounds T of A.
5. Let $P \in F$. The number P is a **greatest lower bound** (also called an **infimum**) of A if P is a lower bound of A, and if $P \geq V$ for all lower bounds V of A. △

As the reader is asked to verify in Exercise 2.3.11, a subset $A \subseteq \mathbb{R}$ is bounded if and only if there is some $M \in \mathbb{R}$ such that $|x| \leq M$ for all $x \in A$; it is always possible to choose M so that $M > 0$. The proof of this fact has to wait until Section 2.3 because that is where we define absolute value.

Whereas having upper bounds and lower bounds of subsets of an ordered field is not remarkable, it is the existence of least upper bounds and greatest lower bounds that characterizes the real numbers. As the reader has most likely encountered, and as we will prove rigorously in Theorem 2.6.11, the number $\sqrt{2}$ is a real number but not a rational number. If we look at the set of all rational numbers that are less than $\sqrt{2}$, that set certainly has an upper bound (for example 2), but it does not have a least upper bound (intuitively because the only possible candidate for such a least upper bound would be $\sqrt{2}$, and that is not in the set of rational numbers). The existence of such problems is what renders the set of rational numbers unfit to be the basis for calculus. No such problem exists in the set of real numbers, and it is this difference between the real numbers and the rational numbers that is the basis for the following definition, and for our axiomatic characterization of the real numbers. Of course, not every subset of the real numbers has a least upper bound, for example a set that has no upper bound, but what truly characterizes the real numbers is that if a non-empty set has an upper bound, then it must have a least upper bound.

Definition 2.2.3. Let F be an ordered field. The ordered field F satisfies the **Least Upper Bound Property** if every non-empty subset of F that is bounded above has a least upper bound. △

We are now ready for our axiomatic characterization of the real numbers, which combines algebraic properties together with the Least Upper Bound Property.

Axiom 2.2.4 (Axiom for the Real Numbers). *There exists an ordered field* $\mathbb{R}$ *that satisfies the Least Upper Bound Property.*

Observe that the Axiom for the Real Numbers (Axiom 2.2.4) does not say that the real numbers are unique, though in fact that turns out to be true, as will be proved in Section 2.7.

Although it may not seem very impressive upon first encounter, we will see in Section 2.6, and throughout this text, just how powerful the Least Upper Bound Property of the real numbers is. Indeed, virtually all the major theorems in this text, concerning such topics as continuous functions, derivatives, integrals, sequences and series, rely upon the Least Upper Bound Property.

Reflections

In the modern approach to mathematics, as seen for example by anyone who has taken a course in abstract algebra, it is standard to define new objects of study axiomatically. It should therefore not surprise the reader that one of the ways, and in fact the most common way, of studying real analysis is by starting with axioms for the real numbers. What is surprising is that the real numbers can be characterized by such a relatively simple set of axioms. The axioms for an ordered field are entirely straightforward and not unexpected, and the Least Upper Bound Property, while not as intuitively simple as the axioms for an ordered field, is also not complicated, and is not intuitively unreasonable in retrospect. And yet, though easily stated, the Least Upper Bound Property is sufficient to distinguish between the real numbers and all other ordered fields, and is sufficient to make all of calculus work. The fact that mathematicians, at the very end of the lengthy period of time during which calculus was developed, were able to figure out precisely what makes the real numbers work as they do should be viewed as a remarkable solution to a very major intellectual puzzle.

The reader might wonder whether it would be possible to replace the Least Upper Bound Property with some other (perhaps more familiar) axiom, and still obtain the real numbers. There are, in fact, a number of possible replacements for the Least Upper Bound Property, some of which should be familiar to the reader. As we will see in Theorem 3.5.4 and Theorem 8.3.17, there are a number of important theorems in calculus, such as the Extreme Value Theorem (Theorem 3.5.1) and the Intermediate Value Theorem (Theorem 3.5.2), which are equivalent to the Least Upper Bound Property; any of the theorems that are equivalent to the Least Upper Bound Property could substitute for the latter in the axioms for the real numbers. However, the Least Upper Bound Property is by far easier to state than any of these other results (which require the definitions of concepts such as continuous functions), and so it makes sense to adopt the Least Upper Bound Property as the axiom. Moreover, if we were to adopt one of these equivalent theorems as an axiom for the real numbers, we would immediately have had to derive the Least Upper Bound Property from that axiom in order to prove the other theorems of calculus, and so it is more efficient to start with the Least Upper Bound Property. Of course, the Greatest Lower Bound Property would be just as good an axiom as the Least Upper Bound Property, and it is an arbitrary, though by now standard, decision to take the latter rather than the former axiomatically.

2.3 Algebraic Properties of the Real Numbers

Whether you have read Section 1.7, in which the real numbers were constructed and their basic properties were proved, or whether you started reading from Section 2.2, where we assumed these basic properties as axioms for the real numbers, from now on we will be making use of the properties of the real numbers only, and not how the real numbers are constructed. In general, what counts in mathematics is how objects behave, not "what they are." Ultimately, the real numbers are numbers that behave in a certain way.

In this section, and in the remaining sections of this chapter, we will explore various aspects of the real numbers, and we will show that they all follow from the axiomatic properties of the real numbers. In the present section we will discuss various useful algebraic properties of the real numbers, by which we mean properties that follow strictly from those aspects of the real numbers that were proved in Theorem 1.7.6 (for those readers who have read Section 1.7), and were stated as the axiom for an ordered field in Definition 2.2.1 (for those readers who have read Section 2.2). The majority of the facts that we prove in the present section are certainly very familiar to the reader, and as such require no motivation, but we need to prove these facts nonetheless, because we will use them later on, and we want to make sure that everything we prove about real analysis throughout this text is ultimately derived from nothing more than the axioms for the real numbers.

In order to show exactly how important the axiomatic properties of the real numbers are, and to make sure that we are not using any assumptions about the real numbers other than those we have stated as axiomatic, we will explicitly refer to such properties in the present section (including exercises) whenever they are used. Such explicit references to the axiomatic properties of the real numbers are admittedly tedious, and we will not do so after the present section, other than references to the Least Upper Bound Property, which is always worth mentioning.

For convenience, we need the following terminology and notation.

Definition 2.3.1.

1. The binary operation $-$ on $\mathbb{R}$ is defined by $a-b=a+(-b)$ for all $a,b\in\mathbb{R}$. The binary operation $\div$ on $\mathbb{R}-\{0\}$ is defined by $a\div b=ab^{-1}$ for all $a,b\in\mathbb{R}-\{0\}$; we also let $0\div s=0\cdot s^{-1}=0$ for all $s\in\mathbb{R}-\{0\}$. The number $a\div b$ is also denoted $\frac{a}{b}$.
2. Let $a\in\mathbb{R}$. The **square** of a, denoted a^2, is defined by $a^2=a\cdot a$.
3. The relation $\leq$ on $\mathbb{R}$ is defined by $x\leq y$ if and only if $x<y$ or $x=y$, for all $x,y\in\mathbb{R}$.
4. The number $2\in\mathbb{R}$ is defined by $2=1+1$. △

Observe that if $b\in\mathbb{R}$ then $0-b=-b$, and if $b\neq 0$ then $\frac{1}{b}=b^{-1}$.

We now see some basic algebraic properties of the real numbers involving addition and multiplication. The reader who has studied some abstract algebra can safely skip the proof of the following lemma, having already seen many such proofs. As is usual, we will write "$-ab$" when we mean "$-(ab)$."

Lemma 2.3.2. *Let $a,b,c\in\mathbb{R}$.*

1. *If $a+c=b+c$ then $a=b$ (Cancellation Law for Addition).*
2. *If $a+b=a$ then $b=0$.*
3. *If $a+b=0$ then $b=-a$.*
4. *$-(a+b)=(-a)+(-b)$.*
5. *$-0=0$.*
6. *If $ac=bc$ and $c\neq 0$, then $a=b$ (Cancellation Law for Multiplication).*
7. *$0\cdot a=0=a\cdot 0$.*
8. *If $ab=a$ and $a\neq 0$, then $b=1$.*
9. *If $ab=1$ then $b=a^{-1}$.*
10. *If $a\neq 0$ and $b\neq 0$, then $(ab)^{-1}=a^{-1}b^{-1}$.*
11. *$(-1)\cdot a=-a$.*
12. *$(-a)b=-ab=a(-b)$.*
13. *$-(-a)=a$.*
14. *$(-1)^2=1$ and $1^{-1}=1$.*
15. *If $ab=0$, then $a=0$ or $b=0$ (No Zero Divisors Law).*
16. *If $a\neq 0$ then $(a^{-1})^{-1}=a$.*
17. *If $a\neq 0$ then $(-a)^{-1}=-a^{-1}$.*

Proof. We will prove Parts (1), (5), (7), (9), (11) and (15), leaving the rest to the reader in Exercise 2.3.1.

(1) Suppose that $a+c=b+c$. Then $(a+c)+(-c)=(b+c)+(-c)$. By the Associative Law for Addition it follows that $a+(c+(-c))=b+(c+(-c))$, and hence by the Inverses Law for Addition we see that $a+0=b+0$, and therefore by the Identity Law for Addition we conclude that $a=b$.

(5) By the Identity Law for Addition we know that $0+0=0$, and it follows from Part (3) of this lemma that $0=-0$.

(7) By the Identity Law for Addition we know that $0+0=0$, and hence $a(0+0)=a\cdot 0$. By the Distributive Law it follows that $a\cdot 0+a\cdot 0=a\cdot 0$, and using the Identity and Commutative Laws for Addition we deduce that $a\cdot 0+a\cdot 0=0+a\cdot 0$. It now follows from Part (1) of this lemma that $a\cdot 0=0$. The Commutative Law for Multiplication then implies that $0\cdot a=0$.

(9) Suppose that $ab=1$. It cannot be the case that $a=0$, because if a were 0, then we would have $ab=0$ by Part (7) of this lemma, which would then imply that $0=1$, which is a contradiction to Non-Triviality. Hence a^{-1} exists, and then using the Identity, Associative, Commutative and Inverses Laws for Multiplication we see that $a^{-1}=a^{-1}\cdot 1=a^{-1}(ab)=(a^{-1}a)b=1\cdot b=b$.

(11) By the Identity Law for Multiplication, the Distributive Law, the Inverses Law for Addition and Part (7) of this lemma we see that $a+a\cdot(-1)=a\cdot 1+a\cdot(-1)=a[1+(-1)]=a\cdot 0=0$. Part (3) of this lemma then implies that $(-1)\cdot a=-a$.

(15) Suppose that $ab=0$ and that $a\neq 0$. Hence a^{-1} exists, and Part (7) of this lemma, and the Commutative, Associative, Inverses and Identity Laws for Multiplication, imply that $0=a^{-1}\cdot 0=a^{-1}(ab)=(a^{-1}a)b=1\cdot b=b$. □

Observe that Lemma 2.3.2 (2) shows that 0 is unique, Part (8) of the lemma shows that 1 is unique, Part (3) of the lemma shows that $-a$ is unique and Part (9) of the lemma shows that a^{-1} is unique.

We now turn to some properties of the real numbers involving the relations less than, and less than or equal to.

Lemma 2.3.3. *Let $a,b,c,d\in\mathbb{R}$.*

1. *If $a\leq b$ and $b\leq a$, then $a=b$.*
2. *If $a\leq b$ and $b\leq c$, then $a\leq c$. If $a\leq b$ and $b<c$, then $a<c$. If $a<b$ and $b\leq c$, then $a<c$.*
3. *If $a\leq b$ then $a+c\leq b+c$.*
4. *If $a<b$ and $c<d$, then $a+c<b+d$; if $a\leq b$ and $c\leq d$, then $a+c\leq b+d$.*
5. *$a>0$ if and only if $-a<0$, and $a<0$ if and only if $-a>0$; also $a\geq 0$ if and only if $-a\leq 0$, and $a\leq 0$ if and only if $-a\geq 0$.*
6. *$a<b$ if and only if $b-a>0$ if and only if $-b<-a$. Also $a\leq b$ if and only if $b-a\geq 0$ if and only if $-b\leq -a$.*
7. *If $a\neq 0$ then $a^2>0$.*

8. $-1 < 0 < 1$.
9. $a < a + 1$.
10. *If* $a \leq b$ *and* $c > 0$*, then* $ac \leq bc$.
11. *If* $0 \leq a < b$ *and* $0 \leq c < d$*, then* $ac < bd$*; if* $0 \leq a \leq b$ *and* $0 \leq c \leq d$*, then* $ac \leq bd$.
12. *If* $a < b$ *and* $c < 0$*, then* $ac > bc$.
13. *If* $a > 0$ *then* $a^{-1} > 0$.
14. *If* $a > 0$ *and* $b > 0$*, then* $a < b$ *if and only if* $b^{-1} < a^{-1}$ *if and only if* $a^2 < b^2$.

Proof. We will prove Parts (1), (3), (5), (7), (8), (11) and (12), leaving the rest to the reader in Exercise 2.3.2.

(1) Suppose that $a \leq b$ and $b \leq a$. By the definition of the relation $\leq$ we see that $a < b$ or $a = b$, and that $b < a$ or $b = a$. First, suppose that $a < b$. By the Trichotomy Law it cannot be the case that $a = b$ or $a > b$, leading to a contradiction. A similar contradiction occurs if we assume that $b < a$. The only possibility that remains is that $a = b$.

(3) Suppose that $a \leq b$. Then $a < b$ or $a = b$. There are now two cases. First, suppose that $a < b$. Then by the Addition Law for Order we know that $a + c < b + c$. Second, suppose that $a = b$. Then clearly $a + c = b + c$. Hence $a + c < b + c$ or $a + c = b + c$, which means that $a + c \leq b + c$.

(5) Suppose that $a > 0$. Then by the Addition Law for Order we see that $a + (-a) > 0 + (-a)$, and then using the Commutative, Identity and Inverses Laws for Addition, we deduce that $0 > -a$. Similarly, if $-a < 0$, then $(-a) + a < 0 + a$, and it follows that $0 < a$. The proofs of the other three parts are similar, and we omit the details.

(7) Suppose that $a \neq 0$. Then by the Trichotomy Law we know that $a > 0$ or $a < 0$. First, suppose that $a > 0$. Then by the Multiplication Law for Order (using $a > 0$ in the role of both inequalities in that law), we deduce that $a \cdot a > 0 \cdot a$. It then follows from Lemma 2.3.2 (7) that $a^2 > 0$.

Second, suppose that $a < 0$. By Part (5) of this lemma we know that $-a > 0$. Using the previous paragraph we deduce that $(-a)^2 > 0$. Applying Lemma 2.3.2 (12) twice we see that $(-a)^2 = (-a)(-a) = -[a(-a)] = -[-a^2]$, and by Part (13) of that lemma we deduce that $(-a)^2 = a^2$. It follows that $a^2 > 0$.

(8) By Non-Triviality we know that $1 \neq 0$. By Part (7) of this lemma it follows that $1^2 > 0$, and by the Identity Law for Multiplication we deduce that $1 > 0$.

From Part (5) of this lemma we now see that $-1 < -0$, and by Lemma 2.3.2 (5) we deduce that $-1 < 0$.

(11) Suppose that $0 \leq a < b$ and $0 \leq c < d$. By Part (2) of this lemma we see that $b > 0$ and $d > 0$.

There are now two cases. First, suppose that $a = 0$. By Lemma 2.3.2 (7) we then deduce that $ac = 0$. Because $b > 0$ and $d > 0$, it follows from the Multiplication Law for Order and Lemma 2.3.2 (7) again that $0 = 0 \cdot d < bd$. Hence $ac < bd$.

Second, suppose that $a > 0$. We then use the Multiplication Law for Order and the Commutative Law for Multiplication to deduce that $ac < ad$. Because $d > 0$, we use the Multiplication Law for Order again to deduce that $ad < bd$. By the Transitive Law it follows that $ac < bd$.

The proof of the other part is similar, and we omit the details.

(12) Suppose that $a < b$ and that $c < 0$. By Part (5) of this lemma we know that $-c > 0$. By the Multiplication Law for Order it follows that $a(-c) < b(-c)$. We then use Lemma 2.3.2 (12) to see that $-ac < -bc$. It now follows from Part (6) of this lemma that $bc < ac$. □

We now have a definition and lemma concerning positive and negative real numbers.

Definition 2.3.4. Let $a \in \mathbb{R}$. The number a is **positive** if $a > 0$; the number a is **negative** if $a < 0$; and the number a is **non-negative** if $a \geq 0$. △

Lemma 2.3.5. *Let $a, b, c, d \in \mathbb{R}$.*

1. *If $a > 0$ and $b > 0$, then $a + b > 0$. If $a > 0$ and $b \geq 0$, then $a + b > 0$. If $a \geq 0$ and $b \geq 0$, then $a + b \geq 0$.*
2. *If $a < 0$ and $b < 0$, then $a + b < 0$. If $a < 0$ and $b \leq 0$, then $a + b < 0$. If $a \leq 0$ and $b \leq 0$, then $a + b \leq 0$.*
3. *If $a > 0$ and $b > 0$, then $ab > 0$. If $a > 0$ and $b \geq 0$, then $ab \geq 0$. If $a \geq 0$ and $b \geq 0$, then $ab \geq 0$.*
4. *If $a < 0$ and $b < 0$, then $ab > 0$. If $a < 0$ and $b \leq 0$, then $ab \geq 0$. If $a \leq 0$ and $b \leq 0$, then $ab \geq 0$.*
5. *If $a < 0$ and $b > 0$, then $ab < 0$. If $a < 0$ and $b \geq 0$, then $ab \leq 0$. If $a \leq 0$ and $b > 0$, then $ab \leq 0$. If $a \leq 0$ and $b \geq 0$, then $ab \leq 0$.*

Proof. We will prove Parts (1) and (5), leaving the rest to the reader in Exercise 2.3.4.

(1) First, suppose that $a > 0$ and $b > 0$. Then by Lemma 2.3.3 (4) and the Identity Law for Addition we see that $a + b > 0 + 0 = 0$.

Second, suppose that $a > 0$ and $b \geq 0$. There are now two subcases. First, suppose that $b > 0$. Then by the previous paragraph we know that $a + b > 0$. Second, suppose that $b = 0$. Then by the Identity Law for Addition we see that $a + b = a + 0 = a > 0$.

Third, suppose that $a \geq 0$ and $b \geq 0$. There are now two subcases. First, suppose that $a > 0$. Then by the previous paragraph we know that $a + b > 0$, which implies that $a + b \geq 0$. Second, suppose that $a = 0$. Then by the Commutative and Identity Laws for Addition we see that $a + b = 0 + b = b + 0 = b \geq 0$.

(5) First, suppose that $a < 0$ and $b > 0$. By the Multiplicative Law for Order and Lemma 2.3.2 (7) we deduce that $ab < 0 \cdot b = 0$.

Second, suppose that $a < 0$ and $b \geq 0$. There are now two subcases. First, suppose that $b > 0$. Then by the previous paragraph we know that $ab < 0$, which implies that $ab \leq 0$. Second, suppose that $b = 0$. Then by Lemma 2.3.2 (7) we see that $ab = a \cdot 0 = 0$, and hence $ab \leq 0$.

The proofs of the other two parts are similar, and we omit the details. □

Observe that Lemma 2.3.5 (1) implies that the positive numbers are closed under addition, and Part (3) of the lemma implies that the positive numbers are closed under multiplication. This use of the term "closed" is employed only informally, and only occasionally, in contrast to the very different, and very important, use of this same word in the following definition.

We now define various types of intervals in the real numbers. The reader has most likely encountered the use of intervals in previous mathematics courses, for example precalculus and calculus, but their importance might not have been evident in those courses. By contrast, the various types of intervals play a fundamental role in real analysis. As the reader will see throughout this text, many of the important theorems in real analysis are stated in terms of intervals (either open or closed, depending upon the situation).

Definition 2.3.6. An **open bounded interval** is a set of the form

$$(a,b) = \{x \in \mathbb{R} \mid a < x < b\},$$

where $a,b \in \mathbb{R}$ and $a \leq b$. A **closed bounded interval** is a set of the form

$$[a,b] = \{x \in \mathbb{R} \mid a \leq x \leq b\},$$

where $a,b \in \mathbb{R}$ and $a \leq b$. A **half-open interval** is a set of the form

$$[a,b) = \{x \in \mathbb{R} \mid a \leq x < b\} \quad \text{or} \quad (a,b] = \{x \in \mathbb{R} \mid a < x \leq b\},$$

where $a,b \in \mathbb{R}$ and $a \leq b$. An **open unbounded interval** is a set of the form

$$(a,\infty) = \{x \in \mathbb{R} \mid a < x\} \quad \text{or} \quad (-\infty,b) = \{x \in \mathbb{R} \mid x < b\} \quad \text{or} \quad (-\infty,\infty) = \mathbb{R},$$

where $a,b \in \mathbb{R}$. A **closed unbounded interval** is a set of the form

$$[a,\infty) = \{x \in \mathbb{R} \mid a \leq x\} \quad \text{or} \quad (-\infty,b] = \{x \in \mathbb{R} \mid x \leq b\},$$

where $a,b \in \mathbb{R}$.

An **open interval** is either an open bounded interval or an open unbounded interval. A **closed interval** is either a closed bounded interval or a closed unbounded interval. A **right unbounded interval** is any interval of the form (a,∞), $[a,\infty)$ or $(-\infty,\infty)$. A **left unbounded interval** is any interval of the form $(-\infty,b)$, $(-\infty,b]$ or $(-\infty,\infty)$. A **non-degenerate interval** is any interval of the form (a,b), $(a,b]$, $[a,b)$ or $[a,b]$ where $a < b$, or any unbounded interval. The number a in intervals of the form $[a,b)$, $[a,b]$ or $[a,\infty)$ is called the **left endpoint** of the interval. The number b in intervals of the form $(a,b]$, $[a,b]$ or $(-\infty,b]$ is called the **right endpoint** of the interval. An **endpoint** of an interval is either a left endpoint or a right endpoint. The **interior** of an interval is everything in the interval other than its endpoints (if it has any). △

We note that there are no intervals that are "closed" at ∞ or $-\infty$ (for example, there is no interval of the form $[a,\infty]$), because ∞ is not a real number, and therefore it cannot be included in an interval contained in the real numbers. The symbol "∞" is simply a shorthand way of saying that an interval "goes on forever."

In the following lemma, and in many places throughout this text (and in virtually all real analysis texts), we will make use of the following convenient, but not entirely proper, phraseology. The symbols ε and δ (and other Greek letters) are often used to denote positive real numbers, which we often think of intuitively as being very small, though in principle they could be any size. Because this situation is so common, rather than quantifying such ε and δ properly by saying "for all $\varepsilon \in \mathbb{R}$ such that $\varepsilon > 0$" or "there is some $\delta \in \mathbb{R}$ such that $\delta > 0$," we will simply say "for all $\varepsilon > 0$" or "there is some $\delta > 0$." In other words, it will always be assumed that "ε" and "δ" in such situations are real numbers.

The following lemma about intervals is simple but very useful. The second part of the lemma is essentially what characterizes open intervals. The reader will learn more about "openness" in an introductory course in point set topology; see [Mun00].

Lemma 2.3.7. *Let $I \subseteq \mathbb{R}$ be an interval.*

1. *If $x, y \in I$ and $x \le y$, then $[x, y] \subseteq I$.*
2. *If I is an open interval, and if $x \in I$, then there is some $\delta > 0$ such that $[x - \delta, x + \delta] \subseteq I$.*

Proof. Left to the reader in Exercise 2.3.6. □

A very useful concept in real analysis is the absolute value of real numbers, which is defined as follows.

Definition 2.3.8. Let $a \in \mathbb{R}$. The **absolute value** of a, denoted $|a|$, is defined by

$$|a| = \begin{cases} a, & \text{if } a \ge 0 \\ -a, & \text{if } a < 0. \end{cases}$$

△

The following lemma states the basic properties of absolute value.

Lemma 2.3.9. *Let $a, b \in \mathbb{R}$.*

1. *$|a| \ge 0$, and $|a| = 0$ if and only if $a = 0$.*
2. *$-|a| \le a \le |a|$.*
3. *$|a| = |b|$ if and only if $a = b$ or $a = -b$.*
4. *$|a| < b$ if and only if $-b < a < b$, and $|a| \le b$ if and only if $-b \le a \le b$.*
5. *$|ab| = |a| \cdot |b|$.*
6. *$|a + b| \le |a| + |b|$ (Triangle Inequality).*
7. *$|a| - |b| \le |a + b|$ and $|a| - |b| \le |a - b|$.*

Proof. We will prove Parts (2), (4), (5) and (6), leaving the rest to the reader in Exercise 2.3.7.

(2) There are two cases. First, suppose that $a \ge 0$. Then $|a| = a$, and hence $|a| \ge 0$. By Lemma 2.3.3 (5) we see that $-|a| \le 0$. It then follows from Lemma 2.3.3 (2) that $-|a| \le a \le |a|$. Second, suppose that $a < 0$. Then $-a = |a|$, and hence by

Lemma 2.3.2 (13) we see that $-|a| = a$. By Part (1) of this lemma we know that $|a| \geq 0$, and Lemma 2.3.3 (2) then implies that $-|a| \leq a \leq |a|$.

(4) Suppose that $|a| < b$. By Part (1) of this lemma we know that $|a| \geq 0$, and hence by Lemma 2.3.3 (2) we deduce that $b > 0$. It follows from Lemma 2.3.3 (5) that $-b < 0$. There are now two cases. First, suppose that $a \geq 0$. Then $|a| = a$, and hence $|a| < b$ implies that $a < b$. We saw that $-b < 0$, and hence by Lemma 2.3.3 (2) we know that $-b < a$. Therefore $-b < a < b$. Second, suppose that $a < 0$. Because $b > 0$, it follows from the Transitive Law that $a < b$. Because $|a| = -a$ and $|a| < b$, we see that $-a < b$. We now use Lemma 2.3.3 (6) and Lemma 2.3.2 (13) to see that $-b < a$. Again, we deduce that $-b < a < b$.

Now suppose that $-b < a < b$. By Lemma 2.3.3 (6) and Lemma 2.3.2 (13) it follows that $-b < -a < b$. Because $|a|$ equals either a or $-a$, we therefore see in either case that $|a| < b$.

The fact that $|a| \leq b$ if and only if $-b \leq a \leq b$ follows from what we have already seen together with Part (3) of this lemma; we omit the details.

(5) There are four cases. First, suppose that $a \geq 0$ and $b \geq 0$. By Lemma 2.3.5 (3) it follows that $ab \geq 0$. Therefore $|a| = a$, and $|b| = b$, and $|ab| = ab$. Hence $|ab| = ab = |a| \cdot |b|$.

Second, suppose that $a \geq 0$ and $b < 0$. Then $a > 0$ or $a = 0$. There are now two subcases. First, suppose that $a > 0$. Then by Lemma 2.3.5 (5) we know that $ab < 0$. Therefore $|a| = a$, and $|b| = -b$, and $|ab| = -ab$. Hence by Lemma 2.3.2 (12) we deduce that $|ab| = -ab = a(-b) = |a| \cdot |b|$. Second, suppose that $a = 0$. Then by Lemma 2.3.2 (7) we see that $ab = 0$. We therefore see that $|a| = a = 0$, and $|b| = -b$, and $|ab| = ab = 0$. Hence, using Lemma 2.3.2 (7) again, we see that $|ab| = 0 = 0 \cdot b = |a| \cdot |b|$.

Third, suppose that $a < 0$ and $b \geq 0$. This case is just like the previous case, and we omit the details.

Fourth, suppose that $a < 0$ and $b < 0$. Then by Lemma 2.3.5 (4) we know that $ab > 0$. We therefore see that $|a| = -a$, and $|b| = -b$, and $|ab| = ab$. Hence, using Lemma 2.3.2 (12) (13) we see that $|ab| = ab = -(-ab) = -[a(-b)] = (-a)(-b) = |a| \cdot |b|$.

(6) By Part (2) of this lemma we know that $-|a| \leq a \leq |a|$ and $-|b| \leq b \leq |b|$. Using Lemma 2.3.3 (4) we deduce that $(-|a|) + (-|b|) \leq a + b \leq |a| + |b|$. By Lemma 2.3.2 (4) we see that $-(|a| + |b|) \leq a + b \leq |a| + |b|$, and it now follows from Part (4) of this lemma that $|a + b| \leq |a| + |b|$. □

It is hard to overstate the importance for real analysis of the Triangle Inequality (Lemma 2.3.9 (6)) and its variants in Lemma 2.3.9 (7). We will use these inequalities repeatedly in proofs throughout this text. Also, the Triangle Inequality can be extended to the sum of more than two numbers, as stated in Exercise 2.5.3, and we will use that version repeatedly as well.

The results that we have proved up till now in the present section should be very familiar (that is, the statements of the results should be familiar, not necessarily the proofs). The following result, by contrast, though quite intuitively reasonable, might be one that the reader has not previously encountered. It happens on occasion in real analysis that we will want to prove that a real number a is equal to zero, but rather than doing so directly, which is sometimes difficult, we do so indirectly by proving that $|a|$ is smaller than any positive real number. The following lemma shows that this type of proof is valid.

Lemma 2.3.10. *Let $a \in \mathbb{R}$.*

1. *$a \leq 0$ if and only if $a < \varepsilon$ for all $\varepsilon > 0$.*
2. *$a \geq 0$ if and only if $a > -\varepsilon$ for all $\varepsilon > 0$.*
3. *$a = 0$ if and only if $|a| < \varepsilon$ for all $\varepsilon > 0$.*

Proof. We will prove Parts (1) and (3), leaving the remaining part to the reader in Exercise 2.3.12.

(1) Suppose that $a \leq 0$. Let $\varepsilon > 0$. Then $a < \varepsilon$ by Lemma 2.3.3 (2).

Now suppose that $a < \delta$ for all $\delta > 0$. We use proof by contradiction. Suppose that $a > 0$. Then $a < a$, which is a contradiction to the Trichotomy Law, because $a = a$. Hence $a > 0$ is false, and by the Trichotomy Law it follows that $a \leq 0$.

(3) Suppose that $a = 0$. Then $|a| = a = 0$. Therefore $|a| \leq 0$. By Part (1) of this lemma we deduce that $|a| < \varepsilon$ for all $\varepsilon > 0$.

Now suppose that $|a| < \delta$ for all $\delta > 0$. It follows from Part (1) of this lemma that $|a| \leq 0$. On the other hand, we know by the first part of Lemma 2.3.9 (1) that $|a| \geq 0$. By Lemma 2.3.3 (1) we deduce that $|a| = 0$. We now use the second part of Lemma 2.3.9 (1) to conclude that $a = 0$. □

Not only is Lemma 2.3.10 technically useful, but it also has a very important philosophical consequence. In the early development of calculus, before the modern concept of a limit was developed, the notion of "infinitesimal" was used, where an infinitesimal is some sort of number that is "infinitely small," which means a non-zero number that has absolute value smaller than every real number (technically, the number zero is also considered to be an infinitesimal, though it is only non-zero infinitesimals that are useful). Although infinitesimals were conceptually important in the early development of calculus, Lemma 2.3.10 implies that there are no non-zero infinitesimals among the real numbers, and as such standard modern treatments of real analysis (such as ours) make no use of infinitesimals. (Infinitesimals can in fact be developed rigorously, but not as numbers that are included in the set of real numbers; rather, they are to be considered as part of a larger set that contains the real numbers, infinitesimals and infinitely large numbers. See [Gol98] for a rigorous treatment of infinitesimals, and see [Kei] or [HK03] for an elementary treatment of calculus using infinitesimals instead of limits.)

Now that we have seen proofs of the basic algebraic properties of the real numbers, we will usually use these properties without reference, to avoid cluttering up difficult proofs with very simple and familiar facts.

Reflections

The various properties of the real numbers proved in this section are intuitively straightforward, and most of them should be familiar to the reader from previous courses; these properties also hold for any other ordered field, such as the rational numbers. However, some of the topics in this section receive more attention in real analysis than in previous courses, for example the use of intervals. Open intervals and closed intervals are used in precalculus and calculus courses, but their importance, and especially the substantial difference between these two types of intervals, is not stressed there. In real analysis, by contrast, intervals have a very important role to play. For example, the Extreme Value Theorem (Theorem 3.5.1), which is an important result about continuous functions, very much depends upon the fact that the domain is a closed bounded interval, and not some other type of interval. A complete understanding of the difference between open intervals and closed intervals awaits the reader in an introductory topology course, where these two types of intervals are seen as special cases of the more general concepts of open sets and closed sets, and where open sets in particular have a starring role.

Exercises

Exercise 2.3.1. **[Used in Lemma 2.3.2.]** Prove Lemma 2.3.2 (2) (3) (4) (6) (8) (10) (12) (13) (14) (16) (17).

Exercise 2.3.2. **[Used in Lemma 2.3.3.]** Prove Lemma 2.3.3 (2) (4) (6) (9) (10) (13) (14).

Exercise 2.3.3. **[Used in Example 4.5.3 and Example 4.6.1.]** For any $a \in \mathbb{R}$, let a^3 denote $a \cdot a \cdot a$.

Let $x, y \in \mathbb{R}$.

(1) Prove that if $x < y$, then $x^3 < y^3$.
(2) Prove that there are $c, d \in \mathbb{R}$ such that $c^3 < x < d^3$.

Exercise 2.3.4. **[Used in Lemma 2.3.5.]** Prove Lemma 2.3.5 (2) (3) (4).

Exercise 2.3.5. **[Used in Exercise 2.3.6 and Exercise 2.8.9.]**

(1) Prove that $1 < 2$.
(2) Prove that $0 < \frac{1}{2} < 1$.
(3) Prove that if $a, b \in \mathbb{R}$ and $a < b$, then $a < \frac{a+b}{2} < b$.

Exercise 2.3.6. **[Used in Lemma 2.3.7.]** Prove Lemma 2.3.7. [Use Exercise 2.3.5 (3).]

Exercise 2.3.7. **[Used in Lemma 2.3.9.]** Prove Lemma 2.3.9 (1) (3) (7).

Exercise 2.3.8. **[Used throughout.]** Let $I \subseteq \mathbb{R}$ be an open interval, let $c \in I$ and let $\delta > 0$. Prove that there is some $x \in I - \{c\}$ such that $|x - c| < \delta$. [Use Exercise 2.3.5 (3).]

Exercise 2.3.9. **[Used in Theorem 10.4.4 and Exercise 10.4.4.]** Let $a \in \mathbb{R}$, let $R \in (0, \infty)$ and let $x \in (a - R, a + R)$. Prove that there is some $P \in (0, R)$ such that $x \in (a - P, a + P)$.

Exercise 2.3.10. **[Used in Theorem 8.4.8.]** Let $I \subseteq \mathbb{R}$ be a non-degenerate open interval, and let $c \in \mathbb{R}$. Prove that there is a non-degenerate open bounded interval $(a,b) \subseteq I$ such that $c \notin [a,b]$.

Exercise 2.3.11. **[Used throughout.]** Let $A \subseteq \mathbb{R}$ be a set. Prove that A is bounded if and only if there is some $M \in \mathbb{R}$ such that $M > 0$ and that $|x| \leq M$ for all $x \in A$.

Exercise 2.3.12. **[Used in Lemma 2.3.10.]** Prove Lemma 2.3.10 (2).

Exercise 2.3.13. **[Used in Exercise 2.5.15.]** Let $a,b,x,y \in \mathbb{R}$. Suppose that $a \leq x \leq b$ and $a \leq y \leq b$. Prove that $|x-y| \leq b-a$.

Exercise 2.3.14. Let $[a,b] \subseteq \mathbb{R}$ be a non-degenerate closed bounded interval, and let $x,y \in [a,b]$. Let $x' = a+b-x$.

(1) Prove that if $x \in \left[a, \frac{a+b}{2}\right]$ then $x' \in \left[\frac{a+b}{2}, b\right]$, and if $x \in \left[\frac{a+b}{2}, b\right]$ then $x' \in \left[a, \frac{a+b}{2}\right]$.
(2) Prove that if $x \in \left[a, \frac{a+b}{2}\right]$ and $y \in \left[\frac{a+b}{2}, b\right]$, then $|x'-y| \leq |x-y|$.

2.4 Finding the Natural Numbers, the Integers and the Rational Numbers in the Real Numbers

Inside the set of real numbers sit three important, and very familiar, sets of numbers: the natural numbers, the integers and the rational numbers. The reader who has read Chapter 1 (starting in either Section 1.2 or 1.4) has already seen a rigorous treatment of these three sets of numbers, and should skip the present section and proceed straight to Section 2.5.

On the other hand, the reader who commenced the study of the real numbers with the axioms for the real numbers given in Section 2.2 has not yet seen a rigorous treatment of the natural numbers, the integers and the rational numbers. The goal of the present section is to show that using only the axioms for the real numbers these three sets of numbers can be found relatively easily inside the set of real numbers, and that these sets of numbers behave just as one would expect. In fact, as was the case in Section 2.3, in order to find the natural numbers, the integers and the rational numbers in the real numbers, we will not need all the properties of the real numbers, but rather only the axiom for an ordered field given in Definition 2.2.1.

We start with the set of natural numbers. What distinguishes this set of numbers from other sets of numbers? One answer might be that the natural numbers are "discrete," in that there is a minimum distance (which is 1) between any two natural numbers. However, the integers are also "discrete" in this same sense, so that property alone does not characterize the natural numbers. One way of viewing the difference between the set of natural numbers and the set of integers is that the former does not contain negative numbers, whereas the latter does. Another intuitive way of differentiating between the natural numbers and the integers is that the former goes to infinity "in one direction" whereas the latter goes to infinity "in two directions." A more formal way of viewing this distinction between the natural numbers and the integers is the ability to do proof by induction in the natural numbers, but not in the integers. (We assume that the reader is familiar, at least informally, with proof by induction, which we will treat in more detail in Section 2.5.) The problem is how to phrase this ability to do proof by induction using only the algebraic properties of the real numbers that we have seen so far. The idea that resolves this problem is that because the natural numbers are tied up with the idea of doing proof by induction, then the natural numbers must certainly contain the number 1, and if a real number n is a natural number, then surely $n+1$ is also a natural number.

Definition 2.4.1. Let $S \subseteq \mathbb{R}$ be a set. The set S is **inductive** if it satisfies the following two properties.

(a) $1 \in S$.
(b) If $a \in S$, then $a+1 \in S$. △

There are many inductive subsets in $\mathbb{R}$, for example the set $\mathbb{R}$ is an inductive subset of itself. On the other hand, the set $\{1\}$ is not an inductive set. As seen in the following definition and lemma, the set of natural numbers is the smallest inductive subset of $\mathbb{R}$.

Definition 2.4.2. The set of **natural numbers**, denoted $\mathbb{N}$, is the intersection of all inductive subsets of $\mathbb{R}$. △

Lemma 2.4.3.

1. *$\mathbb{N}$ is inductive.*
2. *If $A \subseteq \mathbb{R}$ and A is inductive, then $\mathbb{N} \subseteq A$.*
3. *If $n \in \mathbb{N}$ then $n \geq 1$.*

Proof.

(1) Let $\mathcal{A}$ be the collection of all inductive subsets of $\mathbb{R}$. Then $\mathbb{N} = \bigcap_{X \in \mathcal{A}} X$. Because X is inductive for all $X \in \mathcal{A}$, we know that $1 \in X$ for all $X \in \mathcal{A}$. Hence $1 \in \bigcap_{X \in \mathcal{A}} X$. Next, let $b \in \bigcap_{X \in \mathcal{A}} X$. Then $b \in X$ for all $X \in \mathcal{A}$. Therefore $b+1 \in X$ for all $X \in \mathcal{A}$, and it follows that $b+1 \in \bigcap_{X \in \mathcal{A}} X$. We deduce that $\bigcap_{X \in \mathcal{A}} X$ is inductive.

(2) This fact follows immediately from the definition of $\mathbb{N}$ as the intersection of all inductive subsets of $\mathbb{R}$.

(3) First, we verify that the interval $[1,\infty)$ is an inductive set. Clearly $1 \in [1,\infty)$. Let $x \in [1,\infty)$. Then $x \geq 1$. We know that $x < x+1$, and therefore $x+1 > 1$. It follows that $x+1 \in [1,\infty)$, and therefore $[1,\infty)$ is inductive. We now use Part (2) of this lemma to deduce that $\mathbb{N} \subseteq [1,\infty)$. It follows immediately that if $n \in \mathbb{N}$ then $n \geq 1$. □

Our next task is to show that the Peano Postulates for the natural numbers can be recovered from Definition 2.4.2. For the reader who is familiar with the Peano Postulates (for example, if the reader has read Chapter 1, where these postulates were given as Axiom 1.2.1 in Section 1.2, and as Theorem 1.4.8 in Section 1.4), we now see that the set of natural numbers that we have just located inside the real numbers behaves in the exact same way as the set of natural numbers that the reader has previously encountered. For the reader who has not previously encountered the Peano Postulates, it is worth seeing them now, not only because these postulates can be used as axioms for the natural numbers, and they therefore shed some light on what makes the natural numbers what they are, but in particular because Part (c) of the Peano Postulates is the formal statement that proof by induction works for the natural numbers. Hence, the ability to do proof by induction ultimately follows from the axioms for the real numbers.

Theorem 2.4.4 (Peano Postulates). *Let $s\colon \mathbb{N} \to \mathbb{N}$ be defined by $s(n) = n+1$ for all $n \in \mathbb{N}$.*

a. *There is no $n \in \mathbb{N}$ such that $s(n) = 1$.*
b. *The function s is injective.*
c. *Let $G \subseteq \mathbb{N}$ be a set. Suppose that $1 \in G$, and that if $g \in G$ then $s(g) \in G$. Then $G = \mathbb{N}$.*

Proof.

(a) Suppose to the contrary that there is some $x \in \mathbb{N}$ such that $s(x) = 1$. Then $x+1 = 1$, which implies that $x = 0$. However, we know by Lemma 2.4.3 (3) that $x \geq 1$, which is a contradiction to the fact that $0 < 1$.

(b) Suppose that $s(n) = s(m)$ for some $n, m \in \mathbb{N}$. Then $n+1 = m+1$, and hence $n = m$. Therefore s is injective.

(c) By hypothesis, we are assuming that G is inductive. Lemma 2.4.3 (2) then implies that $\mathbb{N} \subseteq G$. Because $G \subseteq \mathbb{N}$, we deduce that $G = \mathbb{N}$. □

The following lemma, which is a nice application of proof by induction, shows that the natural numbers are closed under addition and multiplication; this proof is written in the style of Part (c) of the Peano Postulates (Theorem 2.4.4), rather than in the more familiar style usually used for proofs by induction, in order to emphasize that we are using nothing other than what we have proved so far.

Lemma 2.4.5. *Let $a, b \in \mathbb{N}$. Then $a+b \in \mathbb{N}$ and $ab \in \mathbb{N}$.*

Proof. Let

$$G = \{x \in \mathbb{N} \mid x+y \in \mathbb{N} \text{ for all } y \in \mathbb{N}\}.$$

Clearly $G \subseteq \mathbb{N}$. We will prove that $G = \mathbb{N}$ using Part (c) of the Peano Postulates (Theorem 2.4.4), and it will then follow that $x+y \in \mathbb{N}$ for all $x, y \in \mathbb{N}$.

We first show that $1 \in G$. Let $d \in \mathbb{N}$. Then $d+1 \in \mathbb{N}$, because $\mathbb{N}$ is inductive by Lemma 2.4.3 (1). Hence $1+d \in \mathbb{N}$. It follows that $1 \in G$. Now suppose that $e \in G$. We need to show that $e+1 \in G$. Let $c \in \mathbb{N}$. By the definition of G, we know that $e+c \in \mathbb{N}$. Because $\mathbb{N}$ is inductive, it follows that $(e+c)+1 \in \mathbb{N}$. Hence $(e+1)+c = (e+c)+1 \in \mathbb{N}$. We deduce that $e+1 \in G$. Hence G satisfies the hypotheses of Part (c) of the Peano Postulates, and it follows that $G = \mathbb{N}$.

Next, let

$$H = \{x \in \mathbb{N} \mid xy \in \mathbb{N} \text{ for all } y \in \mathbb{N}\}.$$

Clearly $H \subseteq \mathbb{N}$. We will show that $H = \mathbb{N}$, which will then imply that $xy \in \mathbb{N}$ for all $x, y \in \mathbb{N}$.

Let $u \in \mathbb{N}$. Then $1 \cdot u = u \in \mathbb{N}$. Hence $1 \in H$. Now let $v \in H$. We will show that $v+1 \in H$. Let $g \in \mathbb{N}$. By the definition of H, we know that $vg \in \mathbb{N}$. Using what we have already seen in this proof about addition in $\mathbb{N}$, we deduce that $(v+1)g = vg + g \in \mathbb{N}$. It follows that $v+1 \in H$. Hence H satisfies the hypotheses of Part (c) of the Peano Postulates, and we conclude that $H = \mathbb{N}$. □

We now prove a very important fact about the natural numbers known as the Well-Ordering Principle, which is essentially an alternative view of what characterizes the natural numbers. The Well-Ordering Principle (which was also stated as Theorem 1.2.10, and was taken axiomatically as part of Axiom 1.4.4) captures the intuitive notion that the natural numbers are "discrete" and go to infinity "in one direction" by asserting that every non-empty subset of $\mathbb{N}$ has a smallest element (though not necessarily a largest one).

Theorem 2.4.6 (Well-Ordering Principle). *Let $G \subseteq \mathbb{N}$ be a non-empty set. Then there is some $m \in G$ such that $m \leq g$ for all $g \in G$.*

Proof. Let

$$B = \{a \in \mathbb{N} \mid a \leq g \text{ for all } g \in G\}.$$

By Lemma 2.4.3 (3) we know that $1 \in B$, and hence $B \neq \emptyset$.

As a first step we show that B is not inductive. Suppose to the contrary that B is inductive. We know by definition that $B \subseteq \mathbb{N}$, and it then follows from Theorem 2.4.4 (c) that $B = \mathbb{N}$. Hence $G \subseteq B$. Because $G \neq \emptyset$, there is some $p \in G$, and hence $p \in B$. Because B is inductive it follows that $p+1 \in B$. By the definition of B we know $p+1 \leq g$ for all $g \in G$, and hence in particular $p+1 < p$, which is a contradiction. We deduce that B is not inductive.

Because $1 \in B$, the fact that B is not inductive implies that there is some $m \in B$ such that $m+1 \notin B$. By the definition of B, we know that $m \leq g$ for all $g \in G$.

We will show that $m \in G$. Suppose to the contrary that $m \notin G$. Then $m < g$ for all $g \in G$. Because $m+1 \notin B$, there is some $w \in G$ such that $m+1 \not\leq w$, which means that $w < m+1$. We can then use Theorem 2.4.10 (1) to deduce that $w+1 \leq m+1$ (we have not yet reached that Theorem because it is stated in terms of $\mathbb{Z}$, which we do to avoid having to prove it separately for each of $\mathbb{N}$ and $\mathbb{Z}$, but the reader can be assured that we do not use the present theorem in the proof of Theorem 2.4.10, so there is no circular reasoning here). This last inequality implies that $w \leq m$. We now have a contradiction to the fact that $m < g$ for all $g \in G$. It follows that $m \in G$, and the proof is complete. □

We now turn to the integers, which consist of the natural numbers, their negatives and zero.

Definition 2.4.7. Let

$$-\mathbb{N} = \{x \in \mathbb{R} \mid x = -n \text{ for some } n \in \mathbb{N}\}.$$

The set of **integers**, denoted $\mathbb{Z}$, is defined by

$$\mathbb{Z} = -\mathbb{N} \cup \{0\} \cup \mathbb{N}. \qquad \triangle$$

The following lemma shows the relation between the sets $\mathbb{N}$ and $\mathbb{Z}$.

Lemma 2.4.8.

1. $\mathbb{N} \subseteq \mathbb{Z}$.
2. $a \in \mathbb{N}$ *if and only if* $a \in \mathbb{Z}$ *and* $a > 0$.
3. *The three sets* $-\mathbb{N}$, $\{0\}$ *and* $\mathbb{N}$ *are mutually disjoint.*

Proof.

(1) This part of the lemma follows immediately from the definition of $\mathbb{Z}$.

(2) Because of Part (1) of this lemma, it is sufficient to show that if $a \in \mathbb{Z}$, then $a \in \mathbb{N}$ if and only if $a > 0$.

Let $a \in \mathbb{Z}$. First, suppose that $a \in \mathbb{N}$. Lemma 2.4.3 (3) then implies that $a \geq 1 > 0$. Second, suppose that $a > 0$. By the definition of $\mathbb{Z}$ we know that $a \in \mathbb{N}$ or $a \in \{0\}$ or $a \in -\mathbb{N}$. First, suppose that $a \in \mathbb{N}$. Then there is nothing to prove. Second, suppose that $a \in -\mathbb{N}$. Then $a = -n$ for some $n \in \mathbb{N}$. Hence $-a = n$, and therefore $-a \in \mathbb{N}$. By Lemma 2.4.3 (3) we deduce that $-a \geq 1$, which implies that $a \leq -1$. Hence $a < 0$, which is a contradiction to the fact that $a > 0$. Third, suppose that $a \in \{0\}$. Then $a = 0$. Again we have a contradiction to the fact that $a > 0$. We conclude that $a \in \mathbb{N}$.

(3) It follows from Part (2) of this lemma that $a \in -\mathbb{N}$ if and only if $a \in \mathbb{Z}$ and $a < 0$. The Trichotomy Law then implies that the three sets $-\mathbb{N}$, $\{0\}$ and $\mathbb{N}$ are mutually disjoint. □

The following lemma shows that the integers are closed under addition, multiplication and negation.

Lemma 2.4.9. *Let* $a, b \in \mathbb{Z}$. *Then* $a + b \in \mathbb{Z}$, *and* $ab \in \mathbb{Z}$, *and* $-a \in \mathbb{Z}$.

Proof. We start by showing that $a + b \in \mathbb{Z}$ and $ab \in \mathbb{Z}$. There are five cases.

First, suppose that $a = 0$ or $b = 0$. Then $a + b = a$ or $a + b = b$, and in either case $a + b \in \mathbb{Z}$. We also have $ab = 0 \in \mathbb{Z}$. Second, suppose that $a \in \mathbb{N}$ and $b \in \mathbb{N}$. By Lemma 2.4.5 we see that $a + b \in \mathbb{N} \subseteq \mathbb{Z}$ and $ab \in \mathbb{N} \subseteq \mathbb{Z}$. Third, suppose that $a \in -\mathbb{N}$ and $b \in -\mathbb{N}$. Then $a = -n$ and $b = -m$ for some $n, m \in \mathbb{N}$. By Lemma 2.4.5 we deduce that $a + b = (-n) + (-m) = -(n + m) \in -\mathbb{N} \subseteq \mathbb{Z}$, and that $ab = (-n)(-m) = -[-nm] = nm \in \mathbb{N} \subseteq \mathbb{Z}$. Fourth, suppose that $a \in \mathbb{N}$ and $b \in -\mathbb{N}$. Then $b = -m$ for some $m \in \mathbb{N}$. If $a > m$, then by Exercise 2.4.1 we see that $a + b = a + (-m) = a - m \in \mathbb{N}$. If $a < m$, then using Exercise 2.4.1 again we see that $a + b = (-(-a)) + (-m) = -[(-a) + m] = -[m - a] \in -\mathbb{N}$. If $a = m$, then $a + b = a + (-m) = a + (-a) = 0 \in \mathbb{Z}$. Regardless of whether a or m is larger, we use Lemma 2.4.5 to see that $ab = a(-m) = -am \in -\mathbb{N} \subseteq \mathbb{Z}$. Fifth, suppose that $a \in -\mathbb{N}$ and $b \in \mathbb{N}$. This case is similar to the fourth case, and we omit the details.

Finally, we show that $-a \in \mathbb{Z}$. If $a = 0$, then $-a = -0 = 0 \in \mathbb{Z}$. If $a \in \mathbb{N}$, then

$-a \in -\mathbb{N} \subseteq \mathbb{Z}$. If $a \in -\mathbb{N}$, then $a = -n$ for some $n \in \mathbb{N}$, and therefore $-a = -(-n) = n \in \mathbb{N} \subseteq \mathbb{Z}$. □

Each of the three parts of the following theorem is a version of what it means when we say informally that the integers are "discrete."

Theorem 2.4.10. *Let $a, b \in \mathbb{Z}$.*

1. *If $a < b$ then $a+1 \leq b$.*
2. *There is no $c \in \mathbb{Z}$ such that $a < c < a+1$.*
3. *If $|a-b| < 1$ then $a = b$.*

Proof.

(1) Suppose that $a < b$. Then $b - a > 0$. By Lemma 2.4.9 we know that $b - a = b + (-a) \in \mathbb{Z}$. It follows from Lemma 2.4.8 (2) that $b - a \in \mathbb{N}$. We now use Lemma 2.4.3 (3) to deduce that $b - a \geq 1$. Hence $a + 1 \leq b$.

(2) Suppose that there is some $c \in \mathbb{Z}$ such that $a < c < a+1$. By applying Part (1) of this theorem to the left-hand inequality we deduce that $a+1 \leq c$, which is a contradiction to the fact that $c < a+1$.

(3) Suppose that $|a-b| < 1$. Then $-1 < a-b < 1$, and hence $b-1 < a < b+1$. We know that precisely one of $a < b$ or $a = b$ or $a > b$ holds. If $a > b$, we deduce that $b < a < b+1$, which is a contradiction to Part (2) of this theorem. If $a < b$, then we deduce that $b-1 < a < b$, which means $b-1 < a < (b-1)+1$, which again is a contradiction. We conclude that $a = b$. □

We now turn to the rational numbers (which are also called fractions). In going from the integers to the rational numbers, we lose the notion of "discreteness," but we gain the use of multiplicative inverses for non-zero rational numbers.

Definition 2.4.11. The set of **rational numbers**, denoted $\mathbb{Q}$, is defined by

$$\mathbb{Q} = \{x \in \mathbb{R} \mid x = \frac{a}{b} \text{ for some } a, b \in \mathbb{Z} \text{ such that } b \neq 0\}.$$

The set of **irrational numbers** is the set $\mathbb{R} - \mathbb{Q}$. △

If $x \in \mathbb{Q}$, then by definition $x = \frac{a}{b}$ for some $a, b \in \mathbb{Z}$ such that $b \neq 0$, though, as the reader is aware, the integers a and b are not unique. The reader is also aware, at least informally, that there exist irrational numbers—if there did not, it would have been rather silly of us to have defined the term—though it turns out that it is non-trivial to prove that irrational numbers exist, and we will have to wait until Theorem 2.6.11 to see that.

The following lemma shows the relation between the sets $\mathbb{Z}$ and $\mathbb{Q}$.

Lemma 2.4.12.

1. *$\mathbb{Z} \subseteq \mathbb{Q}$.*
2. *$q \in \mathbb{Q}$ and $q > 0$ if and only if $q = \frac{a}{b}$ for some $a, b \in \mathbb{N}$.*

Proof. We will prove Part (1), leaving the remaining part to the reader in Exercise 2.4.7.

(1) Let $z \in \mathbb{Z}$. By Lemma 2.4.3 (1) we know that $\mathbb{N}$ is inductive, and in particular that $1 \in \mathbb{N}$. Hence $1 \in \mathbb{Z}$. Because $1 \neq 0$, we can define the fraction $\frac{z}{1}$, and we then see that $z = z \cdot 1 = z \cdot 1^{-1} = \frac{z}{1} \in \mathbb{Q}$. Therefore $\mathbb{Z} \subseteq \mathbb{Q}$. □

The following lemma demonstrates that the way we learn to manipulate fractions in elementary school is indeed valid.

Lemma 2.4.13. *Let $a, b, c, d \in \mathbb{Z}$. Suppose that $b \neq 0$ and $d \neq 0$.*

1. $\frac{a}{b} = 0$ *if and only if* $a = 0$.
2. $\frac{a}{b} = 1$ *if and only if* $a = b$.
3. $\frac{a}{b} = \frac{c}{d}$ *if and only if* $ad = bc$.
4. $\frac{a}{b} + \frac{c}{d} = \frac{ad+bc}{bd}$.
5. $-\frac{a}{b} = \frac{-a}{b} = \frac{a}{-b}$.
6. $\frac{a}{b} \cdot \frac{c}{d} = \frac{ac}{bd}$.
7. *If* $a \neq 0$, *then* $\left(\frac{a}{b}\right)^{-1} = \frac{b}{a}$.

Proof. We will prove Parts (1), (4) and (7), leaving the rest to the reader in Exercise 2.4.8.

(1) If $\frac{a}{b} = 0$, then $ab^{-1} = 0$, and therefore $a = a(b^{-1}b) = (ab^{-1})b = 0 \cdot b = 0$. If $a = 0$, then $\frac{a}{b} = 0 \cdot b^{-1} = 0$.

(4) We compute

$$\begin{aligned} \frac{a}{b} + \frac{c}{d} &= ab^{-1} + cd^{-1} = ab^{-1}dd^{-1} + cd^{-1}bb^{-1} \\ &= (ad+bc)(b^{-1}d^{-1}) = (ad+bc)(bd)^{-1} = \frac{ad+bc}{bd}. \end{aligned}$$

(7) Suppose that $a \neq 0$. We then use Parts (2) and (6) of this lemma to compute

$$\frac{a}{b} \cdot \frac{b}{a} = \frac{ab}{ba} = 1.$$

Hence $\left(\frac{a}{b}\right)^{-1} = \frac{b}{a}$. □

The following corollary is an immediate consequence of Lemma 2.4.13, and we omit the proof.

Corollary 2.4.14. *Let $a, b \in \mathbb{Q}$. Then $a + b \in \mathbb{Q}$, and $ab \in \mathbb{Q}$, and $-a \in \mathbb{Q}$, and if $a \neq 0$ then $a^{-1} \in \mathbb{Q}$.*

There are many other algebraic properties of $\mathbb{N}$, $\mathbb{Z}$ and $\mathbb{Q}$ that one might think to prove, though some of these properties, such as the Commutative Law for Addition, are trivially true, because we are assuming these properties for addition and multiplication of real numbers, and $\mathbb{N}$, $\mathbb{Z}$ and $\mathbb{Q}$ are subsets of the real numbers with the same addition, multiplication, negation and multiplicative inverse. We have stated all the distinctive facts about $\mathbb{N}$, $\mathbb{Z}$ and $\mathbb{Q}$ that we will need for later use, and so we will not state any additional properties.

Reflections

The material in this section might seem unnecessary upon first encounter. The real numbers were defined axiomatically in Section 2.2, and given that the natural numbers, the integers and the rational numbers all sit inside the real numbers, what more needs to be said about these three sets of numbers? The answer is that in order to provide rigorous proofs of various results from calculus, we need proofs of some of the properties of these three sets of numbers—properties that do not hold for all real numbers. It is not possible to provide a rigorous proof of something without using precise definitions of the objects under consideration, and so we need precise definitions of exactly which subsets of the real numbers we wish to call the natural numbers, the integers and the rational numbers, and that is what we see in this section.

We have provided proofs of only those properties of the natural numbers, the integers and the rational numbers that are needed subsequently in this text. In particular, the reader might have noticed the absence of any discussion of the cardinality of these sets of numbers. Somewhat surprisingly, although the fact that the set of rational numbers is countable and the set of real numbers is uncountable is very important in some parts of mathematics, this distinction between the cardinalities of the rational numbers and the real numbers is of no direct importance to us in our study of real analysis. We will see a proof that the set of real numbers is uncountable in Section 8.4, but that is simply a nice application of sequences, rather than something useful elsewhere in this text. The difference between the rational numbers and the real numbers that is of interest to us in real analysis is not the size of these sets but the way the elements of the sets are located in relation to each other. More precisely, what concerns us is the fact that the Least Upper Bound Property holds for the real numbers, but not for the rational numbers. An additional distinction between these two sets of numbers that is relevant to real analysis will be seen in Section 5.8, where it is shown that the rational numbers have measure zero, and the real numbers do not.

Exercises

Exercise 2.4.1. **[Used in Lemma 2.4.9 and Exercise 2.7.2.]** Let $a, b \in \mathbb{N}$. Prove that $a > b$ if and only if $a - b \in \mathbb{N}$ if and only if there is some $d \in \mathbb{N}$ such that $b + d = a$.

Exercise 2.4.2. **[Used in Theorem 10.5.2.]** Let $x \in \mathbb{R}$. Prove that at most one of $(x - \frac{1}{2}, x)$ and $(x, x + \frac{1}{2})$ contains an integer.

Exercise 2.4.3. **[Used in Theorem 2.5.4 and Exercise 2.5.13.]** Let $n \in \mathbb{N}$. Suppose that $n \neq 1$. Prove that there is some $b \in \mathbb{N}$ such that $b + 1 = n$.

Exercise 2.4.4. Let $a, b \in \mathbb{Z}$. Prove that if $ab = 1$, then $a = 1$ and $b = 1$, or $a = -1$ and $b = -1$.

Exercise 2.4.5. [Used in Section 2.6 and Exercise 2.6.12.] Prove that there is no $n \in \mathbb{Z}$ such that $n^2 = 2$.

Exercise 2.4.6. [Used in Theorem 2.6.13.] Let $q \in \mathbb{Q}$ and $x \in \mathbb{R} - \mathbb{Q}$.

(1) Prove that $q + x \in \mathbb{R} - \mathbb{Q}$.
(2) Prove that if $q \neq 0$ then $qx \in \mathbb{R} - \mathbb{Q}$.

Exercise 2.4.7. [Used in Lemma 2.4.12.] Prove Lemma 2.4.12 (2).

Exercise 2.4.8. [Used in Lemma 2.4.13.] Prove Lemma 2.4.13 (2) (3) (5) (6).

Exercise 2.4.9. [Used in Theorem 2.7.1.] Let $\frac{a}{b}, \frac{c}{d} \in \mathbb{Q}$. Prove that $\frac{a}{b} < \frac{c}{d}$ if and only if either $cb - ad > 0$ and $bd > 0$, or $cb - ad < 0$ and $bd < 0$.

Exercise 2.4.10. [Used in Section 3.5.] Let $a, b \in \mathbb{Q}$. Suppose that $a > 0$. Prove that there is some $n \in \mathbb{N}$ such that $b < na$. Use only the material in Sections 2.3 and 2.4; do not use the Least Upper Bound Property.

2.5 Induction and Recursion in Practice

The reader has already encountered the theoretical role of proof by induction via the Peano Postulates for the natural numbers in Section 1.2, 1.4 or 2.4. In the present section, by contrast, our purpose is to review the practical use of proof by induction, and then to discuss a very important consequence of it, which is Definition by Recursion. We assume that the reader has some previous experience with proof by induction, and hence we will review such proofs only as much as needed for later purposes, and without intuitive motivation. The reader might not be as familiar with Definition by Recursion, and hence we will give it a slightly more detailed treatment.

Proof by induction, often called the "Principle of Mathematical Induction," is a method of proving certain statements involving the natural numbers, and it is quite distinct from the informal concept of "inductive reasoning," which refers to the process of going from specific examples to more general statements, and which is not restricted to mathematics. More precisely, the method of proof by induction is given in the following theorem, which is just a restatement of Part (c) of the Peano Postulates, and so we state it without proof.

Theorem 2.5.1 (Principle of Mathematical Induction). *Let $G \subseteq \mathbb{N}$. Suppose that*

a. $1 \in G$;
b. *if $n \in G$, then $n + 1 \in G$.*

Then $G = \mathbb{N}$.

It is important to note that Part (b) of Theorem 2.5.1 has the form $P \to Q$, and that to show that Part (b) is true, we do not show that either P or Q is true, but only that the conditional statement $P \to Q$ is true. In other words, we do not need to show directly that $n \in G$, nor that $n + 1 \in G$, but only that $n \in G$ implies $n + 1 \in G$.

Although Theorem 2.5.1 involves the use of a set "G," in practice it is customary to avoid mentioning the set G explicitly. Suppose that we are trying to show that the statement $P(n)$ is true for all $n \in \mathbb{N}$. The formal way to proceed would be to let $G = \{n \in \mathbb{N} \mid P(n) \text{ is true}\}$, and then to verify that $G = \mathbb{N}$ by showing that $1 \in G$, and that $n \in G$ implies $n+1 \in G$ for all $n \in \mathbb{N}$. The less cumbersome, but equally valid, way of proceeding is to state that we are trying to prove by induction that $P(n)$ is true for all $n \in \mathbb{N}$, and then to show that $P(1)$ is true, and that if $P(n)$ is true then so is $P(n+1)$ for all $n \in \mathbb{N}$. The latter of these two steps is often called the "inductive step," and the assumption that $P(n)$ is true in the inductive step is often called the "inductive hypothesis." An equivalent version of the inductive step involves showing that if $P(n-1)$ is true then so is $P(n)$ for all $n \in \mathbb{N}$ such that $n \geq 2$.

We now turn to a very standard example of proof by induction; we will use this formula later on in Example 5.2.3 (1).

Proposition 2.5.2. *Let $n \in \mathbb{N}$. Then*

$$1^2 + 2^2 + \cdots + n^2 = \frac{n(n+1)(2n+1)}{6}. \tag{2.5.1}$$

Proof. We prove the result by induction on n. First, suppose that $n = 1$. Then $1^2 + 2^2 + \cdots + n^2 = 1$, and $\frac{n(n+1)(2n+1)}{6} = \frac{1\cdot(1+1)\cdot(2\cdot 1+1)}{6} = 1$. Therefore Equation 2.5.1 is true for $n = 1$. Now let $n \in \mathbb{N}$. Suppose that Equation 2.5.1 is true for n. That is, suppose that

$$1^2 + 2^2 + \cdots + n^2 = \frac{n(n+1)(2n+1)}{6}.$$

We then compute

$$\begin{aligned} 1^2 + 2^2 + \cdots + (n+1)^2 &= \{1^2 + 2^2 + \cdots + n^2\} + (n+1)^2 \\ &= \frac{n(n+1)(2n+1)}{6} + (n+1)^2 \\ &= (n+1)\frac{2n^2 + n + 6n + 6}{6} \\ &= (n+1)\frac{(n+2)(2n+3)}{6} \\ &= \frac{[n+1][(n+1)+1][2(n+1)+1]}{6}. \end{aligned}$$

This last expression is precisely the right-hand side of Equation 2.5.1 with $n+1$ replacing n. Hence we have proved the inductive step. This completes the proof that Equation 2.5.1 is true for all $n \in \mathbb{N}$. □

We note that the use of "$\cdots$" in Proposition 2.5.2 is not completely rigorous, unless we provide a valid definition of "$\cdots$" in expressions of the form "$a_1 + a_2 + \cdots + a_n$." In general, the notation "$\cdots$" can be used rigorously only in those situations where it is an abbreviation for something that has been properly defined. To avoid a digression at this point we will skip over the definition of "$a_1 + a_2 + \cdots + a_n$" for now, but we note that it is found in Exercise 2.5.19, using material discussed subsequently in this section.

There are various alternative versions of proof by induction, each of which is useful in certain situations where Theorem 2.5.1 might not be directly applicable. For example, instead of proving that a statement $P(n)$ is true for all $n \in \mathbb{N}$, it is possible to prove that it is true for all $n \in \mathbb{N}$ such that $n \geq k$, for some given $k \in \mathbb{N}$, which we would do by proving that $P(k)$ is true, and then proving that if $P(n)$ is true then so is $P(n+1)$ for all $n \in \mathbb{N}$ such that $n \geq k$. We will not be using this variant of proof by induction in this text, so we will not state and prove it formally. A variant of Theorem 2.5.1 that we will need starts at $n = 1$, but has a slightly different type of inductive step, as seen in the following theorem. We start with some notation.

Definition 2.5.3. Let $a, b \in \mathbb{Z}$. The set $\{a, \ldots, b\}$ is defined by $\{a, \ldots, b\} = \{x \in \mathbb{Z} \mid a \leq x \leq b\}$. △

Theorem 2.5.4 (Principle of Mathematical Induction—Variant). *Let $G \subseteq \mathbb{N}$. Suppose that*

a. $1 \in G$;
b. *if $n \in \mathbb{N}$ and $\{1, \ldots, n\} \subseteq G$, then $n+1 \in G$.*

Then $G = \mathbb{N}$.

Proof. Suppose that $G \neq \mathbb{N}$. Let $H = \mathbb{N} - G$. Because $H \subseteq \mathbb{N}$ and $H \neq \emptyset$, the Well-Ordering Principle (Theorem 1.2.10, Axiom 1.4.4 or Theorem 2.4.6) implies that there is some $m \in H$ such that $m \leq h$ for all $h \in H$. By hypothesis we know that $1 \in G$, and hence $1 \notin H$, which implies that $m \neq 1$. Using one of Lemma 1.2.3, Exercise 1.4.6 or Exercise 2.4.3, there is some $b \in \mathbb{N}$ such that $b+1 = m$. Let $p \in \{1, \ldots, b\}$. Then $p \leq b < b+1 = m$. Hence $m \not\leq p$, and therefore $p \notin H$, and so $p \in G$. It follows that $\{1, \ldots, b\} \subseteq G$. By the hypothesis on G we deduce that $b+1 \in G$, which means that $m \in G$, which is a contradiction to the fact that $m \in H$. Hence $G = \mathbb{N}$. □

When using Theorem 2.5.4, the inductive step involves showing that if the desired statement is assumed to be true for all values in $\{1, \ldots, n\}$, then it is true for $n+1$. This method contrasts with Theorem 2.5.1, where we showed that if the statement is assumed to be true only for n, then it is true for $n+1$. It might appear as if Theorem 2.5.4 is unfairly making things too easy by allowing a larger hypothesis in order to derive the same conclusion as Theorem 2.5.1, but Theorem 2.5.4 was proved rigorously, so there is no cheating here. (Although the proof of Theorem 2.5.4 uses the Well-Ordering Principle, we note that the Well-Ordering Principle is proved by using proof by induction, and hence Theorem 2.5.4 is ultimately derived from Theorem 2.5.1.)

We will see an example of the use of Theorem 2.5.4 in the proof of Theorem 2.8.2.

We now turn to the issue of definition by recursion, which is related to proof by induction, but is not the same as it. Definition by recursion involves sequences, a topic we will study from the perspective of real analysis in Chapter 8. Informally, a sequence is an infinite list $a_1, a_2, a_3, \ldots$ of elements of a set. Such a method of writing sequences often suffices for many purposes, but it is not rigorous to define something by writing "...," unless the "..." is provided a rigorous meaning in the given situation. In our particular case, we can define a sequence in a set H to be a function $f\colon \mathbb{N} \to H$, where we think of $f(1)$ as the first element of the sequence, of $f(2)$ as the second element of the sequence and so on; we let $a_n = f(n)$ for all $n \in \mathbb{N}$. Hence, the notation "$a_1, a_2, a_3, \ldots$" corresponds to the rigorous definition of a sequence in terms of a function, and so we will feel free to use this notation.

In Chapter 8 we will prove many important facts about sequences, most of which usually start out with the assumption that we are given a sequence or sequences that have already been defined. In the present section, by contrast, we discuss a very useful way of defining sequences.

There are two standard ways of defining sequences in practice. Suppose that we want to give a formula to define the familiar sequence $1, 2, 4, 8, 16, \ldots$. Let a_n denote the nth term of this sequence for all $n \in \mathbb{N}$. The simplest way to define this sequence is by giving an **explicit** description for a_n for all $n \in \mathbb{N}$, which in this case is $a_n = 2^{n-1}$ for all $n \in \mathbb{N}$. However, although it is easy to find such an explicit formula for a_n in the case of this very simple sequence, it is not always possible to find such formulas for more complicated sequences. Another way of describing this sequence is by stating that $a_1 = 1$, and that $a_{n+1} = 2a_n$ for all $n \in \mathbb{N}$. Such a description is called a **recursive** description of the sequence. There are many sequences for which it is much easier to give a recursive description than an explicit one. Moreover, recursion is important not only in mathematics, but also in logic, and the applications of logic to computer science; see [Rob86], [DSW94, Chapter 3] or [End72, Section 1.2] for more about recursion, and see [Rob84, Section 5.1] for various applications of recursion.

If a sequence in a set H is described by an explicit formula for each a_n in terms of n, it can be useful to find a recursive formula, but there is no question that the sequence exists, because the explicit formula defines a function $\mathbb{N} \to H$. Suppose, by contrast, that a sequence in H is given only by a recursive description, but no explicit formula. For example, suppose that we have a sequence $b_1, b_2, b_3, \ldots$ in $\mathbb{R}$ given by the recursive description $b_1 = 5$, and $b_{n+1} = 1 + 3(b_n)^2$ for all $n \in \mathbb{N}$. Is there a sequence satisfying this description? It appears intuitively as if such a sequence exists because we can produce one element at a time, starting with $b_1 = 5$, then computing $b_2 = 1 + 3(b_1)^2 = 1 + 3 \cdot 5^2 = 76$, then $b_3 = 1 + 3(b_2)^2 = 1 + 3 \cdot 76^2 = 17{,}329$ and so on, proceeding "inductively." However, a sequence in $\mathbb{R}$ is defined as a function $f\colon \mathbb{N} \to \mathbb{R}$, and to show that this recursive description actually produces a sequence, we would need to find a function $f\colon \mathbb{N} \to \mathbb{R}$ such that $f(1) = 5$, and $f(2) = 76$, and $f(3) = 17{,}329$ and so on, and it is not at all obvious how to define such a function. In fact, there is such a function, and it is unique, though the justification for this fact, seen below, is not at all trivial.

The method of describing sequences using Definition by Recursion (also called "recursive definition") can be made completely rigorous, as seen in Theorem 2.5.5 below, but simply saying something such as "just continue inductively" is not satisfactory from a rigorous point of view. Proof by induction works for something that is already defined; here we need to prove that our definition actually produces something. Unfortunately, in some texts that discuss induction, not only is no proof given of the validity of definition by recursion, but no mention is even made of the need for such a proof. It is fine to skip a difficult proof, but mention should always be made that the proof is being skipped.

The simplest form of definition by recursion works as follows. Suppose that we are given a non-empty set H, an element $e \in H$ and a function $k\colon H \to H$. We then want to define a sequence $a_1, a_2, \ldots$ in H such that $a_1 = e$, and that $a_{n+1} = k(a_n)$ for all $n \in \mathbb{N}$.

Given that the formal way to define a sequence in H is as a function $f\colon \mathbb{N} \to H$, we can reformulate definition by recursion as follows: given a set H, an element $e \in H$ and a function $k\colon H \to H$, is there a function $f\colon \mathbb{N} \to H$ such that $f(1) = e$, and that $f(n+1) = k(f(n))$ for all $n \in \mathbb{N}$? The following theorem says that such a function can always be found. Not surprisingly, proof by induction is the main tool in the proof of this theorem. The proof of this theorem, which follows [Dea66, Section 3.5], is a bit trickier than might be expected.

Theorem 2.5.5 (Definition by Recursion). *Let H be a set, let $e \in H$ and let $k\colon H \to H$ be a function. Then there is a unique function $f\colon \mathbb{N} \to H$ such that $f(1) = e$, and that $f(n+1) = k(f(n))$ for all $n \in \mathbb{N}$.*

Proof of Theorem 1.2.4 and Theorem 2.5.5. If you are reading this proof for Theorem 1.2.4 in Section 1.2, please skip this first paragraph. Let $s\colon \mathbb{N} \to \mathbb{N}$ be defined by $s(n) = n+1$ for all $n \in \mathbb{N}$. The condition that $f(n+1) = k(f(n))$ for all $n \in \mathbb{N}$ can then be rephrased as $f \circ s = k \circ f$. Recall that the function s satisfies the Peano Postulates (Theorem 2.4.4).

We have to prove both existence and uniqueness; we start with the latter. Suppose that there are functions $g, h\colon \mathbb{N} \to H$ such that $g(1) = e$ and $h(1) = e$, and that $g \circ s = k \circ g$ and $h \circ s = k \circ h$. Let

$$V = \{a \in \mathbb{N} \mid g(a) = h(a)\}.$$

We will prove that $V = \mathbb{N}$; the fact that $g = h$ will follow immediately. Clearly $V \subseteq \mathbb{N}$. We know that $1 \in V$ because $g(1) = e$ and $h(1) = e$. Now let $n \in V$. Then $g(n) = h(n)$. By the hypotheses on g and h we see that

$$\begin{aligned} g(s(n)) &= (g \circ s)(n) = (k \circ g)(n) = k(g(n)) \\ &= k(h(n)) = (k \circ h)(n) = (h \circ s)(n) = h(s(n)). \end{aligned}$$

Hence $s(n) \in V$. It now follows from Part (c) of the Peano Postulates that $V = \mathbb{N}$, and the proof of uniqueness is complete.

We now prove that a function f with the desired properties exists. A crucial aspect of this proof is to use the formal definition of functions in terms of sets; see [Blo10, Definition 4.1.1] for this definition. In particular, we think of functions $\mathbb{N} \to H$ as subsets of $\mathbb{N} \times H$ satisfying certain properties.

Let

$$C = \{W \subseteq \mathbb{N} \times H \mid (1, e) \in W, \text{ and if } (n, y) \in W \text{ then } (s(n), k(y)) \in W\}.$$

We note that C is non-empty, because the set $\mathbb{N} \times H$ is in C. Let $f = \bigcap_{W \in C} W$. Clearly $f \subseteq \mathbb{N} \times H$. Because $(1, e) \in W$ for all $W \in C$, it follows that $(1, e) \in f$. Suppose that

$(n,y) \in f$. Then $(n,y) \in W$ for all $W \in \mathcal{C}$, and therefore $(s(n),k(y)) \in W$ for all W. Hence $(s(n),k(y)) \in f$. We deduce that $f \in \mathcal{C}$. Clearly $f \subseteq W$ for all $W \in \mathcal{C}$.

We now show that if $(n,y) \in f$ and $(n,y) \neq (1,e)$, then there is some $(m,u) \in f$ such that $(n,y) = (s(m),k(u))$. Suppose to the contrary that there is some $(r,t) \in f$ such that $(r,t) \neq (1,e)$, and that there is no $(m,u) \in f$ such that $(r,t) = (s(m),k(u))$. Let $\hat{f} = f - \{(r,t)\}$. Clearly $\hat{f} \subseteq \mathbb{N} \times H$, and $(1,e) \in \hat{f}$. Let $(n,y) \in \hat{f}$. Then $(n,y) \in f$, and therefore $(s(n),k(y)) \in f$, as seen above. We know that $(r,t) \neq (s(n),k(y))$, and therefore $(s(n),k(y)) \in \hat{f}$. It follows that $\hat{f} \in \mathcal{C}$, which is a contradiction to the fact that $f \subseteq W$ for all $W \in \mathcal{C}$. We deduce that if $(n,y) \in f$ and $(n,y) \neq (1,e)$, then there is some $(m,u) \in f$ such that $(n,y) = (s(m),k(u))$.

Next, we show that f is a function $\mathbb{N} \to H$. Let

$$G = \{a \in \mathbb{N} \mid \text{there is a unique } x \in H \text{ such that } (a,x) \in f\}.$$

We will prove that $G = \mathbb{N}$, and it will follow immediately that f is a function. Clearly $G \subseteq \mathbb{N}$. Because $f \in \mathcal{C}$, we know that $(1,e) \in f$. Now suppose that $(1,p) \in f$ for some $p \in H$ such that $p \neq e$. We know from the previous paragraph that there is some $(m,u) \in f$ such that $(1,p) = (s(m),k(u))$. Hence $1 = s(m)$, which is a contradiction to Part (a) of the Peano Postulates. Therefore e is the unique element in H such that $(1,e) \in f$. Hence $1 \in G$.

Let $n \in G$. Then there is a unique $y \in H$ such that $(n,y) \in f$. Because $f \in \mathcal{C}$, we know that $(s(n),k(y)) \in f$. Now suppose that $(s(n),q) \in f$ for some $q \in H$. By Part (a) of the Peano Postulates we know that $s(n) \neq 1$, and hence $(s(n),q) \neq (1,e)$. Using what we proved above we see that there is some $(a,b) \in f$ such that $(s(n),q) = (s(a),k(b))$. Therefore $s(n) = s(a)$ and $q = k(b)$. By Part (b) of the Peano Postulates we know that s is injective, and hence $n = a$. It follows that $(a,b) = (n,b)$, which means that $(n,b) \in f$. By the uniqueness of y we deduce that $b = y$. Hence $q = k(b) = k(y)$, and therefore $k(y)$ is the unique element of H such that $(s(n),k(y)) \in f$. We deduce that $s(n) \in G$. It follows that $G = \mathbb{N}$.

Finally, we show that $f(1) = e$ and $f \circ s = k \circ f$. The first of these properties is equivalent to saying that $(1,e) \in f$, which we have already seen. Let $n \in \mathbb{N}$. Then $(n, f(n)) \in f$, and hence $(s(n),k(f(n))) \in f$. This last statement can be rephrased as $f(s(n)) = k(f(n))$. Because this last statement holds for all $n \in \mathbb{N}$, we deduce that $f \circ s = k \circ f$. □

As an important example of Definition by Recursion (Theorem 2.5.5), we turn to the notion of raising a real number to an integer power. Let $x \in \mathbb{R}$. We defined x^2 in Definition 2.3.1 (2) simply by letting $x^2 = x \cdot x$. For higher powers of x, we intuitively want to define x^n to be $x^n = x \cdot x \cdot \cdots \cdot x$, where x is multiplied with itself n times, for any $n \in \mathbb{N}$. However, writing $x \cdot x \cdot \cdots \cdot x$ is not a rigorous definition. We will use Definition by Recursion to eliminate the $\cdots$. We note that at present we are considering x^n only for $n \in \mathbb{Z}$. It is certainly possible to define x^r for all $r \in \mathbb{R}$, as long as $x > 0$, but we do not yet have the tools to do so; we will see the definition in Section 7.2.

As our first step, we define x^n for all $n \in \mathbb{N}$ using Definition by Recursion.

Definition 2.5.6. Let $x \in \mathbb{R}$. The number $x^n \in \mathbb{R}$ is defined for all $n \in \mathbb{N}$ by letting $x^1 = x$, and $x^{n+1} = x \cdot x^n$ for all $x \in \mathbb{N}$. △

In order to define x^n for $n \in \mathbb{Z} - \mathbb{N}$, we need the following very simple lemma.

Lemma 2.5.7. *Let $x \in \mathbb{R}$. Suppose that $x \neq 0$. Then $x^n \neq 0$ for all $n \in \mathbb{N}$.*

Proof. Left to the reader in Exercise 2.5.9. □

Lemma 2.5.7 allows us to make the following definition. Observe that in this definition, we encounter a somewhat confusing situation where the expression y^{-1} has two meanings, one being the result of raising the number y to the -1 power, and the other being the multiplicative inverse of y. In particular, in the formula "$x^{-n} = (x^n)^{-1}$," the term "x^{-n}" always denotes raising a number to a power, even when $n = 1$, whereas the term "$(x^n)^{-1}$" denotes the multiplicative inverse of x^n. Although this ambiguity of notation might seem confusing at first, it is actually no problem at all, because substituting $n = 1$ in the following definition shows that the two meanings of y^{-1} are actually equal to each other, even if they are conceptually different.

Definition 2.5.8. Let $x \in \mathbb{R}$. Suppose that $x \neq 0$. The number $x^0 \in \mathbb{R}$ is defined by $x^0 = 1$. For each $n \in \mathbb{N}$, the number x^{-n} is defined by $x^{-n} = (x^n)^{-1}$. △

The following lemma states some very familiar properties of x^n, which we can now prove rigorously using the above definitions.

Lemma 2.5.9. *Let $x \in \mathbb{R}$, and let $n, m \in \mathbb{Z}$. Suppose that $x \neq 0$.*

1. $x^n x^m = x^{n+m}$.
2. $\frac{x^n}{x^m} = x^{n-m}$.

Proof. We will prove Part (1), leaving the remaining part to the reader in Exercise 2.5.11.

(1) There are five cases. First, suppose that $n > 0$ and $m > 0$. By Lemma 2.4.8 (2) we know that $n, m \in \mathbb{N}$. This part of the proof is by induction. We will use induction on k to prove that for each $k \in \mathbb{N}$, the formula $x^k x^p = x^{k+p}$ holds for all $p \in \mathbb{N}$.

Let $k = 1$. Let $p \in \mathbb{N}$. Then by the definition of x^p we see that $x^k x^p = x^1 \cdot x^p = x \cdot x^p = x^{p+1} = x^{1+p} = x^{k+p}$. Hence the result is true for $k = 1$.

Now let $k \in \mathbb{N}$. Suppose that the result is true for k. Let $p \in \mathbb{N}$. Then by the definition of x^k we see that $x^{k+1} x^p = (x \cdot x^k) \cdot x^p = x \cdot (x^k \cdot x^p) = x \cdot x^{k+p} = x^{(k+p)+1} = x^{(k+1)+p}$. Hence the result is true for $k + 1$. It follows by induction that $x^k x^p = x^{k+p}$ for all $p \in \mathbb{N}$, for all $k \in \mathbb{N}$.

Second, suppose that $n = 0$ or $m = 0$. Without loss of generality, assume that $n = 0$. Then by the definition of x^m we see that $x^n x^m = x^0 x^m = 1 \cdot x^m = x^m = x^{0+m} = x^{n+m}$.

Third, suppose that $n > 0$ and $m < 0$. Then $-m > 0$, and hence by Lemma 2.4.8 (2) we know that $n \in \mathbb{N}$ and $-m \in \mathbb{N}$. There are now three subcases.

For the first subcase, suppose that $n + m > 0$. By Lemma 2.4.9 we know that $n + m \in \mathbb{Z}$, and therefore by Lemma 2.4.8 (2) we know that $n + m \in \mathbb{N}$. Let $r = n + m$. Then $n = r + (-m)$. Because $r > 0$ and $-m > 0$, then by the first case in this proof, together with the definition of $x^{-(-m)}$, we see that $x^n x^m = x^{r+(-m)} x^m = (x^r x^{-m}) x^{-(-m)} = x^r (x^{-m} x^{-(-m)}) = x^r (x^{-m} (x^{-m})^{-1}) = x^r \cdot 1 = x^r = x^{n+m}$.

For the second subcase, suppose that $n+m=0$. Then $m=-n$. Using the definition of x^{-n} we see that $x^n x^m = x^n x^{-n} = x^n (x^n)^{-1} = 1 = x^0 = x^{n+m}$.

For the third subcase, suppose that $n+m<0$. Then $(-n)+(-m) = -(n+m) > 0$. Because $-m>0$ and $-n<0$, we can use the first of our three subcases, together with the definition of x^{-n}, to see that $x^n x^m = ((x^n)^{-1})^{-1}((x^m)^{-1})^{-1} = (x^{-n})^{-1}(x^{-m})^{-1} = [x^{-n}x^{-m}]^{-1} = [x^{(-n)+(-m)}]^{-1} = [x^{-(n+m)}]^{-1} = [(x^{n+m})^{-1}]^{-1} = x^{n+m}$. We have now completed the proof in the case where $n>0$ and $m<0$.

Fourth, suppose that $n<0$ and $m>0$. This case is similar to the previous case, and we omit the details.

Fifth, suppose that $n<0$ and $m<0$. Then $-n>0$ and $-m>0$, and we can proceed similarly to the third subcase of the third case of this proof; we omit the details. □

Because of Definition 2.5.6, we can now define polynomials. The following definition might seem a bit trickier than necessary, but from the perspective of real analysis, we are interested in viewing polynomials as functions (so that we can take their derivatives and integrals), rather than as expressions made up of sums of powers of a "variable" multiplied by constants.

Definition 2.5.10. Let $A \subseteq \mathbb{R}$ be a set, and let $f\colon A \to \mathbb{R}$ be a function. The function f is a **polynomial function** if there are some $n \in \mathbb{N} \cup \{0\}$ and $a_0, a_1, \ldots, a_n \in \mathbb{R}$ such that $f(x) = a_0 + a_1 x + \cdots + a_n x^n$ for all $x \in A$. △

It is important to point out that, as with all functions, the name of the polynomial defined in Definition 2.5.10 is "f," not "$f(x)$." It would be commonly understood that the notation "$f(c)$" denotes the value of the polynomial f at the element $c \in A$, and so $f(c)$ would be an element of $\mathbb{R}$, which is the codomain of f. Why should "$f(x)$" mean anything different from "$f(c)$," except that c is one choice of element in the domain, and x is another such element? Historically, following Descartes, mathematicians have often used letters such as x, y and z to denote "variables," and letters such as a, b and c to denote "constants," but from a rigorous standpoint there is no such distinction. Every element of a set is a single element of the set, that is, it is a "constant." There is actually no such thing as a "variable" (though at times it is convenient to use the word "variable" informally). In particular, in careful mathematical writing, we will always use the symbol f to denote the name of the polynomial, and more generally the name of the function, and we will always use the symbol $f(x)$ to denote an element of the codomain. A careless approach in this matter can lead to misunderstandings in some tricky situations.

Although we defined polynomials as functions in Definition 2.5.10, in practice it is customary to define a polynomial functions simply by saying "let $a_0 + a_1 x + \cdots + a_n x^n$ be a polynomial," and not even mentioning that the polynomial is a function. From a rigorous point of view, this informal way of defining polynomials is problematic, because, as mentioned above, there is no real distinction between "constants" and "variables," and so the expression "$a_0 + a_1 x + \cdots + a_n x^n$" is not inherently the definition of a function unless it is stated as such explicitly and properly, which would include the domain and the codomain. However, because no real problem arises from this informal way of defining polynomials, for convenience we will stick to standard practice and define define polynomials in this way. The domain and codomain for polynomials will always be assumed to be $\mathbb{R}$, unless otherwise stated.

Returning to the general issue of definition by recursion, we observe that in Theorem 2.5.5 each a_{n+1} was a function of a_n alone, given by a formula of the form $a_{n+1} = k(a_n)$ for all $n \in \mathbb{N}$. However, we sometimes need to express a_{n+1} in terms of n as well as a_n. For example, suppose that we want to define a sequence by setting $a_1 = 1$, and $a_{n+1} = n + a_n$ for all $n \in \mathbb{N}$. Such a recursive definition does appear to produce a unique sequence that starts $1, 2, 4, 7, 11, \ldots$, though formally this definition by recursion is not covered by Theorem 2.5.5. Such situations are handled by the following variant of Theorem 2.5.5.

Theorem 2.5.11. *Let H be a set, let $e \in H$ and let $t\colon H \times \mathbb{N} \to H$ be a function. Then there is a unique function $g\colon \mathbb{N} \to H$ such that $g(1) = e$, and that $g(n+1) = t((g(n), n))$ for all $n \in \mathbb{N}$.*

Proof. We will prove that g exists, leaving the proof of uniqueness to the reader in Exercise 2.5.16.

We can apply Theorem 2.5.5 to the set $H \times \mathbb{N}$, the element $(e, 1) \in H \times \mathbb{N}$ and the function $r\colon H \times \mathbb{N} \to H \times \mathbb{N}$ defined by $r((x, m)) = (t((x, m)), m+1)$ for all $(x, m) \in H \times \mathbb{N}$. Hence there is a unique function $f\colon \mathbb{N} \to H \times \mathbb{N}$ such that $f(1) = (e, 1)$, and that $f(n+1) = r(f(n))$ for all $n \in \mathbb{N}$.

Let $f_1\colon \mathbb{N} \to H$ and $f_2\colon \mathbb{N} \to \mathbb{N}$ be the coordinate functions of f. That is, the functions f_1 and f_2 are the unique functions such that $f(n) = (f_1(n), f_2(n))$ for all $n \in \mathbb{N}$.

Let $g = f_1$. Because $(g(1), f_2(1)) = f(1) = (e, 1)$, then $g(1) = e$.

Let $n \in \mathbb{N}$. Then the equation $f(n+1) = r(f(n))$ can be rewritten in coordinates as

$$(g(n+1), f_2(n+1)) = r((g(n), f_2(n))) = (t((g(n), f_2(n))), f_2(n) + 1).$$

Hence $g(n+1) = t((g(n), f_2(n)))$ and $f_2(n+1) = f_2(n) + 1$. Observe that the second equation is satisfied if we use $f_2 = 1_{\mathbb{N}}$, where $1_{\mathbb{N}}\colon \mathbb{N} \to \mathbb{N}$ is the identity map. Hence, by the uniqueness of f_2, it must be the case that $f_2 = 1_{\mathbb{N}}$. It now follows that $g(n+1) = t((g(n), n))$. □

Example 2.5.12. We want to define a sequence of real numbers $a_1, a_2, a_3, \ldots$ such that $a_1 = 1$, and $a_{n+1} = (n+1)a_n$ for all $n \in \mathbb{N}$. Using Theorem 2.5.11 with $e = 1$, and with $t(x, m) = (m+1)x$ for all $(x, m) \in \mathbb{R} \times \mathbb{N}$, we see that there is a sequence satisfying the given conditions. This sequence starts $1, 2, 6, 24, 120, \ldots$, and consists of the familiar factorial numbers. We use the symbol $n!$ to denote a_n. The reader might wonder whether we could have dispensed with the Definition by Recursion entirely, and just have explicitly defined $a_n = n!$ for all $n \in \mathbb{N}$, but that would be viewing the situation backwards. The symbol $n!$ is informally defined by $n! = n(n-1)(n-2)\cdots 2 \cdot 1$, but this type of definition is not rigorous, because "$\cdots$" is not a rigorous concept. The formal way to define $n!$ without "$\cdots$" is simply to say that it is the value of a_n for the sequence we have defined by recursion. Observe that from Definition by Recursion, we deduce immediately that $(n+1)! = (n+1)n!$ for all $n \in \mathbb{N}$. ◊

We conclude this section with another application of Theorem 2.5.11, again to a situation where something seems intuitively obvious, but we need to use Definition by Recursion to provide a rigorous definition.

Example 2.5.13. We want to show that any finite set of real numbers has a greatest element and a least element. We will look at greatest elements; least elements behave entirely analogously, and we omit the details.

Before we turn to the specific issue of finding the greatest element of finite sets, we define greatest elements in general. Let $A \subseteq \mathbb{R}$ be a set, and let $c \in A$. The number c is a **greatest element** of A if $x \leq c$ for all $x \in A$.

The reader might find the term "maximal element" more familiar than "greatest element," but whereas these two terms coincide in the context of the real numbers, in the more general context of partially ordered sets (also called posets), the concept of a "maximal element" is not the same as the concept of a "greatest element," and so we will make use here of the latter term, because it corresponds to the above definition in all contexts; see [Blo10, Section 7.4] for definitions of these terms for posets.

Not every subset of $\mathbb{R}$ has a greatest element, for example the open interval $(1,2)$. However, if a set has a greatest element then it is unique, as seen in Exercise 2.5.17. Some other properties of greatest elements can be found in Exercise 2.5.18.

It is easy to define the maximum of two real numbers at a time. Let $x, y \in \mathbb{R}$. The **maximum** of x and y, denoted $\max\{x,y\}$, is defined by

$$\max\{x,y\} = \begin{cases} x, & \text{if } x \geq y \\ y, & \text{if } x \leq y. \end{cases}$$

The problem is providing a rigorous definition of the maximum of arbitrarily sized finite sets of real numbers. More precisely, let $p \in \mathbb{N}$, and let $a_1, a_2, \ldots, a_p \in \mathbb{R}$. We want to show that the set $\{a_1, \ldots, a_p\}$ has a greatest element; if such a greatest element exists, then it would be unique, and it would be denoted $\max\{a_1, \ldots, a_p\}$. (We note, however, that this uniqueness is as an element of the set $\{a_1, \ldots, a_p\}$, and that the greatest element need not be represented by a unique number of the form a_i; for example, if $a_1 = 7$, and $a_2 = 0$ and $a_3 = 7$, then $\{a_1, a_2, a_3\} = \{0, 7\}$, and $\max\{a_1, a_2, a_3\} = 7$, which is a unique number, but we observe that 7 is represented by both a_1 and a_3.)

Simply writing the notation "$\max\{a_1, \ldots, a_p\}$" does not alone suffice to show that a greatest element of the set exists, because, as we have stated previously, the notion of $\ldots$ is not rigorous. We avoid this problem by using Definition by Recursion, where intuitively we find the maximum of $a_1, \ldots, a_n$ by first finding the maximum of a_1 and a_2, then finding the maximum of $\max\{a_1, a_2\}$ and a_3, and continuing in this fashion, until we reach the end.

To avoid writing the convenient but informal notation "$a_1, \ldots, a_p$," we use functions. Let $f\colon \mathbb{N} \to \mathbb{R}$ be a function. We think of $f(n)$ as a_n for all $n \in \mathbb{N}$. (If we are given only a finite sequence $a_1, a_2, \ldots, a_p$ to begin with, we can define $f(i) = a_i$ for all $i \in \{1, \ldots, p\}$, and $f(i) = a_1$ for all $i \in \mathbb{N} - \{1, \ldots, p\}$.) If the reader is wondering why we are willing to write the "$\ldots$" in $\{1, \ldots, p\}$ but not in $a_1, \ldots, a_p$, it is because the former was given a rigorous definition in Definition 2.5.3.

Let $t\colon \mathbb{R} \times \mathbb{N} \to \mathbb{R}$ be defined by $t((x,m)) = \max\{x, f(m+1)\}$ for all $(x,m) \in \mathbb{R} \times \mathbb{N}$. By Theorem 2.5.11 there is a unique function $g\colon \mathbb{N} \to \mathbb{R}$ such that $g(1) = f(1)$, and that $g(n+1) = t((g(n),n))$ for all $n \in \mathbb{N}$. It follows that $g(n+1) = \max\{g(n), f(n+1)\}$ for all $n \in \mathbb{N}$. Hence $g(n+1) \geq g(n)$ and $g(n+1) \geq f(n+1)$ for all $n \in \mathbb{N}$.

We claim that $g(n) \in \{f(i) \mid i \in \{1, \ldots, n\}\}$ for all $n \in \mathbb{N}$. We know that $g(1) = f(1)$, so clearly $g(1) \in \{f(i) \mid i \in \{1, \ldots, 1\}\}$. Next, suppose that $g(k) \in \{f(i) \mid i \in \{1, \ldots, k\}\}$ for some $k \in \mathbb{N}$. Then $g(k+1) = \max\{g(k), f(k+1)\}$, and so $g(k+1) = g(k)$ or $g(k+1) = f(k+1)$. In either case we deduce that $g(k+1) \in \{f(i) \mid i \in \{1, \ldots, k+1\}\}$. By induction we conclude that $g(n) \in \{f(i) \mid i \in \{1, \ldots, n\}\}$ for all $n \in \mathbb{N}$.

Next, let $m \in \mathbb{N}$. We claim that $g(m) \geq f(i)$ for all $i \in \{1, \ldots, m\}$. Because $g(n+1) \geq g(n)$ for all $n \in \{1, \ldots, m-1\}$, it follows from Exercise 2.5.4 that $g(m) \geq g(i)$ for all $i \in \{1, \ldots, m\}$. However, because $g(n) \geq f(n)$ for all $n \in \mathbb{N}$, it follows that $g(m) \geq f(i)$ for all $i \in \{1, \ldots, m\}$. Hence $g(m)$ is a greatest element of $\{f(i) \mid i \in \{1, \ldots, m\}\}$. ◊

In Exercise 2.5.19 it is similarly shown that any finite set of real numbers has a sum; that is, that expressions of the form "$a_1 + a_2 + \cdots + a_n$" are defined for arbitrary $n \in \mathbb{N}$, where $a_1, \ldots, a_n \in \mathbb{R}$.

Reflections

This section has a number of lengthy proofs of apparently obvious facts. The reader has used x^n and its properties confidently for years; has never questioned the definition of $\max\{a_1, \ldots, a_p\}$; has a lot of familiarity with the Principle of Mathematical Induction from previous courses; and has, without any hesitation, used Definition by Recursion many times prior to reading this section. Moreover, induction and recursion are algebraic in nature, and the reader might wonder why they are included in such detail in a text on real analysis, which is about calculus rather than algebra. The reason the material in this section is discussed in such detail is simply that all this material is used subsequently in this text for the treatment of topics that belong to the study of real analysis, and if we want to be sure that everything proved in real analysis follows from the basic properties of the real numbers with no hidden assumptions, we need to prove everything we use, even if intuitively obvious, and even if appearing to belong to a different branch of mathematics.

Exercises

Exercise 2.5.1. [Used in Example 10.4.16.] Let $n \in \mathbb{N}$, and let $a \in \mathbb{R}$. Suppose that $a \geq 1$. Prove that $a^n \geq a$ for all $n \in \mathbb{N}$.

Exercise 2.5.2. [Used in Exercise 2.5.3 and Exercise 2.5.4.] There are occasions when we need to do "induction" to prove something about the first p natural numbers, but not beyond that. In this exercise we will see that such a procedure works. Examples of using this method of proof are found in Exercise 2.5.3 and Exercise 2.5.4.

Let $p \in \mathbb{N}$, and let $G \subseteq \mathbb{N}$. Suppose that

a. $1 \in G$;
b. if $n \in G$ and $n \leq p-1$, then $n+1 \in G$.

Prove that $\{1,\ldots,p\} \subseteq G$.

Exercise 2.5.3. [Used throughout.] Let $n \in \mathbb{N}$, and let $a_1, a_2, \ldots, a_n \in \mathbb{R}$. Prove that $|a_1 + a_2 + \cdots + a_n| \leq |a_1| + |a_2| + \cdots + |a_n|$. [Use Exercise 2.5.2.]

Exercise 2.5.4. [Used in Example 2.5.13.] Let $n \in \mathbb{N}$, and let $a_1, a_2, \ldots, a_n \in \mathbb{R}$. Suppose that $a_i \leq a_{i+1}$ for all $i \in \{1,\ldots,n-1\}$. Prove that $a_i \leq a_n$ for all $i \in \{1,\ldots, n\}$. [Use Exercise 2.5.2.]

Exercise 2.5.5. [Used in Example 5.9.7.] Prove that

$$1+2+\cdots+n = \frac{n(n+1)}{2}$$

for all $n \in \mathbb{N}$.

Exercise 2.5.6. [Used in Example 9.2.4.] Prove that

$$\frac{1}{2}+\frac{1}{4}+\frac{1}{8}+\cdots+\frac{1}{2^n} = 1-\frac{1}{2^n}$$

for all $n \in \mathbb{N}$.

Exercise 2.5.7. [Used in Example 9.2.4.] Prove that

$$\frac{1}{1}+\frac{1}{2}+\frac{1}{3}+\frac{1}{4}+\cdots+\frac{1}{2^n} \geq \frac{n+2}{2}$$

for all $n \in \mathbb{N}$.

Exercise 2.5.8. [Used in Example 8.4.10.] Let $n \in \mathbb{N}$, and let $a_1, a_2, \ldots, a_n \in \mathbb{R}$. Suppose that $a_1 \geq a_2 \geq \cdots \geq a_n \geq 0$. Prove that $0 \leq a_1 - a_2 + a_3 - a_4 + \ldots + (-1)^{n-1} a_n \leq a_1$.

Exercise 2.5.9. [Used in Lemma 2.5.7.] Prove Lemma 2.5.7.

Exercise 2.5.10. The sequence $r_1, r_2, \ldots$ is defined by the conditions $r_1 = 1$, and $r_{n+1} = 4r_n + 7$ for all $n \in \mathbb{N}$. Prove that $r_n = \frac{1}{3}\left(10 \cdot 4^{n-1} - 7\right)$ for all $n \in \mathbb{N}$.

Exercise 2.5.11. [Used in Lemma 2.5.9.] Prove Lemma 2.5.9 Part (2).

Exercise 2.5.12. [Used in Theorem 2.8.10, Exercise 2.8.4, Example 9.2.4, Example 10.3.7, Theorem 10.3.8 and Theorem 10.5.2.] Let $n \in \mathbb{N}$.

(1) Let $x, y \in \mathbb{R}$. Prove that

$$y^n - x^n = (y-x)(y^{n-1} + y^{n-2}x + y^{n-3}x^2 + \cdots + x^{n-1}).$$

(2) Let $x,y \in \mathbb{R}$. Prove that if $0 \le x \le y$, then $x^n \le y^n$.
(3) Let $a,r \in \mathbb{R}$. Suppose that $r \ne 1$. Prove that

$$a + ar + ar^2 + \cdots + ar^{n-1} = \frac{a(1-r^n)}{1-r}.$$

Exercise 2.5.13. [Used in Exercise 2.6.13, Lemma 2.8.1, Example 8.2.13 and Example 9.2.4.]

(1) Let $q \in \mathbb{R}$. Suppose that $q > 0$. Prove that $1 + nq \le (1+q)^n$ for all $n \in \mathbb{N}$.
(2) Let $p \in \mathbb{N}$. Suppose that $p > 1$. Prove that $n < p^n$ for all $n \in \mathbb{N}$.

[Use Exercise 2.4.3.]

Exercise 2.5.14. [Used in Exercise 3.4.6.] Let $n \in \mathbb{N}$, and let $[a_1,b_1],\ldots,[a_n,b_n] \subseteq \mathbb{R}$ be closed bounded intervals. Prove that $\bigcup_{i=1}^n [a_i,b_i]$ equals the union of finitely many disjoint closed bounded intervals.

Exercise 2.5.15. [Used in Example 5.8.2.] Let $[x,y] \subseteq \mathbb{R}$ be a non-degenerate closed bounded interval, let $n \in \mathbb{N}$ and let $[a_1,b_1],\cdots,[a_n,b_n] \subseteq \mathbb{R}$ be non-degenerate closed bounded intervals. Prove that if $[x,y] \subseteq \bigcup_{i=1}^n [a_i,b_i]$, then $y - x \le \sum_{i=1}^n (b_i - a_i)$.

This exercise can be simplified as follows. Without loss of generality, assume that $a_1 \le a_2 \le \cdots \le a_{n+1}$; if that is not the case, the indices can be renamed to make that happen. Moreover, if $[a_i,b_i] \subseteq [a_j,b_j]$ for some $i,j \in \{1,\ldots,n+1\}$, then the interval $[a_i,b_i]$ could be dropped without any change to the hypothesis on $[x,y]$. Hence, without loss of generality, assume that none of the intervals of the form $[a_k,b_k]$ is a subset of another such interval. [Use Exercise 2.3.13.]

Exercise 2.5.16. [Used in Theorem 2.5.11.] Prove the uniqueness of the function g in Theorem 2.5.11.

Exercise 2.5.17. [Used in Example 2.5.13 and Lemma 8.3.8.] Let $A \subseteq \mathbb{R}$ be a set. Prove that if A has a greatest element, then that greatest element is unique.

Exercise 2.5.18. [Used in Example 2.5.13 and Lemma 8.3.8.] Let $A,B \subseteq \mathbb{R}$ be sets. Suppose that $A \subseteq B$.

(1) Prove that if A has a greatest element a and B has a greatest element b, then $a \le b$.
(2) Suppose that $B - A$ is a finite set. Prove that if A has a greatest element then B has a greatest element.
(3) Find an example of sets A and B such that $B - A$ is a finite set and B has a greatest element, but that A does not have a greatest element.

Exercise 2.5.19. [Used throughout.] The binary operations addition and multiplication are defined for only two numbers at a time; hence the word "binary." On the other hand, it is very convenient to use expressions of the form "$a_1 + a_2 + \cdots + a_n$" for arbitrary $n \in \mathbb{N}$, where $a_1,\ldots,a_n \in \mathbb{R}$. However, simply writing "$\cdots$" and saying that we add the numbers two at a time is not rigorous, and it is the purpose of this exercise to give this use of "$\cdots$" a rigorous definition. (A similar definition can be given for the product of finitely many numbers, though we omit the details.)

The intuitive idea of this definition is as follows. Suppose we want to compute the finite sum $3+7+2$. Because addition is defined for only two numbers at a time, there are two ways one might go about finding the sum $3+7+2$, which are $(3+7)+2$ and $3+(7+2)$. Fortunately, because of the Associative Law for Addition, these two sums are equal, and hence we can write $3+7+2$ unambiguously. If we can define the sum of three numbers unambiguously, then a similar argument will show that the sum of four numbers can also be defined and so on.

For the formal definition, rather than writing $a_1+a_2+\cdots+a_n$, it is easier to use functions. Let $f\colon \mathbb{N} \to \mathbb{R}$ be a function. Use Theorem 2.5.11 to give a rigorous definition for the expression "$f(1)+f(2)+\cdots+f(n)$" for all $n \in \mathbb{N}$.

Exercise 2.5.20. **[Used in Example 8.4.10.]** Let H be a non-empty set, let $a,b \in H$ and let $p\colon H \times H \to H$ be a function. Prove that there is a unique function $f\colon \mathbb{N} \to H$ such that $f(1)=a$, that $f(2)=b$ and that $f(n+2)=p((f(n),f(n+1)))$ for all $n \in \mathbb{N}$.

The idea is to apply Theorem 2.5.5 to the set $H \times H$, the element (a,b) and the function $k\colon H \times H \to H \times H$ defined by $k((x,y)) = (y, p(x,y))$ for all $(x,y) \in H \times H$, and then to use the result of that step to find the desired function f.

2.6 The Least Upper Bound Property and Its Consequences

We have now reached the essence of the real numbers. The various properties of the real numbers we saw up till now in this chapter were proved using only the properties of an ordered field (Theorem 1.7.6 or Definition 2.2.1). As such, all the properties of the real numbers that were proved in the previous sections of this chapter would hold in any ordered field, for example the set of rational numbers. We now turn to those properties of the real numbers that make use of the Least Upper Bound Property (Theorem 1.7.9 or Definition 2.2.3). The Least Upper Bound Property is the property of the real numbers that distinguishes these numbers from all other ordered fields, and in particular from the rational numbers, and it is the Least Upper Bound Property that ultimately allows us to do calculus, which is what real analysis is about.

Recall from Definition 1.7.7 or Definition 2.2.2 the concepts of bounded above, bounded below, bounded, upper bound, lower bound, least upper bound and greatest lower bound. Recall also Exercise 2.3.11, which states that a subset $A \subseteq \mathbb{R}$ is bounded if and only if there is some $M \in \mathbb{R}$ such that $M > 0$ and that $|x| \le M$ for all $x \in A$. We now discuss these concepts further, starting with some examples.

Example 2.6.1.

(1) Let $A = [3,5)$. Then 10 is an upper bound of A, and -100 is a lower bound. Hence A is bounded above and bounded below, and therefore A is bounded. Clearly 15 is also an upper bound of A, and 0 is a lower bound of A. In fact, there are infinitely many other upper bounds and lower bounds of A. The least upper bound of A is 5, and the greatest lower bound of A is 3. Observe that A does not contain its least upper bound. On the other hand, if we let $B = [3,5]$, then the least upper bound of B is 5, and B does contain its least upper bound.

(2) Let $C = \{\frac{1}{n} \mid n \in \mathbb{N}\}$. Then C is bounded above by 2 and bounded below by -1. The least upper bound of C is 1, and the greatest lower bound of C is 0.

(3) The set $\mathbb{N}$ is bounded below (a lower bound is 0), but it is not bounded above, as the reader is asked to prove in Exercise 2.6.16, using tools that we will develop later in this section. Hence $\mathbb{N}$ is not bounded. The greatest lower bound of $\mathbb{N}$ is 1, but $\mathbb{N}$ has no least upper bound, because it has no upper bound. ◊

Whereas upper bounds and lower bounds of sets are not unique, as seen in Example 2.6.1 (1), the following lemma shows that least upper bounds and greatest lower bounds, when they exist, are unique.

Lemma 2.6.2. *Let $A \subseteq \mathbb{R}$ be a non-empty set.*

1. *If A has a least upper bound, the least upper bound is unique.*
2. *If A has a greatest lower bound, the greatest lower bound is unique.*

Proof. We will prove Part (1); the other part is similar, and we omit the details.

(1) Suppose that $M, N \in \mathbb{R}$ are both least upper bounds of A. That means that M and N are both upper bounds of A, and that $M \le T$ and $N \le T$ for all upper bounds T of A. It follows that $M \le N$ and $N \le M$, and therefore $M = N$. □

Because of Lemma 2.6.2 we can now refer to "the least upper bound" and "the greatest lower bound" of a set, if they exist. Hence the following definition makes sense.

Definition 2.6.3. Let $A \subseteq \mathbb{R}$ be a non-empty set. If A has a least upper bound, it is denoted $\text{lub}A$. If A has a greatest lower bound, it is denoted $\text{glb}A$. △

In addition to the Least Upper Bound Property of the real numbers, there is also a corresponding—and equivalent—property for greatest lower bounds, called the Greatest Lower Bound Property, which we now prove. The reader who has read Section 1.7 has already seen the Greatest Lower Bound Property (Theorem 1.7.8), and can therefore skip the following theorem.

Theorem 2.6.4 (Greatest Lower Bound Property). *Let $A \subseteq \mathbb{R}$ be a set. If A is non-empty and bounded below, then A has a greatest lower bound.*

Proof. Suppose that A is non-empty and bounded below. Let

$$L = \{b \in \mathbb{R} \mid b \text{ is a lower bound of } A\}.$$

Then $L \neq \emptyset$, because A is bounded below. Let $x \in A$. Then $b \le x$ for all $b \in L$, and hence x is an upper bound of L. Because L is non-empty and bounded above, we can apply the Least Upper Bound Property to L to deduce that L has a least upper bound, say m.

If $y \in A$, then $d \le y$ for all $d \in L$. Therefore every element of A is an upper bound of L. Because m is the least upper bound of L, it follows that $m \le y$ for all $y \in A$. Hence m is a lower bound of X. Because m is an upper bound of L, then $d \le m$ for every $d \in L$. We deduce that m is a greatest lower bound of X. □

We now have a number of useful consequences of the Least Upper Bound Property (and the Greatest Lower Bound Property), starting with the following lemma, which we will use repeatedly. Intuitively, this lemma says that if a set has a least upper bound, then it is possible to find elements of the set as close as one wants to the least upper bound. That is, even if a set does not contain its least upper bound, there is no "gap" between the set and the least upper bound, and similarly for greatest lower bounds.

Lemma 2.6.5. *Let $A \subseteq \mathbb{R}$ be a non-empty set, and let $\varepsilon > 0$.*

1. *Suppose that A has a least upper bound. Then there is some $a \in A$ such that* $\mathrm{lub}A - \varepsilon < a \leq \mathrm{lub}A$.
2. *Suppose that A has a greatest lower bound. Then there is some $b \in A$ such that* $\mathrm{glb}A \leq b < \mathrm{glb}A + \varepsilon$.

Proof. We will prove Part (1); the other part is similar, and we omit the details.

(1) Because $\mathrm{lub}A - \varepsilon < \mathrm{lub}A$, then $\mathrm{lub}A - \varepsilon$ cannot be an upper bound of A, because $\mathrm{lub}A$ is the least upper bound. Hence there is some $a \in A$ such that $\mathrm{lub}A - \varepsilon < a$. Because $\mathrm{lub}A$ is an upper bound of A, then $a \leq \mathrm{lub}A$. □

We observe that in Lemma 2.6.5 (1) the inequality $\mathrm{lub}A - \varepsilon < a \leq \mathrm{lub}A$ cannot in general be replaced with $\mathrm{lub}A - \varepsilon < a < \mathrm{lub}A$, because the latter inequality would not hold, for example, when A has a single element; the analogous fact holds for Part (2) of the lemma.

Our next result is another technically useful fact about least upper bounds and greatest lower bounds. This lemma generalizes the well-known Nested Interval Theorem (Theorem 8.4.7), but is in fact simpler than that result, because it does not require the language of sequences. We will use this lemma in the proof of Theorem 5.4.7, which is a very important result about integration, as well as in the proof of the Nested Interval Theorem. The idea of this lemma is that if the elements of one set of real numbers are all greater than or equal to the elements of another set of real numbers, and if elements of the two sets can be found as close to each other as desired, then there is no gap between the two sets.

Lemma 2.6.6 (No Gap Lemma). *Let $A, B \subseteq \mathbb{R}$ be non-empty sets. Suppose that if $a \in A$ and $b \in B$, then $a \leq b$.*

1. *A has a least upper bound and B has a greatest lower bound, and* $\mathrm{lub}A \leq \mathrm{glb}B$.
2. $\mathrm{lub}A = \mathrm{glb}B$ *if and only if for each $\varepsilon > 0$, there are $a \in A$ and $b \in B$ such that $b - a < \varepsilon$.*

Proof.

(1) Because A and B are both non-empty, then A is bounded above by any element of B, and B is bounded below by any element of A. The Least Upper Bound Property and the Greatest Lower Bound Property imply that A has a least upper bound and B has a greatest lower bound.

Suppose that $\mathrm{lub}A > \mathrm{glb}B$. Let $\mu = \frac{\mathrm{lub}A - \mathrm{glb}B}{2}$. Then $\mu > 0$. We know by Lemma 2.6.5 that there is some $a \in A$ such that $\mathrm{lub}A - \mu < a \leq \mathrm{lub}A$, and there is some $b \in B$ such that $\text{glb}\,B \leq b < \text{glb}\,B + \mu$. It follows that

$$b < \text{glb}\,B + \mu = \text{glb}\,B + \frac{\text{lub}\,A - \text{glb}\,B}{2} = \frac{\text{lub}\,A + \text{glb}\,B}{2}$$
$$= \text{lub}\,A - \frac{\text{lub}\,A - \text{glb}\,B}{2} = \text{lub}\,A - \mu < a,$$

which is a contradiction. Hence $\text{lub}\,A \leq \text{glb}\,B$.

(2) By Part (1) of this lemma we know that $\text{lub}\,A \leq \text{glb}\,B$.

Suppose that $\text{lub}\,A = \text{glb}\,B$. Let $\varepsilon > 0$. By Lemma 2.6.5 there are $p \in A$ and $q \in B$ such that $\text{lub}\,A - \frac{\varepsilon}{2} < p \leq \text{lub}\,A$ and $\text{glb}\,B \leq q < \text{glb}\,B + \frac{\varepsilon}{2}$. Then $\text{lub}\,A - \frac{\varepsilon}{2} < p \leq \text{lub}\,A = \text{glb}\,B \leq q < \text{glb}\,B + \frac{\varepsilon}{2} = \text{lub}\,A + \frac{\varepsilon}{2}$, and it follows that $q - p < \varepsilon$.

Now suppose that $\text{lub}\,A \neq \text{glb}\,B$. Let $\eta = \text{glb}\,B - \text{lub}\,A$. Then $\eta > 0$. If $x \in A$ and $y \in B$, then $x \leq \text{lub}\,A$ and $y \geq \text{glb}\,B$, which implies $y - x \geq \text{glb}\,B - \text{lub}\,A = \eta$. Hence it is not the case that for each $\varepsilon > 0$, there are $a \in A$ and $b \in B$ such that $b - a < \varepsilon$, for example if $\varepsilon = \eta/2$. □

The reader has already encountered a number of basic results about the natural numbers in Section 1.2, 1.4 or 2.4. In contrast to those results, which were about the natural numbers in their own right, we now turn to a very important theorem, called the Archimedean Property of the real numbers, that illuminates how the natural numbers are situated inside the larger set of all real numbers. This theorem may seem so obvious that the reader might question the need for proving it, but in fact the proof is non-trivial, making use of the Least Upper Bound Property. To appreciate the value of the Archimedean Property, which we will use regularly, we note that there exist ordered fields for which the Archimedean Property does not hold; see [Olm62, Sections 711–713] for an example of such an ordered field.

Theorem 2.6.7 (Archimedean Property). *Let $a, b \in \mathbb{R}$. Suppose that $a > 0$. Then there is some $n \in \mathbb{N}$ such that $b < na$.*

Proof. First, suppose that $b \leq 0$. Let $n = 1$. It follows that $b \leq 0 < a = na$. Second, suppose that $b > 0$. We use proof by contradiction. Suppose that $na \leq b$ for all $n \in \mathbb{N}$. Let

$$A = \{ka \mid k \in \mathbb{N}\}.$$

Then $A \subseteq \mathbb{R}$. Because $1 \cdot a \in A$, we see that $A \neq \emptyset$. By hypothesis we know that b is an upper bound of A. The Least Upper Bound Property then implies that A has a least upper bound. Let $m \in \mathbb{N}$. Then $m + 1 \in \mathbb{N}$, and hence $(m+1)a \in A$, which implies $(m+1)a \leq \text{lub}\,A$. Therefore $ma + a \leq \text{lub}\,A$, and hence $ma \leq \text{lub}\,A - a$. Because m was arbitrarily chosen, we deduce that $\text{lub}\,A - a$ is an upper bound of A, which is a contradiction to the fact that $\text{lub}\,A$ is the least upper bound of A. □

Corollary 2.6.8. *Let $x \in \mathbb{R}$.*

1. *There is a unique $n \in \mathbb{Z}$ such that $n - 1 \leq x < n$. If $x \geq 0$, then $n \in \mathbb{N}$.*
2. *If $x > 0$, there is some $m \in \mathbb{N}$ such that $\frac{1}{m} < x$.*

Proof.

(1) First we prove uniqueness. Suppose that there are $n, m \in \mathbb{Z}$ such that $n-1 \leq x < n$ and $m-1 \leq x < m$. We proceed by contradiction. Suppose that $n \neq m$. Without loss of generality, assume that $n < m$. By Theorem 2.4.10 (1) we know that $n+1 \leq m$. Hence $n \leq m-1$, and therefore $x < n \leq m-1 \leq x$, which is a contradiction. We deduce that $n = m$.

We now prove existence. First, suppose that $x = 0$. Then we can let $n = 1$.

Second, suppose that $x > 0$. We can then apply the Archimedean Property (Theorem 2.6.7) to 1 and x to deduce that there is some $k \in \mathbb{N}$ such that $x < k \cdot 1$, which means that $x < k$. Let

$$B = \{m \in \mathbb{N} \mid x < m\}.$$

Clearly $B \subseteq \mathbb{N}$. We have just seen that $k \in B$, and hence $B \neq \emptyset$. By the Well-Ordering Principle (Theorem 1.2.10, Axiom 1.4.4 or Theorem 2.4.6) there is some $n \in B$ such that $n \leq m$ for all $m \in B$. The definition of B implies that $x < n$. By the minimality of n we know that $n-1 \notin B$. If $n-1 \in \mathbb{N}$, then $n-1 \notin B$ implies that $n-1 \leq x$. If $n-1 \notin \mathbb{N}$, then $n = 1$, and so $n-1 = 0$, and hence by hypothesis on x we see that $n-1 < x$. Combining the two cases, we dedcue that $n-1 \leq x < n$.

Third, suppose that $x < 0$. Then $-x > 0$. As we have just seen, there is some $m \in \mathbb{N}$ such that $m-1 \leq -x < m$. Then $-m < x \leq -m+1$. There are now two subcases. First, suppose that $x = -m+1$. Then let $n = -m+2$. Therefore $x = n-1$, and hence $n-1 \leq x < n$. Because $m \in \mathbb{N}$, it follows that $n \in \mathbb{Z}$. Second, suppose that $-m < x < -m+1$. Then let $n = -m+1$, and hence $n-1 \leq x < n$. Because $m \in \mathbb{N}$, then once again $n \in \mathbb{Z}$.

(2) Suppose that $x > 0$. Then $\frac{1}{x} > 0$. We know by Part (1) of this corollary that there is some $m \in \mathbb{N}$ such that $m-1 \leq \frac{1}{x} < m$, and it follows that $\frac{1}{m} < x$. □

We are now ready to fulfill a promise made in Section 2.4, where we said that we would show that there exist irrational numbers, a fact that is not obvious from the definition, because the set of irrational numbers is simply defined to be $\mathbb{R} - \mathbb{Q}$, and we have not yet proved that $\mathbb{Q}$ is not all of $\mathbb{R}$. In particular, we will prove that $\sqrt{2}$ exists in $\mathbb{R}$, and after that we will show that $\sqrt{2}$ is not in $\mathbb{Q}$. The reader might be acquainted with the standard proof by contradiction that $\sqrt{2}$ is irrational using fractions in "lowest terms" and even and odd numbers; see [Blo10, Theorem 2.3.4] for this proof. We will use a different approach here, however, for two reasons. First, we have not given a rigorous treatment of fractions in "lowest terms," nor of even and odd numbers (not because these topics are too difficult, but because we do not need them for any other purpose), and hence we do not have these tools available to us now. Second, and more importantly, the standard proof that $\sqrt{2}$ is irrational shows only that there is no rational number x such that $x^2 = 2$, but it does not show that there exists an irrational number x such that $x^2 = 2$, which is what we need in order to show that $\mathbb{R} - \mathbb{Q}$ is not empty. The reader might wonder whether something as powerful as the Least Upper Bound Property is really needed to prove something as apparently simple as showing that the equation $x^2 - 2 = 0$ has a solution in $\mathbb{R}$, but once we leave the realm of the rational numbers, we should not be surprised that we need to make use of the one tool we have available for the real numbers that is not available for the rational numbers, namely, the Least Upper Bound Property.

The following theorem states that any positive real number has a square root; proving that fact takes no more effort than showing that just the number 2 has a square root.

Theorem 2.6.9. *Let $p \in (0,\infty)$. Then there is a unique $x \in (0,\infty)$ such that $x^2 = p$.*

Proof. We follow [Sto01]. We will prove existence in the case that $p > 1$. The rest of the existence proof, and the uniqueness proof, are left to the reader in Exercise 2.6.18.

Suppose that $p > 1$. Then $p^2 > p$. Let

$$S = \{w \in \mathbb{R} \mid w > 0 \text{ and } w^2 < p\}.$$

Because $1^2 < p$, then $1 \in S$, and therefore $S \neq \emptyset$. Let $y \in \mathbb{R}$. Suppose that $y > p$. Then $y > 0$, and hence $y^2 > p^2 > p$. It follows that $y \notin S$. Therefore, if $z \in S$, then $z \leq p$. We deduce that p is an upper bound of S. Therefore S is bounded above. By the Least Upper Bound Property, the set S has a least upper bound. Let $x = \operatorname{lub} S$. Because $1 \in S$, then $x \geq 1 > 0$.

Let

$$t = x + \frac{p - x^2}{p + x}.$$

The reader can verify that

$$t = \frac{p(x+1)}{p+x} \quad \text{and} \quad t^2 = p + \frac{p(p-1)(x^2-p)}{(p+x)^2}.$$

It follows from the first equality that $t > 0$.

We now show that $x^2 = p$. Suppose to the contrary that $x^2 \neq p$. First, suppose that $x^2 > p$. Then $t < x$ and $t^2 > p$. Let $u \in S$. Then $u^2 < p$, and hence $u^2 < t^2$. By Lemma 2.3.3 (14) we deduce that $u < t$. Therefore t is an upper bound of S, which is a contradiction to the fact that $t < x$ and x is the least upper bound of S.

Second, suppose that $x^2 < p$. Then $t > x$ and $t^2 < p$. Therefore $t \in S$, which is a contradiction to the fact that $t > x$ and x is an upper bound of S.

We conclude that $x^2 = p$. □

Definition 2.6.10. Let $p \in (0,\infty)$. The **square root** of p, denoted $\sqrt{p}$, is the unique $x \in (0,\infty)$ such that $x^2 = p$. △

There is a more pleasant, and less ad hoc, proof that every positive real number has an nth root for every $n \in \mathbb{N}$ in Exercise 3.5.6, though this nicer proof requires the Intermediate Value Theorem, which we have not yet proved. Even more generally, in Section 7.2 we will define x^r for all $x \in (0,\infty)$ and all $r \in \mathbb{R}$, which includes the case of the nth root of x by using $r = \frac{1}{n}$; once again, this more general definition requires tools we have not yet developed, for example integration, which is used to define logarithms, which are needed to define x^r. (It should be noted that both of these alternative methods of constructing roots also ultimately rely upon the Least

Upper Bound Property.) However, even though these more slick, and more general, definitions of the square root of positive real numbers exist, we have given the above proof in order to have square roots available to use now, because we will use them on occasion.

We now show that $\sqrt{p}$ is not rational for any natural number p that is not a perfect square. In particular, it will follows from Exercise 2.4.5 that $\sqrt{2}$ is not rational. The following proof is due to Richard Dedekind (1831–1916), who invented what are now called Dedekind cuts (which were discussed in Section 1.6).

Theorem 2.6.11. *Let $p \in \mathbb{N}$. Suppose that there is no $u \in \mathbb{Z}$ such that $p = u^2$. Then $\sqrt{p} \notin \mathbb{Q}$.*

Proof. By Corollary 2.6.8 (1) there is a unique $v \in \mathbb{Z}$ such that $v - 1 \leq \sqrt{p} < v$. By hypothesis we know that $\sqrt{p} \notin \mathbb{Z}$, and hence $v - 1 < \sqrt{p} < v$. Let $s = v - 1$. Then $s \in \mathbb{Z}$ and $s < \sqrt{p} < s + 1$. Because $p \geq 1$, then $\sqrt{p} \geq 1$, and hence $s \geq 1$.

Suppose that $\sqrt{p} \in \mathbb{Q}$. Then there are $a, b \in \mathbb{Z}$ with $b \neq 0$ such that $\sqrt{p} = \frac{a}{b}$. By Lemma 2.4.12 (2) we can assume that $a, b \in \mathbb{N}$.

Let

$$E = \left\{ d \in \mathbb{N} \mid \text{there is some } c \in \mathbb{Z} \text{ such that } \sqrt{p} = \frac{c}{d} \right\}.$$

Clearly $E \subseteq \mathbb{N}$. Because $\sqrt{p} = \frac{a}{b}$, then $b \in E$, so $E \neq \emptyset$. By the Well-Ordering Principle (Theorem 1.2.10, Axiom 1.4.4 or Theorem 2.4.6) there is some $n \in E$ such that $n \leq x$ for all $x \in E$. By the definition of E there is some $m \in \mathbb{Z}$ such that $\sqrt{p} = \frac{m}{n}$. Because $s < \sqrt{p} < s + 1$, it follows that $s < \frac{m}{n} < s + 1$, and hence, with a bit of rearranging, we deduce that $0 < m - sn < n$.

It can be verified that $(np - sm)^2 - (m - sn)^2 p = (s^2 - p)(m^2 - n^2 p)$ by expanding both sides of the equation; the details are left to the reader. Because $\sqrt{p} = \frac{m}{n}$ then $m^2 - n^2 p = 0$, and it follows that $(np - sm)^2 - (m - sn)^2 p = 0$. Hence $\sqrt{p} = \frac{np - sm}{m - sn}$. Because $m, n, s, p \in \mathbb{Z}$ then $m - sn \in \mathbb{Z}$ and $np - sm \in \mathbb{Z}$. Because $m - sn > 0$, then $m - sn \in \mathbb{N}$. It follows that $m - sn \in E$, which is a contradiction to the fact that $n \leq x$ for all $x \in E$, because $m - sn < n$. We conclude that $\sqrt{p} \notin \mathbb{Q}$. □

The following corollary seems intuitively clear, and we now have the tools to prove it.

Corollary 2.6.12. *The ordered field $\mathbb{Q}$ does not satisfy the Least Upper Bound Property.*

Proof. Let

$$A = \{w \in \mathbb{Q} \mid w > 0 \text{ and } w^2 < 2\}.$$

As in the proof of Theorem 2.6.9, we see that $1 \in A$, which means that $A \neq \emptyset$, and that 2 is an upper bound of A, which means that A is bounded above. Suppose that A has a least upper bound. Let $y = \operatorname{lub} A$. As the reader can verify, the proof of Theorem 2.6.9 works in the context of $\mathbb{Q}$, and it follows from that proof that $y^2 = 2$. Hence $y = \sqrt{2}$. We then have a contradiction to Theorem 2.6.11 and Exercise 2.4.5, and we deduce that A does not have a least upper bound. □

The set of integers is "discrete" in the sense that each integer is isolated from its fellow numbers by a distance of at least 1; see Theorem 2.4.10 for details. By contrast, we now show that the set of rational numbers and the set of irrational numbers are as "indiscrete" as possible, in that numbers of each of these two types can be found arbitrarily close to any given real number. More formally, we will prove that between any two real numbers there exists a rational number and an irrational number. This fact about the rational numbers and the irrational numbers, which relies upon Corollary 2.6.8, and hence ultimately on the Least Upper Bound Property, is known as the "density of the rational numbers" and the "density of the irrational numbers."

Theorem 2.6.13. *Let $a, b \in \mathbb{R}$. Suppose that $a < b$.*

1. *There is some $q \in \mathbb{Q}$ such that $a < q < b$.*
2. *There is some $r \in \mathbb{R} - \mathbb{Q}$ such that $a < r < b$.*

Proof.

(1) We know by hypothesis that $b - a > 0$. By Corollary 2.6.8 (2) there is some $n \in \mathbb{N}$ such that $\frac{1}{n} < b - a$. It follows that $an + 1 < bn$. By Corollary 2.6.8 (1) there is some $m \in \mathbb{Z}$ such that $m - 1 \leq an < m$. It follows that $m \leq an + 1 < bn$. Hence $an < m < bn$, and therefore $a < \frac{m}{n} < b$. By the definition of $\mathbb{Q}$, we know that $\frac{m}{n} \in \mathbb{Q}$.

(2) Because $\sqrt{2} > 0$, then $\frac{1}{\sqrt{2}} > 0$, and it follows that $\frac{a}{\sqrt{2}} < \frac{b}{\sqrt{2}}$. By Part (1) of this theorem we know that there is some $q \in \mathbb{Q}$ such that $\frac{a}{\sqrt{2}} < q < \frac{b}{\sqrt{2}}$. We can choose q so that it is not 0 (otherwise, if q were 0, then we could pick another rational number between q and 0, and use that rational number). Then $a < q\sqrt{2} < b$. Because $q \neq 0$, it follows from Exercise 2.4.6 (2) that $q\sqrt{2} \in \mathbb{R} - \mathbb{Q}$. □

We conclude this section with a result known as the Heine–Borel Theorem, which might appear to be a somewhat unmotivated technicality, but which will be very useful in the proofs of two important theorems (Theorem 3.4.4 and Theorem 5.8.5). Often the name "Heine–Borel Theorem" is used to refer to a more general result than what is stated below, but what we have here is the core of the more general version. It would take us too far afield to motivate the Heine–Borel Theorem, other than to say that the combination of closedness and boundedness for intervals in $\mathbb{R}$ is so powerful that it makes closed bounded intervals behave somewhat analogously to finite sets. For the present, the reader should view the Heine–Borel Theorem as simply a technical necessity, though the reader is encouraged to learn more about this theorem, and the concepts involved. This topic may be found in any book on point set topology, as part of the discussion of the concept of "compactness"; see [Mun00, Sections 26 and 27] for details.

Theorem 2.6.14 (Heine–Borel Theorem). *Let $C \subseteq \mathbb{R}$ be a closed bounded interval, let I be a non-empty set and let $\{A_i\}_{i \in I}$ be a family of open intervals in $\mathbb{R}$. Suppose that $C \subseteq \bigcup_{i \in I} A_i$. Then there are $n \in \mathbb{N}$ and $i_1, i_2, \ldots, i_n \in I$ such that $C \subseteq \bigcup_{k=1}^{n} A_{i_k}$.*

Proof. Let $C = [a, b]$. The result is trivial if $a = b$, so suppose that $a < b$.

Let

$$S = \{r \in [a,b] \mid \text{there are } p \in \mathbb{N} \text{ and } i_1, \ldots, i_p \in I \text{ such that } [a,r] \subseteq \bigcup_{k=1}^{p} A_{i_k}\}.$$

Clearly $S \subseteq [a,b]$. It is evident that $a \in S$, and therefore $S \neq \emptyset$. The set S is bounded above by b. The Least Upper Bound Property implies that S has a least upper bound. Let $z = \operatorname{lub} S$. Then $a \le z \le b$.

Because $z \in [a,b]$, there is some $m \in I$ such that $z \in A_m$. The set A_m is an open interval, and hence Lemma 2.3.7 (2) implies that there is some $\delta > 0$ such that $[z-\delta, z+\delta] \subseteq A_m$.

We now show that $z \in S$. If $z = a$, then we already know that $z \in S$. Now suppose that $z > a$. By definition $z = \operatorname{lub} S$, and so Lemma 2.6.5 (1) implies that there is some $w \in S$ such that $z - \delta < w \le z$. It follows that $[w,z] \subseteq A_m$. By the definition of S there are $p \in \mathbb{N}$ and $i_1, i_2, \ldots, i_p \in I$ such that $[a,w] \subseteq \bigcup_{k=1}^{p} A_{i_k}$. Then $[a,z] = [a,w] \cup [w,z] \subseteq \bigcup_{k=1}^{p} A_{i_k} \cup A_m$. Hence $z \in S$.

Next, we show that $z = b$. Assume that $z \neq b$. Then $z < b$. Let $\eta = \min\{\delta, \frac{b-z}{2}\}$. Then $\eta > 0$. Observe that $[z, z+\eta] \subseteq A_m$, and that $z + \eta \in [a,b]$. Then $[a, z+\eta] = [a,z] \cup [z, z+\eta] \subseteq \bigcup_{k=1}^{p} A_{i_k} \cup A_m$. It follows that $z + \eta \in S$, which is a contradiction to the fact that $z = \operatorname{lub} S$. We deduce that $z = b$. Because $z \in S$, we conclude that $b \in S$, which completes the proof. □

The Heine–Borel Theorem (Theorem 2.6.14) is about closed bounded intervals only. As the reader is asked to show in Exercise 2.6.17, the statement analogous to this theorem, but with an interval that is bounded and either open or half-open, or with an interval that is unbounded and closed, is not true.

The proof of the Heine–Borel Theorem (Theorem 2.6.14) relies upon the Least Upper Bound Property. More strongly, it turns out that the Heine–Borel Theorem is equivalent to the Least Upper Bound Property, as is discussed in Section 3.5 and proved in Theorem 3.5.4.

Reflections

Some parts of this section appear to be complicated proofs of obvious facts, for example the Archimedean Property and the existence of $\sqrt{2}$, and other parts appear to be proofs of unmotivated technicalities, for example the No Gap Lemma and the Heine–Borel Theorem. In fact, both types of proofs are quite typical of what is encountered throughout this text, and throughout much of mathematics. Virtually all the main theorems encountered in this text ultimately rely upon the Least Upper Bound Property. Some of those theorems, for example the Intermediate Value Theorem, have intuitively obvious statements, but have complicated proofs that rely directly upon the Least Upper Bound Property, similarly to some parts of the present section. Other theorems in the book, for example Lebesgue's Theorem, have complicated proofs that rely upon unmotivated technical results such as the Heine–Borel Theorem (which in turn relies upon the Least Upper Bound Property). So, in both content and style, the material of the present section is at the heart of what we will encounter throughout this text.

Exercises

Exercise 2.6.1. **[Used in Exercise 2.6.11, Theorem 2.7.1, Lemma 5.4.6, Theorem 5.9.9 and Theorem 5.8.5.]** Let $A, B \subseteq \mathbb{R}$ be non-empty sets. Suppose that $A \subseteq B$.

(1) Suppose that B has a least upper bound. Prove that A has a least upper bound, and $\text{lub} A \leq \text{lub} B$.
(2) Suppose that B has a greatest lower bound. Prove that A has a greatest lower bound, and $\text{glb} B \leq \text{glb} A$.

Exercise 2.6.2. **[Used in Lemma 3.5.3, Theorem 5.4.11, Example 5.9.7 and Exercise 5.9.1.]** Let $A \subseteq \mathbb{R}$ be a non-empty set. Refer to Example 2.5.13 for the definition of a greatest element of a set; the definition and properties of a least element of a set are completely analogous.

(1) Prove that if $a \in A$ is a greatest element of A, then A has a least upper bound and $\text{lub} A = a$.
(2) Prove that if $c \in \mathbb{R}$ is an upper bound of A and $c \in A$, then A has a least upper bound and $c = \text{lub} A$.
(3) Prove that if $b \in A$ is a least element of A, then A has a greatest lower bound and $\text{glb} A = b$.
(4) Prove that if $d \in \mathbb{R}$ is a lower bound of A and $c \in A$, then A has a greatest lower bound and $d = \text{glb} A$.

Exercise 2.6.3. **[Used in Theorem 2.8.10, Theorem 5.9.9, Theorem 5.9.10 and Section 5.9.]** Let $A, B \subseteq \mathbb{R}$ be non-empty sets.

(1) Suppose that A and B are bounded above, and that for each $b \in B$ and $\varepsilon > 0$, there is some $a \in A$ such that $b - \varepsilon \leq a$. Prove that $\text{lub} B \leq \text{lub} A$. In particular, deduce that if for each $b \in B$, there is some $a \in A$ such that $b \leq a$, then $\text{lub} B \leq \text{lub} A$.
(2) Suppose that A and B are bounded below, and that for each $b \in B$ and $\varepsilon > 0$, there is some $a \in A$ such that $a \leq b + \varepsilon$. Prove that $\text{glb} A \leq \text{glb} B$. In particular, deduce that if for each $b \in B$, there is some $a \in A$ such that $a \leq b$, then $\text{glb} A \leq \text{glb} B$.

Exercise 2.6.4. **[Used in Theorem 3.5.3.]** Let $A \subseteq \mathbb{R}$ be a non-empty set. Suppose that A is bounded above. Let

$$U = \{x \in \mathbb{R} \mid x \text{ is an upper bound of } A\},$$

and let $L = \mathbb{R} - U$. Prove that if $x \in L$ and $y \in U$, then $x < y$.

Exercise 2.6.5. **[Used in Exercise 7.4.1.]** Let $A \subseteq \mathbb{R}$ be a non-empty set. Let

$$-A = \{-x \mid x \in A\}.$$

(1) Prove that $-A$ is bounded below if and only if A is bounded above.
(2) Prove that $-A$ is bounded above if and only if A is bounded below.
(3) Prove that $-A$ has a greatest lower bound if and only if A has a least upper bound, and that if A has a least upper bound then $\operatorname{glb}(-A) = -\operatorname{lub} A$.
(4) Prove that $-A$ has a least upper bound if and only if A has a greatest lower bound, and that if A has a greatest lower bound then $\operatorname{lub}(-A) = -\operatorname{glb} A$.
(5) Use previous parts of this exercise to provide an alternative proof of the Greatest Lower Bound Property (Theorem 2.6.4).

Exercise 2.6.6. **[Used throughout.]** Let $A \subseteq \mathbb{R}$ be a non-empty set, and let $b \in \mathbb{R}$ be an upper bound of A. Suppose that for each $\varepsilon > 0$, there is some $a \in A$ such that $b - \varepsilon < a$. Prove that $b = \operatorname{lub} A$. (In practice, it is often more convenient to show that $b - a < \varepsilon$ rather than the equivalent $b - \varepsilon < a$, but we stated the exercise as we did to make it analogous to Lemma 2.6.5 (1).)

Exercise 2.6.7. **[Used in Theorem 2.7.1.]** Let $b \in \mathbb{R}$. Let $D = \{x \in \mathbb{Q} \mid x < b\}$. Prove that D has a least upper bound, and that $\operatorname{lub} D = b$. (This fact seems trivial intuitively, but a proof is needed.) [Use Exercise 2.6.6.]

Exercise 2.6.8. **[Used in Example 5.9.7.]** Let $A, B \subseteq \mathbb{R}$ be non-empty sets, and let $p \in \mathbb{R}$. Suppose that if $a \in A$ and $b \in B$, then $a \leq b$. Suppose that for each $\varepsilon > 0$, there are $a \in A$ and $b \in B$ such that $p - \varepsilon < a \leq p \leq b < p + \varepsilon$. Prove that A has a least upper bound and B has a greatest lower bound, and $p = \operatorname{lub} A = \operatorname{glb} B$. [Use Exercise 2.6.6.]

Exercise 2.6.9. **[Used in Exercise 2.6.10, Theorem 2.7.1, Exercise 2.8.5, Exercise 5.4.16 and Theorem 5.9.10.]** Let $A, B \subseteq \mathbb{R}$ be non-empty sets. Let $A + B$ and AB be defined by

$$A + B = \{a + b \mid a \in A \text{ and } b \in B\} \quad \text{and} \quad AB = \{ab \mid a \in A \text{ and } b \in B\}.$$

(1) For each $z \in \mathbb{R}$, let $C_z = \{y \in \mathbb{Q} \mid y < z\}$. Let $x, y \in \mathbb{R}$. Prove that $C_{x+y} = C_x + C_y$.
(2) For each $z \in (0, \infty)$, let $\bar{C}_z = \{y \in \mathbb{Q} \mid 0 < y < z\}$. Let $x, y \in (0, \infty)$. Prove that $\bar{C}_{xy} = \bar{C}_x \bar{C}_y$.
(3) Prove that if A and B have least upper bounds, then $A + B$ has a least upper bound, and $\operatorname{lub}(A + B) = \operatorname{lub} A + \operatorname{lub} B$.
(4) Prove that if A and B have greatest lower bounds, then $A + B$ has a greatest lower bound, and $\operatorname{glb}(A + B) = \operatorname{glb} A + \operatorname{glb} B$.
(5) Suppose that $A, B \subseteq (0, \infty)$. Prove that if A and B have least upper bounds, then AB has a least upper bound, and $\operatorname{lub} AB = (\operatorname{lub} A)(\operatorname{lub} B)$.
You may use, without proof, the fact that $(\operatorname{lub} A)(\operatorname{lub} B) - [\operatorname{lub} A + \operatorname{lub} B]x + x^2$ can be made as close as desired to $(\operatorname{lub} A)(\operatorname{lub} B)$ by a suitable choice of positive x; this fact will follow from the continuity of polynomials, discussed in Example 3.3.7 (1).

Exercise 2.6.10. **[Used in Exercise 2.6.11, Exercise 5.4.8 and Theorem 5.5.1.]** Let $A \subseteq \mathbb{R}$ be a non-empty set. Suppose that A has a least upper bound and a greatest lower bound.

(1) Prove that $\text{glb}A \le \text{lub}A$.
(2) Let $x, y \in A$. Prove that $|x - y| \le \text{lub}A - \text{glb}A$.
(3) Prove that
$$\text{lub}\{|x-y| \mid x,y \in A\} = \text{lub}A - \text{glb}A.$$
[Use Exercise 2.6.5 and Exercise 2.6.9.]
(4) Prove that $\text{glb}A = \text{lub}A$ if and only if A has a single element.

Exercise 2.6.11. **[Used in Exercise 3.2.18.]** Let $(a,b) \subseteq \mathbb{R}$ be a non-degenerate open bounded interval, and let $\{D_x\}_{x\in(a,b)}$ be a family of subsets of $\mathbb{R}$. Suppose that D_x is non-empty and bounded for all $x \in (a,b)$, and that $s,t \in (a,b)$ and $s < t$ imply $D_s \subseteq D_t$. For each $s \in (a,b)$ let $a_s = \text{glb}D_s$ and $b_s = \text{lub}D_s$. Let $A = \{a_s \mid s \in (a,b)\}$ and $B = \{b_s \mid s \in (a,b)\}$. Prove that A has a least upper bound and B has a greatest lower bound, and that $\text{lub}A \le \text{glb}B$. [Use Exercise 2.6.1 and Exercise 2.6.10 (1).]

Exercise 2.6.12. **[Used in Lemma 7.2.4.]** Let $a, b \in \mathbb{R}$. Suppose that $a > 0$. Prove that there is some $n \in \mathbb{N}$ such that $b \in [-na, na]$.

Exercise 2.6.13. **[Used in Lemma 2.8.5, Theorem 2.8.6, Theorem 2.8.10 and Exercise 8.4.11.]** Let $p \in \mathbb{N}$. Suppose that $p > 1$. Let $x \in (0,\infty)$. Prove that there is some $n \in \mathbb{N}$ such that $\frac{1}{p^n} < x$. [Use Exercise 2.5.13 (2).]

Exercise 2.6.14. **[Used in Lemma 7.3.2, Section 7.3, Lemma 7.3.4, Exercise 7.3.2 and Lemma 10.5.1.]** Let $a, h \in \mathbb{R}$. Suppose that $h > 0$.

(1) Let $x \in \mathbb{R}$. Prove that there is a unique $n \in \mathbb{Z}$ such that $a + (n-1)h \le x < a + nh$.
(2) Let $x, y \in \mathbb{R}$. Suppose that there is no $n \in \mathbb{Z}$ such that $a + nh$ is strictly between x and y. Prove that $|x - y| \le h$.

Exercise 2.6.15. Let $A \subseteq \mathbb{Z}$ be a set. Suppose that A has a least upper bound. Prove that $\text{lub}A \in \mathbb{Z}$.

Exercise 2.6.16. **[Used in Example 2.6.1.]** Prove that $\mathbb{N}$ is not bounded above as a subset of $\mathbb{R}$.

Exercise 2.6.17. **[Used in Section 2.6.]** Give examples to show that the statement analogous to the Heine–Borel Theorem (Theorem 2.6.14), but with an interval C that is bounded and either open or half-open, or that is unbounded and closed, is not true.

Exercise 2.6.18. **[Used in Theorem 2.6.9.]** Fill in the missing details in the proof of Theorem 2.6.9. That is, prove that x exists when $0 < p \le 1$, and prove uniqueness for all p.

2.7 Uniqueness of the Real Numbers

We have seen two approaches to the existence of the real numbers: In Chapter 1 the real numbers were constructed from the rational numbers, and in the present chapter the real numbers were taken axiomatically. In neither approach, however, was the uniqueness of the real numbers discussed. We now show that the set of real numbers, no matter how its existence is arrived at, is in fact unique.

Before we can prove that the set of real numbers is unique, we need to know what "uniqueness" means in this context. In the modern mathematical approach, what we care about in regard to the set of real numbers, or any other mathematical object, is how the object behaves, not "what it is." The set of real numbers, however conceptualized, behaves according to the properties of an ordered field that satisfies the Least Upper Bound Property; that is what we hypothesized in Axiom 2.2.4 when we took the real numbers axiomatically, and that is what we proved in Theorem 1.7.6 and Theorem 1.7.9 when we constructed the real numbers from the rational numbers. To say that the real numbers are unique is to say that any two ordered fields that satisfy the Least Upper Bound Property behave the same way no matter how these ordered fields were defined. Given that the axiom for an ordered field and the definition of the Least Upper Bound Property are stated in terms of two binary operations (called "addition" and "multiplication") and a relation (called "less than"), to say that two ordered fields that satisfy the Least Upper Bound Property behave the same way means that the binary operations addition and multiplication, and the relation less than, in one of the ordered fields correspond exactly to the binary operations addition and multiplication, and the relation less than, in the other ordered field. Such a correspondence is achieved via a bijective function that "preserves" the two binary operations and the relation, as stated in the following theorem. For the reader who is familiar with rings and fields, we note that the type of function we want is called an order preserving ring isomorphism.

Theorem 2.7.1 (Uniqueness of the Real Numbers). *Let R_1 and R_2 be ordered fields that satisfy the Least Upper Bound Property. Then there is a function $f\colon R_1 \to R_2$ that is bijective, and that satisfies the following properties. Let $x, y \in R_1$.*

a. $f(x+y) = f(x) + f(y)$.
b. $f(xy) = f(x)f(y)$.
c. *If $x < y$, then $f(x) < f(y)$.*

It is important to note in the statement of Theorem 2.7.1 that the symbols "+," "·" and "<" are used in two different contexts, and it is necessary to keep track of what these symbols mean at all times. For example, when we write "$f(x+y) = f(x) + f(y)$," the expression "$x + y$" denotes addition in R_1, whereas "$f(x) + f(y)$" denotes addition in R_2. It would be proper to write "$+_1$" and "$+_2$" respectively to denote the addition operations in each of R_1 and R_2, but doing so would make things very difficult to read, and so we prefer to write "+" to mean both addition operations, with the assumption that everything will be clear from context, and similarly for multiplication and less than. Additionally, we will use the same notation 0 and 1 in both R_1 and R_2.

The idea of the proof of Theorem 2.7.1 is as follows. Given that R_1 and R_2 both satisfy the hypotheses of Axiom 2.2.4, it follows that everything that was proved about $\mathbb{R}$ in this chapter prior to the current section also holds for each of R_1 and R_2. In particular, there are analogs of $\mathbb{N}$, $\mathbb{Z}$ and $\mathbb{Q}$ in each of R_1 and R_2, which we will denote N_1, Z_1, Q_1, and N_2, Z_2, Q_2, respectively. In the proof of Theorem 2.7.1, we will use the analogs for N_1, Z_1, Q_1, N_2, Z_2 and Q_2 of the theorems and exercises we have proved for $\mathbb{N}$, $\mathbb{Z}$ and $\mathbb{Q}$. We then define the function f in stages, first on N_1, then on Z_1, then on Q_1 and finally on R_1. This definition by stages makes the proof of the theorem somewhat lengthier than might be expected.

In the proof we will use the concept of the extension of a function. To remind the reader of this concept, let A and B be sets, let $S \subseteq A$ be a subset and let $f: S \to B$ be a function. An extension of f to A is any function $h: A \to B$ such that $h|_S = f$.

Proof of Theorem 2.7.1. We follow [Spi67, Chapter 29] and [Pow94, Appendix] for parts of this proof.

Step 1 Using Definition by Recursion (Theorem 2.5.5) we see that there are functions $g: N_1 \to N_2$ and $p: N_2 \to N_1$ such that $g(1) = 1$, and $g(n+1) = g(n)+1$ for all $n \in N_1$, and that $p(1) = 1$, and $p(m+1) = p(m)+1$ for all $m \in N_2$. Then $(p \circ g)(1) = p(g(1)) = p(1) = 1$, and $(p \circ g)(n+1) = p(g(n+1)) = p(g(n)+1) = p(g(n))+1 = (p \circ g)(n)+1$ for all $n \in N_1$. Observe that $1_{N_1}(1) = 1$, and $1_{N_1}(n+1) = n+1 = 1_{N_1}(n)+1$ for all $n \in N_1$, where $1_{N_1}: N_1 \to N_1$ is the identity map. If follows from the uniqueness in Theorem 2.5.5 that $p \circ g = 1_{N_1}$. A similar argument shows that $g \circ p = 1_{N_2}$. We deduce that g and p are inverses of each other, and hence each is bijective.

We now prove by induction on n that $g(m+n) = g(m)+g(n)$ for all $m,n \in N_1$. First, suppose that $n = 1$. Let $a \in N_1$. Then $g(a+n) = g(a+1) = g(a)+1 = g(a)+g(1) = g(a)+g(n)$. Now let $n \in N_1$, and suppose that $g(m+n) = g(m)+g(n)$ for all $m \in N_1$. Let $b \in N_1$. Then $g(b+(n+1)) = g((b+n)+1) = g(b+n)+1 = g(b)+g(n)+1 = g(b)+g(n+1)$. Hence, by induction on n, we deduce that $g(m+n) = g(m)+g(n)$ for all $m \in N_1$, for all $n \in N_1$. Similar proofs can be used to show that $g(mn) = g(m)g(n)$ for all $m,n \in N_1$, and that if $m < n$ then $g(m) < g(n)$ for all $m,n \in N_1$; the details are left to the reader in Exercise 2.7.2.

Step 2 Let $h: Z_1 \to Z_2$ be defined by $h(n) = g(n)$ for all $n \in N_1$, by $h(0) = 0$ and by $h(n) = -g(-n)$ for all $n \in -N_1$. By definition h is an extension of g to Z_1. Observe that for $n \in Z_1$, it follows from the definition of h that $h(n) \in N_2$ if and only if $n \in N_1$, that $h(n) = 0$ if and only if $n = 0$, and that $h(n) \in -N_2$ if and only $n \in -N_1$.

It is left to the reader in Exercise 2.7.3 to show that $h(m+n) = h(m)+h(n)$ for all $m,n \in Z_1$, that $h(mn) = h(m)h(n)$ for all $m,n \in Z_1$, and that if $m < n$ then $h(m) < h(n)$ for all $m,n \in Z_1$.

Because $m < n$ implies $h(m) < h(n)$ for all $m,n \in Z_1$, it follows that h is injective.

To show that h is surjective, let $b \in Z_2$. If $b \in N_2$, then because g is surjective there is some $a \in N_1$ such that $g(a) = b$, and hence $h(a) = b$. If $b = 0$, then $h(0) = b$. If $b \in -N_2$, then $-b \in N_2$, and hence there is some $c \in N_1$ such that $g(c) = -b$, which implies that $h(-c) = -g(-(-c)) = -g(c) = -(-b) = b$. We deduce that h is surjective, and hence h is bijective.

Step 3 Let $k: Q_1 \to Q_2$ be defined as follows. Let $x \in Q_1$. Then $x = \frac{a}{b}$ for some $a,b \in Z_1$ such that $b \neq 0$. It follows from the construction of h in Step 2 that $h(b) \neq 0$. Then we let $k(x) = \frac{h(a)}{h(b)}$. To show that this definition makes sense, suppose that $\frac{c}{d} = \frac{s}{t}$ for some $c,d,s,t \in Z_1$ such that $d \neq 0$ and $t \neq 0$. It then follows that $ct = sd$, and hence by Step 2 we see that $h(c)h(t) = h(ct) = h(sd) = h(s)h(d)$, which then implies that $\frac{h(c)}{h(d)} = \frac{h(s)}{h(t)}$. Hence k is well-defined.

Let $n \in Z_1$. Then $n = \frac{n}{1}$, and hence $k(n) = \frac{h(n)}{h(1)} = \frac{h(n)}{1} = h(n)$. It follows that k is an extension of h to Q_1.

Let $x, y \in Q_1$. Then $x = \frac{a}{b}$ and $y = \frac{c}{d}$ for some $a, b, c, d \in Z_1$ such that $b \neq 0$ and $d \neq 0$. We know that $x + y = \frac{ad+bc}{bd}$ and $xy = \frac{ac}{bd}$, and hence we can use Step 2 to deduce that

$$k(x+y) = \frac{h(ad+bc)}{h(bd)} = \frac{h(a)h(d)+h(b)h(c)}{h(b)h(d)} = \frac{h(a)}{h(b)} + \frac{h(c)}{h(d)} = k(x) + k(y)$$

and

$$k(xy) = \frac{h(ac)}{h(bd)} = \frac{h(a)h(c)}{h(b)h(d)} = \frac{h(a)}{h(b)} \cdot \frac{h(c)}{h(d)} = k(x)k(y).$$

Suppose that $x < y$. Then $\frac{a}{b} < \frac{c}{d}$. It follows from Exercise 2.4.9 that either $cb - ad > 0$ and $bd > 0$, or $cb - ad < 0$ and $bd < 0$. If $cd - ad > 0$ and $bd > 0$, then by Step 2 we see that $h(cb - ad) > h(0)$ and $h(bd) > h(0)$, which implies that $h(c)h(b) - h(a)h(d) > 0$ and $h(b)h(d) > 0$; if $cb - ad < 0$ and $bd < 0$, it follows similarly that $h(c)h(b) - h(a)h(d) < 0$ and $h(b)h(d) < 0$. By Exercise 2.4.9 again we deduce from both cases that $\frac{h(a)}{h(b)} < \frac{h(c)}{h(d)}$, which implies that $k(x) < k(y)$.

Because $x < y$ implies $k(x) < k(y)$ for all $x, y \in Q_1$, it follows that k is injective. It is left to the reader in Exercise 2.7.4 to show that k is surjective.

Step 4 We start with two observations. First, let $z, w \in R_2$. Suppose that $z < w$. By Theorem 2.6.13 (1) there is some $t \in Q_2$ such that $z < t < w$. By Step 3 the function k is bijective, and so there is some $s \in Q_1$ such that $k(s) = t$. Hence $z < k(s) < w$. Second, let $x, y \in Q_1$. We know by Step 3 that if $x < y$, then $k(x) < k(y)$. Conversely, if $k(x) < k(y)$, then it must be the case that $x < y$, because if $x \not< y$, then $x = y$ or $y < x$, in which case $k(x) = k(y)$ or $k(y) < k(x)$, which would mean that $k(x) \not< k(y)$.

Let $f\colon R_1 \to R_2$ be defined as follows. Let $x \in R_1$. Let $C_x = \{y \in Q_1 \mid y < x\}$. By Theorem 2.6.13 (1) there is some $q \in Q_1$ such that $x - 1 < q < x$, and therefore $C_x \neq \emptyset$. Hence $k(C_x) \neq \emptyset$. By Corollary 2.6.8 (1) there is some $n \in Z_1$ such that $x < n$. Hence n is an upper bound of C_x. Let $u \in k(C_x)$. Then $u = k(v)$ for some $v \in C_x$. Therefore $v < x < n$. It follows from Step 3 that $u = k(v) < k(n)$. Hence $k(n)$ is an upper bound of $k(C_x)$. By the Least Upper Bound Property the set $k(C_x)$ has a least upper bound. We then let $f(x) = \operatorname{lub} k(C_x)$.

Let $w \in Q_1$. Let $z \in k(C_w)$. Then $z = k(p)$ for some $p \in C_w$. Therefore $p < w$. It follows from Step 3 that $z = k(p) < k(w)$. Hence $k(w)$ is an upper bound of $k(C_w)$, which means that $\operatorname{lub} k(C_w) \leq k(w)$. Suppose that $\operatorname{lub} k(C_w) < k(w)$. By the first observation made at the start of this step of the proof, there is some $s \in Q_1$ such that $\operatorname{lub} k(C_w) < k(s) < k(w)$. By the second observation, we see that $s < w$. Hence $s \in C_w$, which implies that $k(s) \in k(C_w)$. Hence $k(s) \leq \operatorname{lub} k(C_w)$, which is a contradiction. We deduce that $\operatorname{lub} k(C_w) = k(w)$, and it follows that $f(w) = k(w)$. Therefore f is an extension of k to R_1.

Let $x, y \in R_1$. In Exercise 2.6.9 we defined $A+B$ and AB for any two sets $A, B \subseteq \mathbb{R}$. By Part (1) of that exercise we know that $C_{x+y} = C_x + C_y$. We then see that $k(C_{x+y}) = k(C_x + C_y) = k(C_x) + k(C_y)$, where the last equality can be deduced from Step 3; the details are left to the reader. It then follows from Exercise 2.6.9 (3) that $f(x+y) = \text{lub}\, k(C_{x+y}) = \text{lub}[k(C_x) + k(C_y)] = \text{lub}\, k(C_x) + \text{lub}\, k(C_y) = f(x) + f(y)$.

Let $u \in R_1$. Suppose that $u > 0$. Let $\bar{C}_u$ by $\bar{C}_u = \{y \in Q_1 \mid 0 < y < u\}$. As before, the Least Upper Bound Property implies that the set $k(\bar{C}_u)$ has a least upper bound; the details are left to the reader. Moreover, because $u > 0$, it can be verified that $\text{lub}\, k(\bar{C}_u) = \text{lub}\, k(C_u) = f(u)$; again, the details are left to the reader.

We now show that $f(xy) = f(x)f(y)$. There are five cases.

First, suppose that $x = 0$ or $y = 0$. Without loss of generality, assume that $y = 0$. Because $0 \in Z_1$, then $f(0) = k(0) = h(0) = 0$. Then $f(x \cdot 0) = f(0) = 0 = f(x) \cdot 0 = f(x)f(0)$.

Second, suppose that $x > 0$ and $y > 0$. By using Exercise 2.6.9 (2), and a similar argument as before, it is seen that $f(xy) = \text{lub}\, k(\bar{C}_{xy}) = \text{lub}[k(\bar{C}_x)k(\bar{C}_y)] = [\text{lub}\, k(\bar{C}_x)] \cdot [\text{lub}\, k(\bar{C}_y)] = f(x)f(y)$; the details are left to the reader.

Third, suppose that $x < 0$ and $y > 0$. Then $-x > 0$. Using Exercise 2.7.5 (2) we then see that $f(xy) = f(-(-x)y) = -f((-x)y) = -f(-x)f(y) = f(x)f(y)$.

There are two other cases, which are when $x > 0$ and $y < 0$, and when $x < 0$ and $y < 0$; they are similar to the previous case, and we omit the details.

Suppose that $x < y$. Then $C_x \subseteq C_y$, and hence $k(C_x) \subseteq k(C_y)$. It follows from Exercise 2.6.1 (1) that $\text{lub}\, k(C_x) \leq \text{lub}\, k(C_y)$. By Theorem 2.6.13 (1) there is some $a \in Q_1$ such that $x < a < y$, and by the same theorem again we see that there is some $b \in Q_1$ such that $x < a < b < y$. Let $w \in C_x$. Then $w < a < b$, and by Step 3 we see that $k(w) < k(a) < k(b)$. Therefore $k(a)$ is an upper bound of $k(C_x)$, and we deduce that $f(x) = \text{lub}\, k(C_x) \leq k(a)$. Because $b \in C_y$, then $k(b) \in k(C_y)$, which implies that $k(b) \leq \text{lub}\, k(C_y) = f(y)$. We deduce that $f(x) \leq k(a) < k(b) \leq f(y)$.

Because $x < y$ implies $f(x) < f(y)$ for all $x, y \in R_1$, it follows that f is injective.

We now show that f is surjective. Let $b \in R_2$. If $b \in Q_2$, then $b = k(c)$ for some $c \in Q_1$, and hence $b = f(c)$. Now, suppose that $b \in R_2 - Q_2$. Let $D = \{y \in Q_2 \mid y < b\}$. By Exercise 2.6.7 we know that $\text{lub}\, D = b$. By a similar argument to that used about sets of the form C_x, we see that $D \neq \emptyset$, and that there is some $m \in Z_2$ that is an upper bound of D. Because k is bijective, it has an inverse function k^{-1}. Let $E = k^{-1}(D)$, and let $p = k^{-1}(m)$. Observe that $E \subseteq Q_1$ and $p \in Z_1$. Then $E \neq \emptyset$, and, by using the second observation made at the start of this step of the proof, we see that p is an upper bound of E. By the Least Upper Bound Property we know that the set E has a least upper bound. Let $a = \text{lub}\, E$.

Suppose that $a \in Q_1$. Then $k(a) \in Q_2$. It cannot be that $k(a) = b$, because $b \notin Q_2$. First, suppose that $k(a) < b$. Then, by the first observation made at the start of this step of the proof, there is some $w \in Q_1$ such that $k(a) < k(w) < b$. Hence $a < w$ by the second observation. Moreover, because $k(w) \in Q_2$, then $k(w) \in D$, and therefore $w \in E$, which is a contradiction to the fact that a is an upper bound of E. Second, suppose that $b < k(a)$. Then there is some $q \in Q_1$ such that $b < k(q) < k(a)$. Hence $q < a$. Let $r \in E$. Then $k(r) \in D$, so $k(r) < b < k(q)$, and therefore $r < q$. It follows that q is an upper bound of E, which is a contradiction to the fact that $a = \text{lub}\,E$. We conclude that $a \notin Q_1$.

We will show that $E = C_a$, and it will then follow that $f(a) = \text{lub}\,k(C_a) = \text{lub}\,k(E) = \text{lub}\,D = b$. Let $s \in E$. Then $s \le \text{lub}\,E = a$. Observe that $s \ne a$, because $s \in Q_1$ and $a \notin Q_1$. Hence $s < a$, and therefore $s \in C_a$. Let $t \in C_a$. Then $t \in Q_1$ and $t < a$. Because $t < a = \text{lub}\,E$, then t is not an upper bound of E, so there is some $v \in E$ such that $t < v$. Then $k(t) < k(v)$. Because $k(v) \in D$, it follows that $k(v) < b$, and therefore $k(t) < b$. Because $k(t) \in Q_2$, then $k(t) \in D$, which implies $t \in E$. We conclude that $E = C_a$, and hence we have proved that f is surjective. □

Reflections

The proof in this section, though long, has no real surprises. The main lesson to be learned from this proof is not the details (though they are worth knowing), but rather the fact that if a proof this long is needed, then the result proved should not be taken for granted. That is, we were right to raise the question of the uniqueness of the real numbers. The reader is encouraged to ask similar questions—concerning uniqueness and other issues as well—about all new mathematical concepts she encounters.

Exercises

Exercise 2.7.1. Find an example of a set that satisfies all the axioms of an ordered field except for the Inverses Law for Multiplication, and that satisfies the Least Upper Bound Property.

Exercise 2.7.2. **[Used in Theorem 2.7.1.]** Complete the missing parts of Step 1 of the proof of Theorem 2.7.1. That is, let $m, n \in N_1$, and prove that $g(mn) = g(m)g(n)$, and that if $m < n$ then $g(m) < g(n)$. [Use Exercise 2.4.1.]

Exercise 2.7.3. **[Used in Theorem 2.7.1.]** Complete the missing parts of Step 2 of the proof of Theorem 2.7.1. That is, let $m, n \in Z_1$, and prove that $h(m+n) = h(m) + h(n)$, that $h(mn) = h(m)h(n)$, and that if $m < n$ then $h(m) < h(n)$.

Exercise 2.7.4. **[Used in Theorem 2.7.1.]** Complete the missing part of Step 3 of the proof of Theorem 2.7.1. That is, prove that k is surjective.

Exercise 2.7.5. **[Used in Theorem 2.7.1.]** Let R_1 and R_2 be ordered fields that satisfy the Least Upper Bound Property, and let $p\colon R_1 \to R_2$ be a function. Suppose that $p(x+y) = p(x) + p(y)$ for all $x, y \in R_1$.

(1) Prove that $p(0) = 0$.
(2) Prove that $p(-x) = -p(x)$ for all $x \in R_1$.

Exercise 2.7.6. Let R_1 and R_2 be ordered fields that satisfy the Least Upper Bound Property. Prove that there is a unique function $f\colon R_1 \to R_2$ that satisfies the properties stated in Theorem 2.7.1.

2.8 Decimal Expansion of Real Numbers

It is very important to distinguish between the real numbers per se and the way we write them. For example, we know that $1 \in \mathbb{R}$, and hence $1+1 \in \mathbb{R}$. We standardly denote the real number $1+1$ by the symbol 2. We then denote the number $1+1+1$ by the symbol 3. We could, in principle, denote each of the numbers

$$1, 1+1, 1+1+1, 1+1+1+1, 1+1+1+1+1, \ldots$$

by a single, distinct symbol. Of course, we would need infinitely many distinct symbols to do so, and that would not be very convenient for practical calculations. To overcome this problem, a number of systems of writing numbers have been developed by various cultures throughout history, for example Roman numerals. Today, the most commonly used system for writing numbers is base 10 notation, also called decimal notation, which allows us to designate all real numbers using only ten distinct symbols, namely, the symbols $0, 1, 2, \ldots, 9$, though we write these symbols in infinitely many different combinations. For example, we denote $1+1+1+1+1+1+1+1+1$ by the symbol 9, and we denote $1+1+1+1+1+1+1+1+1+1$ by the symbol 10. The base 10 system of writing numbers has proved to be very convenient, though we note that the use of the number 10, as opposed to some other number as the base, is quite arbitrary, where the number 10 was presumably chosen simply because we human beings have that many fingers and toes; there is no particular mathematical advantage to the use of the number 10.

We are so used to thinking of the real numbers in terms of how we write them that we often take our system of writing numbers for granted, though in fact there are two very substantial questions that need to be asked about any system for writing numbers: Can every real number can be represented in the given system, and if yes, is the representation unique? It turns out, as we will see in Theorem 2.8.6 below, that every real number can in fact be written in the base 10 system, though it should not be taken as obvious. As for uniqueness in the base 10 system, it is almost true, with the exception of decimals that eventually become the number 9 repeating. Because there is nothing special about base 10, we will prove our results in the more general context of base p, where p is any natural number greater than 1.

The formal definition of the base p representation of a real number will be given in Definition 2.8.7 below, after we prove some preliminary results. Although the use of base 10 notation is something that the reader learned at a very early age, it will turn out that a surprisingly large amount of effort is needed to formulate and prove that everything works out as expected—we will use the Well-Ordering Principle, the Least Upper Bound Property and the Archimedean Property.

Base 10 notation for the real numbers makes use of powers of 10. For example, if we write 235 in base 10 notation, we mean the number $2 \cdot 10^2 + 3 \cdot 10 + 5 \cdot 1$. Similarly, the base p representation of the real numbers makes use of powers of p, and we will therefore need to make use of Definition 2.5.6 and Definition 2.5.8.

Our development of the base p representation of the real numbers will have two stages: first we deal with the natural numbers, and only after that will we turn to real numbers that are not natural numbers. Also, we will restrict our attention to positive numbers, because the base p representation of a negative number is just the negative of the base p representation of its absolute value, and the base p representation of zero is just zero.

The following simple lemma is the essence of what makes the base p representation of natural numbers work.

Lemma 2.8.1. *Let $p \in \mathbb{N}$. Suppose that $p > 1$. Let $n \in \mathbb{N}$. Then there is a unique $k \in \mathbb{N}$ such that $p^{k-1} \leq n < p^k$.*

Proof. Let

$$G = \{c \in \mathbb{N} \mid n < p^c\}.$$

Clearly $G \subseteq \mathbb{N}$. By Exercise 2.5.13 (2) we know that $n < p^n$, and hence $n \in G$. Therefore $G \neq \emptyset$. By the Well-Ordering Principle (Theorem 1.2.10, Axiom 1.4.4 or Theorem 2.4.6), there is some $k \in G$ such that $k \leq g$ for all $g \in G$. Because $k \in G$ we know that $n < p^k$. By the choice of k, we see that $k - 1 \notin G$, which means that $p^{k-1} \leq n$. □

We now see that every natural number can be written uniquely in base p notation. We note that in the base 10 system, we express any real number in terms of the numbers $0, 1, 2, \ldots, 9$ arranged in the appropriate place value system. For base p, we use the numbers $0, 1, 2, \ldots, p-1$.

In the proof of the following theorem, and subsequently, if there is a summation of the form $\sum_{i=r}^{s} a_i$, and if $r > s$, we take the summation to be zero; doing so allows us to avoid special cases.

Theorem 2.8.2. *Let $p \in \mathbb{N}$. Suppose that $p > 1$. Let $n \in \mathbb{N}$. Then there are unique $k \in \mathbb{N}$ and $a_0, a_1, \ldots, a_{k-1} \in \{0, \ldots, p-1\}$ such that $a_{k-1} \neq 0$, and that*

$$n = \sum_{i=0}^{k-1} a_i p^i. \tag{2.8.1}$$

Proof. To prove uniqueness, suppose that there are $t, r \in \mathbb{N}$ and $c_0, c_1, \ldots, c_{t-1} \in \{0, \ldots, p-1\}$ and $b_0, b_1, \ldots, b_{r-1} \in \{0, \ldots, p-1\}$ such that $c_{t-1} \neq 0$ and $b_{r-1} \neq 0$, and that

$$n = \sum_{i=0}^{t-1} c_i p^i \quad \text{and} \quad n = \sum_{j=0}^{r-1} b_j p^j.$$

Without loss of generality, assume that $t \leq r$. The above equations involving n then imply that

$$\sum_{i=0}^{t-1} (b_i - c_i) p^i + \sum_{j=t}^{r-1} b_j p^j = 0.$$

Because $c_0, c_1, \ldots, c_{t-1} \in \{0, \ldots, p-1\}$ and $b_0, b_1, \ldots, b_{r-1} \in \{0, \ldots, p-1\}$, then $|b_i - c_i| \leq p - 1$ for all $i \in \{0, \ldots, t-1\}$, and $|b_j| \leq p - 1$ for all $j \in \{t, \ldots, r-1\}$. It now follows from Exercise 2.8.4 (3) that $b_i - c_i = 0$ for all $i \in \{0, \ldots, t-1\}$, and that $b_j = 0$ for all $j \in \{t, \ldots, r-1\}$. However, we know by hypothesis that $b_{r-1} \neq 0$, and hence it must be the case that $t = r$. It then follows that $b_i = c_i$ for all $i \in \{0, \ldots, t-1\}$, and the proof of uniqueness is complete.

We now prove existence by induction on n, where we use the variant of proof by induction given in Theorem 2.5.4. First, suppose that $n = 1$. Then $n = 1 \cdot p^0$. Let $k = 1$ and $a_0 = 1$, and then Equation 2.8.1 is satisfied when $n = 1$.

Next, let $n \in \mathbb{N}$. Suppose that $n > 1$, and that the desired result holds for all natural numbers less than n. By Lemma 2.8.1 there is some $k \in \mathbb{N}$ such that $p^{k-1} \le n < p^k$. Let

$$S = \{j \in \{0, \ldots, p-1\} \mid jp^{k-1} \le n\}.$$

Clearly $S \subseteq \{0, \ldots, p-1\}$. We know that $1 \in S$, and hence $S \neq \emptyset$. The set S is a finite set of real numbers, and hence it has a greatest element, as proved in Example 2.5.13. Let a_{k-1} denote this greatest element. Then $a_{k-1}p^{k-1} \le n < (a_{k-1}+1)p^{k-1}$. It follows that $0 \le n - a_{k-1}p^{k-1} < p^{k-1}$.

There are now two cases. First, suppose that $n - a_{k-1}p^{k-1} = 0$. Then $n = a_{k-1}p^{k-1}$. We can therefore let $a_0 = a_1 = \cdots = a_{k-2} = 0$, and then Equation 2.8.1 is satisfied.

Second, suppose that $0 < n - a_{k-1}p^{k-1}$. Because each of a_{k-1} and p^{k-1} is an integer, then so is $n - a_{k-1}p^{k-1}$, and hence $n - a_{k-1}p^{k-1} \in \mathbb{N}$. Because $a_{k-1} > 0$ and $p^{k-1} > 0$, then $n - a_{k-1}p^{k-1} < n$. We can therefore apply the inductive hypothesis to $n - a_{k-1}p^{k-1}$, and we deduce that there are $v \in \mathbb{N}$ and $d_0, d_1, \ldots, d_{v-1} \in \{0, \ldots, p-1\}$ such that $d_{v-1} \neq 0$, and that

$$n - a_{k-1}p^{k-1} = \sum_{i=0}^{v-1} d_i p^i. \tag{2.8.2}$$

We claim that $v < k$. Suppose to the contrary that $v \ge k$. It then follows that $d_{v-1}p^{v-1} \ge 1 \cdot p^{k-1}$, and hence $\sum_{i=0}^{v-1} d_i p^i \ge p^{k-1}$, which would contradict the fact that $n - a_{k-1}p^{k-1} < p^{k-1}$. Hence $v < k$. We now let $a_i = d_i$ for all $i \in \{0, \ldots, v-1\}$, and $a_i = 0$ for all $i \in \{v, \ldots, k-2\}$. Equation 2.8.1 is then a rearrangement of Equation 2.8.2. □

We now turn to the much trickier question of the base p representation of arbitrary positive real numbers. When we write a real number in decimal notation, for example $\pi = 3.14159\ldots$, we have an infinite collection of digits after the decimal points. Such an infinite collection of numbers is called a sequence, a concept that we will discuss in detail in Chapter 8. For now, it is sufficient to think of a sequence of numbers as an infinite list of the form $a_1, a_2, a_3, a_4, \ldots$. It is the presence of infinitely many numbers that makes the base p representation of arbitrary real numbers much more complicated than the representation of natural numbers.

When we write $\pi = 3.14159\ldots$, we mean that

$$\pi = 3 + \frac{1}{10} + \frac{4}{10^2} + \frac{1}{10^3} + \frac{5}{10^4} + \frac{9}{10^5} + \cdots.$$

The difficulty with dealing with such an expression is that it involves adding infinitely many numbers at a time. Addition, as we have seen it in the axioms for the real numbers, is defined for only two numbers at a time; using Definition by Recursion it is possible to add any finite set of numbers at a time, as seen in Exercise 2.5.19, but that approach does not extend to adding infinitely many numbers at a time. An infinite sum of the form written for π above is called a series, a concept that we will discuss in detail in Chapter 9. As the reader will see in that chapter, not every series actually has a sum; those that do are called "convergent" series. It turns out that every base p representation of a real number, when viewed as a series, is in fact convergent, though that is not an obvious fact, and involves ideas that will be seen in Chapter 9; see Exercise 9.3.8 (1) for details. However, even though we have not yet discussed the convergence of series, we can nonetheless give a rigorous treatment of base p representation of arbitrary real numbers in the present section because it is possible to replace the use of series with the use of least upper bounds, a concept with which we are by now quite familiar. We follow (with added detail) the treatment via least upper bounds found in [Gor02, Section 1.3] and [Ros68, Section II.3]. After learning some facts about series in Sections 9.2 and 9.3, it will be left to the reader in Exercise 9.3.8 and Exercise 9.3.9 to provide simplified proofs of some results of the present section; that simplicity is deceptive, however, because it relies upon learning some new concepts first.

We start with the following lemma, which will allow us to make use of the Least Upper Bound Property. For convenience, we will sometimes write p^{-n} and at other times write $\frac{1}{p^n}$.

Lemma 2.8.3. *Let $p \in \mathbb{N}$. Suppose that $p > 1$. Let $a_1, a_2, a_3, \ldots \in \{0, \ldots, p-1\}$. Then the set*

$$\left\{ \sum_{i=1}^{n} a_i p^{-i} \mid n \in \mathbb{N} \right\}$$

is bounded below by 0 *and is bounded above by* 1.

Proof. Let $n \in \mathbb{N}$, and let $x = \sum_{i=1}^{n} a_i p^{-i}$. Because $a_i \geq 0$ for all $i \in \{1, \ldots, n\}$, and because $p > 0$, it follows that $x \geq 0$. Hence S is bounded below by 0. By Exercise 2.8.4 (2) we see that $x = \sum_{i=1}^{n} a_i p^{-i} \leq \frac{1}{p^{1-1}} - \frac{1}{p^n} < 1$. Hence S is bounded above by 1. □

Because of Lemma 2.8.3, we can make the following definition by appealing to the Least Upper Bound Property. Although we will use the infinite summation notation that is standardly used for series in the following definition, at this point we are thinking of this infinite summation strictly as formal notation, and we are not actually adding infinitely many things at a time in the following definition.

Definition 2.8.4. Let $p \in \mathbb{N}$. Suppose that $p > 1$. Let $a_1, a_2, a_3, \ldots \in \{0, \ldots, p-1\}$. The sum $\sum_{i=1}^{\infty} a_i p^{-i}$ is defined by

$$\sum_{i=1}^{\infty} a_i p^{-i} = \text{lub} \left\{ \sum_{i=1}^{n} a_i p^{-i} \mid n \in \mathbb{N} \right\}. \qquad \triangle$$

The following lemma gives some basic properties of $\sum_{i=1}^{\infty} a_i p^{-i}$. To obtain an intuitive feel for why these facts are true, the reader should think of the case $p = 10$, and should think of $\sum_{i=1}^{\infty} a_i p^{-i}$ as the infinite decimal $0.a_1a_2a_3\cdots$, although we formally introduce that notation only later in Definition 2.8.7.

Lemma 2.8.5. *Let $p \in \mathbb{N}$. Suppose that $p > 1$. Let $a_1, a_2, a_3, \ldots \in \{0, \ldots, p-1\}$.*

1. $0 \le \sum_{i=1}^{\infty} a_i p^{-i} \le 1$.
2. $\sum_{i=1}^{\infty} a_i p^{-i} = 0$ *if and only if* $a_i = 0$ *for all* $i \in \mathbb{N}$.
3. $\sum_{i=1}^{\infty} a_i p^{-i} = 1$ *if and only if* $a_i = p-1$ *for all* $i \in \mathbb{N}$.
4. *Let $m \in \mathbb{N}$. Suppose that $m > 1$, and that $a_{m-1} \ne p-1$. Then*
$$\sum_{i=1}^{\infty} a_i p^{-i} \le \sum_{i=1}^{m-2} a_i p^{-i} + \frac{a_{m-1}+1}{p^{m-1}},$$
where equality holds if and only if $a_i = p-1$ for all $i \in \mathbb{N}$ such that $i \ge m$.

Proof. We will prove Parts (2) and (4), leaving the rest to the reader in Exercise 2.8.3.

(2) If $a_i = 0$ for all $i \in \mathbb{N}$, then $\sum_{i=1}^{n} a_i p^{-i} = 0$ for all $n \in \mathbb{N}$, and hence $\sum_{i=1}^{\infty} a_i p^{-i} = \operatorname{lub}\{0\} = 0$. If $a_k \ne 0$ for some $k \in \mathbb{N}$, then $\sum_{i=1}^{k} a_i p^{-i} > 0$, and hence $\sum_{i=1}^{\infty} a_i p^{-i} = \operatorname{lub}\{\sum_{i=1}^{n} a_i p^{-i} \mid n \in \mathbb{N}\} > 0$.

(4) Let
$$Q = \sum_{i=1}^{m-2} a_i p^{-i} + \frac{a_{m-1}+1}{p^{m-1}} \quad \text{and} \quad T = \left\{ \sum_{i=1}^{n} a_i p^{-i} \mid n \in \mathbb{N} \right\}.$$

By definition $\sum_{i=1}^{\infty} a_i p^{-i} = \operatorname{lub} T$. We will show that $\operatorname{lub} T \le Q$, and that $\operatorname{lub} T = Q$ if and only if $a_i = p-1$ for all $i \in \mathbb{N}$ such that $i \ge m$.

Let $n \in \mathbb{N}$. First, suppose that $n \le m-2$. Then $\sum_{i=1}^{n} a_i p^{-i} \le \sum_{i=1}^{m-2} a_i p^{-i} \le Q$. Second, suppose that $n > m-2$. Then $n \ge m-1$ by Theorem 2.4.10 (1). Using Exercise 2.8.4 (2) we see that
$$\sum_{i=1}^{n} a_i p^{-i} = \sum_{i=1}^{m-2} a_i p^{-i} + \frac{a_{m-1}}{p^{m-1}} + \sum_{i=m}^{n} a_i p^{-i} \le \sum_{i=1}^{m-2} a_i p^{-i} + \frac{a_{m-1}}{p^{m-1}} + \frac{1}{p^{m-1}} - \frac{1}{p^n} < Q.$$

Combining these two cases, we see that Q is an upper bound of T. Hence $\operatorname{lub} T \le Q$.

We now show that $\operatorname{lub} T = Q$ if and only if $a_i = p-1$ for all $i \in \mathbb{N}$ such that $i \ge m$. Suppose that there is some $r \in \mathbb{N}$ such that $r \ge m$ and $a_r \ne p-1$. Because $a_r \in \{0, \ldots, p-1\}$, it follows that $a_r \le p-2$, and hence $p-1-a_r \ge 1$.

Let $s \in \mathbb{N}$. We will show that $\sum_{i=1}^{s} a_i p^{-i} \le Q - p^{-r}$. It will then follow that $Q - p^{-r}$ is an upper bound of T, which means that $\operatorname{lub} T \le Q - p^{-r} < Q$, which implies that $\operatorname{lub} T \ne Q$. There are two cases.

First, suppose that $s < r$. Using the fact that $a_i \le p-1$ for all $i \in \mathbb{N}$, together with Exercise 2.8.4 (2), we see that
$$Q - \sum_{i=1}^{s} a_i p^{-i} \ge Q - \sum_{i=1}^{r} a_i p^{-i}$$

$$= \left[\sum_{i=1}^{m-2} a_i p^{-i} + \frac{a_{m-1}+1}{p^{m-1}} \right] - \left[\sum_{i=1}^{m-2} a_i p^{-i} + \frac{a_{m-1}}{p^{m-1}} + \sum_{i=m}^{r} a_i p^{-i} \right]$$
$$\geq \frac{1}{p^{m-1}} - \left[\frac{1}{p^{m-1}} - \frac{1}{p^r} \right] = \frac{1}{p^r}. \tag{2.8.3}$$

It follows that $\sum_{i=1}^{s} a_i p^{-i} \leq Q - p^{-r}$.

Second, suppose that $s \geq r$. Hence $s \geq r \geq m$. Using reasoning similar to the previous case, we see that

$$Q - \sum_{i=1}^{s} a_i p^{-i} = \left[\sum_{i=1}^{m-2} a_i p^{-i} + \frac{a_{m-1}+1}{p^{m-1}} \right]$$
$$- \left[\sum_{i=1}^{m-2} a_i p^{-i} + \frac{a_{m-1}}{p^{m-1}} + \sum_{i=m}^{r-1} a_i p^{-i} + \frac{a_r}{p^r} + \sum_{i=r+1}^{s} a_i p^{-i} \right]$$
$$\geq \frac{1}{p^{m-1}} - \left[\frac{1}{p^{m-1}} - \frac{1}{p^{r-1}} \right] - \frac{p-1}{p^r} + \frac{p-1-a_r}{p^r} - \left[\frac{1}{p^r} - \frac{1}{p^s} \right]$$
$$= \frac{p-1-a_r}{p^r} + \frac{1}{p^s} > \frac{1}{p^r},$$

where the last inequality holds because $p-1-a_r \geq 1$. It follows that $\sum_{i=1}^{s} a_i p^{-i} \leq Q - p^{-r}$. Combining the two cases, we deduce that $\text{lub}\, T \neq Q$

Now, suppose that $a_i = p-1$ for all $i \in \mathbb{N}$ such that $i \geq m$. We saw above that Q is an upper bound of T. Let $\varepsilon > 0$. By Exercise 2.6.13 there is some $k \in \mathbb{N}$ such that $\frac{1}{p^k} < \varepsilon$. Taking a larger value of k will not change this inequality, and so we may assume that $k \geq m$. Using the fact that $a_i = p-1$ for all $i \in \mathbb{N}$ such that $i \geq m$ together with the second half of Exercise 2.8.4 (2), we see that the same reasoning used in Equation 2.8.3 shows that in the present case $Q - \sum_{i=1}^{k} a_i p^{-i} = \frac{1}{p^k} < \varepsilon$. It then follows from Exercise 2.6.6 that $Q = \text{lub}\, T$. □

We now come to our main theorem regarding the base p representation of real numbers, which says that every positive real number has such a representation, and that such a representation is unique if we avoid representations that eventually become the number $(p-1)$ repeating, which is the analog of the number 9 repeating in the decimal system. The idea of the proof of this theorem is not difficult, but the details are somewhat lengthy.

Theorem 2.8.6. *Let $p \in \mathbb{N}$. Suppose that $p > 1$. Let $x \in (0, \infty)$.*

1. *There are $k \in \mathbb{N}$, and $b_0, b_1, \ldots, b_{k-1} \in \{0, \ldots, p-1\}$ and $a_1, a_2, a_3 \ldots \in \{0, \ldots, p-1\}$, such that*

$$x = \sum_{j=0}^{k-1} b_j p^j + \sum_{i=1}^{\infty} a_i p^{-i}. \tag{2.8.4}$$

2. *It is possible to choose $k \in \mathbb{N}$, and $b_0, b_1, \ldots, b_{k-1} \in \{0, \ldots, p-1\}$, and $a_1, a_2, a_3 \ldots \in \{0, \ldots, p-1\}$ in Part (1) of this theorem such that there is no $m \in \mathbb{N}$ such that $a_i = p-1$ for all $i \in \mathbb{N}$ such that $i \geq m$.*
3. *If $x > 1$, then it is possible to choose $k \in \mathbb{N}$, and $b_0, b_1, \ldots, b_{k-1} \in \{0, \ldots, p-1\}$, and $a_1, a_2, a_3 \ldots \in \{0, \ldots, p-1\}$ in Part (1) of this theorem such that $b_{k-1} \neq 0$. If $0 < x < 1$, then it is possible to choose $k = 1$, and $b_0 = 0$, and $a_1, a_2, a_3 \ldots \in \{0, \ldots, p-1\}$ in Part (1) of this theorem.*
4. *If the conditions of Parts (2) and (3) of this theorem hold, then the numbers $k \in \mathbb{N}$, and $b_0, b_1, \ldots, b_{k-1} \in \{0, \ldots, p-1\}$, and $a_1, a_2, a_3 \ldots \in \{0, \ldots, p-1\}$ in Part (1) are unique.*

Proof. We will prove Parts (1) and (4), leaving the rest to the reader in Exercise 2.8.6.

(1) By Corollary 2.6.8 (1) there is a unique $n \in \mathbb{N}$ such that $n-1 \leq x < n$. Let $a_0 = n-1$. Hence $a_0 \geq 0$ and $a_0 \leq x < a_0 + 1$.

We start by defining the numbers $k \in \mathbb{N}$ and $b_0, b_1, \ldots, b_{k-1} \in \{0, \ldots, p-1\}$. If $a_0 = 0$, let $k = 1$ and $b_0 = 0$. Then $a_0 = \sum_{j=0}^{k-1} b_j p^j$. Now suppose that $a_0 > 0$. Then $a_0 \in \mathbb{N}$. We can then apply Theorem 2.8.2 to a_0 to deduce that there is a unique $k \in \mathbb{N}$, and unique $b_0, b_1, \ldots, b_{k-1} \in \{0, \ldots, p-1\}$ such that $b_{k-1} \neq 0$, and that $a_0 = \sum_{j=0}^{k-1} b_j p^j$.

Next, we define the numbers $a_1, a_2, a_3, \ldots \in \{0, \ldots, p-1\}$. Actually, we use Definition by Recursion to define numbers $a_1, a_2, a_3, \ldots \in \{0, \ldots, p-1\}$ and $z_1, z_2, z_3, \ldots \in \mathbb{R}$; the numbers $z_1, z_2, z_3, \ldots$ are not of interest per se, but they help us define $a_1, a_2, a_3, \ldots$. (To be precise, we are using Definition by Recursion to define the single sequence $(a_1, z_1), (a_2, z_3), (a_3, z_3), \ldots$ in $\{0, \ldots, p-1\} \times \mathbb{R}$, though for convenience we do not write it that way.)

First, let $z_1 = x - a_0$. Then $0 \leq z_1 < 1$, and hence $0 \cdot p^{-1} \leq z_1 < pp^{-1}$. Let

$$S_1 = \{j \in \{0, \ldots, p-1\} \mid jp^{-1} \leq z_1\}.$$

Then $0 \in S_1$, and so $S_1 \neq \emptyset$. The set S_1 is finite, and hence it has a greatest element. Let a_1 denote this greatest element. Because $a_1 \in S_1$, then $a_1 jp^{-1} \leq z_1$. If $a_1 \neq p-1$, then the maximality of a_1 implies that $z_1 < (a_1+1)p^{-1}$; if $a_1 = p-1$, then $z_1 < pp^{-1} = (a_1+1)p^{-1}$. Hence $a_1 p^{-1} \leq z_1 < (a_1+1)p^{-1}$, and therefore $0 \leq z_1 - a_1 p^{-1} < p^{-1}$.

Next, let $z_2 = z_1 - a_1 p^{-1}$. Then $0 \leq z_2 < p^{-1}$, and hence $0 \cdot p^{-2} \leq z_2 < pp^{-2}$. Let

$$S_2 = \{j \in \{0, \ldots, p-1\} \mid jp^{-2} \leq z_2\}.$$

As before, we know that $0 \in S_2$, and so $S_2 \neq \emptyset$, and that the finite set S_2 has a greatest element. Let a_2 denote this greatest element. As before, it is seen that $a_2 p^{-2} \leq z_2 < (a_2+1)p^{-2}$; we omit the details.

Next, let $z_3 = z_2 - a_2 p^{-2} = z_1 - (a_1 p^{-1} + a_2 p^{-2})$. Then $0 \leq z_3 < p^{-2}$. Similarly to what we did for a_1 and a_2, we can find $a_3 \in \{0, \ldots, p-1\}$ such that $a_3 p^{-3} \leq z_3 < (a_3+1)p^{-3}$.

We continue in this fashion, obtaining numbers $z_1, z_2, z_3, \ldots \in \mathbb{R}$ and $a_1, a_2, a_3, \ldots \in \{0, \ldots, p-1\}$ such that for all $n \in \mathbb{N}$, it is the case that

$$z_{n+1} = z_1 - \sum_{i=1}^{n} a_i p^{-i} \quad \text{and} \quad 0 \leq z_{n+1} < p^{-n}. \tag{2.8.5}$$

Let

$$T = \left\{ \sum_{i=1}^{n} a_i p^{-i} \mid n \in \mathbb{N} \right\}.$$

Let $n \in \mathbb{N}$. Equation 2.8.5 implies that $z_1 - \sum_{i=1}^{n} a_i p^{-i} = z_{n+1} \geq 0$, and therefore $\sum_{i=1}^{n} a_i p^{-i} \leq z_1$. Hence z_1 is an upper bound of T.

Let $\varepsilon > 0$. By Exercise 2.6.13 there is some $m \in \mathbb{N}$ such that $p^{-m} = \frac{1}{p^m} < \varepsilon$. Using Equation 2.8.5 again, we see that $z_1 - \sum_{i=1}^{m} a_i p^{-i} = z_{m+1} < p^{-m} < \varepsilon$. It now follows from Exercise 2.6.6 that $z_1 = \operatorname{lub} T$. Hence $z_1 = \sum_{i=1}^{\infty} a_i p^{-i}$.

Because $z_1 = x - a_0$ and $a_0 = \sum_{j=0}^{k-1} b_j p^j$, we deduce that

$$x = a_0 + z_1 = \sum_{j=0}^{k-1} b_j p^j + \sum_{i=1}^{\infty} a_i p^{-i}.$$

(4) Suppose that there are $k, u \in \mathbb{N}$, and $b_0, b_1, \ldots, b_{k-1} \in \{0, \ldots, p-1\}$ and $c_0, c_1, \ldots, c_{u-1} \in \{0, \ldots, p-1\}$, and $a_1, a_2, a_3 \ldots \in \{0, \ldots, p-1\}$ and $e_1, e_2, e_3 \ldots \in \{0, \ldots, p-1\}$ such that

$$x = \sum_{j=0}^{k-1} b_j p^j + \sum_{i=1}^{\infty} a_i p^{-i} \quad \text{and} \quad x = \sum_{j=0}^{u-1} c_j p^j + \sum_{i=1}^{\infty} e_i p^{-i}.$$

Suppose further that there is no $m \in \mathbb{N}$ such that $a_i = p-1$ for all $i \in \mathbb{N}$ such that $i \geq m$, or that $e_i = p-1$ for all $i \in \mathbb{N}$ such that $i \geq m$. Suppose also that if $x > 1$, then $b_{k-1} \neq 0$, and if $0 < x < 1$, then $k = 0$ and $b_0 = 0$.

From the above hypotheses it follows that

$$\sum_{j=0}^{k-1} b_j p^j - \sum_{j=0}^{u-1} c_j p^j = \sum_{i=1}^{\infty} e_i p^{-i} - \sum_{i=1}^{\infty} a_i p^{-i}. \tag{2.8.6}$$

We know by Lemma 2.8.5 (1) (3) that each of $\sum_{i=1}^{\infty} e_i p^{-i}$ and $\sum_{i=1}^{\infty} a_i p^{-i}$ is in the interval $[0, 1)$. It follows that $\left|\sum_{i=1}^{\infty} e_i p^{-i} - \sum_{i=1}^{\infty} a_i p^{-i}\right| < 1$. Equation 2.8.6 then implies that $\left|\sum_{j=0}^{k-1} b_j p^j - \sum_{j=0}^{u-1} c_j p^j\right| < 1$. Each of $\sum_{j=0}^{k-1} b_j p^j$ and $\sum_{j=0}^{u-1} c_j p^j$ are integers, and it now follows from Theorem 2.4.10 (3) that $\sum_{j=0}^{k-1} b_j p^j = \sum_{j=0}^{u-1} c_j p^j$.

Suppose that $x \geq 1$. Then $\sum_{j=0}^{k-1} b_j p^j = \sum_{j=0}^{u-1} c_j p^j \neq 0$. Therefore $\sum_{j=0}^{k-1} b_j p^j \in \mathbb{N}$ and $\sum_{j=0}^{u-1} c_j p^j \in \mathbb{N}$, and the uniqueness in the statement of Theorem 2.8.2 implies that $k = u$, and that $b_j = c_j$ for all $j \in \{0, \ldots, k-1\}$. Next, suppose that $0 < x < 1$. Then by hypothesis $k = 0$ and $b_0 = 0$, and $u = 0$ and $c_0 = 0$, so that $k = u$ and $b_0 = c_0$.

Because $\sum_{j=0}^{k-1} b_j p^j = \sum_{j=0}^{u-1} c_j p^j$, it follows from Equation 2.8.6 that $\sum_{i=1}^{\infty} a_i p^{-i} = \sum_{i=1}^{\infty} e_i p^{-i}$. We now use Exercise 2.8.7 to conclude that $a_i = e_i$ for all $i \in \mathbb{N}$. □

We are now, finally, ready to make the following definition.

Definition 2.8.7. Let $p \in \mathbb{N}$. Suppose that $p > 1$. Let $x \in (0, \infty)$. A **base p representation** of the number x is an expression of the form $x = b_{k-1} \cdots b_1 b_0.a_1 a_2 a_3 \cdots$, where $k \in \mathbb{N}$ and $b_0, b_1, \ldots, b_{k-1} \in \{0, \ldots, p-1\}$ and $a_1, a_2, a_3 \ldots \in \{0, \ldots, p-1\}$ are such that

$$x = \sum_{j=0}^{k-1} b_j p^j + \sum_{i=1}^{\infty} a_i p^{-i}. \qquad \triangle$$

We can now restate Theorem 2.8.6 by saying that any positive real number has a base p representation, and that such a representation is unique subject to the conditions stated in Parts (2) and (3) of the theorem.

We conclude this section with the one fundamental issue regarding base p representations that we have not yet addressed, which is characterizing the base p representation of rational numbers. It should be familiar to the reader that the decimal expansion of a rational number is either terminating or eventually repeating. (Actually, this characterization is redundant, because a "terminating" decimal expansion is simply one that eventually has repeating zeros, but we will maintain the standard phraseology.) The analogous fact holds for the base p representation of rational numbers for all p, as we will show in Theorem 2.8.10 below. First, however, we need to state and prove the following theorem, which is known as the "Division Algorithm," although it is not an algorithm, but is rather an existence theorem (the name of the theorem is a historical artifact). The Division Algorithm is a very important tool in a number of branches of mathematics (for example number theory), though in this text we will be using it only in the proof of Theorem 2.8.10, and hence we have included it in this section.

To understand the Division Algorithm, think of how one learns to divide natural numbers in elementary school. Suppose that we want to divide 27 by 4. It is seen that 4 goes into 27 six times, so that the quotient is 6, and there is a remainder of 3. In other words, we write $\frac{27}{4} = 6 + \frac{3}{4}$, which for convenience can also be written as $27 = 6 \cdot 4 + 3$. How did we find the quotient and the remainder? The idea is that we wanted to find as many whole copies of 4 in 27 as possible, and we see that there are 6 copies, because $6 \cdot 4$ is less than 27, but $7 \cdot 4$ is greater than 27. The remainder is what was left over when we subtracted $6 \cdot 4$ from 27. As such, we see that the remainder must be less than 4, or else we could have increased the quotient. The Division Algorithm is just a general statement that in all such situations, there is a unique quotient and remainder.

Theorem 2.8.8 (Division Algorithm). *Let $a \in \mathbb{N} \cup \{0\}$ and $b \in \mathbb{N}$. Then there are unique $q, r \in \mathbb{N} \cup \{0\}$ such that $a = bq + r$ and $0 \le r < b$.*

Proof. To prove uniqueness, suppose that there are $q, p, r, s \in \mathbb{Z}$ such that $a = bq + r$ and $a = bp + s$, and that $0 \le r < b$ and $0 \le s < b$. There are two cases.

First, suppose that $q = p$. Because $bq + r = bp + s$, it follows that $r = s$.

Second, suppose that $q \ne p$. Without loss of generality, assume that $q > p$. Then $q + (-p) > 0$, and because $q + (-p)$ is an integer, it follows that $q + (-p) \ge 1$. Because $bq + r = bp + s$, we then have $s = b(q + (-p)) + r \ge b \cdot 1 + 0 = b$, and this inequality contradicts the hypothesis on s. Hence it cannot be the case that $q \ne p$.

We now prove existence. Again, there are two cases. First, suppose that $a = 0$. Let $q = 0$ and $r = 0$, which yields $bq + r = 0 \cdot q + 0 = 0 = a$, and $0 \leq r < b$.

Second, suppose that $a > 0$. We prove the desired result by induction on a. Let $a = 1$. There are now two subcases. First, suppose that $b = 1$. Let $q = 1$ and $r = 0$, which yields $bq + r = 1 \cdot 1 + 0 = 1 = a$, and $0 \leq r < b$. Second, suppose that $b \neq 1$. Therefore $b > 1$. Let $q = 0$ and $r = 1$, which yields $bq + r = b \cdot 0 + 1 = 1 = a$, and $0 \leq r < b$. Hence the result is true when $a = 1$.

Now suppose that the result is true for a, and we will prove that it is true for $a + 1$. By the inductive hypothesis there are $g, v \in \mathbb{Z}$ such that $a = bg + v$ and $0 \leq v < b$. Then $v + 1 \leq b$. There are now two subcases. First, suppose that $v + 1 < b$. Let $q = g$ and $r = v + 1$. Hence $bq + r = bg + (v + 1) = (bg + v) + 1 = a + 1$, and $0 \leq v < v + 1 = r = v + 1 < b$. Second, suppose that $v + 1 = b$. Hence $v = b - 1$. Let $q = g + 1$ and $r = 0$. Then $bq + r = b(g + 1) + 0 = bg + b = bg + (b - 1) + 1 = (bg + v) + 1 = a + 1$, and $0 \leq r < b$. By induction we now see that the result is true for all $a > 0$. □

The final definition and theorem of this section show that a positive real number is rational if and only if it has an eventually repeating base p representation for any p. Again, the proof is somewhat lengthier than might be expected.

Definition 2.8.9. Let $p \in \mathbb{N}$. Suppose that $p > 1$. Let $x \in (0, \infty)$, and let $x = b_{k-1} \cdots b_1 b_0 . a_1 a_2 a_3 \cdots$ be a base p representation of x. This base p representation is **eventually repeating** if there are some $r, s \in \mathbb{N}$ such that $a_j = a_{j+s}$ for all $j \in \mathbb{N}$ such that $j \geq r$; in that case we write

$$x = b_{k-1} \cdots b_1 b_0 . a_1 a_2 a_3 \cdots a_{r-1} \overline{a_r \cdots a_{r+s-1}}.$$

△

Theorem 2.8.10. *Let $p \in \mathbb{N}$. Suppose that $p > 1$. Let $x \in (0, \infty)$. Then $x \in \mathbb{Q}$ if and only if x has an eventually repeating base p representation.*

Proof. First, suppose that x has an eventually repeating base p representation. Hence $x = b_{k-1} \cdots b_1 b_0 . a_1 a_2 a_3 \cdots a_{r-1} \overline{a_r \cdots a_{r+s-1}}$ for some $k, r, s \in \mathbb{N}$ and for some $b_0, b_1, \ldots, b_{k-1}, a_1, a_2, \ldots, a_{r+s-1} \in \{0, \ldots, p-1\}$.

Let

$$B = \sum_{i=r}^{r+s-1} a_i p^{-i} \quad \text{and} \quad W = \left\{ \sum_{i=r}^{u} a_i p^{-i} \mid u \in \mathbb{N} \text{ and } u \geq r \right\}.$$

As a preliminary step, we will prove that $\operatorname{lub} W = \frac{p^s B}{p^s - 1}$. Let $m \in \mathbb{N}$. Using the fact that $a_j = a_{j+s}$ for all $j \in \mathbb{N}$ such that $j \geq r$, together with Exercise 2.5.12 (3), we see that

$$\begin{aligned}
\frac{p^s B}{p^s - 1} - \sum_{i=r}^{r+ms-1} a_i p^{-i} &= \frac{p^s B}{p^s - 1} - \sum_{k=1}^{m} \sum_{i=r+(k-1)s}^{r+ks-1} a_i p^{-i} \\
&= \frac{p^s B}{p^s - 1} - \sum_{k=1}^{m} p^{-(k-1)s} \sum_{i=r+(k-1)s}^{r+ks-1} a_{i-(k-1)s} p^{-[i-(k-1)s]}
\end{aligned}$$

$$= \frac{p^s B}{p^s - 1} - \sum_{k=1}^{m} p^{-(k-1)s} \sum_{i=r}^{r+s-1} a_i p^{-i}$$

$$= \frac{p^s B}{p^s - 1} - \sum_{k=1}^{m} B p^{-(k-1)s} = \frac{p^s B}{p^s - 1} - \frac{B(1-(p^{-s})^m)}{1-p^{-s}}$$

$$= \frac{B}{(p^s - 1)p^{(m-1)s}}.$$

Now let $u \in \mathbb{N}$. Suppose that $u \geq r$. Let $q = u - r + 1$. Then $q \geq 1$, and hence $q \in \mathbb{N}$. Because $s \in \mathbb{N}$, then $s \geq 1$, and hence $qs \geq 1$. It follows that $u \leq r + qs - 1$. Then

$$\sum_{i=r}^{u} a_i p^{-i} \leq \sum_{i=r}^{r+qs-1} a_i p^{-i},$$

which implies that

$$\frac{p^s B}{p^s - 1} - \sum_{i=r}^{u} a_i p^{-i} \geq \frac{p^s B}{p^s - 1} - \sum_{i=r}^{r+qs-1} a_i p^{-i} = \frac{B}{(p^s - 1)p^{(q-1)s}} \geq 0.$$

Therefore $\sum_{i=r}^{u} a_i p^{-i} \leq \frac{p^s B}{p^s - 1}$. We deduce that $\frac{p^s B}{p^s - 1}$ is an upper bound of W.

Let $\varepsilon > 0$. Clearly $B \geq 0$. If $B = 0$, let $w = 1$. Then $\frac{B}{(p^s-1)p^{(w-1)s}} = 0 < \varepsilon$. Now suppose $B > 0$. By Exercise 2.6.13 there is some $e \in \mathbb{N}$ such that $\frac{1}{p^e} < \frac{(p^s-1)\varepsilon}{B}$. Let $w = e + 1$. Because $es \geq e$, we see that

$$\frac{B}{(p^s - 1)p^{(w-1)s}} = \frac{B}{p^s - 1}\frac{1}{p^{es}} \leq \frac{B}{p^s - 1}\frac{1}{p^e} < \frac{B}{p^s - 1}\frac{(p^s - 1)\varepsilon}{B} = \varepsilon.$$

Putting these two cases together, and using a previous calculation, it follows that

$$\frac{p^s B}{p^s - 1} - \sum_{i=r}^{r+ws-1} a_i p^{-i} = \frac{B}{(p^s - 1)p^{(w-1)s}} < \varepsilon.$$

It follows from Exercise 2.6.6 that $\frac{p^s B}{p^s - 1} = \operatorname{lub} W$, which completes the preliminary step.

If $n, m \in \mathbb{N}$ and $n \leq m$, then $\sum_{i=1}^{n} a_i p^{-i} \leq \sum_{i=1}^{m} a_i p^{-i}$. It follows from Exercise 2.6.3 (1) that

$$\operatorname{lub}\left\{\sum_{i=1}^{n} a_i p^{-i} \mid n \in \mathbb{N}\right\} = \operatorname{lub}\left\{\sum_{i=1}^{u} a_i p^{-i} \mid u \in \mathbb{N} \text{ and } u \geq r\right\}.$$

We now use Exercise 2.8.5 to deduce that

$$\sum_{i=1}^{\infty} a_i p^{-i} = \operatorname{lub}\left\{\sum_{i=1}^{n} a_i p^{-i} \mid n \in \mathbb{N}\right\} = \operatorname{lub}\left\{\sum_{i=1}^{u} a_i p^{-i} \mid u \in \mathbb{N} \text{ and } u \geq r\right\}$$

$$= \operatorname{lub}\left\{\sum_{i=1}^{r-1} a_i p^{-i} + \sum_{i=r}^{u} a_i p^{-i} \mid u \in \mathbb{N} \text{ and } u \geq r\right\}$$

$$= \sum_{i=1}^{r-1} a_i p^{-i} + \text{lub}\left\{\sum_{i=r}^{u} a_i p^{-i} \mid u \in \mathbb{N} \text{ and } u \geq r\right\}$$
$$= \sum_{i=1}^{r-1} a_i p^{-i} + \text{lub}\, W = \sum_{i=1}^{r-1} a_i p^{-i} + \frac{p^s B}{p^s - 1}.$$

Therefore

$$x = b_{k-1} \cdots b_1 b_0 . a_1 a_2 a_3 \cdots a_{r-1} \overline{a_r \cdots a_{r+s-1}}$$
$$= \sum_{j=0}^{k-1} b_j p^j + \sum_{i=1}^{\infty} a_i p^{-i} = \sum_{j=0}^{k-1} b_j p^j + \sum_{i=1}^{r-1} a_i p^{-i} + \frac{p^s B}{p^s - 1}$$
$$= \sum_{j=0}^{k-1} b_j p^j + \sum_{i=1}^{r-1} a_i p^{-i} + \frac{p^s}{p^s - 1} \sum_{i=r}^{r+s-1} a_i p^{-i}.$$

This last expression is a rational number, being the sum of rational numbers, and therefore $x \in \mathbb{Q}$.

Now suppose that $x \in \mathbb{Q}$. We will show that x has an eventually repeating base p representation. The key idea is that even though we have already seen a method for finding a base p representation of x in the proof of Theorem 2.8.6 (1), we will now use a different method that works for the special case when x is a rational number, and this alternative method will allow us to show that the resulting base p representation is eventually repeating.

Because $x \in \mathbb{Q}$ and $x > 0$, we know by Lemma 2.4.12 (2) that $x = \frac{c}{d}$ for some $c, d \in \mathbb{N}$. We now use Definition by Recursion to define numbers $e_0, e_1, e_2, \ldots \in \mathbb{N} \cup \{0\}$ and $r_0, r_1, r_2, \ldots \in \mathbb{N} \cup \{0\}$. (As in the proof of Theorem 2.8.6 (1), we are actually using Definition by Recursion to define a single sequence in $\mathbb{N} \cup \{0\} \times \mathbb{N} \cup \{0\}$, though for convenience we do not write it that way.)

Using the Division Algorithm (Theorem 2.8.8), there are unique $e_0, r_0 \in \mathbb{N} \cup \{0\}$ such that $c = de_0 + r_0$ and $0 \leq r_0 < d$. Using the Division Algorithm again, there are unique $e_1, r_1 \in \mathbb{N} \cup \{0\}$ such that $pr_0 = de_1 + r_1$ and $0 \leq r_1 < d$. Similarly, there are unique $e_2, r_2 \in \mathbb{N} \cup \{0\}$ such that $pr_1 = de_2 + r_2$ and $0 \leq r_2 < d$. We continue in this fashion, obtaining numbers $e_0, e_1, e_2, \ldots \in \mathbb{N} \cup \{0\}$ and $r_0, r_1, r_2, \ldots \in \mathbb{N} \cup \{0\}$ such that for all $n \in \mathbb{N}$, it is the case that

$$pr_{n-1} = de_n + r_n \quad \text{and} \quad 0 \leq r_n < d. \tag{2.8.7}$$

It follows from Exercise 2.8.2 (1) that $e_0 \geq 0$, and from Exercise 2.8.2 (2) that $0 \leq e_n < p$ for all $n \in \mathbb{N}$. Hence $e_n \in \{0, \ldots, p-1\}$ for all $n \in \mathbb{N}$.

Let

$$R = \left\{e_0 + \sum_{i=1}^{n} e_i p^{-i} \mid n \in \mathbb{N}\right\}.$$

It follows from Exercise 2.8.5 that R has a least upper bound, and that $\text{lub}\, R = e_0 + \sum_{i=1}^{\infty} e_i p^{-i}$. We will now show that $\text{lub}\, R = \frac{c}{d}$, which will imply that $\frac{c}{d} = e_0 + \sum_{i=1}^{\infty} e_i p^{-i}$.

Let $n \in \mathbb{N}$. Then

$$\begin{aligned}\frac{c}{d} - \left[e_0 + \sum_{i=1}^{n} e_i p^{-i}\right] &= \frac{c - de_0}{d} - \frac{1}{p^n}\sum_{i=1}^{n} \frac{pr_{i-1} - r_i}{d} p^{n-i} \\ &= \frac{1}{dp^n}\left\{p^n r_0 - \sum_{i=1}^{n} p^{n-(i-1)} r_{i-1} + \sum_{i=1}^{n} p^{n-i} r_i\right\} \\ &= \frac{1}{dp^n}\left\{p^n r_0 - \sum_{i=0}^{n-1} p^{n-i} r_i + \sum_{i=1}^{n} p^{n-i} r_i\right\} \\ &= \frac{1}{dp^n}\{p^n r_0 - p^n r_0 + p^0 r_n\} = \frac{r_n}{bp^n} \geq 0.\end{aligned}$$

Therefore $e_0 + \sum_{i=1}^{n} e_i p^{-i} \leq \frac{c}{d}$. We deduce that $\frac{c}{d}$ is an upper bound of R.

Let $\varepsilon > 0$. By Exercise 2.6.13 there is some $u \in \mathbb{N}$ such that $\frac{1}{p^u} < \varepsilon$. From the above calculation, together with the fact that $r_u < d$, we now see that

$$\frac{c}{d} - \left[e_0 + \sum_{i=1}^{u} e_i p^{-i}\right] = \frac{r_u}{dp^u} < \frac{1}{p^u} < \varepsilon.$$

It now follows from Exercise 2.6.6 that $\frac{c}{d} = \text{lub} R$. Hence $\frac{c}{d} = e_0 + \sum_{i=1}^{\infty} e_i p^{-i}$.

To find the base p representation of $\frac{c}{d}$, there are two cases. If $e_0 = 0$, then $\frac{c}{d} = \sum_{i=1}^{\infty} e_i p^{-i}$ is a base p representation of $\frac{c}{d}$. Now suppose that $e_0 > 0$. Hence $e_0 \in \mathbb{N}$. We can therefore apply Theorem 2.8.2 to e_0 to obtain unique $v \in \mathbb{N}$ and $f_0, f_1, \ldots, f_{v-1} \in \{0, \ldots, p-1\}$, such that $f_{v-1} \neq 0$, and that $e_0 = \sum_{i=0}^{v-1} f_i p^i$. It follows that $\frac{c}{d} = \sum_{i=0}^{v-1} f_i p^i + \sum_{i=1}^{\infty} e_i p^{-i}$ is a base p representation of $\frac{c}{d}$.

To complete the proof of this theorem, we will show that the numbers $e_1, e_2, e_3, \ldots$ are eventually repeating. Because $r_0, r_1, r_2, \ldots \in \mathbb{N} \cup \{0\}$, and $0 \leq r_n < d$ for all $n \in \mathbb{N}$, we see that there are d possible values in $\mathbb{N} \cup \{0\}$ that each r_n can take. Consider the $d+1$ numbers $r_1, r_2, \ldots, r_{d+1}$. Because we have $d+1$ numbers, and each one of these numbers can take on one of d possible values, it follows that at least two of the numbers $r_1, r_2, \ldots, r_{d+1}$ are equal to each other. (Formally, we are using a fact known as the Pigeonhole Principle, which is discussed and proved in many texts on combinatorics, for example [Rob84, Section 8.1]; this principle is really just a theorem about maps of finite sets, and does not need any combinatorial ideas for its formulation and proof, as seen in [Blo10, Exercise 6.3.17].) Hence $r_s = r_t$ for some $s, t \in \{1, 2, \ldots, d+1\}$, where $s < t$.

Recall that the numbers $e_0, e_1, e_2, \ldots \in \mathbb{N} \cup \{0\}$ and $r_0, r_1, r_2, \ldots \in \mathbb{N} \cup \{0\}$ were defined using Definition by Recursion, making use of the Division Algorithm (Theorem 2.8.8); these numbers satisfy Equation 2.8.7 for all $n \in \mathbb{N}$. According to the Division Algorithm, if we know the number r_{k-1} for some $k \in \mathbb{N}$, then the numbers e_k and r_k are uniquely determined. Hence, because $r_s = r_t$, then $r_{s+1} = r_{t+1}$, and then $r_{s+2} = r_{t+2}$ and so on. From this we deduce that the numbers $r_0, r_1, r_2, \ldots$ can be written as $r_0, r_1, \ldots, r_{s-1}$ followed by the numbers $r_s, r_{s+1}, \ldots, r_{t-1}$ repeated. Because each number e_k is determined uniquely by r_{k-1} for all $k \in \mathbb{N}$, then the numbers

$e_1, e_2, e_3, \ldots$ can be written as $e_1, e_2, \ldots, e_s$ followed by the numbers $e_{s+1}, e_{s+2}, \ldots, e_t$ repeated. Therefore $\frac{c}{d} = a_{k-1} \cdots a_1 a_0 . e_1 e_2 e_3 \cdots e_s \overline{e_{s+1} \cdots e_{t-1}}$. □

Reflections

The goal of this section is to prove some facts about the real numbers that are so familiar they are usually taken for granted, and it is rather surprising that the proofs in this section are as complicated as they are. The reason for these lengthy and technical proofs is because decimal expansions involve infinitely many numbers after the decimal point, which means that decimal expansions are a type of infinite sum. In contrast to finite sums, infinite sums do not always exist in general, and the existence of the particular infinite sums used for decimal expansions requires the Least Upper Bound Property. The use of this property explains why none of the familiar facts about decimal expansions are proved rigorously when students first learn about decimal expansions in elementary school, or even subsequently in high school or college calculus courses.

Given that decimal expansions are infinite sums, there is a slightly easier way to prove some of the results in the present section than we have seen here, which is by using series rather than least upper bounds. Of course, doing so requires a rigorous treatment of series, which is found in Chapter 9, and a rigorous treatment of series ultimately relies upon the Least Upper Bound Property, so using series to study decimal expansions does not bypass the Least Upper Bound Property, it simply hides that property inside the study of series. There are no free results in mathematics, and in the present case we can either have longer proofs of the properties of decimal expansions while avoiding a preliminary treatment of series, or we can have shorter proofs of the properties of decimal expansions after having studied series rigorously. We chose the former approach in the present section, though it is a judgment call which method is preferable. Once the reader has learned some facts about series in Sections 9.2 and 9.3, it will be left to the reader in Exercise 9.3.8 and Exercise 9.3.9 to provide simplified proofs of some parts of the present section.

Exercises

Exercise 2.8.1. Here is a magic trick that you can perform. First, photocopy and cut out the six cards shown Figure 2.8.1 (or make your own fancier versions of them). Then, ask a volunteer to pick a whole number from 1 to 60. Give the volunteer the six cards, and ask her to select those cards that have the chosen number on them (anywhere from one to five cards will have the chosen number). Take the selected cards, and say some appropriate magic words. While you do that, add up in your head the numbers in the upper left-hand corners of the selected cards, and that sum will be the chosen number, which you should announce to the audience with appropriate fanfare. (As an alternative, you could say that you are going to guess the volunteer's age, and then ask the volunteer to select the cards that have her age on them; make sure the person you select is not over 60.)

The mathematical question is: explain why this trick works. Use a result from this section.

1 3 5 7 9
11 13 15 17 19
21 23 25 27 29
31 33 35 37 39
41 43 45 47 49
51 53 55 57 59

2 3 6 7 10
11 14 15 18 19
22 23 26 27 30
31 34 35 38 39
42 43 46 47 50
51 54 55 58 59

4 5 6 7 12
13 14 15 20 21
22 23 28 29 30
31 36 37 38 39
44 45 46 47 52
53 54 55 60

8 9 10 11 12
13 14 15 24 25
26 27 28 29 30
31 40 41 42 43
44 45 46 47 56
57 58 59 60

16 17 18 19 20
21 22 23 24 25
26 27 28 29 30
31 48 49 50 51
52 53 54 55 56
57 58 59 60

32 33 34 35 36
37 38 39 40 41
42 43 44 45 46
47 48 49 50 51
52 53 54 55 56
57 58 59 60

Fig. 2.8.1.

Exercise 2.8.2. **[Used in Theorem 2.8.10.]** Let $a, b, q, r \in \mathbb{N} \cup \{0\}$. Suppose that $b > 0$, that $a = bq + r$ and that $0 \le r < b$.

(1) Prove that $0 \le q \le \frac{a}{b}$.
(2) Suppose that $a = xy$ for some $x, y \in \mathbb{N} \cup \{0\}$ such that $x < b$. Prove that $0 \le q < y$.

Exercise 2.8.3. **[Used in Lemma 2.8.5.]** Prove Lemma 2.8.5 (1) (3).

Exercise 2.8.4. **[Used in Theorem 2.8.2, Lemma 2.8.3, Lemma 2.8.5 and Theorem 2.8.6.]** Let $p \in \mathbb{N}$. Suppose that $p > 1$. Let $k \in \mathbb{N}$.

(1) Let $a_0, a_1, \ldots, a_{k-1} \in \{0, \ldots, p-1\}$. Prove that $\sum_{i=0}^{k-1} a_i p^i < p^k$.
(2) Let $a_0, a_1, \ldots, a_{k-1} \in \{0, \ldots, p-1\}$, and let $r, s \in \{0, \ldots, k-1\}$. Suppose that $r \le s$. Prove that $\sum_{i=r}^{s} a_i p^{-i} \le \frac{1}{p^{r-1}} - \frac{1}{p^s}$, and that equality holds if and only if $a_i = p-1$ for all $i \in \{0, \ldots, k-1\}$.
(3) Let $b_0, b_1, \ldots, b_k \in \mathbb{Z}$. Suppose that $|b_i| \le p-1$ for all $i \in \{0, \ldots, k\}$. Prove that if $\sum_{i=0}^{k} b_i p^i = 0$, then $b_i = 0$ for all $i \in \{0, \ldots, k\}$.

[Use Exercise 2.5.3 and Exercise 2.5.12.]

Exercise 2.8.5. **[Used in Theorem 2.8.10.]** Let $p \in \mathbb{N}$. Suppose that $p > 1$. Let $b \in \mathbb{R}$ and $a_1, a_2, a_3 \ldots \in \{0, \ldots, p-1\}$. Let $r \in \mathbb{N}$, and let

$$S = \left\{ b + \sum_{i=r}^{u} a_i p^{-i} \mid u \in \mathbb{N} \text{ and } u \ge r \right\}.$$

Prove that S has a least upper bound, and that

$$\operatorname{lub} S = b + \operatorname{lub}\left\{ \sum_{i=r}^{u} a_i p^{-i} \mid u \in \mathbb{N} \text{ and } u \geq r \right\}.$$

In particular, when $r = 1$, it follows that

$$\operatorname{lub} S = b + \sum_{i=1}^{\infty} a_i p^{-i}.$$

[Use Exercise 2.6.9.]

Exercise 2.8.6. [Used in Theorem 2.8.6.] Prove Theorem 2.8.6 (2) (3).

Exercise 2.8.7. [Used in Theorem 2.8.6.] Let $p \in \mathbb{N}$. Suppose that $p > 1$. Let $a_1, a_2, a_3 \ldots \in \{0, \ldots, p-1\}$ and $e_1, e_2, e_3 \ldots \in \{0, \ldots, p-1\}$. Suppose that there is no $m \in \mathbb{N}$ such that $a_i = p - 1$ for all $i \in \mathbb{N}$ such that $i \geq m$, or that $e_i = p - 1$ for all $i \in \mathbb{N}$ such that $i \geq m$. Prove that if $\sum_{i=1}^{\infty} a_i p^{-i} = \sum_{i=1}^{\infty} e_i p^{-i}$, then $a_i = e_i$ for all $i \in \mathbb{N}$.

Exercise 2.8.8. [Used in Exercise 2.8.9.] In Theorem 2.8.8 we stated the Division Algorithm for $a \in \mathbb{N} \cup \{0\}$ and $b \in \mathbb{N}$. In fact, the Division Algorithm works for any $a, b \in \mathbb{Z}$ such that $b \neq 0$. Prove the following statement. Let $a, b \in \mathbb{Z}$. Suppose that $b \neq 0$. Then there are unique $q, r \in \mathbb{Z}$ such that $a = bq + r$ and $0 \leq r < |b|$. (There is no need to prove again what was proved in Theorem 2.8.8; use what was proved in that theorem, and prove only what was not proved in the text.)

Exercise 2.8.9. Let $n \in \mathbb{Z}$. The integer n is **even** if there is some $k \in \mathbb{Z}$ such that $n = 2k$; the integer n is **odd** if there is some $k \in \mathbb{Z}$ such that $n = 2k + 1$.

Prove that every integer is either even or odd, but not both.

[Use Exercise 2.3.5 (2) and Exercise 2.8.8.]

2.9 Historical Remarks

Please see Section 1.8.

3

Limits and Continuity

3.1 Introduction

Having considered the fundamental properties of the real numbers in Chapters 1 and 2, we now commence our study of real analysis proper. The heart of real analysis, and one of the key features that distinguishes real analysis from algebraic and combinatorial branches of mathematics, is the concept of a limit. There are various types of limits, for example limits of functions (which we will discuss in the present section), and limits of sequences (to be discussed in Section 8.2). However, all of these types of limits have similar features, and gaining familiarity with one type of limit will make learning about the other types much easier. The reader has already encountered limits of functions in an intuitive fashion in calculus courses. However, limits take on a much more important role in real analysis than in calculus because in the former we are concerned with rigorous proofs rather than applications, and limits are at the heart of the rigorous treatment of calculus.

We start this chapter with a treatment of limits of functions in Section 3.2, to be followed by the closely related topic of continuity in Section 3.3. In Section 3.4 we discuss the somewhat technical concept of uniform continuity, which is a very useful variant of the more familiar notion of continuity. The discussion of uniform continuity involves the first substantial proof of the chapter (not surprisingly, a proof that relies upon the Least Upper Bound Property). The concluding section of this chapter, Section 3.5, has further substantial proofs, first of the Extreme Value Theorem and the Intermediate Value Theorem, and finally of the fact that the Heine–Borel Theorem, the Extreme Value Theorem and the Intermediate Value Theorem are all logically equivalent to the Least Upper Bound Property.

3.2 Limits of Functions

To understand the need for limits, consider the familiar formula for the definition of derivative of a function f at the number c, which is

$$\lim_{h \to 0} \frac{f(c+h) - f(c)}{h}.$$

The reader is familiar with this formula from calculus courses; we will discuss this formula from a rigorous point of view in detail in Chapter 4. For now, we want to highlight the use of the limit in the definition of derivatives. What this limit tells us is that we want to evaluate the fraction $\frac{f(c+h)-f(c)}{h}$ for values of h that approach 0, but, and this is the important point to note, not when $h = 0$. In this particular case we could not evaluate the fraction at $h = 0$ even if we had wanted to; if we were to try to substitute $h = 0$ into this fraction we would have 0 in each of the numerator and the denominator, which does not yield a real number.

In general, as is often stated informally in calculus courses, the intuitive idea of a limit of a function f as x goes to a number c in the domain of f is that the value of $f(x)$ gets closer and closer to a number L as the value of x gets closer and closer to c. (As always, note the distinction between the name of the function, which is f, and the value of the function at x, which is $f(x)$.) Not every function has a limit at each number c. It is important to stress that if we look at the values of $f(x)$ as x gets closer and closer to c, we never consider the value of f at c; in fact, the function f need not be defined at c for the limit of f to exist as x goes to c.

More formally, when we look at limits of functions, we will most often consider functions of the form $f\colon I - \{c\} \to \mathbb{R}$, where $I \subseteq \mathbb{R}$ is an open interval and $c \in I$ is a number. See Exercise 3.2.17 for a discussion of whether it is necessary to restrict our attention to I being an open interval, or whether more general sets would work; in practice, however, it usually suffices to look at open intervals, and we will therefore do so.

In the informal discussion of limits in calculus courses, we often have the function f defined on the whole interval I, including at c, but to be able to deal with the most general situation involving limits, we consider functions defined on $I - \{c\}$. Of course, if a function is in fact defined on all of I, then by abuse of notation we can also think of the same function as being defined on $I - \{c\}$, so there is no harm in having the function defined at c, but it is important that it is not required that the function be defined at c (for example to allow for the limit used in the definition of derivatives). And if the function is defined at c, then it is very important to note that the value of $f(c)$ plays no role in the limit of f as x goes to c.

The phrase "gets closer and closer," which is used in the informal approach to limits, is not at all usable in a rigorous definition, and we need to replace it by something more precise. In fact, not only is the phrase "getting closer and closer" a fuzzy description of what happens in limits, it is simply incorrect. Consider first the function $h\colon \mathbb{R} - \{0\} \to \mathbb{R}$ defined by $h(x) = x^2 + 3$ for all $x \in \mathbb{R} - \{0\}$. The graph of this function is seen in Figure 3.2.1 (i). To ask whether the limit of h exists at $c = 0$, we observe that indeed $h(x)$ gets closer and closer to 3 as x gets closer and closer to 0, in exactly the sense that the phrase "gets closer and closer" was intuitively intended. Whatever the rigorous definition of limits is, it is certainly the case that the limit of h as x goes to 0 ought to be 3. On the other hand, let $g\colon \mathbb{R} - \{0\} \to \mathbb{R}$ be defined by

$$g(x) = \begin{cases} x^2 + 3, & \text{if } x > 0 \\ -x^2 - 3, & \text{if } x < 0. \end{cases}$$

The graph of g is seen in Figure 3.2.1 (ii). Whatever the rigorous definition of limits is, the graph of this function tells us that the limit of g as x goes to 0 ought not to exist. And yet, it is the case that as x gets closer and closer to 0, the values of $g(x)$ also get closer and closer to 0. The point is that it is not sufficient for the values of $g(x)$ to get closer and closer to a given number, they have to get arbitrarily close to the given number, and it is the measure of arbitrary closeness that is missing from the phrase "gets closer and closer."

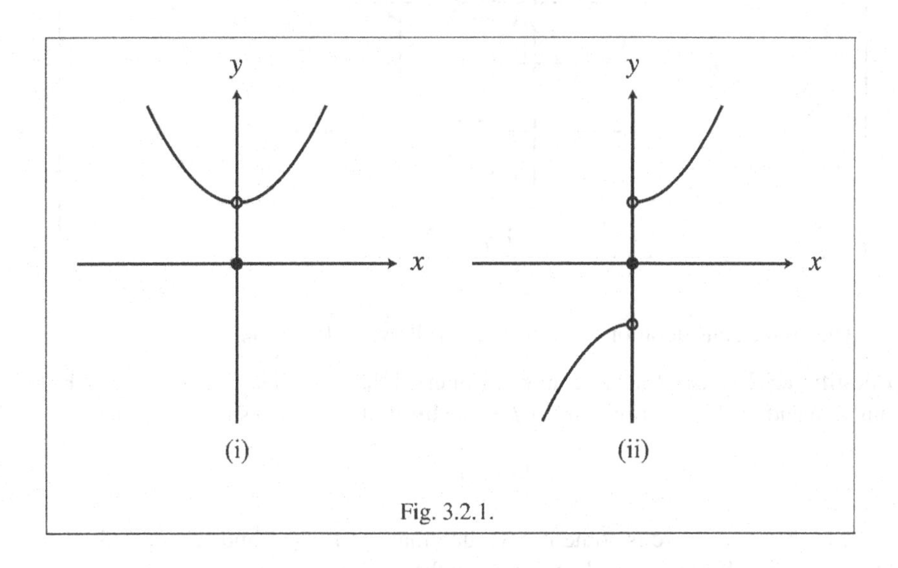

Fig. 3.2.1.

To measure "arbitrary closeness," we use an arbitrarily chosen positive number, often denoted with a symbol such as ε or δ. Let us now rewrite the phrase "the value of $f(x)$ gets closer and closer to a number L as the value of x gets closer and closer to c" using measures of closeness. We start with the first part, relating the closeness of $f(x)$ to L, where we will use ε to denote our measure of closeness. The idea of a limit existing is as follows: if for each possible choice of $\varepsilon > 0$, no matter how small, we can show that for all x sufficiently close to c (though not equal to c), the value of $f(x)$ will be within distance ε of L, then we will say that the limit of f as x goes to c is L. We will use δ to denote the measure of closeness of x to c. Then, if for every possible choice of $\varepsilon > 0$, no matter how small, we can show that there is some $\delta > 0$ such that for all x within distance δ of c (though not including c itself), the value of $f(x)$ will be within distance ε of L, we will say that the limit of f as x goes to c is L. To say that $f(x)$ is within distance ε of L is to say that $|f(x) - L| < \varepsilon$, and to say that x is within distance δ of c, but x is not equal to c, is to say that $x \in I - \{c\}$ and $|x - c| < \delta$. We then see that the rigorous way to say "the value of $f(x)$ gets closer and closer to a number L as the value of x gets closer and closer to c" is to say that for each $\varepsilon > 0$, there is some $\delta > 0$ such that for all $x \in I - \{c\}$ such that $|x - c| < \delta$, it is the case that $|f(x) - L| < \varepsilon$. As seen in Figure 3.2.2, the expression "for all $x \in I - \{c\}$ such that $|x - c| < \delta$, it is the case that $|f(x) - L| < \varepsilon$" can be viewed graphically by saying that $f(x)$ is within a band of width 2ε centered at L whenever x is in $I - \{c\}$ and x is within a band of width 2δ centered at c.

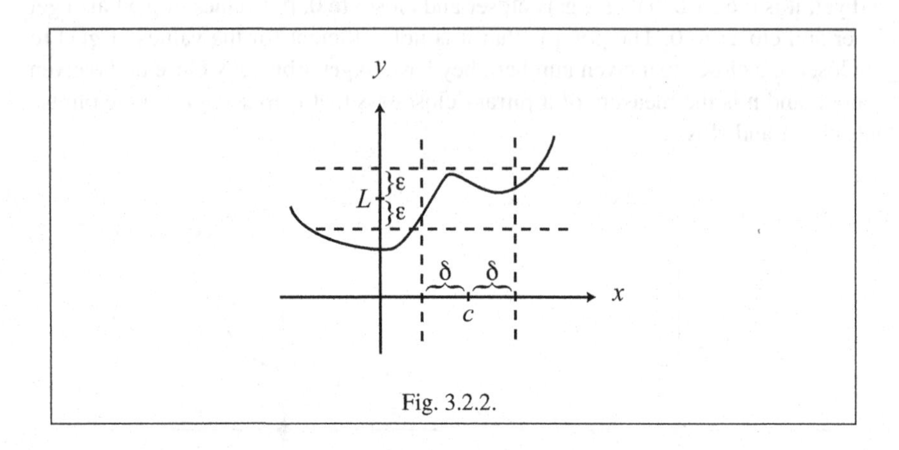

Fig. 3.2.2.

The above considerations lead us to the following definition.

Definition 3.2.1. Let $I \subseteq \mathbb{R}$ be an open interval, let $c \in I$, let $f\colon I - \{c\} \to \mathbb{R}$ be a function and let $L \in \mathbb{R}$. The number L is the **limit** of f as x goes to c, written

$$\lim_{x \to c} f(x) = L,$$

if for each $\varepsilon > 0$, there is some $\delta > 0$ such that $x \in I - \{c\}$ and $|x - c| < \delta$ imply $|f(x) - L| < \varepsilon$. If $\lim_{x \to c} f(x) = L$, we also say that f **converges** to L as x goes to c. If f converges to some real number as x goes to c, we say that $\lim_{x \to c} f(x)$ exists. △

Definition 3.2.1 must be used precisely as stated. First, it is very important that we use a symbol such as ε to measure the closeness of $f(x)$ to L, not a specific numerical value, no matter how small. A symbol such as ε could be any possible positive number, and is not any one specific numerical value. If we find a "δ" for only some specific numerical values of ε, we will not have proved that the limit exists.

Second, the order of the quantifiers in the definition of limits is absolutely crucial. The definition of the limit of a function can be written in logical symbols as

$$(\forall \varepsilon > 0)(\exists \delta > 0)[(x \in I - \{c\} \wedge |x - c| < \delta) \to |f(x) - L| < \varepsilon].$$

The order of the quantifiers cannot be changed. If we want to prove that $\lim_{x \to c} f(x) = L$, the proof must start with choosing an arbitrary $\varepsilon > 0$. Next, after possible argumentation, a value of $\delta > 0$ must be given, where δ may depend upon ε, c and f. We then choose an arbitrary $x \in I - \{c\}$ such that $|x - c| < \delta$. (Observe that saying only

"$|x-c| < \delta$" guarantees neither that $x \in I$ nor that $x \neq c$, and hence we need to say "$x \in I - \{c\}$ and $|x-c| < \delta$" when we are describing x.) Finally, again after possible argumentation, we must deduce that $|f(x) - L| < \varepsilon$. It is important that the arbitrary choices are indeed arbitrary. A typical proof that $\lim_{x \to c} f(x) = L$ must therefore have the following form:

Proof. Let $\varepsilon > 0$.

$\vdots$

(argumentation)

$\vdots$

Let $\delta = \ldots$.

$\vdots$

(argumentation)

$\vdots$

Suppose that $x \in I - \{c\}$ and $|x - c| < \delta$.

$\vdots$

(argumentation)

$\vdots$

Therefore $|f(x) - L| < \varepsilon$. □

Such proofs are often called "ε–δ proofs." Learning to construct correct ε–δ proofs may take some practice, but the effort is very worthwhile, because this type of argument will be used in many places in real analysis.

We will see some examples of limits shortly, but first we need a very important lemma. Although it is not stated in Definition 3.2.1 that the number "L" in the definition is unique, it turns out that if $\lim_{x \to c} f(x) = L$ for some $L \in \mathbb{R}$, then there is only one such number L. In other words, if a function has a limit as x goes to c, that means there is a *single* number L that $f(x)$ is getting closer and closer to; if there is no such number, then there is no limit.

Lemma 3.2.2. *Let $I \subseteq \mathbb{R}$ be an open interval, let $c \in I$ and let $f : I - \{c\} \to \mathbb{R}$ be a function. If $\lim_{x \to c} f(x) = L$ for some $L \in \mathbb{R}$, then L is unique.*

Proof. Suppose that $\lim_{x \to c} f(x) = L_1$ and $\lim_{x \to c} f(x) = L_2$ for some $L_1, L_2 \in \mathbb{R}$ such that $L_1 \neq L_2$. Let $\varepsilon = \frac{|L_1 - L_2|}{2}$. Then $\varepsilon > 0$. By the definition of limits there is some $\delta_1 > 0$ such that $x \in I - \{c\}$ and $|x - c| < \delta_1$ imply $|f(x) - L_1| < \varepsilon$, and there is some $\delta_2 > 0$ such that $x \in I - \{c\}$ and $|x - c| < \delta_2$ imply $|f(x) - L_2| < \varepsilon$. Let $\delta = \min\{\delta_1, \delta_2\}$. Choose some $x \in I - \{c\}$ such that $|x - c| < \delta$; such x exists by Exercise 2.3.8. Then $|x - c| < \delta_1$ and $|x - c| < \delta_2$, and hence

$$|L_1 - L_2| = |L_1 - f(x) + f(x) - L_2| \leq |L_1 - f(x)| + |f(x) - L_2|$$

$$= |f(x) - L_1| + |f(x) - L_2| < \varepsilon + \varepsilon = 2\varepsilon = 2\frac{|L_1 - L_2|}{2} = |L_1 - L_2|,$$

which is a contradiction. We deduce that if $\lim_{x \to c} = L$ for some $L \in \mathbb{R}$, then L is unique. □

Because of Lemma 3.2.2 we can refer to "the" limit of a function at c, if the limit exists.

Example 3.2.3. In the first three parts of this example, we will first do some scratch work prior to the actual proof. It is often necessary to do scratch work as a first step when working with the ε–δ definition of limits, though it is important to avoid confusing the scratch work (which often involves working backwards from the desired conclusion) with the proof.

(1) We will prove that $\lim_{x \to 4} (5x + 1) = 21$. (In principle, we should have stated that the function under consideration is $f\colon \mathbb{R} - \{4\} \to \mathbb{R}$ defined by $f(x) = 5x + 1$ for all $x \in \mathbb{R}$, but that is implicitly clear, and we will not write out the name of the function in other similar situations.)

Scratch Work We work backwards for our scratch work. We want to conclude that $|(5x + 1) - 21| < \varepsilon$, which is the same as $|5x - 20| < \varepsilon$, which is equivalent to $5|x - 4| < \varepsilon$, which in turn is the same as $|x - 4| < \frac{\varepsilon}{5}$. We now see that $\delta = \frac{\varepsilon}{5}$ ought to work, though we will only be sure that it works when we try to write the proof up properly, which means "forwards."

Actual Proof Let $\varepsilon > 0$. Let $\delta = \frac{\varepsilon}{5}$. Suppose that $x \in \mathbb{R} - \{4\}$ and $|x - 4| < \delta$. Then

$$|(5x+1) - 21| = |5x - 20| = 5|x - 4| < 5\delta = 5 \cdot \frac{\varepsilon}{5} = \varepsilon.$$

(2) We will prove that $\lim_{x \to 3} (x^2 - 1) = 8$.

Scratch Work Again, we work backwards for our scratch work. We want $|(x^2 - 1) - 8| < \varepsilon$, which is $|x^2 - 9| < \varepsilon$, which is $|(x - 3)(x + 3)| < \varepsilon$, which is $|x - 3| < \frac{\varepsilon}{|x+3|}$. We cannot take $\delta = \frac{\varepsilon}{|x+3|}$, because δ must be a number, whereas "x" would not have a fixed value at this point in the proof. The number δ can depend upon ε and 3 (which is "c" here), both of which come before δ in the proper order of the quantifiers in the definition of limits, but "x" comes after δ in the definition, and indeed the choice of x depends upon δ, so δ cannot depend upon x. Fortunately, we can define δ properly by using the following trick, which will give us a bound on the possible values of $|x + 3|$. Suppose that $|x - 3| < 1$. Then $-1 < x - 3 < 1$, which implies that $2 < x < 4$, and therefore $5 < x + 3 < 7$, and hence $5 < |x + 3| < 7$. We now choose $\delta = \min\{\frac{\varepsilon}{7}, 1\}$.

Actual Proof Let $\varepsilon > 0$. Let $\delta = \min\{\frac{\varepsilon}{7}, 1\}$. Suppose that $x \in \mathbb{R} - \{3\}$ and $|x - 3| < \delta$. Then $|x - 3| < 1$, which implies that $-1 < x - 3 < 1$, and therefore $2 < x < 4$, and hence $5 < x + 3 < 7$, and we conclude that $5 < |x + 3| < 7$. Then

$$|(x^2 - 1) - 8| = |x^2 - 9| = |x - 3| \cdot |x + 3| < \delta \cdot 7 \le \frac{\varepsilon}{7} \cdot 7 = \varepsilon.$$

(3) We will prove that $\lim_{x \to 0} \frac{1}{x}$ does not exist.

Scratch Work The graph of the function $f\colon \mathbb{R} - \{0\} \to \mathbb{R}$ defined by $f(x) = \frac{1}{x}$ for all $x \in \mathbb{R} - \{0\}$ is seen in Figure 3.2.3 (i), and it is evident that as x goes to 0 from the right-hand side, the values of $f(x)$ go to positive infinity, and that as x goes to 0 from the left-hand side, the values of $f(x)$ go to negative infinity. We now see two intuitive reasons why $\lim_{x \to 0} \frac{1}{x}$ does not exist. First, it is important to stress that when we write "$\lim_{x \to c} f(x) = L$," we always mean that L is a real number. Although we can write the symbol "∞" to represent "infinity," the symbol ∞ is not a real number. If one writes "$\lim_{x \to c} f(x) = \infty$" (which we will not do in the present chapter, but which we will do in Chapter 6), then one is not actually writing a limit in the sense that we are discussing at present; the symbol "∞" in this context means that the value of $f(x)$ grows without bound. The other intuitive reason that $\lim_{x \to 0} \frac{1}{x}$ does not exist is that as x goes to 0 from the right-hand side the values of $f(x)$ do one thing, and as x goes to 0 from the left-hand side the values of $f(x)$ do a different thing. For a limit to exist, the values of $f(x)$ must do the same thing no matter how x approaches c, as will be clarified by Lemma 3.2.17 below.

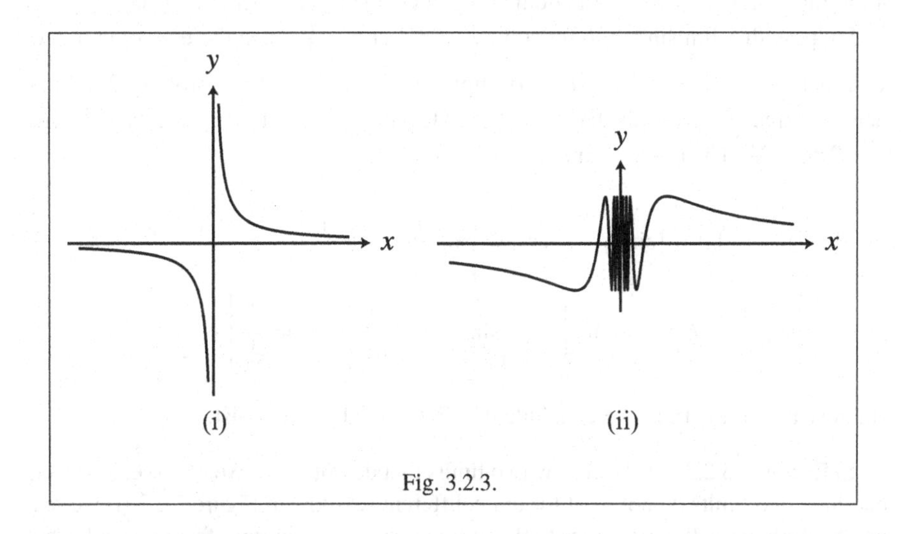

Fig. 3.2.3.

A proof that a limit does not exist requires the ε–δ definition just as much as a proof that a limit does exist. In the present case we will use proof by contradiction. More precisely, we suppose that $\lim_{x \to 0} \frac{1}{x} = L$ for some $L \in \mathbb{R}$. We will then find some $\varepsilon > 0$ for which there is no appropriate δ, which means that for each $\delta > 0$, there is some $x \in \mathbb{R} - \{0\}$ such that $|x - 0| < \delta$ and yet $\left|\frac{1}{x} - L\right| \not< \varepsilon$. Let $\delta > 0$. We now work backwards, and notice that what we need is to have $\frac{1}{x} - L \leq -\varepsilon$ or $\frac{1}{x} - L \geq \varepsilon$. It suffices to find x such that one of these inequalities holds. There are three cases, depending upon whether $L > 0$ or $L = 0$ or $L < 0$. Suppose that $L > 0$. We want

$\frac{1}{x} - L \geq \varepsilon$, which is the same as $\frac{1}{x} \geq L + \varepsilon$, and so we need $x \leq \frac{1}{L+\varepsilon}$. This last step works with any choice of ε because $L > 0$. We also need to have $|x - 0| < \delta$, and so we will choose $x = \min\{\frac{\delta}{2}, \frac{1}{L+\varepsilon}\}$. We will not give the full details of the cases where $L < 0$ and where $L = 0$, except to note that when $L < 0$ we need to choose some $\varepsilon > 0$ so that $L + \varepsilon < 0$, for example we could choose $\varepsilon = \frac{|L|}{2}$, and so we might as well use that ε when $L > 0$ as well; when $L = 0$ we cannot use $\varepsilon = \frac{|L|}{2}$, because we need $\varepsilon > 0$, but any choice of positive ε will work, for example $\varepsilon = 1$.

Actual Proof Suppose that $\lim_{x\to 0} \frac{1}{x} = L$ for some $L \in \mathbb{R}$. Let $\varepsilon = \frac{|L|}{2}$ if $L \neq 0$, and let $\varepsilon = 1$ if $L = 0$. We consider the case when $L > 0$; the other cases are similar, and the details are left to the reader. Let $\delta > 0$. Because $L > 0$, then $L + \varepsilon > 0$. Let $x = \min\{\frac{\delta}{2}, \frac{1}{L+\varepsilon}\}$. Then $x \in (0, \infty)$ and $|x - 0| \leq \frac{\delta}{2} < \delta$. On the other hand, because $x \leq \frac{1}{L+\varepsilon}$, it follows that $L + \varepsilon \leq \frac{1}{x}$, and hence $\frac{1}{x} - L \geq \varepsilon$, which implies that $|\frac{1}{x} - L| \not< \varepsilon$.

(4) We will prove that $\lim_{x\to 0} \sin\frac{1}{x}$ does not exist. We have not yet defined the trigonometric functions rigorously (we will do so in Section 7.3), but we assume that the reader is informally familiar with $\sin x$ and its basic properties, and if we assume these properties, we can give the following proof. See Figure 3.2.3 (ii) for the graph of the function $g\colon \mathbb{R} - \{0\} \to \mathbb{R}$ defined by $g(x) = \sin\frac{1}{x}$ for all $x \in \mathbb{R} - \{0\}$.

Suppose that $\lim_{x\to 0} \sin\frac{1}{x} = L$ for some $L \in \mathbb{R}$. Let $\varepsilon = \frac{1}{2}$. Then there is some $\delta > 0$ such that $x \in I - \{c\}$ and $|x - c| < \delta$ imply $|f(x) - L| < \frac{1}{2}$. By Corollary 2.6.8 (2) there is some $n \in \mathbb{N}$ such that $\frac{1}{n} < 2\delta\pi$. Hence $\frac{1}{2\pi n} < \delta$. Let $x_1 = 2\pi n + \frac{\pi}{2}$ and $x_2 = 2\pi n + \frac{3\pi}{2}$. Then $\frac{1}{x_1} < \delta$ and $\frac{1}{x_2} < \delta$. Therefore

$$2 = |1 - (-1)| = |\sin x_1 - \sin x_2| = \left|\sin\frac{1}{\frac{1}{x_1}} - \sin\frac{1}{\frac{1}{x_2}}\right|$$

$$= \left|\sin\frac{1}{\frac{1}{x_1}} - L + L - \sin\frac{1}{\frac{1}{x_2}}\right| \leq \left|\sin\frac{1}{\frac{1}{x_1}} - L\right| + \left|L - \sin\frac{1}{\frac{1}{x_2}}\right| < \frac{1}{2} + \frac{1}{2} = 1,$$

which is a contradiction. We conclude that $\lim_{x\to 0} \sin\frac{1}{x}$ does not exist. ◊

In Example 3.2.3 (3) (4) we saw two limits that do not exist. We observe, however, that these two limits do not exist for quite different reasons. In Figure 3.2.3 (i) we see that the problem is that as x goes to 0, the function goes to infinity from one side and negative infinity from the other side, whereas in Part (ii) of the figure the function oscillates more and more rapidly with values between 1 and -1 as x gets closer to 0, and hence there is no single value toward which the values of the function get closer as x gets closer to 0.

The ε–δ definition of limits given in Definition 3.2.1 is not always easy to use in practice, and in the rest of this section we will see some lemmas and theorems that allow us to compute limits more easily in certain situations (the proofs of these results make use of the ε–δ approach, but once these results have been proved, we can often apply them without further use of ε–δ). We start with some preliminary results.

Our next result states that if a function has a positive limit at c, then the function must be positive near c, and similarly for a negative limit.

Theorem 3.2.4 (Sign-Preserving Property for Limits). *Let $I \subseteq \mathbb{R}$ be an open interval, let $c \in I$ and let $f: I - \{c\} \to \mathbb{R}$ be a function. Suppose that $\lim_{x \to c} f(x)$ exists.*

1. *If $\lim_{x \to c} f(x) > 0$, then there is some $M > 0$ and some $\delta > 0$ such that $x \in I - \{c\}$ and $|x - c| < \delta$ imply $f(x) > M$.*
2. *If $\lim_{x \to c} f(x) < 0$, then there is some $N < 0$ and some $\delta > 0$ such that $x \in I - \{c\}$ and $|x - c| < \delta$ imply $f(x) < N$.*

Proof. We will prove Part (1); the other part is similar, and we omit the details.

(1) Suppose that $\lim_{x \to c} f(x) > 0$. Let $L = \lim_{x \to c} f(x)$. Let $M = \frac{L}{2}$. Then $M > 0$. By the definition of limits there is some $\delta > 0$ such that $x \in I - \{c\}$ and $|x - c| < \delta$ imply $|f(x) - L| < \frac{L}{2}$. Then $x \in I - \{c\}$ and $|x - c| < \delta$ imply that $-\frac{L}{2} < f(x) - L < \frac{L}{2}$, and hence $\frac{L}{2} < f(x)$, and therefore $f(x) > M$. □

Recall from Definition 1.7.7 (3) or Definition 2.2.2 (3) the concept of a subset of $\mathbb{R}$ being bounded. It was proved in Exercise 2.3.11 that a subset $A \subseteq \mathbb{R}$ is bounded if and only if there is some $M \in \mathbb{R}$ such that $|x| \le M$ for all $x \in A$. We now define what it means for functions to be bounded. Although this definition does not make use of limits, it will be useful in our study of limits, and hence we give it here.

When discussing the boundedness of functions, we will see that the boundedness occurs entirely in the codomain, and hence our definition of bounded functions requires that the codomains of the functions under consideration are sets that are susceptible to the notion of boundedness, which for our purposes means that we need codomains that are subsets of the real numbers. The domain of a bounded function could be any type of set.

Definition 3.2.5. Let A be a set, let $B \subseteq \mathbb{R}$ be a set and let $f: A \to B$ be a function. The function f is **bounded** if the set $f(A)$ is bounded; that is, if there is some $M \in \mathbb{R}$ such that $|f(x)| \le M$ for all $x \in A$. The number M is called a **bound** of f. △

Example 3.2.6. Let $h: [0, 1] \to \mathbb{R}$ be defined by

$$h(x) = \begin{cases} 1, & \text{if } x = 0 \\ \frac{1}{x}, & \text{if } x \in (0, 1]. \end{cases}$$

It is intuitively evident that h is not bounded by looking at its graph, but we need to provide a proof, which we do by showing that for each $M \in \mathbb{R}$, there is some $x \in [0, 1]$ such that $M < |f(x)|$. Let $M \in \mathbb{R}$. There are now two cases. First, suppose that $M \le 0$. Then $M < 1 = |h(0)|$. Second, suppose that $M > 0$. Let $x = M + 1$. Then $x \ge 1$, and hence $\frac{1}{x} \in (0, 1]$. We then see that $|h(\frac{1}{x})| = x > M$. Hence h is not bounded. ◊

Observe that if a function is bounded, the bound of the function is not unique, because any number larger than a bound is also a bound. Clearly any bound of a function is non-negative, and it is always possible to choose a positive bound, which we will do when convenient.

We now have two lemmas that involve limits and boundedness. Clearly a function can have a limit at some number c even if the function is not bounded, for example the function in Example 3.2.3 (1), and a function can be bounded but not have a limit at some number c. There are, however, some relations between limits and boundedness, as we now see.

Lemma 3.2.7. *Let $I \subseteq \mathbb{R}$ be an open interval, let $c \in I$ and let $f\colon I - \{c\} \to \mathbb{R}$ be a function. If $\lim_{x\to c} f(x)$ exists, then there is some $\delta > 0$ such that the restriction of f to $(I - \{c\}) \cap (c-\delta, c+\delta)$ is bounded.*

Proof. Let $L = \lim_{x\to c} f(x)$. Then there is some $\delta > 0$ such that $x \in I - \{c\}$ and $|x - c| < \delta$ imply $|f(x) - L| < 1$. Suppose that $x \in I - \{c\}$ and $|x - c| < \delta$. By Lemma 2.3.9 (7) we see that $|f(x)| - |L| < 1$, and hence $|f(x)| < |L| + 1$. Therefore the restriction of f to $(I - \{c\}) \cap (c-\delta, c+\delta)$ is bounded, with bound $|L| + 1$. □

Our next lemma is a useful result about limits that makes use of boundedness. The reader is asked in Exercise 3.2.4 to show that this hypothesis of boundedness in this lemma cannot be dropped.

Lemma 3.2.8. *Let $I \subseteq \mathbb{R}$ be an open interval, let $c \in I$ and let $f, g\colon I - \{c\} \to \mathbb{R}$ be functions. Suppose that $\lim_{x\to c} f(x) = 0$, and that g is bounded. Then $\lim_{x\to c} f(x)g(x) = 0$.*

Proof. Let $\varepsilon > 0$. Because g is bounded, there is some $M \in \mathbb{R}$ such that $|g(x)| \leq M$ for all $x \in I - \{c\}$; we may assume that $M > 0$. Because $\lim_{x\to c} f(x) = 0$, there is some $\delta > 0$ such that $x \in I - \{c\}$ and $|x - c| < \delta$ imply $|f(x) - 0| < \frac{\varepsilon}{M}$. Suppose that $x \in I - \{c\}$ and $|x - c| < \delta$. Then

$$|f(x)g(x) - 0| = |f(x)g(x)| = |f(x)| \cdot |g(x)| < \frac{\varepsilon}{M} \cdot M = \varepsilon. \qquad \square$$

One of the most convenient situations in which to prove that a function has a limit is when the function is built up out of simpler functions, the limits of which we are more easily able to evaluate. The following definition, which will be useful throughout our study of real analysis, states how to add, subtract, multiply and divide functions. As was the case when we defined bounded functions in Definition 3.2.5, the addition, subtraction, multiplication and division of functions occurs entirely in the codomain, and hence the definition of addition, etc., of functions requires that the codomains of the functions under consideration are sets where we can perform such operations, which for our purposes means that we need codomains that are the real numbers. The domains of such functions could be any type of set.

Definition 3.2.9. Let A, B be sets, let $f\colon A \to \mathbb{R}$ and $g\colon B \to \mathbb{R}$ be functions and let $k \in \mathbb{R}$.

1. The function $f + g\colon A \cap B \to \mathbb{R}$ is defined by $[f + g](x) = f(x) + g(x)$ for all $x \in A \cap B$.

2. The function $f-g\colon A\cap B\to\mathbb{R}$ is defined by $[f-g](x)=f(x)-g(x)$ for all $x\in A\cap B$.
3. The function $kf\colon A\to\mathbb{R}$ is defined by $[kf](x)=kf(x)$ for all $x\in A$.
4. The function $fg\colon A\cap B\to\mathbb{R}$ is defined by $[fg](x)=f(x)\cdot g(x)$ for all $x\in A\cap B$.
5. Let $C=(A\cap B)-\{b\in B\mid g(b)=0\}$. The function $\frac{f}{g}\colon C\to\mathbb{R}$ is defined by $\left[\frac{f}{g}\right](x)=\frac{f(x)}{g(x)}$ for all $x\in C$.
6. The function $|f|\colon A\to\mathbb{R}$ is defined by $|f|(x)=|f(x)|$ for all $x\in A$. △

We now see that limits behave nicely with respect to the addition, subtraction, multiplication and division of functions. The relation of limits to the absolute value of a function is discussed in Exercise 3.2.9.

Theorem 3.2.10. *Let $I\subseteq\mathbb{R}$ be an open interval, let $c\in I$, let $f,g\colon I-\{c\}\to\mathbb{R}$ be functions and let $k\in\mathbb{R}$. Suppose that $\lim_{x\to c}f(x)$ and $\lim_{x\to c}g(x)$ exist.*

1. *$\lim_{x\to c}[f+g](x)$ exists and $\lim_{x\to c}[f+g](x)=\lim_{x\to c}f(x)+\lim_{x\to c}g(x)$.*
2. *$\lim_{x\to c}[f-g](x)$ exists and $\lim_{x\to c}[f-g](x)=\lim_{x\to c}f(x)-\lim_{x\to c}g(x)$.*
3. *$\lim_{x\to c}[kf](x)$ exists and $\lim_{x\to c}[kf](x)=k\lim_{x\to c}f(x)$.*
4. *$\lim_{x\to c}[fg](x)$ exists and $\lim_{x\to c}[fg](x)=[\lim_{x\to c}f(x)]\cdot[\lim_{x\to c}g(x)]$.*
5. *If $\lim_{x\to c}g(x)\neq 0$, then $\lim_{x\to c}\left[\frac{f}{g}\right](x)$ exists and $\lim_{x\to c}\left[\frac{f}{g}\right](x)=\dfrac{\lim_{x\to c}f(x)}{\lim_{x\to c}g(x)}$.*

Proof. We will prove Parts (1), (3) and (4), leaving the rest to the reader in Exercise 3.2.6.

Let $L=\lim_{x\to c}f(x)$ and $M=\lim_{x\to c}g(x)$.

(1) Let $\varepsilon>0$. Then there is some $\delta_1>0$ such that $x\in I-\{c\}$ and $|x-c|<\delta_1$ imply $|f(x)-L|<\frac{\varepsilon}{2}$, and there is some $\delta_2>0$ such that $x\in I-\{c\}$ and $|x-c|<\delta_2$ imply $|g(x)-M|<\frac{\varepsilon}{2}$. Let $\delta=\min\{\delta_1,\delta_2\}$. Suppose that $x\in I-\{c\}$ and $|x-c|<\delta$. Then

$$
\begin{aligned}
|[f+g](x)-(L+M)| &= |(f(x)-L)+(g(x)-M)|\le|f(x)-L|+|g(x)-M|\\
&<\frac{\varepsilon}{2}+\frac{\varepsilon}{2}=\varepsilon.
\end{aligned}
$$

(3) By Exercise 3.2.3 we know that $\lim_{x\to c}[f(x)-L]=0$. Let $h\colon I-\{c\}\to\mathbb{R}$ be defined by $h(x)=k$ for all $x\in I-\{c\}$. Then h is bounded, with bound $|k|$. Hence, Lemma 3.2.8 implies that $\lim_{x\to c}k[f(x)-L]=0$, which then implies that $\lim_{x\to c}[kf(x)-kL]=0$. Using Exercise 3.2.3 again we deduce that $\lim_{x\to c}kf(x)=kL$.

(4) Let $\varepsilon>0$. By Lemma 3.2.7 there is some $\delta_1>0$ such that the restriction of g to $(I-\{c\})\cap(c-\delta_1,c+\delta_1)$ is bounded. Hence there is some $B\in\mathbb{R}$ such that $|g(x)|\le B$ for all $x\in(I-\{c\})\cap(c-\delta_1,c+\delta_2)$. We may assume that $B>0$. Then $B+|L|>0$. There is some $\delta_2>0$ such that $x\in I-\{c\}$ and $|x-c|<\delta_2$ imply

$|f(x) - L| < \frac{\varepsilon}{B+|L|}$, and there is some $\delta_3 > 0$ such that $x \in I - \{c\}$ and $|x - c| < \delta_3$ imply $|g(x) - M| < \frac{\varepsilon}{B+|L|}$. Let $\delta = \min\{\delta_1, \delta_2, \delta_3\}$. Suppose that $x \in I - \{c\}$ and $|x - c| < \delta$. Then

$$\begin{aligned} |[fg](x) - LM| &= |f(x)g(x) - LM| = |f(x)g(x) - g(x)L + g(x)L - LM| \\ &\leq |g(x)| \cdot |f(x) - L| + |L| \cdot |g(x) - M| \\ &< B\frac{\varepsilon}{B+|L|} + |L| \cdot \frac{\varepsilon}{B+|L|} = \varepsilon. \end{aligned}$$ □

Example 3.2.11. Combining Theorem 3.2.10 (3) (4), Example 3.2.3 (2) and Exercise 3.2.1, we see that

$$\begin{aligned} \lim_{x\to 3}(12x^3 + 8x^2 - 12x - 8) &= \lim_{x\to 3}[4(x^2 - 1)(3x + 2)] \\ &= 4 \cdot [\lim_{x\to 3}(x^2 - 1)] \cdot [\lim_{x\to 3}(3x + 2)] = 4 \cdot 8 \cdot 11 = 352. \end{aligned}$$ ◊

The following theorem concerns the relation between limits and the composition of functions.

Theorem 3.2.12. *Let $I, J \subseteq \mathbb{R}$ be open intervals, let $c \in I$, let $d \in J$ and let $g\colon I - \{c\} \to J - \{d\}$ and $f\colon J - \{d\} \to \mathbb{R}$ be functions. Suppose that $\lim_{y\to c} g(y) = d$ and that $\lim_{x\to d} f(x)$ exist. Then $\lim_{y\to c}(f \circ g)(y)$ exists, and $\lim_{y\to c}(f \circ g)(y) = \lim_{x\to d} f(x)$.*

Proof. Left to the reader in Exercise 3.2.10. □

We now turn to the relation of limits to inequalities between functions.

Theorem 3.2.13. *Let $I \subseteq \mathbb{R}$ be an open interval, let $c \in I$ and let $f, g\colon I - \{c\} \to \mathbb{R}$ be functions. Suppose that $f(x) \leq g(x)$ for all $x \in I - \{c\}$. If $\lim_{x\to c} f(x)$ and $\lim_{x\to c} g(x)$ exist, then $\lim_{x\to c} f(x) \leq \lim_{x\to c} g(x)$.*

Proof. Left to the reader in Exercise 3.2.11. □

Our next result provides a convenient way to prove the existence of the limit of a tricky function by "trapping it" between two functions whose limits can be evaluated more easily. The reader may be familiar with the version of this theorem that holds for sequences (which we will see in Theorem 8.2.12); indeed, many of the basic facts about limits of functions have analogs for limits of sequences, as the reader will see in Section 8.2.

Theorem 3.2.14 (Squeeze Theorem for Functions). *Let $I \subseteq \mathbb{R}$ be an open interval, let $c \in I$ and let $f, g, h\colon I - \{c\} \to \mathbb{R}$ be functions. Suppose that $f(x) \leq g(x) \leq h(x)$ for all $x \in I - \{c\}$. If $\lim_{x\to c} f(x) = L = \lim_{x\to c} h(x)$ for some $L \in \mathbb{R}$, then $\lim_{x\to c} g(x)$ exists and $\lim_{x\to c} g(x) = L$.*

Proof. Suppose that $\lim_{x\to c} f(x) = L = \lim_{x\to c} h(x)$ for some $L \in \mathbb{R}$. Let $\varepsilon > 0$. There is some $\delta_1 > 0$ such that $x \in I - \{c\}$ and $|x - c| < \delta_1$ imply $|f(x) - L| < \varepsilon$, and there is some $\delta_2 > 0$ such that $x \in I - \{c\}$ and $|x - c| < \delta_2$ imply $|h(x) - L| < \varepsilon$. Let $\delta = \min\{\delta_1, \delta_2\}$. Suppose that $x \in I - \{c\}$ and $|x - c| < \delta$. Then $|f(x) - L| < \varepsilon$ and $|h(x) - L| < \varepsilon$. It follows that $L - \varepsilon < f(x) < L + \varepsilon$ and $L - \varepsilon < h(x) < L + \varepsilon$. Hence

$$L - \varepsilon < f(x) \leq g(x) \leq h(x) < L + \varepsilon,$$

which implies $|g(x) - L| < \varepsilon$. □

We conclude this section with a brief discussion of "one-sided" limits, which are limits at a number c that involve approaching c from only one side, either from the right (that is, via numbers larger than c) or from the left (that is, via numbers smaller than c). Such one-sided limits are useful at the endpoints of closed intervals. For example, consider the function $g: \mathbb{R} \to \mathbb{R}$ defined at the start of this section, the graph of which is seen in Figure 3.2.1 (ii). Whereas the limit of this function does not exist as x goes to 0, if we restrict our attention to $x > 0$, then the limit of that part of the function as x goes to 0 ought to be 3; that is a right-hand limit. Similarly, if we restrict our attention to $x < 0$, then the limit of that part of the function as x goes to 0 ought to be -3; that is a left-hand limit.

Observe that the interval I in the following definition is not necessarily open.

Definition 3.2.15. Let $I \subseteq \mathbb{R}$ be an interval, let $c \in I$, let $f: I - \{c\} \to \mathbb{R}$ be a function and let $L \in \mathbb{R}$.

1. Suppose that c is not a right endpoint of I. The number L is the **right-hand limit** of f at c, written
$$\lim_{x \to c^+} f(x) = L,$$
if for each $\varepsilon > 0$, there is some $\delta > 0$ such that $x \in I - \{c\}$ and $c < x < c + \delta$ imply $|f(x) - L| < \varepsilon$. If $\lim_{x \to c^+} f(x) = L$, we also say that f **converges** to L as x goes to c from the right. If f converges to some real number as x goes to c from the right, we say that $\lim_{x \to c^+} f(x)$ exists.
2. Suppose that c is not a left endpoint of I. The number L is the **left-hand limit** of f at c, written
$$\lim_{x \to c^-} f(x) = L,$$
if for each $\varepsilon > 0$, there is some $\delta > 0$ such that $x \in I - \{c\}$ and $c - \delta < x < c$ imply $|f(x) - L| < \varepsilon$. If $\lim_{x \to c^-} f(x) = L$, we also say that f **converges** to L as x goes to c from the left. If f converges to some real number as x goes to c from the left, we say that $\lim_{x \to c^-} f(x)$ exists.
3. A **one-sided limit** is either a right-hand limit or a left-hand limit. △

Example 3.2.16. We examine each of $\lim_{x \to 0^+} \frac{|x|}{x}$, and $\lim_{x \to 0^-} \frac{|x|}{x}$ and $\lim_{x \to 0} \frac{|x|}{x}$. The function under consideration here is $f: \mathbb{R} - \{0\} \to \mathbb{R}$ defined by $f(x) = \frac{|x|}{x}$ for all $x \in \mathbb{R} - \{0\}$. The key observation is that

$$\frac{|x|}{x} = \begin{cases} 1, & \text{if } x > 0 \\ -1, & \text{if } x < 0. \end{cases}$$

It then follows that $\lim_{x\to 0+} \frac{|x|}{x} = \lim_{x\to 0+} 1 = 1$, and that $\lim_{x\to 0^-} \frac{|x|}{x} = \lim_{x\to 0^-} -1 = -1$. On the other hand, we see that $\lim_{x\to 0} \frac{|x|}{x}$ does not exist, as follows. Suppose to the contrary that $\lim_{x\to 0} \frac{|x|}{x} = L$ for some $L \in \mathbb{R}$. It cannot be the case that L equals both 1 and -1. Without loss of generality, assume that $L \neq 1$. Let $\varepsilon = \frac{|L-1|}{2}$. Then $\varepsilon > 0$. For any $\delta > 0$, we can choose $y \in (0, \delta)$, and then $f(y) = 1$, which implies $|f(y) - L| = |1 - L| = 2\varepsilon > \varepsilon$. Hence we cannot find the required δ for the given ε, which is a contradiction to the assumption that $\lim_{x\to 0} \frac{|x|}{x} = L$. It follows that $\lim_{x\to 0} \frac{|x|}{x}$ does not exist. ◊

Example 3.2.16 shows that $\lim_{x\to c^+} f(x)$ and $\lim_{x\to c^-} f(x)$ can both exist and be different. We now see what happens when both of these one-sided limits exist and are equal.

Lemma 3.2.17. *Let $I \subseteq \mathbb{R}$ be an open interval, let $c \in I$ and let $f : I - \{c\} \to \mathbb{R}$ be a function. Then $\lim_{x\to c} f(x)$ exists if and only if $\lim_{x\to c^+} f(x)$ and $\lim_{x\to c^-} f(x)$ exist and are equal, and if these three limits exist then they are equal.*

Proof. Suppose that $\lim_{x\to c} f(x)$ exists. Let $L = \lim_{x\to c} f(x)$. Let $\varepsilon > 0$. Then there is some $\delta > 0$ such that $x \in I - \{c\}$ and $|x - c| < \delta$ imply $|f(x) - L| < \varepsilon$. Suppose that $x \in I - \{c\}$ and $c < x < c + \delta$. Then $|x - c| < \delta$, and it follows that $|f(x) - L| < \varepsilon$. We deduce that $\lim_{x\to c^+} f(x) = L$. A similar argument shows that $\lim_{x\to c^-} f(x) = L$, and we omit the details.

Now suppose that $\lim_{x\to c^+} f(x)$ and $\lim_{x\to c^-} f(x)$ exist and are equal. Let $M = \lim_{x\to c^+} f(x) = \lim_{x\to c^-} f(x)$. Let $\varepsilon > 0$. There is some $\delta_1 > 0$ such that $x \in I - \{c\}$ and $c < x < c + \delta_1$ imply $|f(x) - M| < \varepsilon$, and there is some $\delta_2 > 0$ such that $x \in I - \{c\}$ and $c - \delta_2 < x < c$ imply $|f(x) - M| < \varepsilon$. Let $\delta = \min\{\delta_1, \delta_2\}$. Suppose that $x \in I - \{c\}$ and $|x - c| < \delta$. Then $c - \delta < x < c + \delta$ and $x \neq c$, and hence $c - \delta_2 < x < c$ or $c < x < c + \delta_1$. In either case we deduce that $|f(x) - M| < \varepsilon$. It follows that $\lim_{x\to c} f(x) = M$. □

It can be verified that the analogs for one-sided limits of the lemmas and theorems of this section all hold; the proofs are not substantially different in the one-sided case, and we omit the details.

Reflections

The reader has now had her first taste of what real analysis is all about. The ε–δ definition of limits of functions is at the very heart of any real analysis course, for two reasons. First, limits are the conceptual basis for continuity, derivatives, and more, and hence an understanding of limits is crucial for an understanding of much of real analysis. It could be said that it is the limit concept that distinguishes analysis from algebra. Second, the ε–δ definition of limits of functions is the model for a number of other similar definitions, for example the definition of limits of sequences, and even the definition of the Riemann integral of a function, which is a more complicated definition but still involves ε and δ in analogous roles. Hence, mastering ε–δ proofs will serve the student well as preparation for the rest of the material in this text.

Some students find ε–δ proofs a bit confusing upon first encounter, especially in comparison with the material in other junior–senior-level mathematics courses such as abstract algebra. In the author's view, the problem stems from the prominent role of the quantifiers in ε–δ proofs. More generally, in the author's experience teaching a variety of proofs-based undergraduate mathematics courses, problems with quantifiers, whether due to misunderstanding or carelessness, are the source of the majority of errors that students have in the construction of rigorous proofs. In the particular case of the ε–δ definition of limits of functions, it is the existence of two quantifiers—in a particular order—that makes the difficulty in mastering the quantifiers that much greater. Fortunately, the author's experience has also led him to see that if students put in sufficient effort and care practicing ε–δ proofs, they can usually learn to formulate and write such proofs very nicely.

Exercises

Exercise 3.2.1. **[Used throughout.]** Let $m, b, c \in \mathbb{R}$. Using only the definition of limits, prove that $\lim_{x\to c}(mx+b) = mc+b$.

Exercise 3.2.2. Using only the definition of limits, prove that each of the following limits holds.

(1) $\lim_{x\to 1}(x^2+3x+5) = 9$.

(2) $\lim_{x\to 3}\frac{x^2-9}{x-3} = 6$.

Exercise 3.2.3. **[Used in Theorem 3.2.10.]** Let $I \subseteq \mathbb{R}$ be an open interval, let $c \in I$, let $f\colon I-\{c\} \to \mathbb{R}$ be a function and let $L \in \mathbb{R}$. Using only the definition of limits, prove that $\lim_{x\to c} f(x) = L$ if and only if $\lim_{x\to c}[f(x)-L] = 0$.

Exercise 3.2.4. **[Used in Section 3.2.]** Find an example of functions $f, g\colon \mathbb{R}-\{0\} \to \mathbb{R}$ such that $\lim_{x\to 0} f(x) = 0$, and that $\lim_{x\to 0}[fg](x)$ does not exist.

Exercise 3.2.5. **[Used throughout.]** Let $J \subseteq I \subseteq \mathbb{R}$ be open intervals, let $c \in J$ and let $f\colon I-\{c\} \to \mathbb{R}$ be a function. Prove that $\lim_{x\to c} f(x)$ exists if and only if $\lim_{x\to c} f|_J(x)$ exists, and if these limits exist, then they are equal.

Exercise 3.2.6. **[Used in Theorem 3.2.10.]** Prove Theorem 3.2.10 (2) (5).

Exercise 3.2.7. **[Used in Example 4.2.5.]** Let $I \subseteq \mathbb{R}$ be an open interval, let $c \in I$ and let $f, g\colon I-\{c\} \to \mathbb{R}$ be functions. Suppose that $\lim_{x\to c} f(x)$ exists and that $\lim_{x\to c} g(x)$ does not exist.

(1) Prove that $\lim_{x\to c}[f+g](x)$ does not exist.

(2) Prove that if $\lim_{x\to c} f(x) \neq 0$, then $\lim_{x\to c} [fg](x)$ does not exist.

Exercise 3.2.8. [Used in Example 4.6.1.] Let $I \subseteq \mathbb{R}$ be an open interval, let $c \in I$ and let $f : I - \{c\} \to \mathbb{R}$ be a function. Suppose that $f(x) \neq 0$ for all $x \in I - \{c\}$, and that $\lim_{x\to c} f(x) = 0$. Prove that $\lim_{x\to c} \frac{1}{f(x)}$ does not exist. [Use Exercise 3.2.1.]

Exercise 3.2.9. [Used in Section 3.2, Exercise 4.2.4 and Theorem 6.3.3.] Let $I \subseteq \mathbb{R}$ be an open interval, let $c \in I$ and let $f : I - \{c\} \to \mathbb{R}$ be a function.

(1) Let $L \in \mathbb{R}$. Prove that if $\lim_{x\to c} f(x) = L$, then $\lim_{x\to c} |f(x)| = |L|$.
(2) Prove that $\lim_{x\to c} f(x) = 0$ if and only if $\lim_{x\to c} |f(x)| = 0$.
(3) Find an example of a function $g : \mathbb{R} \to \mathbb{R}$ such that $\lim_{x\to c} |g(x)| = 1$, but that $\lim_{x\to c} g(x)$ does not exist.

Exercise 3.2.10. [Used in Theorem 3.2.12.] Prove Theorem 3.2.12.

Exercise 3.2.11. [Used in Theorem 3.2.13.] Prove Theorem 3.2.13.

Exercise 3.2.12. [Used in Lemma 4.4.1 and Theorem 4.5.2.]

(1) Let $I \subseteq \mathbb{R}$ be an open interval, let $c \in I$ and let $f : I - \{c\} \to \mathbb{R}$ be a function. Suppose that $\lim_{x\to c} f(x)$ exists. Prove that if $f(x) \geq 0$ for all $x \in I - \{c\}$, then $\lim_{x\to c} f(x) \geq 0$.
(2) If we were to assume that $f(x) > 0$ for all $x \in I - \{c\}$ as the hypothesis of Part (1) of this exercise, would it be possible to conclude that $\lim_{x\to c} f(x) > 0$? Give a proof or a counterexample.

Exercise 3.2.13. Let $I \subseteq \mathbb{R}$ be an open interval, let $c \in I$ and let $f, g : I - \{c\} \to \mathbb{R}$ be functions. Suppose that $\lim_{x\to c} f(x)$ and $\lim_{x\to c} g(x)$ exist, and that for each $\varepsilon > 0$, there is some $\delta > 0$ such that $x \in I - \{c\}$ and $|x - c| < \delta$ imply $|f(x) - g(x)| < \varepsilon$. Prove that $\lim_{x\to c} f(x) = \lim_{x\to c} g(x)$.

Exercise 3.2.14. [Used in Theorem 5.8.5.] Let $A \subseteq \mathbb{R}$ be a set, and let $f, g, h : A \to \mathbb{R}$ be functions. Suppose that f and h are bounded, and that $f(x) \leq g(x) \leq h(x)$ for all $x \in A$. Prove that g is bounded.

Exercise 3.2.15. [Used in Exercise 5.4.15.] Let $A \subseteq \mathbb{R}$ be a set, let $f : A \to \mathbb{R}$ be a function and let $k \in \mathbb{R}$. Suppose that f is bounded.

(1) Prove that kf is bounded.
(2) Prove that if $k > 0$, then

$$\text{lub}\{[kf](x) \mid x \in A\} = k \cdot \text{lub}\{f(x) \mid x \in A\},$$

and

$$\text{glb}\{[kf](x) \mid x \in A\} = k \cdot \text{glb}\{f(x) \mid x \in A\}.$$

(3) Prove that if $k < 0$, then

$$\text{lub}\{[kf](x) \mid x \in A\} = k \cdot \text{glb}\{f(x) \mid x \in A\},$$

and

$$\text{glb}\{[kf](x) \mid x \in A\} = k \cdot \text{lub}\{f(x) \mid x \in A\}.$$

Exercise 3.2.16. [Used in Exercise 5.4.16.] Let $A \subseteq \mathbb{R}$ be a non-empty set, and let $f, g \colon A \to \mathbb{R}$ be functions. Suppose that f and g are bounded.

(1) Prove that $f + g$ is bounded.
(2) Prove that

$$\text{lub}\{[f+g](x) \mid x \in A\} \le \text{lub}\{f(x) \mid x \in A\} + \text{lub}\{g(x) \mid x \in A\}.$$

(3) Find an example where the inequality in Part (2) of this exercise is strict.
(4) Prove that

$$\text{glb}\{[f+g](x) \mid x \in A\} \ge \text{glb}\{f(x) \mid x \in A\} + \text{glb}\{g(x) \mid x \in A\}.$$

(5) Find an example where the inequality in Part (4) of this exercise is strict.

Exercise 3.2.17. [Used in Section 3.2 and Section 3.3.] In the definition of limits in Definition 3.2.1, we looked at limits of the form $\lim_{x \to c} f(x)$ for functions $f \colon I - \{c\} \to \mathbb{R}$, where I is an open interval. The purpose of this exercise is to discuss whether it would be plausible to define limits for functions with other types of domains; that is, for functions $f \colon I - \{c\} \to \mathbb{R}$ where I is not necessarily an interval.

(1) Let $I = [1,2] \cup \{0\}$, and let $f \colon I - \{0\} \to \mathbb{R}$ be defined by $f(x) = x^2$ for all $x \in I - \{0\}$. If Definition 3.2.1 is used as stated with this function f and with $c = 0$, prove that $\lim_{x \to c} f(x) = r$ for every $r \in \mathbb{R}$. We conclude that Definition 3.2.1 does not work with arbitrary sets I.
(2) Let $A \subseteq \mathbb{R}$ be a non-empty set, and let $c \in \mathbb{R}$. The number c is an **accumulation point** of A if for every $\delta > 0$ there is some $x \in A - \{c\}$ such that $|x - c| < \delta$. Let $I \subseteq \mathbb{R}$ be an open interval, and let $c \in I$. Prove that c is an accumulation point of $I - \{c\}$.
(3) Let $A \subseteq \mathbb{R}$ be a non-empty set, let $c \in \mathbb{R}$ and let $f \colon A - \{c\} \to \mathbb{R}$ be a function. (Note that it is not necessarily the case that $c \in A$.) Suppose that c is an accumulation point of A. Prove that if Definition 3.2.1 is used as stated with this function f and this number c, and that if $\lim_{x \to c} f(x) = L$ for some $L \in \mathbb{R}$, then L is unique. It would therefore be possible to rewrite Definition 3.2.1 with the more general hypotheses that I is an arbitrary non-empty set, and with c an accumulation point of I.
(4) Let $B, D \subseteq \mathbb{R}$ be non-empty sets, let $c \in \mathbb{R}$ and let $f \colon B \cup D - \{c\} \to \mathbb{R}$ be a function. Suppose that c is an accumulation point of each of B and D. Then c is an accumulation point of $B \cup D$. Prove that $\lim_{x \to c} f(x)$ exists if and only if $\lim_{x \to c} f|_{B-\{c\}}(x)$ and $\lim_{x \to c} f|_{D-\{c\}}(x)$ exist and are equal, and if these three limits exist then they are equal. (Observe that Lemma 3.2.17 is a special case of this more general result.)

Exercise 3.2.18. [Used in Exercise 3.4.9 and Section 8.3.] The definition of the limit of a function (Definition 3.2.1) is a very important and very useful definition, but it has one drawback, which is that in order to prove that a function has a limit, one first needs to guess the value of the limit. It would be nice to be able to prove that a limit exists without having to make such a guess. The purpose of this exercise is to prove a result that states, intuitively, that $\lim_{x \to c} f(x)$ exists if and only if $f(x)$ and $f(y)$ get closer and closer to each other as x and y get closer and closer to c. (This characterization is more interesting in theory than it is useful in practice, though there is a much more well-known and widely used analogous characterization of the limits of sequences, called the Cauchy Completeness Theorem (Corollary 8.3.16).) This characterization of limits is as follows.

Let $I \subseteq \mathbb{R}$ be an open interval, let $c \in I$ and let $f \colon I - \{c\} \to \mathbb{R}$ be a function. Then $\lim_{x \to c} f(x)$ exists if and only if for each $\varepsilon > 0$, there is some $\delta > 0$ such that $x, y \in I - \{c\}$ and $|x - c| < \delta$ and $|y - c| < \delta$ imply $|f(x) - f(y)| < \varepsilon$.

We will prove this result in a few steps.

(1) Suppose that $\lim_{x \to c} f(x)$ exists. Prove that for each $\varepsilon > 0$, there is some $\delta > 0$ such that $x, y \in I - \{c\}$ and $|x - c| < \delta$ and $|y - c| < \delta$ imply $|f(x) - f(y)| < \varepsilon$.

(2) For the rest of this exercise, suppose that for each $\varepsilon > 0$, there is some $\delta > 0$ such that $x, y \in I - \{c\}$ and $|x - c| < \delta$ and $|y - c| < \delta$ imply $|f(x) - f(y)| < \varepsilon$. For each $r > 0$, let $A_r = (I - \{c\}) \cap (c - r, c + r)$. Hence, for each $\varepsilon > 0$, there is some $\delta > 0$ such that $x, y \in A_\delta$ implies $|f(x) - f(y)| < \varepsilon$. Prove that there is some $\eta > 0$ such that $f(A_\eta)$ is bounded.

(3) For each $s \in (0, \eta)$, we note that $f(A_s) \subseteq f(A_\eta)$, and hence $f(A_s)$ is bounded. We can therefore define $a_s = \operatorname{glb} f(A_s)$ and $b_s = \operatorname{lub} f(A_s)$. Let $A = \{a_s \mid s \in (0, \eta)\}$ and $B = \{b_s \mid s \in (0, \eta)\}$. By Exercise 2.6.11 we see that A has a least upper bound and B has a greatest lower bound, and that $\operatorname{lub} A \leq \operatorname{glb} B$. Prove that $\operatorname{lub} A = \operatorname{glb} B$. Use the No Gap Lemma (Lemma 2.6.6).

(4) Let $M = \operatorname{lub} A = \operatorname{glb} B$. Prove that $\lim_{x \to c} f(x) = M$.

3.3 Continuity

The idea of a continuous function is quite familiar intuitively, and is often described as a function whose graph $y = f(x)$ can "be drawn without lifting the pencil from the paper." That is, a function is continuous if its graph has no "gaps" or "jumps." The graph seen in Figure 3.3.1 (i) represents a continuous function. The graph seen in Figure 3.3.1 (ii) represents a discontinuous function; this function has only one place at which it is not continuous, namely, at $x = 0$, but that is sufficient for the whole function to be considered discontinuous. Many of the familiar functions treated in calculus, such as polynomials, e^x, $\ln x$, $\sin x$ and $\cos x$, are continuous. On the other hand, we cannot ignore discontinuous functions, because some applications of mathematics in the sciences and engineering require the use of discontinuous functions (for example, the description of an electric circuit that has an open switch up till a certain point in time, at which point the switch is closed). Also, although discontinuous functions are not differentiable, as we will see by Theorem 4.2.4, it will turn out that some (though not all) discontinuous functions are integrable, as we will see in Example 5.2.6 (2) (3) (4), and in more generality in Theorem 5.8.5.

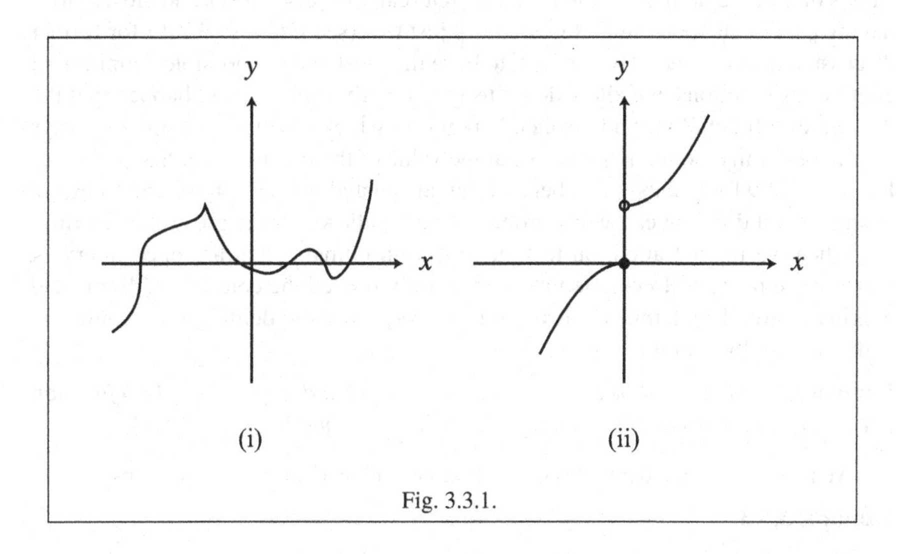

Fig. 3.3.1.

A rigorous treatment of continuity involves the same type of ε–δ arguments that are used in the rigorous treatment of the limits of functions. Indeed, as we will see in Lemma 3.3.2, there is a very close relationship between limits and continuity. There is, however, one fundamental difference between the definitions of limits and continuity. Suppose that we have an open interval $I \subseteq \mathbb{R}$, a number $c \in I$ and a function $f\colon I \to \mathbb{R}$. If we want to find whether the limit of f as x goes to c exists, we do not take into account the value of $f(c)$. Indeed, the function f need not be defined at c for this limit to exist. To study the continuity of f at c, by contrast, we are very much concerned with the value of $f(c)$. To say that f is continuous at c, we need to know that, intuitively, the function "does not jump" at c. More precisely, we need to know that the value of $f(c)$ is just what we would expect it to be if we looked at the values of $f(x)$ as x goes to c. In other words, to say that f is continuous at c, we take the ε–δ of limits given in Definition 3.2.1, and we replace "L" with "$f(c)$."

Definition 3.3.1. Let $A \subseteq \mathbb{R}$ be a set, and let $f\colon A \to \mathbb{R}$ be a function.

1. Let $c \in A$. The function f is **continuous at** c if for each $\varepsilon > 0$, there is some $\delta > 0$ such that $x \in A$ and $|x - c| < \delta$ imply $|f(x) - f(c)| < \varepsilon$. The function f is **discontinuous at** c if f is not continuous at c; in that case we also say that f has a **discontinuity at** c.
2. The function f is **continuous** if it is continuous at every number in A. The function f is **discontinuous** if it is not continuous. △

The reader will have noticed that in the definition of limits we restrict attention to functions with domains that are open intervals with a number removed, whereas in the definition of continuous functions we allow domains that are arbitrary subsets of $\mathbb{R}$. If we are taking the limit at a number $c \in \mathbb{R}$, we need to be sure that there are values of x in the domain of the function that really do get "closer and closer" to c; having c be in an open interval guarantees just that (see Exercise 3.2.17 for further discussion of this issue). By contrast, to have the most general possible definition of continuous functions, we allow domains that are arbitrary. It could happen that the domain of a function has an "isolated" point, in which case the function is always continuous at that point, no matter what the value of the function is at that point; see Exercise 3.3.9 for details. Such behavior at an isolated point might seem somewhat strange, but it does not cause any problems, and it allows for the greatest generality.

When we restrict attention to functions with domains that are open intervals, then we see in the following lemma how closely related the concepts of limits and continuity are. This lemma follows immediately from the definition of limits and continuity, and we omit the proof.

Lemma 3.3.2. *Let $I \subseteq \mathbb{R}$ be an open interval, let $c \in I$ and let $f: I \to \mathbb{R}$ be a function. Then f is continuous at c if and only if $\lim_{x\to c} f(x)$ exists and $\lim_{x\to c} f(x) = f(c)$.*

We now have some examples of continuous and discontinuous functions.

Example 3.3.3.

(1) Let $A \subseteq \mathbb{R}$ be a set, and let $f: A \to \mathbb{R}$ be defined by $f(x) = mx + b$ for all $x \in A$, where $m, b \in \mathbb{R}$. We will prove that f is continuous. It would be possible to do this proof using an ε–δ argument, but we can avoid such an argument by using Exercise 3.2.1. First, suppose that $A = \mathbb{R}$. Then A is an open interval, and it follows from Exercise 3.2.1 and Lemma 3.3.2 that f is continuous. Second, if A is an arbitrary subset of $\mathbb{R}$, we use the previous case together with Exercise 3.3.2 (2) to deduce that f is continuous.

(2) Let $B \subseteq \mathbb{R} - \{0\}$ be a set, and let $p: B \to \mathbb{R}$ be defined by $p(x) = \frac{1}{x}$ for all $x \in B$. We will prove that p is continuous. As was the case with some of our proofs involving limits, we will first do scratch work prior to the actual proof.

Scratch Work We work backwards for our scratch work. Let $c \in B$. There are two cases, depending upon whether $c > 0$ or $c < 0$. We will consider the former case; the latter case is very similar, and the details are left to the reader. We want $\left|\frac{1}{x} - \frac{1}{c}\right| < \varepsilon$, which is $\left|\frac{c-x}{xc}\right| < \varepsilon$, which is $|x - c| < \varepsilon|x|c$. The appearance of "$x$" in the right-hand side of this last inequality is a problem, because "δ" cannot depend upon x. To remedy this situation, we want to impose a positive lower bound on the values of $|x|$, and the only way to do that is by our choice of δ. One way to obtain this lower bound is as follows. Suppose that $|x - c| < \frac{c}{2}$. Then $-\frac{c}{2} < x - c < \frac{c}{2}$, so $\frac{c}{2} < x < \frac{3c}{2}$, and hence $\frac{c}{2} < |x| < \frac{3c}{2}$. When this restriction on x holds, then $\varepsilon|x|c > \frac{\varepsilon c^2}{2}$. We can then use $\delta = \min\{\frac{c}{2}, \frac{\varepsilon c^2}{2}\}$.

Actual Proof Let $c \in B$. We will prove that f is continuous at c. There are two cases. First, suppose that $c > 0$. Let $\varepsilon > 0$. Let $\delta = \min\{\frac{c}{2}, \frac{\varepsilon c^2}{2}\}$. Suppose that $x \in B$

and $|x-c|<\delta$. Then $|x-c|<\frac{c}{2}$, and hence $-\frac{c}{2}<x-c<\frac{c}{2}$, which implies that $\frac{c}{2}<x<\frac{3c}{2}$, and therefore $\frac{c}{2}<|x|$. It then follows that

$$\left|\frac{1}{x}-\frac{1}{c}\right|=\left|\frac{c-x}{xc}\right|=\frac{|x-c|}{|x|c}<\frac{\frac{\varepsilon c^2}{2}}{\frac{c}{2}\cdot c}=\varepsilon.$$

We conclude that p is continuous at c. Second, suppose that $c<0$. This case is similar to the previous case, and we omit the details. Because c was chosen arbitrarily, we conclude that p is continuous.

(3) We assume that the reader is informally familiar with the standard elementary functions (that is, polynomials, power functions, logarithms, exponentials and trigonometric functions). All of these functions are continuous. We have not yet defined these functions rigorously, and so we cannot prove that they are continuous at this point, but for now we will assume the continuity (and other standard properties) of these functions for the sake of examples. (We will treat these functions rigorously in Chapter 7, and in that chapter it will be seen that these functions are indeed continuous, which will be proved by showing that they are differentiable, and then using the fact that differentiable functions are continuous, as seen in Theorem 4.2.4.)

While we are discussing the continuity of the familiar elementary functions from an informal point of view, here is a very interesting function that uses the sine function in its definition. Let $k\colon \mathbb{R}\to\mathbb{R}$ be defined by

$$k(x)=\begin{cases}\sin\frac{1}{x}, & \text{if } x\neq 0\\ 0, & \text{if } x=0.\end{cases}$$

The graph of k is seen in Figure 3.2.3 (ii). We saw in Example 3.2.3 (4) that $\lim_{x\to 0}k(x)$ does not exist. It follows from Lemma 3.3.2 that k is discontinuous at 0. (The function k is continuous at all other real numbers, but we omit the details.)

(4) Is $\tan x$ continuous? A look at the graph of $y=\tan x$ might lead one to think that $\tan x$ is not continuous, because it has vertical asymptotes at $x=\frac{\pi}{2}+n\pi$ for all $n\in\mathbb{Z}$. However, it turns out that $\tan x$ is continuous. The issue is the domain of $\tan x$. The definition of a function formally includes its domain and codomain, and simply writing out a formula for the function (such as "$f(x)=\tan x$") does not rigorously define a function. In the case of some familiar functions, such as $\sin x$ and e^x, we take it as known that the domain and codomain are both $\mathbb{R}$. Observe, however, that $\tan x$ is not defined at $x=\frac{\pi}{2}+n\pi$ for all $n\in\mathbb{Z}$. The correct domain of $\tan x$ is $\cdots\cup\left(-\frac{\pi}{2},\frac{\pi}{2}\right)\cup\left(\frac{\pi}{2},\frac{3\pi}{2}\right)\cup\cdots$, and on this domain, the function $\tan x$ is indeed continuous. A proof of this fact would require first proving rigorously that $\sin x$ and $\cos x$ are continuous on $\mathbb{R}$, and then using the fact that $\tan x=\frac{\sin x}{\cos x}$ for all x in the domain of $\tan x$, combined with Theorem 3.3.5 (5), which will be proved later in this section.

(5) Let $g\colon\mathbb{R}\to\mathbb{R}$ be defined by

$$g(x)=\begin{cases}\frac{|x|}{x}, & \text{if } x\neq 0\\ 0, & \text{if } x=0.\end{cases}$$

The function g is discontinuous at 0, though it is continuous everywhere else. To verify this fact, we observe that

$$g(x) = \begin{cases} 1, & \text{if } x > 0 \\ 0, & \text{if } x = 0 \\ -1, & \text{if } x < 0. \end{cases}$$

To see that g is discontinuous at 0, it would be possible to use the ε–δ definition of continuity directly, but we can save some effort by observing that in Example 3.2.16 we saw that $\lim_{x\to 0} g(x)$ does not exist, and then applying Lemma 3.3.2. It is intuitively clear that g is continuous at all numbers in $\mathbb{R} - \{0\}$; the proof involves an ε–δ argument, which is left to the reader.

(6) Let $r\colon [0,1] \to \mathbb{R}$ be defined by

$$r(x) = \begin{cases} 1, & \text{if } x \in \mathbb{Q} \cap [0,1] \\ 0, & \text{otherwise.} \end{cases}$$

We will prove that r is discontinuous everywhere. Let $c \in [0,1]$. Suppose that r is continuous at c. Then there is some $\delta > 0$ such that $x \in [0,1]$ and $|x-c| < \delta$ imply $|r(x) - r(c)| < \frac{1}{2}$. Because $c \in [0,1]$, there is some $\eta > 0$ such that $(c-\eta, c) \subseteq [0,1]$ or $(c, c+\eta) \subseteq [0,1]$; we cannot be sure that both of these are true, because c might be one of the endpoints of $[0,1]$. Without loss of generality, assume that $(c, c+\eta) \subseteq [0,1]$. Let $\tau = \min\{\delta, \eta\}$.

There are now two cases. First, suppose that c is rational. We know by Theorem 2.6.13 (2) that there is some irrational $y \in \mathbb{R}$ such that $c < y < c + \tau$. It follows that $y \in [0,1]$ and $|y - c| < \delta$. Hence $|r(y) - r(c)| < \frac{1}{2}$. From the definition of the function r it follows that $|0 - 1| < \frac{1}{2}$, which is a contradiction. Second, suppose that c is irrational. A similar contradiction can be obtained, this time using Theorem 2.6.13 (1). We deduce that r is not continuous at c.

(7) For this next example, we need to use the fact that every non-negative rational number can be expressed uniquely as a fraction in "lowest terms," which means as a fraction $\frac{a}{b}$ such that $a \in \mathbb{N} \cup \{0\}$ and $b \in \mathbb{N}$, and that a and b have no common factors other than 1 and -1. (Observe that 0 expressed in lowest terms is $\frac{0}{1}$, because every integer is a factor of 0.) The reader is informally familiar with this fact from years of experience with fractions. This fact can be proved rigorously starting from the basic properties of the integers that we have seen, though working through the details would take us too far afield, and so we will not provide such a proof. For details, the reader can either derive the desired fact about fractions from the Fundamental Theorem of Arithmetic, found for example in [Ros05, Section 3.5], or see a proof in [Olm62, Sections 402 and 404]. We will be using this fact about fractions in lowest terms only in the present example, and in subsequent examples that rely upon this one, but not in any proofs of theorems.

Let $s\colon [0,1] \to \mathbb{R}$ be defined by

$$s(x) = \begin{cases} \frac{1}{q}, & \text{if } x \in \mathbb{Q} \cap [0,1] \text{ and } x = \frac{p}{q} \text{ in lowest terms,} \\ & \quad \text{where } p \in \mathbb{N} \cup \{0\} \text{ and } q \in \mathbb{N} \\ 0, & \text{otherwise.} \end{cases}$$

We will prove that s is discontinuous at every rational number in $[0,1]$, and continuous at every irrational number in $[0,1]$. The strange behavior of this function is somewhat counterintuitive, and it is due to the existence of such strange functions that we need to have rigorous definitions and proofs for concepts such as continuity, because our intuition about such matters might not always be correct.

The proof that s is discontinuous at every rational number in $[0,1]$ is very similar to the proof that the function r in Part (6) of this example is discontinuous everywhere, and the details are left to the reader. The more interesting part of the proof is that s is continuous at every irrational number in $[0,1]$.

Let $c \in [0,1]$. Suppose that c is irrational. Let $\varepsilon > 0$. By Corollary 2.6.8 (2) there is some $m \in \mathbb{N}$ such that $\frac{1}{m} < \varepsilon$. There are only finitely many rational numbers in $[0,1]$ that have denominator m or smaller when expressed in lowest terms; let $q_1, \ldots, q_k \in [0,1]$ be these rational numbers. Let $\delta = \min\{|c - q_1|, \ldots, |c - q_k|\}$. Because c is irrational, then $\delta > 0$.

Suppose that $x \in [0,1]$ and $|x - c| < \delta$. There are now two cases. First, suppose that x is irrational. Then $|s(x) - s(c)| = |0 - 0| < \varepsilon$. Second, suppose that x is rational. Then $x = \frac{a}{b}$ for some $a \in \mathbb{N} \cup \{0\}$ and $b \in \mathbb{N}$, where $\frac{a}{b}$ is in lowest terms. By the choice of δ, we know that $x \neq q_i$ for all $i \in \{1, \ldots, k\}$. Hence $b > m$, and it follows that $|s(x) - s(c)| = |\frac{1}{b} - 0| < \frac{1}{m} < \varepsilon$. Combining these two cases, we deduce that s is continuous at c. ◊

Given the very strange nature of the function s in Example 3.3.3 (7), the reader might wonder whether for any non-degenerate closed bounded interval $C \subseteq \mathbb{R}$, and for any subset $A \subseteq C$, it would be possible to find a function $f\colon C \to \mathbb{R}$ that is continuous at every number in A and discontinuous at every number in $C - A$. The answer turns out to be no. For example, as seen by Exercise 8.4.7, there is no function $g\colon [0,1] \to \mathbb{R}$ that is continuous at every rational number in $[0,1]$, and discontinuous at every irrational number in $[0,1]$. See [TBB01, Section 6.7] for a general discussion of which subsets of $\mathbb{R}$ can be the set of numbers at which a function is continuous.

The following result is the analog for continuous functions of the Sign-Preserving Property for Limits (Theorem 3.2.4).

Theorem 3.3.4 (Sign-Preserving Property for Continuous Functions). *Let $A \subseteq \mathbb{R}$ be a non-empty set, let $c \in A$ and let $f\colon A \to \mathbb{R}$ be a function. Suppose that f is continuous at c.*

1. *If $f(c) > 0$, then there is some $M > 0$ and some $\delta > 0$ such that $x \in A$ and $|x - c| < \delta$ imply $f(x) > M$.*
2. *If $f(c) < 0$, then there is some $N < 0$ and some $\delta > 0$ such that $x \in A$ and $|x - c| < \delta$ imply $f(x) < N$.*

Proof. Left to the reader in Exercise 3.3.5. □

Our next theorem shows that continuity is well behaved with respect to addition, subtraction, multiplication and division of functions. This theorem is very convenient for showing the continuity of functions that are built up out of simpler ones.

Theorem 3.3.5. *Let $A \subseteq \mathbb{R}$ be a non-empty set, let $c \in A$, let $f, g\colon A \to \mathbb{R}$ be functions and let $k \in \mathbb{R}$. Suppose that f and g are continuous at c.*

1. *$f+g$ is continuous at c.*
2. *$f-g$ is continuous at c.*
3. *kf is continuous at c.*
4. *fg is continuous at c.*
5. *If $g(c) \neq 0$, then $\frac{f}{g}$ is continuous at c.*

Proof. If A were an open interval, then this theorem could be deduced immediately by combining Lemma 3.3.2 and Theorem 3.2.10. In the general case, where A is not necessarily an open interval, we cannot use Lemma 3.3.2, but an examination of the details of the proof of Theorem 3.2.10 reveals that that proof can easily be modified to work in the present situation, simply by replacing L with $f(c)$, replacing M with $g(c)$ and replacing $I - \{c\}$ with A. The details are left to the reader. □

The following result is an immediate consequence of Theorem 3.3.5, and we omit the proof.

Corollary 3.3.6. *Let $A \subseteq \mathbb{R}$ be a non-empty set, let $f, g\colon A \to \mathbb{R}$ be functions and let $k \in \mathbb{R}$. Suppose that f and g are continuous. Then $f+g$, $f-g$, kf and fg are continuous, and if $g(x) \neq 0$ for all $x \in I$ then $\frac{f}{g}$ is continuous.*

Example 3.3.7.

(1) Let $A \subseteq \mathbb{R}$ be a non-empty set. For each $n \in \mathbb{N}$, let $f_n\colon A \to \mathbb{R}$ be defined by $f_n(x) = x^n$ for all $x \in A$. We will prove that f_n is continuous for all $n \in \mathbb{N}$ by induction on n. First, let $n = 1$. Then $f_n(x) = f_1(x) = x$ for all $x \in A$, and we have seen that this function is continuous in Example 3.3.3 (1). Now let $n \in \mathbb{N}$, and suppose that f_n is continuous. Then $f_{n+1}(x) = x^{n+1} = x^n \cdot x = f_n(x) f_1(x)$ for all $x \in A$, and hence $f_{n+1} = f_n f_1$. It follows from Theorem 3.3.5 (4) that f_{n+1} is continuous. By induction, we deduce that f_n is continuous for all $n \in \mathbb{N}$.

Example 3.3.3 (1) shows that all constant functions $A \to \mathbb{R}$ are continuous, and it then follows from the previous paragraph together with Theorem 3.3.5 (1) (3), that all polynomial functions are continuous.

(2) Let $p\colon \mathbb{R} - \{0\} \to \mathbb{R}$ be defined by $p(x) = \frac{1}{x}$ for all $x \in \mathbb{R} - \{0\}$. We saw in Example 3.3.3 (2) that p is continuous, using the ε–δ definition of continuity. A much simpler proof can be obtained using Theorem 3.3.5 (5) by observing that the functions $h, k\colon \mathbb{R} - \{0\} \to \mathbb{R}$ defined by $h(x) = 1$ for all $x \in \mathbb{R}$ and $k(x) = x$ for all $x \in \mathbb{R}$ are continuous by Example 3.3.3 (1). ◊

We now see that continuity is also well behaved with respect to the composition of functions.

Theorem 3.3.8. *Let $A, B \subseteq \mathbb{R}$ be non-empty sets, let $c \in A$ and let $g\colon A \to B$ and $f\colon B \to \mathbb{R}$ be functions.*

1. *Suppose that A is an open interval. If $\lim_{x\to c} g(x)$ exists and is in B, and if f is continuous at $\lim_{x\to c} g(x)$, then $\lim_{x\to c} f(g(x)) = f(\lim_{x\to c} g(x))$.*
2. *If g is continuous at c, and if f is continuous at $g(c)$, then $f \circ g$ is continuous at c.*
3. *If g and f are continuous, then $f \circ g$ is continuous.*

Proof.

(1) Suppose that $\lim_{x\to c} g(x)$ exists and is in B, and that f is continuous at $\lim_{x\to c} g(x)$. Let $L = \lim_{x\to c} g(x)$. Let $\varepsilon > 0$. Then there is some $\eta > 0$ such that $y \in B$ and $|y - L| < \eta$ imply $|f(x) - f(L)| < \varepsilon$, and there is some $\delta > 0$ such that $x \in A - \{c\}$ and $|x - c| < \delta$ imply $|g(x) - L| < \eta$. Suppose that $x \in A - \{c\}$ and $|x - c| < \delta$. Then $|g(x) - L| < \eta$, and hence $|f(g(x)) - f(L)| < \varepsilon$, which means $|f(g(x)) - f(\lim_{x\to c} g(x))| < \varepsilon$. It follows that $\lim_{x\to c} f(g(x)) = f(\lim_{x\to c} g(x))$.

(2) Because A is not necessarily an open interval, this part of the theorem cannot be deduced from Part (1) of the theorem. However, the proof of this part of the theorem is very similar to the proof of Part (1), but with $g(c)$ replacing $\lim_{x\to c} g(x)$; the details are left to the reader.

(3) This part of the theorem follows immediately from Part (2) of the theorem. □

Whereas the composition of continuous functions works nicely, as stated in Theorem 3.3.8, the composition of discontinuous functions can behave rather strangely.

Example 3.3.9.

(1) Let $h, k\colon \mathbb{R} \to \mathbb{R}$ be defined by

$$h(x) = \begin{cases} 1, & \text{if } x \neq 0 \\ 0, & \text{if } x = 0, \end{cases} \quad \text{and} \quad k(x) = \begin{cases} 2, & \text{if } x \neq 3 \\ 0, & \text{if } x = 3. \end{cases}$$

It is straightforward to verify that h is discontinuous at 0, but continuous everywhere else, and that k is discontinuous at 3, but continuous everywhere else; the details are left to the reader. Observe that $(k \circ h)(x) = 2$ for all $x \in \mathbb{R}$, so that $k \circ h$ is a constant function, and hence it is continuous by Example 3.3.3 (1). We therefore see that the composition of two discontinuous functions can be continuous.

(2) Let $h\colon \mathbb{R} \to \mathbb{R}$ be the function given in Part (1) of this example, let $r\colon [0,1] \to \mathbb{R}$ be the function given in Example 3.3.3 (6) and let $s\colon [0,1] \to \mathbb{R}$ be the function given in Example 3.3.3 (7). We saw that h is discontinuous at 0, and continuous everywhere else in $[0,1]$, and that s is discontinuous at every rational number in $[0,1]$, and continuous at every irrational number in $[0,1]$. Observe that $h \circ s = r$, and that r was seen to be discontinuous everywhere, which shows that the composition of two discontinuous functions can have "worse" discontinuity than either of the original functions. ◊

Our next result shows that two continuous functions on adjacent closed bounded intervals can be "pasted together" to form a continuous function if they agree on the point common to both domains.

Lemma 3.3.10 (Pasting Lemma). *Let $[a,b] \subseteq \mathbb{R}$ and $[b,c] \subseteq \mathbb{R}$ be non-degenerate closed bounded intervals, and let $f\colon [a,b] \to \mathbb{R}$ and $g\colon [b,c] \to \mathbb{R}$ be functions. Let $h\colon [a,c] \to \mathbb{R}$ be defined by*

$$h(x) = \begin{cases} f(x), & \text{if } x \in [a,b] \\ g(x), & \text{if } x \in [b,c]. \end{cases}$$

If f and g are continuous, and if $f(b) = g(b)$, then h is continuous.

Proof. Left to the reader in Exercise 3.3.10. □

Finally, recall the concept of the extension of a function, as discussed prior to the proof of Theorem 2.7.1. We now see that not every continuous function with domain of the form $A - \{c\}$ can be extended to a continuous function with domain A.

Example 3.3.11. Consider the functions $f, p\colon \mathbb{R} - \{0\} \to \mathbb{R}$ defined by $f(x) = x$ for all $x \in \mathbb{R} - \{0\}$ and $p(x) = \frac{1}{x}$ for all $x \in \mathbb{R} - \{0\}$. Both of these functions are continuous, as we saw in Example 3.3.7 (2). We observe that f can be extended to a continuous function $F\colon \mathbb{R} \to \mathbb{R}$ by defining $F(0) = 0$, so that $F(x) = x$ for all $x \in \mathbb{R}$. On the other hand, the function p cannot be extended to a continuous function $\mathbb{R} \to \mathbb{R}$, as can be seen intuitively by looking at the graph of p, and can be proved by combining Lemma 3.3.2 with the fact that $\lim_{x \to 0} p(x)$ does not exist, which we saw in Example 3.2.3 (3). ◊

Reflections

In contrast to limits, which are often viewed by beginning real analysis students as a necessary technicality at best, the intuitive concept of continuity is one that most people find quite simple and understandable. It is easy to see the intuitive difference between a function that has a graph that can be drawn without lifting one's pencil from the page and a function that has a graph that cannot be drawn that way. Continuity is therefore a concept for which there is a large gap between the intuitive idea, which is simple, and the rigorous definition, which is technical. It is worthwhile to take the time to convince yourself that the ε–δ formulation really captures the idea of drawing a graph without lifting one's pencil.

The main topics of study in real analysis are the central concepts encountered in calculus courses such as derivatives, integrals, sequence, series and the like. In the context of real analysis, the concept of continuity plays a supporting role, though a technically important one. If the reader studies topology, however, then she will see continuity in a starring role. Continuous functions in topology play the analogous role to what homomorphisms play in abstract algebra and linear maps play in linear algebra, in that continuous functions are the type of function that preserves topological structure. An excellent place to learn about the general concept of continuity is the classic introductory topology text [Mun00].

Exercises

Exercise 3.3.1. Using only the definition of continuity, prove that the following functions are continuous.

(1) Let $f\colon \mathbb{R} \to \mathbb{R}$ be defined by $f(x) = x^2 + 1$ for all $x \in \mathbb{R}$.
(2) **[Used in Example 4.2.3.]** Let $A \subseteq \mathbb{R}$ be a set, and let $g\colon A \to \mathbb{R}$ be defined by $g(x) = |x|$ for all $x \in A$.

Exercise 3.3.2. **[Used throughout.]** Let $A \subseteq \mathbb{R}$ be a set, let $c \in A$ and let $f\colon A \to \mathbb{R}$ be a function.

(1) Prove that f is continuous at c if and only if there is some $\delta > 0$ such that $f|_{A\cap(c-\delta,c+\delta)}$ is continuous at c.
(2) Let $B \subseteq A$ be a set. Suppose that $c \in B$. Prove that if f is continuous at c, then $f|_B$ is continuous at c. Deduce that if f is continuous, then $f|_B$ is continuous. Find an example to show that neither of these statements can be made into if and only if statements.

Exercise 3.3.3. Let $I, J \subseteq \mathbb{R}$ be open intervals, and let $f\colon I \to \mathbb{R}$ be a function. Suppose that f is continuous. Let $x \in f^{-1}(J)$. Prove that there is some open interval $K \subseteq \mathbb{R}$ such that $x \in K \cap I \subseteq f^{-1}(J)$.

Exercise 3.3.4. **[Used in Theorem 3.5.2 and Exercise 3.5.8.]** Let $A, B \subseteq \mathbb{R}$ be non-empty sets, and let $f\colon B \to \mathbb{R}$ be a function. Suppose that $A \subseteq B$.

(1) Suppose that A and $f(A)$ have least upper bounds, and that $\operatorname{lub} A \in B$. Prove that if f is continuous at $\operatorname{lub} A$ then $f(\operatorname{lub} A) \leq \operatorname{lub} f(A)$.
(2) Suppose that A and $f(A)$ have greatest lower bounds, and that $\operatorname{glb} A \in B$. Prove that if f is continuous at $\operatorname{glb} A$ then $f(\operatorname{glb} A) \geq \operatorname{glb} f(A)$.

Exercise 3.3.5. **[Used in Theorem 3.3.4.]** Prove Theorem 3.3.4.

Exercise 3.3.6. Theorem 3.3.8 (1) was stated and proved for the case that A is an open interval and B is an arbitrary set. Give a simpler proof of this result in the case where B is also an open interval.

Exercise 3.3.7. **[Used in Exercise 3.3.8.]** Let $A \subseteq \mathbb{R}$ be a set, let $c \in A$, and let $f\colon A \to \mathbb{R}$ be a function. Prove that if f is continuous at c, then there is some $\delta > 0$ such that $f|_{A\cap(c-\delta,c+\delta)}$ is bounded.

Exercise 3.3.8. **[Used in Section 3.4.]** Let $C \subseteq \mathbb{R}$ be a closed bounded interval, and let $f\colon C \to \mathbb{R}$ be a function. Prove that if f is continuous, then f is bounded. [Use Exercise 3.3.7.]

Exercise 3.3.9. **[Used in Section 3.3.]** Let $A \subseteq \mathbb{R}$ be a non-empty set, let $c \in A$ and let $f\colon A \to \mathbb{R}$ be a function. Suppose that there is some $\mu > 0$ such that $(A - \{c\}) \cap (c - \mu, c + \mu) = \emptyset$. Prove that f is continuous at c.

Exercise 3.3.10. **[Used in Lemma 3.3.10.]** Prove Lemma 3.3.10.

Exercise 3.3.11. **[Used in Lemma 7.3.4.]** Let $A \subseteq \mathbb{R}$ be a set, let $f\colon A \to \mathbb{R}$ be a function and let $p \in \mathbb{R}$. Let $A + p$ denote the set $\{a + p \mid a \in A\}$. (The notation "$A + p$" is similar to the notation "$A + B$" used in Exercise 2.6.9, where here we write "p" instead of "$\{p\}$.") Let $g\colon A + p \to \mathbb{R}$ be defined by $g(x) = f(x - p)$ for all $x \in A + p$. Prove that if f is continuous, then g is continuous.

3.4 Uniform Continuity

Continuity is a very important, and intuitively appealing, property of functions, but it turns out that for the proofs of some theorems in real analysis, for example the fact that a continuous function is integrable, a strengthened version of continuity is needed. To understand this strengthened version, let us first review what it means for a function to be continuous. Suppose that $f\colon A \to \mathbb{R}$ is a continuous function for some set $A \subseteq \mathbb{R}$. Then f is continuous at each $c \in A$. Stated informally, to say that f is continuous for each $c \in A$ means that for each choice of $c \in A$, and for each $\varepsilon > 0$, we need to find some $\delta > 0$ such that if $x \in A$ and x is within distance δ of c, then $f(x)$ is within distance ε of $f(c)$. The choice of δ here depends upon both c and ε, and of course upon the function f. For smaller ε we need smaller δ, and so we cannot avoid the fact that δ depends upon ε (other than in exceptional cases, such as constant functions). Could it be the case that for a given ε, we could use the same δ for all values of c? The answer in general is no, though for some functions it is yes.

Let g and f be the functions whose graphs are seen in Figure 3.4.1 and Figure 3.4.2, respectively. If we compare what happens at the two points c and d in the domain of the function g, we see that for the same ε, we need a much smaller δ at c than we do at d (note that δ is not labeled in the figure for lack of space). As we take values of c closer and closer to 0, then we need smaller and smaller values of δ for the same ε. By contrast, we see in the graph of f that for any given ε, it is possible to choose a value of δ that works with respect to this ε for any value of c (intuitively, choose the δ that works where the graph has the largest slope). Of course, we are looking at these graphs just for the intuitive idea; we will see proofs of what we have asserted in Example 3.4.3 (2) and Exercise 4.4.6.

From the above examples, we see that whereas in principle the choice of δ when proving that a function is continuous depends upon c and ε, for some functions the choice of δ depends only upon ε.

To obtain a better understanding of this situation, let us turn to the definition of continuity in terms of quantifiers and logical symbols. Again, suppose that we are given a function $f\colon A \to \mathbb{R}$ for some set $A \subseteq \mathbb{R}$. The condition that f is continuous at each $c \in A$ is expressed by writing

$$(\forall c \in A)[f \text{ is continuous at } c],$$

which can be written completely in symbols as

$$(\forall c \in A)(\forall \varepsilon > 0)(\exists \delta > 0)[(x \in A \wedge |x - c| < \delta) \to |f(x) - f(c)| < \varepsilon].$$

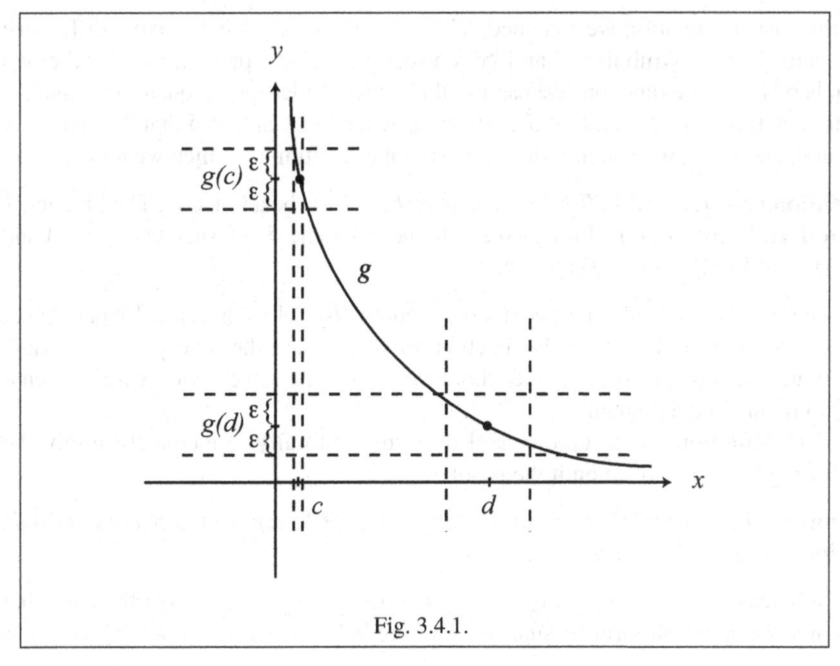

Fig. 3.4.1.

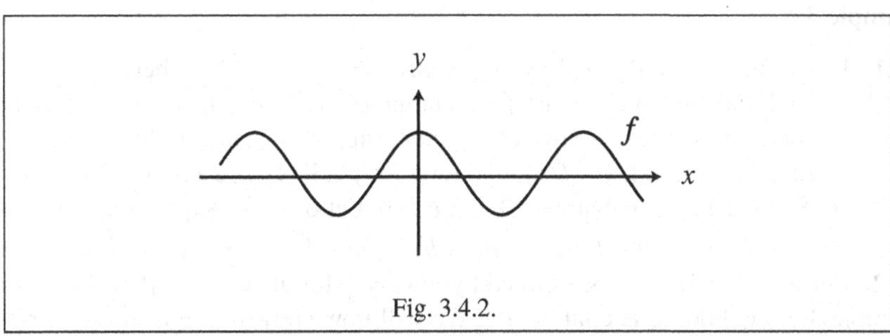

Fig. 3.4.2.

As always, the order of the quantifiers is crucial. Because we are first given c and ε, and we then show that there exists an appropriate δ, the choice of δ can depend upon both c and ε. However, given that we saw above that it could happen for some functions that δ does not depend upon c, even though δ is quantified after c in the given order of quantifiers, it is therefore possible that for some functions we could replace the above statement in logical symbols with

$$(\forall \varepsilon > 0)(\exists \delta > 0)(\forall c \in A)[(x \in A \wedge |x - c| < \delta) \to |f(x) - f(c)| < \varepsilon].$$

In this new formulation, the roles of x and c are in fact equivalent, even if it might not appear as such at first, and we can therefore rewrite this formulation as

$$(\forall \varepsilon > 0)(\exists \delta > 0)(\forall x \in A)(\forall y \in A)[|x - y| < \delta \to |f(x) - f(y)| < \varepsilon].$$

In this last formulation we renamed "c" as "y," which does not make a difference logically, but the symbols "x" and "y" correctly suggest a parallel role for the two numbers. For some functions we can use this revised order for the quantifiers, and for other functions we cannot. In those situations where we can find δ that depends only upon ε, and not c, we obtain a stronger version of continuity, which we now define.

Definition 3.4.1. Let $A \subseteq \mathbb{R}$ be a set, and let $f\colon A \to \mathbb{R}$ be a function. The function f is **uniformly continuous** if for each $\varepsilon > 0$, there is some $\delta > 0$ such that $x, y \in A$ and $|x-y| < \delta$ imply $|f(x) - f(y)| < \varepsilon$. △

Observe that in contrast to the notion of continuity, which is defined separately at each number in the domain of the function, we do not have the concept of "uniformly continuous at a point," because the whole idea is that the same δ works for a given ε for all points in the domain.

The definitions of continuity and uniform continuity immediately imply the following lemma, and we omit the proof.

Lemma 3.4.2. *Let $A \subseteq \mathbb{R}$ be a set, and let $f\colon A \to \mathbb{R}$ be a function. If f is uniformly continuous, then f is continuous.*

Whereas uniform continuity implies continuity, it is not always the case that continuity implies uniform continuity, as we see in Part (2) of the following example.

Example 3.4.3.

(1) Let $f\colon \mathbb{R} \to \mathbb{R}$ be defined by $f(x) = mx + b$ for all $x \in \mathbb{R}$, where $m, b \in \mathbb{R}$. It was shown in Example 3.3.3 (1) that f is continuous, and we will now show that f is uniformly continuous. There are two cases. First, suppose that $m = 0$. In that case f is a constant function, and $|f(x) - f(y)| = 0$ for all $x, y \in \mathbb{R}$. Hence any $\delta > 0$ works for any $\varepsilon > 0$. Second, suppose that $m \neq 0$. Let $\varepsilon > 0$. Let $\delta = \frac{\varepsilon}{|m|}$. Suppose that $x, y \in \mathbb{R}$ and $|x-y| < \delta$. Then $|f(x) - f(y)| = |(mx+b) - (my+b)| = |m| \cdot |x-y| < |m| \cdot \delta = \varepsilon$.

(2) Let $g\colon \mathbb{R} - \{0\} \to \mathbb{R}$ be defined by $g(x) = \frac{1}{x}$ for all $x \in \mathbb{R} - \{0\}$. We saw in Example 3.3.3 (2) that g is continuous. We will now show that g is not uniformly continuous, which corresponds to what we saw intuitively in Figure 3.4.1, which has part of the graph of g. To prove that g is not uniformly continuous, we show that there is some $\varepsilon > 0$ such that for every $\delta > 0$, there are $x, y \in A$ such that $|x - y| < \delta$ and $|g(x) - g(y)| \geq \varepsilon$.

Let $\varepsilon = 1$. Let $\delta > 0$. Let $x = \sqrt{\delta}$ and $y = \frac{\sqrt{\delta}}{\sqrt{\delta}+1}$. Then

$$|x-y| = \left| \sqrt{\delta} - \frac{\sqrt{\delta}}{\sqrt{\delta}+1} \right| = \frac{\delta}{\sqrt{\delta}+1} < \delta,$$

and

$$|g(x) - g(y)| = \left| \frac{1}{x} - \frac{1}{y} \right| = \left| \frac{1}{\sqrt{\delta}} - \frac{\sqrt{\delta}+1}{\sqrt{\delta}} \right| = 1 = \varepsilon.$$

Therefore g is not uniformly continuous.

(3) Let $h\colon (1,\infty) \to \mathbb{R}$ be defined by $h(x) = \frac{1}{x}$ for all $x \in (1,\infty)$. Observe that h is just the restriction to $(1,\infty)$ of the function p in Part (2) of this example. As we now see, the restriction of the domain of p to $(1,\infty)$ yields a uniformly continuous function. Let $\varepsilon > 0$. Let $\delta = \varepsilon$. Suppose that $x, y \in \mathbb{R}$ and $|x-y| < \delta$. Then

$$|h(x) - h(y)| = \left|\frac{1}{x} - \frac{1}{y}\right| = \left|\frac{y-x}{xy}\right| = \frac{|x-y|}{xy} < \frac{\delta}{1 \cdot 1} = \varepsilon. \qquad \Diamond$$

Whereas a comparison of the definitions of uniform continuity and continuity appears to be a matter of epsilons, deltas and quantifiers, a comparison of the various functions examined in Example 3.4.3 yields the intuitive idea that a function is uniformly continuous if it is continuous and if "$f(x)$ does not change too rapidly as x changes." Although the definition of uniform continuity does not have anything to do with differentiability (which we have not yet defined), the intuitive notion of not changing too rapidly is reminiscent of the intuitive notion of derivatives. Indeed, as will be seen in Exercise 4.4.6, if a function is differentiable and has bounded derivative, then it is uniformly continuous. However, it is important to stress that differentiability is not needed for a function to be uniformly continuous, and we mention it here only as an aid to our intuition.

Continuous functions are not always uniformly continuous, but there is one common, and very useful, situation in which continuity does imply uniform continuity, as we see in the following theorem. This theorem is our first truly substantial result involving limits and continuity. In contrast to all the other proofs up till now in the present chapter, which were relatively straightforward, and which relied upon only the algebraic properties of the real numbers, the proof of Theorem 3.4.4, while not long, relies upon the Least Upper Bound Property of the real numbers (via the Heine–Borel Theorem (Theorem 2.6.14), which is proved using the Least Upper Bound Property). Put another way, whereas all the previous results in this chapter would still be true if we considered functions defined on subsets of the rational numbers, Theorem 3.4.4 would not be true in such a situation. For example, let $f\colon [0,2] \cap \mathbb{Q} \to \mathbb{Q}$ be defined by $f(x) = \frac{1}{x^2-2}$ for all $x \in [0,2] \cap \mathbb{Q}$. This function is continuous, because the analogs of all results proved in Section 3.3 would still hold for functions defined on subsets of $\mathbb{Q}$, and because the denominator of $\frac{1}{x^2-2}$ is never zero (because $\sqrt{2} \notin \mathbb{Q}$). However, the function is not uniformly continuous, by an argument similar to that used in Example 3.4.3 (2).

Theorem 3.4.4. *Let $C \subseteq \mathbb{R}$ be a closed bounded interval, and let $f\colon C \to \mathbb{R}$ be a function. If f is continuous, then f is uniformly continuous.*

Proof. Suppose that f is continuous. Let $\varepsilon > 0$. By the definition of continuity, for each $z \in C$ there is there is some $\delta_z > 0$ such that $x \in C$ and $|x-z| < \delta_z$ imply $|f(x) - f(z)| < \frac{\varepsilon}{2}$. We then form the family $\left\{\left(w - \frac{\delta_w}{2}, w + \frac{\delta_w}{2}\right)\right\}_{w \in C}$ of open intervals in $\mathbb{R}$. Because $w \in \left(w - \frac{\delta_w}{2}, w + \frac{\delta_w}{2}\right)$ for all $w \in C$, then $C \subseteq \bigcup_{w \in C}\left(w - \frac{\delta_w}{2}, w + \frac{\delta_w}{2}\right)$. The Heine–Borel Theorem (Theorem 2.6.14) implies that there are $n \in \mathbb{N}$ and $w_1, w_2, \ldots, w_n \in C$ such that $C \subseteq \bigcup_{k=1}^{n} \left(w_k - \frac{\delta_{w_k}}{2}, w_k + \frac{\delta_{w_k}}{2}\right)$. Let $\delta = \min\{\frac{\delta_{w_1}}{2}, \ldots, \frac{\delta_{w_n}}{2}\}$.

Suppose that $x, y \in C$ and $|x - y| < \delta$. Because $y \in C$, there is some $p \in \{1, \ldots, n\}$ such that $y \in \left(w_p - \frac{\delta_{w_p}}{2}, w_p + \frac{\delta_{w_p}}{2}\right)$. Hence $|y - w_p| < \frac{\delta_{w_p}}{2}$. By the definition of δ, we also know that $|x - y| < \frac{\delta_{w_p}}{2}$. It follows that

$$|x - w_p| = |x - y + y - w_p| \le |x - y| + |y - w_p| < \frac{\delta_{w_p}}{2} + \frac{\delta_{w_p}}{2} = \delta_{w_p}.$$

We now deduce from the choice of δ_{w_p} that $|f(y) - f(w_p)| < \frac{\varepsilon}{2}$ and $|f(x) - f(w_p)| < \frac{\varepsilon}{2}$. Therefore

$$\begin{aligned} |f(x) - f(y)| &= |f(x) - f(w_p) + f(w_p) - f(y)| \\ &\le |f(x) - f(w_p)| + |f(w_p) - f(y)| < \frac{\varepsilon}{2} + \frac{\varepsilon}{2} = \varepsilon. \end{aligned}$$ □

Recall the definition of a function being bounded (given in Section 3.2). Is there a relation between continuity or uniform continuity and boundedness? Clearly, a function can be continuous and not bounded, for example the function in Example 3.4.3 (2). A function can also be uniformly continuous and not bounded, for example the function in Example 3.4.3 (1). However, there is something very different about these two examples. For the linear function in Example 3.4.3 (1), the fact that the function is not bounded is due to the fact that the domain is not bounded; that is, if we restrict the function in Example 3.4.3 (1) to any bounded interval, then the restricted function is itself bounded. By contrast, if we restrict the function in Example 3.4.3 (2) to the interval $(0,1)$, then the restricted function is still not bounded, even though the domain of the restricted function is bounded. The difference between these two functions is precisely the difference between uniform continuity and regular continuity. If a function is uniformly continuous, then intuitively $f(x)$ cannot change too much if x does not change too much, and that suggests that the only way a uniformly continuous function can be not bounded is if its domain is not bounded. We now state and prove this fact.

Theorem 3.4.5. *Let $A \subseteq \mathbb{R}$ be a non-empty set, and let $f\colon A \to \mathbb{R}$ be a function. Suppose that A is bounded. If f is uniformly continuous, then f is bounded.*

Proof. Suppose that f is uniformly continuous. Because the set A is bounded, there is some $M \in \mathbb{R}$ such that $|x| \le M$ for all $x \in A$. Hence $A \subseteq [-M, M]$. We may assume that $M > 0$. Because f is uniformly continuous, there is some $\delta > 0$ such that $x, y \in A$ and $|x - y| < \delta$ imply $|f(x) - f(y)| < 1$. By Corollary 2.6.8 (2) there is some $n \in \mathbb{N}$ such that $\frac{1}{n} < \frac{\delta}{2M}$, which implies $\frac{2M}{n} < \delta$.

We now divide the interval $[-M, M]$ into n equal subintervals, which we do by letting $x_0, x_1, \ldots, x_n \in [-M, M]$ be defined by the conditions that $-M = x_0 < x_1 < \cdots < x_n = M$ and that $x_i - x_{i-1} = \frac{2M}{n}$ for all $i \in \{1, \ldots, n\}$.

Let $i \in \{1,\dots,n\}$. Let $E_i \in \mathbb{R}$ be defined as follows. If $A \cap [x_{i-1},x_i] = \emptyset$, then let $E_i = 0$. If $A \cap [x_{i-1},x_i] \neq \emptyset$, then choose some $e_i \in A \cap [x_{i-1},x_i]$ (it does not matter which e_i is chosen), and let $E_i = |f(e_i)| + 1$. Next, let $E = \max\{E_1, E_2, \dots, E_n\}$.

Let $y \in A$. Then $y \in A \cap [x_{k-1},x_k]$ for some $k \in \{1,\dots,n\}$, and therefore $|y - e_k| \leq |x_k - x_{k-1}| = \frac{2M}{n} < \delta$. It follows that $|f(y) - f(e_k)| < 1$, and then Lemma 2.3.9 (7) implies that $|f(y)| < |f(e_k)| + 1 = E_k \leq E$. We deduce that f is bounded, with bound E. □

The following result was proved in Exercise 3.3.8 by a direct use of the Heine–Borel Theorem (Theorem 2.6.14), but a particularly simple proof can be obtained using theorems we have seen in this section. Of course, we cannot really escape the Heine–Borel Theorem here, because it is used in the proof of Theorem 3.4.4, but now that we have proved the latter, we obtain the result stated in Exercise 3.3.8 with no extra work.

Corollary 3.4.6. *Let $C \subseteq \mathbb{R}$ be a closed bounded interval, and let $f : C \to \mathbb{R}$ be a function. If f is continuous, then f is bounded.*

Proof. Suppose that f is continuous. By Theorem 3.4.4 we know that f is uniformly continuous. Because C is bounded, we can apply Theorem 3.4.5 to deduce that f is bounded. □

Reflections

In contrast to the concept of continuity, which from an intuitive point of view is both easy to understand and is familiar from calculus courses, the concept of uniform continuity is neither intuitively simple nor familiar from previous courses. Nonetheless, uniform continuity is a very important technical concept that shows up in the proofs of a number of theorems in real analysis, for example Theorem 5.4.11, which states that continuous functions on non-degenerate closed bounded intervals are integrable.

It is not hard to imagine how someone might have first thought of the concept of continuity at the intuitive level, simply by looking at graphs of functions; the rigorous definition, needless to say, was much harder to think of. We can speculate that uniform continuity, by contrast, might have been first conceptualized not from intuitive considerations but rather as a result of trying to prove theorems such as Theorem 5.4.11, and noticing that something more than the definition of continuity was needed to make the proof work. It would have been subsequently necessary to investigate the relationship between the intuitively familiar concept of continuity and the technically necessary concept of uniform continuity; the culmination of such an investigation would have been Theorem 3.4.4, a theorem that is used in the proof of Theorem 5.4.11. Of course, the actual historical development of mathematics does not always follow the sort of logical order just described, but it is nonetheless useful to think about how mathematical ideas might have developed logically, in order to understand their significance.

The study of uniform continuity should serve to reinforce the importance of quantifiers, because the difference between the definitions of continuity and uniform continuity is precisely in the order of the quantifiers. The formulation of rigorous proofs relies upon a good understanding of the use of quantifiers, and nowhere is this fact more apparent than in proofs involving uniform continuity.

Exercises

Exercise 3.4.1. Using only the definition of uniform continuity, prove that the following functions are uniformly continuous.

(1) Let $f\colon [0,3] \to \mathbb{R}$ be defined by $f(x) = x^2$ for all $x \in [0,3]$.
(2) Let $g\colon [1,2] \to \mathbb{R}$ be defined by $g(x) = \sqrt{x}$ for all $x \in [1,2]$.

Exercise 3.4.2. Using only the definition of uniform continuity, prove that the function $f\colon \mathbb{R} \to \mathbb{R}$ defined by $f(x) = x^2$ for all $x \in \mathbb{R}$ is not uniformly continuous.

Exercise 3.4.3. Let $A \subseteq \mathbb{R}$ be a set, let $f,g\colon A \to \mathbb{R}$ be functions and let $k \in \mathbb{R}$. Suppose that f and g are uniformly continuous.

(1) Prove that $f+g$ is uniformly continuous.
(2) Prove that kf is uniformly continuous.
(3) Find an example to show that fg need not be uniformly continuous.

Exercise 3.4.4. Let $A,B \subseteq \mathbb{R}$ be sets, and let $g\colon A \to B$ and $f\colon B \to \mathbb{R}$ be functions. Suppose that f and g are uniformly continuous. Prove that $f \circ g$ is uniformly continuous.

Exercise 3.4.5. [Used in Exercise 4.4.6, Exercise 4.6.9 and Exercise 10.3.7.] Let $A \subseteq \mathbb{R}$ be a set, and let $f\colon A \to \mathbb{R}$ be a function. The function f satisfies a **Lipschitz condition** if there is some $K \in \mathbb{R}$ such that $|f(x) - f(y)| \le K|x - y|$ for all $x,y \in A$; the number K is called a **Lipschitz constant** for f.

(1) Prove that if f satisfies a Lipschitz condition, then f is uniformly continuous.
(2) Find an example of a function $g\colon [0,\infty) \to \mathbb{R}$ that is uniformly continuous but does not satisfy a Lipschitz condition.

Exercise 3.4.6. [Used in Theorem 5.5.4.] Let $n \in \mathbb{N}$, and let $[a_1,b_1],\ldots,[a_n,b_n] \subseteq \mathbb{R}$ be closed bounded intervals. Let $f\colon [a_1,b_1] \cup \cdots \cup [a_n,b_n] \to \mathbb{R}$ be a function. Prove that if f is continuous, then f is uniformly continuous and bounded.

[Use Exercise 2.5.14.]

Exercise 3.4.7. Find an example of a function $f\colon \mathbb{R} \to \mathbb{R}$ that is continuous and bounded, but that is not uniformly continuous. Be sure to prove that the function is not uniformly continuous.

Exercise 3.4.8. Find an example of two disjoint, non-empty sets $A,B \subseteq \mathbb{R}$ and a function $f\colon A \cup B \to \mathbb{R}$ such that $f|_A$ and $f|_B$ are uniformly continuous, but that f is not uniformly continuous.

Exercise 3.4.9. Let $(a,b) \subseteq \mathbb{R}$ be a non-degenerate open bounded interval, and let $f\colon (a,b) \to \mathbb{R}$ be a function. Prove that f is uniformly continuous if and only if f can be extended to a continuous function $F\colon [a,b] \to \mathbb{R}$.

When proving that if f is uniformly continuous then f can be extended to a continuous function $F\colon [a,b] \to \mathbb{R}$, it suffices to prove that f can be extended to a continuous function $G\colon [a,b) \to \mathbb{R}$; extending G to a continuous function $F\colon [a,b] \to \mathbb{R}$ is completely analogous, and the details can be omitted. To define G, use the one-sided analog of Exercise 3.2.18. [Use Exercise 3.3.2.]

3.5 Two Important Theorems

We are now ready to state and prove two very important theorems concerning continuous functions defined on closed bounded intervals, which are the Extreme Value Theorem and the Intermediate Value Theorem. Both of these theorems are encountered informally in calculus courses, but in real analysis we see their worth more clearly, because they are useful tools in the proofs of important theorems that we will subsequently encounter. For example, the Extreme Value Theorem will be used in the proof of Rolle's Theorem (Lemma 4.4.3), which in turn is used in the proof of the Mean Value Theorem (Theorem 4.4.4), and it is also used in the proof that all continuous functions are integrable (Theorem 5.4.11). The Intermediate Value Theorem is used in the proof that the natural logarithm function is bijective (Lemma 7.2.4), which in turn is used to define the exponential function.

Although both of these theorems are concerned with continuous functions defined on closed bounded intervals, it turns out that they involve very different aspects of closed bounded intervals. This difference can be fully understood only via the study of the concepts of "compactness" and "connectedness," which are treated in an introductory course in point set topology. Indeed, each of the Extreme Value Theorem and the Intermediate Value Theorem can be greatly generalized through the use of these topological concepts. See the classic text [Mun00] for an introduction to point set topology, including connectedness and compactness, and generalized versions of the Extreme Value Theorem and the Intermediate Value Theorem. These generalizations are an instance where greater abstraction actually leads to greater clarity.

Our first theorem, the Extreme Value Theorem, concerns the existence of maximum and minimum values of functions. Must every function have a maximum value and a minimum value? The answer is clearly no, because of functions such as $f\colon \mathbb{R} \to \mathbb{R}$ defined by $f(x) = x$ for all $x \in \mathbb{R}$. Can we find criteria that would guarantee that a function has a maximum value and a minimum value? The function f has a domain that is not bounded, and so it is natural to ask whether functions with bounded domains always have maximum and minimum values, but again the answer is no. The function $g\colon (0,1) \to \mathbb{R}$ defined by $g(x) = \frac{1}{x}$ for all $x \in (0,1)$ has neither maximum value nor minimum value. The apparent problem with the function g is that its domain is an open interval. Would it suffice to restrict our attention to functions with domains that are closed bounded intervals? The answer is still no. The function $h\colon [0,1] \to \mathbb{R}$

defined by

$$h(x) = \begin{cases} \frac{1}{x}, & \text{if } x \in (0,1] \\ 0, & \text{if } x = 0 \end{cases}$$

has a minimum value at $x = 0$, but it does not have a maximum value. The problem with the function h is that it is not continuous. As seen in the following theorem, we have found all possible difficulties, because every continuous function with domain a closed bounded interval has a maximum value and a minimum value. Observe that in the statement of our theorem, we are not concerned with the actual maximum value and minimum value of the function, but only that a maximum value and a minimum value occur somewhere in the domain.

Theorem 3.5.1 (Extreme Value Theorem). *Let $C \subseteq \mathbb{R}$ be a closed bounded interval, and let $f\colon C \to \mathbb{R}$ be a function. Suppose that f is continuous. Then there are $x_{min}, x_{max} \in C$ such that $f(x_{min}) \le f(x) \le f(x_{max})$ for all $x \in C$.*

Proof. By Corollary 3.4.6 we know that f is bounded, which means that the set $f(C)$ is bounded. Because $C \ne \emptyset$, then $f(C) \ne \emptyset$. The Least Upper Bound Property and the Greatest Lower Bound Property imply that $f(C)$ has a least upper bound and a greatest lower bound. We will show that there is some $x_{max} \in C$ such that $f(x_{max}) = \operatorname{lub} f(C)$. It will follow that $f(x) \le f(x_{max})$ for all $x \in C$. A similar proof can be used to find x_{min}, and we omit the details.

For convenience let $M = \operatorname{lub} f(C)$. Then $f(x) \le M$ for all $x \in C$. Suppose that $f(x) < M$ for all $x \in C$. Let $g\colon C \to \mathbb{R}$ be defined by

$$g(x) = \frac{1}{M - f(x)}$$

for all $x \in C$. Because f is continuous, and because the denominator in the definition of g is never zero, it follows from Example 3.3.3 (1) and Theorem 3.3.5 that g is continuous.

By Corollary 3.4.6 again we know that g is bounded. Hence there is some $P \in \mathbb{R}$ such that $|g(x)| \le P$ for all $x \in C$. Observe that $P > 0$. Moreover, because we are assuming that $f(x) < M$ for all $x \in C$, it follows that $g(x) > 0$ for all $x \in C$. Hence $g(x) \le P$ for all $x \in C$, which means that

$$\frac{1}{M - f(x)} \le P$$

for all $x \in C$. Therefore

$$f(x) \le M - \frac{1}{P}$$

for all $x \in C$. We deduce that $M - \frac{1}{P}$ is an upper bound of $f(C)$, which is a contradiction to the fact that $M = \operatorname{lub} f(C)$. It is therefore not the case that $f(x) < M$ for all $x \in C$. Because $f(x) \le M$ for all $x \in C$, there must be some $x_{max} \in C$ such that $f(x_{max}) = M$. □

It is important to note that the numbers x_{min} and x_{max}, whose existence is guaranteed in the statement of the Extreme Value Theorem (Theorem 3.5.1), are not necessarily unique. Moreover, the Extreme Value Theorem does not tell us how to find x_{min} and x_{max}; we are told only that they exist. This theorem is an example of an "existence theorem," as is the Intermediate Value Theorem, to which we now turn.

Suppose that $f\colon [a,b] \to \mathbb{R}$ is a function for some closed bounded interval $[a,b] \subseteq \mathbb{R}$. Must it be the case that f takes on all values between $f(a)$ and $f(b)$? The answer is clearly no. Let $k\colon [0,1] \to \mathbb{R}$ be defined by

$$k(x) = \begin{cases} 1, & \text{if } x \in (0,1] \\ 0, & \text{if } x = 0. \end{cases}$$

Then $k(0) = 0$ and $k(1) = 1$, but k does not take on any values between 0 and 1. Of course, the function k is not continuous, and that is the source of the problem. The following theorem states, as expected, that a continuous function $f\colon [a,b] \to \mathbb{R}$ takes on all values between $f(a)$ and $f(b)$.

Theorem 3.5.2 (Intermediate Value Theorem). *Let $[a,b] \subseteq \mathbb{R}$ be a closed bounded interval, and let $f\colon [a,b] \to \mathbb{R}$ be a function. Suppose that f is continuous. Let $r \in \mathbb{R}$. If r is strictly between $f(a)$ and $f(b)$, then there is some $c \in (a,b)$ such that $f(c) = r$.*

Proof. Suppose that r is strictly between $f(a)$ and $f(b)$. Without loss of generality, assume that $f(a) < r < f(b)$. Let

$$S = \{x \in [a,b] \mid f(x) < r\}.$$

Then $S \subseteq [a,b]$. The set S is non-empty because $a \in S$, and S is bounded above by b. The Least Upper Bound Property implies that S has a least upper bound. Let $c = \operatorname{lub} S$. Because $a \in S$, then $a \le c$, and because b is an upper bound of S, then $c \le b$. Hence $c \in [a,b]$. We will show that $f(c) = r$, which we do by showing that $f(c) \le r$ and that $f(c) \ge r$.

Because $S \ne \emptyset$, then $f(S) \ne \emptyset$. It is evident from the definition of S that $f(S)$ is bounded above by r. The Least Upper Bound Property again implies that $f(S)$ has a least upper bound. By Exercise 3.3.4 (1) we see that $f(\operatorname{lub} S) \le \operatorname{lub} f(S)$. Hence $f(c) \le \operatorname{lub} f(S)$. Because r is an upper bound of $f(S)$, it follows that $\operatorname{lub} f(S) \le r$, and therefore $f(c) \le r$.

Because $f(c) \le r < f(b)$, we see that $c \ne b$. Hence $c < b$. It follows that the interval $(c,b]$ is non-degenerate. Let $B = (c,b]$. Then $f(B) \ne \emptyset$. Clearly $c = \operatorname{glb} B$. Moreover, because $c = \operatorname{lub} S$, it follows that $B \subseteq [a,b] - S$. Hence $f(x) \ge r$ for all $x \in B$. Therefore $f(B)$ is bounded below by r. The Greatest Lower Bound Property implies that $f(B)$ has a greatest lower bound. By Exercise 3.3.4 (2) we see that $f(\operatorname{glb} B) \ge \operatorname{glb} f(B)$. Hence $f(c) \ge \operatorname{glb} f(B)$. Because r is a lower bound of $f(B)$, it follows that $\operatorname{glb} f(B) \ge r$, and therefore $f(c) \ge r$. We deduce that $f(c) = r$.

Finally, because $r \ne f(a)$ and $r \ne f(b)$, it follows that $c \ne a$ and $c \ne b$. Therefore $c \in (a,b)$. □

Similarly to the Extreme Value Theorem (Theorem 3.5.1), the Intermediate Value Theorem (Theorem 3.5.2) is also an existence theorem, in that we are given no information on how to find the number c whose existence is guaranteed by the theorem. Also similarly, the number c is not necessarily unique.

The proofs of the Extreme Value Theorem and the Intermediate Value Theorem both rely upon the Least Upper Bound Property. In fact, we will now show that both of these theorems are equivalent to the Least Upper Bound Property. While we are at it, we will also show that the Heine–Borel Theorem (Theorem 2.6.14) is equivalent to the Least Upper Bound Property.

What we mean by "equivalent" in this context is as follows. As stated in Section 2.2, we have taken as our hypotheses for $\mathbb{R}$ the axiom for an ordered field and the Least Upper Bound Property; from these assumptions we deduce all our results in real analysis. To say that a theorem that we have proved is equivalent to the Least Upper Bound Property means that if an ordered field F is assumed to satisfy this theorem, then the Least Upper Property Property can be proved for F. In other words, for an ordered field, the Least Upper Bound Property and the other theorem each imply the other. The proof of the equivalence of various theorems with the Least Upper Bound Property will be by contrapositive, where we suppose that F is an ordered field that does not satisfy the Least Upper Bound Property, and where we then show that the various theorems do not hold.

Observe that any ordered field, whether or not it satisfies the Least Upper Bound Property, satisfies all the properties of $\mathbb{R}$ that do not rely upon the Least Upper Bound Property, for example all the properties of $\mathbb{R}$ that are proved in Sections 2.3–2.5. Moreover, none of the results concerning limits and continuity that we saw in Sections 3.2 and 3.3 rely upon the Least Upper Bound Property, and hence they hold for all ordered fields.

We start with the following lemma about ordered fields that do not satisfy the Least Upper Bound Property.

Lemma 3.5.3. *Let F be an ordered field. Suppose that F does not satisfy the Least Upper Bound Property. Let $A \subseteq F$ be a non-empty set such that A is bounded above, but A has no least upper bound. Let $a \in A$, and let $b \in F$ be an upper bound of A. Let $Q = \{x \in [a,b] \mid x \text{ is an upper bound of } A\}$ and $P = [a,b] - Q$.*

1. *$P \cup Q = [a,b]$ and $P \cap Q = \emptyset$.*
2. *$a < b$, and $A \cap [a,b] \subseteq P$, and $a \in P$, and $b \in Q$.*
3. *If $x \in P$ and $z \in Q$, then $x < z$.*
4. *If $x \in P$, then there is some $y \in P$ such that $x < y$. If $z \in Q$, then there is some $w \in Q$ such that $w < z$.*
5. *The set P does not have a least upper bound, and the set Q does not have a greatest lower bound.*

Proof.

(1) This part of the lemma is true by the definition of P and Q.

(2) Because $a \in A$ and b is an upper bound of A, it follows that $a \leq b$, and that $b \in Q$. Because A has no least upper bound, then by Exercise 2.6.2 (2) we know that no upper bound of A is in A. Therefore $A \cap Q = \emptyset$, and hence $A \cap [a,b] \subseteq P$. In particular, we see that $a \in P$. By Part (1) of this lemma we know that $P \cap Q = \emptyset$, and therefore $a \neq b$. It follows that $a < b$.

(3) Let $U = \{x \in F \mid x \text{ is an upper bound of } A\}$ and $L = F - Q$. It follows from Exercise 2.6.4 that if $x \in L$ and $y \in U$, then $x < y$; although that exercise was stated for $\mathbb{R}$, its proof did not involve the Least Upper Bound Property, and hence it also holds for the ordered field F. Because $Q \subseteq U$ and $P \subseteq L$, it follows immediately that if $x \in P$ and $z \in Q$, then $x < z$.

(4) Let $x \in P$. Then x is not an upper bound of A by the definition of P, and hence there is some $y \in A$ such that $x < y$. Because b is an upper bound of A, then $y \leq b$, and because $x \in P \subseteq [a,b]$, then $a \leq x < y$. Hence $y \in A \cap [a,b]$, and it follows from Part (2) of this lemma that $y \in P$. Let $z \in Q$. Then z is an upper bound of A. Because A has no least upper bound, there is another upper bound w of A such that $w < z$. Then $a \leq w < z \leq b$, and hence $w \in Q$.

(5) It follows from Part (3) of this lemma that everything in P is a lower bound of Q, and everything in Q is an upper bound of P.

Suppose that Q has a greatest lower bound. Because a is a lower bound of Q and $b \in Q$, then $\operatorname{glb} Q \in [a,b]$. There are now two cases. First, suppose that $\operatorname{glb} Q \in Q$. Then $\operatorname{glb} Q$ is an upper bound of A. Let w be an upper bound of A. Because $a \in A$, then $w \geq a$. If $w \leq b$, then $w \in Q$, and hence $\operatorname{glb} Q \leq w$. If $w \geq b$, then $\operatorname{glb} Q \leq b \leq w$. Therefore $\operatorname{glb} Q$ is the least upper bound of A, which is a contradiction. Second, suppose that $\operatorname{glb} Q \in P$. By Part (4) of this lemma there is some $y \in P$ such that $\operatorname{glb} Q < y$, which is a contradiction to the fact that $\operatorname{glb} Q$ is the greatest lower bound of Q, because y is a lower bound of Q. We conclude that Q does not have a greatest lower bound.

Now suppose that P has a least upper bound. Because $a \in P$ and b is an upper bound of P, then $\operatorname{lub} P \in [a,b]$. Let $v \in A$. Then v is not an upper bound of A, as noted in the proof of Part (2) of this exercise. Because b is an upper bound of A, then $v \leq b$. If $v \geq a$, then $v \in P$, and hence $\operatorname{lub} P \geq v$. If $v \leq a$, then $\operatorname{lub} P \geq a \geq v$. Therefore $\operatorname{lub} P$ is an upper bound of A, which implies that $\operatorname{lub} P \in Q$. By Part (4) of this lemma there is some $w \in Q$ such that $w < \operatorname{lub} P$, which is a contradiction to the fact that $\operatorname{lub} P$ is the least upper bound of P. We conclude that P does not have a least upper bound. □

Theorem 3.5.4. *The following are equivalent.*

a. The Least Upper Bound Property.
b. The Heine–Borel Theorem.
c. The Extreme Value Theorem.
d. The Intermediate Value Theorem.

Proof. We have already seen that the axioms of the real numbers, that is, the axiom for an ordered field together with the Least Upper Bound Property, imply the Heine–Borel Theorem, the Extreme Value Theorem and the Intermediate Value Theorem.

We will now show that each of these theorems, together with the axiom for an ordered field, implies the Least Upper Bound Property, which we will do by letting F be an ordered field that does not satisfy the Least Upper Bound Property, and then deducing that F does not satisfy any of the Heine–Borel Theorem, the Extreme Value Theorem and the Intermediate Value Theorem.

Let a, b, A, P and Q be as in Lemma 3.5.3. By Parts (1) and (2) of that lemma we know that $P \cup Q = [a,b]$, and $P \cap Q = \emptyset$, and $a < b$, and $A \cap [a,b] \subseteq P$, and $a \in P$, and $b \in Q$.

Let $x \in P$. By Lemma 3.5.3 (4) there is some $d_x \in P$ such that $x < d_x$. Let $c_x = x - 1$. Then $x \in (c_x, d_x)$. Let $u \in [a,b] \cap (c_x, d_x)$. Then $u < d_x$. Because $d_x \in P$, it follows from Lemma 3.5.3 (3) that $u \in P$. Hence $[a,b] \cap (c_x, d_x) \subseteq P$. Let $z \in Q$. By a similar argument there is an open interval (s_z, t_z) in $\mathbb{R}$ such that $s_z \in Q$, and $z \in (s_z, t_z)$, and $[a,b] \cap (s_z, t_z) \subseteq Q$.

Because $P \cup Q = [a,b]$, then

$$[a,b] \subseteq \bigcup_{x \in P} (c_x, d_x) \cup \bigcup_{z \in Q} (s_z, t_z).$$

Let $p, q \in \mathbb{N}$, and $x_1, x_2, \ldots, x_p \in P$, and $z_1, z_2, \ldots, z_q \in Q$. We claim that

$$[a,b] \not\subseteq \bigcup_{i=1}^{p} (c_{x_i}, d_{x_i}) \cup \bigcup_{j=1}^{q} (s_{z_j}, t_{z_j}).$$

Let $d = \max\{d_{x_1}, d_{x_2}, \ldots, d_{x_p}\}$. Then $d \in P$. By Lemma 3.5.3 (4) there is some $w \in P$ such that $d < w$. Then

$$w \notin \bigcup_{i=1}^{p} (c_{x_i}, d_{x_i}).$$

By Lemma 3.5.3 (3) we know that $w < s_{z_j}$ for all $j \in \{1, \ldots, q\}$. Then

$$w \notin \bigcup_{j=1}^{q} (s_{z_j}, t_{z_j}).$$

It follows that

$$w \notin \bigcup_{i=1}^{p} (c_{x_i}, d_{x_i}) \cup \bigcup_{j=1}^{q} (s_{z_j}, t_{z_j}),$$

which proves the claim. We have therefore seen that the family $\{(c_x, d_x)\}_{x \in P} \cup \{(s_z, t_z)\}_{z \in Q}$ of open intervals satisfies the hypothesis of the Heine–Borel Theorem, but not the conclusion of the theorem. Therefore the Heine–Borel Theorem does not hold for F.

Next, let $f \colon [a,b] \to \mathbb{R}$ be defined by

$$f(x) = \begin{cases} x, & \text{if } x \in P \\ a-1, & \text{if } x \in Q. \end{cases}$$

Let $v \in [a,b]$. We will show that f is continuous at v. There are two cases. First, suppose that $v \in P$. Then $v \in (c_v, d_v)$, and $[a,b] \cap (c_v, d_v) \subseteq P$. Because (c_v, d_v) is an open interval, by Lemma 2.3.7 (2) there is some $\delta > 0$ such that $(v-\delta, v+\delta) \subseteq (c_v, d_v)$. It follows that $[a,b] \cap (v-\delta, v+\delta) \subseteq P$. Let $g = f|_{[a,b]\cap(v-\delta,v+\delta)}$. Then $g(x) = x$ for all $x \in [a,b] \cap (v-\delta, v+\delta)$. Therefore g is continuous at v by Example 3.3.3 (1) and Exercise 3.3.2 (1), neither of which relies upon the Least Upper Bound Property, and hence both work for F. It now follows from Exercise 3.3.2 (1) that f is continuous at v, where, again, the exercise does not rely upon the Least Upper Bound Property. A similar argument works when $v \in Q$, and we omit the details. We deduce that f is continuous.

If $x \in Q$, then $f(x) = a - 1 < a = f(a)$, and if $x \in P$, then by Lemma 3.5.3 (4) there is some $y \in P$ such that $x < y$, and hence $f(x) = x < y = f(y)$. It follows that there is no $x_{max} \in [a,b]$ such that $f(x) \leq f(x_{max})$ for all $x \in [a,b]$. Hence the Extreme Value Theorem does not hold for F.

Finally, let $g\colon [a,b] \to \mathbb{R}$ be defined by

$$g(x) = \begin{cases} 0, & \text{if } x \in P \\ 1, & \text{if } x \in Q. \end{cases}$$

A similar argument to the one used with the function f shows that g is continuous. On the other hand, by the arguments used in Section 2.4, all of which apply to F, we know that $\mathbb{Q} \subseteq F$, and in particular that $\frac{1}{2} \in F$. Hence the Intermediate Value Theorem does not hold for F. □

Because of Theorem 3.5.4, the reader might jump to the conclusion that all the important theorems of real analysis that make use of the Least Upper Bound Property in their proofs are equivalent to this property, but that is not the case. For example, the Archimedean Property (Theorem 2.6.7) is a very important and useful theorem, and its proof makes use of the Least Upper Bound Property, but the Archimedean Property is in fact not equivalent to the Least Upper Bound Property, for the following reason. It was proved in Exercise 2.4.10 that the Archimedean Property holds for the rational numbers, and it therefore follows that the axiom for an ordered field together with the Archimedean Property cannot imply the Least Upper Bound Property, because the latter property is not satisfied by the rational numbers. However, we note that our use of the Least Upper Bound Property in the proof of the Archimedean Property for the real numbers was necessary, and was not simply a matter of convenience. As mentioned in Section 2.6, there exist ordered fields that do not satisfy the Archimedean Property, and hence any proof of the Archimedean Property for the real numbers must ultimately rely upon some aspect of the real numbers beyond the axiom for an ordered field, and the only axiom for the real numbers other than that of an ordered field is the Least Upper Bound Property.

Reflections

The two theorems referred to in the title of this section, the Extreme Value Theorem and the Intermediate Value Theorem, are models of why we study real analysis. Both theorems are reasonably clear intuitively, and yet they are rather difficult to prove. It would be nice if every proof of every theorem were not only simple, but also provided a clear insight into why the theorem is true. Unfortunately, it happens regularly in mathematics that theorems that seem to be intuitively true have no known simple or direct proof. Of course, we cannot dispense with uninviting proofs when no better ones are available.

The final result in this section, which shows that the Least Upper Bound Property is equivalent to the Extreme Value Theorem, the Intermediate Value Theorem and the Heine–Borel Theorem, is an example of what is known in the mathematical world as a "folk theorem," which is a result that everyone knows is true, but the proof of which is either not written down anywhere, or is written down somewhere but is not widely known. The author has many times over the years told his students in real analysis courses that these three theorems (and some other theorems too, as seen at the end of Section 8.3) are logically equivalent to the Least Upper Bound Property, but it was only during the writing of this text that the author realized that he had never actually seen a proof of this equivalence, and that if he wanted to continue making such a claim, he would need to figure out a proof, which is the one given in this section. The author subsequently found a similar proof in [Olm62, Appendix Sections 1–3].

Exercises

Exercise 3.5.1.

(1) Find an example of a function $f\colon [0,1] \to \mathbb{R}$ such that f is not continuous, but that f satisfies the conclusion of the Extreme Value Theorem.

(2) Find an example of a function $f\colon [0,1] \to \mathbb{R}$ such that f is not continuous, but that f satisfies the conclusion of the Intermediate Value Theorem.

Exercise 3.5.2. Let $[a,b] \subseteq \mathbb{R}$ be a closed bounded interval, let $c \in (a,b)$ and let $f\colon [a,b] \to \mathbb{R}$ be a function. Suppose that $f|_{[a,c]}$ and $f|_{[c,b]}$ both satisfy the conclusion of the Intermediate Value Theorem. Prove that f satisfies the conclusion of the Intermediate Value Theorem.

Exercise 3.5.3. Let $[a,b] \subseteq \mathbb{R}$ be a closed bounded interval, let $k \in \mathbb{R}$ and let $f\colon [a,b] \to \mathbb{R}$ be a function. Suppose that f satisfies the conclusion of the Intermediate Value Theorem. Prove that kf satisfies the conclusion of the Intermediate Value Theorem.

Exercise 3.5.4. Let $[a,b] \subseteq \mathbb{R}$ be a closed bounded interval, and let $f\colon [a,b] \to [a,b]$ be a function. Suppose that f is continuous. Prove that there is some $c \in [a,b]$ such that $f(c) = c$. The number c is called a **fixed point** of f.

Exercise 3.5.5. Let $[a,b] \subseteq \mathbb{R}$ be a closed bounded interval, and let $f\colon [a,b] \to \mathbb{R}$ be a function. Suppose that f is continuous. Prove that $f([a,b])$ is a closed bounded interval.

Exercise 3.5.6. **[Used in Section 2.6, Example 4.6.3, Section 7.1, Exercise 7.2.11 and Example 10.2.8.]** In this exercise we use the Intermediate Value Theorem to prove that every positive real number has an nth root, for every $n \in \mathbb{N}$. More precisely, let $x \in (0,\infty)$, and let $n \in \mathbb{N}$. We will show that there is a unique $z \in (0,\infty)$ such that $z^n = x$. The number z is called the nth **root** of x, and is denoted $\sqrt[n]{x}$. The **square root** of x, denoted $\sqrt{x}$, is another name for $\sqrt[2]{x}$.

If $n = 1$ this result is trivial, so we suppose that $n > 1$.

(1) Let $a,b \in (0,\infty)$. Suppose that $a < b$. Prove that $a^n < b^n$.
(2) Let $a \in (1,\infty)$. Prove that $1 < a < a^n$.
(3) Let $a \in (0,1)$. Prove that $0 < a^n < 1$.
(4) Prove that if $\sqrt[n]{x}$ exists, then it is unique.
(5) Prove that if $x \geq 1$, then $\sqrt[n]{x}$ exists.
(6) Prove that if $0 < x < 1$, then $\sqrt[n]{x}$ exists.

Exercise 3.5.7. Let $p\colon \mathbb{R} \to \mathbb{R}$ be a polynomial function. Suppose that p has odd degree. The purpose of this exercise is to prove that p has a root. This fact is often stated in calculus courses, where it is justified by referring to the Intermediate Value Theorem, and citing the (usually unproved) fact that "if x gets very large then the highest degree term dominates the other terms in the polynomial." The Intermediate Value Theorem (Theorem 3.5.2) is certainly needed here, but we avoid the informal fact cited above as follows.

Suppose that f has the form $f(x) = a_0 + a_1 x + \cdots + a_n x^n$ for all $x \in \mathbb{R}$, for some $n \in \mathbb{N} \cup \{0\}$ and $a_0, a_1, \ldots, a_n \in \mathbb{R}$. Suppose that n is odd, and that $a_n \neq 0$.

(1) Prove that there is some $c \in (0,\infty)$ such that $\left|\frac{p(c)}{a_n c^n} - 1\right| < 1$ and $\left|\frac{p(-c)}{a_n (-c)^n} - 1\right| < 1$.
(2) Use Part (1) of this exercise to prove that there is some $r \in [-c,c]$ such that $f(r) = 0$.

Exercise 3.5.8. **[Used in Exercise 10.4.12.]** Let $[a,b] \subseteq \mathbb{R}$ be a closed bounded interval, let $f\colon [a,b] \to \mathbb{R}$ be a function and let $r \in \mathbb{R}$. Suppose that f is continuous, and that $f(a) < r < f(b)$. Let $S = \{x \in [a,b] \mid f(x) \geq r\}$. Prove that $S \neq \emptyset$, that $\operatorname{glb} S \in (a,b]$ and that $f(\operatorname{glb} S) \geq r$.

The proof may be simplified as follows. Let $g\colon [a,b] \to \mathbb{R}$ be defined by $g(x) = f(x) - r$ for all $x \in [a,b]$. Then $g(a) < 0 < g(b)$, and $S = \{x \in [a,b] \mid g(x) \geq 0\}$, and g is continuous. Hence it suffices to prove the desired result for g instead of f, where we replace r with 0. [Use Exercise 3.3.4 (2).]

3.6 Historical Remarks

From the modern perspective, calculus makes essential use of the concept of the limit of a function and the related concept of a continuous function. Historically, however, the early development of calculus relied upon the somewhat related notion of infinitesimals rather than the later-developed idea of a limit. An infinitesimal is an "infinitely small" but positive number; that is, a positive number that is smaller than any positive real number. In the standard approach to the real numbers and real analysis that is used today, for example as discussed in this text, infinitesimals do not exist (as is proved in Lemma 2.3.10). However, the intuitive notion of infinitesimals was crucial to the development of calculus, whether or not any such thing exists. The idea of an infinitely small number is related to the ideas of infinitely large numbers and infinite processes; hence, our discussion of the history of the concepts of limits and continuity commences with a brief mention of the ancient Greek approach to the infinite, which was influential for many centuries in Europe. The development of calculus, though in part spurred on by ancient Greek successes, is coincident with the waning of the influence of the ancient Greek way of doing mathematics.

Ancient World

Infinitesimals arose, in a non-mathematical way, in ancient Greek thought via the approach of atomists such as Democritus of Abdera (c. 460–c. 370 BCE). The atomist approach was formulated as an attempt to resolve the four arguments of Zeno of Elea (c. 490–c. 425 BCE), who wanted to show that there is no motion. Aristotle (384–322 BCE) rejected infinitely small indivisibles as part of his attempt to refute Zeno's arguments; these arguments are known today via their summary and counter-arguments in Book VI of Aristotle's *Physics*. Aristotle rejected the infinitely large, though he accepted the infinite existing as a potential. One of his arguments against the infinitely large was based upon cardinalities of sets, where he assumed that a proper subset always has smaller cardinality than the original set, and from that he ruled out infinite sets (a correct deduction, but a false assumption).

Archimedes (287–212 BCE) proved various area and volume results using the method of exhaustion, which avoids taking limits or using indivisibles by using a double reductio ad absurdum (proof by contradiction). Such proofs give no clue as to how these results were first discovered. However, in the *Method*, which was found only in 1899, Archimedes explained his method of discovery, which was based upon mechanical ideas. As part of his method of discovery, which he was careful to state was no substitute for a proper proof, Archimedes used the idea that the region of the plane or space is made up of indivisible sections; because this use of indivisibles was stated without explanation, it might have been the case that such an informal use of indivisibles for discovery (though not proof) was known to Archimedes' contemporaries.

Medieval Period

When explaining his graphical representation of certain functions, Nicole Oresme (1323–1382), around 1350, had the view that measurable quantities (excluding whole numbers) vary continuously, and he suggested the idea of a mathematical indivisible (though he acknowledged that physical indivisibles did not exist). Both of these ideas would be used by later mathematicians in the development of calculus.

Nicholas of Cusa (1401–1464) helped bring infinitesimals and the infinite into mathematics. For example, he viewed a circle as a polygon with infinitely many sides, and used that to find the area of a circle, an approach later used by Stevin. He did not contribute any mathematical work that was important per se, but his approach may well have influenced later mathematicians in their use of the infinite, for example Kepler.

The rise of Platonism (helped by Nicholas of Cusa, among others), and the corresponding decline in the influence of Aristotle, allowed for ideas to be established by the intellect, rather than being justified solely on the basis of empirical observation. This idea that mathematics is independent of empirical observation, or possibly prior to it, helped allow for the acceptance of speculative ideas such as the infinite and the infinitesimal, as long as the result of such speculation did not lead to problematic results. Such an approach in general, and the liberation from Aristotle's rejection of the infinite in particular, was crucial for the development of calculus.

Renaissance

Ancient Greek mathematical texts in Latin translation became widely known in Europe in the 16th century, and elicited great interest. One such text was *On the Equilibrium of Planes* by Archimedes, which is about centers of gravity. Simon Stevin (1548–1620) made a significant step forward in the development of limits in his study of centers of gravity in *De Beghinselen der Weeghconst* of 1586. When approximating centers of gravity of curved regions by using inscribed polygons, rather than using the cumbersome (though logical) reductio ad absurdum argument of Archimedes, which was used to avoid taking limits while approximating regions by polygons and polyhedra with ever more sides, Stevin tried to simplify the limit argument by saying that if two quantities differ they do so by a finite amount, and therefore in order to show that two quantities are equal it suffices to show that they differ by less than any finite amount. (This fact, in modern notation, follows from Lemma 2.3.10.)

In his study of hydrostatics *De Beghinselen des Waterwichts* of 1586, Stevin used what he called "proof by means of numbers," where he showed that the average pressure on a vertical square wall of a vessel full of water corresponds to the pressure at the midpoint by working through an example, in which he subdivided the wall into n horizontal strips (first $n = 4$, then 10, then 1000), and he then showed that in each case the answer is between $\frac{1}{2} - \frac{1}{2n}$ and $\frac{1}{2} + \frac{1}{2n}$. He noted that the same idea holds in general, and concluded that the difference between the actual pressure and $\frac{1}{2}$ can be made smaller than any desired quantity, and so the pressure is $\frac{1}{2}$. This approach is similar to a limit, though Stevin did not have the general definition of that concept, and it seems that he did not actually believe in infinite processes. He said that he preferred the ancient Greek approach, and that his method was only an illustration of his results, not a proof. Nonetheless, Stevin's work helped promote the idea of limits, and helped move away from the Archimedean reductio ad absurdum, although the subsequent rise of Cavalieri's use of indivisibles temporarily moved mathematicians away from the limit idea. Stevin's work may have had some influence on Kepler, Cavalieri and Grégoire de Saint-Vincent.

Another person who tried to use some sort of limit idea instead of reductio ad absurdum as part of an attempt to solve some center of gravity problems was Luca Valerio (1552–1618) in 1604. He was probably unaware of the earlier work of Stevin, but had a similar approach in that he did not explicitly think of limits, but rather stated some propositions that allowed him to avoid the details of reductio ad absurdum. Without an adequate notion of functions, however, he could not completely succeed. Cavalieri and Grégoire de Saint-Vincent were familiar with Valeiro's work.

The work of people such as Stevin and Valerio was, in a sense, a continuation of the ancient Greek approach, rather than the new approach that was to be developed in the 17th century. Steven and Valerio, in the course of trying to simplify the Archimedean approach, were responding to that approach, as opposed to subsequent work which was focused more on discovering new results and developing computational techniques than on formulating proofs, and which was less in reference to Archimedes. In the 19th century there was a move to regain rigor once again, though it was entirely different from the ancient Greek approach, and this time was very much based upon limits, as people such as Stevin and Valerio wanted to do, though not at all using their specific methods.

Seventeenth Century

Perhaps the first person to deal explicitly with the type of limit that earlier mathematicians such as Steven and Valerio, and perhaps Archimedes, had in mind but did not state explicitly was Grégoire de Saint-Vincent (1584–1667) in 1647. In contrast to those earlier mathematicians, who subdivided regions only as much as was needed to get the error less than a given amount, which means that only finite subdivisions were used, Grégoire de Saint-Vincent thought of using an actual infinite subdivision, which led him close to the idea of the limit of an infinite process, though he was not rigorous from a modern perspective.

In contrast to Grégoire de Saint-Vincent's informal idea of limits, a more common approach in the 17th century as a replacement for the reductio ad absurdum approach of Archimedes was the use of infinitesimals or indivisibles. These two notions, though used somewhat similarly, are not entirely the same; infinitesimals are infinitely small parts of a region but have the same number of dimensions as the whole region, whereas indivisibles are one dimension lower. One of the first to use this type of approach for finding areas and volumes was Johannes Kepler (1571–1630). In *Nova stereometria doliorum vinariorum* of 1615, Kepler computed volumes of solids of revolution. This work, which was meant to be of practical use for finding volumes of wine casks, focused on getting results rather than using Archimedean niceties, and in it Kepler used infinitesimals freely to obtain his results. For example, Kepler found the area of a circle by cutting it up into infinitely many infinitesimally thin triangles meeting at the center of the circle, and then rearranging them into a single triangle (the idea of viewing the circle as made up of infinitely many triangles was not new, being found, for example, in the work of Nicholas of Cusa and Viète).

Galileo Galilei (1564–1642) never published a work on indivisibles per se, but he used indivisibles and the infinite in his landmark work *Discorsi e dimostrazioni matematiche, intorno a due nuove scienze* of 1638 (often referred to as *Two New Sciences*). He was probably influenced by Kepler. Galileo, who was also influenced by the scholastics in his approach to the infinite, warned in *Two New Sciences* against treating the infinite the same way as the finite.

Bonaventura Cavalieri (1598–1647), a pupil of Galileo, wrote two important works using indivisibles, *Geometria indivisibilibus continuorum nova quadam ratione promota* of 1635 and *Exercitationes geometricae sex* of 1647. The former, which was entirely about the use of indivisibles, and was the first such book, received a lot of attention and was widely discussed. Cavalieri took the concept of indivisible, which is less than clear intuitively, and made it into a workable tool for finding areas and volumes. He viewed planar regions as made up of infinitely many slices by parallel lines, and solid regions as made up of infinitely many slices by parallel planes, and his main idea was that to compare the areas or volumes of two regions, it suffices to compare their slices. Cavalieri's explanations were not very clear, and he seems to have confused indivisibles in the mathematical and physical senses. On the one hand, Cavalieri made progress via his use of indivisibles to obtain many geometric results. On the other hand, Cavalieri's approach, while producing new results, harked back to the indivisibles of Oresme and other medieval scholars, rather than forward to the limit concept that would be developed later.

In contrast to Aristotle, who had denied the existence of the infinitely small though he accepted the idea of the infinitely large as a potential, Blaise Pascal (1623–1662) viewed the infinitely small as complementary to the infinitely large, just as the reciprocal of a very large number is very small.

John Wallis (1616–1703), who was the first person to use the symbol "∞" to denote infinity, looked at parallelograms with thickness "$\frac{1}{\infty}$," which seems to have been both non-zero and zero as needed. He found areas and volumes arithmetically rather than geometrically. He showed, using an unproved analogy, what we write as $\int_0^a x^n\,dx = \frac{a^{n+1}}{n+1}$ for all $n \in \mathbb{N}$, and then claimed that the result was true for all $n \in \mathbb{R} - \{1\}$ by an appeal to "interpolation and induction."

Newton and Leibniz

One of the most remarkable aspects of the invention of calculus is that even though its inventors, Isaac Newton (1643–1727) and Gottfried von Leibniz (1646–1716), did not have our modern tools for mathematical rigor, they got all the basic ideas right. The notion of a function did not yet exist at the time of Newton and Leibniz; they thought about curves in the plane from a geometric point of view. The notion of a limit as we now know it came even later than the notion of a function, and was certainly not available to Newton and Leibniz. Instead of using limits, the originators of calculus used infinitesimals.

The role of infinitesimals in calculus can be seen in the calculation of derivatives. Today we calculate derivatives via the limit $\lim\limits_{h\to 0} \frac{f(c+h)-f(c)}{h}$. If one does not have the notion of limit, it would be possible to evaluate the fraction $\frac{f(c+h)-f(c)}{h}$ by thinking of h as an infinitesimally small but still positive number. Because h is non-zero we can divide by it, but because it is infinitesimally small we can think of it as negligible in comparison to any real number. For example, if we let $f(x) = x^2$, we first obtain $\frac{(x+h)^2-x^2}{h} = \frac{2xh+h^2}{h} = 2x + h$, and we then drop h to obtain $2x$. Such calculations, which lead to the right answers but are on shaky foundational grounds, were done routinely in the early development of calculus.

Newton's initial versions of calculus used infinitesimals, but subsequently his approach turned away from them and toward something approaching the idea of a limit. Indeed, Newton's understanding of infinitesimals and limits was rather sophisticated given the general state of mathematical development of his era. Newton's most important published work was *Philosophiæ Naturalis Principia Mathematica* of 1687 (often referred to as the *Principia*). This work set forth Newton's theory of mechanics and gravitation, and is considered one of the most important texts in the history of science. The *Principia* did not use the calculus that Newton had previously developed, and phrased everything in terms of classical Euclidean geometry, though it is possible to find in the *Principia* an understanding of Newton's approach to infinitesimals and limits. Some people suggest that Newton worked out everything in the *Principia* in terms of fluxions (as he called his version of derivatives), and then redid it in classical terms to avoid controversy; others dispute this conjecture, suggesting that Newton ultimately preferred the classical approach and was uncomfortable with infinitesimals, and also that calculus was not yet sufficiently developed for Newton's purposes in the *Principia*.

Lemma I of Book I of the *Principia* states

> "Quantities, and the ratio's of quantities, which in any finite time converge continually to equality, and before the end of that time approach nearer the one to the other than by any given difference, become ultimately equal."

(Quotes from the *Principia* are from [New].) In this lemma, Newton attempted to formulate the notion of a limit, at least in a special case (see Exercise 3.2.13 for a modern statement of what Newton appears to be suggesting in this lemma). What matters is not the details of this lemma, but Newton's phrase "nearer the one to the other than by any given difference," which has a strong resemblance to the role of ε in the modern definition of limits, though Newton did not give a definition of what he meant by this sort of limit.

Newton stated his philosophical approach to limits and infinitesimals in the Scholium at the end of Section I of Book I of the *Principia*, where we find

> "These lemmas are premised, to avoid the tediousness of deducing perplexed demonstrations *ad absurdum*, according to the method of the ancient geometers. For demonstrations are more contracted by the method of indivisibles: But because the hypothesis of indivisibles seems somewhat harsh, and therefore that method is reckoned less geometrical; I chose rather to reduce the demonstrations of the following propositions to the first and last sums and ratio's of nascent and evanescent quantities, that is, to the limits of those sums and ratio's; and so to premise, as short as I could, the demonstrations of those limits. For hereby the same thing is perform'd as by the method of indivisibles; and now those principles being demonstrated, we may use them with more safety."

Newton understood that philosophical objections might be raised to the above-quoted ideas, and he argued for his view by comparing limits to instantaneous velocity.

Leibniz based his approach to derivatives on the differentials dx and dy, and he appeared to be ambivalent about whether or not to think of dx and dy as infinitesimals. He attempted to avoid the question by saying that if one wanted to, one could take dx and dy to be real numbers that are as small as desired, and then the errors obtained when expressions such as $(dx)^2$ were dropped could be made to be within any given tolerance, which hints at our modern notion of limits; that one could rework everything involving infinitesimals in terms of the method of exhaustion, though he did not actually do so; and that infinitesimals are useful in solving problems, which makes them worthwhile as a useful fiction. However, even though Leibniz was ambivalent about the existence of infinitesimals, he seemed to think that they obeyed certain rules, and could be used properly. Leibniz did not develop an approach resembling limits as did Newton.

Eighteenth Century

In contrast to Leibniz's ambivalence about infinitesimals, some of his important successors, such as Jakob Bernoulli (1654–1705), Johann Bernoulli (1667–1748) and Leonhard Euler (1707–1783), had no qualms about the existence of infinitesimals.

Between the mid-18th century and the first quarter of the 19th century, questions arose as to the requirements of functions being continuous or not, and smooth or not. In Euler's influential textbook *Introductio in analysin infinitorum* of 1748, the only functions considered were those given by single formulas made up from the standard elementary functions by the usual ways of combining functions. However, in response to the question of the permissible initial shape of a string allowed in the study of vibrating strings via partial differential equations, Euler expanded his notion of functions and allowed piecewise smooth ones, though he assumed that functions were continuous, except possibly at isolated points, which were ignored.

Continuity in the 18th century was not just a geometric notion, but a more general idea of going through all intermediate states, or of gradual change. In an effort to clarify what continuity was, Louis Arbogast (1759–1803), in 1791, focused on the inadmissibility of functions that jump abruptly. He invoked the idea of a function obtaining all "the successive values" between two values; that is, he hinted at the Intermediate Value Theorem.

Nineteenth Century

By the early 19th century the previous notion of continuity was challenged. Joseph Fourier (1768–1830), in *Théorie analytique de la chaleur* of 1822, studied what we now call Fourier series, and stated that some discontinous and some non-differentiable functions should be considered. In 1829 Lejeune Dirichlet (1805–1859) described an example of a function that did not satisfy conditions that imply the convergence of Fourier series (this function is given in Example 3.3.3 (6)). All this work stressed the need for a good definition of continuity.

Carl Friedrich Gauss (1777–1855) had a new view of infinitesimals. In the 18th century it was held that infinitesimals behaved similarly to real numbers (except that the cancellation law did not hold), and hence infinitesimals could be manipulated accordingly. Gauss, by contrast, said that caution was needed when using infinite quantities, and that they should be used only if their use can be viewed in terms of limits. However, Gauss' approach was not completely rigorous; for example, he implicitly (and without proof) used the Monotone Convergence Theorem.

The first person to give essentially the modern formulation of continuity, phrased in terms of $f(x+h) - f(x)$ becoming as small as desired if h is sufficiently small (though without the ε–δ formulation), was Bernard Bolzano (1781–1848) in 1817. Bolzano's goal was to prove the Intermediate Value Theorem, and he defined continuity along the way. His approach was not completely rigorous because he did not have the axioms for the real numbers, but it was a significant step forward nonetheless. However, this work, which was privately published, was not widely seen by Bolzano's contemporaries.

The first person to make rigor in these matters important and to have wide influence was Augustin Louis Cauchy (1789–1857), who apparently was not aware of Bolzano's work. Cauchy wrote three important textbooks on analysis, *Cours d'analyse a l'École Royal Polytechnique* of 1821, *Résumé des leçons données a l'École Royal Polytechnique sur le calcul infinitésimal* of 1823 and *Leçons sur le calcul différentiel* of 1829. These texts were the first to promote rigor as an important goal in real analysis. Cauchy tried to replace infinitesimals with functions whose limits are zero. His definition of continuity, phrased in terms of $f(x+h) - f(x)$ decreasing indefinitely with h, is similar to Bolzano's definition, though slightly less satisfactory. Cauchy showed that various elementary functions (for example sine) are continuous, and he gave a proof of the Intermediate Value Theorem using sequences. Cauchy was the first person to use the symbols ε and δ in their now familiar roles, though that was in the proof of a theorem, and not in his definitions of limits and continuity. Cauchy looked at the continuity of a function on an entire interval, not pointwise as we do today. Cauchy's work was a major step forward in the development of the modern approach to limits and continuity, and to rigorous proofs in real analysis in general, though it was not completely rigorous by modern standards. For example, he still referred to infinitesimals as if they were numbers, though no such theory of infinitesimals was developed at the time, and he glossed over the difference between continuity and uniform continuity.

Karl Weierstrass (1815–1897) changed the view of limits from the previous notion of a "variable approaching" something to a static view where the "x" in "$f(x)$" is a member of what we now call a set. Weierstrass gave what amounts to the ε–δ definition of continuity; he used ε as we now use it, though he used an interval in the domain rather than δ. It can be said that Weierstrass ended the use of infinitesimals in real analysis. Eduard Heine (1821–1881), based upon lectures of Weierstrass, was the first person to use the ε–δ definition of continuity as we do now (though he used η rather than δ). Additionally, Heine distinguished in 1872 between continuous and uniformly continuous functions, and showed that a continuous function on a closed bounded interval is uniformly continuous, a result Dirichlet had formulated in 1854 but did not prove. Heine also published the first proof of the Extreme Value Theorem.

Interestingly, whereas the idea of variability had been banished from Greek mathematics because it led to Zeno's paradoxes, it was precisely this concept which, revived in the later Middle Ages and represented geometrically, was a factor in the invention of calculus in the 17th century. On the other hand, once calculus was invented, and an attempt was made in the two centuries following that invention to put calculus on a rigorous basis, the idea of variability was once again banished (see the paragraph following Definition 2.5.10 for a modern approach to the idea of variables).

Twentieth Century

Although infinitesimals were expelled from the rigorous treatment of real analysis by the time of Weierstrass, in 1960 Abraham Robinson (1918–1974) constructed a system of numbers that included the real numbers as well as infinitesimals and infinitely large numbers, and he showed that calculus can in fact be done rigorously via infinitesimals after all. Such an approach, which is known as "non-standard analysis," can be used as an alternative treatment of real analysis that completely avoids limits, though it has not caught on as a popular replacement for the standard approach to real analysis (as found in the present text).

4

Differentiation

4.1 Introduction

Having now finished all of our technical preliminaries (that is, proofs of the needed facts about the real numbers, limits and continuity), we are finally ready for what this text is really about, which is an advanced look at the core material taught in a single-variable calculus course. It has taken us this long to get to material from calculus because limits are at the heart of what makes calculus work, even in theorems that do not explicitly mention limits, and to make limits work, we needed the properties of the real numbers.

It is assumed that the reader has seen derivatives in a calculus course, and so we will go straight to the technical details, without spending time on intuitive motivation, applications or computational examples.

We need one preliminary technical comment before we commence our study of derivatives. In its most basic form, the definition of derivatives is for functions with domains that are non-degenerate *open* intervals in $\mathbb{R}$; such open intervals need not be bounded, and can in fact be all of $\mathbb{R}$. Using one-sided limits it is also possible to define derivatives on closed intervals, and we will do so when needed, but fundamentally derivatives are about open intervals.

4.2 The Derivative

Although it had a surprisingly long historical wait until it appeared, the definition of the derivative is, from our modern point of view, quite straightforward. Intuitively, derivatives are used to resolve two issues: the rate of change of a function, and the slope of the tangent line to a curve. The latter issue is in fact just the geometric version of the former, and so these two issues are really the same.

As the reader has certainly learned in a calculus course, the intuitive idea for finding the slope of a tangent line to a curve is to find the slopes of secant lines to the curve (that is, lines through pairs of distinct points on the curve), and then to take the limit of the slopes of the secant lines as the x-values of the points get closer and closer. If we want to make this intuitive idea rigorous, the question arises as to what the precise definition of the tangent line to a curve is. Although the geometrical idea of a tangent line as the line that "just touches a curve at a given point" is simple to grasp intuitively, as seen in Figure 4.2.1 (i), it is not entirely obvious how to translate that informal notion of a tangent line into a rigorous definition. Moreover, not every curve has a tangent line at every point; see, for example, the origin in Figure 4.2.1 (ii). In fact, as we will see in Section 10.5, there are continuous functions for which no point has a tangent line (the existence of such functions is not at all trivial, and the reader should not expect to be able to define such a function easily). Fortunately, although the notion of a tangent line is important for the intuitive motivation of the notion of the derivative of a function, from a rigorous point of view the material is treated in the opposite order; that is, we first define the derivative, via the standard definition using limits of quotients, and then the tangent line is simply defined to be the unique line through the given point on the curve and with slope equal to the derivative at that point. We will not be making formal use of tangent lines in this text.

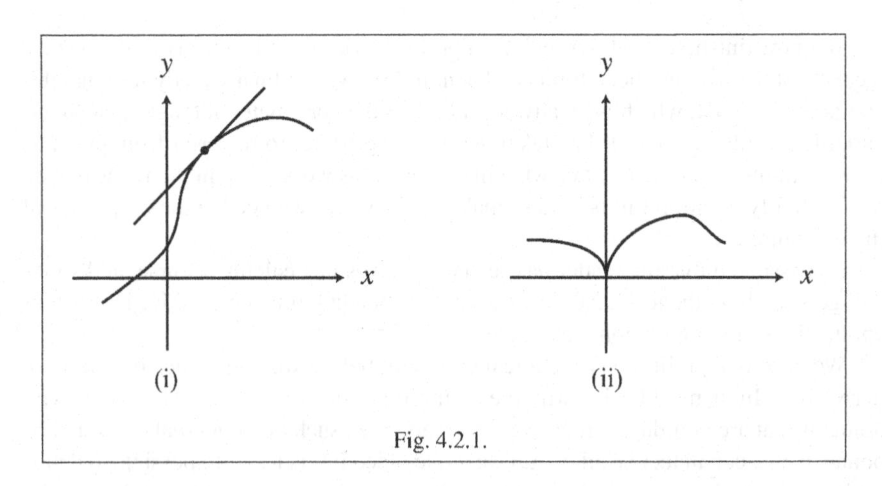

Fig. 4.2.1.

For a rigorous definition of derivatives we need to use limits, and that is why the chapter on limits precedes the present chapter. All of the hard work in making the definition of derivatives rigorous is contained in the rigorous definition of limits, and so our definition of derivatives will be very simple. Indeed, the definition of derivatives is one of the rare places in real analysis where we can be rigorous while appearing to be doing things exactly as they are done in an introductory calculus course.

Suppose that $f\colon I \to \mathbb{R}$ is a function, for some open interval $I \subseteq \mathbb{R}$ be an open interval, and let $c \in I$. To find the slope of the tangent line to the graph at c, if such a tangent line exists, we choose $x \in I - \{c\}$, we compute the slope of the secant line through the points $(c, f(c))$ and $(x, f(x))$, and we then take the limit of this slope as x goes to c. See Figure 4.2.2 for such a secant line and a tangent line. We are therefore led to the following definition.

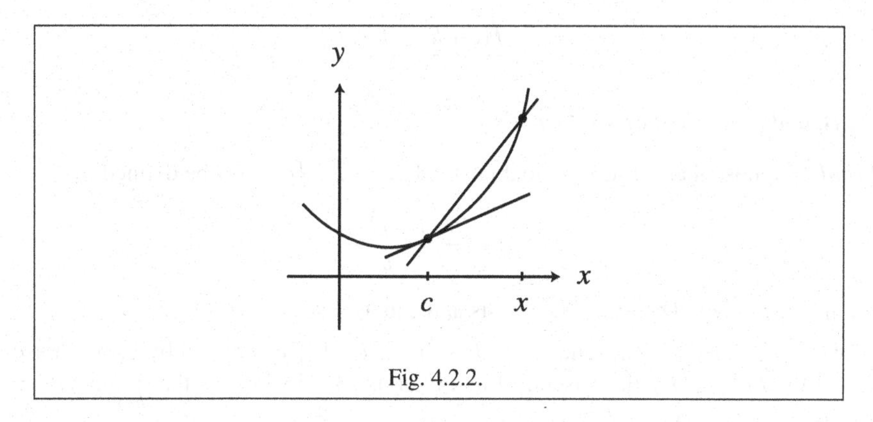

Fig. 4.2.2.

Definition 4.2.1. Let $I \subseteq \mathbb{R}$ be an open interval, let $c \in I$ and let $f\colon I \to \mathbb{R}$ be a function.

1. The function f is **differentiable** at c if

$$\lim_{x \to c} \frac{f(x) - f(c)}{x - c}$$

   exists; if this limit exists, it is called the **derivative** of f at c, and it is denoted $f'(c)$.
2. The function f is **differentiable** if it is differentiable at every number in I. If f is differentiable, the **derivative** of f is the function $f'\colon I \to \mathbb{R}$ whose value at x is $f'(x)$ for all $x \in I$. △

Observe that f' is the name of the derivative function. The notation "$f'(x)$" denotes the value of the derivative function at the point x in the domain of f'. In a calculus course, where it is common to write "$f(x)$" incorrectly as the name of the function, it is also common to write "$f'(x)$" incorrectly as the name of the derivative. In this text we will maintain the correct distinction between "f'" and "$f'(x)$."

In addition to the notation f' for the derivative of f, there are a number of other notations that are used for the derivative, such as $\frac{dy}{dx}$. These other notations exist for historical reasons, and are useful in some circumstances. For our purposes the notation f' is the most appropriate, and we will use it exclusively.

The following lemma gives a standard variant formulation of the definition of the derivative that is often more computationally convenient than the original definition. To see that the limit in this lemma makes sense, suppose that $I \subseteq \mathbb{R}$ is an open interval and that $c \in I$. Then by Lemma 2.3.7 (2) there is some $\delta > 0$ such that $(c - \delta, c + \delta) \subseteq I$, and hence $c + h \in I$ for all $h \in (-\delta, \delta)$. Hence, if $f\colon I \to \mathbb{R}$ is a function, then the function $G\colon (-\delta, \delta) - \{0\} \to \mathbb{R}$ defined by $G(h) = \frac{f(c+h) - f(c)}{h}$ for all $h \in (-\delta, \delta) - \{0\}$ is well-defined, and we can take the limit of this function as h goes to 0.

Lemma 4.2.2. *Let $I \subseteq \mathbb{R}$ be an open interval, let $c \in I$ and let $f\colon I \to \mathbb{R}$ be a function. Then f is differentiable at c if and only if*

$$\lim_{h\to 0} \frac{f(c+h)-f(c)}{h}$$

exists, and if this limit exists it equals $f'(c)$.

Proof. Suppose that f is differentiable at c. Let $F\colon I-\{c\} \to \mathbb{R}$ be defined by

$$F(x) = \frac{f(x)-f(c)}{x-c}$$

for all $x \in I-\{c\}$. Then $\lim_{x\to c} F(x)$ exists and equals $f'(c)$.

Let $J = \{x-c \mid x \in I\}$, and let $g\colon J-\{0\} \to I-\{c\}$ be defined by $g(x) = x+c$ for all $x \in J-\{0\}$. By Exercise 3.2.1 and Exercise 3.2.5 we know that $\lim_{h\to 0} g(h) = c$. It is straightforward to verify that

$$(F \circ g)(h) = \frac{f(c+h)-f(c)}{h}$$

for all $h \in J-\{0\}$. Because $\lim_{x\to c} F(x)$ exists, then Theorem 3.2.12 implies that $\lim_{h\to 0} (F \circ g)(h)$ exists and $\lim_{h\to 0} (F \circ g)(h) = \lim_{x\to c} F(x)$, which is equivalent to saying that

$$\lim_{h\to 0} \frac{f(c+h)-f(c)}{h}$$

exists and equals $f'(c)$.

The other implication is similar, and we omit the details. □

Example 4.2.3.

(1) Let $I \subseteq \mathbb{R}$ be a non-degenerate open interval, and let $f\colon I \to \mathbb{R}$ be defined by $f(x) = mx+b$ for all $x \in I$, where $m, b \in \mathbb{R}$. Let $c \in I$. By Lemma 2.3.7 (2) there is some $\delta > 0$ such that $(c-\delta, c+\delta) \subseteq I$. Then $c+h \in I$ for all $h \in (-\delta, \delta)$, and hence $\frac{f(c+h)-f(c)}{h}$ is defined for all $h \in (-\delta, 0) \cup (0, \delta)$. To find the derivative of f at c, we use Exercise 3.2.1 to see that

$$\lim_{h\to 0} \frac{f(c+h)-f(c)}{h} = \lim_{h\to 0} \frac{[m(c+h)+b]-[mc+b]}{h} = \lim_{h\to 0} \frac{mh}{h} = \lim_{h\to 0} m = m,$$

where we can cancel the h because as we take the limit as h goes to zero, the number h is never equal to zero (which we need to know, because it is not possible to cancel zero).

We therefore see that $f'(x)$ exists and equals m for every $x \in I$. We can abbreviate this derivative by writing $(mx+b)' = m$. In particular, we see that $(x)' = 1$ and $(c)' = 0$.

(2) Let $g\colon \mathbb{R} \to \mathbb{R}$ be defined by $g(x) = x^2$ for all $x \in \mathbb{R}$. We first find the derivative of g at 3 by computing

$$\lim_{h\to 0} \frac{g(3+h)-g(3)}{h} = \lim_{h\to 0} \frac{(3+h)^2-3^2}{h} = \lim_{h\to 0} \frac{6h+h^2}{h} = \lim_{h\to 0} (6+h) = 6,$$

where the limit is found using Exercise 3.2.1. Hence $g'(3)$ exists, and $g'(3) = 6$. We now find the derivative in general by letting $x \in \mathbb{R}$, and computing

$$\lim_{h\to 0} \frac{g(x+h)-g(x)}{h} = \lim_{h\to 0} \frac{(x+h)^2-x^2}{h} = \lim_{h\to 0} \frac{2xh+h^2}{h} = \lim_{h\to 0} (2x+h) = 2x.$$

Hence $g'(x)$ exists and equals $2x$ for all $x \in \mathbb{R}$, which we abbreviate by writing $(x^2)' = 2x$.

(3) Let $k\colon \mathbb{R} \to \mathbb{R}$ be defined by $k(x) = |x|$ for all $x \in \mathbb{R}$. We try to find the derivative of k at 0 by computing

$$\lim_{h\to 0} \frac{k(0+h)-k(0)}{h} = \lim_{h\to 0} \frac{|0+h|-|0|}{h} = \lim_{h\to 0} \frac{|h|}{h}.$$

We saw in Example 3.2.16 that this last limit does not exist. Hence k is not differentiable at 0. This lack of differentiability corresponds to the "corner" in the graph of $y = |x|$ at $x = 0$.

On the other hand, the function k is differentiable at all $x \in \mathbb{R} - \{0\}$, as the graph of $y = |x|$ would suggest. Let $x \in (0, \infty)$. If $h \in \mathbb{R}$ and h is sufficiently close to zero, then $x + h > 0$. Hence

$$\lim_{h\to 0} \frac{k(x+h)-k(x)}{h} = \lim_{h\to 0} \frac{|x+h|-|x|}{h} = \lim_{h\to 0} \frac{(x+h)-x}{h} = \lim_{h\to 0} \frac{h}{h} = \lim_{h\to 0} 1 = 1.$$

Hence $k'(x)$ exists, and $k'(x) = 1$. A similar computation shows that if $x \in (-\infty, 0)$, then $k'(x) = -1$.

By combining the above observations with Exercise 3.3.1 (2), we see that a function can be continuous everywhere but not differentiable everywhere. ◊

We see from Example 4.2.3 (3) that continuity does not imply differentiability. On the other hand, the following theorem states that if a function is differentiable, then it must be continuous. In fact, differentiability at a single point implies continuity at that point. This theorem, though simple, will be useful throughout this text.

Theorem 4.2.4. *Let $I \subseteq \mathbb{R}$ be an open interval, and let $f\colon I \to \mathbb{R}$ be a function. Let $c \in I$. If f is differentiable at c, then f is continuous at c. If f is differentiable, then f is continuous.*

Proof. Suppose that f is differentiable at c. Hence

$$\lim_{x\to c} \frac{f(x)-f(c)}{x-c}$$

exists and equals $f'(c)$. If $x \in I - \{c\}$, then

$$f(x) = \frac{f(x)-f(c)}{x-c}(x-c) + f(c).$$

We now use Theorem 3.2.10 and Exercise 3.2.1 to deduce that

$$\lim_{x \to c} f(x) = \lim_{x \to c} \left[\frac{f(x) - f(c)}{x - c} (x - c) + f(c) \right] = f'(c) \cdot 0 + f(c) = f(c).$$

In particular, $\lim_{x \to c} f(x)$ exists. It now follows from Lemma 3.3.2 that f is continuous at c. □

Suppose that a function f is differentiable. Then f must be somewhat "nicely behaved," for example it is continuous by Theorem 4.2.4. How nicely must f' behave? Must f' also be continuous? And if f' is continuous, is it necessarily differentiable? As seen in the following example, the answer to these last two questions is no.

Example 4.2.5. Our calculations in both parts of this example will be more informal than most of our previous examples, because they require basic formulas for differentiation (such as the Product Rule and Chain Rule, as well as the derivatives of the functions x^r and $\sin x$), with which we assume that the reader is informally familiar, but which we have not yet proved. The Product Rule and Chain Rule will be proved in Section 4.3, the derivative of x^r will be computed in Section 7.2 and the derivative of $\sin x$ will be computed in Section 7.3. Nonetheless, it is nice to see this example now, rather than waiting until we have proved all the details.

(1) Let $f \colon \mathbb{R} \to \mathbb{R}$ be defined by

$$f(x) = \begin{cases} x^2 \sin \frac{1}{x^2}, & \text{if } x \neq 0 \\ 0, & \text{if } x = 0. \end{cases}$$

See Figure 4.2.3 for the graph of f; the parabolas $y = x^2$ and $y = -x^2$ are shown with dashed lines.

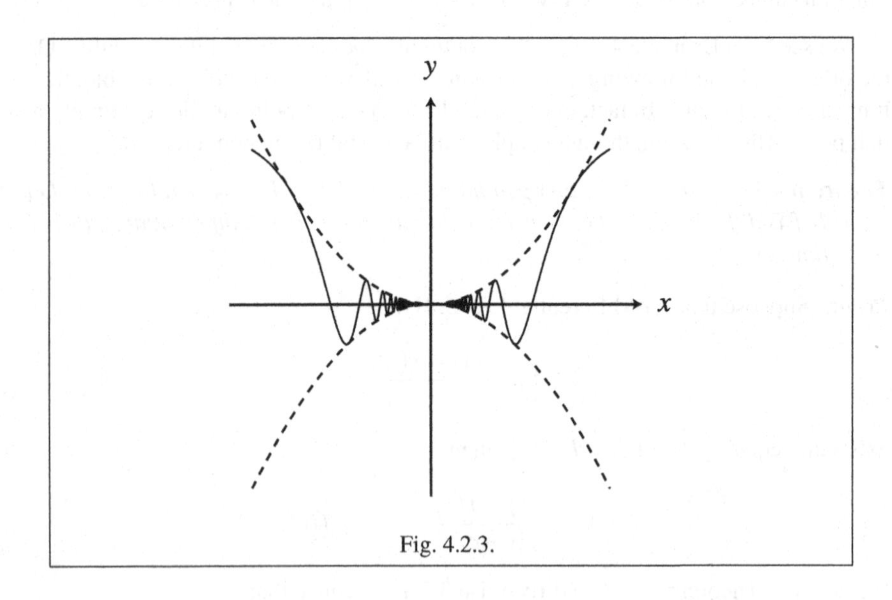

Fig. 4.2.3.

We want to find $f'(x)$ for all $x \in \mathbb{R}$. First, let $x \in \mathbb{R} - \{0\}$. Then $x \in (-\infty, 0)$ or $x \in (0, \infty)$, and in either case by Exercise 4.2.3 (3) we can restrict out attention to the appropriate open interval containing x. We can then find $f'(x)$ using the Product Rule and Chain Rule (stated formally in Theorem 4.3.1 (4) and Theorem 4.3.3). In particular, it is left to the reader to use these rules to verify that

$$f'(x) = 2x\sin\frac{1}{x^2} - \frac{2}{x}\cos\frac{1}{x^2}.$$

Keep in mind that this formula for $f'(x)$ holds only for $x \neq 0$.

To find $f'(0)$, we need to use the definition of derivatives directly. We compute

$$\lim_{h\to 0}\frac{f(0+h)-f(0)}{h} = \lim_{h\to 0}\frac{h^2\sin\frac{1}{h^2} - 0}{h} = \lim_{h\to 0} h\sin\frac{1}{h^2} = 0,$$

where the last equality holds by Lemma 3.2.8, because $\lim_{h\to 0} h = 0$, and $|\sin\frac{1}{h^2}| \leq 1$ for all $h \in \mathbb{R} - \{0\}$.

We have therefore seen that $f'(x)$ exists for all $x \in \mathbb{R}$, and that

$$f'(x) = \begin{cases} 2x\sin\frac{1}{x^2} - \frac{2}{x}\cos\frac{1}{x^2}, & \text{if } x \neq 0 \\ 0, & \text{if } x = 0. \end{cases}$$

We now claim that f' is not continuous at 0 (though it is continuous elsewhere). By Lemma 3.3.2 we need to ask whether or not $\lim_{x\to 0} f'(x)$ exists and $\lim_{x\to 0} f'(x) = f'(0)$. Using our formula for $f'(x)$ when $x \neq 0$, we see that

$$\lim_{x\to 0} f'(x) = \lim_{x\to 0}\left[2x\sin\frac{1}{x^2} - \frac{2}{x}\cos\frac{1}{x^2}\right].$$

Proceeding informally, it can be seen that as x goes to 0 from the right, the value of $\frac{2}{x}$ goes to infinity, and hence $\lim_{x\to 0+} f'(x)$ does not exist; a similar argument holds as x goes to 0 from the left. (For a proof that these limits do not exist, we would need a rigorous treatment of limits to infinity, which will be given in Chapter 6; the details of this proof make use of Exercise 3.2.1, Lemma 3.2.8, Example 6.2.7 (2), Theorem 6.2.8, Exercise 6.2.8 and Exercise 3.2.7 (1), though we omit the details.) We deduce that f' is not continuous at 0. We have therefore found an example of a function f such that f' exists everywhere, but that f' is not continuous.

(2) Let $g\colon \mathbb{R} \to \mathbb{R}$ be defined by

$$g(x) = \begin{cases} x^2, & \text{if } x \geq 0 \\ -x^2, & \text{if } x < 0. \end{cases}$$

A formula for g' can be computed very similarly to Part (1) of this example, by using the formula for the derivative of x^r to compute g' on each of $(-\infty, 0)$ and $(0, \infty)$, and using the definition of the derivative to compute $g'(0)$; the details are left to the reader. The result of such a calculation is that

$$g'(x) = \begin{cases} 2x, & \text{if } x > 0 \\ 0, & \text{if } x = 0 \\ -2x, & \text{if } x < 0. \end{cases}$$

This formula can be condensed into the single formula $g'(x) = 2|x|$ for all $x \in \mathbb{R}$. We know by Exercise 3.3.1 (2) and Theorem 3.3.5 (3) that g' is continuous. However, it follows from Example 4.2.3 (3) that g' is not differentiable; the factor of 2 does not affect differentiability, as the reader knows informally, and as will be proved in Theorem 4.3.1 (3). ♢

Although the derivatives of differentiable functions can be rather poorly behaved, as seen in Example 4.2.5, there are some restrictions on how badly behaved they can be, as we will see in Example 4.4.11.

As strange as the functions in Example 4.2.5 are, we will see even more bizarre functions later on. In Section 5.2 we will see strange examples of integrable and non-integrable functions, and in Section 10.5, our final mathematical section of this text, we will see an example of a function $\mathbb{R} \to \mathbb{R}$ that is continuous everywhere but differentiable nowhere. The strange functions in Example 4.2.5, and the other strange functions we will see subsequently, were constructed to help mathematicians better understand the subtleties of the theoretical underpinnings of calculus. (See [GO03] for even more examples of functions with unusual and surprising properties.) By contrast, most of the standard functions that one encounters in mathematics and its applications, such as polynomials, logarithms, exponentials, sine and cosine, are not only differentiable, but have continuous derivatives, and in fact we can take the derivatives of their derivatives, and the derivatives of those and so on. To state these facts precisely, we need the following terminology.

Definition 4.2.6. Let $I \subseteq \mathbb{R}$ be an open interval, let $c \in I$ and let $f\colon I \to \mathbb{R}$ be a function. Suppose that f is differentiable at c. The function f is **twice differentiable** at c if f' is differentiable at c. If f' is differentiable at c, the derivative $(f')'(c)$ is called the **second derivative** of f at c, and it is denoted $f''(c)$. The function f is **twice differentiable** if it is twice differentiable at every number in I. If f is twice differentiable, the **second derivative** of f is the function $f''\colon I \to \mathbb{R}$ whose value at x is $f''(x)$ for all $x \in I$.

The nth derivative of f for all $n \in \mathbb{N}$ is defined as follows, using Definition by Recursion. If f is differentiable at c, the **first derivative** of f at c is simply the derivative of f at c. Suppose that f is $n-1$ times differentiable at c. The $(n-1)$-st derivative of f at c is denoted $f^{(n-1)}(c)$. The function f is n **times differentiable** at c if $f^{(n-1)}$ is differentiable at c. If $f^{(n-1)}$ is differentiable at c, the derivative $(f^{(n-1)})'(c)$ is called the nth **derivative** of f at c, and it is denoted $f^{(n)}(c)$. The function f is n **times differentiable** if it is n times differentiable at every number in I. If f is n times differentiable, the nth **derivative** of f is the function $f^{(n)}\colon I \to \mathbb{R}$ whose value at x is $f^{(n)}(x)$ for all $x \in I$.

The 0th **derivative** of f is $f^{(0)} = f$. △

We will need the following terminology later on, for example when we discuss Taylor series of functions in Section 10.4.

Definition 4.2.7. Let $I \subseteq \mathbb{R}$ be an open interval, and let $f\colon I \to \mathbb{R}$ be a function. The function f is **continuously differentiable** if f is differentiable and f' is continuous. Let $n \in \mathbb{N}$. The function f is **continuously differentiable of order** n if $f^{(i)}$ exists and is continuous for all $i \in \{1,\ldots,n\}$. The function f is **infinitely differentiable** (also called **smooth**) if $f^{(i)}$ exists all $i \in \mathbb{N}$. △

In general, we take derivatives of functions with domains that are open intervals. However, there are some situations in which it is useful to take derivatives of functions with domains that are other types of non-degenerate intervals. For example, in both versions of the Fundamental Theorem of Calculus (given in Section 5.6), it is necessary to consider derivatives of functions with domains that are non-degenerate closed bounded intervals. At all points of a non-degenerate interval other than the endpoints, we take derivatives as usual, because any interval with its endpoints removed is an open interval. At the endpoints of a non-degenerate interval we simply use one-sided limits, as defined in Section 3.2, instead of ordinary limits in the definition of the derivative.

Definition 4.2.8. Let $I \subseteq \mathbb{R}$ be a non-degenerate interval, let $c \in I$ and let $f\colon I \to \mathbb{R}$ be a function.

1. Suppose that c is a left endpoint of I. The function f is **differentiable** at c if the limit
$$\lim_{x \to c^+} \frac{f(x) - f(c)}{x - c} = \lim_{h \to 0^+} \frac{f(c+h) - f(c)}{h}$$
exists; if this limit exists, it is called the **one-sided derivative** of f at c, and it is denoted $f'(c)$.
2. Suppose that c is a right endpoint of I. The function f is **differentiable** at c if the limit
$$\lim_{x \to c^-} \frac{f(x) - f(c)}{x - c} = \lim_{h \to 0^-} \frac{f(c+h) - f(c)}{h}$$
exists; if this limit exists, it is called the **one-sided derivative** of f at c, and it is denoted $f'(c)$.
3. The function f is **differentiable** if the restriction of f to the interior of I is differentiable in the usual sense, and if f is differentiable at the endpoints of I in the sense of Parts (1) and (2) of this definition if there are endpoints. △

Reflections

In contrast to some of the material in earlier chapters, the basic concepts in the present section (with the possible exception of Example 4.2.5) should be familiar to anyone who has taken a calculus course. It would be a mistake, however, to deduce from this familiarity that the material in the present section, which is the concept of the derivative and some basic facts about this concept, is somehow simpler than the previous material in this text. Rather, the technical difficulties in the concept of the derivative are hidden in the use of limits, which were already dealt with rigorously in Chapter 3. In a typical introductory calculus course, the material on limits is dealt with in an intuitive fashion, but once the basic properties of limits are stated, then the definition of derivatives in a calculus course is precisely the same as our definition. Of course, we now know that our treatment of derivatives is rigorous, because our treatment of limits is.

One fact about differentiable functions that is given more prominence in a real analysis course than in a calculus course is Theorem 4.2.4, which says that if a function is differentiable then it is continuous. This fact might seem intuitively obvious, and it is not particularly useful for the computational aspects of differentiation and integration that are stressed in a calculus course, but it is very important for theoretical purposes, and makes its way into many proofs throughout this text.

Something else found in this section that is not usually found in a calculus course are the weird examples in Example 4.2.5. An introductory calculus course aims to provide students with computational tools that are useful in a broad variety of applications, and hence the focus is on taking derivatives of the sorts of functions that arise in real-world situations, which are usually nicely behaved functions. In a real analysis course, by contrast, the goal is to obtain a better understanding of the rigorous foundations of calculus, and we therefore need weird examples, which do not necessarily arise in any practical application of calculus, to help us determine the range of possible behaviors of functions. Specifically, nice functions tend to have nice derivatives, but we want to know if all differentiable functions have nice derivatives, and the functions in Example 4.2.5 tell us that the answer to this question is no. We will also see some very strange examples of functions in Example 5.2.6 as part of our discussion of integrals.

Exercises

Exercise 4.2.1. Using only the definition of derivatives and Lemma 4.2.2, find the derivative of each of the following functions.

(1) Let $f\colon \mathbb{R} \to \mathbb{R}$ be defined by $f(x) = 3x - 8$ for all $x \in \mathbb{R}$.
(2) Let $g\colon \mathbb{R} \to \mathbb{R}$ be defined by $g(x) = x^3$ for all $x \in \mathbb{R}$.
(3) Let $h\colon (0,\infty) \to \mathbb{R}$ be defined by $h(x) = \frac{1}{x}$ for all $x \in (0,\infty)$.
(4) Let $k\colon (0,\infty) \to \mathbb{R}$ be defined by $k(x) = \sqrt{x}$ for all $x \in (0,\infty)$.

Exercise 4.2.2. Let $f\colon \mathbb{R} \to \mathbb{R}$ be defined by

$$f(x) = \begin{cases} x^3, & \text{if } x \in \mathbb{Q} \\ x^2, & \text{if } x \in \mathbb{R} - \mathbb{Q}. \end{cases}$$

Using only the definition of derivatives and Lemma 4.2.2, determine whether f is differentiable at 0. If it is, find $f'(0)$; if it is not, show why not.

Exercise 4.2.3. **[Used throughout.]** Let $I \subseteq \mathbb{R}$ be a non-degenerate interval, let $c \in I$ and let $f\colon I \to \mathbb{R}$ be a function.

(1) Suppose that c is in the interior of I. Prove that f is differentiable at c if and only if there is some $\delta > 0$ such that $f|_{I\cap(c-\delta,c+\delta)}$ is differentiable at c.

(2) State and prove the analog of Part (1) of this exercise when c is an endpoint of I. (There are two cases, depending upon whether c is a right endpoint or a left endpoint, and it is sufficient to do only one of the cases.)

(3) Let $J \subseteq \mathbb{R}$ be a non-degenerate interval, and let $g\colon J \to \mathbb{R}$ be a function. Suppose that there is some open interval $U \subseteq I \cap J$ such that $c \in U$, and that $f(x) = g(x)$ for all $x \in U$. Prove that g is differentiable at c if and only if f is differentiable at c, and if they are differentiable at c then $g'(c) = f'(c)$.

(4) Let $J \subseteq \mathbb{R}$ be a non-degenerate interval, and let $h\colon J \to \mathbb{R}$ be a function. Suppose that c is an endpoint of J, and that there is some half-open interval $D \subseteq I \cap J$ such that c is the endpoint of D, and that $f(x) = h(x)$ for all $x \in D$. Prove that if f is differentiable at c then h is differentiable at c, and if they are differentiable at c then $h'(c) = f'(c)$. Find an example to show that this result cannot be made into an if and only if statement. (There are two cases, depending upon whether c is a right endpoint or a left endpoint, and it is sufficient to do only one of the cases.)

(5) Let $K \subseteq I$ be a non-degenerate interval. Prove that if f is differentiable, then $f|_K$ is differentiable, and $(f|_K)'(x) = f'(x)$ for all $x \in K$.

Exercise 4.2.4. Let $I \subseteq \mathbb{R}$ be an open interval, let $c \in I$ and let $f\colon I \to \mathbb{R}$ be a function. Suppose that $|f(x)| \le (x-c)^2$ for all $x \in I$. Prove that f is differentiable at c and $f'(c) = 0$. (The function f in Example 4.2.5 (1) is a special case of this exercise, where $c = 0$.)

Exercise 4.2.5. Let $I \subseteq \mathbb{R}$ be an open interval, let $c \in I$ and let $f, g\colon I \to \mathbb{R}$ be functions. Suppose that $f(c) = g(c)$, and that $f(x) \le g(x)$ for all $x \in I$. Prove that if f and g are differentiable at c then $f'(c) = g'(c)$. This result might seem counterintuitive at first, but a sketch shows that to the left of c the secant lines of f through c appear to have larger slope than those of g, whereas to the right of c the secant lines of f through c appear to have smaller slope.

Exercise 4.2.6. Let $I \subseteq \mathbb{R}$ be an open interval, let $c \in I$ and let $f\colon I \to \mathbb{R}$ be a function. Prove that f is differentiable at c if and only if there is some $D \in \mathbb{R}$ such that

$$\lim_{x \to c} \frac{f(x) - f(c) - D(x-c)}{x - c} = 0,$$

and that if there is such a number D, then $D = f'(c)$.

Exercise 4.2.7. Let $I \subseteq \mathbb{R}$ be an open interval, let $c \in I$ and let $f\colon I \to \mathbb{R}$ be a function. The function f is **symmetrically differentiable** at c if

$$\lim_{h \to 0} \frac{f(c+h) - f(c-h)}{2h}$$

exists; if this limit exists, it is called the **symmetric derivative** of f at c.

(1) Prove that if f is differentiable at c, then it is symmetrically differentiable at c, and the symmetric derivative of f at c equals the derivative of f at c.

(2) If f is symmetrically differentiable at c, is it necessarily differentiable at c? Give a proof or a counterexample.

Exercise 4.2.8. Let $[a,b] \subseteq \mathbb{R}$ be a non-degenerate closed bounded interval, let $c \in (a,b)$ and let $f\colon [a,b] \to \mathbb{R}$ be a function. Prove that f is differentiable at c if and only if $f|_{[a,c]}$ and $f|_{[c,b]}$ are both differentiable at c (as one-sided derivatives) and $(f|_{[a,c]})'(c) = (f|_{[c,b]})'(c)$, and if f is differentiable at c then $f'(c) = (f|_{[a,c]})'(c) = (f|_{[c,b]})'(c)$.

Exercise 4.2.9. Let $f\colon (0,\infty) \to \mathbb{R}$ be a function. Suppose that $f(\frac{x}{y}) = f(x) - f(y)$ for all $x, y \in (0,\infty)$, and that $f(1) = 0$.

(1) Prove that f is continuous on $(0,\infty)$ if and only if f is continuous at 1.
(2) Prove that f is differentiable on $(0,\infty)$ if and only if f is differentiable at 1.
(3) Prove that if f is differentiable at 1, then $f'(x) = \frac{f'(1)}{x}$ for all $x \in (0,\infty)$. (It turns out that if $f'(1) = 1$, then f equals the natural logarithm function, though this fact is not needed for this exercise.)

4.3 Computing Derivatives

Defining derivatives is one thing, computing them is another. Although in principle the definition of derivatives applies to all functions, in practice computing derivatives of all but the simplest functions using only the definition would be so cumbersome and time-consuming that it would be impossible in practice to calculate derivatives when they are needed for applications. We now state and prove the standard rules for computing derivatives. These rules, combined with a knowledge of the derivatives of the elementary functions (that is, polynomials, power functions, logarithms, exponentials and trigonometric functions), allow us to take the derivative of virtually any function that can be built up out of elementary functions using sums, differences, products, quotients and compositions. We will see a rigorous treatment of the derivatives of the elementary functions in Chapter 7, though for now it is assumed that the reader knows these derivatives from a calculus course, and we will use such derivatives in examples.

We start with the following theorem, which shows how derivatives work with respect to the addition, subtraction, multiplication and division of functions. Whereas from a computational point of view what is important about this theorem is what the derivative of a sum, difference, product or quotient actually equals, from a theoretical point of view what is important is that the sum, difference, product or quotient of two differentiable functions is itself differentiable (with the usual caveat about not dividing by zero).

Theorem 4.3.1. *Let $I \subseteq \mathbb{R}$ be an open interval, let $c \in I$, let $f, g\colon I \to \mathbb{R}$ be functions and let $k \in \mathbb{R}$. Suppose that f and g are differentiable at c.*

1. *$f + g$ is differentiable at c and $[f+g]'(c) = f'(c) + g'(c)$.*
2. *$f - g$ is differentiable at c and $[f-g]'(c) = f'(c) - g'(c)$.*

3. *kf is differentiable at c and $[kf]'(c) = kf'(c)$.*
4. ***(Product Rule)** fg is differentiable at c and $[fg]'(c) = f'(c)g(c) + f(c)g'(c)$.*
5. ***(Quotient Rule)** If $g(c) \neq 0$, then $\frac{f}{g}$ is differentiable at c and*

$$\left[\frac{f}{g}\right]'(c) = \frac{f'(c)g(c) - f(c)g'(c)}{[g(c)]^2}.$$

Proof. We will prove Parts (1), (4) and (5), leaving the rest to the reader in Exercise 4.3.1.

The following proofs make use of the definition of derivatives, the definition of sums, differences, products and quotients of functions (given in Definition 3.2.9), and Theorem 3.2.10, Theorem 3.3.8 and Theorem 4.2.4. In all parts of the proof, we show that the derivative exists and prove the formula for the derivative simultaneously.

(1) We compute

$$\begin{aligned}
\lim_{h\to 0} & \frac{[f+g](c+h) - [f+g](c)}{h} \\
&= \lim_{h\to 0} \frac{[f(c+h) + g(c+h)] - [f(c) + g(c)]}{h} \\
&= \lim_{h\to 0} \frac{[f(c+h) - f(c)] + [g(c+h) - g(c)]}{h} \\
&= \lim_{h\to 0} \left\{ \frac{f(c+h) - f(c)}{h} + \frac{g(c+h) - g(c)}{h} \right\} \\
&= \lim_{h\to 0} \frac{f(c+h) - f(c)}{h} + \lim_{h\to 0} \frac{g(c+h) - g(c)}{h} = f'(c) + g'(c).
\end{aligned}$$

Hence $[f+g]'(c)$ exists and equals $f'(c) + g'(c)$.

(4) We compute

$$\begin{aligned}
\lim_{h\to 0} & \frac{[fg](c+h) - [fg](c)}{h} \\
&= \lim_{h\to 0} \frac{f(c+h)g(c+h) - f(c)g(c)}{h} \\
&= \lim_{h\to 0} \frac{f(c+h)g(c+h) - f(c+h)g(c) + f(c+h)g(c) - f(c)g(c)}{h} \\
&= \lim_{h\to 0} \left\{ f(c+h)\frac{g(c+h) - g(c)}{h} + \frac{f(c+h) - f(c)}{h} g(c) \right\} \\
&= \lim_{h\to 0} f(c+h)\frac{g(c+h) - g(c)}{h} + \lim_{h\to 0} \frac{f(c+h) - f(c)}{h} g(c) \\
&= f(c)g'(c) + f'(c)g(c).
\end{aligned}$$

Hence $[fg]'(c)$ exists and equals $f'(c)g(c) + f(c)g'(c)$.

(5) We compute

$$\begin{aligned}
\lim_{h\to 0} \frac{\left[\frac{f}{g}\right](c+h) - \left[\frac{f}{g}\right](c)}{h}
&= \lim_{h\to 0} \frac{\frac{f(c+h)}{g(c+h)} - \frac{f(c)}{g(c)}}{h} = \lim_{h\to 0} \frac{f(c+h)g(c) - f(c)g(c+h)}{hg(c)g(c+h)} \\
&= \lim_{h\to 0} \frac{f(c+h)g(c) - f(c)g(c) + f(c)g(c) - f(c)g(c+h)}{hg(c)g(c+h)} \\
&= \lim_{h\to 0} \left[\frac{f(c+h)-f(c)}{h}g(c) - f(c)\frac{g(c+h)-g(c)}{h}\right] \frac{1}{g(c)g(c+h)} \\
&= \frac{f'(c)g(c) - f(c)g'(c)}{[g(c)]^2}.
\end{aligned}$$

Hence $\left[\frac{f}{g}\right]'(c)$ exists and equals $\frac{f'(c)g(c)-f(c)g'(c)}{[g(c)]^2}$. □

The following result is an immediate consequence of Theorem 4.3.1, and we omit the proof.

Corollary 4.3.2. *Let $I \subseteq \mathbb{R}$ be an open interval, let $f, g\colon I \to \mathbb{R}$ be functions and let $k \in \mathbb{R}$. If f and g are differentiable, then $f+g$, $f-g$, kf and fg are differentiable, and if $g(x) \neq 0$ for all $x \in I$ then $\frac{f}{g}$ is differentiable.*

As an application of Theorem 4.3.1, the reader is asked in Exercise 4.3.5 to prove the formula for the derivative of $f\colon \mathbb{R} \to \mathbb{R}$ defined by $f(x) = x^n$ for all $x \in \mathbb{R}$, where $n \in \mathbb{N}$. We will see the analogous formula for all powers of x in Section 7.2, after we have defined such power functions rigorously.

An even more fundamental way of combining two functions than sums, differences, products and quotients is composition, and the following theorem shows how to compute the derivatives of compositions of functions.

Theorem 4.3.3 (Chain Rule). *Let $I, J \subseteq \mathbb{R}$ be open intervals, let $c \in I$ and let $f\colon I \to J$ and $g\colon J \to \mathbb{R}$ be functions. Suppose that f is differentiable at c, and that g is differentiable at $f(c)$. Then $g \circ f$ is differentiable at c and $[g \circ f]'(c) = g'(f(c)) \cdot f'(c)$.*

Before we give a proof of the Chain Rule, we want to present an attempted proof of this theorem that takes the most straightforward possible approach, though in this case the straightforward approach has a flaw. The attempted proof is

$$\begin{aligned}
\text{“}\lim_{x\to c} \frac{(g\circ f)(x) - (g\circ f)(c)}{x-c} &= \lim_{x\to c} \frac{g(f(x)) - g(f(c))}{x-c} \\
&= \lim_{x\to c} \frac{g(f(x)) - g(f(c))}{f(x)-f(c)} \, \frac{f(x)-f(c)}{x-c} \\
&= g'(f(c)) \cdot f'(c).\text{”}
\end{aligned}$$

Before reading a valid proof of the Chain Rule, the reader should try to find the flaw in this attempted proof, to understand why we need to take the less than straightforward approach used in the proof given below; the reader is asked to explain the flaw in Exercise 4.3.6.

Proof of Theorem 4.3.3 (Chain Rule). Let $k\colon J \to \mathbb{R}$ be defined by

$$k(y) = \begin{cases} \frac{g(y)-g(f(c))}{y-f(c)}, & \text{if } y \in J - \{f(c)\} \\ g'(f(c)), & \text{if } y = f(c). \end{cases}$$

Because g is differentiable at $f(c)$, we know that

$$\lim_{y \to f(c)} \frac{g(y) - g(f(c))}{y - f(c)}$$

exists and equals $g'(f(c))$. It therefore follows from Lemma 3.3.2 that k is continuous at $f(c)$.

Because f is differentiable at c, we know by Theorem 4.2.4 that f is continuous at c. Theorem 3.3.8 (2) then implies that $k \circ f$ is continuous at c. It follows from Lemma 3.3.2 that $\lim_{x \to c} (k \circ f)(x)$ exists and equals $(k \circ f)(c)$.

By the definition of k we see that if $y \in J - \{f(c)\}$ then

$$k(y)[y - f(c)] = g(y) - g(f(c)). \tag{4.3.1}$$

Equation 4.3.1 also holds when $y = f(c)$, which is seen by simply substituting $y = f(c)$ into both sides of the equation. Hence Equation 4.3.1 holds for all $y \in J$.

Let $x \in I - \{c\}$. Then $f(x) \in J$, and so we can substitute $y = f(x)$ into Equation 4.3.1 to obtain

$$k(f(x))[f(x) - f(c)] = g(f(x)) - g(f(c)).$$

Dividing both sides of this last equation by $x - c$ (which is not zero) yields

$$(k \circ f)(x) \frac{f(x) - f(c)}{x - c} = \frac{(g \circ f)(x) - (g \circ f)(c)}{x - c}.$$

Finally, using the continuity of $k \circ f$ at c, Theorem 3.2.10 (4), the definition of k and the fact that f is differentiable at c, we see that

$$\lim_{x \to c} \frac{(g \circ f)(x) - (g \circ f)(c)}{x - c} = \lim_{x \to c} \left[(k \circ f)(x) \frac{f(x) - f(c)}{x - c} \right]$$
$$= (k \circ f)(c) \cdot f'(c) = k(f(c)) \cdot f'(c) = g'(f(c)) \cdot f'(c).$$

Hence $[g \circ f]'(c)$ exists and equals $g'(f(c)) \cdot f'(c)$. □

The following result is an immediate consequence of the Chain Rule (Theorem 4.3.3).

Corollary 4.3.4. *Let $I, J \subseteq \mathbb{R}$ be open intervals, and let $f\colon I \to J$ and $g\colon J \to \mathbb{R}$ be functions. If f and g are differentiable, then $g \circ f$ is differentiable.*

As we mentioned at the end of Section 4.2, it is possible to take derivatives of functions with domains that are non-degenerate intervals, where we take one-sided derivatives at the endpoints if there are any. Everything we have proved about derivatives in this section for functions defined on open intervals also holds for functions defined on other types of non-degenerate intervals, as long as we take one-sided derivatives at the endpoints.

There is a very important theoretical implication of the rules of differentiation stated in Theorem 4.3.1 and Theorem 4.3.3. Because it is known how to take the derivatives of the various standard elementary functions (polynomials, power functions, logarithms, exponentials and trigonometric functions), a fact that is familiar to the reader from a calculus course and that is treated rigorously in Chapter 7, it then follows from the rules of differentiation that any function that is made up of elementary functions combined via addition, subtraction, multiplication, division and composition is also differentiable. Most of the functions that arise in the applications of mathematics to real-world problems are such functions (or at worst are piecewise such functions), and therefore most of the functions found in applications are differentiable (or piecewise differentiable), and their derivatives are themselves made up of elementary functions that are combined via addition, subtraction, multiplication, division and composition. This remarkable fact for derivatives is in contrast to the situation for integrals, where it is not possible to take every function that is made up of elementary functions combined via addition, subtraction, multiplication, division and composition, and express its indefinite integral as such a function; and even when it is possible to express the indefinite integral as such a function in principle, it is not always clear how to do so in practice. See the references given in the paragraph following Corollary 5.6.3 for details.

Reflections

The theorems in this section are very familiar to anyone who has taken a calculus course—too familiar, perhaps, because in some introductory calculus courses too much emphasis is placed upon the computing of derivatives at the result of rushing as quickly as possible to the Product Rule, Quotient Rule and Chain Rule, and not enough emphasis is placed upon what the derivative means intuitively, and how it is to be applied. We have therefore separated Section 4.2, which is concerned with the definition and basic properties of derivatives, from the present section, which is about formulas for computing derivatives, in order to emphasize that understanding how derivatives are defined is quite separate from knowing how they are to be computed in practice.

Of course, just as it is problematic when a calculus course neglects the intuitive meaning of derivatives in favor of too rapidly focusing upon how to compute them, it is no less problematic if a calculus course, in its attempt to include intuition and application, neglects basic computational skills. A balance is needed between computing, intuition and application, and, fortunately, some recent calculus texts have attempted to find such a balance. In a real analysis course, by contrast, none of these three aspects of differentiation is central, and the focus is purely theoretical. Even a very familiar theorem such as Theorem 4.3.1, which in a calculus course is viewed as being important for computational purposes, will be seen to be useful in a number of proofs in this text.

Not only should the statement of Theorem 4.3.1 be familiar to the reader from calculus courses, but the proof of this theorem is one of the few proofs in this text that the reader might have already encountered, virtually identically, in a calculus course. All of the hard work for the proof of Theorem 4.3.1 was done in our treatment of limits in Section 3.2, and once the properties of limits have been rigorously established, then the proof of Theorem 4.3.1 found in a calculus course is now seen to be rigorous as well. By contrast, the proof of the Chain Rule (Theorem 4.3.3) is a bit trickier than might be expected, and it is not the same as the informal sketch of a proof that would typically be seen in a calculus course.

Though we have discussed the most common differentiation rules in the present section, there is one method for finding derivatives that is taught in introductory calculus courses that we do not discuss in the present text, namely, implicit differentiation. There are two reasons that we do not deal with this topic. First, implicit differentiation is not really a separate method of differentiation, but rather it is simply an application of the Chain Rule. It is unfortunate that implicit differentiation is sometimes taught as a rote method where one simply inserts the symbol $\frac{dy}{dx}$ in certain places because "that is how it is done," rather than emphasizing the role of the Chain Rule. Second, to give a completely rigorous treatment of implicit differentiation, it would be necessary to prove the Implicit Function Theorem, which says that under the right circumstances an equation of the form $F(x,y) = 0$ can be viewed as locally describing y as a function of x, the right circumstances being essentially that the curve described by $F(x,y) = 0$ is smooth and the tangent line is not vertical at the given point on the curve. To do that rigorously, however, would require functions of two variables and partial derivatives, which we do not treat in this text.

Exercises

Exercise 4.3.1. **[Used in Theorem 4.3.1.]** Prove Theorem 4.3.1 (2) (3).

Exercise 4.3.2. Let $I \subseteq \mathbb{R}$ be an open interval, let $c \in I$, let $n \in \mathbb{N}$ and let $f_1, \ldots, f_n : I \to \mathbb{R}$ be functions. Suppose that f_i is differentiable at c for all $i \in \{1, \ldots, n\}$. Prove that $f_1 f_2 \cdots f_n$ is differentiable at c, and find (and prove) a formula for $(f_1 f_2 \cdots f_n)'(c)$ in terms of $f_1(c), \ldots, f_n(c)$ and $f_1'(c), \ldots, f_n'(c)$.

Exercise 4.3.3. Let $I \subseteq \mathbb{R}$ be an open interval, let $c \in I$ and let $f : I \to \mathbb{R}$ be a function. Suppose that f is differentiable at c. Let $f^2 = f \cdot f$. Using only the definition of derivatives and Lemma 4.2.2, prove that $[f^2]'(c) = 2f(c)f'(c)$. Do not use any other theorems about differentiation, such as Theorem 4.3.1 or Theorem 4.3.3.

Exercise 4.3.4. Let $I \subseteq \mathbb{R}$ be an open interval, let $c \in I$ and let $f : I \to \mathbb{R}$ be a function. Suppose that f is differentiable at c, and that $f'(c) \neq 0$. Using only the definition of derivatives and Lemma 4.2.2, prove that $\left[\frac{1}{f}\right]'(c) = -\frac{f'(c)}{[f(c)]^2}$. Do not use any other theorems about differentiation, such as Theorem 4.3.1.

Exercise 4.3.5. **[Used throughout.]** Let $I \subseteq \mathbb{R}$ be a non-degenerate open interval, and let $n \in \mathbb{N}$. Let $f\colon I \to \mathbb{R}$ be defined by $f(x) = x^n$ for all $x \in I$. Prove that f is differentiable and $f'(x) = nx^{n-1}$ for all $x \in I$. (Strictly speaking, when $n = 1$ the expression x^{n-1} is not defined for $x = 0$, but for convenience we abuse notation and think of the function $g\colon \mathbb{R} \to \mathbb{R}$ defined by "$g(x) = x^0$ for all $x \in \mathbb{R}$" to be the same as the function defined by $g(x) = 1$ for all $x \in \mathbb{R}$.)

Exercise 4.3.6. **[Used in Section 4.3.]** Explain the flaw in the attempted proof of the Chain Rule (Theorem 4.3.3) that is given prior to the correct proof. Restate the Chain Rule with modified hypotheses that would make the attempted proof into a valid proof.

Exercise 4.3.7. **[Used in Lemma 7.3.4 and Theorem 7.3.12.]** Let $[a,b] \subseteq \mathbb{R}$ and $[b,c] \subseteq \mathbb{R}$ be non-degenerate closed bounded intervals, and let $f\colon [a,b] \to \mathbb{R}$ and $g\colon [b,c] \to \mathbb{R}$ be functions. Suppose that $f(b) = g(b)$. Let $h\colon [a,c] \to \mathbb{R}$ be defined by

$$h(x) = \begin{cases} f(x), & \text{if } x \in [a,b] \\ g(x), & \text{if } x \in [b,c]. \end{cases}$$

(1) Suppose that f and g are differentiable, and that $f'(b) = g'(b)$, where $f'(b)$ and $g'(b)$ are one-sided derivatives. Prove that h is differentiable.

(2) Suppose that $b - a = c - b$, that $g(x) = f(a + c - x)$ for all $x \in [b,c]$, and that $f'(b) = 0$, where $f'(b)$ is a one-sided derivative. Prove that h is differentiable, and that $h'(x) = -f'(a + c - x)$ for all $x \in [b,c]$.

4.4 The Mean Value Theorem

We take the derivative of a function to learn more about the function. For example, the reader is familiar with the fact that a positive derivative means the function is increasing. We will see a proof of this fact in Section 4.5. In order to give a rigorous treatment of such theorems, we first need to prove a very important tool that relates functions to their derivatives, which is the Mean Value Theorem (Theorem 4.4.4). As is the case with many important theorems in real analysis, the Mean Value Theorem relies upon the Least Upper Bound Property of the real numbers.

We start with the following two lemmas, which are really the essence of the Mean Value Theorem. The first of our lemmas gives a rigorous statement of the intuitively evident fact that if a differentiable function has a maximum value or a minimum value at a point, then the derivative must be zero at that point. (We use the terms "maximum value" and "minimum value" informally here, and we will not need these terms in the statements of lemmas and theorems.)

Lemma 4.4.1. *Let $[a,b] \subseteq \mathbb{R}$ be a non-degenerate closed bounded interval, let $c \in (a,b)$ and let $f\colon [a,b] \to \mathbb{R}$ be a function. Suppose that f is differentiable at c. If either $f(c) \geq f(x)$ for all $x \in [a,b]$ or $f(c) \leq f(x)$ for all $x \in [a,b]$, then $f'(c) = 0$.*

Proof. Suppose that $f(c) \geq f(x)$ for all $x \in [a,b]$; the other case is similar, and we omit the details.

Because f is differentiable at c, then $\lim_{x \to c} \frac{f(x)-f(c)}{x-c}$ exists and equals $f'(c)$. It follows from Lemma 3.2.17 that $\lim_{x \to c^-} \frac{f(x)-f(c)}{x-c}$ and $\lim_{x \to c^+} \frac{f(x)-f(c)}{x-c}$ exist and are equal to $f'(c)$.

Let $x \in (a,c)$. Because $f(c) \geq f(x)$, then $\frac{f(x)-f(c)}{x-c} \geq 0$. Therefore, using the analog for one-sided limits of Exercise 3.2.12 (1), we deduce that

$$\lim_{x \to c^-} \frac{f(x)-f(c)}{x-c} \geq 0.$$

It follows that $f'(c) \geq 0$. A similar argument shows that

$$\lim_{x \to c^+} \frac{f(x)-f(c)}{x-c} \leq 0,$$

and hence that $f'(c) \leq 0$. We conclude that $f'(c) = 0$. □

The following example shows that Lemma 4.4.1 cannot be made into an if and only if statement.

Example 4.4.2. Let $f\colon [-1,1] \to \mathbb{R}$ be defined by $f(x) = x^3$ for all $x \in [-1,1]$. It can be verified using the definition of derivatives that $f'(0) = 0$; the details are left to the reader. On the other hand, it is certainly not the case that $f(0) \geq f(x)$ for all $x \in [-1,1]$, or that $f(0) \leq f(x)$ for all $x \in [-1,1]$. ◊

The Extreme Value Theorem (Theorem 3.5.1) states that every continuous function defined on a non-degenerate closed bounded interval has both a maximum value and a minimum value. If such a function is differentiable, must these values (or at least one of them) occur where the derivative is zero? Lemma 4.4.1 might appear to imply that the answer is yes, but consider the graph in Figure 4.4.1 (i), where we see that the maximum value and minimum value of a differentiable function with domain that is a non-degenerate closed bounded interval can both occur at the endpoints of the interval, which would not allow us to apply Lemma 4.4.1 to deduce the existence of a point in the interior of the interval with zero derivative. As is formalized in our next lemma, this problem can be avoided if the function has equal values at the two endpoints of the closed interval; see Figure 4.4.1 (ii). This lemma, known as Rolle's Theorem, is actually a special case of the Mean Value Theorem, but it is easier to prove directly than the Mean Value Theorem, and it will be used in the proof of the latter. (In spite of its historical name, we call Rolle's Theorem a lemma because its main role is to prove other more important results, such as the Mean Value Theorem and Taylor's Theorem (Theorem 4.4.6).)

Lemma 4.4.3 (Rolle's Theorem). *Let $[a,b] \subseteq \mathbb{R}$ be a non-degenerate closed bounded interval, and let $f\colon [a,b] \to \mathbb{R}$ be a function. Suppose that f is continuous on $[a,b]$ and differentiable on (a,b). If $f(a) = f(b)$, then there is some $c \in (a,b)$ such that $f'(c) = 0$.*

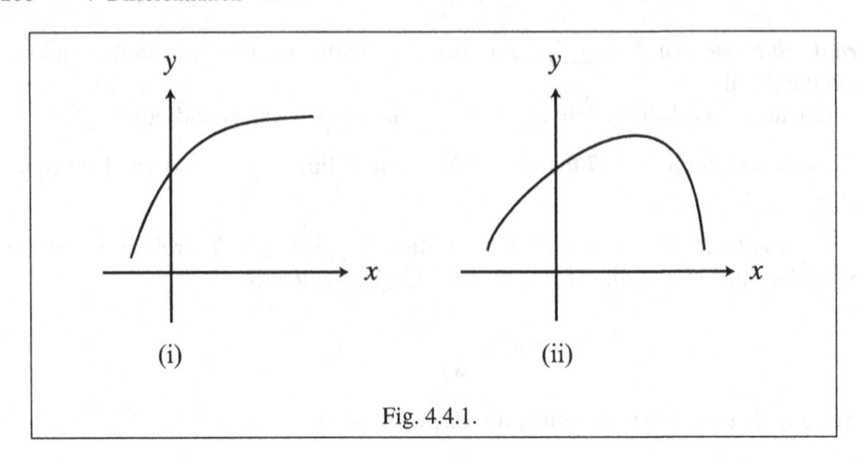

Fig. 4.4.1.

Proof. Suppose that $f(a) = f(b)$. By the Extreme Value Theorem (Theorem 3.5.1) there are $x_{min}, x_{max} \in [a,b]$ such that $f(x_{min}) \leq f(x) \leq f(x_{max})$ for all $x \in [a,b]$. If it were the case that $f(x_{min}) = f(x_{max})$, then f would be a constant function, and therefore by Example 4.2.3 (1) it would follow that $f'(x) = 0$ for all $x \in (a,b)$, and so we could let c be anything in (a,b). Now suppose that $f(x_{min}) < f(x_{max})$. It must therefore be the case that at least one of $f(x_{min})$ and $f(x_{max})$ does not equal $f(a) = f(b)$. Without loss of generality, assume that $f(x_{max}) \neq f(a) = f(b)$; the other case is similar, and we omit the details. Let $c = x_{max}$. Clearly $c \in (a,b)$. Hence f is differentiable at c, and it now follows from Lemma 4.4.1 that $f'(c) = 0$. □

The Mean Value Theorem is a generalization of Rolle's Theorem (Lemma 4.4.3) to the situation where $f(a)$ is not necessarily equal to $f(b)$. As is often the case in mathematics, generalizing a theorem successfully depends upon changing the way one views the theorem. If the goal of Rolle's Theorem is to find a point $c \in (a,b)$ such that $f'(c) = 0$, then as we saw in Figure 4.4.1 (i) it is not possible to drop the requirement that $f(a) = f(b)$. However, we observe that when $f(a) = f(b)$, the line through the points $(a, f(a))$ and $(b, f(b))$ has slope zero, and hence that line is parallel to the tangent line at the point $c \in (a,b)$ such that $f'(c) = 0$. The Mean Value Theorem says that even when $f(a)$ is not necessarily equal to $f(b)$, there is nonetheless always a point $c \in (a,b)$ such that the tangent line at c is parallel to the line through the points $(a, f(a))$ and $(b, f(b))$. See Figure 4.4.2. In addition to the above geometric way of thinking about the Mean Value Theorem, another intuitive way of thinking about this theorem is that if a car is driven on a straight road from time $t = a$ to $t = b$, then at some point during the trip the instantaneous velocity of the car will equal its average velocity for the duration of the trip.

Theorem 4.4.4 (Mean Value Theorem). *Let $[a,b] \subseteq \mathbb{R}$ be a non-degenerate closed bounded interval, and let $f\colon [a,b] \to \mathbb{R}$ be a function. Suppose that f is continuous on $[a,b]$ and differentiable on (a,b). Then there is some $c \in (a,b)$ such that*

$$f'(c) = \frac{f(b) - f(a)}{b - a}.$$

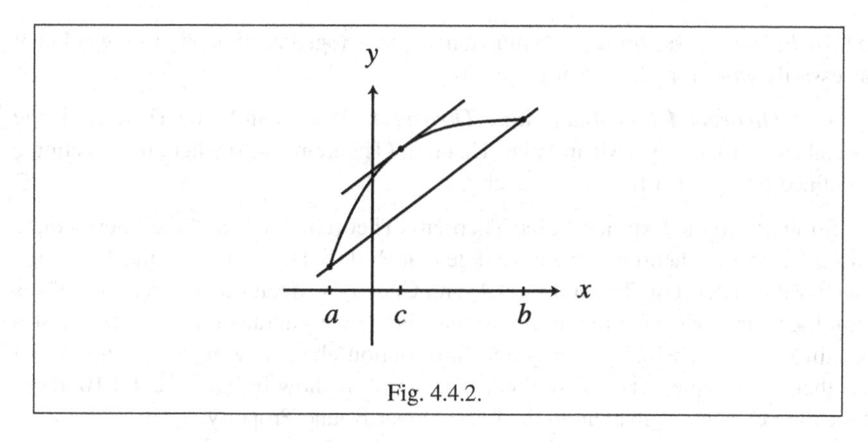

Fig. 4.4.2.

Rather than proving the Mean Value Theorem directly, we will first prove a generalization of it known as Cauchy's Mean Value Theorem, and we will then deduce the Mean Value Theorem as a corollary of this generalization (we will need the generalization later on in any case, and so it is more efficient to proceed as we are doing).

Theorem 4.4.5 (Cauchy's Mean Value Theorem). *Let $[a,b] \subseteq \mathbb{R}$ be a non-degenerate closed bounded interval, and let $f,g\colon [a,b] \to \mathbb{R}$ be functions. Suppose that f and g are continuous on $[a,b]$ and differentiable on (a,b). Then there is some $c \in (a,b)$ such that*

$$[f(b)-f(a)]g'(c) = [g(b)-g(a)]f'(c).$$

Proof. Let $h\colon [a,b] \to \mathbb{R}$ be defined by

$$h(x) = [f(b)-f(a)]g(x) - [g(b)-g(a)]f(x)$$

for all $x \in [a,b]$. We know that constant functions are continuous on $[a,b]$ by Example 3.3.3 (1), and differentiable on (a,b) by Example 4.2.3 (1). We also know that f and g are continuous on $[a,b]$ and differentiable on (a,b), and it then follows from Theorem 3.3.5 and Theorem 4.3.1 that h is continuous on $[a,b]$ and differentiable on (a,b). Moreover, it is seen that $h(a) = f(b)g(a) - g(b)f(a) = h(b)$. We can therefore apply Rolle's Theorem (Lemma 4.4.3) to h, and we deduce that there is some $c \in (a,b)$ such that $h'(c) = 0$. Applying Theorem 4.3.1 to the definition of h we see that $h'(x) = [f(b)-f(a)]g'(x) - [g(b)-g(a)]f'(x)$. The equation $h'(c) = 0$ therefore yields $[f(b)-f(a)]g'(c) - [g(b)-g(a)]f'(c) = 0$, which is what we needed to show. □

Observe that Cauchy's Mean Value Theorem does not follow directly from the Mean Value Theorem, because if we tried to find a "c" for each of the functions f and g individually, we would obtain one number, say $d \in (a,b)$, for the function f, and another number, say $e \in (a,b)$, for the function g, where d and e are not necessarily equal, and then we would deduce that $[f(b)-f(a)]g'(e) = [g(b)-g(a)]f'(d)$, which is not as nice as Cauchy's Mean Value Theorem, which says that you can get one

"$c \in (a,b)$" that works for the two functions f and g together (though this c would not necessarily work for either f or g alone).

Proof of Theorem 4.4.4 (Mean Value Theorem). The Mean Value Theorem is the special case of Cauchy's Mean Value Theorem (Theorem 4.4.5) where the function g is defined by $g(x) = x$ for all $x \in [a,b]$. □

Similarly to the Extreme Value Theorem (Theorem 3.5.1) and the Intermediate Value Theorem (Theorem 3.5.2), we note that Rolle's Theorem (Lemma 4.4.3), the Mean Value Theorem (Theorem 4.4.4) and Cauchy's Mean Value Theorem (Theorem 4.4.5) are existence theorems, in that they each guarantee the existence of a certain number c, without giving any information about how to find c, nor about whether c is unique. Also, as the reader is asked to show in Exercise 4.4.10, these three theorems are equivalent to the Least Upper Bound Property.

We now turn to a very useful extension of the Mean Value Theorem known as Taylor's Theorem; we use the version of the theorem due to Joseph-Louis Lagrange (1736–1813). Taylor's Theorem is useful for dealing with Taylor polynomials and Taylor series, as we will see in Section 10.4, but the proper way to view Taylor's Theorem is as a generalized version of the Mean Value Theorem that uses higher derivatives. As the reader can verify, the Mean Value Theorem is a special case of Taylor's Theorem, obtained by substituting $n = 0$, and $c = a$, and $x = b$ in the latter.

Recall from Definition 4.2.6 that $f^{(0)} = f$ for any function f.

Theorem 4.4.6 (Taylor's Theorem). *Let $[a,b] \subseteq \mathbb{R}$ be a non-degenerate closed bounded interval, let $c \in (a,b)$, let $f\colon [a,b] \to \mathbb{R}$ be a function and let $n \in \mathbb{N} \cup \{0\}$. Suppose that $f^{(k)}$ exists and is continuous on $[a,b]$ for each $k \in \{0,\ldots,n\}$, and that $f^{(n+1)}$ exists on (a,b). Let $x \in [a,b]$. Then there is some p strictly between x and c (except that $p = c$ when $x = c$) such that*

$$f(x) = \sum_{k=0}^{n} \frac{f^{(k)}(c)}{k!}(x-c)^k + \frac{f^{(n+1)}(p)}{(n+1)!}(x-c)^{n+1}.$$

Proof. First, suppose that $x = c$. Let $p = c$. Then the theorem holds in this case, as the reader may verify.

Now suppose that $x \neq c$. Then there is a unique $B \in \mathbb{R}$ such that the following equation holds (simply solve for B):

$$f(x) = \sum_{k=0}^{n} \frac{f^{(k)}(c)}{k!}(x-c)^k + B(x-c)^{n+1}. \tag{4.4.1}$$

To prove the theorem, we will show that there is some p strictly between x and c such that

$$B = \frac{f^{(n+1)}(p)}{(n+1)!}. \tag{4.4.2}$$

Let $F\colon [a,b] \to \mathbb{R}$ be defined by

$$F(z) = \sum_{k=0}^{n} \frac{f^{(k)}(z)}{k!}(x-z)^k + B(x-z)^{n+1} \tag{4.4.3}$$

for all $z \in [a,b]$. Because $f^{(k)}$ exists and is continuous on $[a,b]$ and differentiable on (a,b) for each $k \in \{0,\ldots,n\}$, it follows from standard rules for differentiation (found in Example 4.2.3 (1), Theorem 4.3.1, Theorem 4.3.3 and Exercise 4.3.5) that F is continuous on $[a,b]$ and differentiable on (a,b). Because the closed interval from x to c is contained in $[a,b]$, it follows from Exercise 3.3.2 (2) and Exercise 4.2.3 (5) that F is continuous on this closed interval, and differentiable on the open interval from x to c.

It follows from Equation 4.4.1 and Equation 4.4.3 that $F(c) = f(x)$, and from Equation 4.4.3 alone that $F(x) = f(x)$. We can therefore apply Rolle's Theorem (Lemma 4.4.3) to F on the closed interval from x to c, and we deduce that there is some p strictly between x and c such that $F'(p) = 0$. Using the Product Rule (Theorem 4.3.1 (4)) and some algebraic manipulation, it is left to the reader to verify that

$$F'(z) = \frac{f^{(n+1)}(z)}{n!}(x-z)^n - B(n+1)(x-z)^n$$

for all $z \in (a,b)$. The fact that $F'(p) = 0$ can then be rewritten as

$$0 = \frac{f^{(n+1)}(p)}{n!}(x-p)^n - B(n+1)(x-p)^n,$$

which implies Equation 4.4.2. □

It is important to observe that the number p in Taylor's Theorem (Theorem 4.4.6) depends upon x; the formula for $f(x)$ in the theorem is for a single value of $x \in [a,b]$, and is not a general formula for all $x \in [a,b]$ with the same p.

We conclude this section with an application of the Mean Value Theorem to antiderivatives, a concept we will define after the following lemma. Intuitively, the second part of this lemma states the rather obvious fact that if two horses in a horse race run at the same speed, then they will maintain a constant distance between them.

Lemma 4.4.7. *Let $I \subseteq \mathbb{R}$ be a non-degenerate interval, and let $f,g\colon I \to \mathbb{R}$ be function. Suppose that f and g are continuous on I and differentiable on the interior of I.*

1. *$f'(x) = 0$ for all x in the interior of I if and only if f is constant on I.*
2. *$f'(x) = g'(x)$ for all x in the interior of I if and only if there is some $C \in \mathbb{R}$ such that $f(x) = g(x) + C$ for all $x \in I$.*

Proof.

(1) Suppose that f is constant on I. It then follows from Example 4.2.3 (1) that $f'(x) = 0$ for all x in the interior of I.

Next, suppose that $f'(x) = 0$ for all x in the interior of I. Let $p,q \in I$. Suppose that $p \neq q$. Without loss of generality, assume that $p < q$. Then $[p,q] \subseteq I$. The Mean Value Theorem (Theorem 4.4.4) applied to $f|_{[p,q]}$ implies that there is some $c \in (p,q)$ such that

$$f'(c) = \frac{f(q) - f(p)}{q - p}.$$

Because $c \in (p,q)$ then c is in the interior of I, and it follows that $f(q) - f(p) = f'(c)(q-p) = 0$. We deduce that f is constant on I.

(2) This part of the lemma follows from Part (1) of this lemma applied to $f - g$; we omit the details. □

We now turn to the definition of antiderivatives. The crucial thing to keep in mind when considering antiderivatives is that, although they turn out (via the Fundamental Theorem of Calculus) to be intimately related to integrals, antiderivatives are defined strictly in relation to derivatives. The relation between antiderivatives and integrals is an amazing theorem that requires proof, and not just a matter of definition.

Definition 4.4.8. Let $I \subseteq \mathbb{R}$ be an open interval, and let $f: I \to \mathbb{R}$ be a function. An **antiderivative** of f is a function $F: I \to \mathbb{R}$ such that F is differentiable and $F' = f$. △

For a given function $f: I \to \mathbb{R}$, where $I \subseteq \mathbb{R}$ is an open interval, we ask whether it has an antiderivative, and if it does, whether the antiderivative unique. We start with the latter question, which is the simpler of the two. Of course, as the reader knows from calculus courses, if a function has an antiderivative, then it will have more than one antiderivative. For example, both x^2 and $x^2 + 7$ are antiderivatives of $2x$. Hence antiderivatives are not unique. However, as stated in the following lemma, the next best thing holds, which is that on an open interval any two antiderivatives differ by a constant. The following lemma is an immediate consequence of Lemma 4.4.7 (2), and we omit the proof.

Corollary 4.4.9. *Let $I \subseteq \mathbb{R}$ be a non-degenerate open interval, and let $f: I \to \mathbb{R}$ be a function. If $F, G: I \to \mathbb{R}$ are antiderivatives of f, then there is some $C \in \mathbb{R}$ such that $F(x) = G(x) + C$ for all $x \in I$.*

Does every function $f: I \to \mathbb{R}$, where $I \subseteq \mathbb{R}$ is an open interval, have an antiderivative? In other words, is every such function the derivative of a function? It turns out that every continuous function has an antiderivative, as we will see using Corollary 5.6.3, which is an immediate consequence of the Fundamental Theorem of Calculus Version I (Theorem 5.6.2). However, as we now show, not every function in general has an antiderivative. Our example of a function that does not have an antiderivative makes use of the following theorem, which relies upon ideas developed earlier in this section.

Theorem 4.4.10 (Intermediate Value Theorem for Derivatives). *Let $I \subseteq \mathbb{R}$ be an open interval, and let $f: I \to \mathbb{R}$ be a function. Suppose that f is differentiable. Let $a, b \in I$, and suppose that $a < b$. Let $r \in \mathbb{R}$. If r is strictly between $f'(a)$ and $f'(b)$, then there is some $c \in (a,b)$ such that $f'(c) = r$.*

Proof. Suppose that r is strictly between $f'(a)$ and $f'(b)$. Without loss of generality, assume that $f'(a) < r < f'(b)$.

Let $g: I \to \mathbb{R}$ be defined by $g(x) = f(x) - rx$ for all $x \in I$. Because f is differentiable, it follows from Example 4.2.3 (1) and Theorem 4.3.1 (2) that g is differentiable, and that $g'(x) = f'(x) - r$ for all $x \in I$. By Theorem 4.2.4 we know that g is continuous.

The Extreme Value Theorem (Theorem 3.5.1) applied to $g|_{[a,b]}$ implies that there is some $c \in [a,b]$ such that $g(c) \leq g(x)$ for all $x \in [a,b]$.

The definition of derivatives, combined with Lemma 3.2.17, imply that

$$g'(a) = \lim_{x \to a^+} \frac{g(x) - g(a)}{x - a}.$$

Because $g(x) = f(x) - rx$ for all $x \in I$, it follows that $g'(a) < 0 < g'(b)$. We then use the analog for one-sided limits of Theorem 3.2.4 (2) to deduce that there is some $N < 0$ and some $\delta > 0$ such that $x \in I - \{a\}$ and $x \in (a, a+\delta)$ imply

$$\frac{g(x) - g(a)}{x - a} < N.$$

Because I is an open interval, we use Lemma 2.3.7 (2) to see that by taking a smaller value of δ if necessary, we may assume that $(a, a+\delta) \subseteq I$. Let $y \in (a, a+\delta)$. Then $y - a > 0$, and hence $g(y) - g(a) < 0$, which implies that $g(y) < g(a)$. Therefore $c \neq a$. A similar argument shows that $c \neq b$, and we omit the details. Hence $c \in (a,b)$.

Lemma 4.4.1 applied to $g|_{(a,b)}$ implies that $g'(c) = 0$. Hence $f'(c) - r = 0$, and it follows that $f'(c) = r$. □

What makes the Intermediate Value Theorem for Derivatives (Theorem 4.4.10) interesting is that even though the derivative of a differentiable function need not be continuous, as seen in Example 4.2.5 (1), it turns out that even a discontinuous derivative must satisfy the property given in the Intermediate Value Theorem (Theorem 3.5.2). Hence, as we see in the following simple example, although derivatives need not be continuous, not every discontinuous function is the derivative of something.

Example 4.4.11. Let $g \colon \mathbb{R} \to \mathbb{R}$ be defined by

$$g(x) = \begin{cases} 1, & \text{if } x \leq 1 \\ 2, & \text{if } x > 1. \end{cases}$$

Then g is not the derivative of any function, because it does not satisfy the conclusion of the Intermediate Value Theorem for Derivatives (Theorem 4.4.10). ◊

Reflections

The Mean Value Theorem is often treated very cursorily in introductory calculus courses, or is not treated at all, which is understandable due to the applied and computational focus of such courses. From our present perspective, by contrast, the Mean Value Theorem is a crucial tool used to relate the behavior of the derivative of a function to the behavior of the original function, as will be seen, for example, in the proof of Theorem 4.5.2.

The concept of antiderivatives is introduced in this section, though in principle this concept could have been defined in Section 4.2, because nothing more than the definition of derivatives is needed. However, it is only in the present section that we are able to say something interesting about antiderivatives, and so we have delayed the definition of this concept till here. There is no problem delaying the definition of antiderivatives by a few sections as we have done, but there is a problem delaying that definition, as do some calculus books, until the chapter on integration. It is certainly true, as the reader knows from calculus courses, that it is in the calculation of definite integrals that the concept of antiderivatives become spectacularly useful; it is also true that antiderivatives, when written as "indefinite integrals," use a notation that is extremely similar to the notation for "definite integrals." Nonetheless, locating the definition of antiderivatives in the chapter on integrals can cause students to lose sight of the very important facts, stated above but worth stressing again, that antiderivatives are defined strictly in terms of derivatives, in spite of their notation, and that the close relation between antiderivatives and integrals (meaning the "definite" kind) is an amazing theorem and is not true simply by virtue of definition or notation.

Exercises

Exercise 4.4.1. Find an example of a function $f\colon [a,b] \to \mathbb{R}$ for some non-degenerate closed bounded interval $[a,b] \subseteq \mathbb{R}$ such that f is continuous on $[a,b]$, that f is differentiable on (a,b) except at one point and that f does not satisfy the conclusion of the Mean Value Theorem.

Exercise 4.4.2. Find an example of a function $f\colon [a,b] \to \mathbb{R}$ for some non-degenerate closed bounded interval $[a,b] \subseteq \mathbb{R}$ such that f is continuous and differentiable on (a,b), and that f does not satisfy the conclusion of the Mean Value Theorem.

Exercise 4.4.3. Does Corollary 4.4.9 hold if I is not a single non-degenerate open interval, but is rather the union of finitely many such intervals? Give a proof or a counterexample.

Exercise 4.4.4. Prove that $\sqrt{1+4x} < 2x+1$ for all $x \in (0,\infty)$. You may use standard rules for differentiation, even if we have not yet proved them.

Exercise 4.4.5. Let $f\colon \mathbb{R} \to \mathbb{R}$ be a function. Suppose that f is differentiable, that $f(0) = 1$ and that $|f'(x)| \leq 1$ for all $x \in \mathbb{R}$. Prove that $|f(x)| \leq |x| + 1$ for all $x \in \mathbb{R}$.

Exercise 4.4.6. [Used in Section 3.4 and Lemma 10.5.1.] This exercise refers to Exercise 3.4.5. Let $[a,b] \subseteq \mathbb{R}$ be a non-degenerate closed bounded interval, and let $f\colon [a,b] \to \mathbb{R}$ be a function. Suppose that f is continuous on $[a,b]$ and differentiable on (a,b), and that there is some $M \in \mathbb{R}$ such that $|f'(x)| \leq M$ for all $x \in (a,b)$. Prove that f satisfies a Lipschitz condition with Lipschitz constant M. It follows from Exercise 3.4.5 (1) that f is uniformly continuous.

Exercise 4.4.7. [Used in Exercise 7.3.6.] Let $[a,b] \subseteq \mathbb{R}$ be a non-degenerate closed bounded interval, and let $f\colon [a,b] \to \mathbb{R}$ be a function. Suppose that f is continuous on $[a,b]$ and differentiable on (a,b). Prove that if $\lim_{x\to b^-} f'(x)$ exists, then the one-sided derivative $f'(b)$ exists and equals $\lim_{x\to b^-} f'(x)$.

Exercise 4.4.8. Let $I \subseteq \mathbb{R}$ be an open interval, and let $f\colon I \to \mathbb{R}$ be a function. Suppose that f is twice differentiable. Suppose that there are $x, y, z \in I$ such that $x < y < z$, and $f(x) > f(y)$ and $f(y) < f(z)$. Prove that there is some $c \in I$ such that $f''(c) > 0$.

Exercise 4.4.9. **[Used in Theorem 4.6.4 and Theorem 6.3.5.]** Let $I \subseteq \mathbb{R}$ be a non-degenerate open interval, and let $f\colon I \to \mathbb{R}$ be a function. Suppose that f is differentiable, and that $f'(x) \neq 0$ for all $x \in I$.

(1) Prove that f is injective.
(2) Prove that either $f'(x) > 0$ for all $x \in I$, or that $f'(x) < 0$ for all $x \in I$.

Exercise 4.4.10. **[Used in Section 4.4.]** Using the ideas in the proof of Theorem 3.5.4, prove that Rolle's Theorem (Lemma 4.4.3), the Mean Value Theorem (Theorem 4.4.4) and Cauchy's Mean Value Theorem (Theorem 4.4.5) are equivalent to the Least Upper Bound Property.

4.5 Increasing and Decreasing Functions, Part I: Local and Global Extrema

Now that we have the Mean Value Theorem (Theorem 4.4.4) at our disposal, we turn to one of the main reasons why we are interested in derivatives, which is that the derivative of a function yields geometric information about the original function, for example whether it is increasing or decreasing. Much of the original motivation for looking at such geometric properties of functions was to help people graph functions, which was no mean feat before graphing calculators and computers. However, even though modern technology makes graphing functions easy, concepts such as increasing and decreasing are also useful in other aspects of mathematics and its applications, such as optimization problems. Moreover, doing a little bit of graphing by hand, even when computing technology is available, helps us develop a better intuitive feel for functions and their graphs.

As will be seen below, a number of key geometric concepts such as increasing, decreasing, local maximum, local minimum and others are not about calculus per se; they will be *defined* without reference to differentiability. However, it turns out that even though calculus is not part of the definition of these concepts, there are some questions involving these concepts that are difficult to solve without calculus, and which can be solved easily with calculus—that is one of the reasons calculus is so great.

Our most fundamental geometric definition is the following.

Definition 4.5.1. Let $A \subseteq \mathbb{R}$ be a set, and let $f\colon A \to \mathbb{R}$ be a function.

1. The function f is **increasing** if $x < y$ implies $f(x) \leq f(y)$ for all $x, y \in A$.
2. The function f is **strictly increasing** if $x < y$ implies $f(x) < f(y)$ for all $x, y \in A$.
3. The function f is **decreasing** if $x < y$ implies $f(x) \geq f(y)$ for all $x, y \in A$.
4. The function f is **strictly decreasing** if $x < y$ implies $f(x) > f(y)$ for all $x, y \in A$.

5. The function f is **monotone** if it is either increasing or decreasing.
6. The function f is **strictly monotone** if it is either strictly increasing or strictly decreasing. △

Some books use the terms "non-decreasing" and "increasing" to mean what we call "increasing" and "strictly increasing," respectively, and similarly for decreasing. There is no definitive terminology here, and in any book that discusses these issues, it is worth checking the precise definitions that are used.

It is often difficult to compute directly from the definition when a function is increasing or decreasing, but in the differentiable case there is a very simple way to show that a function is increasing or decreasing.

Theorem 4.5.2. *Let $I \subseteq \mathbb{R}$ be a non-degenerate interval, and let $f\colon I \to \mathbb{R}$ be a function. Suppose that f is continuous on I and differentiable on the interior of I.*

1. *$f'(x) \geq 0$ for all x in the interior of I if and only if f is increasing on I.*
2. *If $f'(x) > 0$ for all x in the interior of I, then f is strictly increasing on I.*
3. *$f'(x) \leq 0$ for all x in the interior of I if and only if f is decreasing on I.*
4. *If $f'(x) < 0$ for all x in the interior of I, then f is strictly decreasing on I.*

Proof. We will prove Part (1), leaving the rest to the reader in Exercise 4.5.6.

(1) Suppose that $f'(x) \geq 0$ for all x in the interior of I. Let $p, q \in I$. Suppose that $p < q$. Then $[p,q] \subseteq I$. The Mean Value Theorem (Theorem 4.4.4) applied to $f|_{[p,q]}$ implies that there is some $c \in (p,q)$ such that

$$f'(c) = \frac{f(q) - f(p)}{q - p}.$$

Because $c \in (p,q)$ then c is in the interior of I, and it follows that $f(q) - f(p) = f'(c)(q-p) \geq 0$. Therefore $f(p) \leq f(q)$. We deduce that f is increasing.

Now suppose that f is increasing. Let c be in the interior of I. By Lemma 2.3.7 (2) there is some $\delta > 0$ such that $(c - \delta, c + \delta) \subseteq I$. Let $x \in (c, c+\delta)$. Then $f(c) \leq f(x)$, because f is increasing. Hence $\frac{f(x)-f(c)}{x-c} \geq 0$. Using the analog for one-sided limits of Exercise 3.2.12 (1) we deduce that

$$\lim_{x \to c^+} \frac{f(x) - f(c)}{x - c} \geq 0.$$

Because $f'(c)$ exists, then by Lemma 3.2.17 we know that $f'(c)$ must equal the limit in the above equation. It follows that $f'(c) \geq 0$. □

We now see in the following example that Theorem 4.5.2 (2) (4) cannot be made into "if and only if" statements.

Example 4.5.3. Let $f\colon \mathbb{R} \to \mathbb{R}$ be defined by $f(x) = x^3$ for all $x \in \mathbb{R}$. The function f is strictly increasing, as seen by Exercise 2.3.3 (1); that exercise does not make use of derivatives. However, we know by Exercise 4.3.5 that $f'(x) = 3x^2$ for all $x \in \mathbb{R}$, and hence $f'(0) = 0$. Therefore Theorem 4.5.2 (2) cannot be made into an "if and only if" statement. A similar example shows that Theorem 4.5.2 (4) cannot be made into an "if and only if" statement. ◊

Our next issue involves finding maximum values and minimum values of a function. Again, the definition does not involve derivatives.

Definition 4.5.4. Let $A \subseteq \mathbb{R}$ be a set, let $c \in A$ and let $f: A \to \mathbb{R}$ be a function.

1. The number c is a **local maximum** of f if there is some $\delta > 0$ such that $x \in A$ and $|x - c| < \delta$ imply $f(x) \leq f(c)$.
2. The number c is a **local minimum** of f if there is some $\delta > 0$ such that $x \in A$ and $|x - c| < \delta$ imply $f(x) \geq f(c)$.
3. The number c is a **local extremum** of f if it is either a local maximum or a local minimum.
4. The number c is a **global maximum** of f if $f(x) \leq f(c)$ for all $x \in A$.
5. The number c is a **global minimum** of f if $f(x) \geq f(c)$ for all $x \in A$.
6. The number c is a **global extremum** of f if it is either a global maximum or a global minimum. △

There is, once again, no definitive terminology here. Some books use the terms "relative maximum" and "absolute maximum," respectively, to mean what we call "local maximum" and "local minimum," and similarly for minima. Note that the plurals of "maximum," "minimum" and "extremum" are "maxima," "minima" and "extrema," respectively.

Observe in Definition 4.5.4 that the actual maximum values and minimum values of a function are not discussed, but only where such values occur, if they exist.

A global maximum is always a local maximum, but not vice versa, and similarly for minima. A number c is a local maximum of a function if and only if it is a global maximum of the restriction of the function to some open interval containing c, and again similarly for minima. Additionally, observe that we use $\leq$ and $\geq$ in the definition of local and global maxima and minima, rather than $<$ and $>$, respectively. Our definition is completely standard, and is very convenient, even though it does mean that local and global maxima and minima are not necessarily unique. For example, any point in the domain of a constant function is both a global maximum and a global minimum, which may not sound right at first glance, but is true according to Definition 4.5.4.

As the reader is familiar from an introductory calculus course, there are functions with various combinations of global extrema and local extrema. For example, the function $f: \mathbb{R} \to \mathbb{R}$ defined by $f(x) = 3x$ for all $x \in \mathbb{R}$ has no local extrema or global extrema of any kind. By contrast, the function $g: [0,7] \to \mathbb{R}$ defined by $f(x) = 3x$ for all $x \in [0,7]$ has a global (and hence local) maximum and a global (and hence local) minimum. The function $h: \mathbb{R} \to \mathbb{R}$ defined by $h(x) = |x|$ for all $x \in \mathbb{R}$ has a global minimum, but no local or global maximum. The function $k: \mathbb{R} \to \mathbb{R}$ defined by $k(x) = \sin x$ for all $x \in \mathbb{R}$ has infinitely many global maxima and global minima. The function $p: \mathbb{R} \to \mathbb{R}$ defined by $p(x) = x^3 - x$ for all $x \in \mathbb{R}$ has a local maximum and a local minimum, but no global extrema.

For many real-world applications, the goal is to find global extrema of functions. It is easier, however, to find local extrema, and finding them, when they exist, helps us locate global extrema. Hence, we examine local extrema first. The following lemma gives a very simple method for finding local extrema. Similarly to Definition 4.5.4, this lemma does not involve differentiability.

Lemma 4.5.5. *Let $A \subseteq \mathbb{R}$ be a set, let $c \in A$ and let $f: A \to \mathbb{R}$ be a function.*

1. *If there is some $\delta > 0$ such that $f|_{A\cap(c-\delta,c]}$ is increasing and $f|_{A\cap[c,c+\delta)}$ is decreasing, then c is a local maximum of f.*
2. *If there is some $\delta > 0$ such that $f|_{A\cap(c-\delta,c]}$ is decreasing and $f|_{A\cap[c,c+\delta)}$ is increasing, then c is a local minimum of f.*

Proof. Left to the reader in Exercise 4.5.7. □

The reader is asked in Exercise 4.5.8 to show that not every local extremum of a function satisfies the hypotheses of either part of Lemma 4.5.5.

When a function is differentiable, there is a nice way to locate local extrema. We will need the following definition and lemma

Definition 4.5.6. Let $I \subseteq \mathbb{R}$ be an open interval, let $c \in I$ and let $f: I \to \mathbb{R}$ be a function. The number c is a **critical point** of f if either f is differentiable at c and $f'(c) = 0$, or f is not differentiable at c. △

Lemma 4.5.7. *Let $I \subseteq \mathbb{R}$ be an open interval, let $c \in I$ and let $f: I \to \mathbb{R}$ be a function. If c is a local extremum of f, then c is a critical point of f.*

Proof. Suppose that c is a local extremum of f. We assume that c is a local maximum; the case where c is a local minimum is similar, and we omit the details. By the definition of local maxima, there is some $\delta > 0$ such that $x \in I$ and $|x - c| < \delta$ imply $f(x) \le f(c)$. Because I is an open interval, we use Lemma 2.3.7 (2) to see that by taking a smaller value of δ if necessary, we may assume that $[c-\delta, c+\delta] \subseteq I$. Then $f(x) \le f(c)$ for all $x \in [c-\delta, c+\delta]$.

If f is not differentiable at c, then there is nothing to prove, so suppose that f is differentiable at c. By Exercise 4.2.3 (3) we know that $f|_{[c-\delta,c+\delta]}$ is differentiable at c and $(f|_{[c-\delta,c+\delta]})'(c) = f'(c)$. It follows from Lemma 4.4.1 applied to $f|_{[c-\delta,c+\delta]}$ that $(f|_{[c-\delta,c+\delta]})'(c) = 0$, and hence $f'(c) = 0$. □

It is important to recognize that critical points need not be local extrema, as seen in the following example.

Example 4.5.8. Let $f: [-1,1] \to \mathbb{R}$ be defined by $f(x) = x^3$ for all $x \in [-1,1]$. Because $f'(x) = 3x^2$ for all $x \in \mathbb{R}$, then $f'(0) = 0$, and hence 0 is a critical point of f. However, as remarked in Example 4.5.3, the function f is strictly increasing, and therefore 0 is neither a local maximum nor a local minimum of f. ◊

Even though not all critical points of a function are local extrema, if we want to find the local extrema of a function, the standard approach is to find all the critical points first, and then identify which, if any, of the critical points are actually local extrema. The following theorem provides a good way to tell which critical points are local extrema.

Theorem 4.5.9 (First Derivative Test). *Let $I \subseteq \mathbb{R}$ be an open interval, let $c \in I$ and let $f: I \to \mathbb{R}$ be a function. Suppose that c is a critical point of f, and that f is continuous on I and differentiable on $I - \{c\}$.*

1. *Suppose that there is some $\delta > 0$ such that $x \in I$ and $c - \delta < x < c$ imply $f'(x) \geq 0$, and that $x \in I$ and $c < x < c + \delta$ imply $f'(x) \leq 0$. Then c is a local maximum of f.*
2. *Suppose that there is some $\delta > 0$ such that $x \in I$ and $c - \delta < x < c$ imply $f'(x) \leq 0$, and that $x \in I$ and $c < x < c + \delta$ imply $f'(x) \geq 0$. Then c is a local minimum of f.*
3. *Suppose that there is some $\delta > 0$ such that $x \in I - \{c\}$ and $|x - c| < \delta$ imply $f'(x) > 0$, or that $x \in I - \{c\}$ and $|x - c| < \delta$ imply $f'(x) < 0$. Then c is not a local extremum of f.*

Proof. We will prove Part (1); the other parts are similar, and we omit the details.

(1) Because I is an open interval, we use Lemma 2.3.7 (2) to see that by taking a smaller value of δ if necessary, we may assume that $[c - \delta, c + \delta] \subseteq I$. Let $p = c - \delta$ and $q = c + \delta$. Then $[p, q] \subseteq I$. Because f is continuous on I and differentiable on $I - \{c\}$, it follows from Exercise 3.3.2 (2) and Exercise 4.2.3 (5) that f is continuous on $[p, c]$ and $[c, q]$, and differentiable on (p, c) and (c, q). By hypothesis we know that $f'(x) \geq 0$ for all $x \in (p, c)$ and $f'(x) \leq 0$ for all $x \in (c, q)$. Using Theorem 4.5.2 (1) (3) we see that $f|_{[p,c]}$ is increasing and $f|_{[c,q]}$ is decreasing. Lemma 4.5.5 (1) now implies that c is a local maximum. □

It might appear that the three parts of the First Derivative Test (Theorem 4.5.9) do not cover all possible cases, because the third part has $>$ and $<$ rather than $\geq$ and $\leq$. However, if the critical point c is isolated, which means that there is some open interval containing c with no other critical points, then the three parts of the First Derivative Test do cover all possible cases. In practice, most critical points encountered in the applications of these methods are isolated.

In addition to the First Derivative Test, there is another widely used test for finding local maxima and minima, namely, the Second Derivative Test. In principle, it would be possible to live without the Second Derivative Test, and use only the First Derivative Test, because the latter is usable in all cases, whereas the former is not. However, the Second Derivative Test is sometimes easier to use in practice than the First Derivative Test, and so it is worth knowing.

Theorem 4.5.10 (Second Derivative Test). *Let $I \subseteq \mathbb{R}$ be an open interval, let $c \in I$ and let $f: I \to \mathbb{R}$ be a function. Suppose that f is differentiable, that $f'(c) = 0$ and that f is twice differentiable at c.*

1. *If $f''(c) > 0$, then c is a local minimum of f.*
2. *If $f''(c) < 0$, then c is a local maximum of f.*

Proof. We will prove Part (1); the other part is similar, and we omit the details.

(1) Suppose that $f''(c) > 0$. By the definition of derivatives, we know that

$$f''(c) = \lim_{x \to c} \frac{f'(x) - f'(c)}{x - c}.$$

Because $f'(c) = 0$ and $f''(c) > 0$, it follows that

$$\lim_{x \to c} \frac{f'(x)}{x - c} > 0.$$

By Theorem 3.2.4 (1) we know that there is some $M > 0$ and some $\delta > 0$ such that $x \in I - \{c\}$ and $|x - c| < \delta$ imply $\frac{f'(x)}{x-c} > M$. If $x \in I$ and $c - \delta < x < c$, then $x - c < 0$, and hence $\frac{f'(x)}{x-c} > 0$ implies $f'(x) < 0$. If $x \in I$ and $c < x < c + \delta$, then a similar argument shows that $f'(x) > 0$. Part (2) of the First Derivative Test (Theorem 4.5.9) now implies that c is a local minimum of f. □

The Second Derivative Test (Theorem 4.5.10) does not say anything about what happens when $f''(c) = 0$. In such a situation, it turns out that c could be a local maximum, a local minimum or neither, as seen in the first part of the following example. Moreover, there are situations where the Second Derivative Test does not work, for example when f is not differentiable at c, but where the First Derivative Test (Theorem 4.5.9) can still be used, as seen in the second part of the following example.

Example 4.5.11.

(1) Let $f, g\colon \mathbb{R} \to \mathbb{R}$ be defined by $f(x) = x^3$ and $g(x) = x^4$ for all $x \in \mathbb{R}$. It is straightforward to verify that $f'(0) = 0$ and $g'(0) = 0$, and that $f''(0) = 0$ and $g''(0) = 0$. Because $x^4 = (x^2)^2 \geq 0$ for all $x \in \mathbb{R}$, then 0 is a local (and also global) minimum of g. As noted in Example 4.5.8, the number 0 is not a local extremum of f.

(2) Let $k\colon \mathbb{R} \to \mathbb{R}$ be defined by $k(x) = |x|$ for all $x \in \mathbb{R}$. We saw in Example 4.2.3 (3) that k is not differentiable at 0, and hence 0 is a critical point of k. We also saw that $k'(x) = -1$ for all $x \in (-\infty, 0)$, and $k'(x) = 1$ for all $x \in (0, \infty)$. Because k is not differentiable at 0, we cannot apply the Second Derivative Test (Theorem 4.5.10) to k at 0. However, the First Derivative Test (Theorem 4.5.9) can still be applied, and we see that 0 is a local minimum of k, which is just what we would expect by looking at the graph of k. ◊

We now turn to global extrema. Not every function has a global maximum or a global minimum. However, there are two very useful situations where we can guarantee the existence of global extrema.

The first situation concerns continuous functions of the form $f\colon [a, b] \to \mathbb{R}$, where $[a, b]$ is a non-degenerate closed bounded interval. We know by the Extreme Value Theorem (Theorem 3.5.1) that such a function f has a global maximum and a global minimum, and so the question in this situation is not the existence of global extrema, but rather how to find them in practice (recall that the Extreme Value Theorem provides no such information). The key observation, which is really very simple, is that a global extremum must also be a local extremum. Where are the local extrema of our function? On the interval (a, b), we know by Lemma 4.5.7 that the local extrema must be critical points. The only other possibility for local extrema are the endpoints of the interval $[a,b]$. Hence, given a continuous function $f\colon [a,b] \to \mathbb{R}$ on a non-degenerate closed bounded interval, the global extrema can be found by first finding the critical points of f, then computing the value of f at the critical points and endpoints, and then comparing these values—the largest value occurs at a global maxima, and the smallest value occurs at a global minima. There is no need for the First Derivative Test or the Second Derivative Test in this situation.

The other situation where it is easy to find global extrema concerns continuous functions of the form $f\colon I \to \mathbb{R}$, where I is a non-degenerate open interval, and when we have the added condition that f has only one critical point. The following theorem, which is rarely stated explicitly in calculus texts, is often used implicitly in optimization word problems, and so we prove it here.

Theorem 4.5.12. *Let $I \subseteq \mathbb{R}$ be an open interval, let $c \in I$ and let $f\colon I \to \mathbb{R}$ be a function. Suppose that f is continuous, and that c is the only critical point of f.*

1. *If c is a local maximum, then it is a global maximum.*
2. *If c is a local minimum, then it is a global minimum.*

Proof. We will prove Part (1); the other part is similar, and we omit the details.

(1) Suppose that c is a local maximum. Suppose further that c is not a global maximum. Hence, there is some $d \in I$ such that $f(c) < f(d)$. Without loss of generality, assume that $c < d$.

Because c is a local maximum of f, there is some $\delta > 0$ such that $x \in I$ and $|x-c| < \delta$ imply $f(x) \leq f(c)$. By choosing δ sufficiently small, we may suppose that $[c, c+\delta) \subseteq [c,d]$. Then $x \in [c, c+\delta)$ implies $f(x) \leq f(c)$.

Because f is continuous, then $f|_{[c,d]}$ is continuous by Exercise 3.3.2 (2). The Extreme Value Theorem (Theorem 3.5.1) applied to $f|_{[c,d]}$ implies that there are $x_{min}, x_{max} \in [c,d]$ such that $f(x_{min}) \leq f(x) \leq f(x_{max})$ for all $x \in [c,d]$. Because $f(c) < f(d)$, it cannot be the case that $x_{min} = d$.

There are now two cases. First, suppose that $x_{min} = c$. It follows that $f(c) \leq f(x)$ for all $x \in [c, c+\delta)$. However, we saw above that $f(x) \leq f(c)$ for all $x \in [c, c+\delta)$, and we deduce that f is constant on $[c, c+\delta)$. It follows that f is differentiable at x and $f'(x) = 0$ for all $x \in (c, c+\delta)$. Therefore every number in $(c, c+\delta)$ is a critical point of f, which is a contradiction to the fact that c is the only critical point of f. Second, suppose that $x_{min} \neq c$. Therefore $x_{min} \in (c,d)$. By Lemma 4.5.7 applied to $f|_{(c,d)}$ we know that x_{min} must be a critical point of $f|_{(c,d)}$, and hence x_{min} must be a critical point of f by Exercise 4.2.3 (3), again a contradiction to the fact that c is the only critical point of f. We conclude that c is a global maximum. □

Reflections

The material in this section affords us the opportunity to consider the interplay between intuitive concepts and rigorous definitions. For example, we all have an intuitive idea of what it means for a function to be increasing—the graph "goes up" as we move to the right. The actual rigorous definition of this concept given in Definition 4.5.1 (1) appears to capture that idea of "going up" quite nicely, and so most of us would not hesitate to use this rigorous definition. However, there are some intuitive concepts for which it does not appear possible to find a rigorous definition that so self-evidently captures the intuitive concept. The definition of area in Section 5.9 is such an example. We all have an intuitive idea of what the area of a region in the plane means, but the rigorous definition of this concept is somewhat tricky, and took mathematicians a very long time to figure out.

Ultimately, the problem is that we cannot *prove* that what has been defined rigorously is the same as the idea about which we have an intuitive understanding, because intuitive ideas are not susceptible to rigorous proofs. Intuitive ideas and rigorous definitions exist in separate worlds, but they are worlds that, if we are careful with our choice of rigorous definitions, will nicely correlate with each other. Even if one cannot prove that a rigorous definition faithfully captures an intuitive idea, it is possible in many cases to be reasonably certain that a rigorous definition truly captures the intuitive idea if one can verify that the concept that has been defined rigorously behaves the way the intuitive concept is supposed to behave. It is the behavior of mathematical objects, not how they are defined, that provides evidence for a correlation between the world of rigor and the world of intuition.

Exercises

Exercise 4.5.1. **[Used in Exercise 4.6.4.]** Let $A \subseteq \mathbb{R}$ be a set, and let $f\colon A \to \mathbb{R}$ be a function. Prove that if f is monotone and injective, then f is strictly monotone.

Exercise 4.5.2.

(1) Find an example of a continuous increasing function $f\colon (0,1) \to \mathbb{R}$ such that $f((0,1))$ is a non-degenerate closed bounded interval.

(2) Can there be a strictly increasing function $f\colon (0,1) \to \mathbb{R}$ such that $f((0,1))$ is a closed bounded interval? Either give an example to show that there is such a function, or give a proof that there cannot be one.

Exercise 4.5.3. Let $[a,b] \subseteq \mathbb{R}$ and $[b,c] \subseteq \mathbb{R}$ be non-degenerate closed bounded intervals, and let $f\colon [a,b] \to \mathbb{R}$ and $g\colon [b,c] \to \mathbb{R}$ be functions. Let $h\colon [a,c] \to \mathbb{R}$ be defined by

$$h(x) = \begin{cases} f(x), & \text{if } x \in [a,b] \\ g(x), & \text{if } x \in [b,c]. \end{cases}$$

Suppose that $f(b) = g(b)$. Prove that if f and g are increasing, then h is increasing.

Exercise 4.5.4. Let $A \subseteq \mathbb{R}$ be a set, and let $f, g\colon A \to \mathbb{R}$ be functions. Suppose that f and g are increasing.

(1) Prove that $f + g$ is increasing.

(2) Is $f - g$ necessarily either increasing or decreasing? Give a proof or a counterexample.

Exercise 4.5.5. **[Used in Exercise 4.6.4.]** Let $I \subseteq \mathbb{R}$ be a non-degenerate open bounded interval, let $c \in I$ and let $f: I \to \mathbb{R}$ be a function. Suppose that f is differentiable at c.

(1) Prove that if $f'(c) > 0$, then there is some $\delta > 0$ such that $(c-\delta, c+\delta) \subseteq I$, that $x \in (c-\delta, c]$ implies $f(x) < f(c)$, and that $x \in [c, c+\delta)$ implies $f(x) > f(c)$.

(2) Prove that if $f'(c) < 0$, then there is some $\delta > 0$ such that $(c-\delta, c+\delta) \subseteq I$, that $x \in (c-\delta, c]$ implies $f(x) > f(c)$, and that $x \in [c, c+\delta)$ implies $f(x) < f(c)$.

Exercise 4.5.6. **[Used in Theorem 4.5.2.]** Prove Theorem 4.5.2 (2) (3) (4).

Exercise 4.5.7. **[Used in Lemma 4.5.5.]** Prove Lemma 4.5.5.

Exercise 4.5.8. **[Used in Section 4.5.]** Find an example of a function $f: \mathbb{R} \to \mathbb{R}$ such that f has a local minimum at 0, but that f does not satisfy the hypotheses of Lemma 4.5.5 (2). Defining the function by sketching its graph is sufficient.

Exercise 4.5.9. **[Used in Exercise 7.4.1.]** Let $(a,b) \subseteq \mathbb{R}$ be a non-degenerate open interval, let $c \in (a,b)$ and let $f: (a,b) \to \mathbb{R}$ be a function. Suppose that f is increasing and bounded. Prove that $\operatorname{lub} f([c,b)) = \operatorname{lub} f((a,b))$.

Exercise 4.5.10. **[Used in Exercise 5.8.8, Theorem 6.4.11 and Exercise 7.4.1.]** Let $[a,b) \subseteq \mathbb{R}$ be a non-degenerate half-open interval, and let $f, g: [a,b) \to \mathbb{R}$ be functions. Suppose that f is increasing.

(1) Prove that if f is bounded, then $\lim_{x \to b^-} f(x)$ exists and $\lim_{x \to b^-} f(x) = \operatorname{lub} f([a,b))$.

(2) Prove that if $f(x) \leq g(x)$ for all $x \in [a,b)$, and if $\lim_{x \to b^-} g(x)$ exists, then $\lim_{x \to b^-} f(x)$ exists and $\lim_{x \to b^-} f(x) \leq \lim_{x \to b^-} g(x)$.

Exercise 4.5.11. **[Used in Exercise 5.7.5 and Theorem 6.4.12.]** Let $[a,b] \subseteq \mathbb{R}$ be a non-degenerate closed bounded interval, and let $f: [a,b] \to \mathbb{R}$ be a function. Suppose that f is continuous and injective. Prove that f is strictly monotone. [Use Exercise 3.3.2.]

4.6 Increasing and Decreasing Functions, Part II: Further Topics

In this section we discuss two additional topics that are related to the concept of increasing and decreasing functions, the first of which is the differentiability of inverse functions, and the second of which is concave up functions. These two topics are independent of each other; the second topic starts after the proof of Theorem 4.6.4.

For our first topic, rather than using differentiation as a tool to help analyze geometric properties of functions, as we did in Section 4.5, we reverse the approach, and will use the concepts of increasing and decreasing to help us differentiate functions, specifically the inverse functions of bijective differentiable functions. This topic will be useful in our study of exponential functions in Section 7.2 and trigonometric functions in Section 7.3.

We start with a very brief review of some basic facts about inverse functions generally, before turning to the specific question of differentiability of such functions. We assume that the reader is familiar with basic concepts such as injectivity, surjectivity, bijectivity and inverse functions; we mention here only a few key facts for review. See [Blo10, Section 4.4] for a more thorough treatment of this material.

Let A and B be sets, and let $f\colon A \to B$ be a function. Suppose that f is bijective. Then f has an inverse function, which is denoted $f^{-1}\colon B \to A$. By the definition of inverse functions, we know that $f^{-1}(f(x)) = x$ for all $x \in A$ and $f(f^{-1}(x)) = x$ for all $x \in B$. These two equations can be rephrased by saying that $x = f^{-1}(y)$ if and only if $y = f(x)$ for all $x \in A$ and $y \in B$. Whereas this latter formulation is usually less convenient than the former formulation from the point of view of rigorous proofs, we mention it because it might be familiar to the reader from precalculus and calculus courses, for example where the natural logarithm function ln is defined by saying that $y = \ln x$ if and only if $x = e^y$, and similarly for the inverse trigonometric functions. In the particular case of a function of the form $f\colon A \to B$, where $A, B \subseteq \mathbb{R}$, such a function f is bijective if and only if every horizontal line through a number in B intersects the graph of f in precisely one point. In that case the graph of f^{-1} can be obtained by reflecting the graph of f in the line $y = x$.

Injectivity alone is not sufficient to guarantee that a function has an inverse. However, by restricting the codomain of an injective function $f\colon A \to B$, we can also view it as a bijective function $f\colon A \to f(A)$. In principle, changing the codomain of a function changes the function, and we should not really use the same letter "f" to denote both $f\colon A \to B$ and $f\colon A \to f(A)$. However, we will use this abuse of notation because it will make for easier reading, and because no confusion should arise. If $f\colon A \to B$ is injective, then the function $f\colon A \to f(A)$ has an inverse function, which will be denoted $f^{-1}\colon f(A) \to A$.

We now turn to the question of the differentiability of inverse functions. Suppose that $f\colon I \to \mathbb{R}$ is an injective differentiable function, where $I \subseteq \mathbb{R}$ is a non-degenerate open interval; as above we view f as a bijective function $f\colon I \to f(I)$. Is $f^{-1}\colon f(I) \to I$ necessarily differentiable? Indeed, is $f(I)$ necessarily an open interval, which we would want in order to take the derivative of f^{-1}? Although the answer to the latter question is yes, as the reader is asked to prove in Exercise 4.6.4, the answer to the former questions is no, as we see in the following example.

Example 4.6.1. Let $f\colon \mathbb{R} \to \mathbb{R}$ be defined by $f(x) = x^3$ for all $x \in \mathbb{R}$. Intuitively, we know that the function f is bijective, and hence it has an inverse function $f^{-1}\colon \mathbb{R} \to \mathbb{R}$, which we write as $f^{-1}(x) = \sqrt[3]{x}$ for all $x \in \mathbb{R}$. Moreover, we know that the graph of f^{-1} is obtained from the graph of f by reflection in the line $y = x$. Because f has a horizontal tangent line at the origin, then the graph of f^{-1} has a vertical tangent line at $x = 0$, which makes it not differentiable at $x = 0$.

The above intuitive ideas, though correct, do not constitute a rigorous proof, because we cannot rely upon graphical arguments. Moreover, we have not yet seen a rigorous treatment of how to find the derivative of power functions such as $\sqrt[3]{x}$; we will see that in Section 7.2, but it will have to wait until after we have discussed integration in Chapter 5. In the meantime, however, we can offer the following proof that f^{-1} is not differentiable at $x = 0$ using an ad hoc argument based upon only what we have seen so far, together with Lemma 4.6.2, which immediately follows this example, and which does not make use of anything in this example.

As stated in Example 4.5.3, the function f is strictly increasing (and hence injective), and it is differentiable, with derivative $f'(x) = 3x^2$ for all $x \in \mathbb{R}$. Observe that $f'(0) = 0$. By Theorem 4.2.4 we know that f is continuous. Additionally, it follows from Exercise 2.3.3 (2) that $f(\mathbb{R})$ is not bounded above and is not bounded below.

We can now use the various parts of Lemma 4.6.2 below to deduce that f is bijective, that $f(\mathbb{R}) = \mathbb{R}$ and that f^{-1} is continuous and strictly increasing. We will show that f^{-1} is not differentiable at 0.

We start with some preliminary observations. First, we know that $0^3 = 0$, and hence $\sqrt[3]{0} = 0$. Further, because f^{-1} is bijective, then $\sqrt[3]{x} \neq 0$ when $x \neq 0$. Second, the condition $f(f^{-1}(x)) = x$ for all $x \in \mathbb{R}$ can be written as $(\sqrt[3]{x})^3 = x$ for all $x \in \mathbb{R}$. We deduce that $(\sqrt[3]{x})^2 \sqrt[3]{x} = x$ for all $x \in \mathbb{R}$, and hence $\frac{\sqrt[3]{x}}{x} = \frac{1}{(\sqrt[3]{x})^2}$ for all $x \in \mathbb{R}$ such that $x \neq 0$. (Of course, this last fact can be proved more easily using the standard properties of power functions, with which the reader is informally familiar, and which we will prove in Section 7.2.) Third, we note that because f^{-1} is continuous, we can deduce in particular that $\lim_{x\to 0} \sqrt[3]{x} = \sqrt[3]{0} = 0$. It then follows from Theorem 3.2.10 (4) that $\lim_{x\to 0} (\sqrt[3]{x})^2 = 0$. We now use Exercise 3.2.8 to deduce that

$$\lim_{x\to 0} \frac{f^{-1}(x) - f^{-1}(0)}{x - 0} = \lim_{x\to 0} \frac{\sqrt[3]{x} - \sqrt[3]{0}}{x - 0} = \lim_{x\to 0} \frac{1}{(\sqrt[3]{x})^2}$$

does not exist. It follows that f^{-1} is not differentiable at 0. ◊

The reasoning used in Example 4.6.1 applies to any injective differentiable function that has its derivative equal to zero at a point, so that the inverse of any such function will not be differentiable. Fortunately, as seen in Theorem 4.6.4 below, a derivative being zero is the only obstacle to the differentiability of inverse functions. We start with the following lemma, which is about inverses of monotone functions (not necessarily differentiable).

Lemma 4.6.2. *Let $I \subseteq \mathbb{R}$ be a non-degenerate open interval, and let $f : I \to \mathbb{R}$ be a function. Suppose that f is strictly monotone.*

1. *The function $f : I \to f(I)$ is bijective.*
2. *Suppose that f is continuous. Then $f(I)$ is a non-degenerate open interval, and one of the following holds:*
   a. *If the interval $f(I)$ is bounded, then $f(I) = (\mathrm{glb}\, f(I), \mathrm{lub}\, f(I))$.*
   b. *If the interval $f(I)$ is bounded above but is not bounded below, then $f(I) = (-\infty, \mathrm{lub}\, f(I))$.*
   c. *If the interval $f(I)$ is bounded below but is not bounded above, then $f(I) = (\mathrm{glb}\, f(I), \infty)$.*
   d. *If the interval $f(I)$ is not bounded above and is not bounded below, then $f(I) = \mathbb{R}$.*

3. *If f is continuous and strictly increasing (or strictly decreasing), then the function $f^{-1}\colon f(I) \to I$ is continuous and strictly increasing (or strictly decreasing, respectively).*

Proof. We will prove Part (2), leaving the rest to the reader in Exercise 4.6.1.

(2) Suppose that f is strictly increasing; the case where f is strictly decreasing is similar, and we omit the details.

We will prove Item (a) of this part of the lemma, leaving the rest to the reader in Exercise 4.6.2.

Suppose that $f(I)$ is bounded. Because $I \neq \emptyset$, then $f(I) \neq \emptyset$, and therefore the Least Upper Bound Property and the Greatest Lower Bound Property imply that $f(I)$ has a least upper bound and a greatest lower bound.

Let $x \in f(I)$. Then there is some $z \in I$ such that $f(z) = x$. Because I is an open interval, it follows from Lemma 2.3.7 (2) that there are $c, d \in I$ such that $c < z < d$. Because f is strictly increasing, then $f(c) < f(z) < f(d)$, and hence $f(c) < x < f(d)$. Therefore $\operatorname{glb} f(I) \leq f(c) < x < f(d) \leq \operatorname{lub} f(I)$, and hence $x \in (\operatorname{glb} f(I), \operatorname{lub} f(I))$. We deduce that $f(I) \subseteq (\operatorname{glb} f(I), \operatorname{lub} f(I))$.

Let $y \in (\operatorname{glb} f(I), \operatorname{lub} f(I))$. Then $\operatorname{glb} f(I) < y < \operatorname{lub} f(I)$. Let $\varepsilon = \operatorname{lub} f(I) - y$. Then $\varepsilon > 0$. By Lemma 2.6.5 (1) there is some $q \in f(I)$ such that $\operatorname{lub} f(I) - \varepsilon < q \leq \operatorname{lub} f(I)$. Hence $\operatorname{lub} f(I) - (\operatorname{lub} f(I) - y) < q \leq \operatorname{lub} f(I)$, which yields $y < q$. A similar argument shows that there is some $p \in f(I)$ such that $p < y$; we omit the details.

Because $p, q \in f(I)$, there are $s, t \in I$ such that $f(s) = p$ and $f(t) = q$. We know $p \neq q$, and hence $s \neq t$. If $s > t$, then because f is strictly increasing it follows that $f(s) > f(t)$, which means $p > q$, which is a contradiction. Therefore $s < t$.

Because I is an interval, we know that $[s,t] \subseteq I$. It follows from Exercise 3.3.2 (2) that $f|_{[s,t]}$ is continuous. Observe that $f(s) < y < f(t)$. The Intermediate Value Theorem (Theorem 3.5.2) applied to $f|_{[s,t]}$ implies that there is some $r \in (s,t)$ such that $f(r) = y$. Hence $y \in f(I)$. We deduce that $(\operatorname{glb} f(I), \operatorname{lub} f(I)) \subseteq f(I)$. Therefore $f(I) = (\operatorname{glb} f(I), \operatorname{lub} f(I))$. □

Example 4.6.3. We want to show that the square root function is continuous. Let $f\colon (0,\infty) \to \mathbb{R}$ be defined by $f(x) = x^2$ for all $x \in \mathbb{R}$. By Exercise 3.5.6 (1) we see that f is strictly increasing, and by Example 3.3.7 (1) we see that f is continuous. Exercise 3.5.6 implies that $f((0,\infty)) = (0,\infty)$. It then follows from Lemma 4.6.2 (3) that $f^{-1}\colon (0,\infty) \to (0,\infty)$ is continuous and strictly increasing. By Definition 2.6.10 we see that $f^{-1}(x) = \sqrt{x}$ for all $x \in (0,\infty)$. The continuity of this function could also be shown directly by an ε–δ proof, but Lemma 4.6.2 allows us to avoid that. ◊

We are now ready to prove the formula for the derivative of the inverse of a differentiable function, subject to suitable hypotheses. This formula, given in Theorem 4.6.4 (4) below, is often "proved" in a calculus course roughly as follows (though without our way of writing functions). "Let $f\colon I \to f(I)$ be a differentiable function that has an inverse function $f^{-1}\colon f(I) \to I$, which means that $f(f^{-1}(x)) = x$ for all $x \in f(I)$. Taking the derivative of each side of this equation, and making use of the Chain Rule on the left-hand side, we obtain $f'(f^{-1}(x)) \cdot [f^{-1}]'(x) = 1$ for all $x \in f(I)$.

Dividing both sides of this equation by $f'(f^{-1}(x))$ yields the desired formula for $[f^{-1}]'(x)$."

The problem with the above "proof" is that it assumes that f^{-1} is differentiable, which is needed for the Chain Rule to be used, but such an assumption is not justified unless we prove it. The proof that f^{-1} is differentiable turns out to be a bit more tricky than the above alleged proof. Moreover, this proof also shows, with no extra effort, that the formula for $[f^{-1}]'(x)$ holds, and because of that we can skip the above alleged proof entirely.

Theorem 4.6.4. *Let $I \subseteq \mathbb{R}$ be a non-degenerate open interval, and let $f: I \to \mathbb{R}$ be a function. Suppose that f is differentiable, and that $f'(x) \neq 0$ for all $x \in I$.*

1. *The function f is strictly monotone.*
2. *The function $f: I \to f(I)$ is bijective.*
3. *The function $f^{-1}: f(I) \to I$ is differentiable.*
4. *The derivative of f^{-1} is given by*

$$[f^{-1}]'(x) = \frac{1}{f'(f^{-1}(x))}$$

for all $x \in f(I)$.

Proof. Throughout this proof, when we write f, we will think of it as a function $I \to f(I)$.

(1) This part of the theorem follows immediately from Exercise 4.4.9 (2) and Theorem 4.5.2 (2) (4), which is stated for closed intervals, but also holds for open intervals.

(2) The fact that f is bijective follows immediately from Part (1) of this theorem and Lemma 4.6.2 (1).

(3) & (4) We start with some preliminary observations. By Theorem 4.2.4 we know that f is continuous, and by Part (1) of this theorem we know that f is strictly monotone. Lemma 4.6.2 (3) implies that $f^{-1}: f(I) \to I$ is continuous and strictly monotone.

Let $c \in f(I)$. Let $d = f^{-1}(c)$. Let $F: I - \{d\} \to \mathbb{R}$ be defined by

$$F(x) = \frac{x-d}{f(x) - f(d)}$$

for all $x \in I - \{d\}$. By Part (2) of this theorem we know that f is bijective, and hence if $x \in I - \{d\}$ then $f(x) \neq f(d)$, which implies that F is well-defined.

By hypothesis we know that $f'(d) \neq 0$. We can therefore use Theorem 3.2.10 (5), together with the definition of derivatives, to compute

$$\lim_{x \to d} F(x) = \lim_{x \to d} \frac{x-d}{f(x) - f(d)} = \lim_{x \to d} \frac{1}{\frac{f(x)-f(d)}{x-d}} = \frac{1}{f'(d)} = \frac{1}{f'(f^{-1}(c))}.$$

Because f^{-1} is continuous, we know by Lemma 3.3.2 that $\lim_{y\to c} f^{-1}(y) = f^{-1}(c) = d$. Theorem 3.2.12 now implies that $\lim_{y\to c} (F \circ f^{-1})(y)$ exists and $\lim_{y\to c} (F \circ f^{-1})(y) = \lim_{x\to d} F(x)$. Then

$$\begin{aligned}\lim_{y\to c} \frac{f^{-1}(y) - f^{-1}(c)}{y - c} &= \lim_{y\to c} \frac{f^{-1}(y) - d}{f(f^{-1}(y)) - f(d)} \\ &= \lim_{y\to c} (F \circ f^{-1})(y) = \lim_{x\to d} F(x) = \frac{1}{f'(f^{-1}(c))}.\end{aligned}$$

It follows that $[f^{-1}]'(c)$ exists and

$$[f^{-1}]'(c) = \frac{1}{f'(f^{-1}(c))}.$$

□

We now turn to the second topic of this section, which is the idea of functions being concave up or concave down. We follow [Gor02] in part. Similarly to our discussion in Section 4.5, this topic concerns geometric properties of graphs of functions. It is quite simple to visualize concave up and concave down functions intuitively, as in Figure 4.6.1. For a rigorous approach to these concepts, we take as our model the concepts of increasing and decreasing, in that these concepts are defined without reference to calculus, and then we saw in Theorem 4.5.2 that for differentiable functions, increasing and decreasing can be characterized in terms of the derivative. Unfortunately, the analogy between concave up and concave down, and increasing and decreasing, is not perfect. On the one hand, we will have a theorem analogous to Theorem 4.5.2, but with the second derivative replacing the first derivative. On the other hand, the standard definition of concave up and concave down that is given in calculus courses, which is that the derivative of the function is increasing, is quite different in nature from the geometric (and non-calculus-based) definition of increasing and decreasing. It would be nicer to have a non-calculus definition of concave up and concave down, because the idea is inherently geometric. For the sake of brevity, we will restrict our attention to concave up; a treatment of concave down is similar, and we omit the details.

We will, in fact, give two variant definitions of concave up in Theorem 4.6.6 below, though these characterizations of concave up are not as simple as the definition of increasing and decreasing. These characterizations involve the notion of a secant line to a curve, which is a line through pairs of distinct points on the curve, and which is defined in Definition 4.6.5. In Figure 4.6.2 a concave up graph of a function is shown, together with two of its secant lines. The reader will observe that each secant line is above the curve in the interval that is between the points where the line intersects the curve; that is the intuitive idea of the first characterization of concave up. The reader will also observe that of the two secant lines, the one that is to the right has a larger slope; that is the intuitive idea of the second characterization of concave up.

Definition 4.6.5. Let $I \subseteq \mathbb{R}$ be an open interval, let $a, b \in I$ and let $f: I \to \mathbb{R}$ be a function. Suppose that $a < b$. The **secant line** through $(a, f(a))$ and $(b, f(b))$ is the function $S_{a,b}: \mathbb{R} \to \mathbb{R}$ defined by

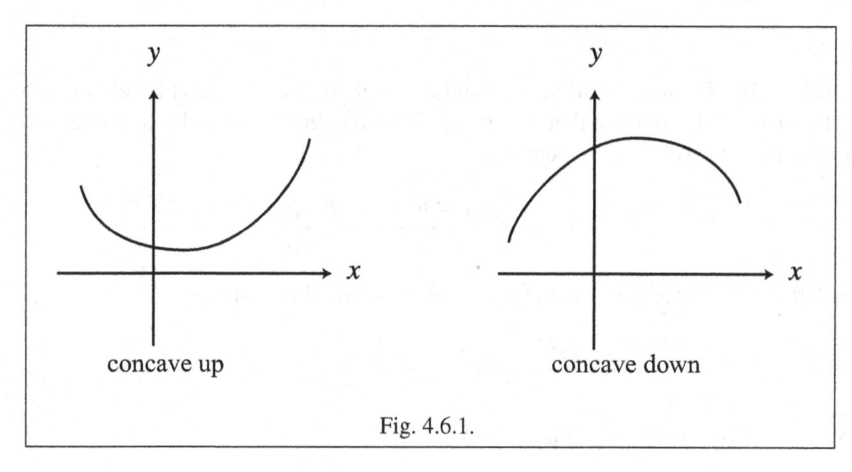

Fig. 4.6.1.

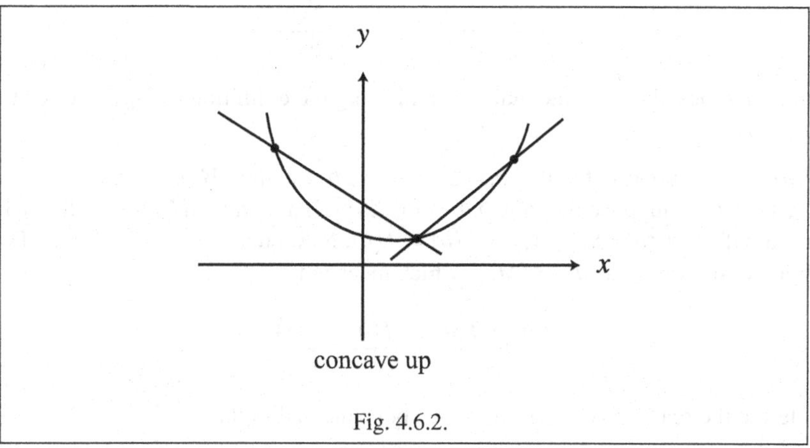

Fig. 4.6.2.

$$S_{a,b}(x) = f(a)\frac{b-x}{b-a} + f(b)\frac{x-a}{b-a}$$

for all $x \in \mathbb{R}$. The **slope of the secant line** through $(a, f(a))$ and $(b, f(b))$, denoted $M_{a,b}$, is defined by

$$M_{a,b} = \frac{f(b) - f(a)}{b-a}. \qquad \triangle$$

It is left to the reader to verify that the formula for $S_{a,b}(x)$ given in Definition 4.6.5 is indeed the straight line through the points $(a, f(a))$ and $(b, f(b))$.

Theorem 4.6.6. *Let $I \subseteq \mathbb{R}$ be an open interval, and let $f\colon I \to \mathbb{R}$ be a function. The following are equivalent.*

a. *If $a, b \in I$ and $a < b$, then $f(x) \le S_{a,b}(x)$ for all $x \in [a, b]$ (Function Lies Below Its Secant Lines).*

b. *If $a, b, c \in I$ and $a < b < c$, then $M_{a,b} \le M_{b,c}$ (Function Has Increasing Secant Line Slopes).*

Proof.

(a) ⇒ (b) Suppose that if $a, b \in I$ and $a < b$, then $f(x) \leq S_{a,b}(x)$ for all $x \in [a, b]$.

Let $a, b, c \in I$. Suppose that $a < b < c$. Then $b \in [a, c]$, and so by hypothesis we know $f(b) \leq S_{a,c}(b)$, which means that

$$f(b) \leq f(a)\frac{c-b}{c-a} + f(c)\frac{b-a}{c-a}.$$

It is left to the reader to deduce from the above inequality that

$$\frac{f(b)-f(a)}{b-a} \leq \frac{f(c)-f(a)}{c-a}.$$

A similar calculation shows that

$$\frac{f(c)-f(a)}{c-a} \leq \frac{f(c)-f(b)}{c-b}.$$

Combining these last two inequalities, and using the definition of $M_{a,b}$, we see that $M_{a,b} \leq M_{b,c}$.

(b) ⇒ (a) Suppose that if $a, b, c \in I$ and $a < b < c$, then $M_{a,b} \leq M_{b,c}$.

Let $a, b \in I$. Suppose that $a < b$. Let $x \in [a, b]$. If $x = a$, then $S_{a,b}(x) = S_{a,b}(a) = f(a)$, and if $x = b$ then $S_{a,b}(x) = S_{a,b}(b) = f(b)$. Now suppose that $a < x < b$. Then by hypothesis we know $M_{a,x} \leq M_{x,b}$, which means that

$$\frac{f(x)-f(a)}{x-a} \leq \frac{f(b)-f(x)}{b-x}.$$

It is left to the reader to deduce from the above inequality that

$$f(x) \leq f(a)\frac{b-x}{b-a} + f(b)\frac{x-a}{b-a},$$

which means that $f(x) \leq S_{a,b}(x)$. □

We now use Theorem 4.6.6 as the basis for the following definition.

Definition 4.6.7. Let $I \subseteq \mathbb{R}$ be an open interval, and let $f\colon I \to \mathbb{R}$ be a function. The function f is **concave up** if either of the two conditions in Theorem 4.6.6 hold. △

In the differentiable case, we can use both the first derivative and second derivative to characterize when a function is concave up; a similar result holds for concave down. The first part of the following theorem shows that the usual definition of concave up given in calculus courses is equivalent to the geometric approach of Theorem 4.6.6.

Theorem 4.6.8. *Let $I \subseteq \mathbb{R}$ be an open interval, and let $f\colon I \to \mathbb{R}$ be a function.*

1. *Suppose that f is differentiable. Then the two conditions in Theorem 4.6.6 hold if and only if f' is increasing on I.*

2. *Suppose that f is twice differentiable. Then the two conditions in Theorem 4.6.6 hold if and only if $f''(x) \geq 0$ for all $x \in I$.*

Proof.

(1) Because of Theorem 4.6.6, it will suffice to prove that f' is increasing if and only if Theorem 4.6.6 (b) holds.

Suppose that f' is increasing. Let $a,b,c \in I$. Suppose that $a < b < c$. Because f is differentiable, then it is continuous by Theorem 4.2.4. The Mean Value Theorem (Theorem 4.4.4) applied to each of $f|_{[a,b]}$ and $f|_{[b,c]}$ implies that there are $p \in (a,c)$ and $q \in (c,b)$ such that $f'(p) = M_{a,c}$ and $f'(q) = M_{c,b}$. Because f' is increasing, we see that $f'(p) \leq f'(q)$, and hence $M_{a,b} \leq M_{b,c}$. It follows that Theorem 4.6.6 (b) holds.

Now suppose that Theorem 4.6.6 (b) holds. Let $x,y \in I$. Suppose that $x < y$. Because I is open there are $w,z \in I$ such that $w < x$ and $y < z$. Then by hypothesis we know that $M_{w,x} \leq M_{x,y}$ and $M_{x,y} \leq M_{y,z}$. By the definition of derivatives, combined with Lemma 3.2.17, we see that $f'(x) = \lim_{w \to x^-} M_{w,x}$ and $f'(y) = \lim_{z \to y^+} M_{y,z}$. Using the analogs for one-sided limits of Theorem 3.2.13 and Exercise 3.2.1, we deduce that $f'(x) \leq M_{x,y} \leq f'(y)$. It follows that f' is increasing.

(2) This part of the theorem follows immediately from Part (1) of this theorem together with Theorem 4.5.2 (1) applied to f'. □

Reflections

The concepts of concave up and concave down are partly analogous to the concepts of increasing and decreasing, though with second derivatives instead of first derivatives. The reader might have noticed, however, that we did not take this analogy as far as it can go, in that we did not discuss the second derivative analog of local extrema. Of course, there is such an analog, namely, inflection points, which the reader has seen in calculus courses. This analogy is almost, but not entirely, complete. Recall from the First Derivative Test (Theorem 4.5.9) that a local extremum occurs where the function changes from increasing to decreasing or vice versa. Whereas the First Derivative Test is a theorem about local extrema, and is not the definition of this concept, the analog for second derivatives of the idea in the First Derivative Test is taken as the definition of inflection points, which are numbers where the function changes from concave up to concave down or vice versa. Moreover, in contrast to the distinction made between the two types of local extrema, namely, local maxima and local minima, no distinction is made between the type of inflection point where the function changes from concave up to concave down, and the type of inflection point where the function changes from concave down to concave up.

The analogy between local extrema and inflection points is, nonetheless, a very useful one. To find local extrema, we defined critical points, and then proved that those are the only places we need to look for local extrema, though not all critical points are local extrema. Similarly, we could define something that might be called "second critical points." That is, a point c is a second critical point if either f is twice differentiable at c and $f''(c) = 0$, or f is not twice differentiable at c. Then the second critical points would be the only places we need to look for inflection points, though not all second critical points are inflection points; the analog of Theorem 4.5.9 would then say that a second critical point is an inflection point if and only if $f''(x)$ changes from positive to negative or vice versa at the second critical point. No one seems to use the term "second critical point," though perhaps it would be a nice idea. For the sake of leaving room for more important topics, we do not give a rigorous treatment of inflection points in this text.

Exercises

Exercise 4.6.1. [Used in Lemma 4.6.2.] Prove Lemma 4.6.2 (1) (3).

Exercise 4.6.2. [Used in Lemma 4.6.2.] Prove Items (b)–(d) of Lemma 4.6.2 (2).

Exercise 4.6.3. [Used in Exercise 4.6.5, Exercise 5.7.5, Theorem 6.4.12, Lemma 7.3.7 and Exercise 7.3.5.] Let $[a,b] \subseteq \mathbb{R}$ be a non-degenerate closed bounded interval, and let $f\colon [a,b] \to \mathbb{R}$ be a function. Suppose that f is continuous and strictly monotone.

(1) Prove that if f is strictly increasing, then $f([a,b]) = [f(a), f(b)]$ and the function $f\colon [a,b] \to [f(a), f(b)]$ is bijective; and that if f is strictly decreasing, then $f([a,b]) = [f(b), f(a)]$ and the function $f\colon [a,b] \to [f(b), f(a)]$ is bijective.

(2) Prove that $f^{-1}\colon f([a,b]) \to [a,b]$ is continuous, and is strictly increasing or strictly decreasing if f is strictly increasing or strictly decreasing, respectively.

Exercise 4.6.4. [Used in Section 4.6.] Let $I \subseteq \mathbb{R}$ be a non-degenerate open interval, and let $f\colon I \to \mathbb{R}$ be a function.

(1) Suppose that f is continuous. Let $x, p, q, y \in I$. Suppose that $x < p < q < y$, and that $f(x) < f(p)$ and $f(q) > f(y)$. Prove that f is not injective.

(2) Suppose that f is continuous. Let $c, d \in I$. Suppose that f is differentiable at c and d, and that $f'(c) > 0$ and $f'(d) < 0$. Prove that f is not injective. [Use Exercise 4.5.5.]

(3) Suppose that f is differentiable and injective. Prove that $f'(x) \geq 0$ for all $x \in I$, or that $f'(x) \leq 0$ for all $x \in I$. Deduce that f is monotone.

(4) Suppose that f is differentiable and injective. Prove that $f(I)$ is a non-degenerate open interval. [Use Exercise 4.5.1.]

Exercise 4.6.5. [Used in Exercise 7.4.1.] Let $(a,b) \subseteq \mathbb{R}$ be a non-degenerate open bounded interval, and let $f\colon (a,b) \to \mathbb{R}$ be a function. Suppose that f is continuous, strictly increasing and bounded. Let $F\colon [a,b] \to \mathbb{R}$ be defined by

$$F(x) = \begin{cases} \operatorname{glb} f((a,b)), & \text{if } x = a \\ f(x), & \text{if } a < x < b \\ \operatorname{lub} f((a,b)), & \text{if } x = b. \end{cases}$$

(1) Prove that F is continuous.

(2) Prove that F is strictly increasing.
(3) Prove that $F([a,b]) = [\text{glb}\, f((a,b)), \text{lub}\, f((a,b))]$. [Use Exercise 4.6.3 (1).]

Exercise 4.6.6. [Used in Theorem 6.3.9 and Theorem 6.4.12.] Let $(a,b] \subseteq \mathbb{R}$ be a non-degenerate half-open interval, and let $f\colon (a,b] \to \mathbb{R}$ be a function. Suppose that f is strictly increasing.

(1) Prove that one of the following holds.
a. If $f(I)$ is bounded below, then $f((a,b]) = (\text{glb}\, f((a,b]), f(b)]$.
b. If $f(I)$ is not bounded below, then $f((a,b]) = (-\infty, f(b)]$.
(2) Suppose that f is continuous at b. Prove that $f^{-1}\colon f((a,b]) \to (a,b]$ is continuous at $f(b)$. [Use the one-sided analog of Exercise 2.3.8.]

Exercise 4.6.7. Let $I \subseteq \mathbb{R}$ be an open interval, and let $f\colon I \to \mathbb{R}$ be a function. Suppose that f is concave up. Prove that if $a,b,c \in I$ and $a < b < c$, then $M_{a,b} \le M_{a,c} \le M_{b,c}$.

Exercise 4.6.8. Let $[a,b] \subseteq \mathbb{R}$ be a non-degenerate closed bounded interval, and let $f\colon [a,b] \to \mathbb{R}$ be a function. Suppose that if $r,s \in (a,b)$ and $r < s$, then $M_{r,b} \le M_{s,b}$. Is $f|_{(a,b)}$ necessarily concave up? Give a proof or a counterexample.

Exercise 4.6.9. Let $I \subseteq \mathbb{R}$ be a non-degenerate open interval, and let $f\colon I \to \mathbb{R}$ be a function.

(1) Let $[a,b] \subseteq I$ be a non-degenerate closed bounded interval. Prove that if the set $\{M_{p,q} \mid p,q \in [a,b] \text{ and } p < q\}$ is bounded, then $f|_{[a,b]}$ is uniformly continuous. [Use Exercise 3.4.5.]
(2) Prove that if f is concave up, then f is continuous. [Use Exercise 3.3.2.]
(3) Is Part (2) of this exercise true if I is a closed interval? Give a proof or a counterexample.

Exercise 4.6.10. Let $I \subseteq \mathbb{R}$ be an open interval, and let $f\colon I \to \mathbb{R}$ be a function. The function f is **convex** if $a,b \in I$ and $a < b$ imply $f(ta + (1-t)b) \le tf(a) + (1-t)f(b)$ for all $t \in [0,1]$.

Prove that f is concave up if and only if f is convex.

4.7 Historical Remarks

Of the two fundamental topics in calculus, differentiation and integration, the former appears to the modern student to be much simpler than the latter, a view that will be maintained when the reader encounters the treatment of integration in Chapter 5. As such, it is no surprise that in today's calculus and real analysis texts, differentiation is almost always taught before integration (a notable exception being the classic text [Apo67]). Historically, however, differentiation has a much less rich history than integration, the latter having strong roots in the ancient world, and the former having to wait until the 17th century for substantial treatment.

The late historical development of differentiation in contrast to integration should come as no surprise. The basic question that gave rise to the study of integration is the computation of areas of regions of the plane and volumes of regions of space, and such questions certainly arose very early in human civilization, for example in architecture, farming, art, commerce and the like. By contrast, the notion of the rate of change of a function, which is the essence of derivatives, requires the concept of a function, which is a much later historical development than the notion of area, with functions as we think of them today making their appearance only in very preliminary form in the 14th century, and in a more developed form in the 18th century. Prior to the invention of analytic geometry in the 17th century, curves in the plane were viewed as geometric objects, rather than the graphs of functions, and the question of rate of change is not nearly as natural a question for curves as it is for graphs of functions. If one views a curve as the motion of a particle, then one can consider its velocity vector, but that too came relatively late historically. It is possible to consider tangent lines to curves from a strictly geometric viewpoint, but it is difficult to come up with a precise geometric definition of what tangent lines are. In ancient Greece, and until close to when calculus was invented, a tangent line to a curve was viewed as a line that touched the curve in one point, or that satisfied some other similar geometric definition. Such an approach is not strictly correct, as can be seen by Example 4.2.5 (1), where the function crosses its tangent line at $x = 0$ infinitely many times as 0 is approached. However, it is hard to come up with a better definition of tangent lines without the function concept, or without thinking of a curve as representing the motion of a particle.

Ancient World

Euclid (c. 325–c. 265 BCE) discussed tangent lines to circles in Book III of the *Elements*, Archimedes (287–212 BCE) discussed tangent lines to what is now called the Archimedean spiral, and Apollonius (c. 262 BC–c. 190 BC) discussed tangent lines to conic sections. The ancient Greeks, and the rest of the ancient world, did not appear to know much more than that about tangent lines to curves.

Medieval Period

Bhaskara II (1114–1185), also known as Bhaskaracharya, appeared to have conceived of the basic ideas of differentiation in *Siddhanta Siromani* of 1150, which was primarily about astronomy, but also contained some mathematics. He had the idea of locating maxima and minima where the derivative is zero, had a version of Rolle's Theorem and had the equivalent of the fact that the derivative of sine is cosine.

Nicole Oresme (1323–1382) observed, via his graphical representation of variation, a special case of the idea of locating maxima and minima where the rate of change is zero.

Seventeenth Century

The modern study of tangent lines started in the first half of the 17th century, a time when mathematics was in general advancing very rapidly. The approach to geometry in this period was very different from the impressive but restrictive ancient Greek approach; algebra, which was underdeveloped in ancient Greece, had shown considerable development in the meantime; analytic geometry had recently been developed, and many new curves were studied, giving further impetus to the need to find tangent lines to curves; the idea of a function was starting to take shape (though the fully modern approach to functions was yet to be developed). In general, in keeping with the spirit of the times, a practical problem-solving approach to mathematics had developed by the 17th century.

Johannes Kepler (1571–1630), in *Nova stereometria doliorum vinariorum* of 1615, wanted to provide practical methods for finding volumes of wine casks. In particular, he found maximum volumes using an experimental approach, listing volumes for given dimensions, and then selecting the best. In the process, he noticed that as one got closer to the maximum volume, the amount that the volume changed for a given amount of change in the dimensions of the solid decreased until it is negligible, which is essentially a recognition that the maximum is found when the rate of change is zero.

A major step forward in the study of tangent lines, not long before the invention of calculus, was due to Pierre de Fermat (1601–1665), who in the late 1620s found the maximum and minimum values of curves by considering what we write as $\frac{f(x+e)-f(x)}{e}$, dividing as if e is non-zero, and then dropping e at some point in order to get the answer to the problem. Fermat also used this method to find tangent lines to curves. Fermat's argument resembles the use of infinitesimals, though he did not clearly explain if that is how he understood it. Moreover, whereas the idea of the derivative is implicit in Fermat's method, he did not appear to have recognized the concept of the derivative per se.

René Descartes (1596–1650), in the appendix *La Géométrie* of the philosophical work *Discours de la méthode pour bien conduire sa raison et chercher la vérité dans les sciences* of 1637, found normal lines (which are perpendicular to tangent lines) to some curves by intersecting them with circles, and having the two points of intersection get closer and closer, which lead to a double root of an equation when the points are thought of as having merged.

In the 1630s and 1640s Evangelista Torricelli (1608–1647) and Gilles de Roberval (1602–1675) helped advance the notion of viewing a curve in the plane as the path of a moving object (so that each of x and y is a function of time), which allowed for tangent lines to be viewed as lines of instantaneous motion. In the 1650s and early 1660s Johann Hudde (1628–1704), René de Sluse (1622–1685) and Christiaan Huygens (1629–1695) independently discovered algorithmic rules for computing the slopes of the tangent lines of arbitrary algebraic curves. Whereas today we always approach tangent line computations via derivatives, because that is certainly the most convenient way to do so, the method of Hudde and Sluse was not based upon the ideas that eventually became calculus. Calculus is so convenient that we tend to forget that some (though not all) of the problems that are now solved with calculus can also be solved without it.

A number of mathematicians, including James Gregory (1638–1675), Isaac Barrow (1630–1677) and Blaise Pascal (1623–1662), used a "differential triangle," which is an infinitesimal right triangle with hypotenuse that is tangent to the curve, and with sides parallel to the coordinate axes. In contrast to the others, who used this triangle for tangent problems, Pascal used it as part of an area problem, but it was from Pascal's work of 1658 that Leibniz learned of this idea, which was important to his understanding of the derivative.

Further progress in the development of the derivative was due to Isaac Barrow (1630–1677), who was Newton's predecessor as the Lucasian Professor of Mathematics at Cambridge. In his lectures in the mid-1660s, which might have been attended by Newton, Barrow computed slopes of tangent lines by implicitly using the idea of approximating tangent lines with secant lines, and dropping higher powers of infinitesimals.

Newton and Leibniz

The next step in the development of the derivative was part of the larger invention of calculus by Isaac Newton (1643–1727) and Gottfried von Leibniz (1646–1716). On the one hand, Newton and Leibniz (independently, and in some ways rather complementarily) moved our understanding of the derivative forward substantially. On the other hand, Newton and Leibniz invented calculus not in a vacuum but rather in the context of the substantial mathematical activity that preceded them. What sets Newton and Leibniz apart from their predecessors is that they went beyond solving particular tangent or area problems, and recognized that behind the various particular cases was a general method for treating these sorts of problems, and they then worked out many of the details of this general method.

The first to conceive of what we now call calculus was Newton, who worked out the basics of his version of it in the period 1665–1666. In his first (and unpublished) paper on calculus, referred to as the *October 1666 Tract on Fluxions*, Newton studied the tangent problem by thinking of a point moving along a curve given by an equation of the form $f(x,y) = 0$, and considering what he later wrote as $\dot{x}$ and $\dot{y}$ (and we write as $\frac{dx}{dt}$ and $\frac{dy}{dt}$). The derivative as we know it today was $\frac{\dot{y}}{\dot{x}}$. Newton calculated the derivative of any algebraic curve by using infinitesimals and implicit differentiation. He essentially worked out the Chain Rule, and showed how to take derivatives of products and quotients, though he did not explicitly formulate the Product Rule and the Quotient Rule. He also recognized the importance of what we call antidifferentiation. In his unpublished *Tractatus de methodis serierum et fluxionum* of 1671, Newton found maxima and minima by setting the derivative equal to zero and solving.

Leibniz, who worked out his version of calculus in the period 1675–1677, had a very different conceptual approach than Newton. Rather than thinking of derivatives in terms of a point moving along a curve and the rates of change of its x and y coordinates, Leibniz considered infinitesimal changes in x and y, which he denoted dx and dy, and thought of the tangent line to a curve as a secant line connecting two infinitesimally close points on the curve. The derivative was the ratio of the differentials dy and dx. Leibniz worked out the derivative of power functions, as well as the Product Rule and Quotient Rule, though he wrote everything in terms of differentials rather than derivatives, for example $d(xy) = xdy + ydx$. He observed that dv is positive when v is increasing, and analogously for decreasing, and hence that local extrema occur only when $dv = 0$, and that inflection points occur only when $d(dv) = 0$. As an application,

Leibniz gave a derivation of Snell's Law of Refraction (which was already known at the time), just as we would do today in a calculus course.

The development of calculus is not only about derivatives, and it is not possible to see the full range of Newton's and Leibniz's accomplishments without seeing their contributions to other aspects of calculus as well, especially integration; see the historical discussion for other chapters for details. It is not worth dwelling on the famous dispute about whether Newton or Leibniz should be given priority in the invention of calculus. The modern view is that each formulated his version of calculus independently of the other; Newton's work came first, though Leibniz was the first to publish. Newton was the better mathematician of the two—he is viewed by many as one of the three greatest mathematicians of all time, together with Archimedes and Gauss. On the other hand, Leibniz's approach to calculus, and in particular his notation, had a larger impact on the immediate development of the subject. They both deserve to share the credit.

Eighteenth Century

Newton and Leibniz were, in keeping with the level of mathematical rigor of their era, not overly careful with the use of infinitesimals (though Newton's views on infinitesimals changed over time). A rigorous treatment of the real numbers was not developed until the 19th century, and it is only with a rigorous foundation for the real numbers that all the theorems of calculus can be proved. Nonetheless, questions about the use of infinitesimals were raised much earlier, notably by George Berkeley (1685–1753) in his essay *The Analyst; or, A Discourse Addressed to an Infidel Mathematician* of 1734. Berkeley pointed out some philosophical and logical problems with both Newton's and Leibniz's approach to derivatives. For example, Berkeley correctly pointed out the logical problem that occurs when computing $\frac{f(c+h)-f(c)}{h}$ using the approach of his day (before the invention of limits), where one first assumes that h is non-zero in order to divide by it, and one then subsequently assumes that h is zero in order to drop terms containing h from consideration. Of course, even with the dubious use of infinitesimals, calculus right from the beginning proved to be very useful, and so the response to such criticisms was not to abandon calculus, but rather to find better foundations for it.

One response to Berkeley was by Jean d'Alembert (1717–1783), in the article *Différentiel* of 1754, which was published in the influential French *Encyclopédie*, of which d'Alembert was an editor. In that article d'Alembert proposed that the derivative be viewed as $\lim_{\Delta x \to 0} \frac{\Delta y}{\Delta x}$, rather than as Newton's ratio of fluxions or Leibniz's ratio of differentials. He did not have a rigorous definition of limits, but his approach was nonetheless a step forward in the development of the derivative as we now know it.

Joseph-Louis Lagrange (1736–1813), in *Théorie des fonctions analytiques* of 1797, attempted to avoid both infinitesimals and limits by viewing all functions as power series, and then picking off the derivative as a certain coefficient in such series. Lagrange also introduced the term "derivative" ("fonction dérivée" in the original) and the notation "$f'(x)$." It was subsequently shown by Cauchy that not every differentiable function can be written as a power series, and hence Lagrange's approach to avoiding infinitesimals and limits was not satisfactory.

Nineteenth Century

It was Augustin Louis Cauchy (1789–1857) who brought the derivative, and much of calculus, into the modern form with which we are familiar today. Cauchy's approach to the derivative was influenced by Sylvestre François Lacroix (1765–1843), who wrote some calculus textbooks starting in 1797 that were widely used, but which Cauchy found not entirely rigorous; both Lacroix and Cauchy taught at the École Polytechnique, and wrote their texts to be able to teach calculus in ways they viewed as satisfactory. Cauchy's work on calculus is found in three important textbooks he wrote in the 1820s. Cauchy's predecessors (with the exception of d'Alembert and Lagrange) took the derivative as the starting point of calculus, whereas Cauchy started with limits as the fundamental concept, and computed the derivative by the familiar formula $\lim_{h \to 0} \frac{f(c+h)-f(c)}{h}$. Cauchy introduced the Chain Rule as we know it, although he gave the mistaken proof mentioned right after the statement of Theorem 4.3.3. Cauchy was also the first person to give the Mean Value Theorem its now central role, although the current approach of proving the Mean Value Theorem from Rolle's Theorem is due later to Pierre Bonnet (1819–1892).

5

Integration

5.1 Introduction

Having looked at differentiation in Chapter 4, we now turn to the other main part of calculus, namely, integration. It is important to understand that, although calculus is unified by the Fundamental Theorem of Calculus, each of differentiation and integration has its own motivation and technical details, and in principle can be treated separately up until the Fundamental Theorem of Calculus, and studied in either order.

When we refer to integration at this point in the text we mean "definite integration." Indefinite integration is another name for antidifferentiation, and is defined solely in terms of differentiation. The "real" integration is definite integration, and the *definition* of this type of integral has nothing to do with differentiation per se. The Fundamental Theorem of Calculus, which relates definite integration and differentiation, is an amazing and surprising fact, and is not simply a matter of definition.

In our treatment of integration we will be using the terms "Riemann sum" and "Riemann integral." As the reader might guess from this terminology, there are other kinds of integrals as well, the most well-known being the "Lebesgue integral." We will not treat these other types of integrals, but it is worth knowing that they exist. All of these types of integrals agree on continuous functions, but sometimes differ on more complicated functions. If we just say "integral," we mean the Riemann integral. See [Str00, Chapter 14] for the Lebesgue integral in $\mathbb{R}$, and see [Bar96] for the Henstock–Kurzweil integral (also known as the generalized Riemann integral or the gauge integral).

As was the case for derivatives, it is assumed that the reader has seen integrals in a calculus course, and so we will spend little time on intuitive motivation, and not discuss applications or computational examples at all.

5.2 The Riemann Integral

The geometric motivation for integration is to find the area of curved regions in the plane. It is easy to find the area of simple shapes such as rectangles and triangles, and from there it is also possible to find the area of any polygon by cutting it up into rectangles and triangles. By contrast, the area of curved regions is much harder to compute.

The simplest type of curved region in the plane is the region under the graph of a function, and that is the type of area directly addressed by integration. The fundamental idea of the Riemann integral is to approximate the area under the graph of a function by approximating the region with rectangles (the areas of which are easy to find), then adding up the areas of the rectangles, and finally taking the limit as the widths of the rectangles get thinner and thinner. Of course, as with any other type of limit, not all limits of this type exist, and that corresponds to when the function is not integrable, which means geometrically that the function is so wildly behaved that we cannot assign a number in a meaningful way to the area of the region beneath the graph of the function.

There are a variety of ways that the region under the graph of a function might be approximated with rectangles. Two standard ways are to use right-hand sums and left-hand sums, as seen respectively in Figure 5.2.1 (i) (ii). Suppose that $f\colon A \to \mathbb{R}$ is a function, where $A \subseteq \mathbb{R}$ is a set, and we want to find the area under the graph of f and above an interval $[a,b] \subseteq A$. To compute a right-hand sum, we divide the interval $[a,b]$ into smaller subintervals (in the figure there are seven such subintervals), and we then form a rectangle above each subinterval, where the height of the rectangle equals the value of the function f at the right endpoint of the subinterval; we then add up the areas of the rectangles, to obtain an approximation of the area under the graph of the function. Left-hand sums are similar to right-hand sums with the obvious modification. Intuitively, to find the exact area under the graph of the function, we then need to take some sort of limit as the subintervals get smaller and smaller.

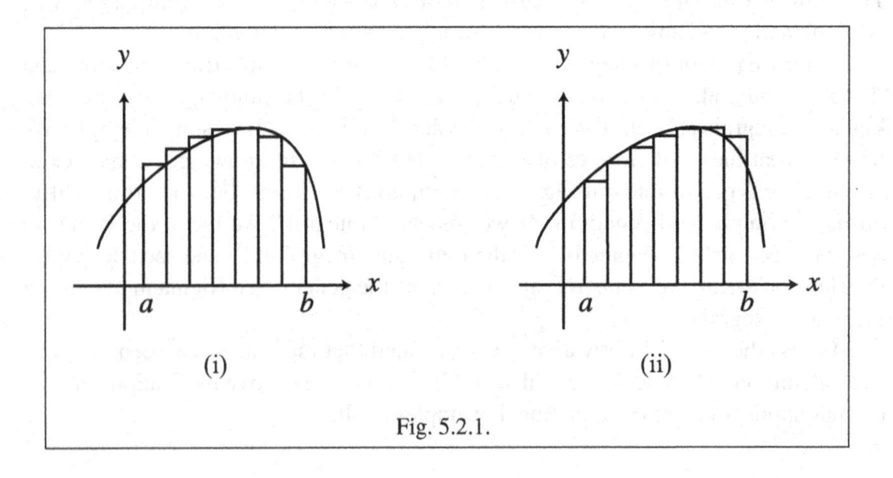

Fig. 5.2.1.

The Riemann integral is based upon a generalization of right-hand sums and left-hand sums. Although it is convenient for computational purposes to subdivide the interval $[a,b]$ into subintervals of equal length, and to use something uniform such as right endpoints or left endpoints, from a theoretical point of view we want to make sure that we do not miss any bad behavior of the function, and so we need to look at sums of rectangles where the subintervals are not necessarily of equal length, and where the height of each rectangle is the value of the function at some point in the corresponding subinterval, but not necessarily endpoints, and not necessarily in the "same location" in each subinterval. A more general sum of this sort is illustrated in Figure 5.2.2. The notation used in this figure will be explained in the following definition, after which we will give the definition of these more general sums.

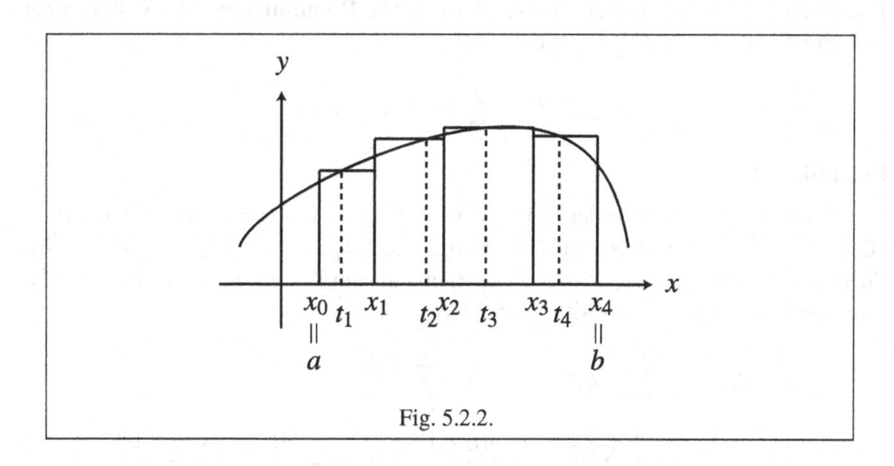

Fig. 5.2.2.

Definition 5.2.1. Let $[a,b] \subseteq \mathbb{R}$ be a non-degenerate closed bounded interval.

1. A **partition** of $[a,b]$ is a set $P = \{x_0, x_1, \ldots, x_n\}$ such that $a = x_0 < x_1 < \cdots < x_n = b$, for some $n \in \mathbb{N}$.
2. If $P = \{x_0, x_1, \ldots, x_n\}$ is a partition of $[a,b]$, the **norm** (also called the **mesh**) of P, denoted $||P||$, is defined by

$$||P|| = \max\{x_1 - x_0, x_2 - x_1, \ldots, x_n - x_{n-1}\}.$$

3. If $P = \{x_0, x_1, \ldots, x_n\}$ is a partition of $[a,b]$, a **representative set** of P is a set $T = \{t_1, t_2, \ldots, t_n\}$ such that $t_i \in [x_{i-1}, x_i]$ for all $i \in \{1, \ldots, n\}$. △

We note that the definition of representative sets in Definition 5.2.1, while intuitively correct, has a slight technical problem. It could happen that $t_{i-1} = t_i$ for some $i \in \{1, \ldots, n\}$, in which case writing "$T = \{t_1, t_2, \ldots, t_n\}$" would lead to the set T having fewer than n elements, because a single element is never written twice in a set, and yet we want T to have one element for each of the n subintervals of $[a,b]$. Therefore, the technically correct way to define a representative set would be as a function $T\colon \{1, \ldots, n\} \to [a,b]$ such that $T(i) \in [x_{i-1}, x_i]$ for all $i \in \{1, \ldots, n\}$. However, because no problem will arise, we will use the more convenient notation of Definition 5.2.1 (3) and write a representative set as $T = \{t_1, t_2, \ldots, t_n\}$, where we think of $t_i = T(i)$ for all $i \in \{1, \ldots, n\}$.

The term "partition" as used here does not have the same meaning as the term "partition" that is used in the context of equivalence relations (as discussed in [Blo10, Section 5.3]); both uses of the term "partition" are quite standard, and the proper meaning is easy to tell from the context. The term "representative set," by contrast, is not standard, but there does not appear to be a universally accepted term for this concept (some books do not even give this concept a name).

Definition 5.2.2. Let $[a,b] \subseteq \mathbb{R}$ be a non-degenerate closed bounded interval, let $f\colon [a,b] \to \mathbb{R}$ be a function, let $P = \{x_0, x_1, \ldots, x_n\}$ be a partition of $[a,b]$ and let $T = \{t_1, t_2, \ldots, t_n\}$ be a representative set of P. The **Riemann sum** of f with respect to P and T, denoted $S(f,P,T)$, is defined by

$$S(f,P,T) = \sum_{i=1}^{n} f(t_i)(x_i - x_{i-1}). \qquad \triangle$$

Example 5.2.3.

(1) Let $f\colon [0,2] \to \mathbb{R}$ be defined by $f(x) = x^2$ for all $x \in [0,2]$. Let $n \in \mathbb{N}$. Let $P_n = \{0, \frac{2}{n}, \frac{4}{n}, \ldots, \frac{2n}{n}\}$. Then P_n is a partition of $[0,2]$, and $\|P_n\| = \frac{2}{n}$. Let $T_n = \{\frac{2}{n}, \frac{4}{n}, \ldots, \frac{2n}{n}\}$. Then T_n is a representative set of P_n. The Riemann sum $S(f,P_n,T_n)$ is an example of a right-hand sum. Using Proposition 2.5.2 we see that

$$\begin{aligned} S(f,P_n,T_n) &= \sum_{i=1}^{n} f(t_i)(x_i - x_{i-1}) = \sum_{i=1}^{n} \left(\frac{2i}{n}\right)^2 \frac{2}{n} \\ &= \frac{8}{n^3} \sum_{i=1}^{n} i^2 = \frac{8}{n^3} \frac{n(n+1)(2n+1)}{6} = \frac{4(n+1)(2n+1)}{3n^2}. \end{aligned}$$

We were able to compute an explicit formula for $S(f,P_n,T_n)$ in terms of n only because the function f was so simple that we had the convenient formula from Proposition 2.5.2 available; for more complicated functions it is rarely possible to find such explicit formulas for Riemann sums.

(2) Let $r\colon [0,1] \to \mathbb{R}$ be defined by

$$r(x) = \begin{cases} 1, & \text{if } x \in \mathbb{Q} \cap [0,1] \\ 0, & \text{otherwise.} \end{cases}$$

Let $P = \{x_0, x_1, \ldots, x_n\}$ be a partition of $[0,1]$. By Theorem 2.6.13 (1) it is possible to choose a representative set $T = \{t_1, t_2, \ldots, t_n\}$ of P such that $t_1, t_2, \ldots, t_n$ are all rational numbers. Then

$$S(r,P,T) = \sum_{i=1}^{n} f(t_i)(x_i - x_{i-1}) = \sum_{i=1}^{n} 1 \cdot (x_i - x_{i-1}) = x_n - x_0 = 1.$$

On the other hand, by Theorem 2.6.13 (2) it is possible to choose a representative set $S = \{s_1, s_2, \ldots, s_n\}$ of P such that $s_1, s_2, \ldots, s_n$ are all irrational numbers. It is then seen that $S(r,P,S) = 0$. It is also possible to choose a representative set $U = \{u_1, u_2, \ldots, u_n\}$ of P such that some of $u_1, u_2, \ldots, u_n$ are rational and some are irrational, in which case $0 < S(r,P,U) < 1$. We therefore see that, at least for some functions, the choice of representative set can make a big difference when computing Riemann sums. $\Diamond$

We now turn to the definition of the Riemann integral. The intuitive idea is that the integral exists if there is some real number that all Riemann sums get closer and closer to, as the widths of the rectangles in the Riemann sums get thinner and thinner. The Riemann integral is a type of limit, and similarly to the definition of limits of functions in Definition 3.2.1, we use an ε–δ-type approach for the definition of the Riemann integral. If the ε–δ formulation of the Riemann integral appears somewhat more complicated than the ε–δ formulation of limits of functions, that is because for integrals we need to take into account all possible partitions, and all possible representative sets of each partition.

Definition 5.2.4. Let $[a,b] \subseteq \mathbb{R}$ be a non-degenerate closed bounded interval, let $f\colon [a,b] \to \mathbb{R}$ be a function and let $K \in \mathbb{R}$. The number K is the **Riemann integral** of f, written

$$\int_a^b f(x)\,dx = K,$$

if for each $\varepsilon > 0$, there is some $\delta > 0$ such that if P is a partition of $[a,b]$ with $\|P\| < \delta$, and if T is a representative set of P, then $|S(f,P,T) - K| < \varepsilon$. If the Riemann integral of f exists, we say that f is **Riemann integrable**. △

It is important to stress that in Definition 5.2.4, when it says "if T is a representative set of P," that means that T can be any representative set of P; it is not sufficient to show that if P is a partition of $[a,b]$ with $\|P\| < \delta$, then $|S(f,P,T) - K| < \varepsilon$ for some choice of representative set T. We will see the importance of this fact in Example 5.2.6 (3).

The definition of the Riemann integral of a function f makes an implicit assumption, which is that if f is Riemann integrable, then the number K in the definition of Riemann integrability is unique; if that were not the case, then there would not be a single number that would be called "the Riemann integral" of the function on the given interval. Fortunately, as we see in the following lemma, this assumption of uniqueness is justified.

Lemma 5.2.5. *Let $[a,b] \subseteq \mathbb{R}$ be a non-degenerate closed bounded interval, and let $f\colon [a,b] \to \mathbb{R}$ be a function. If f is Riemann integrable, then there is a unique $K \in \mathbb{R}$ such that $\int_a^b f(x)\,dx = K$.*

Proof. Suppose that f is Riemann integrable. Suppose further that $\int_a^b f(x)\,dx = K_1$ and $\int_a^b f(x)\,dx = K_2$ for some $K_1, K_2 \in \mathbb{R}$ such that $K_1 \neq K_2$. Let $\varepsilon = \frac{|K_2 - K_1|}{2}$. Then $\varepsilon > 0$. There is some $\delta_1 > 0$ such that if P is a partition of $[a,b]$ with $\|P\| < \delta_1$, and if T is a representative set of P, then $|S(f,P,T) - K_1| < \varepsilon$, and there is some $\delta_2 > 0$ such that if Q is a partition of $[a,b]$ with $\|Q\| < \delta_2$, and if S is a representative set of Q, then $|S(f,Q,S) - K_2| < \varepsilon$. Let $\delta = \min\{\delta_1, \delta_2\}$. Let R be a partition of $[a,b]$ with $\|R\| < \delta$, which exists by Exercise 5.2.1, and let V be a representative set of R. Then

$$\begin{aligned}|K_2 - K_1| &= |K_2 - S(f,R,V) + S(f,R,V) - K_1| \\ &\leq |K_2 - S(f,R,V)| + |S(f,R,V) - K_1| < \varepsilon + \varepsilon = |K_2 - K_1|,\end{aligned}$$

which is a contradiction. Hence, if f is Riemann integrable, then there is a unique $K \in \mathbb{R}$ such that $\int_a^b f(x)\,dx = K$. □

We will usually drop the word "Riemann" and just say "integrable" and "integral," because we will not be dealing with any other type of integral in this text. Also, although we will not be using the term "definite integral" in this text, we will always mean "definite integral" when we say "integral," unless it is clear from the context that we mean "indefinite integral," a term that will be defined in Section 5.6.

Although we use the standard notation $\int_a^b f(x)\,dx$ to denote the Riemann integral, the reader might well ask why we write "$f(x)$," given that the name of the function is actually "f" (recall the brief discussion of this matter following Definition 2.5.10), and why we have the "dx" at all. Those would be very sensible questions. In fact, there are texts that simply write $\int_a^b f$ to denote the Riemann integral of the function f on the interval $[a,b]$, and doing so is quite reasonable. We use the traditional notation $\int_a^b f(x)\,dx$ simply for the sake of familiarity with what the reader has already seen in calculus (and other) courses. It is important to note, however, that the "x" in this notation is a "dummy variable," and that it has no intrinsic meaning. We could just as well write $\int_a^b f(y)\,dy$ to mean the same integral. The standard notation for integrals that we use, due to Gottfried von Leibniz (1646–1716), is meant to remind us of Riemann sums, which have the form $\sum_{i=1}^n f(t_i)(x_i - x_{i-1})$. The symbol "$\int$" is simply an elongated letter "S," and it stands for sum, as does the Greek letter "Σ"; the "dx" is meant to remind us of $(x_i - x_{i-1})$, which is sometimes abbreviated "Δx_i," though we will not be using that notation. In practive, the "dx" is convenient because it helps us keep track of things when we do substitution.

In most cases, it is tricky (or virtually impossible) to show that a function is integrable using only the definition of integrability. Nonetheless, we now see a few examples where the definition of integrability can be used directly.

Example 5.2.6.

(1) Let $c \in \mathbb{R}$, and let $f\colon [a,b] \to \mathbb{R}$ be defined by $f(x) = c$ for all $x \in [a,b]$. We will show that f is integrable, and that $\int_a^b f(x)\,dx = c(b-a)$. Let $P = \{x_0, x_1, \ldots, x_n\}$ be a partition of $[a,b]$, and let $T = \{t_1, t_2, \ldots, t_n\}$ be a representative set of P. Then

$$S(f,P,T) = \sum_{i=1}^{n} f(t_i)(x_i - x_{i-1}) = c\sum_{i=1}^{n}(x_i - x_{i-1}) = c(x_n - x_0) = c(b-a).$$

Given that all Riemann sums of this function have the same value, it is evident that f is integrable and that $\int_a^b f(x)\,dx = c(b-a)$.

(2) Let $g\colon [0,1] \to \mathbb{R}$ be defined by

$$g(x) = \begin{cases} 7, & \text{if } x = 0 \\ 0, & \text{if } x \in (0,1]. \end{cases}$$

We will show that g is integrable, and that $\int_0^1 g(x)\,dx = 0$. Let $\varepsilon > 0$. Let $\delta = \frac{\varepsilon}{7}$. Let $P = \{x_0, x_1, \ldots, x_n\}$ be a partition of $[0,1]$ with $\|P\| < \delta$, and let $T = \{t_1, t_2, \ldots, t_n\}$ be a representative set of P. Then $g(t_1)$ might equal 7 or it might equal 0, which means $|g(t_1)| \leq 7$, and $g(t_i) = 0$ for $i \in \{2,3,\ldots,n\}$. Therefore

$$\begin{aligned} |S(g,P,T) - 0| &= \left| \sum_{i=1}^{n} g(t_i)(x_i - x_{i-1}) \right| = |g(t_1)(x_1 - x_0)| \\ &= |g(t_1)| \cdot |x_1 - x_0| < 7\delta = \varepsilon. \end{aligned}$$

It follows that g is integrable and that $\int_0^1 g(x)\,dx = 0$.

(3) Let $r\colon [0,1] \to \mathbb{R}$ be defined by

$$r(x) = \begin{cases} 1, & \text{if } x \in \mathbb{Q} \cap [0,1] \\ 0, & \text{otherwise.} \end{cases}$$

It was shown in Example 3.3.3 (6) that r is discontinuous everywhere. We will now show that this function is not integrable. Suppose to the contrary that r is integrable. Let $\varepsilon = \frac{1}{2}$. We will obtain a contradiction by showing that there is no δ that "works" for this ε. Let $\delta > 0$. Let $P = \{x_0, x_1, \ldots, x_n\}$ be a partition of $[0,1]$ with $\|P\| < \delta$. There are now two cases. First, suppose that $\int_0^1 r(x)\,dx < \frac{1}{2}$. By Theorem 2.6.13 (1) we can choose a representative set $T = \{t_1, t_2, \ldots, t_n\}$ of P such that $t_1, t_2, \ldots, t_n$ are all rational numbers. It was seen in Example 5.2.3 (2) that $S(r,P,T) = 1$. Hence

$$S(r,P,T) - \int_0^1 r(x)\,dx > 1 - \frac{1}{2} = \frac{1}{2},$$

which implies that

$$\left| S(r,P,T) - \int_0^1 r(x)\,dx \right| > \frac{1}{2}.$$

Second, suppose that $\int_0^1 r(x)\,dx \geq \frac{1}{2}$. By Theorem 2.6.13 (2) we can choose a representative set $T = \{t_1, t_2, \ldots, t_n\}$ of P such that $t_1, t_2, \ldots, t_n$ are all irrational numbers. It was seen in Example 5.2.3 (2) that $S(r,P,T) = 0$. Hence

$$\left| S(r,P,T) - \int_0^1 r(x)\,dx \right| = \left| \int_0^1 r(x)\,dx \right| \geq \frac{1}{2}.$$

We have therefore shown that for any $\delta > 0$, there is some partition P of $[0,1]$, and some representative set T of P, such that

$$\left| S(r,P,T) - \int_0^1 r(x)\,dx \right| \not< \varepsilon,$$

which is a contradiction to our assumption that r is integrable.

(4) Let $s\colon [0,1] \to \mathbb{R}$ be defined by

$$s(x) = \begin{cases} \frac{1}{q}, & \text{if } x \in \mathbb{Q} \cap [0,1] \text{ and } x = \frac{p}{q} \text{ in lowest terms,} \\ & \qquad \text{where } p \in \mathbb{N} \cup \{0\} \text{ and } q \in \mathbb{N} \\ 0, & \text{otherwise.} \end{cases}$$

The use of the expression “lowest terms” was discussed in Example 3.3.3 (7), where it was shown that s is discontinuous at every rational number in $[0,1]$, but it is continuous at every irrational number in $[0,1]$. It is not entirely evident intuitively whether or not the function s is integrable, in that it has more discontinuities than the function g in Part (2) of this example, but fewer discontinuities than the function r in Part (3) of this example. It turns out that s is integrable and that $\int_a^b s(x)\,dx = 0$, though we need a slightly trickier argument than for any of the previous parts of this example.

Let $\varepsilon > 0$. We will choose our δ soon, but first we need a preliminary step. By Corollary 2.6.8 (1) there is some $q_0 \in \mathbb{N}$ such that $q_0 > \max\{\frac{2}{\varepsilon}, 2\}$. Then $\frac{1}{q_0} < \frac{\varepsilon}{2}$. Let

$$A = \left\{ x \in [0,1] \mid s(x) > \frac{1}{q_0} \right\}.$$

Then $x \in A$ if and only if $x \in \mathbb{Q} \cap [0,1]$ and $x = \frac{p}{q}$ in lowest terms, where $p \in \mathbb{N} \cup \{0\}$ and $q \in \mathbb{N}$ and $q < q_0$. Because $1 \in A$ we know that $A \neq \emptyset$, and it is seen that A is finite. Let M be the number of elements in A. Then $M \geq 1$.

Let $\delta = \frac{\varepsilon}{2M}$. Then $\delta > 0$. Let $P = \{x_0, x_1, \ldots, x_n\}$ be a partition of $[0,1]$ with $\|P\| < \delta$, and let $T = \{t_1, t_2, \ldots, t_n\}$ be a representative set of P. Let $i \in \{1, \ldots, n\}$. If $t_i \notin \mathbb{Q}$ then $s(t_i) = 0$; if $t_i \in A$ then $\frac{1}{q_0} < s(t_i) \leq 1$; and if $t_i \in \mathbb{Q} - A$ then $0 < s(t_i) \leq \frac{1}{q_0}$. Using Exercise 2.5.3, we see that

$$\begin{aligned}
|S(s,P,T) - 0| &= \left| \sum_{i=1}^{n} s(t_i)(x_i - x_{i-1}) \right| \leq \sum_{i=1}^{n} |s(t_i)| \cdot |x_i - x_{i-1}| \\
&= \sum_{\substack{i \in \{1,\ldots,n\} \\ t_i \notin \mathbb{Q}}} |s(t_i)| \cdot |x_i - x_{i-1}| + \sum_{\substack{i \in \{1,\ldots,n\} \\ t_i \in A}} |s(t_i)|(x_i - x_{i-1}) \\
&\quad + \sum_{\substack{i \in \{1,\ldots,n\} \\ t_i \in \mathbb{Q} - A}} |s(t_i)| \cdot |x_i - x_{i-1}| \\
&\leq \sum_{\substack{i \in \{1,\ldots,n\} \\ t_i \notin \mathbb{Q}}} 0 \cdot |x_i - x_{i-1}| + \sum_{\substack{i \in \{1,\ldots,n\} \\ t_i \in A}} 1 \cdot |x_i - x_{i-1}| \\
&\quad + \sum_{\substack{i \in \{1,\ldots,n\} \\ t_i \in \mathbb{Q} - A}} \frac{1}{q_0} |x_i - x_{i-1}| \\
&< 0 + M \cdot 1 \cdot \delta + \frac{1}{q_0} \cdot \sum_{i=1}^{n} |x_i - x_{i-1}| < M \cdot \frac{\varepsilon}{2M} + \frac{\varepsilon}{2} \cdot \sum_{i=1}^{n} (x_i - x_{i-1}) \\
&= \frac{\varepsilon}{2} + \frac{\varepsilon}{2} \cdot (x_n - x_0) = \varepsilon.
\end{aligned}$$

(5) Let $v\colon [0,1] \to \mathbb{R}$ be defined by

$$v(x) = \begin{cases} 0, & \text{if } x = 0 \\ 1, & \text{if } x \in (0,1]. \end{cases}$$

The function v is integrable, as the reader is asked to show in Exercise 5.2.3. Let $s\colon [0,1] \to \mathbb{R}$ be the function given in Part (4) of this example; we saw that s is integrable. By abuse of notation we can view s as a function $s\colon [0,1] \to [0,1]$, and hence we can form the composition $v \circ s$. Then

$$(v \circ s)(x) = \begin{cases} 1, & \text{if } x \in \mathbb{Q} \cap [0,1] \\ 0, & \text{otherwise.} \end{cases}$$

Hence $v \circ s = r$, where r is the function given in Part (3) of this example. It was shown that r is not integrable, and we therefore see that the composition of two integrable functions can be a non-integrable function. This situation contrasts with the fact that the composition of continuous functions is continuous (Theorem 3.3.8 (3)), and the composition of differentiable functions is differentiable (Corollary 4.3.4). ◊

There are two basic questions that arise about integrals: which functions are integrable, and how do we integrate the integrable ones? Over the course of this chapter we will deal with the first question thoroughly and the second question partially. For derivatives, the intuitive picture is quite clear—a function is differentiable intuitively if its graph has no "corners." It is much harder to get an intuitive picture of integrable functions, because, as we saw in Example 5.2.6 (4), a function can be rather strange, and in particular it can be discontinuous at many points, and yet still be integrable. The precise nature of "how badly discontinuous" a function can be and yet still be integrable will be clarified in Section 5.8. As for computing integrals of integrable functions, again the situation is much more complicated than for derivatives. Not all rules for computing derivatives have exact analogs for integrals, for example the Product Rule, the Quotient Rule and the Chain Rule. In principle, as we will see in Theorem 5.4.11, any continuous function defined on a closed bounded interval is integrable. However, there is no guarantee that such an integral can be computed in practice. Some such integrals are very hard to compute, and others impossible to compute exactly, and only numerical approximations can be obtained; see Section 5.6, after the statement of Corollary 5.6.3, for some references.

Reflections

It is customary in most real analysis texts to have the chapter on integrals (by which we mean "definite integrals") follow the chapter on derivatives, for the simple reason that integrals are harder to define, and harder to prove theorems about, than derivatives. However, because the Fundamental Theorem of Calculus allows us to evaluate integrals by using antiderivatives, and because of the similarity of notation between definite integrals and indefinite integrals (which are simply antiderivatives), some beginning students mistakenly think that the definition of integrals is related to the definition of derivatives, which is most certainly not the case. In fact, the well-known calculus textbook [Apo67] treats integrals before derivatives, both for historical reasons and to clarify the relation between derivatives and integrals.

There are two standard ways of defining the Riemann integral that are found in introductory textbooks on real analysis; one approach uses Riemann sums, and the other uses upper integrals and lower integrals. These two methods are completely equivalent, and the choice of method used in any text is simply a decision about what to take as the definition, and what to prove using the chosen definition. We use the Riemann sum approach in this text because that approach will be familiar to the reader from introductory calculus courses, where the informal discussion is always based upon Riemann sums. The definition of upper integrals and lower integrals is found in Section 5.4, and the proof that the Riemann sum approach is equivalent to the upper integral and lower integral approach is given in Theorem 5.4.10. We will use upper integrals and lower integrals in the rigorous treatment of area in Section 5.9.

The definition of the Riemann integral in terms of Riemann sums uses an ε and a δ in a way that is reminiscent of the ε–δ definition of limits of functions (given in Definition 3.2.1). In spite of this resemblance, however, the definition of the Riemann integral is not, strictly speaking, the limit of a function, in contrast to the definition of the derivative, which is precisely the limit of a function (and hence the definition of the derivative avoids direct mention of ε–δ, because that is subsumed in the use of limits). The reason that the definition of the Riemann integral is not the limit of a function is that we need to take into account all possible partitions of the given closed bounded interval, and all possible representative sets of each partition, and the collection of all partitions and representative sets are not contained in the real numbers, and more importantly cannot be arranged in a linear order, in contrast to the real numbers. It would be nice if there were some way to generalize the notion of limits of functions so that the generalized notion contains as special cases both limits of functions and the type of ε–δ construction used in the definition of integrals, and in fact there is a way of doing that, using the idea of directed sets. See [Bea97] for an introductory treatment of real analysis using this approach. The advantage of using limits defined in the general context of directed sets is that a number of different definitions, for example limits of functions and Riemann integrals, are special cases of one general type of limit, and various analogous theorems that are proved separately in the standard approach can be proved only once in the context of the more general type of limit. The disadvantages of using directed sets are that doing so makes it more difficult to develop an intuitive understanding of limits of functions, Riemann integrals and the like, and that because students do not have to see similar definitions and proofs in different contexts, they are deprived of the chance to have their newly acquired skills at ε–δ proofs reinforced by constant practice.

Exercises

Exercise 5.2.1. **[Used throughout.]** Let $[a,b] \subseteq \mathbb{R}$ be a non-degenerate closed bounded interval, and let $\varepsilon > 0$. Prove that there is a partition R of $[a,b]$ such that $\|R\| < \varepsilon$.

Exercise 5.2.2. **[Used in Theorem 10.2.11.]** Let $[a,b] \subseteq \mathbb{R}$ be a non-degenerate closed bounded interval, and let $f,g\colon [a,b] \to \mathbb{R}$ be functions. Suppose that there is some $M \in \mathbb{N}$ such that $|f(x) - g(x)| \leq M$ for all $x \in [a,b]$. Let P be a partition of $[a,b]$, and let T be a representative set of P. Prove that $|S(f,P,T) - S(g,P,T)| \leq M(b-a)$.

Exercise 5.2.3. **[Used in Example 5.2.6.]** Let $v\colon [0,1] \to \mathbb{R}$ be the function defined in Example 5.2.6 (5). Prove that v is integrable, using only the definition of integrability.

Exercise 5.2.4. Let $f\colon [0,3] \to \mathbb{R}$ be defined by

$$f(x) = \begin{cases} 5, & \text{if } x \in [0,1] \\ 0, & \text{if } x \in (1,3]. \end{cases}$$

Using only the definition of integrability, prove that f is integrable.

Exercise 5.2.5. [Used in Example 5.6.1 and Example 5.6.5.] Let $h\colon [0,2] \to \mathbb{R}$ be defined by

$$h(x) = \begin{cases} 1, & \text{if } x \in [0,1] \\ 2, & \text{if } x \in (1,2]. \end{cases}$$

Using only the definition of integrability, prove that h is integrable.

Exercise 5.2.6. [Used in Example 5.5.2 and Exercise 5.5.1.] Let $[a,b] \subseteq \mathbb{R}$ be a non-degenerate closed bounded interval, and let $f\colon [a,b] \to \mathbb{R}$ be defined by $f(x) = x$ for all $x \in [a,b]$. Using only the definition of integrability, prove that f is integrable and $\int_a^b f(x)\,dx = \frac{b^2-a^2}{2}$. Use the fact that

$$\sum_{i=1}^{n} \frac{x_i + x_{i-1}}{2}(x_i - x_{i-1}) = \frac{b^2-a^2}{2},$$

which you should verify.

Exercise 5.2.7. Given an example of a function $f\colon [0,1] \to \mathbb{R}$ such that f is not integrable, but that $|f|$ is integrable.

Exercise 5.2.8. Let $[a,b] \subseteq \mathbb{R}$ be a non-degenerate closed bounded interval, and let $f,g,h\colon [a,b] \to \mathbb{R}$ be functions. Suppose that f and h are integrable, and that $\int_a^b f(x)\,dx = \int_a^b h(x)\,dx$. Prove that if $f(x) \le g(x) \le h(x)$ for all $x \in [a,b]$, then g is integrable and $\int_a^b g(x)\,dx = \int_a^b f(x)\,dx$.

Exercise 5.2.9. Let $[a,b] \subseteq \mathbb{R}$ be a non-degenerate closed bounded interval, let $c \in \mathbb{R}$ and let $f\colon [a,b] \to \mathbb{R}$ be a function. Let $h\colon [a+c,b+c] \to \mathbb{R}$ be defined by $h(x) = f(x-c)$ for all $x \in [a+c,b+c]$. Prove that h is integrable if and only if f is integrable, and if they are integrable then $\int_{a+c}^{b+c} h(x)\,dx = \int_a^b f(x)\,dx$.

Exercise 5.2.10. [Used in Exercise 5.5.4.] Let $[a,b] \subseteq \mathbb{R}$ be a non-degenerate closed bounded interval, and let $f\colon [a,b] \to \mathbb{R}$ be a function. Let $g\colon [-b,-a] \to \mathbb{R}$ be defined by $g(x) = f(-x)$ for all $x \in [-b,-a]$. Prove that g is integrable if and only if f is integrable, and if they are integrable then $\int_{-b}^{-a} g(x)\,dx = \int_a^b f(x)\,dx$.

Exercise 5.2.11. [Used in Exercise 5.3.6, Exercise 5.3.7 and Exercise 5.5.11.] Let $[a,b] \subseteq \mathbb{R}$ be a non-degenerate closed bounded interval, and let $f,g\colon [a,b] \to \mathbb{R}$ be functions. Suppose that g is increasing. If $P = \{x_0, x_1, \ldots, x_n\}$ is a partition of $[a,b]$, and if $T = \{t_1, t_2, \ldots, t_n\}$ is a representative set of P, the **Riemann–Stieltjes sum** of f with respect to P, T and g, denoted $S(f,P,T,g)$, is defined by

$$S(f,P,T,g) = \sum_{i=1}^{n} f(t_i)(g(x_i) - g(x_{i-1})).$$

Let $K \in \mathbb{R}$. The number K is the **Riemann–Stieltjes integral** of f with respect to g, written

$$\int_a^b f(x)\,dg = K$$

(also written $\int_a^b f(x)\,dg(x) = K$), if for each $\varepsilon > 0$, there is some $\delta > 0$ such that if P is a partition of $[a,b]$ with $\|P\| < \delta$, and if T is a representative set of P, then $|S(f,P,T,g) - K| < \varepsilon$. If the Riemann–Stieltjes integral of f exists, we say that f is **Riemann–Stieltjes integrable** with respect to g.

(1) Let $c \in (a,b)$. Suppose that g is defined by

$$g(x) = \begin{cases} 0, & \text{if } x \in [a,c) \\ 1, & \text{if } x \in [c,b]. \end{cases}$$

Prove that if f is continuous at c, then f is Riemann–Stieltjes integrable with respect to g.

(2) Let $c \in (a,b)$, and let g be as in Part (1) of this exercise. Let $f = g$. Prove that f is not Riemann–Stieltjes integrable with respect to g.

See Exercise 5.5.11 for a clarification of the relation of the Riemann integral and the Riemann–Stieltjes integral, see [Sto01, Section 6.5] for a general discussion of the Riemann–Stieltjes integral and see [Ros80] for a rethinking of the definition of this type of integral.

5.3 Elementary Properties of the Riemann Integral

Having defined the concept of the integral in Section 5.1, we now turn to some elementary properties of integrals that follow directly from the definition. We will see some additional properties of integrals in Section 5.5, after we look more closely at the meaning of integrability in Section 5.4, which will provide us with a powerful tool to prove some results about integrals that would be very hard to prove directly from the definition of integration.

Theorem 5.3.1. *Let $[a,b] \subseteq \mathbb{R}$ be a non-degenerate closed bounded interval, let $f,g\colon [a,b] \to \mathbb{R}$ be functions and let $k \in \mathbb{R}$. Suppose that f and g are integrable.*

1. *$f+g$ is integrable and $\int_a^b [f+g](x)\,dx = \int_a^b f(x)\,dx + \int_a^b g(x)\,dx$.*
2. *$f-g$ is integrable and $\int_a^b [f-g](x)\,dx = \int_a^b f(x)\,dx - \int_a^b g(x)\,dx$.*
3. *kf is integrable and $\int_a^b [kf](x)\,dx = k\int_a^b f(x)\,dx$.*
4. *$\int_a^b k\,dx = k(b-a)$.*

Proof. We prove Part (1); Part (2) is very similar to Part (1), and we omit the details. Part (3) is left to the reader in Exercise 5.3.1. Part (4) was proved in Example 5.2.6 (1), and is stated here simply for ease of reference.

(1) Let $\varepsilon > 0$. There is some $\delta_1 > 0$ such that if P is a partition of $[a,b]$ with $\|P\| < \delta_1$, and if T is a representative set of P, then $|S(f,P,T) - \int_a^b f(x)\,dx| < \frac{\varepsilon}{2}$, and there is some $\delta_2 > 0$ such that if Q is a partition of $[a,b]$ with $\|Q\| < \delta_2$, and if W is a representative set of Q, then $|S(g,Q,W) - \int_a^b g(x)\,dx| < \frac{\varepsilon}{2}$. Let $\delta = \min\{\delta_1, \delta_2\}$.

Let $R = \{x_0, x_1, \ldots, x_n\}$ be a partition of $[a,b]$ with $\|R\| < \delta$, and let $V = \{v_1, v_2, \ldots, v_n\}$ be a representative set of P. Then

$$\begin{aligned} S(f+g,R,V) &= \sum_{i=1}^{n}[f+g](v_i)(x_i - x_{i-1}) = \sum_{i=1}^{n}[f(v_i)+g(v_i)](x_i - x_{i-1}) \\ &= \sum_{i=1}^{n} f(v_i)(x_i - x_{i-1}) + \sum_{i=1}^{n} g(v_i)(x_i - x_{i-1}) = S(f,R,V) + S(g,R,V). \end{aligned}$$

Hence

$$\begin{aligned} &\left| S(f+g,R,V) - \left[\int_a^b f(x)\,dx + \int_a^b g(x)\,dx\right]\right| \\ &\qquad = \left| S(f,R,V) - \int_a^b f(x)\,dx + S(g,R,V) - \int_a^b g(x)\,dx \right| \\ &\qquad \le \left| S(f,R,V) - \int_a^b f(x)\,dx \right| + \left| S(g,R,V) - \int_a^b g(x)\,dx \right| \\ &\qquad < \frac{\varepsilon}{2} + \frac{\varepsilon}{2} = \varepsilon. \end{aligned}$$

□

We note that whereas we wrote "$\int_a^b [f+g](x)\,dx$" in Theorem 5.3.1 (1), it is quite common to write "$\int_a^b [f(x)+g(x)]\,dx$" to mean the same thing. The latter notation is not entirely proper, because the name of the function being integrated is "$f+g$," and not "$f(x)+g(x)$," but fortunately this notation causes no harm.

The reader will have noticed that missing from Theorem 5.3.1 is a statement concerning the integrability of the product or quotient of integrable functions. As we will see in Section 5.5, it is true that the product and quotient of integrable functions are integrable (under suitable hypotheses for quotients), but the proof of that fact requires more tools than we presently have at our disposal. The proof of the integrability of the sum of integrable functions was simple because of the formula $S(f+g,R,V) = S(f,R,V) + S(g,R,V)$, but there is no comparable formula for the Riemann sum of a product or quotient of functions. Correspondingly, although we will prove in Section 5.5 that products and quotients of appropriate integrable functions are integrable, there are no nice formulas for the integrals of such products or quotients.

We now turn to some useful results concerning integrals and inequalities. The idea behind the third part of the following theorem is illustrated in Figure 5.3.1; the area under the curve is greater than the area of the more heavily shaded rectangle, which is $m(b-a)$, and is less than the areas of the two shaded rectangles together, which is $M(b-a)$.

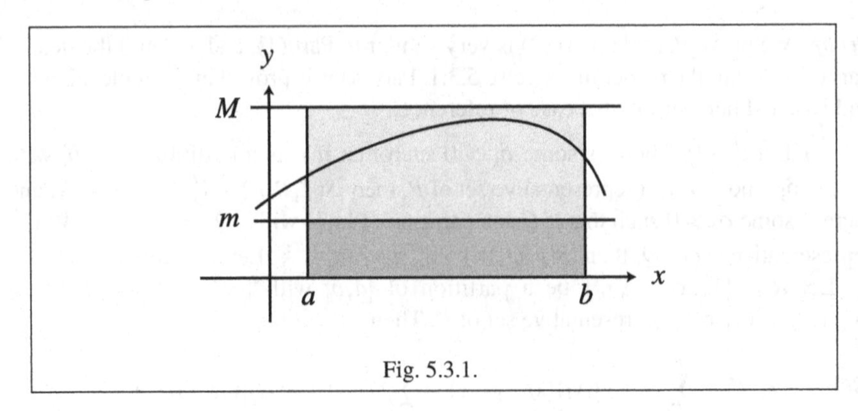

Fig. 5.3.1.

Theorem 5.3.2. *Let $[a,b] \subseteq \mathbb{R}$ be a closed bounded interval, and let $f,g\colon [a,b] \to \mathbb{R}$ be functions. Suppose that f and g are integrable.*

1. *If $f(x) \geq 0$ for all $x \in [a,b]$, then $\int_a^b f(x)\,dx \geq 0$.*
2. *If $f(x) \geq g(x)$ for all $x \in [a,b]$, then $\int_a^b f(x)\,dx \geq \int_a^b g(x)\,dx$.*
3. *Let $m,M \in \mathbb{R}$. If $m \leq f(x)$ for all $x \in [a,b]$, then $m(b-a) \leq \int_a^b f(x)\,dx$, and if $f(x) \leq M$ for all $x \in [a,b]$, then $\int_a^b f(x)\,dx \leq M(b-a)$.*

Proof. We will prove Part (1); each of the other parts will then follow from the previous part together with Theorem 5.3.1.

(1) Suppose that $f(x) \geq 0$ for all $x \in [a,b]$. Let $\varepsilon > 0$. Then there is some $\delta > 0$ such that if P is a partition of $[a,b]$ with $\|P\| < \delta$, and if T is a representative set of P, then $|S(f,P,T) - \int_a^b f(x)\,dx| < \varepsilon$. Let R be a partition of $[a,b]$ with $\|R\| < \delta$, and let V be a representative set of R. Then $|S(f,R,V) - \int_a^b f(x)\,dx| < \varepsilon$, which implies that $\int_a^b f(x)\,dx > S(f,R,V) - \varepsilon$. Because $f(x) \geq 0$ for all $x \in [a,b]$, then clearly $S(f,R,V) \geq 0$. Hence $\int_a^b f(x)\,dx > -\varepsilon$. By Lemma 2.3.10 (2) we deduce that $\int_a^b f(x)\,dx \geq 0$. □

Recall the notion of a function being bounded, as defined in Definition 3.2.5. Is there a relation between integrability and boundedness? The function in Example 5.2.6 (3) is bounded but not integrable, so boundedness alone does not imply integrability. Does integrability imply boundedness? That is, must a function be bounded to be integrable? The reader who is familiar with the concept of an "improper integral," as studied informally in many calculus courses, might suppose that the answer to this question is no, due to previously encountered examples of improper integrals. For example, the improper integral $\int_0^1 \frac{1}{\sqrt{x}}\,dx$ equals 2, as will be seen in Example 6.4.8 (1), even though the function being integrated is not bounded. We will discuss improper integrals in detail in Section 6.4, but for now we clarify that an improper integral is not evaluated directly as a Riemann integral on a closed bounded interval, but rather is evaluated as the limit of Riemann integrals on closed bounded subintervals of the original interval. For example, the improper integral $\int_0^1 \frac{1}{\sqrt{x}}\,dx$ is evaluated as a limit of integrals of the form $\int_s^1 \frac{1}{\sqrt{x}}\, dx$, where $s \in (0,1]$, and the function being integrated is bounded when restricted to each of the subintervals $[s,1]$. As seen in the following theorem, if we restrict our attention to actual Riemann integrals on closed bounded intervals, then integrable functions are always bounded.

Theorem 5.3.3. *Let $[a,b] \subseteq \mathbb{R}$ be a non-degenerate closed bounded interval, and let $f\colon [a,b] \to \mathbb{R}$ be a function. If f is integrable, then f is bounded.*

Proof. Suppose that f is integrable. Then there is some $\delta > 0$ such that if P is a partition of $[a,b]$ with $\|P\| < \delta$, and if T is a representative set of P, then $|S(f,P,T) - \int_a^b f(x)\, dx| < \frac{1}{2}$.

Let $Q = \{x_0, x_1, \ldots, x_q\}$ be a partition of $[a,b]$ with $\|Q\| < \delta$, which exists by Exercise 5.2.1. Let $V = \{x_1, x_2, \ldots, x_q\}$, which is the "right-hand" representative set of Q. Let

$$M = \max\left\{|f(x_1)| + \frac{1}{x_1 - x_0}, \ldots, |f(x_q)| + \frac{1}{x_q - x_{q-1}}\right\}.$$

Let $x \in [a,b]$. Then there is some $k \in \{1,\ldots,q\}$ such that $x \in [x_{k-1}, x_k]$. The number k will not be unique if x happens to be one of $x_1, x_2, \ldots, x_{q-1}$, but in that case we choose one of the values of k that works. Let $W = \{s_1, s_2, \ldots, s_n\}$ be the representative set of Q defined by $s_i = x_i$ if $i \neq k$, and $s_k = x$. Then V and W differ in at most one place, namely, at x_k and s_k. Hence

$$|S(f,Q,W) - S(f,Q,V)| = |[f(x) - f(x_k)](x_k - x_{k-1})| = |f(x) - f(x_k)|(x_k - x_{k-1}).$$

On the other hand, we see that

$$\begin{aligned} |S(f,Q,W) - S(f,Q,V)| &= \left|S(f,Q,W) - \int_a^b f(x)\,dx + \int_a^b f(x)\,dx - S(f,Q,V)\right| \\ &\leq \left|S(f,Q,W) - \int_a^b f(x)\,dx\right| + \left|\int_a^b f(x)\,dx - S(f,Q,V)\right| \\ &< \frac{1}{2} + \frac{1}{2} = 1. \end{aligned}$$

It follows that

$$|f(x) - f(x_k)|(x_k - x_{k-1}) < 1,$$

and hence

$$|f(x) - f(x_k)| < \frac{1}{x_k - x_{k-1}}.$$

Using Lemma 2.3.9 (7) we deduce that

$$|f(x)| < |f(x_k)| + \frac{1}{x_k - x_{k-1}} \leq M.$$

Hence f is bounded. □

Because of Theorem 5.3.3, there will be no loss of generality in restricting our attention to bounded functions from now on in our study of integration. The assumption of boundedness will allow for some very useful technicalities that we will see in Section 5.4.

Reflections

Given that the definition of the Riemann integral is a bit tricky, what is striking about the present section is how straightforward, relatively speaking, the proofs are. Of course, in mathematics straightforward does not necessarily mean easy; in this context it means that the proofs follow from the definition without any additional concepts or devious tricks. The reader should not, however, be fooled into thinking that all proofs involving integrals are straightforward. Indeed, the reason this section is so short is that we have included only those theorems about integrals that have straightforward proofs, and there are not very many of those. This section should be viewed as a warm up, with the main action involving integrals about to start in the next section.

Exercises

Exercise 5.3.1. **[Used in Theorem 5.3.1.]** Prove Theorem 5.3.1 (3).

Exercise 5.3.2. **[Used in Exercise 5.5.6.]** Let $[a,b] \subseteq \mathbb{R}$ be a non-degenerate closed bounded interval, and let $f\colon [a,b] \to \mathbb{R}$ be a function. Suppose that f is integrable. Prove that if $|f(x)| \le M$ for all $x \in [a,b]$, for some $M \in \mathbb{R}$, then $\left|\int_a^b f(x)\,dx\right| \le M(b-a)$.

Exercise 5.3.3. **[Used in Exercise 5.3.4, Example 5.5.2, Section 6.4, Exercise 6.4.3 and Example 10.2.4.]** Let $[a,b] \subseteq \mathbb{R}$ be a non-degenerate closed bounded interval, and let $f,g\colon [a,b] \to \mathbb{R}$ be functions.

(1) Suppose that f is zero except at one point. Prove that f is integrable and that $\int_a^b f(x)\,dx = 0$.
(2) Suppose that f is zero except at finitely many points. Prove that f is integrable and that $\int_a^b f(x)\,dx = 0$.
(3) Suppose that f and g are equal except at finitely many points. Prove that f is integrable if and only if g is integrable, and if they are integrable then $\int_a^b f(x)\,dx = \int_a^b g(x)\,dx$.

Exercise 5.3.4. **[Used in Exercise 5.5.7 and Theorem 5.8.5.]** Let $[a,b] \subseteq \mathbb{R}$ be a non-degenerate closed bounded interval, and let $f\colon [a,b] \to \mathbb{R}$ be a function.

(1) Suppose that there is some non-degenerate closed bounded interval $[c,d] \subseteq [a,b]$, and some $k \in \mathbb{R}$, such that $f(x) = k$ if $x \in (c,d)$ and $f(x) = 0$ if $x \in [a,b] - [c,d]$. It does not matter what values f has at $x = c$ and $x = d$. Prove that f is integrable and that $\int_a^b f(x)\,dx = k(d-c)$.

(2) The function f is a **step function** if there is a partition $P = \{x_0, x_1, \ldots, x_n\}$ of $[a,b]$, and numbers $k_1, \ldots, k_n \in \mathbb{R}$, such that for each $i \in \{1, \ldots, n\}$, if $x \in (x_{i-1}, x_i)$ then $f(x) = k_i$. It does not matter what values f has at the elements of P. See Figure 5.3.2. Prove that if f is a step function, then f is integrable and $\int_a^b f(x)\,dx = \sum_{i=1}^n k_i(x_i - x_{i-1})$. [Use Exercise 5.3.3 (3).]

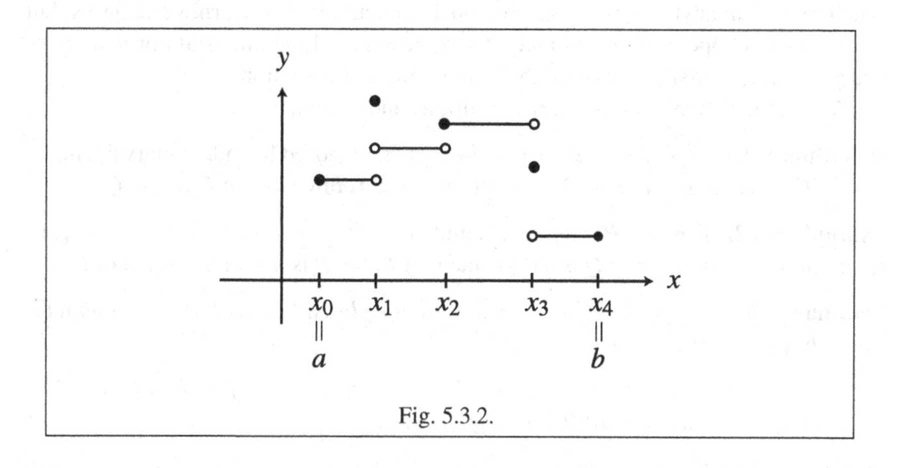

Fig. 5.3.2.

Exercise 5.3.5. [Used in Exercise 5.4.8.] Let $[a,b] \subseteq \mathbb{R}$ be a non-degenerate closed bounded interval, and let $f\colon [a,b] \to \mathbb{R}$ be a function. Suppose that there is some $M > 0$ and some $\delta > 0$ such that if P and Q are partitions of $[a,b]$ with $\|P\| < \delta$ and $\|Q\| < \delta$, and if T is a representative set of P, and V is a representative set of Q, then $|S(f,P,T) - S(f,Q,V)| < M$. Prove that f is bounded.

Exercise 5.3.6. This exercise makes use of Exercise 5.2.11.

(1) State and prove the analog of Theorem 5.3.1 (1) for Riemann–Stieltjes integrals.
(2) State and prove the analog of Theorem 5.3.1 (4) for Riemann–Stieltjes integrals.

Exercise 5.3.7. This exercise makes use of Exercise 5.2.11.

(1) Let $[a,b] \subseteq \mathbb{R}$ be a non-degenerate closed bounded interval, and let $f, g : [a,b] \to \mathbb{R}$ be functions. Suppose that g is strictly increasing, and that f is Riemann–Stieltjes integrable with respect to g. Prove that f is bounded.
(2) Give examples of functions $f, g\colon [0,1] \to \mathbb{R}$ such that g is increasing, that f is Riemann–Stieltjes integrable with respect to g and that f is not bounded.

5.4 Upper Sums and Lower Sums

Although the definition of the Riemann integral via Riemann sums is intuitively appealing in that it corresponds to the way integrals are treated in calculus courses, from a technical point of view Riemann sums are not always easy to work with. In particular, there are some useful properties of integrals, such as those that will be proved in Section 5.5, that would be quite difficult to prove directly using the definition of the Riemann integral. Fortunately, these results can be proved more easily using an alternative characterization of integrability that is given in Theorem 5.4.7 below. The material in this section is somewhat technical, and the proof of Theorem 5.4.7 is lengthier than anything we have previously encountered concerning integrals, but there does not appear to be any nice way of proving all the important properties of integrals without first going through the material in this section.

We start with some preliminary definitions and lemmas.

Definition 5.4.1. Let $[a,b] \subseteq \mathbb{R}$ be a non-degenerate closed bounded interval, and let P and Q be partitions of $[a,b]$. The partition Q is a **refinement** of P if $P \subseteq Q$. △

Example 5.4.2. The sets $P = \{0, \frac{1}{2}, 1\}$, and $Q = \{0, \frac{1}{4}, \frac{1}{2}, \frac{3}{4}, 1\}$ and $R = \{0, \frac{1}{3}, \frac{2}{3}, 1\}$ are partitions of $[0,1]$. Then Q is a refinement of P, but R is not a refinement of P. ◇

Lemma 5.4.3. *Let $[a,b] \subseteq \mathbb{R}$ be a non-degenerate closed bounded interval, and let P and Q be partitions of $[a,b]$.*

1. *$P \cup Q$ is a partition of $[a,b]$, and $P \cup Q$ is a refinement of each of P and Q.*
2. *If Q is a refinement of P, then $\|Q\| \leq \|P\|$.*

Proof. Left to the reader in Exercise 5.4.1. □

The essence of the alternative characterization of integrability, to be given in Theorem 5.4.7, is that for a function to be integrable, its values cannot vary too much when restricted to sufficiently small subintervals of its domain. For example, the values of a continuous function (which we will see later by Theorem 5.4.11 must be integrable) will not vary much on a sufficiently small interval—that is what the ε–δ definition of continuity says. On the other hand, the values of the function r given in Example 5.2.6 (3), which is not integrable, vary the same amount on any subinterval, no matter how small. We saw in Example 5.2.6 (2) (4) that it is possible for discontinuous functions to be integrable, so the obstacle to the integrability of the function r given in Example 5.2.6 (3) is not its discontinuity per se, but must be something else, and that something is precisely how the values of the function vary in small subintervals.

The idea of measuring precisely how much the values of a function vary on an interval can be made precise (see [TBB01, Section 6.7]), but we will not need this concept, and instead we proceed as follows. For a bounded function defined on a non-degenerate closed bounded interval, we will define the "upper sum" and "lower sum" of the function with respect to each partition of the interval, where these sums are similar to Riemann sums, but where the height of each rectangle represents, intuitively, the highest and lowest possible values respectively of the function on each subinterval of the partition. We then capture the idea of whether or not a function varies too much on sufficiently small subintervals by looking at the difference between the upper sum and lower sum when partitions with smaller and smaller norms are used. More specifically, Theorem 5.4.7 will say that a function is integrable if and only if the difference between the upper sum and lower sum can be made as small as desired if we choose partitions with sufficiently small norm.

Before we can state and prove Theorem 5.4.7, we need another definition and a lemma. Recall that by Theorem 5.3.3, it is no restriction in our study of integration that we consider only bounded functions for the rest of this section.

Definition 5.4.4. Let $[a,b] \subseteq \mathbb{R}$ be a non-degenerate closed bounded interval, let $f\colon [a,b] \to \mathbb{R}$ be a function and let $P = \{x_0, x_1, \ldots, x_n\}$ be a partition of $[a,b]$. Suppose that f is bounded.

1. For each $i \in \{1, \ldots, n\}$, let

$$M_i(f) = \operatorname{lub} f([x_{i-1}, x_i]) \quad \text{and} \quad m_i(f) = \operatorname{glb} f([x_{i-1}, x_i]).$$

If it is necessary to indicate the partition being used, we will write $M_i^P(f)$ and $m_i^P(f)$.

2. The **upper sum** of f with respect to P, denoted $U(f,P)$, is defined by

$$U(f,P) = \sum_{i=1}^{n} M_i(f)(x_i - x_{i-1}),$$

and the **lower sum** of f with respect to P, denoted $L(f,P)$, is defined by

$$L(f,P) = \sum_{i=1}^{n} m_i(f)(x_i - x_{i-1}). \qquad \triangle$$

Observe in Definition 5.4.4 that each set of the form $f([x_{i-1}, x_i])$ is non-empty, and it is bounded because the function f is bounded, and therefore the Least Upper Bound Property and the Greatest Lower Bound imply that $f([x_{i-1}, x_i])$ has a least upper bound and a greatest lower bound. The function f in this definition need not be continuous, however, and hence the functions of the form $f|_{[x_{i-1}, x_i]}$ need not have maximum values or minimum values (which is why we need to use least upper bounds and greatest lower bounds in the definition). Because the numbers $M_i(f)$ and $m_i(f)$ need not equal the value of f at any number in $[x_{i-1}, x_i]$, we observe that upper sums and lower sums are not necessarily Riemann sums.

Example 5.4.5.

(1) Let $f\colon [-1,1] \to \mathbb{R}$ be defined by $f(x) = x^2$ for all $x \in [-1,1]$. Let $P = \{-1, -\frac{1}{2}, 0, \frac{1}{2}, 1\}$, which is a partition of $[-1,1]$. Then

$$U(f,P) = (-1)^2 \cdot \frac{1}{2} + \left(-\frac{1}{2}\right)^2 \cdot \frac{1}{2} + \left(\frac{1}{2}\right)^2 \cdot \frac{1}{2} + 1^2 \cdot \frac{1}{2} = \frac{5}{4},$$

and

$$L(f,P) = \left(-\frac{1}{2}\right)^2 \cdot \frac{1}{2} + 0^2 \cdot \frac{1}{2} + 0^2 \cdot \frac{1}{2} + \left(\frac{1}{2}\right)^2 \cdot \frac{1}{2} = \frac{1}{4}.$$

(2) Let $g\colon [0,1] \to \mathbb{R}$ be defined by

$$g(x) = \begin{cases} 7, & \text{if } x = 0 \\ 0, & \text{if } x \in (0,1]. \end{cases}$$

It was shown in Example 5.2.6 (2) that g is integrable. Let $P = \{x_0, x_1, \ldots, x_n\}$ be a partition of $[0,1]$. Then $U(f,P) = 7(x_1 - x_0)$ and $L(f,P) = 0$.

(3) Let $r\colon [0,1] \to \mathbb{R}$ be defined by

$$r(x) = \begin{cases} 1, & \text{if } x \in \mathbb{Q} \cap [0,1] \\ 0, & \text{otherwise.} \end{cases}$$

It was shown in Example 5.2.6 (3) that r is not integrable. Let $P = \{x_0, x_1, \ldots, x_n\}$ be a partition of $[0,1]$. Then $U(r,P) = 1$ and $L(r,P) = 0$. ◊

Although upper sums and lower sums are not necessarily Riemann sums themselves, we see in Part (1) of the following lemma that every Riemann sum for a given partition is squeezed between the upper sum and lower sum for that partition. That fact is what makes upper sums and lower sums so useful.

Lemma 5.4.6. *Let $[a,b] \subseteq \mathbb{R}$ be a non-degenerate closed bounded interval, let $f\colon [a,b] \to \mathbb{R}$ be a function and let P be a partition of $[a,b]$. Suppose that f is bounded.*

1. *If T is a representative set of P, then $L(f,P) \le S(f,P,T) \le U(f,P)$.*
2. *If R is a refinement of P, then $L(f,P) \le L(f,R) \le U(f,R) \le U(f,P)$.*
3. *If Q is a partition of $[a,b]$, then $L(f,P) \le U(f,Q)$.*

Proof.

(1) Let T be a representative set of P. Suppose that $P = \{x_0, x_1, \ldots, x_n\}$ and $T = \{t_1, t_2, \ldots, t_n\}$. For each $i \in \{1, \ldots, n\}$, we know that $t_i \in [x_{i-1}, x_i]$, and therefore

$$m_i(f) = \operatorname{glb} f([x_{i-1}, x_i]) \le f(t_i) \le \operatorname{lub} f([x_{i-1}, x_i]) = M_i(f).$$

Hence

$$\sum_{i=1}^{n} m_i(f)(x_i - x_{i-1}) \le \sum_{i=1}^{n} f(t_i)(x_i - x_{i-1}) \le \sum_{i=1}^{n} M_i(f)(x_i - x_{i-1}),$$

which means that $L(f,P) \le S(f,P,T) \le U(f,P)$.

(2) Let R be a refinement of P. Suppose that $P = \{x_0, x_1, \ldots, x_n\}$ and $R = \{y_0, y_1, \ldots, y_k\}$, where $\{x_0, x_1, \ldots, x_n\} \subseteq \{y_0, y_1, \ldots, y_k\}$.

Let $i \in \{1, \ldots, n\}$. Then there are $s, t \in \{1, \ldots, k\}$ such that $x_{i-1} = y_{s-1}$ and $x_i = y_t$. Hence $[x_{i-1}, x_i] = [y_{s-1}, y_s] \cup [y_s, y_{s+1}] \cup \cdots \cup [y_{t-1}, y_t]$. If $j \in \{s, \ldots, t\}$, then $f([y_{j-1}, y_j]) \subseteq f([x_{i-1}, x_i])$, and therefore by Exercise 2.6.1 (2) we see that $m_i^P(f) = \operatorname{glb} f([x_{i-1}, x_i]) \le \operatorname{glb} f([y_{j-1}, y_j]) = m_j^R(f)$. It follows that

$$m_i^P(f)(x_i - x_{i-1}) = m_i^P(f)[(y_s - y_{s-1}) + \cdots + (y_t - y_{t-1})]$$
$$\leq m_s^R(f)(y_s - y_{s-1}) + \cdots + m_t^R(f)(y_t - y_{t-1}),$$

and hence

$$L(f,P) = \sum_{i=1}^{n} m_i^P(f)(x_i - x_{i-1}) \leq \sum_{j=1}^{k} m_j^R(f)(y_j - y_{j-1}) = L(f,R).$$

A similar argument shows that $U(f,R) \leq U(f,P)$, and we omit the details. By Part (1) of this lemma we know that $L(f,R) \leq U(f,R)$, and the proof is complete.

(3) By Lemma 5.4.3 (1) we know that $P \cup Q$ is a refinement of each of P and Q, and hence Part (2) of this lemma implies

$$L(f,P) \leq L(f,P \cup Q) \leq U(f,P \cup Q) \leq U(f,Q). \qquad \square$$

We are now ready to state and prove the major result in this section, Theorem 5.4.7, which gives an alternative characterization of what it means for a function to be integrable. There are actually two variants of this characterization given in the theorem, where the characterization given in Part (c) is the more useful, but we need to prove the other variant, given in Part (b), in order to prove Part (c). The proof of this theorem is somewhat lengthy, but the value of the theorem, which we will see later in this section and in Section 5.5, makes the long proof worth the effort.

The definition of the integral, given in Definition 5.2.4, is relatively straightforward from an intuitive point of view, but it has one major disadvantage. In order to prove that a given function is integrable via that definition, it is necessary first to guess what the integral actually equals (that is the number K in Definition 5.2.4). In some situations, however, we might want prove that a function is integrable in principle even though we have no idea what the actual value of the integral is. For example, we will prove that all continuous functions on non-degenerate closed bounded intervals are integrable, even though we have no hope of finding a general formula for the integrals of all continuous functions. The beauty of Theorem 5.4.7 is that it gives a characterization of integrability in terms of upper sums and lower sums getting closer and closer to each other, without having to specify what number (which would be the value of the integral) the upper sums and lower sums are getting closer to.

Theorem 5.4.7. *Let $[a,b] \subseteq \mathbb{R}$ be a non-degenerate closed bounded interval, and let $f\colon [a,b] \to \mathbb{R}$ be a function. Suppose that f is bounded. The following are equivalent.*

a. The function f is integrable.
b. For each $\varepsilon > 0$, there is some $\delta > 0$ such that if P is a partition of $[a,b]$ with $\|P\| < \delta$, then $U(f,P) - L(f,P) < \varepsilon$.
c. For each $\varepsilon > 0$, there is some partition P of $[a,b]$ such that $U(f,P) - L(f,P) < \varepsilon$.

Proof.

(a) ⇒ (b) Suppose that f is integrable.

Let $\varepsilon > 0$. By the definition of integrability, there is some $\delta > 0$ such that if R is a partition of $[a,b]$ with $\|R\| < \delta$, and if T is a representative set of R, then $|S(f,R,T) - \int_a^b f(x)\,dx| < \frac{\varepsilon}{4}$.

Let $P = \{z_0, z_1, \ldots, z_k\}$ be a partition of $[a,b]$ with $\|P\| < \delta$. Let $i \in \{1, \ldots, k\}$. By the definition of $M_i(f)$ and $m_i(f)$, together with Lemma 2.6.5, there are $c_i, d_i \in [z_{i-1}, z_i]$ such that

$$m_i(f) \le f(d_i) < m_i(f) + \frac{\varepsilon}{4\delta k} \quad \text{and} \quad M_i(f) - \frac{\varepsilon}{4\delta k} < f(c_i) \le M_i(f).$$

Hence

$$|m_i(f) - f(d_i)| < \frac{\varepsilon}{4\delta k} \quad \text{and} \quad |M_i(f) - f(c_i)| < \frac{\varepsilon}{4\delta k}.$$

Let $C = \{c_1, c_2, \ldots, c_k\}$ and $D = \{d_1, d_2, \ldots, d_k\}$. Then C and D are representative sets of P.

Using Exercise 2.5.3, we see that

$$\begin{aligned} |U(f,P) - S(f,P,C)| &= \left| \sum_{i=1}^{k} M_i(f)(z_i - z_{i-1}) - \sum_{i=1}^{k} f(c_i)(z_i - z_{i-1}) \right| \\ &\le \sum_{i=1}^{k} |M_i(f) - f(c_i)|(z_i - z_{i-1}) < \sum_{i=1}^{k} \frac{\varepsilon}{4\delta k}\delta = \frac{\varepsilon}{4}. \end{aligned}$$

A similar argument shows that

$$|S(f,P,D) - L(f,P)| < \frac{\varepsilon}{4}.$$

Therefore

$$\begin{aligned} U(f,P) - L(f,P) &= |U(f,P) - L(f,P)| \\ &= \left| U(f,P) - S(f,P,C) + S(f,P,C) - \int_a^b f(x)\,dx + \int_a^b f(x)\,dx \right. \\ &\qquad \left. - S(f,P,D) + S(f,P,D) - L(f,P) \right| \\ &\le |U(f,P) - S(f,P,C)| + \left| S(f,P,C) - \int_a^b f(x)\,dx \right| \\ &\qquad + \left| \int_a^b f(x)\,dx - S(f,P,D) \right| + |S(f,P,D) - L(f,P)| \\ &< \frac{\varepsilon}{4} + \frac{\varepsilon}{4} + \frac{\varepsilon}{4} + \frac{\varepsilon}{4} = \varepsilon. \end{aligned}$$

(b) ⇒ (a) Suppose that for each $\varepsilon > 0$, there is some $\delta > 0$ such that if P is a partition of $[a,b]$ with $\|P\| < \delta$, then $U(f,P) - L(f,P) < \varepsilon$.

Let

$$\mathcal{U} = \{U(f,P) \mid P \text{ is a partition of } [a,b]\},$$

and

$$\mathcal{L} = \{L(f,P) \mid P \text{ is a partition of } [a,b]\}.$$

Then $\mathcal{U}$ and $\mathcal{L}$ are non-empty subsets of $\mathbb{R}$, because there exist partitions of $[a,b]$, for example $P = \{a,b\}$. By Lemma 5.4.6 (3), we know that if $L(f,Q) \in \mathcal{L}$ and $U(f,R) \in \mathcal{U}$, then $L(f,Q) \leq U(f,R)$. Let $\mu > 0$. By hypothesis, there is some $\beta > 0$ such that if P is a partition of $[a,b]$ with $\|P\| < \beta$, then $U(f,P) - L(f,P) < \mu$. Let P be a partition of $[a,b]$ with $\|P\| < \beta$, which exists by Exercise 5.2.1. Then $U(f,P) - L(f,P) < \mu$. We now see that $\mathcal{L}$ and $\mathcal{U}$ satisfy the hypotheses of both parts of the No Gap Lemma (Lemma 2.6.6), and therefore $\mathcal{L}$ has a least upper bound and $\mathcal{U}$ has a greatest lower bound, and $\operatorname{lub}\mathcal{L} = \operatorname{glb}\mathcal{U}$.

Let $K = \operatorname{lub}\mathcal{L} = \operatorname{glb}\mathcal{U}$. Let $\varepsilon > 0$. By hypothesis, there is some $\delta > 0$ such that if P is a partition of $[a,b]$ with $\|P\| < \delta$, then $U(f,P) - L(f,P) < \varepsilon$.

Now let W be a partition of $[a,b]$ with $\|W\| < \delta$, and let T be a representative set of W. Then $U(f,W) - L(f,W) < \varepsilon$. By Lemma 5.4.6 (1) we know that $L(f,W) \leq S(f,W,T) \leq U(f,W)$. We also know that $K = \operatorname{lub}\mathcal{L} = \operatorname{glb}\mathcal{U}$, and hence $L(f,W) \leq K \leq U(f,W)$. It follows that $|S(f,W,T) - K| \leq U(f,W) - L(f,W)$. Because $U(f,W) - L(f,W) < \varepsilon$, it follows that $|S(f,W,T) - K| < \varepsilon$. Hence f is integrable, with the integral of f equal to K.

(b) ⇒ (c) This implication is trivial because of Exercise 5.2.1.

(c) ⇒ (b) Suppose that for each $\varepsilon > 0$, there is some partition P of $[a,b]$ such that $U(f,P) - L(f,P) < \varepsilon$.

Let $\varepsilon > 0$. By hypothesis, there is some partition $Q = \{x_0, x_1, \ldots, x_n\}$ of $[a,b]$ such that $U(f,Q) - L(f,Q) < \frac{\varepsilon}{2}$. Because f is bounded, there is some $B \in \mathbb{R}$ such that $|f(x)| \leq B$ for all $x \in [a,b]$. We may assume that $B > 0$. Let $\delta = \frac{\varepsilon}{4Bn}$. Then $\delta > 0$.

Let $Z = \{z_0, z_1, \ldots, z_k\}$ be a partition of $[a,b]$ with $\|Z\| < \delta$. Let

$$W = \{i \in \{1,\ldots,k\} \mid x_j \in (z_{i-1}, z_i) \text{ for some } j \in \{1,\ldots,n\}\}.$$

For each $j \in \{1,\ldots,n\}$, let

$$V_j = \{i \in \{1,\ldots,k\} \mid [z_{i-1}, z_i] \subseteq [x_{j-1}, x_j]\}.$$

Then $W \cup V_1 \cup \cdots \cup V_n = \{1,\ldots,k\}$, and the sets $W, V_1, \ldots, V_n$ are pairwise disjoint. Because $x_0 = a$ and $x_n = b$, it follows that W has at most $n-1$ elements.

For each $i \in \{1,\ldots,k\}$, it follows from Exercise 5.4.9 (4) that $M_i^Z(f) - m_i^Z(f) \leq 2B$. For each $j \in \{1,\ldots,n\}$, observe that $\sum_{i \in V_j}(z_i - z_{i-1}) \leq x_j - x_{j-1}$, and that if $i \in V_j$ then $M_i^Z(f) - m_i^Z(f) \leq M_j^Q(f) - m_j^Q(f)$.

Then

$$U(f,Z) - L(f,Z) = \sum_{i=1}^{k} M_i^Z(f)(z_i - z_{i-1}) - \sum_{i=1}^{k} m_i^Z(f)(z_i - z_{i-1})$$

$$
\begin{aligned}
&= \sum_{i=1}^{k} [M_i^Z(f) - m_i^Z(f)](z_i - z_{i-1}) \\
&= \sum_{i\in W} [M_i^Z(f) - m_i^Z(f)](z_i - z_{i-1}) + \sum_{j=1}^{n} \sum_{i\in V_j} [M_i^Z(f) - m_i^Z(f)](z_i - z_{i-1}) \\
&\le \sum_{i\in W} 2B(z_i - z_{i-1}) + \sum_{j=1}^{n} [M_j^Q(f) - m_j^Q(f)] \sum_{i\in V_j} (z_i - z_{i-1}) \\
&\le (n-1)2B\delta + \sum_{j=1}^{n} [M_j^Q(f) - m_j^Q(f)](x_j - x_{j-1}) \\
&= (n-1)2B\frac{\varepsilon}{4Bn} + [U(f,Q) - L(f,Q)] < \frac{\varepsilon}{2} + \frac{\varepsilon}{2} = \varepsilon.
\end{aligned}
$$
□

It would have been nicer to have proved Theorem 5.4.7 by proving (a) ⇒ (b) ⇒ (c) ⇒ (a), but unfortunately there does not appear to be a direct way of going from (c) to (a).

The proof of (b) ⇒ (a) in Theorem 5.4.7 suggests yet another characterization of integrability, based upon the following definition.

Definition 5.4.8. Let $[a,b] \subseteq \mathbb{R}$ be a non-degenerate closed bounded interval, and let $f\colon [a,b] \to \mathbb{R}$ be a function. Suppose that f is bounded. The **upper integral** of f, denoted $\overline{\int_a^b} f(x)\,dx$, is defined by

$$\overline{\int_a^b} f(x)\,dx = \operatorname{glb}\{U(f,P) \mid P \text{ is a partition of } [a,b]\},$$

and the **lower integral** of f, denoted $\underline{\int_a^b} f(x)\,dx$, is defined by

$$\underline{\int_a^b} f(x)\,dx = \operatorname{lub}\{L(f,P) \mid P \text{ is a partition of } [a,b]\}.$$
△

Observe that $\overline{\int_a^b} f(x)\,dx = \operatorname{glb} \mathcal{U}$ and $\underline{\int_a^b} f(x)\,dx = \operatorname{lub} \mathcal{L}$ in the notation of the proof of (b) ⇒ (a) in Theorem 5.4.7. The following lemma is derived immediately from the arguments used in that proof, though using only Part (1) of the No Gap Lemma (Lemma 2.6.6); we omit the details.

Lemma 5.4.9. *Let $[a,b] \subseteq \mathbb{R}$ be a non-degenerate closed bounded interval, and let $f\colon [a,b] \to \mathbb{R}$ be a function. Suppose that f is bounded. Then the upper integral and lower integral of f always exist, and $\underline{\int_a^b} f(x)\,dx \le \overline{\int_a^b} f(x)\,dx$.*

We are now ready to state our additional characterization of integrability.

Theorem 5.4.10. *Let $[a,b] \subseteq \mathbb{R}$ be a non-degenerate closed bounded interval, and let $f\colon [a,b] \to \mathbb{R}$ be a function. Suppose that f is bounded. Then f is integrable if and only if $\underline{\int_a^b} f(x)\,dx = \overline{\int_a^b} f(x)\,dx$, and if this equality holds then $\int_a^b f(x)\,dx = \underline{\int_a^b} f(x)\,dx = \overline{\int_a^b} f(x)\,dx$.*

Proof. Left to the reader in Exercise 5.4.13. □

Some real analysis texts define the integrability of a function f by saying that f is integrable if $\underline{\int_a^b} f(x)\,dx = \overline{\int_a^b} f(x)\,dx$, and then prove as a theorem what we take as the definition of integrability. The advantage of that alternative approach is that the proofs of some theorems about integrals can be reached rapidly, though at the price of a definition of integrability that is less intuitively meaningful, and less familiar to students who have taken calculus.

We conclude this section with the following important result, which is a nice application of Theorem 5.4.7. In contrast to differentiable functions, which are very well behaved (for example, we saw in Theorem 4.2.4 that all differentiable functions are continuous), integrable functions can be rather strange (for example, we saw in Example 5.2.6 (2) that integrable functions need not be continuous). However, as we now see, all continuous functions on non-degenerate closed bounded intervals are integrable. The reason we need to use Theorem 5.4.7 in the proof of the following theorem is that for an arbitrary continuous function there is no way to guess the value of the integral, and so we have no candidate for the number K in the original definition of integrals.

Theorem 5.4.11. *Let $[a,b] \subseteq \mathbb{R}$ be a non-degenerate closed bounded interval, and let $f\colon [a,b] \to \mathbb{R}$ be a function. If f is continuous, then f is integrable.*

Proof. Suppose that f is continuous. Let $\varepsilon > 0$. By Theorem 3.4.4 we see that f is uniformly continuous. Hence there is some $\delta > 0$ such that $x, y \in [a,b]$ and $|x-y| < \delta$ imply $|f(x) - f(y)| < \frac{\varepsilon}{b-a}$.

Let $P = \{x_0, x_1, \ldots, x_n\}$ be a partition of $[a,b]$ with $\|P\| < \delta$. Let $i \in \{1, \ldots, n\}$. By Exercise 3.3.2 (2) the function $f|_{[x_{i-1},x_i]}$ is continuous. The Extreme Value Theorem (Theorem 3.5.1) applied to $f|_{[x_{i-1},x_i]}$ implies that there are $x^i_{max}, x^i_{min} \in [x_{i-1}, x_i]$ such that $f(x^i_{min}) \le f(x) \le f(x^i_{max})$ for all $x \in [x_{i-1}, x_i]$. By Exercise 2.6.2 we see that $M_i(f) = f(x^i_{max})$ and $m_i(f) = f(x^i_{min})$. Because $\|P\| < \delta$ we know that $|x_i - x_{i-1}| < \delta$, and hence $|x^i_{max} - x^i_{min}| < \delta$. Therefore $|f(x^i_{max}) - f(x^i_{min})| < \frac{\varepsilon}{b-a}$, and it follows that $M_i(f) - m_i(f) < \frac{\varepsilon}{b-a}$. Then

$$\begin{aligned} U(f,P) - L(f,P) &= \sum_{i=1}^{n} [M_i(f) - m_i(f)](x_i - x_{i-1}) \\ &< \frac{\varepsilon}{b-a} \sum_{i=1}^{n} (x_i - x_{i-1}) = \frac{\varepsilon}{b-a}(b-a) = \varepsilon. \end{aligned}$$

By Theorem 5.4.7 (b) we conclude that f is integrable. □

It follows from Theorem 5.4.11 that many familiar functions, such as polynomials, exponentials, logarithms, sine and cosine, are all integrable on any non-degenerate closed bounded interval. A more detailed discussion of which functions are integrable is given in Section 5.8.

As with the other main theorems of real analysis, Theorem 5.4.11 ultimately relies upon the Least Upper Bound Property. Observe that this theorem would not be true for continuous functions of the form $f\colon [a,b]\cap\mathbb{Q}\to\mathbb{R}$. For example, the function $f\colon [0,2]\cap\mathbb{Q}\to\mathbb{Q}$ be defined by $f(x)=\frac{1}{x^2-2}$ for all $x\in[0,2]\cap\mathbb{Q}$ is continuous but not bounded, and hence not integrable. The reader is encouraged to locate exactly where the Least Upper Bound Property is used in our proof of Theorem 5.4.11 (it is used in more than one place).

Reflections

The transition from the previous section to this one is akin to the transition from a pleasant stroll to a hike up a steep hill; whereas the previous section had relatively straightforward theorems and proofs about integrals, this section has technical material the use of which will be apparent only in the subsequent section, and has a lengthier proof than anything we have previously encountered concerning integrals. In contrast to derivatives, where the proofs of the basic properties are not particularly difficult, there appears to be no way to prove some of the basic (and intuitive) properties of integrals without first going through some tricky technicalities. In a calculus course, where the properties of integrals are not given rigorous proofs, this tricky material can be glossed over, but not here. The nature of mathematics is such that, if it is done properly, we are required to be satisfied, at times, with delayed gratification.

Exercises

Exercise 5.4.1. [Used in Lemma 5.4.3.] Prove Lemma 5.4.3.

Exercise 5.4.2. Find the upper sum and lower sum for each of the following functions with respect to the given partition.

(1) Let $f\colon [1,3]\to\mathbb{R}$ be defined by $f(x)=\frac{1}{x}$ for all $x\in[1,3]$, and let $P=\{1,1.4,1.8,2.2,2.6,3\}$.
(2) Let s be the function given in Example 5.2.6 (4), and let $Q=\{0,\frac{1}{5},\frac{2}{5},\frac{3}{5},\frac{4}{5},1\}$.

Exercise 5.4.3. [Used in Theorem 5.5.6 and Theorem 5.5.7.] Let $[a,b]\subseteq\mathbb{R}$ be a non-degenerate closed bounded interval, let $c\in(a,b)$, let $f\colon [a,b]\to\mathbb{R}$ be a function, let P be a partition of $[a,c]$ and let Q be a partition of $[c,b]$. Then $P\cup Q$ is a partition of $[a,b]$. Prove that

$$U(f,P\cup Q)=U(f|_{[a,c]},P)+U(f|_{[c,b]},Q)$$

and

$$L(f,P\cup Q)=L(f|_{[a,c]},P)+L(f|_{[c,b]},Q).$$

Exercise 5.4.4.

(1) Let $[a,b]\subseteq\mathbb{R}$ be a non-degenerate closed bounded interval, let $f\colon [a,b]\to\mathbb{R}$ be a function and let P be a partition of $[a,b]$. Suppose that f is continuous. Prove that there are representative sets S and T of P such that $S(f,P,S)=L(f,P)$ and $S(f,P,T)=U(f,P)$.

(2) Find an example of a bounded function $g\colon [-1,1] \to \mathbb{R}$, and a partition Q of $[-1,1]$, such that $U(g,Q)$ is not equal to any Riemann sum of g with respect to Q.

Exercise 5.4.5. Let $[a,b] \subseteq \mathbb{R}$ be a non-degenerate closed bounded interval, let $f\colon [a,b] \to \mathbb{R}$ be a function and let P be a partition of $[a,b]$. Suppose that there are $m, M \in \mathbb{R}$ such that $m \le f(x) \le M$ for all $x \in [a,b]$. Prove that $m(b-a) \le L(f,P) \le U(f,P) \le M(b-a)$.

Exercise 5.4.6. **[Used in Theorem 5.8.5.]** Let $[a,b] \subseteq \mathbb{R}$ be a non-degenerate closed bounded interval, let $f,g\colon [a,b] \to \mathbb{R}$ be functions and let P be a partition of $[a,b]$. Suppose that f and g are bounded, and that $f(x) \le g(x)$ for all $x \in [a,b]$. Prove that $L(f,P) \le L(g,P)$ and $U(f,P) \le U(g,P)$.

Exercise 5.4.7. **[Used in Exercise 5.4.8.]** Let $[a,b] \subseteq \mathbb{R}$ be a non-degenerate closed bounded interval, let $f\colon [a,b] \to \mathbb{R}$ be a function and let P be a partition of $[a,b]$. Suppose that f is bounded. Prove that

$$U(f,P) = \mathrm{lub}\{S(f,P,T) \mid T \text{ is a representative set of } P\},$$

and

$$L(f,P) = \mathrm{glb}\{S(f,P,T) \mid T \text{ is a representative set of } P\}.$$

[Use Exercise 2.6.6.]

Exercise 5.4.8. Let $[a,b] \subseteq \mathbb{R}$ be a non-degenerate closed bounded interval, and let $f\colon [a,b] \to \mathbb{R}$ be a function. Prove that f is integrable if and only if for each $\varepsilon > 0$, there is some $\delta > 0$ such that if P and Q are partitions of $[a,b]$ with $\|P\| < \delta$ and $\|Q\| < \delta$, and if T is a representative set of P, and V is a representative set of Q, then $|S(f,P,T) - S(f,Q,V)| < \varepsilon$. [Use Exercise 2.6.10, Exercise 5.3.5 and Exercise 5.4.7.]

Exercise 5.4.9. **[Used in Theorem 5.4.7, Theorem 5.5.1, Theorem 5.8.5 and Lemma 6.4.9.]** Let $[a,b] \subseteq \mathbb{R}$ be a non-degenerate closed bounded interval, let $f\colon [a,b] \to \mathbb{R}$ be a function and let $P = \{x_0, x_1, \ldots, x_n\}$ be a partition of $[a,b]$. Suppose that f is bounded. Let $i \in \{1, \ldots, n\}$.

(1) Let $y, z \in [x_{i-1}, x_i]$. Prove that $|f(y) - f(z)| \le M_i(f) - m_i(f)$.
(2) Prove that

$$M_i(f) - m_i(f) = \mathrm{lub}\{|f(y) - f(z)| \mid y, z \in [x_{i-1}, x_i]\}.$$

(3) Suppose that there is some $P \in \mathbb{R}$ such that $|f(x) - f(y)| \le P$ for all $x, y \in [x_{i-1}, x_i]$. Prove that $M_i(f) - m_i(f) \le P$.
(4) Suppose that there is some $M \in \mathbb{R}$ such that $|f(x)| \le M$ for all $x \in [x_{i-1}, x_i]$. Prove that $M_i(f) - m_i(f) \le 2M$.

Exercise 5.4.10. **[Used in Exercise 5.4.11 and Theorem 5.5.7.]** Let $[a,b] \subseteq \mathbb{R}$ be a non-degenerate closed bounded interval, let $f\colon [a,b] \to \mathbb{R}$ be a function and let P be a partition of $[a,b]$. Suppose that f integrable. Prove that

$$L(f,P) \le \int_a^b f(x)\,dx \le U(f,P).$$

Exercise 5.4.11. **[Used in Theorem 5.8.5.]** Let $[a,b] \subseteq \mathbb{R}$ be a non-degenerate closed bounded interval, and let $f,g\colon [a,b] \to \mathbb{R}$ be functions. Suppose that f integrable, that g is bounded, and that for each $\varepsilon > 0$, there is a partition P of $[a,b]$ such that $U(f,P) - L(f,P) < \varepsilon$ and $L(f,P) \leq L(g,P) \leq U(g,P) \leq U(f,P)$.

(1) Prove that g is integrable.
(2) Prove that $\int_a^b g(x)\,dx = \int_a^b f(x)\,dx$. [Use Exercise 5.4.10.]

Exercise 5.4.12. **[Used in Exercise 5.5.3 and Theorem 9.3.6.]** Let $[a,b] \subseteq \mathbb{R}$ be a non-degenerate closed bounded interval, and let $f\colon [a,b] \to \mathbb{R}$ be a function. Prove that if f is monotone, then f is integrable.

Exercise 5.4.13. **[Used in Theorem 5.4.10.]** Prove Theorem 5.4.10. You may cite parts of the proof of Theorem 5.4.7 without repeating them.

Exercise 5.4.14. Find the upper integral and lower integral for each of the following functions.

(1) Let g be the function given in Example 5.2.6 (2).
(2) Let r be the function given in Example 5.2.6 (3).

Exercise 5.4.15. Let $[a,b] \subseteq \mathbb{R}$ be a non-degenerate closed bounded interval, let $k \in \mathbb{R}$ and let $f\colon [a,b] \to \mathbb{R}$ be a function. Suppose that f is bounded.

(1) Prove that if $k > 0$, then $\overline{\int_a^b}[kf](x)\,dx = k\overline{\int_a^b} f(x)\,dx$.
(2) Prove that if $k < 0$, then $\overline{\int_a^b}[kf](x)\,dx = k\underline{\int_a^b} f(x)\,dx$.

[Use Exercise 3.2.15.]

Exercise 5.4.16. Let $[a,b] \subseteq \mathbb{R}$ be a non-degenerate closed bounded interval, and let $f,g\colon [a,b] \to \mathbb{R}$ be functions. Suppose that f and g are bounded.

(1) Prove that $\overline{\int_a^b}[f+g](x)\,dx \leq \overline{\int_a^b} f(x)\,dx + \overline{\int_a^b} g(x)\,dx$, and give an example where the inequality is strict.
(2) Prove that $\underline{\int_a^b}[f+g](x)\,dx \geq \underline{\int_a^b} f(x)\,dx + \underline{\int_a^b} g(x)\,dx$, and give an example where the inequality is strict.

[Use Exercise 2.6.9 and Exercise 3.2.16.]

5.5 Further Properties of the Riemann Integral

We now discuss some additional properties of the Riemann integral, the proofs of which rely upon Theorem 5.4.7. The first property involves the integrability of the composition of functions. Recall from Example 5.2.6 (5) that the composition of integrable functions need not be integrable. The following theorem, which will be useful to us shortly, uses uniform continuity to circumvent the strange behavior seen in that example.

In this theorem, we have functions $f\colon [a,b] \to \mathbb{R}$ and $g\colon D \to \mathbb{R}$, where $f([a,b]) \subseteq D$, and we wish to form the composition $g \circ f$. Because the codomain of f is not equal to the domain of g, then technically we would first need to change f into a function $[a,b] \to D$ before forming the composition, but to avoid cumbersome writing we abuse notation and simply write $g \circ f$, which should cause no confusion.

Theorem 5.5.1. *Let $[a,b] \subseteq \mathbb{R}$ be a non-degenerate closed bounded interval, let $D \subseteq \mathbb{R}$ be a set and let $f\colon [a,b] \to \mathbb{R}$ and $g\colon D \to \mathbb{R}$ be functions. Suppose that f is integrable, and that $f([a,b]) \subseteq D$.*

1. *If g is uniformly continuous and bounded, then $g \circ f$ is integrable.*
2. *If D is a non-degenerate closed bounded interval and g is continuous, then $g \circ f$ is integrable.*

Proof.

(1) Suppose that g is uniformly continuous and bounded. We will show that $g \circ f$ is integrable by showing that it satisfies the criterion given in Theorem 5.4.7 (c).

Let $\varepsilon > 0$. Because g is bounded, there is some $N \in \mathbb{R}$ such that $|g(x)| \le N$ for all $x \in D$. Observe that $N \ge 0$. Let $\eta = \frac{\varepsilon}{2(b-a+2N)}$. Then $\eta > 0$. Because g is uniformly continuous, there is some $\delta > 0$ such that $x,y \in D$ and $|x-y| < \delta$ imply $|g(x) - g(y)| < \eta$. By taking a smaller value of δ if necessary, we may suppose that $\delta < \eta$. Because f is integrable, we know by Theorem 5.4.7 (c) that there is some partition $P = \{x_0, x_1, \ldots, x_n\}$ of $[a,b]$ such that $U(f,P) - L(f,P) < \delta^2$.

Let

$$W = \{i \in \{1,\ldots,n\} \mid M_i(f) - m_i(f) < \delta\}$$

and

$$V = \{i \in \{1,\ldots,n\} \mid M_i(f) - m_i(f) \ge \delta\}.$$

Then $W \cup V = \{1,\ldots,n\}$ and $W \cap V = \emptyset$.

Let $j \in W$. If $y,z \in [x_{j-1}, x_j]$, then by Exercise 2.6.10 (2) we see that $|f(y) - f(z)| \le M_j(f) - m_j(f) < \delta$, and therefore $|(g \circ f)(y) - (g \circ f)(z)| = |g(f(y)) - g(f(z))| < \eta$. By Exercise 5.4.9 (3) we deduce that $M_j(g \circ f) - m_j(g \circ f) \le \eta$. Then

$$\sum_{i \in W} [M_i(g \circ f) - m_i(g \circ f)](x_i - x_{i-1}) \le \eta \sum_{i \in W} (x_i - x_{i-1}) \le \eta(b-a).$$

Let $k \in V$. Then $M_k(f) - m_k(f) \ge \delta$, and so $\frac{M_k(f) - m_k(f)}{\delta} \ge 1$. Because $|g(x)| \le N$ for all $x \in D$, we see that if $s,t \in D$, then $|g(s) - g(t)| \le |g(s)| + |g(t)| \le 2N$. It follows that if $y,z \in [x_{k-1}, x_k]$, then $|(g \circ f)(y) - (g \circ f)(z)| = |g(f(y)) - g(f(z))| \le 2N$. Hence, using Exercise 5.4.9 (3) again, we deduce that $M_k(g \circ f) - m_k(g \circ f) \le 2N$. Then

$$\begin{aligned}
\sum_{i \in V} [M_i(g \circ f) - m_i(g \circ f)](x_i - x_{i-1}) &\le 2N \sum_{i \in V} (x_i - x_{i-1}) \\
&\le 2N \sum_{i \in V} \frac{M_i(f) - m_i(f)}{\delta} (x_i - x_{i-1}) \\
&\le \frac{2N}{\delta} \sum_{i=1}^{n} [M_i(f) - m_i(f)](x_i - x_{i-1}) \\
&= \frac{2N}{\delta} [U(f,P) - L(f,P)] < \frac{2N}{\delta} \delta^2 = 2N\delta < 2N\eta.
\end{aligned}$$

Putting the above calculations together we see that

$$\begin{aligned}
U(g\circ f,P)-L(g\circ f,P) &= \sum_{i=1}^{n}[M_i(g\circ f)-m_i(g\circ f)](x_i-x_{i-1}) \\
&= \sum_{i\in W}[M_i(g\circ f)-m_i(g\circ f)](x_i-x_{i-1}) \\
&\quad + \sum_{i\in V}[M_i(g\circ f)-m_i(g\circ f)](x_i-x_{i-1}) \\
&\leq \eta(b-a)+2N\eta = \eta(b-a+2N) = \frac{\varepsilon}{2} < \varepsilon.
\end{aligned}$$

Hence $g\circ f$ satisfies the criterion given in Theorem 5.4.7 (c).

(2) This part of the theorem follows immediately from Part (1) of this theorem together with Theorem 3.4.4 and Corollary 3.4.6. □

We now turn to the integrability of the product and quotient of integrable functions. There is no problem with products of integrable functions, as we will see below, but the situation is slightly trickier for quotients. We saw in Theorem 3.3.5 (5) that the quotient of continuous functions is continuous as long as the denominator is not zero, and we saw the analogous result for differentiability in Theorem 4.3.1 (5) (the Quotient Rule). Unfortunately, as we see in the following example, just knowing that the denominator is not zero is not sufficient to guarantee that the quotient of integrable functions is integrable.

Example 5.5.2. Let $f,g\colon [0,1]\to\mathbb{R}$ be defined by $f(x)=1$ for all $x\in[0,1]$, and

$$g(x)=\begin{cases}1, & \text{if } x=0\\ x, & \text{if } x\in(0,1].\end{cases}$$

Then

$$\left[\frac{f}{g}\right](x)=\begin{cases}1, & \text{if } x=0\\ \frac{1}{x}, & \text{if } x\in(0,1].\end{cases}$$

We know by Example 5.2.6 (1) that f is integrable. The function g is also integrable, as can be seen by combining Exercise 5.2.6 and Exercise 5.3.3 (3). However, even though $g(x)\neq 0$ for all $x\in[0,1]$, the function $\frac{f}{g}$ is not integrable, because integrable functions are bounded by Theorem 5.3.3, and yet $\frac{f}{g}$ is not bounded, a fact that is evident by looking at the graph of $\frac{f}{g}$, and is proved in Example 3.2.6. ◊

The following definition allows us to avoid the problem seen in Example 5.5.2.

Definition 5.5.3. Let $A\subseteq\mathbb{R}$ be a set, and let $f\colon A\to\mathbb{R}$ be a function. The function f is **bounded away from zero** if there is some $P>0$ such that $|f(x)|\geq P$ for all $x\in A$. △

If a function $f\colon A\to\mathbb{R}$ is bounded away from zero, then clearly $f(x)\neq 0$ for all $x\in A$.

Theorem 5.5.4. *Let $[a,b] \subseteq \mathbb{R}$ be a non-degenerate closed bounded interval, and let $f,g\colon [a,b] \to \mathbb{R}$ be functions. Suppose that f and g are integrable.*

1. *f^n is integrable for all $n \in \mathbb{N}$.*
2. *fg is integrable.*
3. *If g is bounded away from zero, then $\frac{f}{g}$ is integrable.*

Proof. Because f and g are integrable, then by Theorem 5.3.3 we know that f and g are bounded. Hence there are $M_1, M_2 \in \mathbb{R}$ such that $|f(x)| \leq M_1$ and $|g(x)| \leq M_2$ for all $x \in [a,b]$. We may suppose that $M_1 > 0$ and $M_2 > 0$. Let $M = \max\{M_1, M_2\}$. Then $f([a,b]) \subseteq [-M,M]$ and $g([a,b]) \subseteq [-M,M]$.

(1) Let $n \in \mathbb{N}$. Let $h\colon [-M,M] \to \mathbb{R}$ be defined by $h(x) = x^n$ for all $x \in [-M,M]$. Then h is continuous by Example 3.3.7 (1). It now follows from Theorem 5.5.1 (2) that $f^n = h \circ f$ is integrable.

(2) Because f and g are integrable, then $f+g$ is integrable by Theorem 5.3.1 (1). By Part (1) of this theorem we know that f^2, and g^2 and $(f+g)^2$ are all integrable. Observe that

$$fg = \frac{1}{2}\left[(f+g)^2 - f^2 - g^2\right].$$

It follows from Theorem 5.3.1 (2) (3) that fg is integrable.

(3) Suppose that g is bounded away from zero. Hence there is some $P > 0$ such that $|g(x)| \geq P$ for all $x \in [a,b]$. It follows that $g([a,b]) \subseteq [-M,-P] \cup [M,P]$. Let $k\colon [-M,-P] \cup [M,P] \to \mathbb{R}$ be defined by $k(x) = \frac{1}{x}$ for all $x \in [-M,-P] \cup [M,P]$. Then k is continuous by Example 3.3.3 (2). It follows from Exercise 3.4.6 that k is uniformly continuous and bounded. Using Theorem 5.5.1 (1) we see that $\frac{1}{g} = k \circ g$ is integrable. We then use Part (2) of this theorem to deduce that $\frac{f}{g} = f \cdot \frac{1}{g}$ is integrable. □

Our next result concerns the integrability of the absolute value of a function.

Theorem 5.5.5. *Let $[a,b] \subseteq \mathbb{R}$ be a non-degenerate closed bounded interval, and let $f\colon [a,b] \to \mathbb{R}$ be a function. If f is integrable, then $|f|$ is integrable and*

$$\left| \int_a^b f(x)\, dx \right| \leq \int_a^b |f(x)|\, dx.$$

Proof. Suppose that f is integrable. By Theorem 5.3.3 we know that f is bounded. Hence there is some $M \in \mathbb{R}$ such that $|f(x)| \leq M$. We may assume that $M > 0$. Hence $f([a,b]) \subseteq [-M,M]$. Let $h\colon [-M,M] \to \mathbb{R}$ be defined by $h(x) = |x|$ for all $x \in [-M,M]$. We know by Exercise 3.3.1 (2) that h is continuous. It follows from Theorem 5.5.1 (2) that $|f| = h \circ f$ is integrable.

Observe that $-|f(x)| \leq f(x) \leq |f(x)|$ for all $x \in [a,b]$. It now follows from Theorem 5.3.1 (3) and Theorem 5.3.2 (2) that

$$-\int_a^b |f(x)|\, dx \leq \int_a^b f(x)\, dx \leq \int_a^b |f(x)|\, dx.$$

Because $|f(x)| \geq 0$ for all $x \in [a,b]$, then Theorem 5.3.2 (1) implies that $\int_a^b |f(x)|\,dx \geq 0$. Hence

$$\left| \int_a^b f(x)\,dx \right| \leq \int_a^b |f(x)|\,dx. \qquad \square$$

Theorem 5.5.5 can be viewed as an analog for integrals of the Triangle Inequality (Lemma 2.3.9 (6)) and its extension to finite sums in Exercise 2.5.3, where instead of the sum of finitely many numbers we have an integral, which for the sake of the analogy can be thought of intuitively as the infinite sum of all the values of the function f (though it is not really such a sum). As with the Triangle Inequality, we cannot in general replace the inequality in Theorem 5.5.5 with equality because if f takes on both positive and negative values, then there will be cancellation in $\int_a^b f(x)\,dx$, whereas there will be no cancellation in $\int_a^b |f(x)|\,dx$.

We now turn to the restriction of an integrable function to a subinterval of its domain. It might seem obvious that such a restriction of an integrable function is integrable, but the proof is non-trivial, making use of Theorem 5.4.7. What makes the proof less straightforward than might be expected is the fact that if one starts with a partition P of an interval $[a,b]$ and a representative set T of P, and if $[c,d] \subseteq [a,b]$, the numbers c and d might not be in P, and the numbers in T might not be in $[c,d]$.

Theorem 5.5.6. *Let $D \subseteq C \subseteq \mathbb{R}$ be non-degenerate closed bounded intervals, and let $f\colon C \to \mathbb{R}$ be a function. If f is integrable, then $f|_D$ is integrable.*

Proof. Suppose that f is integrable. Let $C = [a,b]$ and $D = [c,d]$. If $c = a$ and $d = b$ then there is nothing to prove, so suppose that at least one of these equalities is false. Without loss of generality, assume that $a < c$. Because $c < d$ then $a < d$. We will first show that $f|_{[a,d]}$ is integrable. If $d = b$ then there is nothing to prove, so we suppose that $d < b$. Let $\varepsilon > 0$. By Theorem 5.4.7 (c) there is some partition P of $[a,b]$ such that $U(f,P) - L(f,P) < \varepsilon$. Let $Q = P \cup \{d\}$, let $R = Q \cap [a,d]$ and let $Z = Q \cap [d,b]$. Then Q is a partition of $[a,b]$ that is a refinement of P, and R is a partition of $[a,d]$, and Z is a partition of $[d,b]$, and $Q = R \cup Z$. By Exercise 5.4.3 we know that

$$U(f,Q) = U(f|_{[a,d]},R) + U(f|_{[d,b]},Z)$$

and

$$L(f,Q) = L(f|_{[a,d]},R) + L(f|_{[d,b]},Z).$$

Lemma 5.4.6 (2) implies that $L(f,P) \leq L(f,Q) \leq U(f,Q) \leq U(f,P)$. Hence $U(f,Q) - L(f,Q) < \varepsilon$. By Lemma 5.4.6 (1) we see that $U(f|_{[d,b]},Z) - L(f|_{[d,b]},Z) \geq 0$. Then

$$\begin{aligned} &U(f|_{[a,d]},R) - L(f|_{[a,d]},R) \\ &\qquad \leq [U(f|_{[a,d]},R) - L(f|_{[a,d]},R)] + [U(f|_{[d,b]},Z) - L(f|_{[d,b]},Z)] \\ &\qquad = U(f,Q) - L(f,Q) < \varepsilon. \end{aligned}$$

Therefore $f|_{[a,d]}$ satisfies the criterion in Theorem 5.4.7 (c), and hence $f|_{[a,d]}$ is integrable.

A similar argument shows that $f|_{[c,d]} = (f|_{[a,d]})|_{[c,d]}$ is integrable, and we omit the details. $\square$

Our next theorem is illustrated in Figure 5.5.1. The intuitive idea is that if we want to integrate a function on an interval $[a,b]$, we can break up the interval into two subintervals $[a,c]$ and $[c,b]$, integrate on each of the two subintervals and add these two integrals. The proof, once again, is not as simple as might be expected.

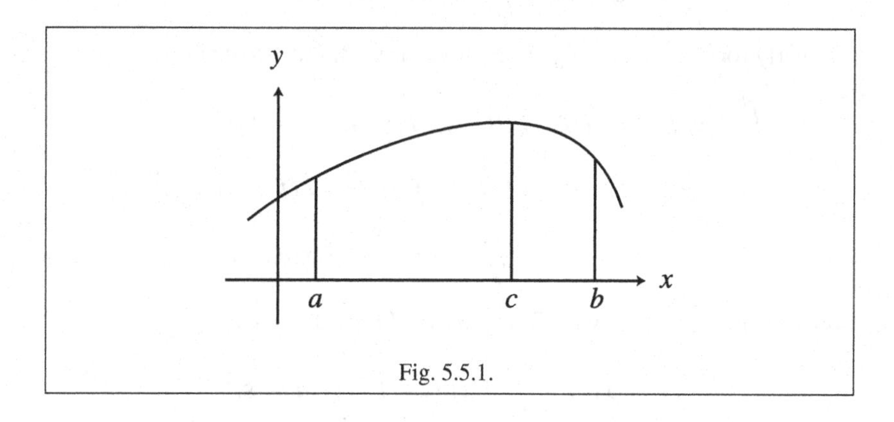

Fig. 5.5.1.

Theorem 5.5.7. *Let $[a,b] \subseteq \mathbb{R}$ be a non-degenerate closed bounded interval, let $c \in (a,b)$ and let $f\colon [a,b] \to \mathbb{R}$ be a function.*

1. *f is integrable if and only if $f|_{[a,c]}$ and $f|_{[c,b]}$ are integrable.*
2. *If f is integrable, then*

$$\int_a^b f(x)\,dx = \int_a^c f(x)\,dx + \int_c^b f(x)\,dx.$$

Proof.

(1) Suppose that f is integrable. It then follows immediately from Theorem 5.5.6 that $f|_{[a,c]}$ and $f|_{[c,b]}$ are integrable.

Now suppose that $f|_{[a,c]}$ and $f|_{[c,b]}$ are integrable. Let $\varepsilon > 0$. By applying the criterion in Theorem 5.4.7 (c) to each of $f|_{[a,c]}$ and $f|_{[c,b]}$, we know that there is a partition P_1 of $[a,c]$ such that $U(f|_{[a,c]},P_1) - L(f|_{[a,c]},P_1) < \frac{\varepsilon}{2}$ and a partition P_2 of $[c,b]$ such that $U(f|_{[c,b]},P_2) - L(f|_{[c,b]},P_2) < \frac{\varepsilon}{2}$. Let $P = P_1 \cup P_2$. Then P is a partition of $[a,b]$. Using Exercise 5.4.3 we see that

$$\begin{aligned} U(f,P) - L(f,P) &= [U(f|_{[a,c]},P_1) + U(f|_{[c,b]},P_2)] - [L(f|_{[a,c]},P_1) + L(f|_{[c,b]},P_2)] \\ &= [U(f|_{[a,c]},P_1) - L(f|_{[a,c]},P_1)] + [U(f|_{[c,b]},P_2) - L(f|_{[c,b]},P_2)] \\ &< \frac{\varepsilon}{2} + \frac{\varepsilon}{2} = \varepsilon. \end{aligned}$$

Therefore f satisfies the criterion in Theorem 5.4.7 (c), and hence f is integrable.

(2) Suppose that f is integrable. Let $\varepsilon > 0$. Let P_1, P_2 and P be as in the proof of Part (1) of this theorem. Recall that $U(f|_{[a,c]},P_1) - L(f|_{[a,c]},P_1) < \frac{\varepsilon}{2}$ and

$U(f|_{[c,b]}, P_2) - L(f|_{[c,b]}, P_2) < \frac{\varepsilon}{2}$. Therefore $U(f|_{[a,c]}, P_1) < L(f|_{[a,c]}, P_1) + \frac{\varepsilon}{2}$ and $U(f|_{[c,b]}, P_2) < L(f|_{[c,b]}, P_2) + \frac{\varepsilon}{2}$.

By Exercise 5.4.10 we know that

$$L(f,P) \leq \int_a^b f(x)\,dx \leq U(f,P),$$

and similarly for $f|_{[a,c]}$ and $f|_{[c,b]}$. Using Exercise 5.4.3 we see that

$$\begin{aligned}\int_a^b f(x)\,dx \leq U(f,P) &= U(f|_{[a,c]}, P_1) + U(f|_{[c,b]}, P_2)\\ &< L(f|_{[a,c]}, P_1) + \frac{\varepsilon}{2} + L(f|_{[c,b]}, P_2) + \frac{\varepsilon}{2}\\ &\leq \int_a^c f(x)\,dx + \int_c^b f(x)\,dx + \varepsilon.\end{aligned}$$

A similar argument starting with $\int_a^b f(x)\,dx \geq L(f,P)$ shows that

$$\int_a^b f(x)\,dx \geq \int_a^c f(x)\,dx + \int_c^b f(x)\,dx - \varepsilon.$$

Hence

$$\left[\int_a^c f(x)\,dx + \int_c^b f(x)\,dx\right] - \varepsilon < \int_a^b f(x)\,dx < \left[\int_a^c f(x)\,dx + \int_c^b f(x)\,dx\right] + \varepsilon,$$

which is equivalent to

$$\left|\int_a^b f(x)\,dx - \left[\int_a^c f(x)\,dx + \int_c^b f(x)\,dx\right]\right| < \varepsilon.$$

Because ε was arbitrarily chosen, it now follows from Lemma 2.3.10 (3) that

$$\int_a^b f(x)\,dx - \left[\int_a^c f(x)\,dx + \int_c^b f(x)\,dx\right] = 0. \qquad \square$$

In Theorem 5.5.7 it is assumed that c is between a and b, but in practice it is sometimes useful to allow c to be outside $[a,b]$ as well; we will see such a need in the proof of the Fundamental Theorem of Calculus Version I (Theorem 5.6.2). To be able to handle that situation, we first need to consider integrals of the form $\int_p^q f(x)\,dx$, where q is not greater than p. The following definition is precisely what is needed to extend Theorem 5.5.7 to the case where c is not necessarily between a and b.

Definition 5.5.8. Let $[a,b] \subseteq \mathbb{R}$ be a non-degenerate closed bounded interval, and let $f\colon [a,b] \to \mathbb{R}$ be a function. Suppose that f is integrable. Let $\int_b^a f(x)\,dx$ be defined by

$$\int_b^a f(x)\,dx = -\int_a^b f(x)\,dx,$$

and let $\int_a^a f(x)\,dx$ be defined by

$$\int_a^a f(x)\,dx = 0. \qquad \triangle$$

An intuitive way to think of the definition of $\int_a^b f(x)\,dx$ when $b < a$ versus when $a < b$ is by thinking of an integral as giving a positive value if, when you travel on the x-axis from a to b, the function is to your left.

Corollary 5.5.9. *Let $C \subseteq \mathbb{R}$ be a closed bounded interval, and let $f\colon C \to \mathbb{R}$ be a function. Let $a, b, c \in C$. If f is integrable, then*

$$\int_a^b f(x)\,dx = \int_a^c f(x)\,dx + \int_c^b f(x)\,dx.$$

Proof. Left to the reader in Exercise 5.5.2. □

Reflections

On first encounter, the properties of the Riemann integral presented in this section do not appear, from their statements, to be of a different nature than the properties presented in Section 5.3; the properties in both sections seem intuitively reasonable, and are clearly useful. The sole reason for dividing the properties of the Riemann integral into these two sections is that the proofs of the properties in this section require the technicalities about upper sums and lower sums proved in Section 5.4, specifically Theorem 5.4.7, whereas the properties in Section 5.3 can be proved using only the definition of the Riemann integral. What is disturbing intuitively is that it was not obvious ahead of time which properties of the Riemann integral would turn out to have straightforward proofs and which would rely upon a technical result such as Theorem 5.4.7. When trying to prove a new theorem, it is always worth trying the most straightforward approach first, but, unfortunately, it is only with hindsight that one can know if such an approach will work.

Exercises

Exercise 5.5.1. Use Theorem 5.5.1 (2) to give an alternative proof of Theorem 5.4.11. Make sure you do not indirectly use Theorem 5.4.11 in your proof.

[Use Exercise 5.2.6.]

Exercise 5.5.2. **[Used in Corollary 5.5.9.]** Prove Corollary 5.5.9.

Exercise 5.5.3. **[Used in Exercise 7.2.3.]** Let $[a,b] \subseteq \mathbb{R}$ be a non-degenerate closed bounded interval, and let $f\colon [a,b] \to \mathbb{R}$ be a function. Suppose that f is decreasing. Exercise 5.4.12 implies that f is integrable. Prove that if f is strictly decreasing, then $f(b)(b-a) < \int_a^b f(x)\,dx < f(a)(b-a)$.

Exercise 5.5.4. Let $a \in (0,\infty)$, and let $f\colon [-a,a] \to \mathbb{R}$ be a function. Suppose that f is integrable.

(1) Suppose that $f(-x) = f(x)$ for all $x \in [-a,a]$; such a function is called an **even** function. Prove that $\int_{-a}^a f(x)\,dx = 2\int_0^a f(x)\,dx$.

(2) Suppose that $f(-x) = -f(x)$ for all $x \in [-a,a]$; such a function is called an **odd** function. Prove that $\int_{-a}^a f(x)\,dx = 0$.

[Use Exercise 5.2.10.]

Exercise 5.5.5. [Used in Theorem 5.9.17.] Let $[a,b] \subseteq \mathbb{R}$ be a non-degenerate closed bounded interval, and let $f\colon [a,b] \to \mathbb{R}$ be a function. Suppose that f is bounded.

(1) Suppose that f is continuous except possibly at a single point in $[a,b]$. Prove that f is integrable.

(2) Suppose that f is continuous except possibly at finitely many points in $[a,b]$. Prove that f is integrable.

Exercise 5.5.6. [Used in Section 6.4 and Lemma 6.4.9.] Let $[a,b] \subseteq \mathbb{R}$ be a non-degenerate closed bounded interval, and let $f\colon [a,b] \to \mathbb{R}$ be a function. Suppose that f is integrable. Theorem 5.5.6 implies that $f|_{[a,t]}$ is integrable for each $t \in (a,b)$. Prove that $\lim_{t \to b^-} \int_a^t f(x)\,dx = \int_a^b f(x)\,dx$. [Use Exercise 5.3.2.]

Exercise 5.5.7. [Used in Section 5.8, Theorem 7.4.5 and Exercise 7.4.3.] Let $[a,b] \subseteq \mathbb{R}$ be a non-degenerate closed bounded interval, and let $f\colon [a,b] \to \mathbb{R}$ be a function. Suppose that $f(x) \geq 0$ for all $x \in [a,b]$, and that f is continuous. Prove that if $\int_a^b f(x)\,dx = 0$, then $f(x) = 0$ for all $x \in [a,b]$. Example 5.2.6 (2) shows that the requirement of continuity cannot be dropped. [Use Exercise 5.3.4.]

Exercise 5.5.8. Let $[a,b] \subseteq \mathbb{R}$ be a non-degenerate closed bounded interval, and let $f\colon [a,b] \to \mathbb{R}$ be a function. Suppose that f is continuous. Prove that if $\int_a^c f(x)\,dx = 0$ for all $c \in [a,b]$, then $f(x) = 0$ for all $x \in [a,b]$.

Exercise 5.5.9. Let $[a,b] \subseteq \mathbb{R}$ be a non-degenerate closed bounded interval, and let $f,g\colon [a,b] \to \mathbb{R}$ be functions. Suppose that f and g are integrable, and that at least one of $\int_a^b f^2(x)\,dx$ or $\int_a^b g^2(x)\,dx$ is not zero. Prove that

$$\left[\int_a^b [fg](x)\,dx\right]^2 \leq \left(\int_a^b f^2(x)\,dx\right)\left(\int_a^b g^2(x)\,dx\right).$$

(The integrals in this inequality all exist by Theorem 5.5.4.) This result is known as the Cauchy–Schwarz Inequality for Integrals. The hypothesis that at least one of $\int_a^b f^2(x)\,dx$ or $\int_a^b g^2(x)\,dx$ is not zero is not actually necessary to prove this inequality, but it makes the problem simpler.

To prove the inequality, without loss of generality assume that $\int_a^b g^2(x)\,dx \neq 0$, and observe that $(f - cg)^2$ is integrable for any $c \in \mathbb{R}$ by Theorem 5.3.1 and Theorem 5.5.4. At some point in the proof choose a useful value of c.

Exercise 5.5.10. Let $[a,b] \subseteq \mathbb{R}$ be a non-degenerate closed bounded interval, and let $f,g\colon [a,b] \to \mathbb{R}$ be functions. Suppose that f is continuous, that g is integrable and that $g(x) \geq 0$ for all $x \in [a,b]$. Theorem 5.4.11 implies that f is integrable, and Theorem 5.5.4 (2) then implies that fg is integrable.

(1) By the Extreme Value Theorem (Theorem 3.5.1) there are $x_{min}, x_{max} \in [a,b]$ such that $f(x_{min}) \leq f(x) \leq f(x_{max})$ for all $x \in [a,b]$. Prove that

$$f(x_{min}) \int_a^b g(x)\,dx \leq \int_a^b [fg](x)\,dx \leq f(x_{max}) \int_a^b g(x)\,dx.$$

(2) Prove that there is some $c \in [a,b]$ such that $\int_a^b [fg](x)\,dx = f(c)\int_a^b g(x)\,dx$. This result is known as the Generalized Mean Value Theorem for Integrals.

Exercise 5.5.11. This exercise makes use of Exercise 5.2.11. Let $[a,b] \subseteq \mathbb{R}$ be a non-degenerate closed bounded interval, and let $f,g\colon [a,b] \to \mathbb{R}$ be functions.

(1) Suppose that f is integrable, and that g is increasing and continuously differentiable. Prove that f is Riemann–Stieltjes integrable with respect to g, and that $\int_a^b f(x)\,dg = \int_a^b f(x)g'(x)\,dx$.
(2) Find an example to show that if the requirement that g is continuously differentiable in Part (1) of this exercise is changed to differentiable, it will not necessarily be the case that fg' is integrable, and therefore it will not be possible to express $\int_a^b f(x)\,dg$ in terms of $\int_a^b f(x)g'(x)\,dx$.

5.6 Fundamental Theorem of Calculus

The Fundamental Theorem of Calculus does two things at once: It shows that there is a relation between derivatives and integrals (by definition these two concepts are quite distinct—recall that by the term "integral" we mean "definite integral"), and it gives us a method for calculating integrals (doing so directly from the definition of integrals is very difficult in all but the simplest cases). If calculus is to mathematics as Shakespeare is to English literature, then the Fundamental Theorem of Calculus is Hamlet or King Lear. Mathematically, it does not get much better than this. It is probably fair to say that without the Fundamental Theorem of Calculus, integrals—one of the most important mathematical tools in science and technology—would not be very usable, and in that case much of modern science and technology would not exist.

There are actually two versions of the Fundamental Theorem of Calculus. The two versions are equivalent to each other, and the order in which they are discussed does not matter; we will follow the more customary order. Both versions essentially say that differentiation and integration are inverse operations. In the first version we do integration first and then differentiation, and in the other version we do the reverse. In our discussion we will make use of the concept of an antiderivative, as defined in Definition 4.4.8; note that antiderivatives are defined strictly in terms of derivatives. If we can establish a connection between integrals and antiderivatives, then we will have established a connection between integrals and derivatives.

How are integrals and derivatives related? As defined, these two concepts are somewhat hard to compare, because derivatives take functions and yield functions, whereas integrals take functions (and closed bounded intervals) and yield numbers. One way to compare integrals with derivatives is modify integrals to obtain functions from them. More specifically, the idea is to let the "b" in $\int_a^b f(x)\,dx$ vary. That is, we want to consider integrals of the form $\int_a^x f(t)\,dt$, where we think of x as a "variable." (Observe that we wrote $f(t)\,dt$ in the above integral rather than $f(x)\,dx$, to emphasize that the symbol x in the integral $\int_a^x f(t)\,dt$ is not the same as the "dummy variable" x in the integral $\int_a^b f(x)\,dx$.)

Example 5.6.1.

(1) Let $f\colon [0,2] \to \mathbb{R}$ be defined by $f(x) = x$ for all $x \in [0,2]$. Let $F\colon [0,2] \to \mathbb{R}$ be defined by

$$F(x) = \int_1^x f(t)\,dt$$

for all $x \in [0,2]$. We want to find an explicit formula for F; recall that we do not yet have the Fundamental Theorem of Calculus at our disposal—we are looking at this example as motivation for that theorem. Because the function f is so simple, a formula for F can be found using some basic geometry. Let $x \in [0,2]$. As seen in Figure 5.6.1, the value of $F(x)$ represents the area of the shaded trapezoid when $x > 1$; a similar argument holds when $x \leq 1$, and we omit the details. The two bases of the trapezoid have lengths 1 and x, and the height of the trapezoid is $x - 1$. Using the formula for the area of a trapezoid, it follows that

$$F(x) = \frac{1+x}{2} \cdot (x-1) = \frac{1}{2}x^2 - \frac{1}{2}.$$

It is evident that F is differentiable, and that $F' = f$. We have therefore seen a concrete example of the Fundamental Theorem of Calculus Version I (Theorem 5.6.2), which we will prove after this example.

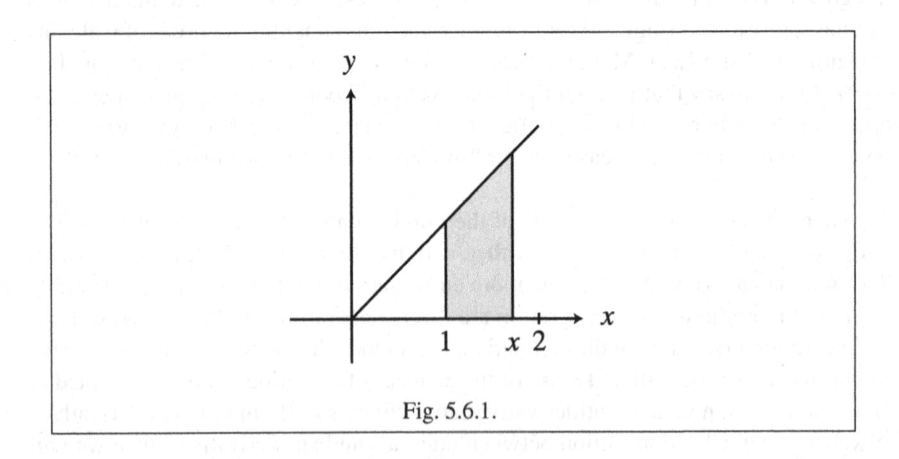

Fig. 5.6.1.

(2) Let $h\colon [0,2] \to \mathbb{R}$ be defined by

$$h(x) = \begin{cases} 1, & \text{if } x \in [0,1] \\ 2, & \text{if } x \in (1,2]. \end{cases}$$

By Exercise 5.2.5 we know that h is integrable. Let $H\colon [0,2] \to \mathbb{R}$ be defined by

$$H(x) = \int_0^x h(t)\,dt$$

for all $x \in [0,2]$. Using Figure 5.6.2, it is left to the reader to verify that

$$H(x) = \begin{cases} x, & \text{if } x \in [0,1] \\ 2x-1, & \text{if } x \in [1,2], \end{cases}$$

and that H is not differentiable at $x = 1$.

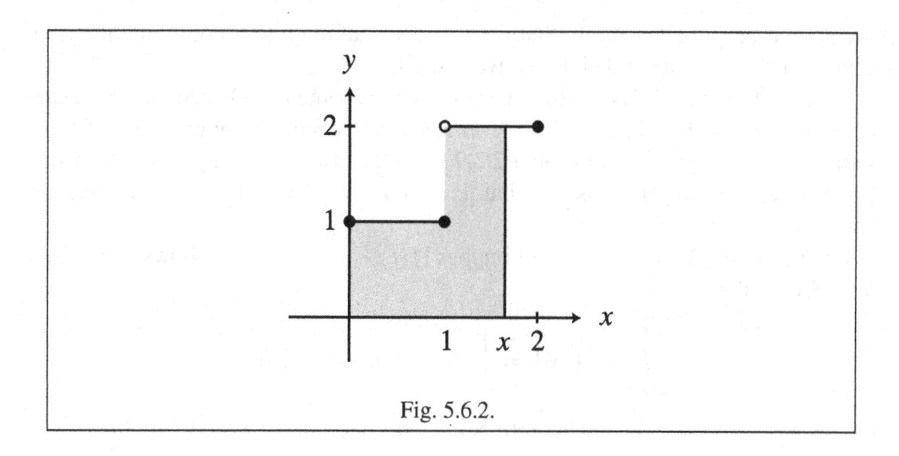

Fig. 5.6.2.

Hence, it is not always the case that a function of the form $\int_a^x h(t)\,dt$ is differentiable, and, of course, when $\int_a^x h(t)\,dt$ is not differentiable, then it makes no sense to assert that its derivative is h. In fact, combining Example 4.4.11 with Exercise 4.2.3, we see that the function h is not the derivative of any function. ◊

As we saw in the two parts of Example 5.6.1, sometimes functions of the form $\int_a^x f(t)\,dt$ are differentiable and sometimes they are not. As we now see in the first version of the Fundamental Theorem of Calculus, if a function $f\colon [a,b] \to \mathbb{R}$ is continuous then $\int_a^x f(t)\,dt$ is differentiable, and it is an antiderivative of f. We note, however, that continuity is not necessary for $\int_a^x f(t)\,dt$ to be differentiable; see Exercise 5.6.2. In general, a function of the form $\int_a^x f(t)\,dt$ is better behaved than the function f, even when f is not continuous; it is seen in Exercise 5.6.5 that if f is integrable, then $\int_a^x f(t)\,dt$ is uniformly continuous.

Theorem 5.6.2 (Fundamental Theorem of Calculus Version I). *Let $I \subseteq \mathbb{R}$ be a non-degenerate interval, let $a \in I$ and let $f\colon I \to \mathbb{R}$ be a function. Suppose that $f|_C$ is integrable for every non-degenerate closed bounded interval $C \subseteq I$. Let $F\colon I \to \mathbb{R}$ be defined by*

$$F(x) = \int_a^x f(t)\,dt$$

for all $x \in I$. Let $c \in I$. If f is continuous at c, then F is differentiable at c and $F'(c) = f(c)$. If f is continuous, then F is differentiable and $F' = f$.

Proof. Suppose that f is continuous at c. Suppose further that c is not a right endpoint of I. We will show that

$$\lim_{h \to 0^+} \frac{F(c+h) - F(c)}{h} = f(c). \tag{5.6.1}$$

A similar argument, the details of which we omit, shows that if c is not a left endpoint of I, then

$$\lim_{h \to 0^-} \frac{F(c+h) - F(c)}{h} = f(c).$$

These two cases together imply that $F'(c) = f(c)$, taking into account that if c is an endpoint of I then a one-sided derivative is used.

Let $\varepsilon > 0$. Because f is continuous at c, there is some $\delta > 0$ such that $w \in I$ and $|w - c| < \delta$ imply that $|f(w) - f(c)| < \frac{\varepsilon}{2}$. Because c is not a right endpoint of I, we use the one-sided analog of Lemma 2.3.7 (2) to see that by taking a smaller value of δ if necessary, we may assume that $[c, c+\delta) \subseteq I$. Hence $w \in [c, c+\delta)$ implies $|f(w) - f(c)| < \frac{\varepsilon}{2}$.

Let $h \in (0, \delta)$. Then $t \in [c, c+h]$ implies $|f(t) - f(c)| < \frac{\varepsilon}{2}$. It follows from Theorem 5.3.1 (4) that

$$\frac{1}{h} \int_c^{c+h} f(c)\, dt = \frac{1}{h} f(c)(c+h-c) = f(c).$$

We now use Corollary 5.5.9, Theorem 5.5.5 and Theorem 5.3.2 (3) to see that

$$\begin{aligned}
\left| \frac{F(c+h) - F(c)}{h} - f(c) \right| &= \left| \left[\frac{1}{h} \int_a^{c+h} f(t)\, dt - \frac{1}{h} \int_a^c f(t)\, dt \right] - f(c) \right| \\
&= \left| \frac{1}{h} \int_c^{c+h} f(t)\, dt - \frac{1}{h} \int_c^{c+h} f(c)\, dt \right| \\
&= \left| \frac{1}{h} \int_c^{c+h} [f(t) - f(c)]\, dt \right| \leq \frac{1}{h} \int_c^{c+h} |f(t) - f(c)|\, dt \\
&\leq \frac{1}{h} \frac{\varepsilon}{2} (c+h-c) = \frac{\varepsilon}{2} < \varepsilon.
\end{aligned}$$

It then follows from the definition of one-sided limits that Equation 5.6.1 holds. □

Observe that in the statement of the Fundamental Theorem of Calculus Version I (Theorem 5.6.2), if the interval I is closed and bounded, then it is sufficient to assume that f is integrable, because of Theorem 5.5.6.

The Fundamental Theorem of Calculus Version I, together with Exercise 3.3.2 (2) and Theorem 5.4.11, immediately imply the following important fact.

Corollary 5.6.3. *Let $I \subseteq \mathbb{R}$ be a non-degenerate interval, and let $f \colon I \to \mathbb{R}$ be a function. If f is continuous, then f has an antiderivative.*

Whereas Corollary 5.6.3 says that in principle every continuous function has an antiderivative, it says nothing about how to compute such antiderivatives in practice. Indeed, as the reader has seen in calculus courses, computing antiderivatives can be quite tricky. In contrast to differentiation, where we have a short list of rules that allow us to take the derivative of virtually any function that can be built up out of elementary functions using sums, differences, products, quotients and compositions, there is no simple list of techniques for finding antiderivatives of all such functions. Some antiderivatives are very difficult to compute, requiring ad hoc techniques, and some cannot be expressed at all as nice formulas made up out of elementary functions. For example, let $f\colon \mathbb{R} \to \mathbb{R}$ be defined by $f(x) = e^{-x^2}$ for all $x \in \mathbb{R}$. The Fundamental Theorem of Calculus Version I implies that the function $g\colon \mathbb{R} \to \mathbb{R}$ defined by $g(x) = \int_0^x e^{-t^2}\,dt$ for all $x \in \mathbb{R}$ is an antiderivative for f, but this antiderivative is not very useful in practice, and in fact there is no simple formula for this antiderivative, which is a great pity, because this particular antiderivative is useful in a variety of application of mathematics, for example probability and statistics. See [Kas80] or [MZ94] for general discussion and historical remarks about which antiderivatives can be written "in finite terms" (which means as an appropriate type of formula involving elementary functions), and see [Ros72] for a precise statement and detailed proof of such a result (the proof involves ideas from both complex analysis and abstract algebra). We will return to the question of finding antiderivatives later in this section, but for now we continue with our discussion of the Fundamental Theorem of Calculus.

In Version I of the Fundamental Theorem of Calculus we first integrated a function (albeit with the "variable" x instead of b), and then differentiated the result of the integration, and we obtained the function with which we started. In Version II of the Fundamental Theorem of Calculus we reverse the order of differentiation and integration. The basic idea is to start with a function $f\colon [a,b] \to \mathbb{R}$, take its derivative, and then integrate the derivative, which yields $\int_a^b f'(x)\,dx$. We cannot directly compare this integral with the original function, because the integral is a number, and not a function, but this problem can be avoided, as suggested by the following intuitive idea.

Suppose that $s\colon [a,b] \to \mathbb{R}$ represents the position of an object on the x-axis as a function of time. We want to compute the average velocity of the object from time $t = a$ to time $t = b$. One way to calculate average velocity is to divide the total distance by the total time, which yields $\frac{s(b)-s(a)}{b-a}$. On the other hand, there is a general way to calculate the average value of an integrable function $g\colon [a,b] \to \mathbb{R}$, which is $\frac{1}{b-a}\int_a^b g(x)\,dx$. We assume that the reader is informally familiar with this formula for average value from a calculus course; it is discussed in detail in Example 8.4.6. Hence, we can calculate the average velocity of our position function as $\frac{1}{b-a}\int_a^b s'(t)\,dt$, assuming that s' is integrable. If the world is as nice as one would hope, then these two methods of computing average velocity ought to give the same result, which means that $\frac{1}{b-a}\int_a^b s'(t)\,dt = \frac{s(b)-s(a)}{b-a}$, and hence that $\int_a^b s'(t)\,dt = s(b) - s(a)$. What we see is that the integral $\int_a^b s'(t)\,dt$, which is a number, is related to the function s by looking at the values of s at the endpoints $[a,b]$, and that is how we solve the problem of comparing numbers and functions in Version II of the Fundamental Theorem of Calculus. Indeed, the formula $\int_a^b s'(t)\,dt = s(b) - s(a)$ is the same as the Fundamental Theorem of Calculus Version II, except that we need to rename s' by f, and then s is an antiderivative of f, which we call F. Of course, this intuitive argument is not a proof, but it lends plausibility.

Theorem 5.6.4 (Fundamental Theorem of Calculus Version II). *Let $[a,b] \subseteq \mathbb{R}$ be a non-degenerate closed bounded interval, and let $f\colon [a,b] \to \mathbb{R}$ be a function. Suppose that f is integrable and f has an antiderivative. If $F\colon [a,b] \to \mathbb{R}$ is an antiderivative of f, then*

$$\int_a^b f(x)\,dx = F(b) - F(a).$$

Proof. Let $F\colon [a,b] \to \mathbb{R}$ be an antiderivative of f. Let $\varepsilon > 0$. Because f is integrable, there is some $\delta > 0$ such that if P is a partition of $[a,b]$ with $\|P\| < \delta$, and if T is a representative set of P, then $|S(f,P,T) - \int_a^b f(x)\,dx| < \varepsilon$. Let $R = \{x_0, x_1, \ldots, x_n\}$ be a partition of $[a,b]$ with $\|R\| < \delta$, which exists by Exercise 5.2.1.

Because F is differentiable (its derivative is f), then F is continuous by Theorem 4.2.4. Let $i \in \{1,\ldots,n\}$. The Mean Value Theorem (Theorem 4.4.4) applied to $F|_{[x_{i-1},x_i]}$ implies that there is some $s_i \in (x_{i-1}, x_i)$ such that

$$F'(s_i) = \frac{F(x_i) - F(x_{i-1})}{x_i - x_{i-1}}.$$

Hence $F(x_i) - F(x_{i-1}) = F'(s_i)(x_i - x_{i-1}) = f(s_i)(x_i - x_{i-1})$. Let $S = \{s_1, s_2, \ldots, s_n\}$. Then S is a representative set of R. Therefore

$$\begin{aligned} S(f,R,S) &= \sum_{i=1}^{n} f(s_i)(x_i - x_{i-1}) = \sum_{i=1}^{n} [F(x_i) - F(x_{i-1})] \\ &= F(x_n) - F(x_0) = F(b) - F(a). \end{aligned}$$

Because S is a representative set of R, then $|S(f,R,S) - \int_a^b f(x)\,dx| < \varepsilon$, and it follows that $|[F(b) - F(a)] - \int_a^b f(x)\,dx| < \varepsilon$. The desired result now follows from Lemma 2.3.10 (3). □

The expression "$F(b) - F(a)$" in the Fundamental Theorem of Calculus Version II (Theorem 5.6.4) is so frequently used in calculations of integrals that it is often written in the compact notation "$[F(x)]_a^b$," or some variant of that. We will not make use of such notation in the statements of theorems or in proofs, though for convenience we will use it in some examples.

The reader might wonder whether the hypotheses of the Fundamental Theorem of Calculus Version II (Theorem 5.6.4), which are that the function f is integrable and has an antiderivative, are redundant, in that these two criteria might appear to be related. However, even though the two versions of the Fundamental Theorem of Calculus show a relationship between integration and antidifferentiation, the relationship is not as straightforward as one might expect. As we see in Example 5.6.5, a function can have an antiderivative and yet not be integrable, and a function can be integrable and yet not have an antiderivative.

Example 5.6.5.

(1) Let $f\colon \mathbb{R} \to \mathbb{R}$ be defined by

$$f(x) = \begin{cases} x^2 \sin \frac{1}{x^2}, & \text{if } x \neq 0 \\ 0, & \text{if } x = 0. \end{cases}$$

It was shown in Example 4.2.5 (1) that f is differentiable. However, it was also noted in that example that the values of f' go to infinity as x goes to 0 from the right, and hence f' is not bounded on any closed bounded interval containing 0 in its interior. Therefore f' cannot be integrable on any closed bounded interval containing 0 in its interior by Theorem 5.3.3. Viewed another way, the function f' has an antiderivative, namely, the function f, and yet it is not integrable on any closed bounded interval that contains 0 in its interior.

(2) Let $h\colon [0,2] \to \mathbb{R}$ be defined by

$$h(x) = \begin{cases} 1, & \text{if } x \in [0,1] \\ 2, & \text{if } x \in (1,2]. \end{cases}$$

The function h is integrable by Exercise 5.2.5. However, as noted in Example 5.6.1 (2), the function h is not the derivative of any function, which is another way of saying that h does not have an antiderivative. ◊

We know by Theorem 5.4.11 and Corollary 5.6.3 that a continuous function is both integrable and has an antiderivative, and therefore the Fundamental Theorem of Calculus Version II applies to any continuous function. Are there functions that are both integrable and have antiderivatives, but are not continuous? If not, then we could just as well have stated the Fundamental Theorem of Calculus Version II with the simpler hypothesis that f is continuous. It turns out, however, as the reader is asked to show in Exercise 5.8.7, that there are such functions, which is why we stated the hypotheses of the Fundamental Theorem of Calculus Version II as we did.

Although our initial motivation for Version II of the Fundamental Theorem of Calculus was that we wanted to reverse the order of integration and differentiation from what was done in Version I, in fact Version II turns out to be by far the more useful of the two versions in the applications of calculus. The reader has undoubtedly computed many integrals via the Fundamental Theorem of Calculus Version II in calculus courses, and so we will provide only one such example, together with an example of how not to use the Fundamental Theorem of Calculus.

Example 5.6.6.

(1) Let $g\colon [0,2] \to \mathbb{R}$ be defined by $g(x) = x^2$ for all $x \in [0,2]$. We know by Example 3.3.7 (1) that g is continuous. As remarked above, it follows that g satisfies the hypotheses of the Fundamental Theorem of Calculus Version II (Theorem 5.6.4). Let $G\colon [0,2] \to \mathbb{R}$ be defined by $G(x) = \frac{x^3}{3}$ for all $x \in [0,2]$. Then G is an antiderivative of g, as the reader can easily verify. The Fundamental Theorem of Calculus Version II now implies that

$$\int_0^2 x^2\,dx = \int_0^2 g(x)\,dx = G(2) - G(0) = \frac{2^3}{3} - \frac{0^3}{3} = \frac{8}{3}.$$

(2) As useful as the Fundamental Theorem of Calculus Version II is, it is important not to get carried away using this wonderful theorem when it is not applicable. When faced with the integral $\int_{-1}^{1} \frac{1}{x^2}\,dx$, a student in a calculus course who has just learned the Fundamental Theorem of Calculus Version II might try the calculation

$$\text{“}\int_{-1}^{1} \frac{1}{x^2}\,dx = \int_{-1}^{1} x^{-2}\,dx = \left[-\frac{1}{x}\right]_{-1}^{1} = \left(-\frac{1}{1}\right) - \left(-\frac{1}{-1}\right) = -2.\text{”}$$

This calculation cannot be correct, because the function given by $f(x) = \frac{1}{x^2}$ is always positive, so its integral cannot be negative by Theorem 5.3.2 (1). The problem is that the hypotheses of the Fundamental Theorem of Calculus Version II are not satisfied by the function f. Indeed, this function f is not defined on the whole interval $[-1,1]$. It would be possible to extend f to the interval $[-1,1]$ by giving it an arbitrary value at $x = 0$, but the extended function is not bounded, and hence it is not integrable by Theorem 5.3.3; the extended function also does not have an antiderivative, because the function given by the formula $f(x) = -\dfrac{1}{x}$, which is the antiderivative on $[-1,0) \cup (0,1]$, cannot be extended to a continuous function on $[-1,1]$, and therefore it cannot be extended to a differentiable function by Theorem 4.2.4. Hence, it is crucial to make sure that the Fundamental Theorem of Calculus Version II is applicable before trying to use it.

We will see the proper way of dealing with the integral $\int_{-1}^{1} \frac{1}{x^2}\,dx$ in Example 6.4.10, after we have defined improper integrals. ◊

In contrast to the proof of the Fundamental Theorem of Calculus Version I (Theorem 5.6.2), which required results that were proved in Section 5.5, the proof of the Fundamental Theorem of Calculus Version II (Theorem 5.6.4) required nothing about integrals beyond their definition given in Section 5.2. The Fundamental Theorem of Calculus Version II could therefore have been stated and proved in that earlier section, but we delayed stating and proving it till now in order to keep both versions of the Fundamental Theorem of Calculus together.

The relation between the two versions of the Fundamental Theorem of Calculus can be seen more easily by looking at the case of continuous functions. Although it is not required that f be continous in the statement of Version II of the Fundamental Theorem of Calculus, as noted above continuous functions always satisfy the hypotheses of this version. We now see that in the continuous case it is possible to use the Fundamental Theorem of Calculus Version I to give a simpler proof of Version II. Suppose that $f\colon [a,b] \to \mathbb{R}$ is a continuous function, where $[a,b]$ is a non-degenerate closed bounded interval, and suppose that $F\colon [a,b] \to \mathbb{R}$ is an antiderivative of f. Let $G\colon [a,b] \to \mathbb{R}$ be defined by

$$G(x) = \int_a^x f(t)\,dt$$

for all $x \in [a,b]$. Then G is an antiderivative of f by the Fundamental Theorem of Calculus Version I, because we are assuming that f is continuous. Lemma 4.4.7 (2) implies that there is some $C \in \mathbb{R}$ such that $F(x) = G(x) + C$ for all $x \in [a,b]$. Then

$$F(b) - F(a) = [G(b) + C] - [G(a) + C] = \int_a^b f(t)\,dt - \int_a^a f(t)\,dt = \int_a^b f(x)\,dx.$$

Reflections

Our two versions of the Fundamental Theorem of Calculus are the same as the versions seen in introductory calculus courses, except for the fact that we are now more careful with stating their precise hypotheses. In our statement of the Fundamental Theorem of Calculus Version I it should be noted that the continuity of f, and the corresponding differentiability of F, occur at single numbers, not necessarily on the whole interval. Even more interesting is the statement of the Fundamental Theorem of Calculus Version II, where it is required that f is integrable and that f has an antiderivative. Both of these conditions hold if f is continuous, and that is usually assumed when the Fundamental Theorem of Calculus Version II is stated in an introductory calculus course. However, it is important to note that for non-continuous functions, being integrable is a distinct condition from having an antiderivative, as noted in Example 5.6.5, where there is a function that is integrable and yet does not have an antiderivative, and there is a function that has an antiderivative and yet is not integrable.

The Fundamental Theorem of Calculus Version II is stated for functions of a single variable, but it has generalizations to higher dimensions. Observe that under suitable hypotheses on f, this theorem can be restated as $\int_a^b f'(x)\,dx = f(b) - f(a)$. In other words, the "total amount" of the derivative of f over the whole interval $[a,b]$ can be found by evaluating the function f in an appropriate way on the endpoints of the interval. The key to generalizing this result to higher dimensions is to think of the endpoints of a non-degenerate closed bounded interval as the boundary of the interval. Theorems such as Green's Theorem, the Divergence Theorem and Stokes' Theorem, which one encounters in a multivariable calculus course, all relate the integral of some sort of derivative of a function on a region (in the plane or space) to the integral of the function on the boundary of the region (where the boundary is one dimension lower than the whole region). Such theorems are direct generalizations of the Fundamental Theorem of Calculus Version II. An even more general approach to this matter, which includes the above-mentioned theorems as special cases, is the generalized Stokes' Theorem on smooth manifolds; see the classic text [Spi65] for details.

Exercises

Exercise 5.6.1. Find the derivative of each of the following functions.

(1) Let $F\colon [0,5] \to \mathbb{R}$ be defined by

$$F(x) = \int_1^x t^2\,dt$$

for all $x \in [0,5]$.

(2) Let $G\colon [1,2] \to \mathbb{R}$ be defined by

$$G(x) = \int_1^{x^2} (t^5 + 7)\, dt$$

for all $x \in [1,2]$.

Exercise 5.6.2. [Used in Section 5.6.] Find an example of a function $f\colon [a,b] \to \mathbb{R}$ for some non-degenerate closed bounded interval $[a,b] \subseteq \mathbb{R}$ such that f is integrable but not continuous, and that the function $F\colon (a,b) \to \mathbb{R}$ defined by

$$F(x) = \int_a^x f(t)\, dt$$

for all $x \in [a,b]$ is differentiable. For your function f, does $F' = f$?

Exercise 5.6.3. Let $a \in (0,\infty)$, and let $g\colon [-a,a] \to \mathbb{R}$ be a function. Suppose that g is integrable. Let $G\colon [-a,a] \to \mathbb{R}$ be defined by

$$G(x) = \int_{-x}^x g(t)\, dt$$

for all $x \in [-a,a]$. Prove that if g is continuous, then G is differentiable, and $G'(x) = g(x) + g(-x)$ for all $x \in [-a,a]$.

Exercise 5.6.4. [Used in Theorem 6.4.11 and Theorem 9.3.6.] Let $I \subseteq \mathbb{R}$ be a non-degenerate interval that has the form $[a,b)$ or $[a,\infty)$ for some $a,b \in \mathbb{R}$, and let $f\colon I \to \mathbb{R}$ be a function. Suppose that $f(x) \geq 0$ for all $x \in I$, and that $f|_{[a,t]}$ is integrable for every $t \in I$. Let $F\colon I \to \mathbb{R}$ be defined by

$$F(x) = \int_a^x f(t)\, dt$$

for all $x \in I$. Prove that F is increasing.

Exercise 5.6.5. [Used in Section 5.6.] Let $[a,b] \subseteq \mathbb{R}$ be a closed bounded interval, and let $f\colon [a,b] \to \mathbb{R}$ be a function. Suppose that f is integrable. Let $F\colon [a,b] \to \mathbb{R}$ be defined by

$$F(x) = \int_a^x f(t)\, dt$$

for all $x \in [a,b]$. Prove that F is uniformly continuous.

Exercise 5.6.6. [Used in Theorem 5.7.4.] Let $[a,b] \subseteq \mathbb{R}$ be a non-degenerate closed bounded interval, and let $f\colon [a,b] \to \mathbb{R}$ be a function. Suppose that f is integrable and f has an antiderivative. Let $F\colon [a,b] \to \mathbb{R}$ be an antiderivative of f. Prove that if $s,t \in [a,b]$, then

$$\int_s^t f(x)\, dx = F(t) - F(s).$$

5.7 Computing Antiderivatives

The Fundamental Theorem of Calculus Version II (Theorem 5.6.4) shows that to compute definite integrals, it is necessary to be able to compute antiderivatives. We saw the definition of antiderivatives in the latter part of Section 4.4; now that we know how important they are, we discuss some methods for computing them.

Unfortunately, although the concept of an antiderivative is simple to define in principle, in practice it is not always easy to compute antiderivatives of complicated functions. Moreover, as we saw in Example 4.4.11, not every function has an antiderivative. By Corollary 5.6.3 we know that every continuous function has an antiderivative. Some discontinuous functions also have antiderivatives, as we saw in Example 4.2.5 (1), where there is an example of a differentiable function with a discontinuous derivative, and therefore that derivative is a discontinuous function with an antiderivative.

If a function has an antiderivative, then it will have more than one, though fortunately such antiderivatives cannot be very different from one another, because Corollary 4.4.9 states that any two antiderivatives of a given function defined on an open interval differ by a constant. Hence, once we know one antiderivative of a function, we know all others by adding constants to the first antiderivative. In other words, once we find a single antiderivative, we can obtain the most general antiderivative by adding "$+C$" to the single antiderivative, where C is an arbitrary real number. We now give a name to this general antiderivative.

Definition 5.7.1. Let $I \subseteq \mathbb{R}$ be an open interval, and let $f\colon I \to \mathbb{R}$ be a function. Suppose that f has an antiderivative. The **indefinite integral** of f, denoted $\int f(x)\,dx$, is the most general antiderivative of f. If F is any antiderivative of f, then

$$\int f(x)\,dx = F(x) + C,$$

where C is an arbitrary real number. △

Given that an indefinite integral is the most general antiderivative, the "$+C$" in indefinite integration is crucial. By contrast, an antiderivative is a single function, and does not include the "$+C$."

It is worth stressing that what we called an "integral" without any adjective prior to Definition 5.7.1 is what is referred to in calculus courses as a "definite integral." In general, we will continue to use the word "integral" without an adjective in that meaning, and if we mean "indefinite integral" we will refer to it as such, unless it is clear from the context that the word "integral" alone means "indefinite integral."

The notation $\int f(x)\,dx$ for indefinite integrals, though very standard, is rather unfortunate, in that it looks very similar to the notation $\int_a^b f(x)\,dx$ for "definite integrals." It is true that these two types of integrals are ultimately related by the Fundamental Theorem of Calculus, but it is important to keep in mind that definite integrals and indefinite integrals are very different constructions, and mean very different things—that these two concepts are related is a remarkable theorem, not an obvious consequence of their definitions. As such, it would be less confusing if these two concepts had names and notations that were not so similar, but we are stuck with the standard notation.

We conclude this section with some basic facts about indefinite integrals, starting with the most elementary properties, which are analogous to the corresponding properties of derivatives and definite integrals.

Theorem 5.7.2. *Let $I \subseteq \mathbb{R}$ be an open interval, let $f, g\colon I \to \mathbb{R}$ be functions and let $k \in \mathbb{R}$. Suppose that f and g have antiderivatives.*

1. *$f+g$ is has an antiderivative and $\int [f+g](x)\,dx = \int f(x)\,dx + \int g(x)\,dx$.*
2. *$f-g$ is has an antiderivative and $\int [f-g](x)\,dx = \int f(x)\,dx - \int g(x)\,dx$.*
3. *kf has an antiderivative and $\int [kf](x)\,dx = k\int f(x)\,dx$.*

Proof. We will prove Part (1), leaving the rest to the reader in Exercise 5.7.1.

(1) Let $F, G\colon I \to \mathbb{R}$ be antiderivatives of f and g, respectively. By Theorem 4.3.1 (1) we know that $[F+G]' = F' + G' = f + g$. Hence $F+G$ is an antiderivative of $f+g$. It follows that

$$\int [f+g](x)\,dx = F(x) + G(x) + C,$$

where C is an arbitrary constant. We also know that

$$\int f(x)\,dx = F(x) + D \quad \text{and} \quad \int g(x)\,dx = G(x) + E,$$

where D and E are arbitrary constants. However, if D and E are arbitrary constants, then so is $D+E$, and hence we can write

$$\int [f+g](x)\,dx = F(x) + G(x) + D + E.$$

It follows immediately that

$$\int [f+g](x)\,dx = \int f(x)\,dx + \int g(x)\,dx. \qquad \square$$

If there is one single most useful technique of integration, it is the one given in the following theorem, which is based upon the Chain Rule.

Theorem 5.7.3 (Integration by Substitution). *Let $I, J \subseteq \mathbb{R}$ be open intervals, and let $g\colon I \to J$ and $f\colon J \to \mathbb{R}$ be functions. Suppose that g is differentiable. If $F\colon J \to \mathbb{R}$ is an antiderivative of f, then*

$$\int f(g(x))g'(x)\,dx = F(g(x)) + C$$

for all $x \in I$.

Proof. By definition we know that F is differentiable and $F' = f$. Hence, by the Chain Rule (Theorem 4.3.3) we deduce that $F \circ g$ is differentiable and $[F \circ g]'(x) = F'(g(x)) \cdot g'(x) = f(g(x)) \cdot g'(x)$ for all $x \in I$. Therefore $F \circ g$ is an antiderivative of $(f \circ g)g'$. $\square$

The reader might recall from a calculus course a slightly different formulation of Integration by Substitution than the one we have given in Theorem 5.7.3. The calculus course version is usually stated as

$$\int f(g(x))g'(x)\,dx = \int f(u)\,du,$$

where the substitution $u = g(x)$ is used. We do not use that formulation in Theorem 5.7.3, both because the formulation that we used is more in keeping with our discussion up till now, and because we have not dealt with "dx" other than as a formal symbol. The calculus course version, which is equivalent to the one we use, can be made rigorous if one deals properly with "dx" and "du," though doing so would take us too far afield. The symbols "dx" and "du are examples of differential forms; see [Spi65, "Fields and Forms"] for a rigorous treatment of this subject.

Theorem 5.7.3 is for indefinite integrals. There is also a definite integral version of Integration by Substitution, which we now state and prove. Although to students in a calculus course it might appear that the definite integral version of Integration by Substitution is simply the result of putting the "a" and "b" on the integrals in the indefinite integral version, it is not rigorous to use "proof by similarity of notation." In fact, because integrability and antidifferentiability are not the same, as we saw in Example 5.6.5, the following theorem is proved quite differently from Theorem 5.7.3, though both theorems ultimately rely upon the Chain Rule.

Theorem 5.7.4 (Integration by Substitution for Definite Integrals). *Let $[a,b]$, $[c,d] \subseteq \mathbb{R}$ be non-degenerate closed bounded intervals, and let $g\colon [a,b] \to [c,d]$ and $f\colon [c,d] \to \mathbb{R}$ be functions. Suppose that f is continuous, that g is differentiable and that g' is integrable. Then $(f \circ g)g'$ is integrable and*

$$\int_a^b f(g(x))g'(x)\,dx = \int_{g(a)}^{g(b)} f(x)\,dx.$$

Proof. By Theorem 4.2.4 we know that g is continuous, and hence $f \circ g$ is continuous by Theorem 3.3.8 (3). It follows from Theorem 5.4.11 that $f \circ g$ is integrable, and therefore $(f \circ g)g'$ is integrable by Theorem 5.5.4 (2).

Let $F\colon [c,d] \to \mathbb{R}$ be defined by

$$F(x) = \int_c^x f(t)\,dt$$

for all $x \in [c,d]$. The Fundamental Theorem of Calculus Version I (Theorem 5.6.2) implies that F is differentiable and $F' = f$. By the Chain Rule (Theorem 4.3.3) we see that $[F \circ g]' = (F' \circ g)g' = (f \circ g)g'$. Hence $[F \circ g]'$ is integrable by the previous paragraph. Observe also that $[F \circ g]'$ has an antiderivative, which is $F \circ g$. Because f is continuous then f is integrable by Theorem 5.4.11, and we also know that f has an antiderivative, which is F. By the Fundamental Theorem of Calculus Version II (Theorem 5.6.4) applied to $[F \circ g]'$, and Exercise 5.6.6 applied to f, we see that

$$\int_a^b f(g(x))g'(x)\,dx = \int_a^b [F \circ g]'(x)\,dx = F(g(b)) - F(g(a)) = \int_{g(a)}^{g(b)} f(x)\,dx. \quad \square$$

Here is another very useful method for computing antiderivatives, this one based upon the Product Rule. Once again, we have both an indefinite integral version and a definite integral version.

Theorem 5.7.5 (Integration by Parts). *Let $I \subseteq \mathbb{R}$ be an open interval, and let $f, g \colon I \to \mathbb{R}$ be functions. Suppose that f and g are continuously differentiable. Then $f'g$ and fg' have antiderivatives and*

$$\int f(x)g'(x)\,dx = f(x)g(x) - \int f'(x)g(x)\,dx$$

for all $x \in I$.

Proof. Because f and g are differentiable, then they are continuous by Theorem 4.2.4. Because f and g are continuously differentiable, then by definition we know that f' and g' are continuous. It follows from Theorem 3.3.5 (4) that $f'g$ and fg' are continuous, and hence by Corollary 5.6.3 we know that $f'g$ and fg' have antiderivatives. By the Product Rule (Theorem 4.3.1 (4)) we know that fg is differentiable and $[fg]' = f'g + fg'$. Therefore fg is an antiderivative of $f'g + fg'$, and hence

$$\int [f'(x)g(x) + f(x)g'(x)]\,dx = f(x)g(x) + C.$$

Using Theorem 5.7.2 (1) we deduce that

$$\int f(x)g'(x)\,dx = f(x)g(x) - \int f'(x)g(x)\,dx + C.$$

However, we can now drop the "$+C$," because it is included in the indefinite integral on each side of the equals sign. □

Theorem 5.7.6 (Integration by Parts for Definite Integrals). *Let $[a,b] \subseteq \mathbb{R}$ be a non-degenerate closed bounded interval, and let $f, g \colon [a,b] \to \mathbb{R}$ be functions. Suppose that f and g are differentiable, and that f' and g' are integrable. Then $f'g$ and fg' are integrable and*

$$\int_a^b f(x)g'(x)\,dx = [f(b)g(b) - f(a)g(a)] - \int_a^b f'(x)g(x)\,dx.$$

Proof. Because f and g are differentiable, then they are continuous by Theorem 4.2.4, and hence they are integrable by Theorem 5.4.11. It follows from Theorem 5.5.4 (2) that $f'g$ and fg' are integrable. By the Product Rule (Theorem 4.3.1 (4)) we know that $[fg]' = f'g + fg'$. Using Theorem 5.3.1 (1) we see that

$$\begin{aligned}\int_a^b [f(x)g(x)]'\,dx &= \int_a^b [f'(x)g(x) + f(x)g'(x)]\,dx \\ &= \int_a^b f'(x)g(x)\,dx + \int_a^b f(x)g'(x)\,dx.\end{aligned}$$

Because fg is an antiderivative of $[fg]'$, the Fundamental Theorem of Calculus Version II (Theorem 5.6.4) implies that

$$f(b)g(b) - f(a)g(a) = \int_a^b f'(x)g(x)\,dx + \int_a^b f(x)g'(x)\,dx,$$

and the desired result follows immediately. □

The reader might be more familiar with the formulation of Integration by Parts that is written $\int u\,dv = uv - \int v\,du$, rather than the formulation we used in Theorem 5.7.5. These two formulations of Integration by Parts are equivalent, but we chose the formulation we used for the same reasons that we chose our formulation of Integration by Substitution in Theorem 5.7.3. For the sake of familiarity and computational simplicity, however, we will feel free to use the more common formulation of Integration by Parts using du and dv in a few computations, for example in the proof of Theorem 7.4.4, which says that the area of a circle of radius r is πr^2.

There are other techniques of integration in addition to Integration by Substitution and Integration by Parts, for example partial fractions and trigonometric substitution. Each of these techniques is useful for a particular type of indefinite integral, and by using all of these techniques it is possible to evaluate many, though certainly not all, of the integrals one encounters in applications. These techniques of integration are also used by various computer algebra systems, which can evaluate many integrals very effectively. We will not give the details of any of these other techniques of integration here. See [BML, Section III.11] for a treatment of the algebra that underlies partial fractions.

Reflections

Although we defined the concept of antiderivatives in Section 4.4, we did not provide a notation for this concept at the time, and instead waited to define the notation $\int f(x)\,dx$ for indefinite integrals until the present section, which is after the notation for definite integrals was defined, lest the reader mistakenly think that definite integrals are defined in terms of indefinite integrals. As previously observed, the similarity of the notations for indefinite integrals and for definite integrals causes some confusion, because these two types of integrals are very different conceptually, and are related to one another not by definition, but by the Fundamental Theorem of Calculus (both versions). Though unfortunate, we are stuck with this notation for the two types of integrals for historical reasons. Somewhat surprisingly, the notation for indefinite integrals (the less fundamental concept) was not created by taking the notation for definite integrals (the more fundamental concept) and removing the "a" and "b," but rather the other way around. The notation for indefinite integrals was due to Leibniz, right at the birth of calculus, and it proved very useful (especially in comparison with Newton's notation, which did not include the "dx," and which therefore did not work as nicely with substitutions); the notation for definite integrals was due to Fourier over a century after Leibniz.

In addition to the difference in meaning between the notation $\int f(x)\,dx$ and the notation $\int_a^b f(x)\,dx$, another important difference between the two is that the "x" in the latter is a "dummy variable" whereas the x in the former is not. That is, if we change x to u in $\int_a^b f(x)\,dx$ we obtain $\int_a^b f(u)\,du$, which has the same numerical value as $\int_a^b f(x)\,dx$. By contrast, the indefinite integral $\int f(x)\,dx$ is a function with "variable" x, whereas $\int f(u)\,du$ is a function with "variable" u, and hence these two functions, though related, are not identical. Of course, the expressions $\int f(x)\,dx$ and $\int f(u)\,du$ are equally meaningful; the practice of some introductory calculus texts of writing their tables of integrals (which are indefinite integrals) with u rather than x, as if that somehow makes the table of integrals more general, is quite silly, because a formula for $\int f(x)\,dx$ and a formula for $\int f(u)\,du$ tell us the same information.

Exercises

Exercise 5.7.1. [Used in Theorem 5.7.2.] Prove Theorem 5.7.2 (2) (3).

Exercise 5.7.2. Prove that there do not exist numbers $a, b \in \mathbb{R}$ such that $\int [fg](x)\,dx = ag(x)\int f(x)\,dx + bf(x)\int g(x)\,dx$ for all functions $f, g\colon \mathbb{R} \to \mathbb{R}$ such that f, g and fg have antiderivatives.

Exercise 5.7.3. Find an example of functions $f, g\colon \mathbb{R} \to \mathbb{R}$ such that f and g do not have antiderivatives, but that fg has an antiderivative.

Exercise 5.7.4. Explain the flaw in the following attempted "proof" that $0 = 1$: "Using Integration by Parts (Theorem 5.7.5) with $u = \frac{1}{x}$ and $dv = dx$, we see that

$$0 + \int \frac{1}{x}\,dx = \int \frac{1}{x}\,dx = \frac{1}{x}\cdot x - \int \left(-\frac{1}{x^2}\right) x\,dx = 1 + \int \frac{1}{x}\,dx.$$

By canceling we deduce that $0 = 1$."

Exercise 5.7.5. Let $[a, b] \subseteq \mathbb{R}$ be a non-degenerate closed bounded interval, and let $f\colon [a, b] \to \mathbb{R}$ be a function. Suppose that f is differentiable, that f' is integrable and that f is injective. We view f as a function $f\colon [a, b] \to f([a, b])$, and hence $f^{-1}\colon f([a, b]) \to [a, b]$ exists. Prove that

$$\int_a^b f(x)\,dx + \int_{f(a)}^{f(b)} f^{-1}(x)\,dx = bf(b) - af(a).$$

Make sure to verify that the hypotheses for any theorems used are satisfied.

[Use Exercise 4.5.11 and Exercise 4.6.3 (2).]

Exercise 5.7.6. [Used in Exercise 9.3.11 and Exercise 10.4.10.]

(1) Let $p \in \mathbb{N} \cup \{0\}$, and let $a_0, a_2, \ldots, a_p, b_0, b_2, \ldots, b_p \in \mathbb{R}$. Prove that if $u, r, s \in \{0, \ldots, p\}$ and $u \le r \le s$, then

$$\sum_{i=r}^{s} a_i b_i = a_s \left(\sum_{k=u}^{s} b_k\right) - a_r \left(\sum_{k=u}^{r-1} b_k\right) - \sum_{i=r}^{s-1} (a_{i+1} - a_i) \left(\sum_{k=u}^{i} b_k\right),$$

where any summation of the form $\sum_c^d$ with $c > d$ is taken to be zero.

(2) The formula in Part (1) of this exercise is known as Abel's Formula.Of which theorem in this section is Abel's Formula the discrete analog?

Exercise 5.7.7. This exercise discusses the **beta function**.
Let $B\colon [1,\infty)\times[1,\infty)\to\mathbb{R}$ be defined by

$$B(x,y)=\int_0^1 t^{x-1}(1-t)^{y-1}\,dt$$

for all $x,y\in[1,\infty)$.

(1) Prove that $B(y,x)=B(x,y)$ for all $x,y\in[1,\infty)$.
(2) Prove that $B(x,y)=\frac{x-1}{y}B(x-1,y+1)$ for all $x\in[2,\infty)$ and $y\in[1,\infty)$.
(3) Prove that if $n,m\in\mathbb{N}$, then $B(n,m)=\frac{(n-1)!(m-1)!}{(n+m-1)!}$. (If the reader is familiar with the gamma function, as discussed in Exercise 6.4.13, then it will be observed that this equality can be rephrased as $B(n,m)=\frac{\Gamma(n)\Gamma(m)}{\Gamma(n+m)}$ for all $n,m\in\mathbb{N}$. In fact, it can be verified that $B(x,y)=\frac{\Gamma(x)\Gamma(y)}{\Gamma(x+y)}$ for all $x,y\in[1,\infty)$; see [Wad00, Theorem 12.69(i)] for a proof.)

Exercise 5.7.8. The purpose of this exercise is to provide an alternative version of Taylor's Theorem (Theorem 4.4.6). Let $[a,b]\subseteq\mathbb{R}$ be a non-degenerate closed bounded interval, let $c\in(a,b)$, let $f\colon[a,b]\to\mathbb{R}$ be a function and let $n\in\mathbb{N}\cup\{0\}$. Suppose that $f^{(k)}$ exists and is continuous on $[a,b]$ for each $k\in\{0,\ldots,n+1\}$. Let $x\in[a,b]$. Prove that

$$f(x)=\sum_{k=0}^{n}\frac{f^{(k)}(c)}{k!}(x-c)^k+\int_c^x\frac{f^{(n+1)}(t)}{n!}(x-t)^n\,dt.$$

Do not try to deduce this result from Taylor's Theorem (Theorem 4.4.6); prove it on its own, by induction on n.

5.8 Lebesgue's Theorem

We saw in Theorem 5.4.11 that all continuous functions are integrable. Non-continuous functions may or may not be integrable; the functions in Example 5.2.6 (2) (4) are not continuous and are integrable, and the function in Example 5.2.6 (3) is not continuous and not integrable. Intuitively, for a function to be integrable, it can be discontinuous, but the set of numbers where the function is discontinuous cannot be "too large." This fact can be made precise, as we will see in Theorem 5.8.5 below, known as Lebesgue's Theorem. We start with some preliminaries, beginning with a definition that characterizes what we mean when we say that a set that is "not too large" from the point of view of integration. Our treatment of this material follows [Sto01, Section 6.7].

To understand the material in this section, the reader must be familiar with basic properties of countable and uncountable sets, and with the cardinalities of the standard sets of numbers, such as the rational numbers and the real numbers; see [Blo10, Sections 6.5–6.7] for details.

If the functions in Example 5.2.6 (2) (4) are compared with the function in Example 5.2.6 (3), the reader will notice that the sets of numbers at which the functions are discontinuous in Example 5.2.6 (2) (4) are countable (recall that a finite set is considered countable, and the rational numbers are countable), whereas the set of numbers at which the function is discontinuous in Example 5.2.6 (3) is uncountable (any non-degenerate interval in $\mathbb{R}$ is uncountable). It would be tempting to conjecture from these examples that a function is integrable if and only if the set of numbers at which it is discontinuous is countable, but such a conjecture turns out to be false, as the reader is asked to show in Exercise 8.4.11. The issue of countability versus uncountability is not the correct way to characterize sets that are "not too large" from the point of view of integration.

Rather than looking at the cardinality of sets, we need to look at the "size" of subsets of $\mathbb{R}$. Although a complete characterization of such size, known as Lebesgue measure, would take us too far afield, for our present purpose we need to characterize only those subsets of $\mathbb{R}$ that have "size zero," known formally as "measure zero," to be defined below. See [Str00, Chapter 14] for a discussion of Lebesgue measure.

As with many other aspects of calculus, the basic idea in determining the size of subsets of $\mathbb{R}$ is to approximate more complicated subsets with simpler ones, and in particular with sets for which it is easier to compute the size. The subsets of $\mathbb{R}$ for which size is particularly easy to compute are bounded intervals. For any bounded interval, which has the form (a,b), $[a,b)$, $(a,b]$ or $[a,b]$, its length is $b-a$. For our present purposes, it turns out that we can restrict our attention to open bounded intervals. The essential idea of a set having "measure zero" is that, rather than dealing with zero directly, the set is shown to be "smaller" than any given $\varepsilon > 0$, where "smaller" is measured in terms of being a subset of an appropriate collection of open intervals, the lengths of which add up to less than ε.

Whatever "measure zero" is defined to be, certainly any finite subset of $\mathbb{R}$ should have measure zero. Clearly, for any finite set, no matter how large, we can always find a finite collection of open intervals in $\mathbb{R}$ that contains the finite set, and that the sum of the lengths of the open intervals is less than any given positive number. It is this last fact that implies that every finite set has measure zero. More generally, any subset of $\mathbb{R}$ that has this property, namely, that for any $\varepsilon > 0$, there is a finite collection of open intervals that contains the subset, and the sum of the lengths of the open intervals is less than ε, should be considered as having measure zero. However, and this is the subtle insight of Lebesgue's Theorem, it turns out that to characterize sets as having measure zero, it does not suffice to consider only finite collections of open intervals, but rather countable collections of open intervals as well.

A countable collection of open intervals in $\mathbb{R}$ can be viewed as a sequence of open intervals, and hence, for the following definition, we need the notions of sequences and series. These concepts will be discussed in detail in Sections 8.2 and 9.2, respectively, but we assume that the reader is familiar with these concepts informally, and that the reader will agree to take on faith whatever facts we need about sequences and series for now until we deal with them rigorously later on, and in that way we can examine Lebesgue's Theorem in the chapter on integration, where it belongs.

Definition 5.8.1. Let $A \subseteq \mathbb{R}$ be a set. The set A has **measure zero** if for each $\varepsilon > 0$, there is a sequence $\{(a_n, b_n)\}_{n=1}^{\infty}$ of open bounded intervals in $\mathbb{R}$ such that $A \subseteq \bigcup_{n=1}^{\infty} (a_n, b_n)$ and $\sum_{n=1}^{\infty} (b_n - a_n) < \varepsilon$. △

We note that the open intervals of the form (a_n, b_n) used in Definition 5.8.1 are allowed to be degenerate, that is, we allow $a_n = b_n$, which means that such intervals are empty. Hence, although we use sequences of the form $\{(a_n, b_n)\}_{n=1}^{\infty}$, in those cases where all but finitely many of these open intervals are degenerate, then we really have finite collections of non-degenerate open intervals. We have phrased the definition as we did to avoid having to consider two separate cases, namely, finite collections of non-degenerate open intervals and sequences of non-degenerate open intervals.

Example 5.8.2.

(1) The empty set has measure zero. Let $\varepsilon > 0$. Let $(a_1, b_1) = \left(0, \frac{\varepsilon}{2}\right)$, and let $(a_n, b_n) = (0, 0)$ for all $n \in \mathbb{N}$ such that $n > 1$. Then $\emptyset \subseteq \bigcup_{n=1}^{\infty} (a_n, b_n)$ and $\sum_{n=1}^{\infty} (b_n - a_n) = \frac{\varepsilon}{2} < \varepsilon$.

(2) Any finite subset of $\mathbb{R}$ has measure zero. The reader is asked to prove this fact in Exercise 5.8.1.

(3) A non-degenerate interval does not have measure zero. Let $I \subseteq \mathbb{R}$ be a non-degenerate interval. We consider the case where I is a closed bounded interval; the case where I is not a closed bounded interval is left to the reader in Exercise 5.8.2.

Because we are assuming that I is a non-degenerate closed bounded interval, then $I = [c, d]$ for some $c, d \in \mathbb{R}$ such that $c < d$. Suppose that $[c, d]$ has measure zero. Then there is a sequence $\{(a_n, b_n)\}_{n=1}^{\infty}$ of open bounded intervals in $\mathbb{R}$ such that $[c, d] \subseteq \bigcup_{n=1}^{\infty} (a_n, b_n)$ and $\sum_{n=1}^{\infty} (b_n - a_n) < d - c$. By the Heine–Borel Theorem (Theorem 2.6.14) there are $n \in \mathbb{N}$ and $i_1, i_2, \ldots, i_n \in \mathbb{N}$ such that $[c, d] \subseteq \bigcup_{k=1}^{n} (a_{i_k}, b_{i_k})$. It follows from Exercise 2.5.15 that

$$d - c \leq \sum_{k=1}^{n} (b_{i_k} - a_{i_k}) \leq \sum_{n=1}^{\infty} (b_n - a_n) < d - c,$$

which is a contradiction. Therefore I does not have measure zero. (The Heine–Borel Theorem might seem stronger than necessary to prove this result, but it allowed us to avoid dealing with series more than absolutely necessary.)

(4) Any countable subset of $\mathbb{R}$ has measure zero. Let $C \subseteq \mathbb{R}$ be a countable set. By definition a countable set is either finite or countably infinite. Finite sets were treated in Part (2) of this example, so suppose that C is countably infinite. That is, suppose that $C = \{c_1, c_2, c_3, \ldots\}$, where $c_1, c_2, \ldots \subseteq \mathbb{R}$. Let $\varepsilon > 0$. For each $n \in \mathbb{N}$, let $(a_n, b_n) = \left(c_n - \frac{\varepsilon}{2^{n+2}}, c_n + \frac{\varepsilon}{2^{n+2}}\right)$. Then $c_n \in (a_n, b_n)$ for all $n \in \mathbb{N}$, and hence $A \subseteq \bigcup_{n=1}^{\infty} (a_n, b_n)$. Because $b_n - a_n = \frac{\varepsilon}{2^{n+1}}$ for all $n \in \mathbb{N}$, then

$$\sum_{n=1}^{\infty} (b_n - a_n) = \sum_{n=1}^{\infty} \frac{\varepsilon}{2^{n+1}} = \varepsilon \left(\frac{1}{4} + \frac{1}{8} + \frac{1}{16} + \cdots \right).$$

The series $\frac{1}{4} + \frac{1}{8} + \frac{1}{16} + \cdots$ is a geometric series. The reader is likely to be familiar with geometric series from calculus courses or before; such series will be treated rigorously in Example 9.2.4 (4). From that example it is seen that $\frac{1}{4}+\frac{1}{8}+\frac{1}{16}+\cdots$ is convergent, and that $\frac{1}{4}+\frac{1}{8}+\frac{1}{16}+\cdots=\frac{1}{2}$. It follows that $\sum_{n=1}^{\infty}(b_n-a_n)=\frac{\varepsilon}{2}<\varepsilon$. Hence A has measure zero.

(5) There are uncountable subsets of $\mathbb{R}$ that have measure zero, though it is hard to imagine such sets intuitively. A famous example of an uncountable subset of $\mathbb{R}$ that has measure zero is the Cantor set, which will be discussed in detail in Example 8.4.9, when we have more tools involving sequences at our disposal. ◊

It is evident that if a subset $A\subseteq\mathbb{R}$ has measure zero, and if $B\subseteq A$, then B has measure zero; we omit the details. A less evident, though very useful, fact about sets of measure zero is given in the following lemma.

Lemma 5.8.3. *Let $\{A_n\}_{n=1}^{\infty}$ be a sequence of subsets of $\mathbb{R}$. Suppose that A_n has measure zero for all $n\in\mathbb{N}$. Then $\bigcup_{n=1}^{\infty}A_n$ has measure zero.*

Proof. Let $\varepsilon>0$. Let $n\in\mathbb{N}$. Then there is a sequence $\left\{\left(a_k^n,b_k^n\right)\right\}_{k=1}^{\infty}$ of open bounded intervals in $\mathbb{R}$ such that $A_n\subseteq\bigcup_{k=1}^{\infty}\left(a_k^n,b_k^n\right)$ and $\sum_{k=1}^{\infty}(b_k^n-a_k^n)<\frac{\varepsilon}{2^{n+1}}$. We saw in Example 5.8.2 (4) that $\sum_{n=1}^{\infty}\frac{\varepsilon}{2^{n+1}}=\frac{\varepsilon}{2}$.

Let $f\colon\mathbb{N}\to\mathbb{N}\times\mathbb{N}$ be a bijective function. Such a function exists because $\mathbb{N}\times\mathbb{N}$ is countably infinite, which is a standard fact about the cardinality of the number systems; see [Blo10, Sections 6.5–6.7] for details. Let $\{(c_n,d_n)\}_{n=1}^{\infty}$ be defined as follows. For each $i\in\mathbb{N}$, we have $f(i)=(n_i,k_i)$ for some $n_i,k_i\in\mathbb{N}$, and then let $c_i=a_{k_i}^{n_i}$ and $d_i=b_{k_i}^{n_i}$. Because f is surjective, it follows that $\bigcup_{n=1}^{\infty}A_n\subseteq\bigcup_{n=1}^{\infty}(c_n,d_n)$. We now use Exercise 9.3.6 to deduce that $\sum_{n=1}^{\infty}(d_n-c_n)$ is convergent and $\sum_{n=1}^{\infty}(d_n-c_n)\le\sum_{k=1}^{\infty}\frac{\varepsilon}{2^{k+1}}=\frac{\varepsilon}{2}<\varepsilon$. □

Although Lemma 5.8.3 is stated in terms of sequence of subsets of $\mathbb{R}$ of measure zero, the lemma also applies to any finite collection of subsets of measure zero. If $A_1,A_2,\ldots,A_p$ is a finite collection of subsets of $\mathbb{R}$ of measure zero, we can let $A_i=\{0\}$ for all $i\in\mathbb{N}$ such that $i>p$, and then $\{A_n\}_{n=1}^{\infty}$ satisfies the hypotheses of the lemma, and so $\bigcup_{n=1}^{\infty}A_n$ has measure zero, and hence $\bigcup_{n=1}^{p}A_n$ has measure zero.

The following lemma, which relates integration and sets of measure zero, is the first step toward Lebesgue's Theorem.

Lemma 5.8.4. *Let $[a,b]\subseteq\mathbb{R}$ be a non-degenerate closed bounded interval, and let $f\colon[a,b]\to\mathbb{R}$ be a function. Suppose that f is integrable, that $f(x)\ge 0$ for all $x\in[a,b]$ and that $\int_a^b f(x)\,dx=0$. Then the set $\{x\in[a,b]\mid f(x)>0\}$ has measure zero.*

Proof. Let $E=\{x\in[a,b]\mid f(x)>0\}$. For each $n\in\mathbb{N}$, let $E_n=\{x\in[a,b]\mid f(x)\ge\frac{1}{n}\}$. Clearly $E_n\subseteq E$ for all $n\in\mathbb{N}$. Let $x\in E$. Then $x\in[a,b]$ and $f(x)>0$. By Corollary 2.6.8 (2) there is some $m\in\mathbb{N}$ such that $\frac{1}{m}<f(x)$. Hence $x\in E_m$. It follows that $E=\bigcup_{n=1}^{\infty}E_n$. We will show that for each $n\in\mathbb{N}$, the set E_n has measure zero. It will then follow from Lemma 5.8.3 that E has measure zero.

Let $n\in\mathbb{N}$. Let $\varepsilon>0$. Because $\int_a^b f(x)\,dx=0$, there is some $\delta>0$ such that if P is a partition of $[a,b]$ with $\|P\|<\delta$, and if T is a representative set of P, then $|S(f,P,T)-0|<\frac{\varepsilon}{2n}$.

Let $R = \{x_0, x_1, \ldots, x_n\}$ be a partition of $[a,b]$ such that $\|R\| < \delta$, which exists by Exercise 5.2.1. Let $V = \{i \in \{1,\ldots,n\} \mid E_n \cap [x_{i-1}, x_i] \neq \emptyset\}$. It is evident from the definition of V that

$$E_n \subseteq \bigcup_{i \in V} [x_{i-1}, x_i] \subseteq \bigcup_{i \in V} (x_{i-1}, x_i) \cup \bigcup_{j=0}^{n} \left(x_j - \frac{\varepsilon}{4(n+1)}, x_j + \frac{\varepsilon}{4(n+1)} \right).$$

Let $U = \{u_1, \ldots, u_n\}$ be the representative set of R defined as follows. If $i \in V$, let u_i be an arbitrary element of $E_n \cap [x_{i-1}, x_i]$; if $i \in \{1,\ldots,n\} - V$, let u_i be an arbitrary element of $[x_{i-1}, x_i]$. Then $f(u_i) \geq \frac{1}{n}$ for all $i \in V$, and $f(u_i) \geq 0$ for all $i \in \{1,\ldots,n\} - V$. Then

$$\begin{aligned} \frac{1}{n} \sum_{i \in V} (x_i - x_{i-1}) \leq \sum_{i \in V} f(u_i)(x_i - x_{i-1}) &\leq \sum_{i=1}^{n} f(u_i)(x_i - x_{i-1}) \\ &= |S(f,R,U) - 0| < \frac{\varepsilon}{2n}. \end{aligned}$$

It follows that $\sum_{i \in V} (x_i - x_{i-1}) < \frac{\varepsilon}{2}$. It is straightforward to verify that

$$\sum_{j=0}^{n} \left[\left(x_j - \frac{\varepsilon}{4(n+1)} \right) - \left(x_j - \frac{\varepsilon}{4(n+1)} \right) \right] = \frac{\varepsilon}{2}.$$

Therefore

$$\sum_{i \in V} (x_i - x_{i-1}) + \sum_{j=0}^{n} \left[\left(x_j - \frac{\varepsilon}{4(n+1)} \right) - \left(x_j - \frac{\varepsilon}{4(n+1)} \right) \right] < \frac{\varepsilon}{2} + \frac{\varepsilon}{2} = \varepsilon.$$

We deduce that E_n has measure zero. □

Lemma 5.8.4 is interesting precisely because there are discontinuous functions that are integrable; in Exercise 5.5.7 it was seen that if the function f in Lemma 5.8.4 is continuous, then $\{x \in [a,b] \mid f(x) > 0\}$ is the empty set, which has measure zero as noted in Example 5.8.2 (1).

We are now ready to prove the following remarkable characterization of integrable functions.

Theorem 5.8.5 (Lebesgue's Theorem). *Let $[a,b] \subseteq \mathbb{R}$ be a non-degenerate closed bounded interval, and let $f\colon [a,b] \to \mathbb{R}$ be a function. Then f is integrable if and only if f is bounded and the set of numbers at which f is discontinuous has measure zero.*

Proof. Suppose that f is integrable. Then by Theorem 5.3.3 we know that f is bounded.

Let $\{Q_n\}_{n=1}^{\infty}$ be the sequence of partitions of $[a,b]$ defined as follows. For each $k \in \mathbb{N}$, it follows from Theorem 5.4.7 (c) that there is a partition Q_k of $[a,b]$ such that $U(f,Q_k) - L(f,Q_k) < \frac{1}{k}$; there is more than one such partition Q_k, so we choose one. Let $\{P_n\}_{n=1}^{\infty}$ be the sequence of partitions of $[a,b]$ defined by $P_k = \bigcup_{i=1}^{k} Q_i$ for all $k \in \mathbb{N}$.

Let $n \in \mathbb{N}$. Then P_n is a refinement of both Q_n and P_{n-1}. It follows that if $m \in \mathbb{N}$ and $m > n$, then P_m is a refinement of P_n; the proof of this fact requires induction, and the details are left to the reader. By Lemma 5.4.6 (2) we see that $L(f,Q_n) \leq L(f,P_n) \leq U(f,P_n) \leq U(f,Q_n)$, and hence $U(f,P_n) - L(f,P_n) < \frac{1}{n}$. We denote the elements of P_n by $P_n = \{x_0^n, x_1^n, \ldots, x_{p_n}^n\}$.

Let $u_n, l_n \colon [a,b] \to \mathbb{R}$ be defined by

$$u_n(x) = \begin{cases} M_i^{P_n}(f), & \text{if } x \in (x_{i-1}^n, x_i^n) \text{ for some } i \in \{1,\ldots,p_n\} \\ f(x), & \text{if } x \in P_n \end{cases}$$

and

$$l_n(x) = \begin{cases} m_i^{P_n}(f), & \text{if } x \in (x_{i-1}^n, x_i^n) \text{ for some } i \in \{1,\ldots,p_n\} \\ f(x), & \text{if } x \in P_n. \end{cases}$$

Then u_n and l_n are step functions, as defined in Exercise 5.3.4 (2).

It is seen from the definition of u_n and l_n that $l_n(x) \leq f(x) \leq u_n(x)$ for all $x \in [a,b]$. Let $i \in \{1,\ldots,p_n\}$. We can compute $M_i^{P_n}(u_n)$ and $m_i^{P_n}(l_n)$, because step functions are bounded. Observe that $l_n(x_{i-1}^n) = f(x_{i-1}^n) = u_n(x_{i-1}^n)$ and $l_n(x_i^n) = f(x_i^n) = u_n(x_i^n)$, and that if $x \in (x_{i-1}^n, x_i^n)$, then $l_n(x) = m_i^{P_n}(f)$ and $u_n(x) = M_i^{P_n}(f)$. It follows that $m_i^{P_n}(l_n) = \min\{f(x_{i-1}^n), m_i^{P_n}(f), f(x_i^n)\} = m_i^{P_n}(f)$, and similarly $M_i^{P_n}(u_n) = M_i^{P_n}(f)$. We deduce that

$$U(u_n,P_n) = \sum_{j=1}^{p_n} M_j^{P_n}(u_n)(x_j^n - x_{j-1}^n) = \sum_{j=1}^{p_n} M_j^{P_n}(f)(x_j^n - x_{j-1}^n) = U(f,P_n),$$

and similarly $L(l_n,P_n) = L(f,P_n)$.

Let $m \in \mathbb{N}$. Suppose that $m > n$. Let $x \in [a,b] - P_m$. Then $x \in (x_{t-1}^m, x_t^m)$ for some $t \in \{1,\ldots,p_m\}$. Because P_m is a subdivision of P_n, then $[x_{t-1}^m, x_t^m] \subseteq [x_{h-1}^n, x_h^n]$ for some $h \in \{1,\ldots,p_n\}$. It then follows from Exercise 2.6.1 (1) that $u_m(x) = M_t^{P_m}(f) = \operatorname{lub} f([x_{t-1}^m, x_t^m]) \leq \operatorname{lub} f([x_{h-1}^n, x_h^n]) = M_h^{P_n}(f) = u_n(x)$. A similar argument shows that $l_n(x) \leq l_m(x)$.

Let $s,t \in \mathbb{N}$. Let $y \in [a,b]$. We will show that $l_s(y) \leq u_t(y)$. First, suppose that $s = t$. Then we have already noted that $l_s(y) \leq u_t(y)$. Second, suppose that $s > t$. If $y \in [a,b] - P_s$, then combining various facts mentioned above we see that $l_s(y) \leq u_s(y) \leq u_t(y)$; if $y \in P_s$, then $l_s(y) = f(y) \leq u_t(y)$. Third, suppose that $s < t$. This case is similar to the previous case, and we omit the details.

Let $u, l \colon [a,b] \to \mathbb{R}$ be defined as follows. Let $x \in [a,b]$. Let

$$\mathcal{L}_x = \{l_n(x) \mid n \in \mathbb{N}\} \quad \text{and} \quad \mathcal{U}_x = \{u_n(x) \mid n \in \mathbb{N}\}.$$

Clearly $\mathcal{L}_x$ and $\mathcal{U}_x$ are both non-empty subsets of $\mathbb{R}$. By the previous paragraph, we know that $\mathcal{L}_x$ and $\mathcal{U}_x$ satisfy the hypotheses of Part (1) of the No Gap Lemma (Lemma 2.6.6), and hence $\mathcal{L}_x$ has a least upper bound and $\mathcal{U}_x$ has a greatest lower bound, and $\operatorname{lub} \mathcal{L}_x \leq \operatorname{glb} \mathcal{U}_x$. Let $u(x) = \operatorname{glb} \mathcal{U}_x$ and $l(x) = \operatorname{glb} \mathcal{L}_x$. We have therefore defined the functions u and l, and we see that $l(x) \leq u(x)$ for all $x \in [a,b]$.

We now show that u and l are integrable, and that $\int_a^b u(x)\,dx = \int_a^b f(x)\,dx = \int_a^b l(x)\,dx$. First, observe that $l_1(x) \leq l(x) \leq u(x) \leq u_1(x)$ for all $x \in [a,b]$, and that l_1 and u_1 are bounded. It follows from Exercise 3.2.14 that u and l are bounded. Let $\varepsilon > 0$. By Corollary 2.6.8 (2) there is some $m \in \mathbb{N}$ such that $\frac{1}{m} < \varepsilon$. Then $U(f,P_m) - L(f,P_m) < \frac{1}{m} < \varepsilon$. We know that $l_m(x) \leq l(x) \leq u(x) \leq u_m(x)$ for all $x \in [a,b]$, and then by Exercise 5.4.6 and Lemma 5.4.6 (1) we deduce that $L(l_m,P_m) \leq L(l,P_m) \leq U(l,P_m) \leq U(u_m,P_m)$. We saw above that $L(l_m,P_m) = L(f,P_m)$ and $U(u_m,P_m) = U(f,P_m)$, and hence $L(f,P_m) \leq L(l,P_m) \leq U(l,P_m) \leq U(f,P_m)$. A similar argument shows that $L(f,P_m) \leq L(u,P_m) \leq U(u,P_m) \leq U(f,P_m)$. It now follows from Exercise 5.4.11 that u and l are integrable, and that $\int_a^b u(x)\,dx = \int_a^b f(x)\,dx = \int_a^b l(x)\,dx$.

By Theorem 5.3.1 (2) we know that $u - l$ is integrable, and that $\int_a^b [u-l](x)\,dx = \int_a^b u(x)\,dx - \int_a^b l(x)\,dx = 0$. We also know that $(u-l)(x) \geq 0$ for all $x \in [a,b]$. It now follows from Lemma 5.8.4 that the set $\{x \in [a,b] \mid (u-l)(x) > 0\}$ has measure zero.

Let

$$F = \{x \in [a,b] \mid (u-l)(x) > 0\} \cup \bigcup_{n=1}^{\infty} P_n.$$

For each $n \in \mathbb{N}$, the set P_n is finite, and hence it has measure zero by Example 5.8.2 (2). It follows from Lemma 5.8.3 that F has measure zero.

We now show that f is continuous at all numbers in $[a,b] - F$. Let $c \in [a,b] - F$. Let $\varepsilon > 0$. By the definition of F we know that $(u-l)(c) = 0$, which means that $u(c) = l(c)$. Because $l(c) = \operatorname{lub} \mathcal{L}_c$, we can use Lemma 2.6.5 (1) to deduce that there is some $v \in \mathbb{N}$ such that $l(c) - \frac{\varepsilon}{2} < l_v(c) \leq l(c)$. Similarly, there is some $k \in \mathbb{N}$ such that $u(c) \leq u_k(c) < u(c) + \frac{\varepsilon}{2}$. Suppose that $v \geq k$; the case where $v < k$ is similar, and we omit the details. Because $c \notin \bigcup_{n=1}^{\infty} P_n$ then $c \in [a,b] - P_v$, and, as noted previously, it follows that $u(c) \leq u_v(c) \leq u_k(c)$. Because $l(c) = u(c)$, we deduce that $u(c) - \frac{\varepsilon}{2} < l_v(c) < u(c) < u_v(c) < u(c) + \frac{\varepsilon}{2}$. Therefore $u_v(c) - l_v(c) < \varepsilon$.

Because $c \in [a,b] - P_v$, it follows that $c \in (x_{i-1}^v, x_i^v)$ for some $i \in \{1, \ldots, p_v\}$. By Lemma 2.3.7 (2) there is some $\delta > 0$ such that $(c-\delta, c+\delta) \subseteq (x_{i-1}^v, x_i^v)$.

Suppose that $x \in [a,b]$ and $|x - c| < \delta$. Then $x \in (x_{i-1}^v, x_i^v)$, and it follows from the definition of u_v and l_v as step functions that $u_v(x) = u_v(c)$ and $l_v(x) = l_v(c)$. We know that $l_v(c) \leq f(c) \leq u_v(c)$ and $l_v(x) \leq f(x) \leq u_v(x)$, and hence $l_v(c) \leq f(x) \leq u_v(c)$. It follows that $|f(x) - f(c)| \leq u_v(c) - l_v(c) < \varepsilon$. We deduce that f is continuous at c.

Let G be the set of numbers at which f is discontinuous. Because f is continuous at all numbers in $[a,b] - F$, it follows that $G \subseteq F$. Because F has measure zero, then G has measure zero, and that completes this part of the proof.

Now suppose that f is bounded and that the set of numbers at which f is discontinuous has measure zero.

Let $\varepsilon > 0$. Because f is bounded, there is some $M \in \mathbb{R}$ such that $|f(x)| \leq M$ for all $x \in [a,b]$; we may assume that $M > 0$. Let D be the set of numbers at which f is discontinuous. Then D has measure zero. Hence there is a sequence $\{(a_n,b_n)\}_{n=1}^{\infty}$ of open bounded intervals in $\mathbb{R}$ such that $D \subseteq \bigcup_{n=1}^{\infty} (a_n,b_n)$ and $\sum_{n=1}^{\infty} (b_n - a_n) < \frac{\varepsilon}{4M}$.

Let $x \in [a,b] - D$. (We note that $[a,b] - D \neq \emptyset$, because $[a,b]$ does not have measure zero by Example 5.8.2 (3), and hence $D \neq [a,b]$.) Then f is continuous at x. Hence there is some $\delta_x > 0$ such that $y \in [a,b]$ and $|y-x| < \delta_x$ imply $|f(y) - f(x)| < \frac{\varepsilon}{4(b-a)}$. Suppose that $y,z \in [a,b] \cap (x-\delta_x, x+\delta_x)$. Then $|y-x| < \delta_x$ and $|z-x| < \delta_x$, and it follows that

$$\begin{aligned}|f(z)-f(y)| &= |f(z)-f(x)+f(x)-f(y)| \leq |f(z)-f(x)| + |f(x)-f(y)| \\ &< \frac{\varepsilon}{4(b-a)} + \frac{\varepsilon}{4(b-a)} = \frac{\varepsilon}{2(b-a)}.\end{aligned}$$

Because $D \subseteq \bigcup_{n=1}^{\infty} (a_n, b_n)$, then

$$[a,b] = ([a,b]-D) \cup D \subseteq \bigcup_{x \in [a,b]-D} \left(x - \frac{\delta_x}{2}, x + \frac{\delta_x}{2}\right) \cup \bigcup_{n=1}^{\infty} (a_n, b_n).$$

The Heine–Borel Theorem (Theorem 2.6.14) implies that there are $p,q \in \mathbb{N}$, and $x_1, x_2, \ldots, x_p \in [a,b] - D$, and $n_1, n_2, \ldots, n_q \in \mathbb{N}$ such that

$$[a,b] \subseteq \bigcup_{j=1}^{p} \left(x_j - \frac{\delta_{x_j}}{2}, x_j + \frac{\delta_{x_j}}{2}\right) \cup \bigcup_{i=1}^{q} (a_{n_i}, b_{n_i}).$$

Let

$$Q = \{a, b, x_1 - \frac{\delta_{x_1}}{2}, x_1 + \frac{\delta_{x_1}}{2}, \ldots, x_p - \frac{\delta_{x_p}}{2}, x_p + \frac{\delta_{x_p}}{2}, a_{n_1}, b_{n_1}, a_{n_2}, b_{n_2}, \ldots, a_{n_q}, b_{n_q}\}.$$

Let $P = Q \cap [a,b]$. Arrange the elements of P in increasing order, and rename them $y_0, y_1, \ldots, y_n$. Then $P = \{y_0, y_1, \ldots, y_n\}$ is a partition of $[a,b]$. Let $i \in \{1, \ldots, n\}$. Then (y_{i-1}, y_i) is a subset of at least one interval of the form $\left(x_j - \frac{\delta_{x_j}}{2}, x_j + \frac{\delta_{x_j}}{2}\right)$ or of the form (a_{n_k}, b_{n_k}). Let

$$V = \{i \in \{1, \ldots, n\} \mid (y_{i-1}, y_i) \subseteq (a_{n_k}, b_{n_k}) \text{ for some } k \in \{1, \ldots, q\}\},$$

and let $W = \{1, \ldots, n\} - V$.

Let $k \in W$. Then $(y_{k-1}, y_k) \subseteq \left(x_j - \frac{\delta_{x_j}}{2}, x_j + \frac{\delta_{x_j}}{2}\right)$ for some $j \in \{1, \ldots, p\}$. It follows that $[y_{k-1}, y_k] \subseteq (x_j - \delta_{x_j}, x_j + \delta_{x_j})$. If $z, w \in [y_{k-1}, y_k]$, then $z, w \in [a,b] \cap (x_j - \delta_{x_j}, x_j + \delta_{x_j})$, and, as we saw above, it follows that $|f(z) - f(w)| < \frac{\varepsilon}{2(b-a)}$. By Exercise 5.4.9 (3) we see that $M_k(f) - m_k(f) \leq \frac{\varepsilon}{2(b-a)}$. Therefore

$$\begin{aligned}\sum_{i \in W} [M_i(f) - m_i(f)](y_i - y_{i-1}) &\leq \frac{\varepsilon}{2(b-a)} \sum_{i \in W} (y_i - y_{i-1}) \leq \frac{\varepsilon}{2(b-a)} \sum_{i=1}^{n} (y_i - y_{i-1}) \\ &= \frac{\varepsilon}{2(b-a)} (b-a) = \frac{\varepsilon}{2}.\end{aligned}$$

Let $r \in V$. Because $|f(x)| \leq M$ for all $x \in [a,b]$, it follows from Exercise 5.4.9 (4) that $M_r(f) - m_r(f) \leq 2M$. Therefore

$$\sum_{i\in V}[M_i(f)-m_i(f)](y_i-y_{i-1}) \le 2M\sum_{i\in V}(y_i-y_{i-1}) \le 2M\sum_{n=1}^{\infty}(b_n-a_n) < 2M\frac{\varepsilon}{4M} = \frac{\varepsilon}{2}.$$

Putting the above calculations together we conclude that

$$\begin{aligned} U(f,P)-L(f,P) &= \sum_{i=1}^{n}[M_i(f)-m_i(f)](x_i-x_{i-1}) \\ &= \sum_{i\in V}[M_i(f)-m_i(f)](x_i-x_{i-1}) + \sum_{i\in W}[M_i(f)-m_i(f)](x_i-x_{i-1}) \\ &< \frac{\varepsilon}{2}+\frac{\varepsilon}{2} = \varepsilon. \end{aligned}$$

Therefore f satisfies the criterion given in Theorem 5.4.7 (c), and hence f is integrable. □

The following corollary is an immediate consequence of Lebesgue's Theorem (Theorem 5.8.5) combined with Example 5.8.2 (4).

Corollary 5.8.6. *Let $[a,b]\subseteq\mathbb{R}$ be a non-degenerate closed bounded interval, and let $f\colon[a,b]\to\mathbb{R}$ be a function. If f is bounded and is discontinuous at countably many numbers, then f is integrable.*

Corollary 5.8.6 immediately implies that the function in Example 5.2.6 (4) is integrable, though of course proving this corollary, which requires Lebesgue's Theorem (Theorem 5.8.5), requires more work than going through the details of that example, so no real effort has been saved. However, from now on, we can easily treat any similar such examples. Moreover, Lebesgue's Theorem can be used to give alternative, and often simpler, proofs of various theorems that we have already seen, for example Theorem 5.5.1 (1) and Theorem 5.5.4 (2); the reader is asked to provide such proofs in Exercise 5.8.4 and Exercise 5.8.5, respectively.

Finally, we mention that Lebesgue's Theorem refers to Riemann integration only, not to other types of integration. For example, although Lebesgue integration agrees with Riemann integration for continuous functions, there are some very discontinuous functions that are Lebesgue integrable even though they are not Riemann integrable, for example the function given in Example 5.2.6 (3). See [Str00, Chapter 14] for a discussion of Lebesgue integration. We will not refer to Lebesgue integration further in this text.

Reflections

The proof of Lebesgue's Theorem is the lengthiest, and possibly the trickiest, proof in this book, and it is certainly acceptable to skip this proof upon first reading. Indeed, one can have a solid understanding of introductory real analysis without knowing Lebesgue's Theorem at all. However, it is hard to skip the statement of this theorem and feel that one has a good grasp of the nature of integrable functions. In contrast to differentiable functions, where we have a simple intuitive idea of differentiability in terms of graphs of functions not having "corners" or vertical tangent lines, there is no correspondingly simple picture of what makes a function integrable. Lebesgue's Theorem provides the closest thing we have to an intuitive understanding of integrable functions. As for the proof of Lebesgue's Theorem, though skipping it is understandable, the reader is encouraged to work through it, to appreciate the cleverness involved, and to see precisely how the notion of measure zero is used in the proof. Moreover, though real analysis would be a very tedious subject if all the proofs were as long as that of Lebesgue's Theorem, working through a long and difficult proof on occasion is a worthwhile endeavor somewhat similar to challenging physical exercise—getting in shape sometimes requires us to push ourselves to the limit of what we had thought possible.

Exercises

Exercise 5.8.1. **[Used in Example 5.8.2.]** Prove that any finite subset of $\mathbb{R}$ has measure zero.

Exercise 5.8.2. **[Used in Example 5.8.2.]** Complete the proof of Example 5.8.2 (3). That is, treat the case where I is not a closed bounded interval.

Exercise 5.8.3. Let $[a,b] \subseteq \mathbb{R}$ be a non-degenerate closed bounded interval, and let $f\colon [a,b] \to \mathbb{R}$ be a function. Suppose that f is integrable, that $f(x) \geq 0$ for all $x \in [a,b]$ and that the set $\{x \in [a,b] \mid f(x) > 0\}$ does not have measure zero. Prove that $\int_a^b f(x)\,dx > 0$.

Exercise 5.8.4. **[Used in Section 5.8.]** Use Lebesgue's Theorem (Theorem 5.8.5) to give an alternative (and shorter) proof of Theorem 5.5.1 (1).

Exercise 5.8.5. **[Used in Section 5.8.]** Use Lebesgue's Theorem (Theorem 5.8.5) to give an alternative (and shorter) proof of Theorem 5.5.4 (2).

Exercise 5.8.6. Use Lebesgue's Theorem (Theorem 5.8.5) to give an alternative (and shorter) proof of the first part of Theorem 5.5.5, which is the fact that if f is integrable, then $|f|$ is integrable. (Do not try to prove the inequality in Theorem 5.5.5 using Lebesgue's Theorem.)

Exercise 5.8.7. **[Used in Section 5.6.]** Find an example of a function with domain a non-degenerate closed bounded interval that is integrable and has an antiderivative, but that is not continuous. It is acceptable if the lack of continuity is asserted without proof, but a proof must be provided for the other two properties. One way to construct an example is to modify Example 4.2.5 (1).

Exercise 5.8.8. Let $[a,b] \subseteq \mathbb{R}$ be a non-degenerate closed bounded interval, and let $f\colon [a,b] \to \mathbb{R}$ be a function. Suppose that f is monotone. The purpose of this exercise is to prove that the set of numbers at which f is discontinuous is countable. It will then follow from Example 5.8.2 (4) and Lebesgue's Theorem (Theorem 5.8.5) that a monotone function is integrable. (A previous proof that a monotone function is integrable was given in Exercise 5.4.12, using Theorem 5.4.7 rather than Lebesgue's Theorem; the present proof, though a bit longer, yields more information, because we learn about the set of numbers at which a monotone function is discontinuous.)

Suppose that f is increasing; the other case is similar, and we omit the details. If $f(a) = f(b)$, then the function is constant, and hence continuous. Now suppose that $f(a) < f(b)$.

For convenience, if $c \in [a,b]$ we will write $f_-(c)$ to denote $\lim_{x \to c^-} f(x)$, and $f_+(c)$ to denote $\lim_{x \to c^+} f(x)$, where we replace $\lim_{x \to c^-} f(x)$ with $f(a)$ when $c = a$, and we replace $\lim_{x \to c^+} f(x)$ with $f(b)$ when $c = b$.

(1) Let $c \in [a,b]$. Prove that $f_-(c)$ exists and $f_-(c) \leq f(c)$; prove that $f_+(c)$ exists and $f(c) \leq f_+(c)$. [Use Exercise 4.5.10 (1).]

(2) Let $c \in [a,b]$. Prove that f is discontinuous at c if and only if $f_-(c) < f_+(c)$.

(3) Let $c,d \in [a,b]$. Suppose that $c < d$. Prove that $f_+(c) \leq f_-(d)$.

(4) Let $E = \{r \in [a,b] \mid f \text{ is discontinuous at } r\}$. Let $h\colon E \to \mathbb{Q}$ be defined as follows. Let $r \in E$. By Part (2) of this exercise we know that $f_-(c) < f_+(c)$. It follows from Theorem 2.6.13 (1) that there is some $q \in \mathbb{Q}$ such that $f_-(c) < q < f_+(c)$. We then let $h(r) = q$. There will always be more than one possible value of q, but we choose one such value arbitrarily. Prove that h is injective. Because $\mathbb{Q}$ is countable, it follows from standard facts about countable sets that E is also countable; see [Blo10, Sections 6.5 and 6.6] for information about countable sets.

5.9 Area and Arc Length

The geometric motivation for integration is the need to find the area of curved regions of the plane. And yet, in the discussion of area via integration in calculus courses, a substantial aspect of the study of area is always skipped over. In calculus courses it is taken for granted that the integral of a non-negative function equals the area under the graph of the function, but a rigorous study of area requires a proof of this fact. Of course, such a proof requires that we start with a rigorous definition of the concept of area of subsets of the plane, and such a definition is precisely what is glossed over in calculus courses—because it involves technicalities not available in such courses, but which are available to us.

In Section 4.5 we saw the definition of some geometric properties of graphs of functions, for example increasing and decreasing. These properties were *defined* without reference to calculus, though it was then proved that for differentiable functions, it is possible to use derivatives to provide an easier way to verify whether or not a function satisfies these properties. We now have an analogous situation involving integrals. More specifically, we discuss the concepts of area and arc length, both of which will be defined without reference to calculus, and both of which can be computed much more easily using calculus in the case of integrable functions. In contrast to concepts such as increasing and decreasing, which are quite easy to define geometrically, the geometric definitions of area and arc length are rather tricky, making use of least upper bounds and greatest lower bounds.

We start with a discussion of area. Suppose that $A \subseteq \mathbb{R}^2$ is a set. We would like to associate to the set A a number called the area of A. It will turn out that it is not possible to do so for every set A, as will be seen in Example 5.9.7 (2). The basic idea for finding the area of the set A is to try to approximate A with polygons, the areas of which are easy to compute, and then take some sort of limit. However, it would be possible to do such an approximation in one of two ways, using either polygons that are contained in A, or polygons that contain A; in the former case the "limit" would be computed via a least upper bound, and in the latter case via a greatest lower bound. To make sure that there are polygons that contain the set A, we restrict our attention to bounded subsets of $\mathbb{R}^2$, as will be defined below. If the set A is to have something that we would want to call area, then it ought to be the case that the same result is obtained using polygons contained in A and polygons that contain A, and we will define the area of A only when we have such equality. To make things as simple as possible technically, and in order to avoid having to define polygons in general, we restrict our attention to those polygons that are made up out of rectangles that have edges that are parallel to the x-axis and the y-axis.

Definition 5.9.1. Let $S \subseteq \mathbb{R}^2$.

1. The set S is **bounded** if there are closed bounded intervals $[a,b],[c,d] \subseteq \mathbb{R}$ such that $S \subseteq [a,b] \times [c,d]$.
2. The set S is a **rectangle** if $S = [a,b] \times [c,d]$ for some closed bounded intervals $[a,b],[c,d] \subseteq \mathbb{R}$.
3. Suppose that S is a rectangle. Then $S = [a,b] \times [c,d]$ for some closed bounded intervals $[a,b],[c,d] \subseteq \mathbb{R}$. The **interior** of S is the set $(a,b) \times (c,d)$. The rectangle S is **non-degenerate** if $[a,b]$ and $[c,d]$ are both non-degenerate intervals. △

Our use of the term "rectangle" here is restricted, for convenience, to those rectangles in the plane that have edges that are parallel to the x-axis and y-axis; we will not use any other type of rectangle.

Observe that the closed bounded intervals in the definition of rectangles are allowed to be degenerate, meaning single points, which therefore means that rectangles are themselves allowed to be degenerate, meaning vertical line segments of the form $[a,a] \times [c,d]$, horizontal line segments of the form $[a,b] \times [c,c]$ and single points of the form $[a,a] \times [c,c]$. Degenerate rectangles have empty interiors. Although line segments and points are not normally considered rectangles, viewing them as such allows us to avoid some special cases.

Definition 5.9.2. A **special polygon** is a collection of finitely many rectangles in $\mathbb{R}^2$ such that the interiors of any two of the rectangles are disjoint. If S is a special polygon, the **underlying space** of S, denoted $U(S)$, is the union of the rectangles in S. △

A rectangle in $\mathbb{R}^2$ can be thought of as a special polygon that has one element.

See Figure 5.9.1 for an example of a special polygon that has four rectangles, one of which is degenerate. Observe that the rectangles that make up a special polygon are not required to touch each other, so that a special polygon need not be "connected" (a term we have not defined, and will not be using rigorously).

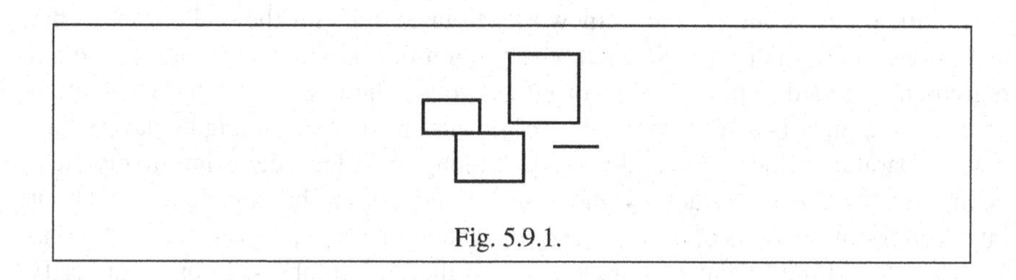

Fig. 5.9.1.

We now turn to the areas of rectangles and special polygons, which are very simple to define, but which are the basis for finding the areas of more complicated subsets of $\mathbb{R}^2$.

Definition 5.9.3. Let $S \subseteq \mathbb{R}^2$.

1. Suppose that S is a rectangle. Then $S = [a,b] \times [c,d]$ for some closed bounded intervals $[a,b],[c,d] \subseteq \mathbb{R}$. The **area** of S, denoted $A(S)$, is defined by $A(S) = (b-a)(d-c)$.
2. Suppose that S is a special polygon. Then S is a collection of finitely many rectangles in $\mathbb{R}^2$ such that the interiors of any two of the rectangles are disjoint. The **area** of S, denoted $A(S)$, is the sum of the areas of the rectangles in S. $\triangle$

In order to avoid being distracted by some elementary, but tedious, proofs involving rectangles in the plane, we will state without proof some properties of special polygons that we will need. Suppose that $P,Q \subseteq \mathbb{R}^2$ are special polygons. Properly speaking, if we write $P \cup Q$, that does not define a special polygon, because some of the rectangles in P and Q might have interiors that are not disjoint. However, it can be shown (by dividing up the rectangles in each of P and Q) that there is a special polygon $W \subseteq \mathbb{R}^2$ such that $U(W) = U(P) \cup U(Q)$. In order to avoid cumbersome notation, we will abuse notation and write "$P \cup Q$" to mean the special polygon W. Hence $U(P \cup Q) = U(P) \cup U(Q)$. A similar idea holds for "$P \cap Q$" and "$P - Q$." Also, if $x \in U(Q) - U(P)$, it can be shown that there is a non-degenerate rectangle $R \subseteq \mathbb{R}^2$ such that $x \in R \subseteq U(Q) - U(P)$. Additionally, it can be shown that the expected area formulas for special polygons hold, for example $A(P \cup Q) = A(P) + A(Q) - A(P \cap Q)$, and if $U(P) \subseteq U(Q)$, then $A(P) \leq A(Q)$ and $A(P - Q) = A(P) - A(Q)$.

The basic idea for proving the above facts about special polygons is that if the rectangles in a special polygon are broken up into subrectangles by intersecting the original rectangles with a horizontal or vertical line, then the special polygon consisting of the smaller rectangles will have the same underlying space and the same area as the original special polygon. Then, when we wish to take the union, intersection or set difference of two special polygons, we first break up the two special polygons using the lines containing the edges of all the rectangles in the two special polygons, and then the intersection of the underlying spaces of the original special polygons consists of a (possibly empty) union of rectangles that are in both of the new special polygons.

The reader might be concerned that the use of such geometric facts about rectangles in the plane is outside the framework of our axioms for the real numbers, but in fact even here the proofs ultimately rely upon our axioms. There are two main ingredients needed to make such a proof rigorous. The first is that the intersection of two rectangles is also a rectangle (recall that all of our rectangles have edges that are parallel to the x-axis and y-axis). Although this fact seems intuitively clear geometrically, it is really not a geometric statement at all, but a general fact about intersections of products of sets; see [Blo10, Theorem 3.3.12] for this fact about sets.

The second ingredient is more specific to the real numbers. In order to verify the above statements about breaking up the rectangles in a special polygon, we would need to proceed one step at a time, in which we take a single rectangle, and break it up into two rectangles by intersecting it with a horizontal or vertical line, and verifying that the two new rectangles have the same combined area as the original rectangle. For example, suppose that $S = [a,b] \times [c,d]$ for some closed bounded intervals $[a,b], [c,d] \subseteq \mathbb{R}$, and we wish to break up S into two rectangles using the vertical line $x = p$. If $[a,b]$ is degenerate, or if p is not in the interior of $[a,b]$, then there is nothing to do, so suppose that $[a,b]$ is non-degenerate and that $p \in (a,b)$. We break up the rectangle S into two rectangles, and form a new special polygon $T = \{[a,p] \times [c,d], [p,b] \times [c,d]\}$. We then need the Trichotomy Law to prove that $[a,p] \cup [p,b] = [a,b]$, which in turn is needed to prove that $U(T) = ([a,p] \times [c,d]) \cup ([p,b] \times [c,d]) = ([a,p] \cup [p,b]) \times [c,d] = [a,b] \times [c,d] = U(S)$, and we need the Distributive Law and the Associative, Commutative, Identity and Inverses Laws for Addition to prove that $A(T) = (p-a)(d-c) + (b-p)(d-c) = [(p-a) + (b-p)](d-c) = (b-a)(d-c) = A(S)$. What appears to be a geometric fact about rectangles in the plane is really an application of the properties of the real numbers.

Assuming the above facts about special polygons, we now find the area of a more general subset of $\mathbb{R}^2$ by looking at the least upper bound of the areas of the special polygons that have underlying spaces contained in the set, and the greatest lower bound of the areas of the special polygons that have underlying spaces containing the set. The following definition and lemma are the first step in this process.

Definition 5.9.4. Let $S \subseteq \mathbb{R}^2$ be a non-empty set. Suppose that S is bounded. Let

$$I_S = \{A(P) \mid P \text{ is a special polygon in } \mathbb{R}^2 \text{ such that } U(P) \subseteq S\}$$

and

$$O_S = \{A(Q) \mid Q \text{ is a special polygon in } \mathbb{R}^2 \text{ such that } S \subseteq U(Q)\},$$

The **inner content** of S, denoted $IC(S)$, is defined by $IC(S) = \text{lub}\, I_S$, and the **outer content** of S, denoted $OC(S)$, is defined by $OC(S) = \text{glb}\, O_S$. △

The first part of the following lemma shows that Definition 5.9.4 makes sense.

Lemma 5.9.5. *Let $S \subseteq \mathbb{R}^2$ be a non-empty set. Suppose that S is bounded.*

1. *I_S has a least upper bound, and O_S has a greatest lower bound.*

2. $IC(S) \leq OC(S)$.
3. $IC(S) = OC(S)$ *if and only if for each* $\varepsilon > 0$, *there are special polygons* P *and* Q *in* $\mathbb{R}^2$ *such that* $U(P) \subseteq S \subseteq U(Q)$ *and* $A(Q) - A(P) < \varepsilon$.

Proof. We prove the three parts of the lemma together. Because $S \neq \emptyset$, there is some $(x,y) \in S$. Hence the special polygon $V = \{[x,x] \times [y,y]\}$ has the property $U(V) \subseteq S$, and therefore $A([x,x] \times [y,y]) = 0$ is in I_S. Because S is bounded, there is some rectangle $[a,b] \times [c,d] \subseteq \mathbb{R}^2$ such that $S \subseteq [a,b] \times [c,d]$. Hence the special polygon $W = \{[a,b] \times [c,d]\}$ has the property $S \subseteq U(W)$, and therefore $A([a,b] \times [c,d]) = (b-a)(d-c)$ is in O_S. Therefore I_S and O_S are non-empty. Let $P, Q \subseteq \mathbb{R}^2$ be special polygons such that $U(P) \subseteq S \subseteq U(Q)$. Then $A(P) \leq A(Q)$. Hence, if $a \in I_S$ and $b \in O_S$, then $a \leq b$. The three parts of the lemma now follow immediately from the No Gap Lemma (Lemma 2.6.6). □

We are now ready to give the definition of the area of subsets of $\mathbb{R}^2$, when such areas exist.

Definition 5.9.6. Let $S \subseteq \mathbb{R}^2$ be a non-empty set. The set S is **squarable** if S is bounded and $IC(S) = OC(S)$. If S is squarable, the **area** of S, denoted $A(S)$, is defined by $A(S) = IC(S) = OC(S)$. △

Observe that the term "squarable" is analogous to "differentiable" and "integrable," whereas the term "area" is analogous to "derivative" and "integral." (There would be a closer analogy if we used the term "area-able" rather than "squarable," but the former term does not, fortunately, appear in the literature, whereas "squarable" does.) The concept of squarability is the 2-dimensional case of the more general concept of Jordan measurable, which is defined for subsets of $\mathbb{R}^n$ for all $n \in \mathbb{N}$. Although the concept of squarability is the most elementary way to define area of subsets of $\mathbb{R}^2$, and hence we are using it here, the more general concept of Jordan measure is not widely studied today, because it has been superseded by the more powerful, though slightly more difficult to define, concept of Lebesgue measure. This latter type of measure is used as the basis for the definition of the Lebesgue integral, which is widely used in more advanced treatments of real analysis. See [Str00, Chapter 14] for an exposition of Lebesgue measure and integration.

It is hard to evaluate the squarability of most subsets of $\mathbb{R}^2$ directly from the definition, and hence the following examples, where we can make such an evaluation, might seem to be either uninteresting or a lot of work for nothing, but it is the best we can do.

Example 5.9.7.

(1) Let $S \subseteq \mathbb{R}^2$ be the triangle with vertices $(0,0)$, $(1,0)$ and $(0,1)$. The reader is familiar with the formula for the area of a triangle, and according to that formula the area of this triangle is $\frac{1}{2}$. However, we have not given a proof of that area formula using the definition of area in Definition 5.9.6; we need to show that S is squarable using only what we have stated so far regarding squarability. The reader might then suggest the following argument, which is really just the proof of the area formula for triangles in our particular case: we know that the area of the unit square is 1, which

holds because a square is a rectangle, and we know that the area of the triangle S is half the area of the unit square, and therefore the area of S must be $\frac{1}{2}$. However, such reasoning requires a number of assumptions that, while ultimately correct, would need to be proved, including the fact that the triangle is squarable (not every subset of a squarable set is squarable, as will be seen in Part (2) of this example); the fact that congruent subsets of the plane have equal areas; and the fact that the area of the union of two subsets of the plane that do not have any overlap of their interiors is the sum of the areas of the two parts. Proving all of those facts would take more effort than showing directly in the present case that S is squarable.

For each $\varepsilon > 0$, we will construct special polygons $P, Q \subseteq \mathbb{R}^2$ such that $U(P) \subseteq S \subseteq U(Q)$, and that $\frac{1}{2} - \varepsilon < A(P) \leq \frac{1}{2} \leq A(Q) < \frac{1}{2} + \varepsilon$. It will then follow from Exercise 2.6.8, which is a variant of the No Gap Lemma (Lemma 2.6.6), that

$$IC(S) = OC(S) = \frac{1}{2},$$

and we will then deduce that S is squarable and that $A(S) = \frac{1}{2}$.

Let $\varepsilon > 0$. There are many ways to construct the desired special polygons P and Q, and we will show one such construction. By Corollary 2.6.8 (2) there is some $n \in \mathbb{N}$ such that $\frac{1}{n} < \varepsilon$. In Figure 5.9.2 we see a way of constructing P and Q (the figure shows the case $n = 4$). In general, the special polygons P and Q each have n rectangles (one of the rectangles in P is degenerate, and so it is not visible in the figure), where all the rectangles have height $\frac{1}{n}$, and where the rectangles in P have widths $0, \frac{1}{n}, \ldots, \frac{n-1}{n}$, respectively, and the rectangles in Q have widths $\frac{1}{n}, \frac{2}{n}, \ldots, 1$, respectively. By using the well-known formula for the sum of the first n integers, found in Exercise 2.5.5, it is seen that $A(P) = \frac{1}{2} - \frac{1}{2n}$ and $A(Q) = \frac{1}{2} + \frac{1}{2n}$. Then clearly $\frac{1}{2} - \varepsilon < A(P) \leq \frac{1}{2} \leq A(Q) < \frac{1}{2} + \varepsilon$.

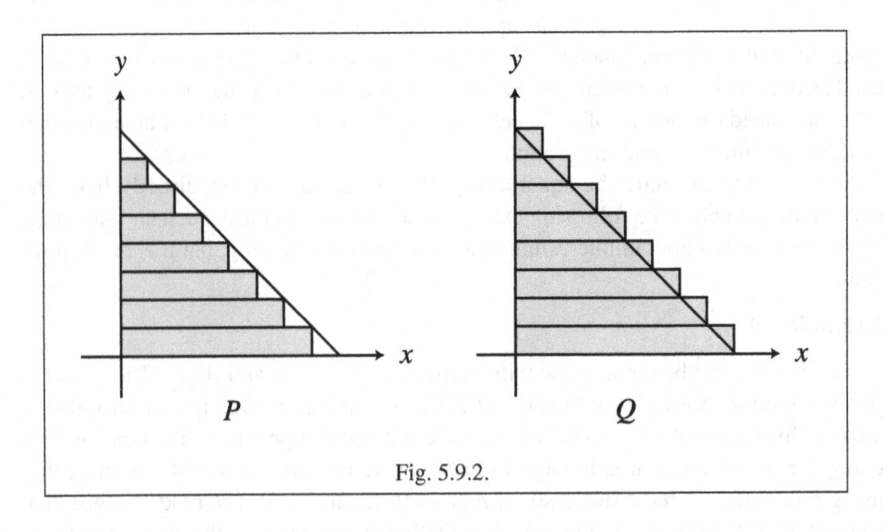

Fig. 5.9.2.

(2) Let $T = ([0,1] \times [0,1]) \cap (\mathbb{Q} \times \mathbb{Q}) \subseteq \mathbb{R}^2$. We will show that T is not squarable.

Let $[a,b] \times [c,d] \subseteq \mathbb{R}^2$ be a rectangle. If both $[a,b]$ and $[c,d]$ are non-degenerate, then each contains an irrational number by Theorem 2.6.13 (2), and it follows that $[a,b] \times [c,d]$ could not be a subset of T. Hence, if $P \subseteq \mathbb{R}^2$ is a special polygon such that $U(P) \subseteq T$, then P is the union of finitely many degenerate rectangles, and hence $A(P) = 0$. It follows that $IC(T) = 0$.

Let $Q \subseteq \mathbb{R}^2$ be a special polygon such that $T \subseteq U(Q)$. Suppose that $[0,1] \times [0,1] \not\subseteq Q$. Then there is some $x \in ([0,1] \times [0,1]) - Q$. The rectangle $[0,1] \times [0,1]$ is a special polygon, and then using a remark made earlier in this section there is a non-degenerate rectangle $R \subseteq \mathbb{R}^2$ such that $x \in R \subseteq U([0,1] \times [0,1]) - U(T)$. By Theorem 2.6.13 (1) there exists a point with rational coordinates in R, which is a contradiction to the fact that $T \subseteq U(Q)$. Hence $[0,1] \times [0,1] \subseteq Q$. It now follows that $1 = A([0,1] \times [0,1]) \leq A(Q)$. Hence 1 is a least element of $\mathcal{O}_T$, and it follows from Exercise 2.6.2 (3) that $OC(T) = \operatorname{glb} \mathcal{O}_T = 1$.

Because $IC(T) \neq OC(T)$, then T is not squarable. ◊

Before proceeding, we note that there is one slight problem with the definition of the area of squarable sets that needs to be clarified. Let $S \subseteq \mathbb{R}^2$ be a rectangle. Then the area of S was defined to be $A(S) = (b-a)(d-c)$ in Definition 5.9.3 (1). On the other hand, we can think of S simply as a subset of $\mathbb{R}^2$, and as such we can ask whether it is squarable, and if it is, whether its area using Definition 5.9.6 equals $(b-a)(d-c)$. If these two approaches to the area of a rectangle do not yield the same result, then our definition of area would be very questionable. Fortunately, as seen in Exercise 5.9.1, everything works out as one would hope.

Although it is hard to evaluate the squarability of most subsets of $\mathbb{R}^2$ directly from the definition, in the special case of the region under the graph of a non-negative integrable function, we can use integration to compute area, as seen in Theorem 5.9.9 below. We start with the following definition.

Definition 5.9.8. Let $f, g\colon [a,b] \to \mathbb{R}$ be functions. Suppose that $f(x) \leq g(x)$ for all $x \in [a,b]$. The **region between the graphs** of f and g, denoted $R_a^b(f,g)$, is defined by

$$R_a^b(f,g) = \{(x,y) \in \mathbb{R}^2 \mid a \leq x \leq b \text{ and } f(x) \leq y \leq g(x)\}.$$

If $f(x) \geq 0$ for all $x \in [a,b]$, the **region under the graph** of f, denoted $R_a^b(f)$, is defined by

$$R_a^b(f) = \{(x,y) \in \mathbb{R}^2 \mid a \leq x \leq b \text{ and } 0 \leq y \leq f(x)\}.$$ △

For the proof of the following theorem, recall the concept of upper integral and lower integral defined in Definition 5.4.8.

Theorem 5.9.9. *Let $[a,b] \subseteq \mathbb{R}$ be a non-degenerate closed bounded interval, and let $f\colon [a,b] \to \mathbb{R}$ be a function. Suppose that f is bounded, and that $f(x) \geq 0$ for all $x \in [a,b]$. Then $R_a^b(f)$ is squarable if and only if f is integrable, and if f is integrable then $A(R_a^b(f)) = \int_a^b f(x)\,dx$.*

Proof. We will show that $IC(R_a^b(f)) = \underline{\int_a^b} f(x)\,dx$. A similar argument shows that $OC(R_a^b(f)) = \overline{\int_a^b} f(x)\,dx$, and we omit the details. It will then follow that $R_a^b(f)$ is squarable if and only if $\underline{\int_a^b} f(x)\,dx = \overline{\int_a^b} f(x)\,dx$, and by Theorem 5.4.10 we know that the latter condition holds if and only if f is integrable, and that if f is integrable, then $\int_a^b f(x)\,dx = \underline{\int_a^b} f(x)\,dx = \overline{\int_a^b} f(x)\,dx = IC(R_a^b(f)) = OC(R_a^b(f)) = A(R_a^b(f))$.

Let

$$\mathcal{L} = \{L(f,P) \mid P \text{ is a partition of } [a,b]\}.$$

By definition we know that $IC(R_a^b(f)) = \operatorname{lub} I_{R_a^b(f)}$ and $\underline{\int_a^b} f(x)\,dx = \operatorname{lub} \mathcal{L}$. We will prove that (1) $\mathcal{L} \subseteq I_{R_a^b(f)}$, and that (2) for each $x \in I_{R_a^b(f)}$, there is some $y \in \mathcal{L}$ such that $x \le y$. It will then follow from (1) together with Exercise 2.6.1 (1) that $\operatorname{lub} \mathcal{L} \le \operatorname{lub} I_{R_a^b(f)}$, and it will follow from (2) together with Exercise 2.6.3 (1) that $\operatorname{lub} I_{R_a^b(f)} \le \operatorname{lub} \mathcal{L}$, which together imply that $\operatorname{lub} I_{R_a^b(f)} = \operatorname{lub} \mathcal{L}$, which means that $IC(R_a^b(f)) = \underline{\int_a^b} f(x)\,dx$.

Let $V = \{x_0, x_1, \ldots, x_u\}$ be a partition of $[a,b]$. Let $\hat{V}$ be the special polygon defined by

$$\hat{V} = \left\{[x_{i-1}, x_i] \times \left[0, m_i^P(f)\right]\right\}_{i=1}^{u}. \tag{5.9.1}$$

It is evident that $A(\hat{V}) = L(f,V)$. It follows that $\mathcal{L} \subseteq I_{R_a^b(f)}$, and so (1) has been proved.

Let $x \in I_{R_a^b(f)}$. Then $x = A(P)$ for some special polygon $P \subseteq \mathbb{R}^2$ such that $U(P) \subseteq R_a^b(f)$. Then $P = \{[a_i, b_i] \times [c_i, d_i]\}_{i=1}^n$, where $[a_i, b_i] \times [c_i, d_i]$ is a rectangle for each $i \in \{1, \ldots, n\}$, and where the interiors of any two of these rectangles are disjoint. See Figure 5.9.3 (i).

Let $k \in \{1, \ldots, n\}$. Because $[a_k, b_k] \times [c_k, d_k] \subseteq R_a^b(f)$, it follows that $0 \le c_k \le d_k \le f(x)$ for all $x \in [a_k, b_k]$. We deduce that $[a_k, b_k] \times [c_k, d_k] \subseteq [a_k, b_k] \times [0, d_k] \subseteq R_a^b(f)$. Hence

$$U(P) = \bigcup_{i=1}^{n} [a_i, b_i] \times [c_i, d_i] \subseteq \bigcup_{i=1}^{n} [a_i, b_i] \times [0, d_i].$$

See Figure 5.9.3 (ii). It might be the case that the rectangles of the form $[a_i, b_i] \times [0, d_i]$ overlap in their interiors. However, by subdividing these rectangles vertically if necessary, we can find new rectangles $\left\{[p_j, q_j] \times [0, s_j]\right\}_{j=1}^{m}$, for some $m \in \mathbb{N}$, such that

$$U(P) \subseteq \bigcup_{i=1}^{n} [a_i, b_i] \times [0, d_i] \subseteq \bigcup_{j=1}^{m} [p_j, q_j] \times [0, s_j] \subseteq R_a^b(f),$$

and that the interiors of any two of these new rectangles are disjoint. See Figure 5.9.3 (iii). By renumbering the rectangles in $\left\{[p_j, q_j] \times [0, s_j]\right\}_{j=1}^{m}$ if necessary, we may suppose that $a \le p_1 \le q_1 \le p_2 \le q_2 \le \cdots \le p_m \le q_m \le b$. Because the numbers $a, p_1, q_1, \ldots, p_m, q_m, b$ might not all be distinct, we let $y_0, y_1, \ldots, y_v$ be the same list of numbers, in the same order, but with each number listed only once. Then $Q = \{y_0, y_1, \ldots, y_v\}$ is the partition of $[a,b]$. For each $r \in \{1, \ldots, v\}$, we form the rectangle $[y_{r-1}, y_r] \times [0, t_r]$, where

$$t_r = \begin{cases} s_j, & \text{if } [y_{r-1}, y_r] = [p_j, q_j] \text{ for some } j \in \{1, \ldots, m\} \\ 0, & \text{otherwise.} \end{cases}$$

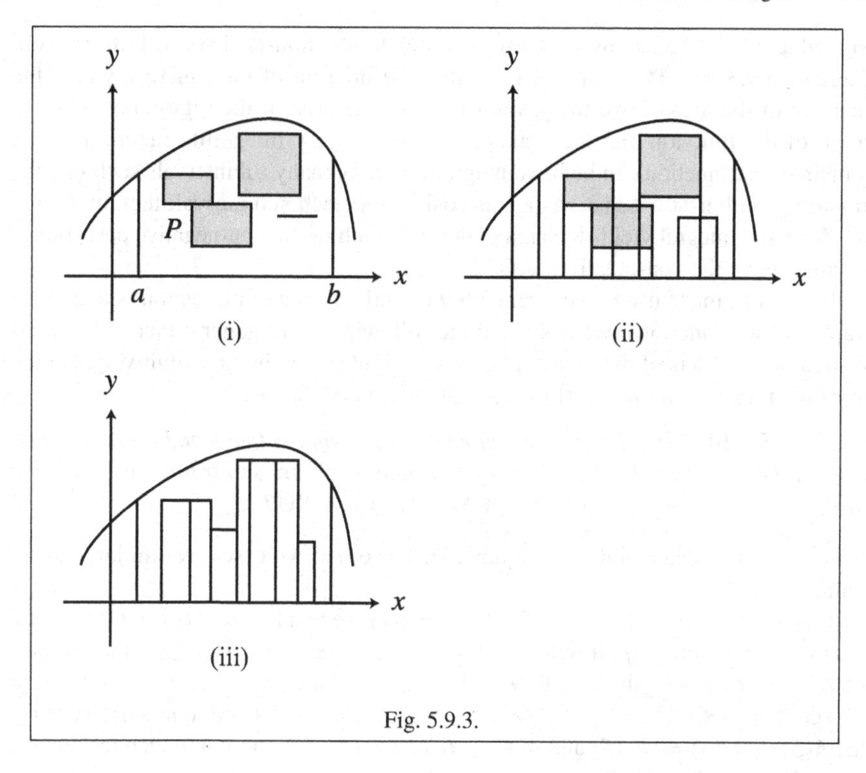

Fig. 5.9.3.

By definition, the collection of rectangles $\{[y_{r-1}, y_r] \times [0, t_r]\}_{r=1}^{v}$ consists of all rectangles of the form $[p_j, q_j] \times [0, s_j]$ for $j \in \{1, \ldots, m\}$, with possibly some additional degenerate rectangles located in the x-axis. It follows that

$$U(P) \subseteq \bigcup_{j=1}^{m} [p_j, q_j] \times [0, s_j] \subseteq \bigcup_{r=1}^{v} [y_{r-1}, y_r] \times [0, t_r] \subseteq R_a^b(f).$$

Let $r \in \{1, \ldots, v\}$. Then $[y_{r-1}, y_r] \times [0, t_r] \subseteq R_a^b(f)$. It follows that $t_r \leq f(x)$ for all $x \in [y_{r-1}, y_r]$. We deduce that $t_r \leq m_r^Q(f)$. Let $\hat{Q}$ be the special polygon defined analogously to Equation 5.9.1. Then

$$U(P) \subseteq \bigcup_{r=1}^{v} [y_{r-1}, y_r] \times [0, t_r] \subseteq U(\hat{Q}) \subseteq R_a^b(f).$$

We deduce that $x = A(P) \leq A(\hat{Q}) = L(f, Q)$. If we let $y = L(f, Q)$, then $y \in \mathcal{L}$. We have therefore shown that there is some $y \in \mathcal{L}$ such that $x \leq y$, and so (2) has been proved. □

It is important to stress that Theorem 5.9.9 applies only to non-negative integrable functions. For arbitrary integrable functions, it is not the case that integration yields the area between the graph of the function and the x-axis. Rather, integration gives "signed area," in the following sense. If a function is non-positive, it follows from Exercise 5.9.3 and Theorem 5.3.1 (3) that the integral of the function yields the negative of the area above the graph (meaning the area of the region between the graph of the function and the x-axis). For an arbitrary integrable function, if the domain of the function can be broken up into finitely many subintervals such that the function is either non-negative or non-positive on each subinterval, then by Corollary 5.5.9 the integral yields the area under the graph of the non-negative parts minus the area above the non-positive parts.

In order to make use of Theorem 5.9.9 to find the area of the region between the graphs of two functions, we first need the following more general fact about areas of squarable subsets of the plane. This result might seem obvious intuitively, but the proof requires a bit more effort than might at first be assumed.

Theorem 5.9.10. *Let $S,T \subseteq \mathbb{R}^2$ be non-empty sets. Suppose that S and T are bounded, and that $OC(S\cap T) = 0$. If any two of S, T and $S\cup T$ are squarable then so is the third, and if they are squarable then $A(S)+A(T) = A(S\cup T)$.*

Proof. Suppose that S and T are squarable. The other two cases are similar, and we omit the details.

Using the notation of Exercise 2.6.9, we will prove that $\text{lub}(I_S + I_T) = \text{lub}\, I_{S\cup T}$ and $\text{glb}(O_S + O_T) = \text{lub}\, O_{S\cup T}$. It will then follow from Exercise 2.6.9 (3) (4) that $\text{lub}\, I_S + \text{lub}\, I_T = \text{lub}\, I_{S\cup T}$ and $\text{glb}\, O_S + \text{glb}\, O_T = \text{lub}\, O_{S\cup T}$, and therefore that $IC(S)+IC(T) = IC(S\cup T)$ and $OC(S)+OC(T) = OC(S\cup T)$. Because S and T are squarable, we know that $A(S) = IC(S) = OC(S)$ and $A(T) = IC(T) = OC(T)$, and it will then follow that $IC(S\cup T) = IC(S)+IC(T) = OC(S)+OC(T) = OC(S\cup T)$, which means that $S\cup T$ is squarable, and that $A(S\cup T) = IC(S\cup T) = IC(S)+IC(T) = A(S)+A(T)$.

Let $z \in O_S + O_T$. Then $z = x+y$ for some $x \in O_S$ and $y \in O_T$. Then $x = A(P)$ and $y = A(Q)$ for some special polygons $P,Q \subseteq \mathbb{R}^2$ such that $S \subseteq U(P)$ and $T \subseteq U(Q)$. Then $S\cup T \subseteq U(P\cup Q)$, and $A(P\cup Q) = A(P)+A(Q)-A(P\cap Q) \le A(P)+A(Q)$. Hence there is some $w \in O_{S\cup T}$, specifically, the number $w = A(P\cup Q)$, such that $w \le x+y = z$. It then follows from Exercise 2.6.3 (2) that $\text{glb}\, O_{S\cup T} \le \text{glb}(O_S + O_T)$.

Let $a \in O_{S\cup T}$. Let $\varepsilon > 0$. Then $a = A(G)$ for some special polygon $G \subseteq \mathbb{R}^2$ such that $S\cup T \subseteq U(G)$. Because $OC(S\cap T) = 0$, it follows from Lemma 2.6.5 (2) that there is some $b \in O_{S\cap T}$ such that $b < \frac{\varepsilon}{2}$. Then $b = A(C)$ for some special polygon $C \subseteq \mathbb{R}^2$ such that $S\cap T \subseteq U(C)$. Without loss of generality, we may assume that $U(C) \subseteq U(G)$, because if not, we could replace C with $C\cap G$, and this replacement for C would have the same properties as the original C.

Because S is squarable, then by the definition of squarability together with Lemma 5.9.5 (3) there are special polygons $V,W \subseteq \mathbb{R}^2$ such that $U(V) \subseteq S \subseteq U(W)$ and $A(W)-A(V) < \frac{\varepsilon}{2}$. Without loss of generality, we may assume that $U(W) \subseteq U(G)$, because if not, we could replace W with $W\cap G$, and this replacement for W would have the same properties as the original W. Observe that $U(V) \subseteq U(G)$. Let $B = [G-V]\cup C$. Then $A(B) \le [A(G)-A(V)]+A(C)$. It is left to the reader to verify that $T \subseteq U(B)$, and that $U(W\cap B) \subseteq [U(W)-U(V)]\cup U(C)$. Then $A(W\cap B) \le [A(W)-A(V)]+A(C) < \frac{\varepsilon}{2}+\frac{\varepsilon}{2} = \varepsilon$. It can also be verified that $U(W\cup B) = U(G)$. Therefore $A(G) = A(W)+A(B)-A(W\cap B) > A(W)+A(B)-\varepsilon$, which implies that

$A(W)+A(B) < A(G)+\varepsilon$. Let $r = A(W)$ and $s = A(B)$. Then $r+s \le a+\varepsilon$. Because $r+s \in O_S + O_T$, it follows from Exercise 2.6.3 (2) that $\text{glb}(O_S + O_T) \le \text{glb}\, O_{S\cup T}$. We deduce that $\text{glb}(O_S + O_T) = \text{glb}\, O_{S\cup T}$.

Let $e \in I_S + I_T$. Then $e = f+g$ for some $f \in I_S$ and $g \in I_T$. Then $f = A(H)$ and $g = A(K)$ for some special polygons $H, K \subseteq \mathbb{R}^2$ such that $U(H) \subseteq S$ and $U(K) \subseteq T$. Then $U(H \cup K) \subseteq S \cup T$, and $A(H\cup K) = A(H)+A(K)-A(H\cap K)$, and $U(H\cap K) \subseteq S\cap T$. Because $OC(S\cap T) = 0$, then Lemma 5.9.5 (2) implies that $IC(S\cap T) = 0$, and it follows that $A(H\cap K) = 0$. Therefore $A(H\cup K) = A(H)+A(K)$. Hence there is some $m \in I_{S\cup T}$, specifically, the number $m = A(H\cup K)$, such that $m = f+g = e$. It then follows from Exercise 2.6.3 (1) that $\text{lub}\, I_{S\cup T} \ge \text{lub}(I_S + I_T)$.

Let $k \in I_{S\cup T}$. Let $\varepsilon > 0$. Then $k = A(L)$ for some special polygon $L \subseteq \mathbb{R}^2$ such that $U(L) \subseteq S\cup T$. Let b and C be as above. Then $U(L-C) \subseteq (S-T)\cup(T-S)$. Let M be the collection of those rectangles in $L-C$ that are contained in $S-T$, and let N be the collection of those rectangles in $L-C$ that are contained in $T-S$. We see that $L-C = M\cup N$, and $M\cap N = \emptyset$, and $U(M) \subseteq S$ and $U(N) \subseteq T$. Then $A(M)+A(N) = A(M\cap N) \ge A(L)-A(C) > A(L)-\varepsilon$. Let $v = A(M)$ and $w = A(N)$. Then $v+w \ge k-\varepsilon$. Because $v+w \in I_S + I_T$, it follows from Exercise 2.6.3 (1) that $\text{lub}(I_S + I_T) \ge \text{lub}\, I_{S\cup T}$. We deduce that $\text{lub}(I_S + I_T) = \text{lub}\, I_{S\cup T}$. □

The following theorem is proved using Theorem 5.9.9 and Theorem 5.9.10.

Theorem 5.9.11. *Let $[a,b] \subseteq \mathbb{R}$ be a non-degenerate closed bounded interval, and let $f,g\colon [a,b] \to \mathbb{R}$ be functions. Suppose that $f(x) \le g(x)$ for all $x \in [a,b]$, and that f and g are integrable. Then the region between the graphs of f and g is squarable and $A(R_a^b(f,g)) = \int_a^b [f-g](x)\,dx$.*

Proof. Left to the reader in Exercise 5.9.7. □

As mentioned above, we have defined area using only special polygons, and not arbitrary polygons. It is technically simpler to use only special polygons, but the reader would have good reason to ask whether we would have obtained a different definition of area had we used more general polygons. Fortunately, it turns out that computing inner content and outer content using arbitrary polygons always gives the same result as using only special polygons. We will not go through a proof of this fact, because doing so requires a rigorous definition of arbitrary polygons (which for the purpose of computing area must include degenerate polygons); one of the advantages of using special polygons is precisely that it allows us to avoid such a definition. The intuitive idea of such a proof, however, is seen in Figure 5.9.4. Suppose that we have an arbitrary polygon contained in a subset of the plane, as in Figure 5.9.4 (i). If the polygon has an edge that is not parallel to the x-axis or y-axis, then we can replace it with a "staircase" and break up the modified polygon into rectangles, as in Figure 5.9.4 (ii), yielding a special polygon. Although the new special polygon has smaller area than the original polygon, the special polygon can be chosen to have area within any $\varepsilon > 0$ of the area of the original polygon. It then follows from Exercise 2.6.1 (1) and Exercise 2.6.3 (1) that the least upper bound of the areas of special polygons contained in the region of the plane equals the least upper bound of the areas of all polygons contained in the region. Hence the definition of inner content would not change had we used arbitrary polygons instead of just special polygons. A similar result holds for outer content, and so we might as well stick with special polygons because they are easier to work with.

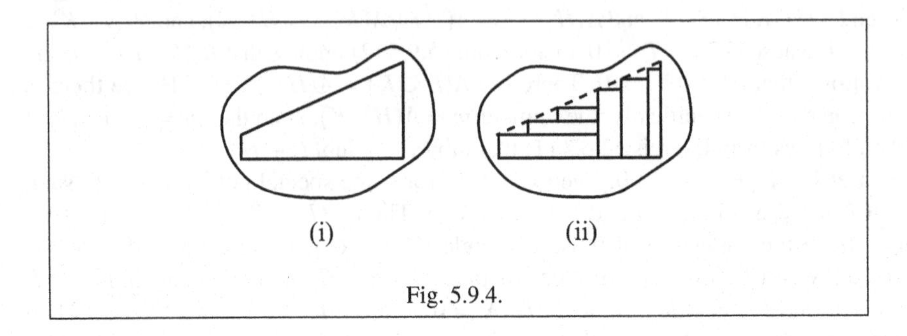

Fig. 5.9.4.

We now turn to the arc length of graphs of functions. It is also possible to find the arc length of more general curves in the plane, but doing so requires using functions $\mathbb{R} \to \mathbb{R}^2$ rather than functions $\mathbb{R} \to \mathbb{R}$, and hence is outside the scope of this book.

As was the case in our discussion of area, we start with a geometric definition of arc length, which does not involve differentiation and integration, and only after that do we show that for nicely behaved functions we can use integration to compute arc length. The intuitive idea is that we calculate the arc length of a graph of a function by approximating the graph by polygonal arcs made up of straight line segments between points on the graph, and then taking some sort of limit. As was the case with our definition of area, the "limit" will be computed by finding a least upper bound. In contrast to the definition of area, however, where we approximated a subset of the plane from both the inside and the outside, and hence we needed both least upper bounds and greatest lower bounds, for the arc length of the graph of a function, there is no "inside" or "outside," and we will need only least upper bounds, because the length of a polygonal approximation of the graph of a function is always less than or equal to the length of the graph itself.

Just as we based our discussion of the area of arbitrary subsets of the plane on the formula for the area of a rectangle, and we then added up such areas to approximate the area of a more complicated set, for arc length we base our discussion on the formula for the length of a line segment, and we then add up such lengths to approximate the arc length of a more complicated curve. We start with the following definition, which uses partitions of closed bounded intervals to determine polygonal approximations to graphs of functions, and the Pythagorean Theorem to compute lengths of line segments. See Figure 5.9.5 (i) for the graph of a function, and Part (ii) of that figure for a polygonal arc that approximates the graph.

Definition 5.9.12. Let $[a,b] \subseteq \mathbb{R}$ be a non-degenerate closed bounded interval, let $f\colon [a,b] \to \mathbb{R}$ be a function and let $P = \{x_0, x_1, \ldots, x_n\}$ be a partition of $[a,b]$. The **polygonal sum** of f with respect to P, denoted $C(f,P)$, is defined by

$$C(f,P) = \sum_{i=1}^{n} \sqrt{[x_i - x_{i-1}]^2 + [f(x_i) - f(x_{i-1})]^2}. \qquad \triangle$$

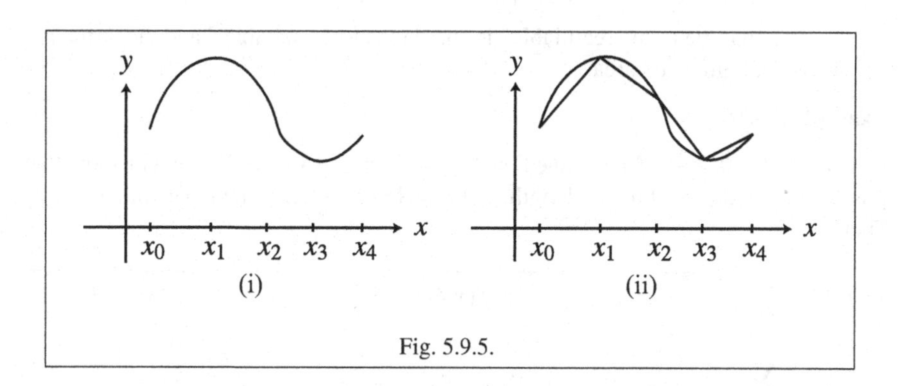

Fig. 5.9.5.

Example 5.9.13. Let $f\colon [0,1] \to \mathbb{R}$ be defined by $f(x) = x^{\frac{3}{2}}$ for all $x \in [0,1]$, and let $P = \{0, \frac{1}{4}, \frac{1}{2}, \frac{3}{4}, 1\}$. Then

$$\begin{aligned} C(f,P) &= \sqrt{\left[\tfrac{1}{4} - 0\right]^2 + \left[\left(\tfrac{1}{4}\right)^{\frac{3}{2}} - 0^{\frac{3}{2}}\right]^2} + \sqrt{\left[\tfrac{1}{2} - \tfrac{1}{4}\right]^2 + \left[\left(\tfrac{1}{2}\right)^{\frac{3}{2}} - \left(\tfrac{1}{4}\right)^{\frac{3}{2}}\right]^2} \\ &\quad + \sqrt{\left[\tfrac{3}{4} - \tfrac{1}{2}\right]^2 + \left[\left(\tfrac{3}{4}\right)^{\frac{3}{2}} - \left(\tfrac{1}{2}\right)^{\frac{3}{2}}\right]^2} + \sqrt{\left[1 - \tfrac{3}{4}\right]^2 + \left[1^{\frac{3}{2}} - \left(\tfrac{3}{4}\right)^{\frac{3}{2}}\right]^2} \\ &\approx 1.436. \end{aligned}$$

The actual length of the curve will be computed in Example 5.9.18. ◊

The following lemma about polygonal sums is quite reasonable intuitively if we think about lengths of edges of triangles, though the proof, which uses some ideas from linear algebra, is a bit more complicated than might be expected. These ideas are sketched out in an exercise, with no assumption that the reader is familiar with linear algebra.

Lemma 5.9.14. *Let $[a,b] \subseteq \mathbb{R}$ be a non-degenerate closed bounded interval, let $f\colon [a,b] \to \mathbb{R}$ be a function and let P and Q be partitions of $[a,b]$. If Q is a refinement of P, then $C(f,Q) \geq C(f,P)$.*

Proof. Left to the reader in Exercise 5.9.9. □

We are now ready to give the definition of the arc length of graphs of function, when such arc lengths exist.

Definition 5.9.15. Let $[a,b] \subseteq \mathbb{R}$ be a non-degenerate closed bounded interval, and let $f\colon [a,b] \to \mathbb{R}$ be a function. Let

$$\mathcal{A}_f = \{C(f,P) \mid P \text{ is a partition of } [a,b]\}.$$

The function f is **rectifiable** if $\mathcal{A}_f$ is bounded above. If f is rectifiable, the **arc length** of f, denoted $L_a^b(f)$, is defined by $L_a^b(f) = \text{lub}\,\mathcal{A}_f$. △

Observe that the term "rectifiable" is analogous to "squarable," and the term "arc length" is analogous to "area."

Example 5.9.16.

(1) Let $f\colon [0,2] \to \mathbb{R}$ be defined by $f(x) = 3x$ for all $x \in [0,2]$. We will show that f is rectifiable and find its arc length. Let $P = \{x_0, x_1, \ldots, x_n\}$ be a partition of $[0,2]$. Then

$$\begin{aligned} C(f,P) &= \sum_{i=1}^{n} \sqrt{[x_i - x_{i-1}]^2 + [f(x_i) - f(x_{i-1})]^2} = \sum_{i=1}^{n} \sqrt{[x_i - x_{i-1}]^2 + [3x_i - 3x_{i-1}]^2} \\ &= \sum_{i=1}^{n} \sqrt{10}(x_i - x_{i-1}) = \sqrt{10}(x_n - x_0) = 2\sqrt{10}. \end{aligned}$$

Hence, all polygonal sums of f are equal to $2\sqrt{10}$, and therefore $\mathcal{A}_f = \{2\sqrt{10}\}$. Clearly $\mathcal{A}_f$ is bounded above, and hence f is rectifiable. Moreover, we see that $L_0^2(f) = \text{lub}\,\mathcal{A}_f = 2\sqrt{10}$. Given that the graph of f is a straight line, a quick calculation with the Pythagorean Theorem shows that the arc length that we computed with polygonal sums is the expected answer.

(2) Let $r\colon [0,1] \to \mathbb{R}$ be defined by

$$r(x) = \begin{cases} 1, & \text{if } x \in \mathbb{Q} \cap [0,1] \\ 0, & \text{otherwise.} \end{cases}$$

It was shown in Example 3.3.3 (6) that r is discontinuous everywhere, and in Example 5.2.6 (3) that r is not integrable. We will now show that this function is not rectifiable. Let $M \in \mathbb{R}$. We will show that there is a partition Q of $[0,1]$ such that $C(r,Q) > M$. It will follow that $\mathcal{A}_r$ is not bounded above, which means that r is not rectifiable.

By Corollary 2.6.8 (1) there is some $p \in \mathbb{Z}$ such that $M < p$. By taking a larger value of p if necessary, we may assume that $p > 0$ and that p is even. Then $p = 2n$ for some $n \in \mathbb{N}$. Let $Q = \{x_0, x_1, \ldots, x_{2n}\}$ be a partition of $[0,1]$ defined as follows. For each even number $i \in \{0,1,\ldots,2n\}$, let $x_i = \frac{i}{2n}$. For each odd number $i \in \{0,1,\ldots,2n\}$, let x_i be an irrational number such that $x_{i-1} < x_i < x_{i+1}$, where such an x_i can be found by Theorem 2.6.13 (2). Then

$$\begin{aligned} C(r,Q) &= \sum_{i=1}^{2n} \sqrt{[x_i - x_{i-1}]^2 + [f(x_i) - f(x_{i-1})]^2} \\ &= \sum_{i=1}^{n} \sqrt{[x_{2i-1} - x_{2i-2}]^2 + [f(x_{2i-1}) - f(x_{2i-2})]^2} \end{aligned}$$

$$+\sum_{i=1}^{n}\sqrt{[x_{2i}-x_{2i-1}]^2+[f(x_{2i})-f(x_{2i-1})]^2}$$
$$=\sum_{i=1}^{n}\sqrt{[x_{2i-1}-x_{2i-2}]^2+[0-1]^2}+\sum_{i=1}^{n}\sqrt{[x_{2i}-x_{2i-1}]^2+[1-0]^2}$$
$$>\sum_{i=1}^{n}1+\sum_{i=1}^{n}1=2n=p>M.$$

Whereas the discontinuous function r is not rectifiable, we note that some discontinuous functions are rectifiable, as the reader is asked to prove in Exercise 5.9.10.

(3) In Part (2) of this example we saw a function that is discontinuous and not rectifiable. The reader might wonder whether there is an example of a continuous function that is not rectifiable. There are such functions, though they are not as simple to describe as the function in Part (2) of this example. The intuitive idea is that we need a function defined on a non-degenerate closed bounded interval that is continuous but has a graph that is infinitely wiggly, which makes it have infinite length. An example of such a function is seen in Exercise 10.5.3; this function is a special case of the continuous but nowhere differentiable functions discussed in Section 10.5. ◊

It is hard to evaluate the rectifiability of most functions directly from the definition. See [Tri95, Chapter 7] for a thorough geometric discussion of rectifiable curves, and elsewhere in that book for the relation of non-rectifiable curves to fractal curves. Fortunately, if a function is continuously differentiable, the situation is much easier, as we see in the following theorem. We note that in contrast to Theorem 5.9.9, the following theorem is not an "if and only if" result, because it is possible for the graph of a non-differentiable function to be rectifiable, for example the graph of the absolute value function.

Theorem 5.9.17. *Let $[a,b]\subseteq\mathbb{R}$ be a non-degenerate closed bounded interval, and let $f\colon [a,b]\to\mathbb{R}$ be a function. Suppose that f is continuous on $[a,b]$ and continuously differentiable on (a,b), and that f' is bounded on (a,b). Then f is rectifiable and*

$$L_a^b(f)=\int_a^b\sqrt{1+[f'(x)]^2}\,dx.$$

Proof. Let $g\colon [a,b]\to\mathbb{R}$ be defined by

$$g(x)=\begin{cases}\sqrt{1+[f'(x)]^2}, & \text{if } x\in(a,b)\\ 0, & \text{otherwise.}\end{cases}$$

Because f' is bounded on (a,b), there is some $M\in\mathbb{R}$ such that $|f'(x)|\le M$ for all $x\in(a,b)$; we may assume that $M>0$. Then $f'((a,b))\subseteq[-M,M]$. Let $h\colon [-M,M]\to\mathbb{R}$ be defined by $h(x)=\sqrt{1+x^2}$ for all $x\in[-M,M]$. It is left to the reader to verify that h is continuous on $[-M,M]$ and differentiable on $(-M,M)$, and that $|h'(x)|\le 1$ for all $x\in(-M,M)$; in addition to using facts we have already seen, the reader will need to use Theorem 7.2.13 (1). It then follows from Exercise 4.4.6 that h is uniformly continuous. By Theorem 3.4.5 we see that h is bounded. It follows immediately that $h|_{f'((a,b))}$ is bounded, and it follows from Lemma 3.4.2 and Exercise 3.3.2 (2) that $h|_{f'((a,b))}$ is continuous. By abuse of notation we can think of f' as a function $(a,b) \to f'((a,b))$. Then $g|_{(a,b)} = h|_{f'((a,b))} \circ f'$. We deduce immediately that $g|_{(a,b)}$ is bounded, and it follows from Theorem 3.3.8 (3) that $g|_{(a,b)}$ is continuous. Then g is bounded, and g is continuous except possibly at a and b. It now follows from Exercise 5.5.5 (2) that g is integrable.

As a preliminary step, we show that if P is a partition of $[a,b]$, then there is a representative set S of P such that $C(f,P) = S(g,P,S)$. Let $P = \{x_0, x_1, \ldots, x_n\}$ be a partition of $[a,b]$. Let $i \in \{1,\ldots,n\}$. Because f is continuous on $[x_{i-1}, x_i]$ and differentiable on (x_{i-1}, x_i), the Mean Value Theorem (Theorem 4.4.4) implies that there is some $s_i \in (x_{i-1}, x_i)$ such that

$$f'(s_i) = \frac{f(x_i) - f(x_{i-1})}{x_i - x_{i-1}}.$$

Hence $f(x_i) - f(x_{i-1}) = f'(s_i)(x_i - x_{i-1})$. Let $S = \{s_1, \ldots, s_n\}$. Then S is a representative set of Z. We now see that

$$\begin{aligned} C(f,P) &= \sum_{i=1}^{n} \sqrt{[x_i - x_{i-1}]^2 + [f(x_i) - f(x_{i-1})]^2} \\ &= \sum_{i=1}^{n} \sqrt{[x_i - x_{i-1}]^2 + [f'(s_i)(x_i - x_{i-1})]^2} \\ &= \sum_{i=1}^{n} \sqrt{1 + [f'(s_i)]^2}(x_i - x_{i-1}) = S(g,P,S). \end{aligned}$$

We now show that $\int_a^b g(x)\,dx$ is an upper bound of $\mathcal{A}_f$. Suppose to the contrary that there is some partition X of $[a,b]$ such that $C(f,X) > \int_a^b g(x)\,dx$. Let $\varepsilon = C(f,X) - \int_a^b g(x)\,dx$. Then $\varepsilon > 0$. Because g is integrable, there is some $\delta > 0$ such that if Q is a partition of $[a,b]$ with $\|Q\| < \delta$, and if T is a representative set of Q, then $\left|S(g,Q,T) - \int_a^b g(x)\,dx\right| < \varepsilon$. Let Y be a partition of $[a,b]$ with $\|Y\| < \delta$, which exists by Exercise 5.2.1. Let $Z = X \cup Y$. By Lemma 5.4.3 we know that Z is a refinement of both X and Y, and that $\|Z\| \le \|Y\| < \delta$. It follows from Lemma 5.9.14 that $C(f,Z) \ge C(f,X) > \int_a^b g(x)\,dx$. Hence

$$\left|C(f,Z) - \int_a^b g(x)\,dx\right| = C(f,Z) - \int_a^b g(x)\,dx \ge C(f,X) - \int_a^b g(x)\,dx = \varepsilon.$$

By the preliminary step there is a representative set H of Z such that $C(f,Z) = S(g,Z,H)$. Hence $\left|S(g,Z,H) - \int_a^b g(x)\,dx\right| \ge \varepsilon$. On the other hand, because $\|Z\| < \delta$, we know that $\left|S(g,Z,H) - \int_a^b g(x)\,dx\right| < \varepsilon$, which is a contradiction. We conclude that $\int_a^b g(x)\,dx$ is an upper bound of $\mathcal{A}_f$, and it follows that f is rectifiable.

We now show that $\int_a^b g(x)\,dx = \operatorname{lub} \mathcal{A}_f$, which means that $\int_a^b g(x)\,dx = L_a^b(f)$. Let $\eta > 0$. Because g is integrable, there is some $\beta > 0$ such that if Q is a partition of $[a,b]$

with $\|Q\| < \beta$, and if T is a representative set of Q, then $\left|S(g,Q,T) - \int_a^b g(x)\,dx\right| < \eta$. Let W be a partition of $[a,b]$ with $\|W\| < \beta$, which exists by Exercise 5.2.1. Again using the preliminary step, there is a representative set D of W such that $C(f,W) = S(g,W,D)$. Then $\left|C(f,W) - \int_a^b g(x)\,dx\right| < \eta$. Because $\int_a^b g(x)\,dx$ is an upper bound of $\mathcal{A}_f$, it follows that $C(f,W) \le \int_a^b g(x)\,dx$. Hence $\int_a^b g(x)\,dx - C(f,W) < \eta$. Exercise 2.6.6 now implies that $\int_a^b g(x)\,dx = \operatorname{lub} \mathcal{A}_f$. □

Example 5.9.18. Let $f\colon [0,1] \to \mathbb{R}$ be defined by $f(x) = x^{\frac{3}{2}}$ for all $x \in [0,1]$. In Example 5.9.13 we computed that the arc length of f was approximately 1.436. We now compute the exact arc length. As the reader knows informally from calculus, and as we will prove in Theorem 7.2.13 (1), the function f is continuously differentiable, and $f'(x) = \frac{3}{2}x^{\frac{1}{2}}$ for all $x \in [0,1]$. By Theorem 5.9.17 the arc length of f is

$$L_0^1(f) = \int_0^1 \sqrt{1 + \left[\frac{3}{2}x^{\frac{1}{2}}\right]^2}\,dx = \int_0^1 \sqrt{1 + \frac{9x}{4}}\,dx = \frac{1}{2}\int_0^1 (4+9x)^{\frac{1}{2}}\,dx$$

$$= \frac{1}{27}\left[(4+9x)^{\frac{3}{2}}\right]_0^1 = \frac{1}{27}\left[13^{\frac{3}{2}} - 4^{\frac{3}{2}}\right] \approx 1.440. \qquad \Diamond$$

Although the integral in Example 5.9.18 was easy to compute, in practice it is difficult to compute the integral in Theorem 5.9.17 for most simple functions, though numerical approximations are always possible.

Reflections

Whereas today it seems self-evident that we measure the area of regions in the plane by associating a number—called the area—to each such region, this idea of associating a number to represent the area of a region of the plane is in fact not obvious. The ancient Greeks did not associate a number called area to each region, but rather discussed the equality of areas of different regions by proving that one region could be broken up and rearranged into the other; if two regions had different areas then the two regions could be compared by using the ratio of their areas, without actually finding the areas of the individual regions. For example, there is no formula in Euclid's *Elements* for the area of a circle in terms of the radius; rather, what Euclid proved about the areas of circles was that for any two circles, the ratio of their areas equals the ratio of the squares of their diameters. To the ancient Greeks, geometric objects (such as circles) could be compared only with similar types of objects, though it was possible to compare the ratio of two objects of one type with the ratio of two objects of another type.

Even though today we have the notion of area, volume, arc length and the like as numerical values, and indeed these notions seem very clear intuitively, it turns out that providing rigorous definitions of these concepts is not at all trivial if we want to find the area of subsets of the plane more complicated than polygons, and similarly for the other concepts. Historically, people used concepts such as area and volume long before they were rigorously defined. The definition of area stated in the present section, which is the 2-dimensional version of the more general concept of Jordan measure, is from the 19th century.

Jordan measure is not the most common method used today to define the measure of subsets of Euclidean space; the more common method, known as Lebesgue measure, is better behaved than Jordan measure, but it is also trickier to define, and its use would take us too far afield to be included in this text.

The discussion of area in this section made use of some elementary properties of rectangles, including the fact that if a rectangle with edges parallel to the coordinate axes is broken up into finitely many subrectangles by lines parallel to the coordinate axes, then the sum of the areas of the smaller rectangles will equal the area of the larger rectangle. This fact is simple to prove for such rectangles, but it raises the following more general question about polygons in the plane. If a polygon is broken up into finitely many smaller polygons, and the smaller polygons are rearranged into a new polygon, then the new polygon will have the same area as the original polygon, but is the converse true? That is, if two polygons have the same area, can one polygon be obtained from the other by breaking it up into finitely many smaller polygons and rearranging? The answer, as intuitively expected, is yes; that result is called the Bolyai–Gerwien Theorem. David Hilbert, as one of his famous 23 problems from 1900, asked whether the analogous result is true for polyhedra in $\mathbb{R}^3$, and, rather surprisingly, Max Dehn showed soon thereafter that the answer is no. That is, there are two polyhedra in $\mathbb{R}^3$ that have the same volume, but one cannot be obtained from the other by breaking one up into finitely many smaller polyhedra and rearranging. See [Bol78] or [Pak, Chapters 15–17] for details in both the 2-dimensional and 3-dimensional cases.

Exercises

Exercise 5.9.1. [Used in Section 5.9.] Let $S \subseteq \mathbb{R}^2$ be a rectangle. Then $S = [a,b] \times [c,d]$ for some closed bounded intervals $[a,b], [c,d] \subseteq \mathbb{R}$. Prove that S is squarable, and that its area using Definition 5.9.6 equals $(b-a)(d-c)$. [Use Exercise 2.6.2.]

Exercise 5.9.2. Let $K = \{(0,0)\} \cup \{(\frac{1}{n},0) \mid n \in \mathbb{N}\} \subseteq \mathbb{R}^2$. Either prove that K is squarable and find its area, or prove that K is not squarable.

Exercise 5.9.3. [Used in Section 5.9 and Theorem 7.4.4.] Let $[a,b] \subseteq \mathbb{R}$ be a non-degenerate closed bounded interval, and let $f\colon [a,b] \to \mathbb{R}$ be a function. Suppose that $f(x) \geq 0$ for all $x \in [a,b]$. Prove that the region between the graphs of $-f$ and the x-axis is squarable if and only if the region under the graph of f is squarable, and if they are squarable then the areas of these two regions are equal.

Exercise 5.9.4. [Used in Exercise 5.9.7 and Theorem 7.4.4.] Let $[a,b] \subseteq \mathbb{R}$ be a non-degenerate closed bounded interval, let $f,g\colon [a,b] \to \mathbb{R}$ be functions and let $k \in \mathbb{R}$. Let $\hat{f},\hat{g}\colon [a,b] \to \mathbb{R}$ be the functions defined by $\hat{f}(x) = f(x)+k$ and $\hat{g}(x) = g(x)+k$ for all $x \in [a,b]$. Suppose that $f(x) \leq g(x)$ for all $x \in [a,b]$. Prove that the region between the graphs of f and g is squarable if and only if the region between the graphs of $\hat{f}$ and $\hat{g}$ is squarable, and if they are squarable then $A(R_a^b(\hat{f},\hat{g})) = A(R_a^b(f,g))$.

Exercise 5.9.5. **[Used in Theorem 7.4.4.]** Let $[a,b] \subseteq \mathbb{R}$ be a non-degenerate closed bounded interval, and let $f,g\colon [a,b] \to \mathbb{R}$ be functions. Let $\tilde{f},\tilde{g}\colon [-b,-a] \to \mathbb{R}$ be the functions defined by $\tilde{f}(x) = f(-x)$ and $\tilde{g}(x) = g(-x)$ for all $x \in [-b,-a]$. Suppose that $f(x) \le g(x)$ for all $x \in [a,b]$. Prove that the region between the graphs of f and g is squarable if and only if the region between the graphs of $\tilde{f}$ and $\tilde{g}$ is squarable, and if they are squarable then $A(R_{-b}^{-a}(\tilde{f},\tilde{g})) = A(R_a^b(f,g))$.

Exercise 5.9.6. **[Used in Theorem 7.4.4.]** Let $[a,b] \subseteq \mathbb{R}$ be a non-degenerate closed bounded interval, let $f,g\colon [a,b] \to \mathbb{R}$ be functions and let $c \in \mathbb{R}$. Let the functions $\bar{f},\bar{g}\colon [a+c,b+c] \to \mathbb{R}$ be defined by $\bar{f}(x) = f(x-c)$ and $\bar{g}(x) = g(x-c)$ for all $x \in [a+c,b+c]$. Suppose that $f(x) \le g(x)$ for all $x \in [a,b]$. Prove that the region between the graphs of f and g is squarable if and only if the region between the graphs of $\bar{f}$ and $\bar{g}$ is squarable and if they are squarable then $A(R_{a+c}^{b+c}(\bar{f},\bar{g})) = A(R_a^b(f,g))$.

Exercise 5.9.7. **[Used in Theorem 5.9.11.]** Prove Theorem 5.9.11. When you use Theorem 5.9.9 and Theorem 5.9.10, make sure that all of the hypotheses of each are satisfied. [Use Exercise 5.9.4.]

Exercise 5.9.8. In Theorem 5.9.11 it is assumed that the functions f and g are integrable. Is that necessary, or would it have sufficed to assume that $g - f$ is integrable? In other words, is it possible to have functions $f,g\colon [a,b] \to \mathbb{R}$ such that $g - f$ is integrable, but the region between the graphs of f and g is not squarable? If yes, give an example. If not, prove why not.

Exercise 5.9.9. **[Used in Lemma 5.9.14 and Exercise 5.9.10.]** Let $[a,b] \subseteq \mathbb{R}$ be a non-degenerate closed bounded interval, let $f\colon [a,b] \to \mathbb{R}$ be a function and let P and Q be partitions of $[a,b]$. The purpose of this exercise is to prove that if Q is a refinement of P, then $C(f,Q) \ge C(f,P)$. We start with some preliminaries, with the actual proof given in Part (5) of this exercise.

(1) This part of the exercise is a special case of the Cauchy–Schwarz Inequality for vectors in an inner product space. Let $(x_1,x_2),(y_1,y_2) \in \mathbb{R}^2$. Prove that

$$[x_1y_1 + x_2y_2]^2 \le [(x_1)^2 + (x_2)^2] \cdot [(y_1)^2 + (y_2)^2].$$

To prove the inequality, the first step is to let $p\colon \mathbb{R} \to \mathbb{R}$ be the function defined by

$$p(t) = [(x_1)^2 + (x_2)^2]t^2 + 2[x_1y_1 + x_2y_2]t + [(y_1)^2 + (y_2)^2]$$

for all $t \in \mathbb{R}$, and to prove that $p(t) \ge 0$ for all $t \in \mathbb{R}$.

(2) This part of the exercise is a special case of the Triangle Inequality for vectors in an inner product space. Let $(x_1,x_2),(y_1,y_2) \in \mathbb{R}^2$. Prove that

$$\sqrt{[x_1+y_1]^2 + [x_2+y_2]^2} \le \sqrt{[x_1]^2 + [x_2]^2} + \sqrt{[y_1]^2 + [y_2]^2}.$$

(3) Let $(u_1,u_2),(v_1,v_2),(w_1,w_2) \in \mathbb{R}^2$. Prove that

$$\sqrt{[w_1 - u_1]^2 + [w_2 - u_2]^2}$$
$$\leq \sqrt{[v_1 - u_1]^2 + [v_2 - u_2]^2} + \sqrt{[w_1 - v_1]^2 + [w_2 - u_2]^2}.$$

(4) Let $x, y, z \in [a,b]$. Suppose that $x < y < z$. Prove that

$$\sqrt{[z-x]^2 + [f(z) - f(x)]^2}$$
$$\leq \sqrt{[y-x]^2 + [f(y) - f(x)]^2} + \sqrt{[z-y]^2 + [f(z) - f(y)]^2}.$$

(5) Prove that if Q is a refinement of P, then $C(f,Q) \geq C(f,P)$.

Exercise 5.9.10. **[Used in Example 5.9.16.]** Let $g\colon [0,3] \to \mathbb{R}$ be defined by

$$g(x) = \begin{cases} x, & \text{if } x \neq 1 \\ 2, & \text{if } x = 1. \end{cases}$$

Prove that g is rectifiable, and find $L_0^3(g)$. It is acceptable to use geometric reasoning, such as the fact that the length of one edge of a triangle is always less than or equal to the sum of the lengths of the other two edges (this particular fact was proved in Exercise 5.9.9 (3)).

Exercise 5.9.11. **[Used in Theorem 7.4.3.]** Let $[a,b] \subseteq \mathbb{R}$ be a non-degenerate closed bounded interval, and let $f\colon [a,b] \to \mathbb{R}$ be a function. Prove that $-f$ is rectifiable if and only if f is rectifiable, and if they are rectifiable then $L_a^b(-f) = L_a^b(f)$.

Exercise 5.9.12. **[Used in Theorem 7.4.3.]** Let $[a,b] \subseteq \mathbb{R}$ be a non-degenerate closed bounded interval, let $f\colon [a,b] \to \mathbb{R}$ be a function and let $k \in \mathbb{R}$. Let $g\colon [a,b] \to \mathbb{R}$ be defined by $g(x) = f(x) + k$ for all $x \in [a,b]$. Prove that g is rectifiable if and only if f is rectifiable, and if they are rectifiable then $L_a^b(g) = L_a^b(f)$.

Exercise 5.9.13. **[Used in Theorem 7.4.3.]** Let $[a,b] \subseteq \mathbb{R}$ be a non-degenerate closed bounded interval, and let $f\colon [a,b] \to \mathbb{R}$ be a function. Let $g\colon [-b,-a] \to \mathbb{R}$ be defined by $g(x) = f(-x)$ for all $x \in [-b,-a]$. Prove that g is rectifiable if and only if f is rectifiable, and if they are rectifiable then $L_{-b}^{-a}(g) = L_a^b(f)$.

Exercise 5.9.14. **[Used in Theorem 7.4.3.]** Let $[a,b] \subseteq \mathbb{R}$ be a non-degenerate closed bounded interval, let $f\colon [a,b] \to \mathbb{R}$ be a function and let $c \in \mathbb{R}$. Let $h\colon [a+c, b+c] \to \mathbb{R}$ be defined by $h(x) = f(x-c)$ for all $x \in [a+c, b+c]$. Prove that h is rectifiable if and only if f is rectifiable, and if they are rectifiable then $L_{a+c}^{b+c}(h) = L_a^b(f)$.

5.10 Historical Remarks

The problem that motivated the development of integration, which is the need to find areas and volumes of non-rectilinear shapes, is the oldest aspect of calculus, having solid roots in the ancient world. It is evident that in many ancient cultures, activities such as art, architecture, farming, commerce, astronomy and more, would lead to a practical need to find areas and volumes of a variety of shapes. Areas of rectangles, triangles, trapezoids and even circles were discussed in the ancient world in a variety of cultures; finding areas of more complicated shapes, however, required the development of sophisticated mathematical techniques.

Ancient World

The most important method for finding areas and volumes of complicated regions in the ancient world was the method of exhaustion, developed in ancient Greece. The work of Antiphon the Sophist (480–411 BCE), who found the areas of circles by using inscribed polygons with successive doubling of the number of edges, may have been the inspiration for the method of exhaustion. Bryson of Heraclea (c. 450–c. 390 BCE) took this idea one step closer to the method of exhaustion by considering both inscribed and circumscribed polygons for circles. A classic example of the method of exhaustion is the proof of Proposition 2 of Book XII of Euclid's *Elements*, which is also about the areas of circles. The idea is similar to the approach of Antiphon the Sophist, in that polygons and polyhedra with ever-increasing numbers of sides are used to approximate the region of interest, but to make the proof complete, rather than taking a limit to infinity (as we would do today but was never done in ancient Greece), or even using infinitesimals (as was done in the early development of calculus), the proof is based upon a double proof by contradiction (referred to as reductio ad absurdum), making use of the Axiom of Exhaustion, which is Proposition 1 of Book X of the *Elements*. This method is attributed to Eudoxus of Cnidus (408–355 BCE). The term "exhaustion" is due to Grégoire de Saint-Vincent in 1647.

The culmination of the method of exhaustion was the work of Archimedes (287–212 BCE), who found areas and volumes of a variety of regions, for example the area inside a parabola. The method of exhaustion was never formulated as a general method, however, and was used in an ad hoc fashion for each particular case. Archimedes' use of the method of exhaustion was extremely clever and sophisticated, but it was also very tedious. Moreover, this method, which is a way of proving that a given area or volume is correct, hid any intuitive understanding of how the area or volume was first arrived at. When Archimedes' *The Method* was found in 1906, it was a rare ancient Greek text that discussed how results were arrived at. In this particular case, Archimedes' method of discovery was by viewing areas as made up of line elements. Archimedes hinted that Democritus may have used similar ideas. Hence, it might have been the case that some ancient Greeks thought intuitively in terms of infinitesimals, though then wrote up their proofs in the standard ancient Greek style. However, whereas Archimedes' methods have some resemblance to certain aspects of calculus, for example the use of upper sums and lower sums, it is not correct to say that Archimedes was essentially doing calculus—he did not have the general idea of derivatives or integrals, nor the relation between tangent problems and area problems, and he did not provide broadly applicable computational tools.

Medieval Period

The Arab mathematicians in the Middle Ages understood the method of exhaustion as described in Euclid's *Elements*, though European mathematicians at the time did not. In particular, Abu Ali al-Hasan ibn al-Haytham (965–1039), also known as Alhazen, calculated volumes of some solids of revolution. Archimedes had looked at rotating a parabolic segment about its axis, but Alhazen rotated it about some other lines as well.

Nicole Oresme (1323–1382) had the idea of studying variation via representation by coordinates, a precursor to the subsequent invention of analytical geometry. He appears to have understood, in special cases, that if the velocity of an object is graphed, then the area under the graph represents the distance traveled, which is the essential idea of the Fundamental Theorem of Calculus. Oresme also had the idea of mathematical indivisibles, which were used for area and volume calculations a few centuries later.

Renaissance

As mathematics in Europe started to become more sophisticated, in part due to their importation of ideas from the Arab world, the work of Archimedes became known and appreciated, though such knowledge had a mixed impact. On the one hand, Archimedes' calculations of some areas and volumes provided an inspiration to compute even more areas and volumes. On the other hand, the method of exhaustion, based upon the use of reductio ad absurdum to avoid infinitesimals or limits, became an unbearably heavy burden in all but the simplest cases.

One of the people who helped promote a more computationally friendly method for computing areas and volumes than the strict Archimedean approach was Simon Stevin (1548–1620). As an engineer focused on getting results, Stevin, in *De Beghinselen der Weeghconst* of 1586, tried to modify the method of exhaustion by replacing the reductio ad absurdum with some simplifying ideas, for example the fact that if two quantities differ they do so by a finite amount, and therefore in order to show that two quantities are equal it suffices to show that they differ by less than any finite amount. (This fact, in modern notation, follows from Lemma 2.3.10.)

Seventeenth Century

Mathematicians in the 17th century continued to abandon the method of exhaustion, and started to make use of infinitesimals and indivisibles to find areas and volumes (limits in their modern form were not used for another two centuries). One of the first to do so was Johannes Kepler (1571–1630), who is most known for his three laws of planetary motion, but who also did work in geometry. As part of a calculation concerning planetary motion, Kepler showed what we write as $\int_0^a \sin x\,dx = 1 - \cos a$ by dividing the region into infinitely many small parts. In *Nova stereometria doliorum vinariorum* of 1615, which was meant to be of practical use for finding volumes of wine casks, Kepler computed volumes of solids of revolution by using infinitesimals freely to obtain his results.

An important use of indivisibles for finding areas and volumes was the work of Bonaventura Cavalieri (1598–1647), who was influenced by the approach of his teacher Galileo Galilei (1564–1642). Cavalieri's *Geometria indivisibilibus continuorum nova quadam ratione promota* of 1635 was the first book entirely about the use of indivisibles, and *Exercitationes geometricae sex* of 1647 elaborated upon his approach; these books were very influential. Although Kepler and Cavalieri both found areas and volumes by breaking up regions into infinitely small pieces, their approaches were different, in that Cavalieri did not directly add up the infinitely many pieces as did Kepler, but rather found areas and volumes by comparing the sizes of the slices of different shapes. More specifically, Cavalieri used what we now call "Cavalieri's Principle," which says that if two regions in the plane are located between two parallel lines, and if every line parallel to these two lines intersects the two regions in line segments of equal lengths, then the two regions have equal area; the analogous result holds for regions in space. The use of indivisibles, which harked back to Oresme and other medieval scholars, essentially hid the limits involved in finding areas and volumes. Cavalieri's Principle was touched upon by Heron of Alexandria (c. 10–c. 75) and known to Galileo, but it was Cavalieri who took this idea and made it into a workable tool for finding areas and volumes. Cavalieri calculated, though not rigorously, what we write as $\int_0^a x^n\,dx = \frac{a^{n+1}}{n+1}$ for all $n \in \mathbb{N}$. He used clever reasoning going back and forth among dimensions, for example viewing what we write as $\int_0^a x^2\,dx$ as both the area under a parabola and the volume of a pyramid with square cross section. He argued the result up to $n = 9$ on an ad hoc basis, and inferred the general result.

In addition to Cavalieri, who published it first in 1639, other mathematicians in the period 1635–1655 independently came up with $\int_0^a x^n\,dx = \frac{a^{n+1}}{n+1}$ for all $n \in \mathbb{N}$, including Torricelli, Roberval, Pascal, Fermat and Wallis. Rather than using the geometric approach of Cavalieri, this integral was justified by Fermat, Pascal and Roberval by using upper sums and lower sums, and using the limit $\lim_{n\to\infty} \frac{1^k+2^k+\cdots+n^k}{n^{k+1}} = \frac{1}{k+1}$ for all $k \in \mathbb{N}$, though the idea of a limit was not yet rigorous. Such calculations by algebraic manipulation were a step forward toward calculus.

Pierre de Fermat (1601–1665), though a lawyer by profession, was in correspondence with many contemporary mathematicians such as Mersenne, Roberval and Descartes, and his mathematical work included both tangent problems and area problems. He correctly evaluated the integral $\int_0^a x^{\frac{p}{q}}\,dx$, using a subdivision of the interval $[0,a]$ by geometric series. He was among the first to notice, though only in special cases, a link between tangent problems and area problems. Fermat appears to have anticipated the invention of calculus more than most of his contemporaries, in that he had some of the ingredients of both derivatives and integrals, but he did not appear to recognize either the derivative or the integral as a concept in its own right, and he cannot be considered as having invented calculus.

Gilles de Roberval (1602–1675) was less concerned with rigor than Fermat, but he was very ingenious, obtaining, for example, various trigonometric integrals. Roberval also saw a relation, in some special cases, between tangent problems and area problems.

The work of Blaise Pascal (1623–1662) on areas and volumes was not as original as some of his contemporaries, but he was interested in exposition, and his writing on this matter influenced Leibniz.

Evangelista Torricelli (1608–1647) solved a variety of tangent and area problems, including the area under the curve $(\frac{y}{a})^n(\frac{x}{b})^m = 1$, which was proposed by Fermat. He made use of the idea of having inscribed and circumscribed figures differ by less than a given magnitude in order to find areas (which is similar to the idea of Lemma 5.9.5 (3)). Torricelli had an informal understanding of the relation between tangent problems and area problems, though he did not explictly state the Fundamental Theorem of Calculus as we know it.

René Descartes (1596–1650), in *La Géométrie* of 1637, stated that he thought it would not be possible for the human mind to determine the exact arc lengths of curves. He was proved wrong by a number of people, including William Neile (1637–1670), who found the arc length of the curve $y = cx^{\frac{3}{2}}$ in 1657, the architect and mathematician Christopher Wren (1632–1723), who found the arc length of the cycloid in 1658, and Hendrik van Heuraet (1634–1660), who found the geometric equivalent of the general formula for arc length in 1659. All three used the same approach we use today, which is to approximate the curve by polygons, and then use either an infinitesimal or limit argument. Some of the ideas of van Heuraet, including the differential triangle and the association of the area under a new curve with the arc length of the original curve, helped lead to the subsequent discovery of the general relation between tangent problems and area problems.

In *Geometriae pars universalis* of 1668, James Gregory (1638–1675) compiled the known tangent, area and volume calculations from a variety of writers such as Cavalieri, Torricelli and others. Gregory's writing, though verbal and geometric rather than analytical, helped put the focus on general methods by separating them from special cases, thereby eliminating a lot of the repetition found in the work of earlier mathematicians. He was aware of a special case of the inverse relation between tangent problems and area problems, in the context of a discussion of arc length of curves. *Geometriae pars universalis* could be considered to have the first published version of the Fundamental Theorem of Calculus, albeit in a special case, and in a geometric form as a relation between tangents and areas, and not in terms of differentiation and integration of functions.

John Wallis (1616–1703) introduced x^k where k is negative and/or rational in *Arithmetica infinitorum* of 1656. This work had an influence on Newton. Wallis conjectured the correct result for $\int_0^a x^{\frac{p}{q}}\,dx$, and proved it for $\frac{1}{q}$. Wallis' approach was to use algebraic, rather than geometric, methods.

Isaac Barrow (1630–1677), like Gregory, worked in a geometric vein, and his work, while not yet calculus as we know it, can be thought of as the end of the development of the geometric approach prior to the invention of calculus. Barrow, Newton's predecessor as the Lucasian Professor of Mathematics at Cambridge, possibly influenced the young Newton's mathematical development; conversely, Newton made some suggestions for Barrow's book *Lectiones geometricae* of 1670. In this book, Barrow, who thought in terms of time and motion, and who built upon the ideas of the medievalists, as well as Galileo, Cavalieri, Torricelli and Roberval, among others, had geometric statements of both versions of the Fundamental Theorem of Calculus. These results were expressed in terms of areas under curves and tangent lines to curves, rather than in terms of integrals and derivatives. However, though he had a definite understanding of the Fundamental Theorem of Calculus, Barrow did not exploit this understanding to provide a method for computing areas under curves, and in general did not transform the various available ideas about tangents and areas into a practical computational tool; as such, Barrow cannot be said to have invented calculus, though he was close.

Newton and Leibniz

Though it took two more centuries until calculus was brought into the form we know it today, the essence of calculus was first thought of by Isaac Newton (1643–1727) in the period 1665–1666 while home from Cambridge University because of the plague. In the unpublished *October 1666 Tract on Fluxions* Newton used an intuitive argument to show what we would phrase by saying that if A is the area under the curve $y = f(x)$ then $\frac{dA}{dx} = y$, thereby establishing the Fundamental Theorem of Calculus Version I. Newton computed a table of antiderivatives, in part using Integration by Substitution, where some of the antiderivatives were given explicitly and others were reduced to hyperbolic or circular functions, which Newton then handled with the binomial series (which he had previously discovered). He then used these ideas to solve some area problems essentially as we do today, and that was the birth of calculus.

In the unpublished *Tractatus de methodis serierum et fluxionum* of 1671, Newton further elaborated upon what amounts to Integration by Substitution, implicitly used the equivalent of Integration by Parts, and worked out the arc length formula and computed some examples, such as $y = \frac{x^3}{a} + \frac{a}{12x}$ and $y = ax^{\frac{2}{3}}$; for more complicated functions, he worked out the arc lengths in terms of power series.

Gottfried von Leibniz (1646–1716) started working toward his version of calculus, presumably independently of Newton in spite of all the controversy, in late 1673 or early 1674, when he discovered a version of the inverse relation of tangent problems and area problems, by using something like Cavalieri's Principle to show that an area under one curve is equivalent to the area under a certain associated curve, the construction of which involves tangents. In modern terms, he ended up doing something like a special case of Integration by Parts. He used this method to solve some area problems, and to obtain the series $\frac{\pi}{4} = 1 - \frac{1}{3} + \frac{1}{5} - \frac{1}{7} + \cdots$. Leibniz worked out his version of calculus in the latter part of 1675, as recorded in a series of unpublished notes, by the end of which he had the $\int$ and dx notations, and he had solved some non-trivial problems. Leibniz thought of dx and dy as infinitely small changes in x and y, respectively; he seemed to think of curves as polygons with infinitesimal edges. He viewed the notation $\int y\,dx$ (which we write as $\int_a^b y(x)\,dx$) as meaning the sum of infinitesimal rectangles with heights y and widths dx. He stated the Fundamental Theorem of Calculus Version I, and he had the formula for arc length. Leibniz published his calculus in three papers in 1684, 1686 and 1693, after

Newton had written some manuscripts, but before he published his version of calculus in 1704.

Newton and Leibniz had different approaches to integrals. Leibniz viewed integrals as a type of sum, and he viewed differentials as a type of difference; by analogy with the finite case, Leibniz thought of integrals as inverse to differentials. Hence, Leibniz viewed integrals as separately defined from derivatives. For Newton integrals are what we call indefinite integrals, though he solved area problems with them, by looking at rates of change of what we write as $\int_a^x f(t)\,dt$. Ultimately, both Newton and Leibniz arrived at the inverse relation of area problems and tangent problems, and they exploited this relation to provide simple methods for solving a variety of problems.

Eighteenth Century

Although integrals (in the guise of areas) were originally viewed as some sort of infinite sum, after Newton and Johann Bernoulli (1667–1748) the view switched to viewing integrals as antiderivatives, which resulted in some confusion about the relation of what we now call definite integrals and indefinite integrals. Because integrals were viewed as antiderivatives, derivatives became the central object of calculus. The idea of area at the time was rather intuitive, though perhaps because Newton's approach seemed to work for nice functions, and poorly behaved functions were generally avoided, no one felt compelled to look any further at the meaning of the concept of area.

By the time of Joseph Fourier (1768–1830), however, it was realized that integrals of more complicated functions, which arose in real-world situations, were needed. Moreover, Fourier, who wanted to calculate what we call Fourier coefficients, changed the focus from indefinite integrals back to definite integrals. Fourier introduced the notation $\int_a^b f(x)\,dx$; Leibniz used only the notation $\int f(x)\,dx$. Cauchy was influenced by Fourier in focusing on definite integrals, though Cauchy's definition of such integrals was precise, rather than Fourier's informal notion of area.

Nineteenth Century

Augustin Louis Cauchy (1789–1857) was the first person to provide a rigorous treatment of integrals, meaning definite integrals, separately from their relation to derivatives. Cauchy used left-hand sums sums to define integrals; he did not invent such sums (Euler and others had used them to approximate integrals), but Cauchy was the first person to use them in the definition of integrals. Cauchy aimed to prove that definite integrals of continuous functions always exist. He used partitions and refinements of partitions, just as we do today. Cauchy had some logical gaps in his argument, glossing over the difference between continuity and uniform continuity, and implicitly using results that could only be proved with the axioms for the real numbers, but the core ideas he used are familiar to modern students of calculus. Cauchy gave the first rigorous statements and proofs of the two versions of the Fundamental Theorem of Calculus; his arguments, which are still used, work for continuous, and also piecewise continuous, functions. Because Cauchy restricted his attention to nicely behaved functions, however, he did not give a general characterization of integrability.

The next major step forward in the development of integration was due to Georg Friedrich Bernhard Riemann (1826–1866). In his Habilitationschrift *Über die Darstellbarkeit einer Function durch eine trigonometrische Reihe* of 1854 (published in 1867), Riemann looked at the representation of functions as Fourier series, and in the course of his study, where he wanted to look at functions that were not necessarily continuous, he recognized the need to give a more precise definition of integrability to accommodate functions that are not continuous (because Fourier coefficients are computed via integrals). Riemann then defined what we now call Riemann sums, and essentially gave our modern definition of integrals in terms of such sums (though without the ε–δ formulation). Riemann gave criteria that are equivalent to integrability for bounded functions, including the equivalent of looking at the difference between upper sums and lower sums, though upper sums and lower sums were introduced only in the 1870s by several mathematicians, including Gaston Darboux (1842–1917). Riemann also gave an example of a function with a dense set of discontinuities that is integrable. Similarly to Cauchy, Riemann ignored issues that today we know rely upon the axiomatic properties of the real numbers.

In 1875 Darboux looked at upper integrals and lower integrals, though using limits rather than Least Upper Bounds, and he proved what we call the Fundamental Theorem of Calculus Version I. Giuseppe Peano (1858–1932) pointed out that it would be possible to use greatest lower bounds and least upper bounds instead of limits in the definition of upper integrals and lower integrals, as we do now.

The first person to give a rigorous definition of area was Peano, in 1887. He considered the inner area, denoted $a_i(S)$, of a planar region S as the least upper bound of the areas of all polygons contained in the region, and similarly for the outer area, denoted $a_o(S)$. Clearly $a_i(S) \leq a_o(S)$. The area was said to exist if and only if $a_i(S) = a_o(S)$, and if equality held this number was defined to be the area. Peano showed that $a_i(S) = \underline{\int_a^b} f(x)\,dx$ and $a_o(S) = \overline{\int_a^b} f(x)\,dx$ for the region below the graph of a non-negative function; it then followed that the region under the graph of a non-negative function has an area if and only if the function is integrable, in which case the area is the integral. Peano's definition of area is called Jordan content today, due to the further development of the subject by Camille Jordan (1838–1922) in 1893. Jordan used only polygons with horizontal and vertical sides, which is equivalent to what we do in this text.

Twentieth Century

A major step forward in the study of integration was due to Henri Lebesgue (1875–1941) in the early 20th century. He proved what we now call Lebesgue's Theorem, which is a characterization of which functions are Riemann integrable, and he defined what we call the Lebesgue integral, which equals the Riemann integral for Riemann integrable functions, but which exists for many functions that are not Riemann integrable (for example the Dirichlet function defined in Example 3.3.3 (6)), and which has some very convenient properties not obeyed by the Riemann integral. The

Lebesgue integral is defined in terms of Lebesgue measure, which supersedes the use of Jordan content. Other types of integrals were also defined in the 20th century, for example the Henstock–Kurzweil integral.

6

Limits to Infinity

6.1 Introduction

When we studied limits of functions in Chapter 3, we often considered expressions of the form "$\lim_{x \to c} f(x) = L$," where the symbols c and L both denoted real numbers. However, it is also possible to consider limits involving not only real numbers, but limits to "infinity" and "negative infinity," which are written $\lim_{x \to \infty} f(x) = L$, and $\lim_{x \to -\infty} f(x) = L$, and $\lim_{x \to c} f(x) = \infty$, and $\lim_{x \to c} f(x) = -\infty$. It is also possible to combine these two types of limits, for example $\lim_{x \to \infty} f(x) = \infty$. In all of the types of limits that involve ∞ and $-\infty$, it is important to recognize that the symbols "∞" and "$-\infty$" are not real numbers, but are rather a shorthand way of indicating that something is growing without bound either in the positive direction or in the negative direction.

We will define the different types of limits to infinity in Section 6.2. In Sections 6.3 and 6.4 we discuss two very useful topics involving limits to infinity, both of which appear in calculus courses, namely, l'Hôpital's Rule and improper integrals. Limits to infinity have many applications; for example, we will use such limits, and improper integrals, in our discussion of trigonometric functions and π in Sections 7.3 and 7.4. Other useful applications include Laplace transforms, which in turn are used for solving differential equations, and continuous probability; such topics are beyond the scope of this book. See [BD09, Chapter 6] for Laplace transforms as used for differential equations, and see [Ros10, Chapter 5] for continuous probability.

As was the case in previous chapters, here too we assume that the reader is informally familiar with the standard elementary functions and their basic properties, such as continuity and differentiability, in order to have sufficiently many functions to see interesting examples of the material in this chapter. We will see a rigorous treatment of the elementary functions in Chapter 7; the proofs in that chapter, though making use of some of the general ideas in the present chapter, will not make use of the examples in the present chapter, and there is no circular reasoning.

6.2 Limits to Infinity

We define two types of limits to infinity. The first type, which we will refer to as Type 1 limits to infinity, and which is denoted $\lim_{x\to\infty} f(x) = L$ or $\lim_{x\to-\infty} f(x) = L$, has x go to infinity or negative infinity, but has the value of $f(x)$ go to a real number L. The second type, which we will refer to as Type 2 limits to infinity, and which is denoted $\lim_{x\to c} f(x) = \infty$ or $\lim_{x\to c} f(x) = -\infty$, has x go to a real number c, but has the value of $f(x)$ go to infinity or negative infinity. Type 1 limits to infinity correspond to horizontal asymptotes of graphs of functions, as seen in Figure 6.2.1 (i), and Type 2 limits correspond to vertical asymptotes, as seen in the Part (ii) of the figure.

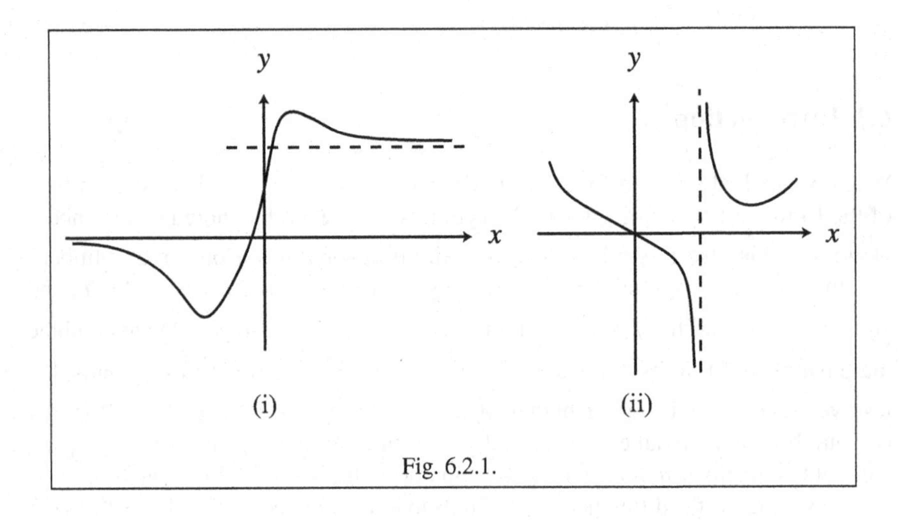

Fig. 6.2.1.

We start our discussion with Type 1 limits to infinity. In the ordinary type of limit, denoted $\lim_{x\to c} f(x) = L$, we mean intuitively that $f(x)$ gets closer and closer to a number L as the value of x gets closer and closer to a number c. By contrast, in a limit of the form $\lim_{x\to\infty} f(x) = L$, the idea is that $f(x)$ gets closer and closer to a number L as the value of x gets larger and larger, which is symbolically denoted by "$x \to \infty$," though there is no real number "∞" that the number x is getting closer and closer to. Hence, in our definition of this type of limit, we replace the expression "$|x - c| < \delta$," which is thought of as x being near c, with the expression "$x > M$," which is thought of as x being large.

For the following definition, recall the definition of right unbounded interval and left unbounded interval given in Definition 2.3.6.

Definition 6.2.1.

1. Let $I \subseteq \mathbb{R}$ be a right unbounded interval, let $f\colon I \to \mathbb{R}$ be a function and let $L \in \mathbb{R}$. The number L is the **limit** of f as x goes to infinity, written

$$\lim_{x\to\infty} f(x) = L,$$

if for each $\varepsilon > 0$, there is some $M \in \mathbb{R}$ such that $x \in I$ and $x > M$ imply $|f(x) - L| < \varepsilon$. If $\lim_{x \to \infty} f(x) = L$, we also say that f **converges** to L as x goes to infinity. If f converges to some real number as x goes to infinity, we say that $\lim_{x \to \infty} f(x)$ exists.

2. Let $J \subseteq \mathbb{R}$ be a left unbounded interval, let $f\colon J \to \mathbb{R}$ be a function and let $L \in \mathbb{R}$. The number L is the **limit** of f as x goes to negative infinity, written

$$\lim_{x \to -\infty} f(x) = L,$$

if for each $\varepsilon > 0$, there is some $P \in \mathbb{R}$ such that $x \in J$ and $x < P$ imply $|f(x) - L| < \varepsilon$. If $\lim_{x \to -\infty} f(x) = L$, we also say that f **converges** to L as x goes to negative infinity. If f converges to some real number as x goes to negative infinity, we say that $\lim_{x \to -\infty} f(x)$ exists. △

As was the case for ordinary limits, here too we need to prove that if $\lim_{x \to \infty} f(x) = L$ or $\lim_{x \to -\infty} f(x) = L$ for some $L \in \mathbb{R}$, then there is only one such number L. The analog of the following lemma for limits to negative infinity also holds, though for the sake of brevity we will not state it.

Lemma 6.2.2. *Let $I \subseteq \mathbb{R}$ be a right unbounded interval, and let $f\colon I \to \mathbb{R}$ be a function. If $\lim_{x \to \infty} f(x) = L$ for some $L \in \mathbb{R}$, then L is unique.*

Proof. Left to the reader in Exercise 6.2.2. □

Because of Lemma 6.2.2 we can refer to "the" limit of a function as x goes to infinity, if the limit exists, and similarly for limits to negative infinity.

Example 6.2.3.

(1) We will prove that $\lim_{x \to \infty} \frac{5x+4}{2x+3} = \frac{5}{2}$. (In principle, we should have stated that the function under consideration is $f\colon (-\frac{3}{2}, \infty) \to \mathbb{R}$ defined by $f(x) = \frac{5x+4}{2x+3}$ for all $x \in (-\frac{3}{2}, \infty)$, but that is implicitly clear, and we will not write out the name of the function in other similar situations.)

Let $\varepsilon > 0$. Let $M = \frac{7}{4\varepsilon}$. Suppose that $x \in (-\frac{3}{2}, \infty)$ and $x > M$. Then $x > \frac{7}{4\varepsilon}$, which means that $x > 0$, and hence $\frac{7}{4x} < \varepsilon$. Then

$$\left| \frac{5x+4}{2x+3} - \frac{5}{2} \right| = \left| \frac{7}{4x+6} \right| = \frac{7}{4x+6} < \frac{7}{4x} < \varepsilon.$$

(2) We will prove that $\lim_{x \to \infty} x^2$ does not exist. Suppose that $\lim_{x \to \infty} x^2 = L$ for some $L \in \mathbb{R}$. Let $\varepsilon = 1$. Let $M \in \mathbb{R}$. Let $x = \max\{M+1, L+1, 1\}$. Then $x > M$. There are now two cases. First, suppose that $L \geq 0$. Because $x \geq L+1 \geq 1$, it follows that $x^2 \geq (L+1)^2 \geq L+1$. Hence $x^2 - L \geq 1$, and therefore $|x^2 - L| \geq 1 = \varepsilon$. Second, suppose that $L < 0$. Because $x \geq 1$, then $x^2 \geq 1$, and it follows that $x^2 - L \geq 1$, which again implies that $|x^2 - L| \geq 1 = \varepsilon$. Putting the two cases together, we see that $|x^2 - L| \not< \varepsilon$, which is a contradiction to the hypothesis that $\lim_{x \to \infty} x^2 = L$. ◊

Many of the lemmas and theorems that were proved for ordinary limits in Section 3.2 have analogs for Type 1 limits to infinity. We will not state and prove all such results, but as a useful example we will prove the following theorem; the analog of this theorem also holds for limits to negative infinity, though again for the sake of brevity we will not state it. The reader is urged to observe the exact analogy between not only the statements of Theorem 3.2.10 and Theorem 6.2.4, but also the proofs of these theorems.

Theorem 6.2.4. *Let $I \subseteq \mathbb{R}$ be a right unbounded interval, let $f, g\colon I \to \mathbb{R}$ be functions and let $k \in \mathbb{R}$. Suppose that $\lim_{x\to\infty} f(x)$ and $\lim_{x\to\infty} g(x)$ exist.*

1. *$\lim_{x\to\infty} [f+g](x)$ exists and $\lim_{x\to\infty} [f+g](x) = \lim_{x\to\infty} f(x) + \lim_{x\to\infty} g(x)$.*
2. *$\lim_{x\to\infty} [f-g](x)$ exists and $\lim_{x\to\infty} [f-g](x) = \lim_{x\to\infty} f(x) - \lim_{x\to\infty} g(x)$.*
3. *$\lim_{x\to\infty} [kf](x)$ exists and $\lim_{x\to\infty} [kf](x) = k \lim_{x\to\infty} f(x)$.*
4. *$\lim_{x\to\infty} [fg](x)$ exists and $\lim_{x\to\infty} [fg](x) = [\lim_{x\to\infty} f(x)] \cdot [\lim_{x\to\infty} g(x)]$.*
5. *If $\lim_{x\to\infty} (g)(x) \neq 0$, then $\lim_{x\to\infty} \left[\frac{f}{g}\right](x)$ exists and $\lim_{x\to\infty} \left[\frac{f}{g}\right](x) = \dfrac{\lim_{x\to\infty} f(x)}{\lim_{x\to\infty} g(x)}$.*

Proof. We will prove Parts (1) and (4), leaving the rest to the reader in Exercise 6.2.3. Let $L = \lim_{x\to\infty} f(x)$ and $M = \lim_{x\to\infty} g(x)$.

(1) Let $\varepsilon > 0$. Then there is some $P \in \mathbb{R}$ such that $x \in I$ and $x > P$ imply $|f(x) - L| < \frac{\varepsilon}{2}$, and there is some $Q \in \mathbb{R}$ such that $x \in I$ and $x > Q$ imply $|g(x) - M| < \frac{\varepsilon}{2}$. Let $R = \max\{P, Q\}$. Suppose that $x \in I$ and $x > R$. Then

$$\begin{aligned} |[f+g](x) - (L+M)| &= |(f(x)-L) + (g(x)-M)| \le |f(x)-L| + |g(x)-M| \\ &< \frac{\varepsilon}{2} + \frac{\varepsilon}{2} = \varepsilon. \end{aligned}$$

(4) Let $\varepsilon > 0$. There is some $P \in \mathbb{R}$ such that $x \in I$ and $x > P$ imply $|g(x) - M| < 1$. Using Lemma 2.3.9 (7) we see that $x \in I$ and $x > P$ imply $|g(x)| - |M| < 1$, and hence $|g(x)| < |M| + 1$. Observe that $|L| + |M| + 1 > 0$. There is some $Q \in \mathbb{R}$ such that $x \in I$ and $x > Q$ imply $|f(x) - L| < \frac{\varepsilon}{|L|+|M|+1}$, and there is some $R \in \mathbb{R}$ such that $x \in I$ and $x > R$ imply $|g(x) - M| < \frac{\varepsilon}{|L|+|M|+1}$. Let $S = \max\{P, Q, R\}$. Suppose that $x \in I$ and $x > S$. Then

$$\begin{aligned} |[fg](x) - LM| &= |f(x)g(x) - LM| = |f(x)g(x) - g(x)L + g(x)L - LM| \\ &\le |g(x)| \cdot |f(x) - L| + |L| \cdot |g(x) - M| \\ &< (|M|+1)\frac{\varepsilon}{|L|+|M|+1} + |L| \cdot \frac{\varepsilon}{|L|+|M|+1} = \varepsilon. \end{aligned}$$ □

We now turn to Type 2 limits to infinity, including one-sided limits of this type. On the one hand, there is a certain technical similarity between Type 1 and Type 2 limits to infinity, though with the roles of the "x" and "y" coordinates reversed. Conceptually, however, there is a substantial difference between these two types of limits to infinity.

For a Type 1 limit to infinity, when we write "$\lim_{x\to\infty} f(x) = L$," we say that f *converges* to L as x goes to infinity. By contrast, for a Type 2 limit to infinity, when we write "$\lim_{x\to c} f(x) = \infty$," we mean that $f(x)$ is growing larger and larger as x gets closer and closer to c, which implies in particular that there is no $L \in \mathbb{R}$ that the values of $f(x)$ are approaching. Hence, when we write "$\lim_{x\to c} f(x) = \infty$," we say that f *diverges* to infinity as x goes to c. It might seem strange to say that f "diverges to" something, because divergence means a lack of convergence, and convergence means getting closer and closer to something. However, there are a variety of ways in which a function f can be divergent as x goes to a number c. Compare, for example, the behavior near 0 of the functions $f, g\colon \mathbb{R} - \{0\} \to \mathbb{R}$ defined by $f(x) = \frac{1}{x^2}$ for all $x \in \mathbb{R} - \{0\}$, seen in Figure 6.2.2 (i), and $g(x) = \sin\frac{1}{x}$ for all $x \in \mathbb{R} - \{0\}$, seen in Part (ii) of the figure. The function f diverges as x goes to 0 because the values of $f(x)$ are getting larger and larger, whereas the function g diverges as x goes to 0 because the values of $g(x)$ oscillate more and more frequently. In both cases the values of $f(x)$ are not approaching a single real number, and hence both $\lim_{x\to 0} f(x)$ and $\lim_{x\to 0} g(x)$ do not exist, but the reasons for the non-existence are quite different. Hence, when we say that "f diverges to infinity as x goes to 0," we are saying something about the type of divergence.

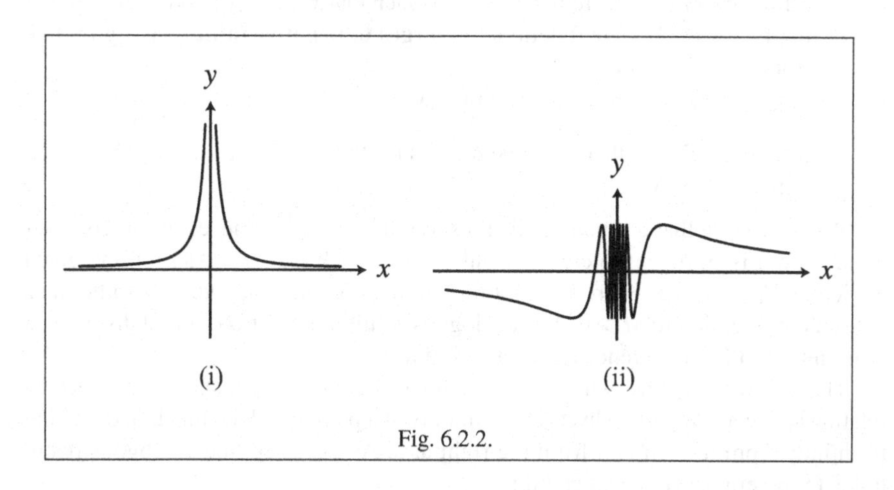

Fig. 6.2.2.

Definition 6.2.5. Let $I \subseteq \mathbb{R}$ be an interval, let $c \in I$ and let $f\colon I - \{c\} \to \mathbb{R}$ be a function.

1. Suppose that I is an open interval. The function f **diverges to infinity** as x goes to c, written
$$\lim_{x\to c} f(x) = \infty,$$
if for each $M \in \mathbb{R}$, there is some $\delta > 0$ such that $x \in I - \{c\}$ and $|x - c| < \delta$ imply $f(x) > M$. The function f **diverges to negative infinity** as x goes to c, written

$$\lim_{x \to c} f(x) = -\infty,$$

if for each $N \in \mathbb{R}$, there is some $\delta > 0$ such that $x \in I - \{c\}$ and $|x - c| < \delta$ imply $f(x) < N$.

2. Suppose that c is not a right endpoint of I. The function f **diverges to infinity** as x goes to c from the right, written

$$\lim_{x \to c^+} f(x) = \infty,$$

if for each $M \in \mathbb{R}$, there is some $\delta > 0$ such that $x \in I - \{c\}$ and $c < x < c + \delta$ imply $f(x) > M$. The function f **diverges to negative infinity** as x goes to c from the right, written

$$\lim_{x \to c^+} f(x) = -\infty,$$

if for each $N \in \mathbb{R}$, there is some $\delta > 0$ such that $x \in I - \{c\}$ and $c < x < c + \delta$ imply $f(x) < N$.

3. Suppose that c is not a left endpoint of I. The function f **diverges to infinity** as x goes to c from the left, written

$$\lim_{x \to c^-} f(x) = \infty,$$

if for each $M \in \mathbb{R}$, there is some $\delta > 0$ such that $x \in I - \{c\}$ and $c - \delta < x < c$ imply $f(x) > M$. The function f **diverges to negative infinity** as x goes to c from the left, written

$$\lim_{x \to c^-} f(x) = -\infty,$$

if for each $N \in \mathbb{R}$, there is some $\delta > 0$ such that $x \in I - \{c\}$ and $c - \delta < x < c$ imply $f(x) < N$. △

When proving that a function diverges to infinity, it is not necessary to consider all $M \in \mathbb{R}$; it is sufficient to consider only $M \in \mathbb{R}$ such that $M > M_0$, for any given choice of $M_0 \in \mathbb{R}$. See Exercise 6.2.9 for details. For example, we will sometimes restrict our attention to $M > 0$. The analogous result holds for one-sided divergence to infinity, and for divergence to negative infinity.

The following lemma, which is the analog of Lemma 3.2.17, shows the expected relation between a function diverging to infinity at a point c and the function diverging to infinity from the left and from the right at c. As expected, the analogous result holds for divergence to negative infinity.

Lemma 6.2.6. *Let $I \subseteq \mathbb{R}$ be an open interval, let $c \in I$ and let $f\colon I - \{c\} \to \mathbb{R}$ be a function. Then $\lim_{x \to c} f(x) = \infty$ if and only if $\lim_{x \to c^+} f(x) = \infty$ and $\lim_{x \to c^-} f(x) = \infty$.*

Proof. Left to the reader in Exercise 6.2.12. □

Example 6.2.7.

(1) We will prove that $\lim_{x \to 0} \frac{1}{x^2} = \infty$. Let $M \in \mathbb{R}$. We may assume that $M > 0$. Let $\delta = \frac{1}{\sqrt{M}}$. Suppose that $x \in \mathbb{R} - \{0\}$ and $|x - 0| < \delta$. Then $|x| < \delta$, and hence $|x| < \frac{1}{\sqrt{M}}$,

which implies $x^2 = |x|^2 < \frac{1}{M}$. Because $x^2 > 0$ and $M > 0$, we deduce that $\frac{1}{x^2} > M$.

(2) Let $f\colon \mathbb{R} - \{0\} \to \mathbb{R}$ be defined by $f(x) = \frac{1}{x}$ for all $x \in \mathbb{R} - \{0\}$. We will prove that f does not diverge to infinity or to negative infinity as x goes to 0. First, we will prove that $\lim_{x \to 0^+} \frac{1}{x} = \infty$. Let $M \in \mathbb{R}$. We may assume that $M > 0$. Let $\delta = \frac{1}{M}$. Suppose that $x \in \mathbb{R} - \{0\}$ and $0 < x < 0 + \delta$. Then $0 < x < \frac{1}{M}$, and hence $\frac{1}{x} > M$. It follows that $\lim_{x \to 0^+} \frac{1}{x} = \infty$. A similar argument shows that $\lim_{x \to 0^-} \frac{1}{x} = -\infty$. It now follows from Lemma 6.2.6 that f does not diverge to infinity or to negative infinity as x goes to 0. ◊

Our next theorem is the Type 2 analog of parts of Theorem 6.2.4. Observe that Theorem 6.2.8 is missing a treatment of differences and quotients of functions, an absence that will be explained in Section 6.3. Even those parts of Theorem 6.2.4 that do have analogs in Theorem 6.2.8 are more complicated in the latter than in the former. In general, Type 2 limits to infinity are not as well behaved as Type 1 limits to infinity, which should not be too surprising, given that Type 1 limits to infinity involve convergence to real numbers, and so we can use addition, subtraction, multiplication and division with such limits, whereas Type 2 limits to infinity involve divergence, and so we cannot use these four operations.

In the following theorem, when we say "$\lim_{x \to c} h(x)$ exists" we mean it as an ordinary limit, not as a Type 2 limit to infinity, because the latter type of limit does not exist as a real number, which is the only type of number with which we are working.

Theorem 6.2.8. *Let $I \subseteq \mathbb{R}$ be an open interval, let $c \in I$, let $f, g, h\colon I - \{c\} \to \mathbb{R}$ be functions and let $k \in \mathbb{R}$.*

1. *Suppose that $\lim_{x \to c} f(x) = \infty$ and $\lim_{x \to c} g(x) = \infty$, and that $\lim_{x \to c} h(x)$ exists. Then $\lim_{x \to c} [f+g](x) = \infty$ and $\lim_{x \to c} [f+h](x) = \infty$.*
2. *Suppose that $\lim_{x \to c} f(x) = -\infty$ and $\lim_{x \to c} g(x) = -\infty$, and that $\lim_{x \to c} h(x)$ exists. Then $\lim_{x \to c} [f+g](x) = -\infty$ and $\lim_{x \to c} [f+h](x) = -\infty$.*
3. *Suppose that $\lim_{x \to c} f(x) = \infty$ and $\lim_{x \to c} g(x) = -\infty$. If $k > 0$, then $\lim_{x \to c} [kf](x) = \infty$ and $\lim_{x \to c} [kg](x) = -\infty$. If $k < 0$, then $\lim_{x \to c} [kf](x) = -\infty$ and $\lim_{x \to c} [kg](x) = \infty$.*
4. *Suppose that $\lim_{x \to c} f(x) = \infty$ and $\lim_{x \to c} g(x) = \infty$, and that $\lim_{x \to c} h(x)$ exists. Then $\lim_{x \to c} [fg](x) = \infty$. If $\lim_{x \to c} h(x) > 0$, then $\lim_{x \to c} [fh](x) = \infty$. If $\lim_{x \to c} h(x) < 0$, then $\lim_{x \to c} [fh](x) = -\infty$.*
5. *Suppose that $\lim_{x \to c} f(x) = -\infty$ and $\lim_{x \to c} g(x) = -\infty$, and that $\lim_{x \to c} h(x)$ exists. Then $\lim_{x \to c} [fg](x) = \infty$. If $\lim_{x \to c} h(x) > 0$, then $\lim_{x \to c} [fh](x) = -\infty$. If $\lim_{x \to c} h(x) < 0$, then $\lim_{x \to c} [fh](x) = \infty$.*
6. *Suppose that $\lim_{x \to c} f(x) = \infty$ and $\lim_{x \to c} g(x) = -\infty$. Then $\lim_{x \to c} [fg](x) = -\infty$.*

Proof. We will prove Parts (1) and (4), leaving the rest to the reader in Exercise 6.2.13.

(1) We start with $f + g$. Let $M \in \mathbb{R}$. We may assume that $M > 0$. There is some

$\delta_1 > 0$ such that $x \in I - \{c\}$ and $|x - c| < \delta_1$ imply $f(x) > M$, and there is some $\delta_2 > 0$ such that $x \in I - \{c\}$ and $|x - c| < \delta_2$ imply $g(x) > M$. Let $\delta = \min\{\delta_1, \delta_2\}$. Suppose that $x \in I - \{c\}$ and $|x - c| < \delta$. Then

$$[f+g](x) = f(x) + g(x) > M + M > M.$$

We now turn to $f + h$. Let $Q = \lim_{x \to c} h(x)$. Let $N \in \mathbb{R}$. There is some $\eta_1 > 0$ such that $x \in I - \{c\}$ and $|x - c| < \eta_1$ imply $f(x) > N - Q + 1$, and there is some $\eta_2 > 0$ such that $x \in I - \{c\}$ and $|x - c| < \eta_2$ imply $|h(x) - Q| < 1$. Let $\eta = \min\{\eta_1, \eta_2\}$. Suppose that $x \in I - \{c\}$ and $|x - c| < \eta$. Then $|h(x) - Q| < 1$, which implies that $-1 < h(x) - Q < 1$, and hence $Q - 1 < h(x)$. Then

$$(f+h)(x) = f(x) + h(x) > (N - Q + 1) + (Q - 1) = N.$$

(4) We start with fg. Let $M \in \mathbb{R}$. We may assume that $M > 1$. There is some $\delta_1 > 0$ such that $x \in I - \{c\}$ and $|x - c| < \delta_1$ imply $f(x) > M$, and there is some $\delta_2 > 0$ such that $x \in I - \{c\}$ and $|x - c| < \delta_2$ imply $g(x) > M$. Let $\delta = \min\{\delta_1, \delta_2\}$. Suppose that $x \in I - \{c\}$ and $|x - c| < \delta$. Then

$$[fg](x) = f(x)g(x) > M \cdot M > M.$$

We now turn to fh. Suppose that $\lim_{x \to c} h(x) > 0$. By the Sign-Preserving Property for Limits (Theorem 3.2.4) there is some $Q > 0$ and some $\eta_1 > 0$ such that $x \in I - \{c\}$ and $|x - c| < \eta_1$ imply $h(x) > Q$.

Let $N \in \mathbb{R}$. We may assume that $N > 0$. There is some $\eta_2 > 0$ such that $x \in I - \{c\}$ and $|x - c| < \eta_2$ imply $f(x) > \frac{N}{Q}$. Let $\eta = \min\{\eta_1, \eta_2\}$. Suppose that $x \in I - \{c\}$ and $|x - c| < \eta$. Then

$$[fh](x) = f(x)h(x) > \frac{N}{Q} \cdot Q = N.$$

The case where $\lim_{x \to c} h(x) < 0$ is similar, and we omit the details. □

The analog of Theorem 6.2.8 for right-hand limits and left-hand limits also holds, though we omit the details.

Having discussed Type 1 and Type 2 limits to infinity separately, we note that it is also possible to combine these two types of limits, and consider limits of the form $\lim_{x \to \infty} f(x) = \infty$, and also with $-\infty$ replacing one or both occurrences of ∞. This topic is left to the reader in Exercise 6.2.15.

Finally, we note that because many of the properties of limits to infinity (especially Type 1 limits to infinity) are analogous to properties of ordinary limits (as discussed in Section 3.2), some authors combine various aspects of limits to infinity and ordinary limits by making use of the "extended real numbers," which is the set $\mathbb{R}^* = \mathbb{R} \cup \{-\infty, \infty\}$, where "$-\infty$" and "$\infty$" are two symbols not in the real numbers. The operations addition and multiplication can then partially be extended to include $-\infty$ and ∞, for example by stating $\infty + \infty = \infty$, and $x + \infty = \infty$ for all $x \in \mathbb{R}$. These last two properties are meant to reflect Theorem 6.2.8 (1), and other parts of that theorem also lead to properties of the extended real numbers. It is also possible to define an order relation on the extended real numbers by setting $-\infty < x < \infty$ for all $x \in \mathbb{R}$. However, not all properties of the real numbers have analogs in the extended real numbers. For example, as will be seen in Exercise 6.3.2 and Example 6.3.1, it is not possible to say that $\infty + (-\infty)$ equals 0, or that $\frac{\infty}{\infty}$ equals 1. Formally, the extended real numbers are not an ordered field, as defined in Definition 2.2.1. The symbols "$-\infty$" and "∞" are not real numbers, and do not behave as do real numbers, and hence care is needed when using the extended real numbers. Having said that, the extended real numbers can be defined quite rigorously, for example using the method of Dedekind cuts, which we used to construct the ordinary real numbers in Sections 1.6 and 1.7. The advantage of using the extended real numbers is efficiency, for example by combining the statements of Theorem 3.2.10 and Theorem 6.2.4 into a single theorem by looking at limits of the form $\lim_{x \to c} f(x)$, where $c \in \mathbb{R}^*$. For our purposes, however, the efficiency gained by using the extended real numbers is not worth the price of having to offer a rigorous treatment of the extended real numbers, and hence we will not be using this concept.

Reflections

The reader might wonder why this section was not included in Chapter 3, which is devoted to limits and continuity. The definitions of limits in the present section are indeed just variants of the definition of limits given in Chapter 3, and technically it would have made sense to include the present section in that earlier chapter. There are, however, a few pedagogical reasons for arranging the material as we have. First, learning to use the ε–δ definition of limits given in Section 3.2 is initially tricky for some students, and there is no advantage in making matters even more difficult by introducing a few variants of the original definition right at the start. Second, even though limits to infinity are formally rather similar to regular limits, there is a substantial conceptual difference between the limits discussed in Section 3.2 and those discussed in the present section, because there is no number called "infinity," and hence the notion of converging to a number is quite distinct from converging to infinity. It is therefore helpful to make this distinction clear by separating the discussion of the two types of limits. Third, limits to infinity show up in two standard places in a typical calculus course (and introductory real analysis course), namely, l'Hôpital's Rule and improper integrals, and it is convenient to cluster these applications of limits to infinity together with the definition of such limits, rather than having the definition a few chapters before it is first used. The author organizes the material involving limits to infinity when he teaches introductory calculus the same way as is found in the present chapter.

Exercises

Exercise 6.2.1. Using only the definition of Type 1 limits to infinity, prove that each of the following limits holds.

(1) $\lim_{x \to \infty} \frac{1}{x} = 0$.

(2) $\lim\limits_{x\to\infty} \frac{2x+7}{5x+1} = \frac{2}{5}$.

(3) $\lim\limits_{x\to\infty} \frac{x^2+1}{3x^2+x} = \frac{1}{3}$.

Exercise 6.2.2. [Used in Lemma 6.2.2.] Prove Lemma 6.2.2.

Exercise 6.2.3. [Used in Theorem 6.2.4.] Prove Theorem 6.2.4 (2) (3) (5).

Exercise 6.2.4. Let $I \subseteq \mathbb{R}$ be a right unbounded interval, let $f: I \to \mathbb{R}$ be a function and let $L \in \mathbb{R}$. Prove that $\lim\limits_{x\to\infty} f(x) = L$ if and only if $\lim\limits_{x\to-\infty} f(-x) = L$.

Exercise 6.2.5. Let $I \subseteq \mathbb{R}$ be a right unbounded interval, and let $f, g: I \to \mathbb{R}$ be functions. Suppose that $\lim\limits_{x\to\infty} f(x) = 0$, and that g is bounded. Prove that $\lim\limits_{x\to\infty} [fg](x) = 0$.

Exercise 6.2.6. Let $I \subseteq \mathbb{R}$ be a right unbounded interval, and let $f, g: I \to \mathbb{R}$ be functions. Suppose that $f(x) \leq g(x)$ for all $x \in I$. Prove that if $\lim\limits_{x\to\infty} f(x)$ and $\lim\limits_{x\to\infty} g(x)$ exist, then $\lim\limits_{x\to\infty} f(x) \leq \lim\limits_{x\to\infty} g(x)$.

Exercise 6.2.7. Let $I \subseteq \mathbb{R}$ be an interval, let $c \in I$ and let $f: I - \{c\} \to \mathbb{R}$ be a function. Suppose that for each $M \in \mathbb{R}$, there is some $x \in I - \{c\}$ such that $f(x) > M$.

(1) Prove that $\lim\limits_{x\to c} f(x)$ does not exist.

(2) Is it necessarily true that $\lim\limits_{x\to c} f(x) = \infty$? Give a proof or a counterexample.

Exercise 6.2.8. [Used in Example 4.2.5.] Let $I \subseteq \mathbb{R}$ be an interval, let $c \in I$ and let $f, g: I - \{c\} \to \mathbb{R}$ be functions. Suppose that $\lim\limits_{x\to c} f(x) = \infty$. Suppose that there is some $q \in \mathbb{R}$ such that $q \neq 0$, and that for each $\delta > 0$, there is some $x \in I - \{c\}$ such that $|x - c| < \delta$ and $g(x) = q$. Prove that $\lim\limits_{x\to c} f(x)g(x)$ does not exist.

Exercise 6.2.9. [Used in Section 6.2.] Let $I \subseteq \mathbb{R}$ be an interval, let $c \in I$, let $f: I - \{c\} \to \mathbb{R}$ be a function and let $M_0 \in \mathbb{R}$. Prove that $\lim\limits_{x\to c} f(x) = \infty$ if and only if for each $M \in \mathbb{R}$ such that $M > M_0$, there is some $\delta > 0$ such that $x \in I - \{c\}$ and $|x - c| < \delta$ imply $f(x) > M$.

Exercise 6.2.10. Let $J \subseteq I \subseteq \mathbb{R}$ be open intervals, let $c \in J$ and let $f: I - \{c\} \to \mathbb{R}$ be a function. Prove that $\lim\limits_{x\to c} f(x) = \infty$ if and only if $\lim\limits_{x\to c} f|_J(x) = \infty$.

Exercise 6.2.11. Using only the definition of Type 2 limits to infinity, prove that $\lim\limits_{x\to 3^+} \frac{x}{x-3} = \infty$.

Exercise 6.2.12. [Used in Lemma 6.2.6.] Prove Lemma 6.2.6.

Exercise 6.2.13. [Used in Theorem 6.2.8.] Prove Theorem 6.2.8 (2) (3) (5) (6).

Exercise 6.2.14. Let $I, J \subseteq \mathbb{R}$ be open intervals, let $c \in I$, let $d \in J$ and let $g: I - \{c\} \to J - \{d\}$ and $f: J - \{d\} \to \mathbb{R}$ be functions. Suppose that $\lim\limits_{y\to c} g(y) = d$ and that $\lim\limits_{x\to d} f(x) = \infty$. Prove that $\lim\limits_{y\to c} (f \circ g)(y) = \infty$.

Exercise 6.2.15. [Used in Section 6.2, Exercise 6.2.16, Example 6.4.3, Exercise 7.2.16 and Exercise 8.2.10.]

(1) Let $I \subseteq \mathbb{R}$ be a right unbounded interval, and let $f: I \to \mathbb{R}$ be a function. Give a precise definition of what it would mean to say that the function f **diverges to infinity** as x goes to infinity, written

$$\lim_{x \to \infty} f(x) = \infty.$$

(2) Using only the definition you gave in Part (1) of this exercise, prove that $\lim_{x \to \infty} x^2 = \infty$.

(3) Using only the definition you gave in Part (1) of this exercise, prove that $\lim_{x \to \infty} \sqrt{x} = \infty$.

(4) Using only the definition you gave in Part (1) of this exercise, prove that $\lim_{x \to \infty} \frac{x^2}{x+1} = \infty$.

Exercise 6.2.16. This exercise makes use of Exercise 6.2.15. Let $I, J \subseteq \mathbb{R}$ be right unbounded intervals, and let $f: I \to \mathbb{R}$ and $g: J \to I$ be functions. Suppose that $\lim_{x \to \infty} f(x) = L$ for some $L \in \mathbb{R}$, and that $\lim_{x \to \infty} g(x) = \infty$. Prove that $\lim_{x \to \infty} (f \circ g)(x) = L$.

6.3 Computing Limits to Infinity

Type 2 limits to infinity are not as well behaved as Type 1 limits to infinity. This difference can be seen comparing Theorem 6.2.4, which is for Type 1 limits to infinity, with Theorem 6.2.8, which is for Type 2. The fact that the latter theorem is missing a treatment of differences and quotients of functions, an absence we will explain shortly, makes Type 2 limits to infinity harder to compute in practice. In the present section we discuss a few methods that help us compute specific cases of Type 2 limits to infinity, the most well-known of which is l'Hôpital's Rule, and related topics.

To see what is difficult about certain categories of Type 2 limits to infinity, let us recall Theorem 6.2.8, which shows us those aspects of such limits that do work nicely. The statement of Theorem 6.2.8 (1), for example, can be summarized by writing "$\infty + \infty = \infty$," and "$\infty + c = \infty$" for all $c \in \mathbb{R}$. Of course, we do not literally mean addition of real numbers in these expressions, because the symbols "∞" and "$-\infty$" are not real numbers, and the notion of addition that we have for real numbers does not apply to these two symbols. However, these two expressions are useful in that they suggest in very concise terms the result stated in Theorem 6.2.8 (1), and we can use such expressions as long as we do not take them to be more than just suggestive notation.

Using the above notation, we observe that whereas Theorem 6.2.8 treats expressions such as $\infty + \infty$ and $\infty \cdot \infty$, missing from that theorem are expressions such as $\infty - \infty$ and $\frac{\infty}{\infty}$. The following example shows us why $\frac{\infty}{\infty}$ is missing from Theorem 6.2.8. The reader is asked in Exercise 6.3.2 to supply similar examples for $\infty - \infty$, which show why $\infty + (-\infty)$ is missing from the theorem.

Example 6.3.1. Let $f,g,h\colon \mathbb{R}-\{0\}\to\mathbb{R}$ be defined by $f(x)=\frac{1}{x^2}$ and $g(x)=\frac{1}{x^2}$ and $h(x)=\frac{1}{x^4}$ for all $x\in\mathbb{R}-\{0\}$. We saw in Example 6.2.7 (1) that $\lim_{x\to 0} f(x)=\lim_{x\to 0} g(x)=\infty$, and it follows from Theorem 6.2.8 (4) that $\lim_{x\to 0} h(x)=\lim_{x\to 0}[fg](x)=\infty$. Hence the three limits $\lim_{x\to 0}\left[\frac{f}{g}\right](x)$, and $\lim_{x\to 0}\left[\frac{f}{h}\right](x)$, and $\lim_{x\to 0}\left[\frac{h}{f}\right](x)$ all have the form $\frac{\infty}{\infty}$.

On the other hand, we observe that $\left[\frac{f}{g}\right](x)=1$ and $\left[\frac{f}{h}\right](x)=x^2$ and $\left[\frac{h}{f}\right](x)=\frac{1}{x^2}$ for all $x\in\mathbb{R}-\{0\}$. Therefore $\lim_{x\to 0}\left[\frac{f}{g}\right](x)=1$, and $\lim_{x\to 0}\left[\frac{f}{h}\right](x)=0$, and $\lim_{x\to 0}\left[\frac{h}{f}\right](x)=\infty$. Hence there is no single value, either a real number or infinity, shared by all limits of the form $\frac{\infty}{\infty}$. ◊

Because of Exercise 6.3.2 and Example 6.3.1, we refer to the expressions $\infty-\infty$ and $\frac{\infty}{\infty}$ as "indeterminate forms." By contrast, the expressions $\infty+\infty$ and $\infty\cdot\infty$ are not indeterminate forms, because they both equal something specific (which happens to be ∞).

Although the above discussion was motivated by the need to compute Type 2 limits to infinity, we note that there are also indeterminate forms for some types of limits that do not involve infinity. In the following example, we see that $\frac{0}{0}$ is also an indeterminate form.

Example 6.3.2. Let $f,g,h\colon \mathbb{R}-\{5\}\to\mathbb{R}$ be defined by $f(x)=x^2-25$ and $g(x)=x-5$ and $h(x)=x^3-15x^2+75x-125$ for all $x\in\mathbb{R}-\{5\}$. Because f, g and h are polynomial functions, then they are continuous, as remarked in Example 3.3.7 (1), and hence it is seen that $\lim_{x\to 5} f(x)=f(5)=0$, and similarly $\lim_{x\to 5} g(x)=0$ and $\lim_{x\to 5} h(x)=0$. Hence the three limits $\lim_{x\to 5}\left[\frac{h}{g}\right](x)$, and $\lim_{x\to 5}\left[\frac{f}{g}\right](x)$, and $\lim_{x\to 5}\left[\frac{g}{h}\right](x)$ all have the form $\frac{0}{0}$.

We cannot apply Theorem 3.2.10 (5) for any of these three limits, because the hypotheses of that theorem are not satisfied. Nonetheless, we can evaluate these three limits as follows. First, we note that $f(x)=(x-5)(x+5)$ and $h(x)=(x-5)^3$ for all $x\in\mathbb{R}-\{5\}$. Next, we observe that as we take the limit of a function as x approaches 5, we never use the value $x=5$ in our function, and hence $x-5$ is never 0, and therefore we can cancel by $x-5$ when needed. Then

$$\lim_{x\to 5}\left[\frac{f}{g}\right](x)=\lim_{x\to 5}\frac{x^2-25}{x-5}=\lim_{x\to 5}\frac{(x-5)(x+5)}{x-5}=\lim_{x\to 5}(x+5)=5+5=10.$$

Similar computations show that $\lim_{x\to 5}\left[\frac{h}{g}\right](x)=0$ and $\lim_{x\to 5}\left[\frac{g}{h}\right](x)=\infty$, where the latter makes use of Example 6.2.7 (1) and Exercise 6.2.14; the details are left to the reader. Hence there is no single value, either a real number or infinity, shared by all limits of the form $\frac{0}{0}$. ◊

Although expressions such as $\frac{0}{0}$ and $\frac{\infty}{\infty}$ are indeterminate forms, it is important to note that not every fraction involving at least one of 0 or ∞ is an indeterminate form. For example, Theorem 3.2.10 (5) tells us that limits of the form $\frac{0}{c}$, where $c\in\mathbb{R}-\{0\}$, are always equal to 0, and hence are not indeterminate.

We now have two theorems that show, when appropriately stated, that the forms $\frac{c}{0}$ and $\frac{c}{\infty}$ are not indeterminate, having values ∞ and 0, respectively. For the first of these theorems, the hypotheses in Part (2) are needed to guarantee that the function does not approach ∞ from one side and $-\infty$ from the other side.

Theorem 6.3.3. *Let $I \subseteq \mathbb{R}$ be an open interval, let $c \in I$ and let $f, g\colon I - \{c\} \to \mathbb{R}$ be functions. Suppose that $\lim_{x\to c} f(x)$ exists and $\lim_{x\to c} f(x) \neq 0$, that $g(x) \neq 0$ for all $x \in I - \{c\}$ and that $\lim_{x\to c} g(x) = 0$.*

1. *$\lim_{x\to c} \left[\frac{f}{g}\right](x)$ does not exist.*
2. *If $\lim_{x\to c} f(x) > 0$ and $g(x) > 0$ for all $x \in I - \{c\}$, or if $\lim_{x\to c} f(x) < 0$ and $g(x) < 0$ for all $x \in I - \{c\}$, then $\lim_{x\to c} \left[\frac{f}{g}\right](x) = \infty$. If $\lim_{x\to c} f(x) > 0$ and $g(x) < 0$ for all $x \in I - \{c\}$, or if $\lim_{x\to c} f(x) < 0$ and $g(x) > 0$ for all $x \in I - \{c\}$, then $\lim_{x\to c} \left[\frac{f}{g}\right](x) = -\infty$.*

Proof. We will prove Part (1), leaving the remaining part to the reader in Exercise 6.3.5.

(1) Let $L = \lim_{x\to c} f(x)$. Exercise 3.2.9 (1) implies that $\lim_{x\to c} |f(x)| = |L|$. By hypothesis $L \neq 0$, and so $|L| > 0$. By the Sign-Preserving Property for Limits (Theorem 3.2.4) applied to the function $|f|$ there is some $M > 0$ and some $\delta_1 > 0$ such that $x \in I - \{c\}$ and $|x - c| < \delta_1$ imply $|f(x)| > M$.

Suppose that $\lim_{x\to c} \left[\frac{f}{g}\right](x)$ exists. Let $P = \lim_{x\to c} \left[\frac{f}{g}\right](x)$. Then there is some $\delta_2 > 0$ such that $x \in I - \{c\}$ and $|x - c| < \delta_2$ imply $\left|\left[\frac{f}{g}\right](x) - P\right| < 1$. Because $\lim_{x\to c} g(x) = 0$, then there is some $\delta_3 > 0$ such that $x \in I - \{c\}$ and $|x - c| < \delta_3$ imply $|g(x) - 0| < \frac{M}{|P|+1}$. Let $\delta = \min\{\delta_1, \delta_2, \delta_3\}$. Suppose that $x \in I - \{c\}$ and $|x - c| < \delta$. Then $|g(x)| = |g(x) - 0| < \frac{M}{|P|+1}$. We also have $\left|\left[\frac{f}{g}\right](x) - P\right| < 1$. By Lemma 2.3.9 (7) it follows that $\left|\left[\frac{f}{g}\right](x)\right| - |P| < 1$, and hence $\frac{|f(x)|}{|g(x)|} < |P| + 1$. We also know that $|f(x)| > M$. Therefore $\frac{M}{|g(x)|} < \frac{|f(x)|}{|g(x)|} < |P| + 1$, and hence $\frac{M}{|P|+1} < |g(x)|$, which is a contradiction. We conclude that $\lim_{x\to c} \left[\frac{f}{g}\right](x)$ does not exist. □

Theorem 6.3.4. *Let $I \subseteq \mathbb{R}$ be an open interval, let $c \in I$ and let $f, g\colon I - \{c\} \to \mathbb{R}$ be functions. Suppose that $\lim_{x\to c} f(x)$ exists, that $g(x) \neq 0$ for all $x \in I - \{c\}$, that $\lim_{x\to c^-} g(x) = \infty$ or $\lim_{x\to c^-} g(x) = -\infty$ and that $\lim_{x\to c^+} g(x) = \infty$ or $\lim_{x\to c^+} g(x) = -\infty$. Then $\lim_{x\to c} \left[\frac{f}{g}\right](x)$ exists and $\lim_{x\to c} \left[\frac{f}{g}\right](x) = 0$.*

Proof. Left to the reader in Exercise 6.3.6. □

The analogs of Theorem 6.3.3 and Theorem 6.3.4 for right-hand limits and left-hand limits also hold, though we omit the details.

We now return to indeterminate forms, and in particular the forms $\frac{0}{0}$ and $\frac{\infty}{\infty}$. In Example 6.3.2 we were able to evaluate some limits of the form $\frac{0}{0}$, but that was only because we were able to cancel the "bad parts" of the numerator and the denominator. However, such cancellation is often not possible in limits of the form $\frac{0}{0}$ and $\frac{\infty}{\infty}$, for example in the limit $\lim_{x\to 0} \frac{e^x-1}{x}$, and it would be helpful to have other ways of computing such limits. A very useful tool for this purpose is l'Hôpital's Rule. We will state and prove two versions of l'Hôpital's Rule, the first for $\frac{0}{0}$, and the second for $\frac{\infty}{\infty}$.

We start with l'Hôpital's Rule for $\frac{0}{0}$, which is somewhat easier to prove than the $\frac{\infty}{\infty}$ case.

Theorem 6.3.5 (l'Hôpital's Rule for $\frac{0}{0}$). *Let $I \subseteq \mathbb{R}$ be an open interval, let $c \in I$ and let $f, g\colon I - \{c\} \to \mathbb{R}$ be functions. Suppose that f and g are differentiable, and that $g'(x) \neq 0$ for all $x \in I - \{c\}$. Suppose that $\lim_{x\to c} f(x) = 0$ and $\lim_{x\to c} g(x) = 0$. If $\lim_{x\to c} \frac{f'(x)}{g'(x)}$ exists, then $\lim_{x\to c} \frac{f(x)}{g(x)}$ exists and*

$$\lim_{x\to c} \frac{f(x)}{g(x)} = \lim_{x\to c} \frac{f'(x)}{g'(x)}.$$

Proof. Suppose that $\lim_{x\to c} \frac{f'(x)}{g'(x)}$ exists. Let $L = \lim_{x\to c} \frac{f'(x)}{g'(x)}$.

Let $\varepsilon > 0$. Then there is some $\delta_1 > 0$ such that $x \in I - \{c\}$ and $|x - c| < \delta_1$ imply $\left|\frac{f'(x)}{g'(x)} - L\right| < \frac{\varepsilon}{2}$.

Let $I = (a,b)$. Because $g|_{(a,c)}$ and $g|_{(c,b)}$ are differentiable, and because $g'(x) \neq 0$ for all $x \in I - \{c\} = (a,c) \cup (c,b)$, it follows from Exercise 4.4.9 (1) that $g|_{(a,c)}$ and $g|_{(c,b)}$ are injective.

Because there is at most one $x \in (a,c)$ such that $g(x) = 0$, and at most one $x \in (c,b)$ such that $g(x) = 0$, there is some $\delta_2 > 0$ such that $x \in I - \{c\}$ and $|x - c| < \delta_2$ imply $g(x) \neq 0$. By Lemma 2.3.7 (2) there is some $\delta_3 > 0$ such that $(c - \delta_3, c + \delta_3) \subseteq I$. Let $\delta = \min\{\delta_1, \delta_2, \delta_3\}$.

Suppose that $w \in I - \{c\}$ and $|w - c| < \delta$. Then $w \in (c - \delta, c)$ or $w \in (c, c + \delta)$. Suppose that $w \in (c - \delta, c)$; the other case is similar, and we omit the details. By the choice of δ we know that $g(w) \neq 0$. Because $\lim_{x\to c} f(x) = 0$ and $\lim_{x\to c} g(x) = 0$, it follows from Exercise 3.2.1 and Theorem 3.2.10 that

$$\lim_{x\to c} \frac{f(w) - f(x)}{g(w) - g(x)} = \frac{f(w) - 0}{g(w) - 0} = \frac{f(w)}{g(w)}.$$

Hence there is some $\eta > 0$ such that $x \in I - \{c\}$ and $|x - c| < \eta$ imply

$$\left|\frac{f(w) - f(x)}{g(w) - g(x)} - \frac{f(w)}{g(w)}\right| < \frac{\varepsilon}{2}.$$

Choose some $e \in (w, c)$ such that $|e - c| < \eta$. By Cauchy's Mean Value Theorem (Theorem 4.4.5) there is some $q \in (w, e)$ such that $[f(e) - f(w)]g'(q) = [g(e) -$

$g(w)]f'(q)$. Because $g|_{(a,c)}$ is injective, it follows that $g(w)-g(e) \neq 0$. We know by hypothesis that $g'(q) \neq 0$. Hence

$$\frac{f(w)-f(e)}{g(w)-g(e)} = \frac{f'(q)}{g'(q)}.$$

Because $q \in (w,e) \subseteq (c-\delta, c)$, then $q \in I - \{c\}$ and $|q-c| < \delta$. Hence $\left|\frac{f'(q)}{g'(q)} - L\right| < \frac{\varepsilon}{2}$, and therefore

$$\left|\frac{f(w)-f(e)}{g(w)-g(e)} - L\right| < \frac{\varepsilon}{2}.$$

Then

$$\begin{aligned}\left|\frac{f(w)}{g(w)} - L\right| &= \left|\frac{f(w)}{g(w)} - \frac{f(w)-f(e)}{g(w)-g(e)} + \frac{f(w)-f(e)}{g(w)-g(e)} - L\right| \\ &\leq \left|\frac{f(w)}{g(w)} - \frac{f(w)-f(e)}{g(w)-g(e)}\right| + \left|\frac{f(w)-f(e)}{g(w)-g(e)} - L\right| < \frac{\varepsilon}{2} + \frac{\varepsilon}{2} = \varepsilon.\end{aligned}$$

We conclude that $\lim_{x\to c} \frac{f(x)}{g(x)}$ exists, and that $\lim_{x\to c} \frac{f(x)}{g(x)} = L = \lim_{x\to c} \frac{f'(x)}{g'(x)}$. □

The above proof of l'Hôpital's Rule for $\frac{0}{0}$ (Theorem 6.3.5) might appear to be more complicated than expected, but that is because we have kept the hypotheses of the theorem to a minimum. A much shorter proof of l'Hôpital's Rule for $\frac{0}{0}$, though with stronger hypotheses, is found in Exercise 6.3.7. This shorter proof, though limited in its applicability, provides the closest thing one could call intuitive motivation for l'Hôpital's Rule for $\frac{0}{0}$; ultimately, what is good about this theorem is its usefulness for computing specific limits, not its intuitive appeal.

There are also variants of l'Hôpital's Rule for $\frac{0}{0}$, where $x \to c$ is replaced with one of $x \to c^+$, or $x \to c^-$, or $x \to \infty$, or $x \to -\infty$; we will not state these variants, because there are no changes of substance between them and the version in Theorem 6.3.5, but we will use them as needed.

Example 6.3.6.

(1) We use l'Hôpital's Rule for $\frac{0}{0}$ (Theorem 6.3.5) to evaluate the limit $\lim_{x\to 0} \frac{e^x-1}{x}$. Because both the numerator and the denominator are continuous functions (by Theorem 7.2.7 (2), Theorem 4.2.4, Example 3.3.3 (1) and Theorem 3.3.5), we see that $\lim_{x\to 0}(e^x - 1) = e^0 - 1 = 0$ and $\lim_{x\to 0} x = 0$. It is left to the reader to verify that the hypotheses of l'Hôpital's Rule for $\frac{0}{0}$ hold for this example, and hence we see that

$$\lim_{x\to 0} \frac{e^x-1}{x} = \lim_{x\to 0} \frac{[e^x-1]'}{[x]'} = \lim_{x\to 0} \frac{e^x}{1} = \frac{e^0}{1} = 1.$$

(2) The limit $\lim_{x\to 0} \frac{\sin x}{x} = 1$ is often encountered in an introductory calculus course. This limit certainly satisfies the hypotheses of l'Hôpital's Rule for $\frac{0}{0}$, and it can be computed very easily by

$$\lim_{x \to 0} \frac{\sin x}{x} = \lim_{x \to 0} \frac{\cos x}{1} = \frac{\cos 0}{1} = \frac{1}{1} = 1.$$

And yet, we need to ask whether this use of l'Hôpital's Rule is legitimate. This limit is not computed using l'Hôpital's Rule in an introductory calculus course, but rather it is usually computed via a geometric argument using the unit circle, and the reader is urged to try to figure out why before reading on.

The reason is that in an introductory calculus course, the limit $\lim_{x \to 0} \frac{\sin x}{x} = 1$ is used in the proof that $\sin' = \cos$, and so it would not be legitimate to use that fact to compute the limit, but that is just what is done when the limit is computed using l'Hôpital's Rule. By contrast, our proof that $\sin' = \cos$, which will be given in Theorem 7.3.12 (1), uses a very different definition of sin than the geometric unit circle definition seen in calculus courses, and in particular does not use the limit $\lim_{x \to 0} \frac{\sin x}{x} = 1$. Hence, in our context, it is fine to use l'Hôpital's Rule for this limit.

(3) l'Hôpital's Rule is so pleasant to use that a problem in its application is that sometimes it is used even when the situation does not permit it. It is left to the reader to find the flaw with the following "calculation" written by an overly eager calculus student:

$$\text{“}\lim_{x \to 0} \frac{\sin x}{x + x^2} = \lim_{x \to 0} \frac{\cos x}{1 + 2x} = \lim_{x \to 0} \frac{-\sin x}{2} = \frac{-\sin 0}{2} = 0.\text{”}$$

(The correct value of the limit is 1.) ◊

We now turn to l'Hôpital's Rule for $\frac{\infty}{\infty}$. Here we make use of Type 2 limits to infinity. The proof for the $\frac{\infty}{\infty}$ case is a bit trickier than for the $\frac{0}{0}$ case, and in contrast to the $\frac{0}{0}$ case, where a shorter proof may be obtained by strengthening the hypotheses, there is no such easy route in the $\frac{\infty}{\infty}$ case.

Theorem 6.3.7 (l'Hôpital's Rule for $\frac{\infty}{\infty}$). *Let $I \subseteq \mathbb{R}$ be an open interval, let $c \in I$ and let $f, g \colon I - \{c\} \to \mathbb{R}$ be functions. Suppose that f and g are differentiable, and that $g'(x) \neq 0$ for all $x \in I - \{c\}$. Suppose that $\lim_{x \to c^-} f(x) = \infty$ or $\lim_{x \to c^-} f(x) = -\infty$, that $\lim_{x \to c^+} f(x) = \infty$ or $\lim_{x \to c^+} f(x) = -\infty$, that $\lim_{x \to c^-} g(x) = \infty$ or $\lim_{x \to c^-} g(x) = -\infty$ and that $\lim_{x \to c^+} g(x) = \infty$ or $\lim_{x \to c^+} g(x) = -\infty$. If $\lim_{x \to c} \frac{f'(x)}{g'(x)}$ exists, then $\lim_{x \to c} \frac{f(x)}{g(x)}$ exists and*

$$\lim_{x \to c} \frac{f(x)}{g(x)} = \lim_{x \to c} \frac{f'(x)}{g'(x)}.$$

Proof. Suppose that $\lim_{x \to c} \frac{f'(x)}{g'(x)}$ exists. Let $L = \lim_{x \to c} \frac{f'(x)}{g'(x)}$.

We will prove that $\lim_{x \to c^-} \frac{f(x)}{g(x)}$ exists and $\lim_{x \to c^-} \frac{f(x)}{g(x)} = L$. A similar argument shows that $\lim_{x \to c^+} \frac{f(x)}{g(x)}$ exists and $\lim_{x \to c^+} \frac{f(x)}{g(x)} = L$, and we omit the details. It will then follow from Lemma 3.2.17 that $\lim_{x \to c} \frac{f(x)}{g(x)}$ exists and equals $\lim_{x \to c} \frac{f(x)}{g(x)} = L$.

Let $\varepsilon > 0$. By Lemma 3.2.17 we see that $\lim_{x \to c^-} \frac{f'(x)}{g'(x)}$ exists and $\lim_{x \to c^-} \frac{f'(x)}{g'(x)} = L$. Then there is some $\delta_1 > 0$ such that $x \in I - \{c\}$ and $c - \delta_1 < x < c$ imply $\left| \frac{f'(x)}{g'(x)} - L \right| < \frac{\varepsilon}{2}$.

Let $I = (a,b)$. Choose some $u \in (a,c)$ such that $|u-c| < \delta_1$. Because $\lim_{x\to c^-} f(x) = \infty$ or $\lim_{x\to c^-} f(x) = -\infty$, there is some $\delta_2 > 0$ such that $x \in I - \{c\}$ and $c - \delta_2 < x < c$ imply $|f(x)| > |f(u)|$. In particular, we see that $x \in I - \{c\}$ and $c - \delta_2 < x < c$ imply $f(x) \neq 0$ and $f(x) \neq f(u)$. Similarly, because $\lim_{x\to c^-} g(x) = \infty$ or $\lim_{x\to c^-} g(x) = -\infty$, it follows that there is some $\delta_3 > 0$ such that $x \in I - \{c\}$ and $x \in c - \delta_3 < x < c$ imply $g(x) \neq 0$ and $g(x) \neq g(u)$.

Let $\eta = \min\{\delta_2, \delta_3, \frac{c-u}{2}\}$. Then $u < c - \eta$. Let $S\colon (c-\eta, c) \to \mathbb{R}$ be defined by

$$S(x) = \frac{1 - \frac{g(u)}{g(x)}}{1 - \frac{f(u)}{f(x)}}$$

for all $x \in (c-\eta, c)$. Because $\lim_{x\to c^-} f(x) = \infty$ or $\lim_{x\to c^-} f(x) = -\infty$, and $\lim_{x\to c^-} g(x) = \infty$ or $\lim_{x\to c^-} g(x) = -\infty$, then it follows from the one-sided analogs of Theorem 6.3.4, Exercise 3.2.1 and Theorem 3.2.10 that $\lim_{x\to c^-} S(x) = 1$. Hence there is some $\delta_4 > 0$ such that $x \in I - \{c\}$ and $c - \delta_4 < x < c$ imply $|S(x) - 1| < \frac{\varepsilon}{2|L|+\varepsilon}$.

Let $\delta = \min\{\delta_2, \delta_3, \delta_4, \frac{c-u}{2}\}$. Suppose that $w \in I - \{c\}$ and $c - \delta < w < c$. Then $u < w < c$, and $g(w) \neq g(u)$, and $|S(w) - 1| < \frac{\varepsilon}{2|L|+\varepsilon}$. Because $g(w) \neq g(u)$, then $S(w) \neq 0$.

By Cauchy's Mean Value Theorem (Theorem 4.4.5) there is some $q \in (u,w)$ such that $[f(w) - f(u)]g'(q) = [g(w) - g(u)]f'(q)$. We know by hypothesis that $g'(q) \neq 0$. Hence, with a bit of rearranging, we see that

$$\frac{f(w)}{g(w)} \frac{1 - \frac{f(u)}{f(w)}}{1 - \frac{g(u)}{g(w)}} = \frac{f'(q)}{g'(q)},$$

and therefore

$$\frac{f(w)}{g(w)} = \frac{f'(q)}{g'(q)} S(w).$$

Because $q \in (u,w)$, it follows that $c - \delta_1 < u < q < w < c$. Hence $q \in I - \{c\}$ and $|q - c| < \delta_1$. Therefore $\left|\frac{f'(q)}{g'(q)} - L\right| < \frac{\varepsilon}{2}$. By Lemma 2.3.9 (7) we see that $\left|\frac{f'(q)}{g'(q)}\right| - |L| < \frac{\varepsilon}{2}$, and therefore $\left|\frac{f'(q)}{g'(q)}\right| < |L| + \frac{\varepsilon}{2}$.

Then

$$\begin{aligned}
\left|\frac{f(w)}{g(w)} - L\right| &= \left|\frac{f'(q)}{g'(q)} S(w) - L\right| = \left|\frac{f'(q)}{g'(q)} S(w) - \frac{f'(q)}{g'(q)} + \frac{f'(q)}{g'(q)} - L\right| \\
&\leq \left|\frac{f'(q)}{g'(q)} S(w) - \frac{f'(q)}{g'(q)}\right| + \left|\frac{f'(q)}{g'(q)} - L\right| \\
&= \left|\frac{f'(q)}{g'(q)}\right| \cdot |S(w) - 1| + \left|\frac{f'(q)}{g'(q)} - L\right|
\end{aligned}$$

$$< \left(|L| + \frac{\varepsilon}{2}\right) \cdot \frac{\varepsilon}{2|L| + \varepsilon} + \frac{\varepsilon}{2} = \varepsilon. \qquad \square$$

There are also variants of l'Hôpital's Rule for $\frac{\infty}{\infty}$ (Theorem 6.3.7), where $x \to c$ is replaced with one of $x \to c^+$, or $x \to c^-$, or $x \to \infty$, or $x \to -\infty$; again, we will not state these variants, but will use them as needed.

Example 6.3.8.

(1) We want to evaluate the limit $\lim_{x \to 0^+} x \ln x$. We cannot evaluate this limit as $[\lim_{x \to 0^+} x] \cdot [\lim_{x \to 0^+} \ln x]$, because the one-sided analog of Theorem 3.2.10 (4) holds only when each of these two limits exists, and yet $\lim_{x \to 0^+} \ln x = -\infty$, which will be proved in Exercise 7.2.5. The limit $\lim_{x \to 0^+} x \ln x$ therefore has the form "$0 \cdot \infty$," which is another type of indeterminate form, as seen in Exercise 6.3.3. However, we can rewrite this limit as $\lim_{x \to 0^+} \frac{\ln x}{\frac{1}{x}}$, which has the form $\frac{\infty}{\infty}$, and using the one-sided variant of l'Hôpital's Rule for $\frac{\infty}{\infty}$ (Theorem 6.3.7) we see that

$$\lim_{x \to 0^+} x \ln x = \lim_{x \to 0^+} \frac{\ln x}{\frac{1}{x}} = \lim_{x \to 0^+} \frac{\frac{1}{x}}{\frac{-1}{x^2}} = \lim_{x \to 0^+} (-x) = 0.$$

(2) We want to evaluate the limit $\lim_{x \to \infty} \frac{x}{\sqrt{5+x^2}}$. This limit has the form $\frac{\infty}{\infty}$, and the $x \to \infty$ variant of l'Hôpital's Rule for $\frac{\infty}{\infty}$ is applicable. However, if we try to use l'Hôpital's Rule we obtain

$$\begin{aligned}\lim_{x \to \infty} \frac{x}{\sqrt{5+x^2}} &= \lim_{x \to \infty} \frac{x}{(5+x^2)^{\frac{1}{2}}} = \lim_{x \to \infty} \frac{1}{2x(5+x^2)^{-\frac{1}{2}}} = \lim_{x \to \infty} \frac{(5+x^2)^{\frac{1}{2}}}{2x} \\ &= \lim_{x \to \infty} \frac{2x(5+x^2)^{-\frac{1}{2}}}{2} = \lim_{x \to \infty} \frac{x}{(5+x^2)^{\frac{1}{2}}} = \cdots.\end{aligned}$$

Having used l'Hôpital's Rule twice, we returned to the original limit; clearly using l'Hôpital's Rule again would simply repeat the process. Hence, although l'Hôpital's Rule is applicable in this case, it is not actually helpful. Fortunately, there is an alternative (and much simpler) way to evaluate this limit, which is

$$\lim_{x \to \infty} \frac{x}{\sqrt{5+x^2}} = \lim_{x \to \infty} \frac{\frac{x}{x}}{\sqrt{\frac{5}{x^2} + \frac{x^2}{x^2}}} = \lim_{x \to \infty} \frac{1}{\sqrt{\frac{5}{x^2} + 1}} = \lim_{x \to \infty} \frac{1}{\sqrt{0+1}} = 1. \qquad \lozenge$$

The final result in this section, though not about indeterminate forms, is about a particular situation involving Type 2 limits to infinity. This theorem, which relates such limits to the derivatives of inverse functions, might appear somewhat technical, but it will be useful when we prove that the sine and cosine functions are differentiable in Section 7.3.

Theorem 6.3.9. *Let $(a,b] \subseteq \mathbb{R}$ be a non-degenerate half-open interval, and let $f\colon (a,b] \to \mathbb{R}$ be a function. Suppose that f is continuous on $(a,b]$ and differentiable on (a,b). Suppose that $f'(x) > 0$ for all $x \in (a,b)$, and that $\lim_{x \to b^-} f'(x) = \infty$. Then the* *function* $f^{-1}\colon f((a,b]) \to (a,b]$ *is differentiable, and* $[f^{-1}]'(f(b)) = 0$*, where this derivative is one-sided.*

Proof. By Theorem 4.5.2 (2) we know that f is strictly increasing. It follows from Exercise 4.6.6 (1) that $f((a,b])$ is an interval of the form either $(\operatorname{glb} f((a,b]), f(b)]$ or $(-\infty, f(b)]$. Because f is strictly increasing, we know f is injective, and hence $f((a,b)) = f((a,b]) - \{f(b)\}$, which means that $f((a,b))$ is an interval of the form either $(\operatorname{glb} f((a,b]), f(b))$ or $(-\infty, f(b))$.

By Theorem 4.6.4 we know that f^{-1} is differentiable on $f((a,b))$, and therefore all that needs to be proved is that f^{-1} is differentiable at $f(b)$ and $[f^{-1}]'(f(b)) = 0$.

Let $F\colon (a,b) \to \mathbb{R}$ be defined by

$$F(x) = \frac{x-b}{f(x)-f(b)}$$

for all $x \in (a,b)$. Because f is injective, then $f(x) \neq f(b)$ for all $x \in (a,b)$, and therefore F is well-defined. We now show that $\lim_{z\to b^-} F(z) = 0$. Let $\varepsilon > 0$. Because $\lim_{x\to b^-} f'(x) = \infty$, there is some $\delta > 0$ such that $x \in (a,b)$ and $b - \delta < x < b$ imply $f'(x) > \frac{1}{\varepsilon}$. Suppose that $z \in (a,b)$ and $b - \delta < z < b$. By the Mean Value Theorem (Theorem 4.4.4) there is some $c \in (z,b)$ such that

$$f'(c) = \frac{f(b)-f(z)}{b-z}.$$

Hence $f(b) - f(z) = f'(c)(b-z)$. Because $c \in (z,b)$, it follows that $c \in (a,b)$ and that $b - \delta < c < b$. Hence $f'(c) > \frac{1}{\varepsilon}$, and therefore $\frac{1}{f'(c)} < \varepsilon$. We now compute

$$|F(z) - 0| = \left|\frac{z-b}{f(z)-f(b)}\right| = \left|\frac{1}{\frac{f(z)-f(b)}{z-b}}\right| = \left|\frac{1}{f'(c)}\right| = \frac{1}{f'(c)} < \varepsilon.$$

Hence $\lim_{z\to b^-} F(z) = 0$.

Given that f is strictly increasing and continuous, Exercise 4.6.6 (2) implies that $f^{-1}\colon f((a,b]) \to (a,b]$ is continuous at $f(b)$. Hence, by the one-sided analog of Lemma 3.3.2, we see that $\lim_{y\to f(b)^-} f^{-1}(y)$ exists and $\lim_{y\to f(b)^-} f^{-1}(y) = f^{-1}(f(b)) = b$. The one-sided analog of Theorem 3.2.12 now implies that $\lim_{y\to f(b)^-} (F \circ f^{-1})(y)$ exists and $\lim_{y\to f(b)^-} (F \circ f^{-1})(y) = \lim_{z\to b^-} F(z)$. Therefore

$$\begin{aligned}\lim_{y\to f(b)^-} \frac{f^{-1}(y) - f^{-1}(f(b))}{y - f(b)} &= \lim_{y\to f(b)^-} \frac{f^{-1}(y) - b}{f(f^{-1}(y)) - f(b)}\\ &= \lim_{y\to f(b)^-} (F \circ f^{-1})(y) = \lim_{z\to b^-} F(z) = 0.\end{aligned}$$

It follows that f^{-1} is differentiable at $f(b)$ and $[f^{-1}]'(f(b)) = 0$. □

Reflections

In the larger scheme of real analysis, l'Hôpital's Rule is not a particularly important result. It is, nonetheless, included in this text because it is well-liked by students in calculus courses, and it is quite useful in various calculations. For example, we will see an application of l'Hôpital's Rule to the number e in Example 8.4.3. Moreover, the proofs of both the $\frac{0}{0}$ and $\frac{\infty}{\infty}$ cases, which are nice applications of Cauchy's Mean Value Theorem (Theorem 4.4.5), are much more complicated than might at first be expected—a good sign that something of interest is occurring.

Exercises

Exercise 6.3.1. [Used in Exercise 6.4.13.] Let $p \in (0, \infty)$.

(1) Prove that $\lim\limits_{x\to\infty} \frac{x^p}{e^x} = 0$. Intuitively, this limit says that the exponential function grows faster than any polynomial as x goes to infinity.

(2) Prove that $\lim\limits_{x\to\infty} \frac{\ln x}{x^p} = 0$. Intuitively, this limit says that the logarithm grows slower than any polynomial as x goes to infinity.

Exercise 6.3.2. [Used in Section 6.2 and Section 6.3.] Find an example of functions $f, g, k \colon (0, \infty) \to \mathbb{R}$ such that $\lim\limits_{x\to c} [f - g](x)$ and $\lim\limits_{x\to c} [f - h](x)$ and $\lim\limits_{x\to c} [f - k](x)$ have the form $\infty - \infty$, that $\lim\limits_{x\to c} [f - g](x) = 0$, that $\lim\limits_{x\to c} [f - h](x)$ is a positive number and that $\lim\limits_{x\to c} [f - k](x) = \infty$.

Exercise 6.3.3. [Used in Example 6.3.8.] In Example 6.3.8 (1) we saw an example of a limit that has the form $0 \cdot \infty$, where the value of the limit was 0. Find an example of functions $f, g, h \colon (0, \infty) \to \mathbb{R}$ such that $\lim\limits_{x\to c} [fg](x)$ and $\lim\limits_{x\to c} [fh](x)$ have the form $0 \cdot \infty$, that $\lim\limits_{x\to c} [fg](x)$ is a positive number and that $\lim\limits_{x\to c} [fh](x) = \infty$.

Exercise 6.3.4. Find an example of functions $f, g \colon \mathbb{R} - \{0\} \to \mathbb{R}$ such that $\lim\limits_{x\to 0} \frac{f(x)}{g(x)} = 0$, but that $\lim\limits_{x\to 0} \frac{f'(x)}{g'(x)}$ does not exist. An informal argument is sufficient.

Exercise 6.3.5. [Used in Theorem 6.3.3.] Prove Theorem 6.3.3 (2).

Exercise 6.3.6. [Used in Theorem 6.3.4.] Prove Theorem 6.3.4.

Exercise 6.3.7. [Used in Section 6.3.] Consider the following "proof" of l'Hôpital's Rule for $\frac{0}{0}$:

$$\text{“}\lim_{x\to c} \frac{f(x)}{g(x)} = \lim_{x\to c} \frac{f(x) - 0}{g(x) - 0} = \lim_{x\to c} \frac{f(x) - f(c)}{g(x) - g(c)}$$
$$= \lim_{x\to c} \frac{\frac{f(x)-f(c)}{x-c}}{\frac{g(x)-g(c)}{x-c}} = \frac{f'(c)}{g'(c)} = \lim_{x\to c} \frac{f'(x)}{g'(x)}.\text{”}$$

This proof is much simpler than the proof we gave for Theorem 6.3.5, but that is because this shorter proof requires stronger hypotheses. Restate l'Hôpital's Rule for $\frac{0}{0}$ with the hypotheses needed to make this shorter proof work.

Exercise 6.3.8. **[Used in Section 7.2.]**

(1) The limit $\lim_{x \to 0+} x^x$ has the form 0^0. Prove that $\lim_{x \to 0+} x^x = 1$.
(2) Find an example of a limit of the form 0^0 such that the limit has value 0.

Exercise 6.3.9. **[Used in Exercise 6.3.10 and Example 8.4.3.]** Let $f \colon \mathbb{R} - \{0\} \to \mathbb{R}$ be a function.

(1) Prove that $\lim_{x \to 0+} f(\frac{1}{x})$ exists if and only if $\lim_{t \to \infty} f(t)$ exists, and if these limits exist then they are equal.
(2) Prove that $\lim_{x \to 0-} f(\frac{1}{x})$ exists if and only if $\lim_{t \to -\infty} f(t)$ exists, and if these limits exist then they are equal.
(3) Prove that $\lim_{x \to 0} f(\frac{1}{x})$ exists if and only if $\lim_{t \to \infty} f(t)$ and $\lim_{t \to -\infty} f(t)$ exist and are equal, and if these three limits exist then they are equal.

Exercise 6.3.10. **[Used in Example 10.4.11.]** Let $p \colon \mathbb{R} \to \mathbb{R}$ be a polynomial function.

(1) Prove that

$$\lim_{x \to 0} p\left(\frac{1}{x}\right) e^{-\frac{1}{x^2}} = 0.$$

[Use Exercise 6.3.9 (3).]

(2) Let $f \colon \mathbb{R} \to \mathbb{R}$ be defined by

$$f(x) = \begin{cases} p\left(\frac{1}{x}\right) e^{-\frac{1}{x^2}}, & \text{if } x \neq 0 \\ 0, & \text{if } x = 0. \end{cases}$$

Prove that f is differentiable, and that there is a polynomial function $r \colon \mathbb{R} \to \mathbb{R}$ such that

$$f'(x) = \begin{cases} r\left(\frac{1}{x}\right) e^{-\frac{1}{x^2}}, & \text{if } x \neq 0 \\ 0, & \text{if } x = 0. \end{cases}$$

(3) Let $h \colon \mathbb{R} \to \mathbb{R}$ be defined by

$$h(x) = \begin{cases} e^{-\frac{1}{x^2}}, & \text{if } x \neq 0 \\ 0, & \text{if } x = 0. \end{cases}$$

Prove that h is infinitely differentiable, and that $h^{(n)}(0) = 0$ for all $n \in \mathbb{N}$.

6.4 Improper Integrals

In our treatment of the Riemann integral in Chapter 5, we stressed that an integral of the form $\int_a^b f(x)\,dx$ was for functions defined on *closed bounded intervals*. We also saw that integrable functions are bounded. However, in various applications of integration, for example as Laplace transforms and continuous probability, it is necessary to look at integrals where either the domain is not bounded (which is often when the function has a horizontal asymptote), or the domain is a half-open interval or an open bounded interval (which is often when the function has a vertical asymptote). Such integrals cannot be evaluated directly as Riemann integrals, but it turns out that they can be evaluated as limits of such integrals. These two types of limits of integrals are called "improper integrals." There are two types of improper integrals, corresponding roughly to the two types of limits to infinity that we saw in Section 6.2.

The main idea in the evaluation of improper integrals is as follows. Suppose that we have a function with domain $[a,b)$. We can think of approximating this interval by closed intervals of the form $[a,t]$, where $t \in (a,b)$, and where t is thought of as getting closer and closer to b. We can then define define the improper integral of the function on $[a,b)$ by evaluating the ordinary integral of the function on each closed integral $[a,t]$, and then taking the limit as t goes to b, if the limit exists. To be sure that this approach is a good one, however, we should ask whether such a limiting process works when our function is in fact defined on a closed bounded interval. That is, let $[a,b] \subseteq \mathbb{R}$ be a non-degenerate closed bounded interval, and let $f\colon [a,b] \to \mathbb{R}$ be a function. Suppose that f is integrable. By Theorem 5.5.6 we know that $f|_{[a,t]}$ is integrable for each $t \in (a,b)$. Is it true that $\lim_{t \to b^-} \int_a^t f(x)\,dx = \int_a^b f(x)\,dx$? Fortunately, it was proved in Exercise 5.5.6 that the answer is yes. Hence, our idea for defining improper integrals is consistent with the definition of integrals for closed bounded intervals. A similar limiting process is used for functions with domains that are not bounded.

We start with the following definition, which is needed for both types of improper integrals.

Definition 6.4.1. Let $I \subseteq \mathbb{R}$ be an interval, and let $f\colon I \to \mathbb{R}$ be a function. The function f is **locally integrable** if $f|_{[a,b]}$ is integrable for every non-degenerate closed bounded interval $[a,b] \subseteq I$. △

It follows from Theorem 5.4.11 that any continuous function is locally integrable.

We now turn to the first type of improper integral, called a Type 1 improper integral, and corresponding to Type 1 limits to infinity.

Definition 6.4.2.

1. Let $[a,\infty) \subseteq \mathbb{R}$ be a closed unbounded interval, and let $f\colon [a,\infty) \to \mathbb{R}$ be a function. Suppose that f is locally integrable. The function f is **improperly integrable** if $\lim_{t \to \infty} \int_a^t f(x)\,dx$ exists. If this limit exists, it is denoted $\int_a^\infty f(x)\,dx$, and it is called the **improper integral** of f. If f is improperly integrable, we also say that the improper integral $\int_a^\infty f(x)\,dx$ is **convergent**; otherwise we say that the improper integral $\int_a^\infty f(x)\,dx$ is **divergent.**
2. Let $(-\infty,b] \subseteq \mathbb{R}$ be a closed unbounded interval, and let $g\colon (-\infty,b] \to \mathbb{R}$ be a function. Suppose that g is locally integrable. The function g is **improperly integrable** if $\lim_{s \to -\infty} \int_s^b g(x)\,dx$ exists. If this limit exists, it is denoted $\int_{-\infty}^b g(x)\,dx$, and it is called the **improper integral** of g. If g is improperly integrable, we

also say that the improper integral $\int_{-\infty}^{b} g(x)\,dx$ is **convergent**; otherwise we say that the improper integral $\int_{-\infty}^{b} g(x)\,dx$ is **divergent**. △

Example 6.4.3.

(1) Let $f\colon [1,\infty) \to \mathbb{R}$ be defined by $f(x) = \frac{1}{x^2}$ for all $x \in [1,\infty)$. Then f is continuous, and therefore it is locally integrable. Hence we can compute

$$\lim_{t\to\infty} \int_1^t \frac{1}{x^2}\,dx = \lim_{t\to\infty} \left[-\frac{1}{x}\right]_1^t = \lim_{t\to\infty} \left[\left(-\frac{1}{t}\right) - \left(-\frac{1}{1}\right)\right] = 0 - (-1) = 1.$$

It follows that the improper integral $\int_1^\infty \frac{1}{x^2}\,dx$ is convergent and $\int_1^\infty \frac{1}{x^2}\,dx = 1$.

(2) Let $g\colon [1,\infty) \to \mathbb{R}$ be defined by $g(x) = \frac{1}{\sqrt{x}}$ for all $x \in [1,\infty)$. Then g is continuous, and therefore it is locally integrable. Hence we can compute

$$\lim_{t\to\infty} \int_1^t \frac{1}{\sqrt{x}}\,dx = \lim_{t\to\infty} \left[2\sqrt{x}\right]_1^t = \lim_{t\to\infty} \left[2\sqrt{t} - 2\sqrt{1}\right] = \infty,$$

where the final equality follows from Exercise 6.2.15 (3) and Theorem 6.2.8. Hence the improper integral $\int_1^t \frac{1}{\sqrt{x}}\,dx$ is divergent. ◊

Type 1 improper integrals behave similarly to regular integrals in some ways, but not all. For example, there are analogs for Type 1 improper integrals of Theorem 5.3.1 (1) (2) (3) and Theorem 5.3.2 (1) (2); see Exercise 6.4.2 for some of these. On the other hand, there are clearly no Type 1 improper integral analogs of Theorem 5.3.1 (4) and Theorem 5.3.2 (3). Moreover, whereas an integrable function must be bounded by Theorem 5.3.3, a function can be Type 1 improperly integrable and yet not bounded, as the reader is asked to show in Exercise 6.4.3.

Definition 6.4.2 deals with Type 1 improper integrals of functions defined on closed unbounded intervals. We now turn to Type 1 improper integrals for functions defined on all $\mathbb{R}$. That is, we want to evaluate improper integrals of the form $\int_{-\infty}^{\infty} f(x)\,dx$. The key observation is that the limits to negative infinity and to infinity must be taken separately, to avoid cancellation due to coincidental symmetry; we will see an example of such cancellation in Example 6.4.6 (2). The simplest way to to treat the limits to negative infinity and to infinity separately is to break up the integral $\int_{-\infty}^{\infty} f(x)\,dx$ into a sum of the form $\int_{-\infty}^{c} f(x)\,dx + \int_{c}^{\infty} f(x)\,dx$. The question then arises as to whether the choice of c makes a difference, though fortunately the following lemma shows that it does not.

Lemma 6.4.4. *Let $f\colon \mathbb{R} \to \mathbb{R}$ be a function, and let $c,d \in \mathbb{R}$. Suppose that f is locally integrable. Then $\int_{-\infty}^{c} f(x)\,dx$ and $\int_{c}^{\infty} f(x)\,dx$ are both convergent if and only if $\int_{-\infty}^{d} f(x)\,dx$ and $\int_{d}^{\infty} f(x)\,dx$ are both convergent, and if these improper integrals are convergent then*

$$\int_{-\infty}^{c} f(x)\,dx + \int_{c}^{\infty} f(x)\,dx = \int_{-\infty}^{d} f(x)\,dx + \int_{d}^{\infty} f(x)\,dx.$$

Proof. Suppose that $\int_{-\infty}^{c} f(x)\,dx$ and $\int_{c}^{\infty} f(x)\,dx$ are convergent. The other implication is similar, and we omit the details.

Let $s,t \in \mathbb{R}$. By Corollary 5.5.9 and Definition 5.5.8 we see that

$$\int_{s}^{d} f(x)\,dx = \int_{s}^{c} f(x)\,dx + \int_{c}^{d} f(x)\,dx$$

and

$$\int_{d}^{t} f(x)\,dx = \int_{d}^{c} f(x)\,dx + \int_{c}^{t} f(x)\,dx = -\int_{c}^{d} f(x)\,dx + \int_{c}^{t} f(x)\,dx.$$

We now use Theorem 6.2.4 and its analog for limits to negative infinity, together with the fact that the limit to infinity or to negative infinity of a constant function is that constant, to deduce that

$$\lim_{s\to-\infty} \int_{s}^{d} f(x)\,dx = \lim_{s\to-\infty} \left\{ \int_{s}^{c} f(x)\,dx + \int_{c}^{d} f(x)\,dx \right\} = \int_{-\infty}^{c} f(x)\,dx + \int_{c}^{d} f(x)\,dx$$

and

$$\lim_{t\to\infty} \int_{d}^{t} f(x)\,dx = \lim_{t\to\infty} \left\{ -\int_{c}^{d} f(x)\,dx + \int_{c}^{t} f(x)\,dx \right\} = -\int_{c}^{d} f(x)\,dx + \int_{c}^{\infty} f(x)\,dx.$$

Hence $\int_{-\infty}^{d} f(x)\,dx$ and $\int_{d}^{\infty} f(x)\,dx$ are convergent, and

$$\int_{-\infty}^{d} f(x)\,dx + \int_{d}^{\infty} f(x)\,dx = \int_{-\infty}^{c} f(x)\,dx + \int_{c}^{\infty} f(x)\,dx. \qquad \square$$

Lemma 6.4.4 allows us to make the following definition.

Definition 6.4.5. Let $f\colon \mathbb{R} \to \mathbb{R}$ be a function. Suppose that f is locally integrable. The function f is **improperly integrable** if for any $c \in \mathbb{R}$ the improper integrals $\int_{-\infty}^{c} f(x)\,dx$ and $\int_{c}^{\infty} f(x)\,dx$ are convergent. If both of these improper integrals are convergent, the sum $\int_{-\infty}^{c} f(x)\,dx + \int_{c}^{\infty} f(x)\,dx$ is denoted $\int_{-\infty}^{\infty} f(x)\,dx$, and it is called the **improper integral** of f. If f is improperly integrable, we also say that the improper integral $\int_{-\infty}^{\infty} f(x)\,dx$ is **convergent**; otherwise we say that the improper integral $\int_{-\infty}^{\infty} f(x)\,dx$ is **divergent**. △

Example 6.4.6.

(1) Let $f\colon \mathbb{R} \to \mathbb{R}$ be defined by $f(x) = |x|e^{-x^2}$ for all $x \in \mathbb{R}$. Then f is continuous, and therefore it is locally integrable. Hence we can compute

$$\begin{aligned}\int_{-\infty}^{0} |x|e^{-x^2}\,dx + \int_{0}^{\infty} |x|e^{-x^2}\,dx &= \lim_{s\to-\infty} \int_{s}^{0} (-x)e^{-x^2}\,dx + \lim_{t\to\infty} \int_{0}^{t} xe^{-x^2}\,dx \\ &= -\lim_{s\to-\infty} \left[\frac{e^{-x^2}}{-2}\right]_{s}^{0} + \lim_{t\to\infty} \left[\frac{e^{-x^2}}{-2}\right]_{0}^{t}\end{aligned}$$

$$= -\lim_{s\to-\infty}\left[\frac{e^{-0^2}}{-2} - \frac{e^{-s^2}}{-2}\right] + \lim_{t\to\infty}\left[\frac{e^{-t^2}}{-2} - \frac{e^{-0^2}}{-2}\right]$$
$$= -\left[-\frac{1}{2}+0\right] + \left[0+\frac{1}{2}\right] = 1,$$

where the equality before last will be proved in Exercise 7.2.8 (2). It follows that the improper integral $\int_{-\infty}^{\infty} |x|e^{-x^2}\,dx$ is convergent and $\int_{-\infty}^{\infty} |x|e^{-x^2}\,dx = 1$.

(2) We want to evaluate the improper integral $\int_{-\infty}^{\infty} x\,dx$. A common mistake in evaluating such integrals is to try to take the limits to negative infinity and to infinity simultaneously, for example by the calculation

$$\int_{-\infty}^{\infty} x\,dx = \lim_{t\to\infty}\int_{-t}^{t} x\,dx = \lim_{t\to\infty}\left[\frac{x^2}{2}\right]_{-t}^{t} = \lim_{t\to\infty}\left[\frac{t^2}{2} - \frac{(-t)^2}{2}\right] = \lim_{t\to\infty} 0 = 0.$$

It would be a mistake to deduce from the above calculation that the improper integral $\int_{-\infty}^{\infty} x\,dx$ is convergent; the limit in the above calculation exists only because of the symmetry of the function, not because of actual convergence of the improper integral. If we evaluate the integral properly, we see that

$$\int_{-\infty}^{0} x\,dx + \int_{0}^{\infty} x\,dx = \lim_{s\to-\infty}\int_{s}^{0} x\,dx + \lim_{t\to\infty}\int_{0}^{t} x\,dx = \lim_{s\to-\infty}\left[\frac{x^2}{2}\right]_{s}^{0} + \lim_{t\to\infty}\left[\frac{x^2}{2}\right]_{0}^{t}$$
$$= \lim_{s\to-\infty}\frac{-s^2}{2} + \lim_{t\to\infty}\frac{t^2}{2}.$$

It can be verified that

$$\lim_{s\to-\infty}\frac{-s^2}{2} = -\infty \quad\text{and}\quad \lim_{t\to\infty}\frac{t^2}{2} = \infty;$$

we omit the details. In particular, neither of these limits exists. Hence each of the improper integrals $\int_{-\infty}^{0} x\,dx$ and $\int_{0}^{\infty} x\,dx$ is divergent, and therefore $\int_{-\infty}^{\infty} x\,dx$ is divergent. It is important to note that $-\infty$ and ∞ do not "cancel each other out"; observe that there is no mention of $\infty + (-\infty)$ in Theorem 6.2.8. ◊

We now turn to the second type of improper integral, called a Type 2 improper integral. For this type of improper integral, rather than looking at functions defined on closed unbounded intervals, we now look at functions defined on open or half-open bounded intervals. The evaluation of Type 1 improper integrals involves Type 1 limits to infinity, and so there is a nice correspondence between these two uses of the term "Type 1." The evaluation of Type 2 improper integrals also involves taking limits, though not necessarily limits to infinity, so the correspondence between the uses of the term "Type 2" for improper integrals and limits to infinity is not immediately evident, though we will see that there is more of a correspondence than is at first apparent after we prove Lemma 6.4.9 below.

Definition 6.4.7.

1. Let $[a,b) \subseteq \mathbb{R}$ be a non-degenerate half-open interval, and let $f\colon [a,b) \to \mathbb{R}$ be a function. Suppose that f is locally integrable. The function f is **improperly integrable** if $\lim_{t\to b^-} \int_a^t f(x)\,dx$ exists. If this limit exists, it is denoted $\int_a^b f(x)\,dx$, and it is called the **improper integral** of f. If f is improperly integrable, we also say that the improper integral $\int_a^b f(x)\,dx$ is **convergent**; otherwise we say that the improper integral $\int_a^b f(x)\,dx$ is **divergent**.
2. Let $(a,b] \subseteq \mathbb{R}$ be a non-degenerate half-open interval, and let $f\colon [a,b) \to \mathbb{R}$ be a function. Suppose that f is locally integrable. The function f is **improperly integrable** if $\lim_{s\to a^+} \int_s^b f(x)\,dx$ exists. If this limit exists, it is denoted $\int_a^b f(x)\,dx$, and it is called the **improper integral** of f. If f is improperly integrable, we also say that the improper integral $\int_a^b f(x)\,dx$ is **convergent**; otherwise we say that the improper integral $\int_a^b f(x)\,dx$ is **divergent**. △

Example 6.4.8.

(1) Let $f\colon (0,1] \to \mathbb{R}$ be defined by $f(x) = \frac{1}{\sqrt{x}}$ for all $x \in (0,1]$. Then f is continuous, and therefore it is locally integrable. Hence we can compute

$$\begin{aligned}\lim_{s\to 0^+} \int_s^1 \frac{1}{\sqrt{x}}\,dx &= \lim_{s\to 0^+} \int_s^1 x^{-\frac{1}{2}}\,dx = \lim_{s\to 0^+} \left[2x^{\frac{1}{2}}\right]_s^1 \\ &= \lim_{s\to 0^+} \left[2 - 2\sqrt{s}\right] = 2 - 2\cdot 0 = 2,\end{aligned}$$

where the penultimate equality follows from Exercise 7.2.11 (3) and the one-sided analog of Theorem 3.2.10. It follows that the improper integral $\int_0^1 \frac{1}{\sqrt{x}}\,dx$ is convergent and $\int_0^1 \frac{1}{\sqrt{x}}\,dx = 2$.

(2) Let $g\colon (0,1] \to \mathbb{R}$ be defined by $g(x) = \frac{1}{x^2}$ for all $x \in (0,1]$. Then g is continuous, and therefore it is locally integrable. Hence we can compute

$$\lim_{s\to 0^+} \int_s^1 \frac{1}{x^2}\,dx = \lim_{s\to 0^+} \left[-\frac{1}{x}\right]_s^1 = \lim_{s\to 0^+} \left[-1 + \frac{1}{s}\right] = \infty,$$

where the final equality follows from Example 6.2.7 (2) and Theorem 6.2.8 (1). Hence the improper integral $\int_0^1 \frac{1}{x^2}\,dx$ is divergent. ◊

For Type 1 improper integrals, where the function is defined on intervals of the form $[a,\infty)$, or $(-\infty,b]$ or $(-\infty,\infty)$, it is evident that the function cannot be integrated in the ordinary (non-improper) way, because the function is not defined on a closed bounded interval. On the other hand, consider the function $f\colon [0,1) \to \mathbb{R}$ defined by $f(x) = x^2$ for all $x \in [0,1)$. In principle, if we wanted to integrate this function, we would need to do so as an improper integral, because the function is not defined as written on a closed bounded interval. Of course, in practice it would be very silly to evaluate the integral $\int_0^1 x^2\,dx$ as an improper integral, because the function f can be extended to a continuous, and hence integrable, function $g\colon [0,1] \to \mathbb{R}$ defined by

$g(x) = x^2$ for all $x \in [0,1]$. The functions for which Type 2 improper integrals are really intended are those defined on open or half-open intervals for which the function cannot be extended to an integrable function at the endpoints of the interval. As we see in the following lemma, the real use of Type 2 improper integrals is when the function is not bounded, because bounded functions can always be dealt with by extending the function to a closed bounded interval. The choice of such extension does not matter, because, as was proved in Exercise 5.3.3 (3), if two functions defined on a non-degenerate closed bounded interval differ at only finitely many points, then one is integrable if and only if the other is, and if they are integrable then their integrals are equal.

In the following lemma, as well as in the other results we will subsequently prove for Type 2 improper integrals, we treat functions defined on intervals of the form $[a,b)$. The analogous results hold for intervals of the form $(a,b]$; for the sake of brevity we will not state such results, though we will use them as needed.

Lemma 6.4.9. *Let $[a,b) \subseteq \mathbb{R}$ be a non-degenerate half-open interval, and let $f\colon [a,b) \to \mathbb{R}$ be a function. Suppose that f is locally integrable and bounded. Then f is improperly integrable if and only if any extension $g\colon [a,b] \to \mathbb{R}$ of f is integrable. If f is improperly integrable, then $\int_a^b f(x)\,dx = \int_a^b g(x)\,dx$ for any extension $g\colon [a,b] \to \mathbb{R}$ of f.*

Proof. First, suppose that f is improperly integrable. Let $g\colon [a,b] \to \mathbb{R}$ be an extension of f. Because f is bounded, there is some $M \in \mathbb{R}$ such that $|f(x)| \leq M$ for all $x \in [a,b)$. We may assume that $M > 0$. Let $N = \max\{M, |g(b)|\}$. Then $N > 0$, and $|g(x)| \leq N$ for all $x \in [a,b]$.

Let $\varepsilon > 0$. Let $t \in (a,b)$. Suppose that $|t-b| < \frac{\varepsilon}{4N}$. Because f is locally integrable, we know that $g|_{[a,t]} = f|_{[a,t]}$ is integrable. By Theorem 5.4.7 (c) there is some partition $P = \{x_0, x_1, \ldots, x_n\}$ of $[a,t]$ such that $U(g|_{[a,t]}, P) - L(g|_{[a,t]}, P) < \frac{\varepsilon}{2}$. Observe that $x_n = t$. Let $x_{n+1} = b$, and let $Q = \{x_0, x_1, \ldots, x_{n+1}\}$. Then Q is a partition of $[a,b]$.

Because $|g(x)| \leq N$ for all $x \in [a,b]$, Exercise 5.4.9 (4) implies that $M_{n+1}(g) - m_{n+1}(g) \leq 2N$. Then

$$\begin{aligned}
U(g,Q)-L(g,Q)& \\
&= \sum_{i=1}^{n+1} [M_i(g) - m_i(g)](x_i - x_{i-1}) \\
&= \sum_{i=1}^{n} [M_i(g) - m_i(g)](x_i - x_{i-1}) + [M_{n+1}(g) - m_{n+1}(g)](x_{n+1} - x_n) \\
&= [U(g|_{[a,t]}, P) - L(g|_{[a,t]}, P)] + [M_{n+1}(g) - m_{n+1}(g)](b-t) \\
&< \frac{\varepsilon}{2} + 2N \cdot \frac{\varepsilon}{4N} = \varepsilon.
\end{aligned}$$

Therefore g is integrable by Theorem 5.4.7 (c). We can now apply Exercise 5.5.6 to g, and we deduce that $\int_a^b g(x)\,dx = \lim_{t \to b^-} \int_a^t g(x)\,dx = \lim_{t \to b^-} \int_a^t f(x)\,dx = \int_a^b f(x)\,dx$.

Second, let $h\colon [a,b] \to \mathbb{R}$ be an extension of f. Suppose that h is integrable. Because $h|_{[a,t]} = f|_{[a,t]}$ for all $t \in (a,b)$, we use Exercise 5.5.6 again to see that

$\lim_{t \to b^-} \int_a^t f(x)\,dx = \lim_{t \to b^-} \int_a^t h(x)\,dx = \int_a^b h(x)\,dx$. Hence f is improperly integrable and $\int_a^b f(x)\,dx = \int_a^b h(x)\,dx$. □

Because of Lemma 6.4.9, if one encounters an integral of the form $\int_a^b f(x)\,dx$, and if the function f is bounded, then the integral is an ordinary one; if the function is not bounded then the integral is improper. The most commonly encountered functions that are not bounded, but for which the domains are bounded intervals, are functions with vertical asymptotes, and hence Type 2 improper integrals are often associated in calculus courses with the notion of vertical asymptotes. Moreover, because Type 2 improper integrals most commonly occur in practice in the case of vertical asymptotes, there is a partial correspondence between the use of the term "Type 2" for improper integrals and the use of that term for limits to infinity.

As was the case for Type 1 improper integrals, we note that Type 2 improper integrals satisfy analogs of some, though not all, properties of ordinary integrals; see Exercise 6.4.7 for analogs of Theorem 5.3.1 (1) (3) and Theorem 5.3.2 (2).

The simplest use of Type 2 improper integrals is when a function is defined everywhere on a non-degenerate closed bounded interval except one of the endpoints. However, it is also possible to look at more complicated situations, for example when a function is undefined at both endpoints of a non-degenerate closed bounded interval, or is undefined at a point (or finitely many points) in the interior of such an interval. In all cases, we break up the domain of the function into finitely many subintervals such that Definition 6.4.7 can be applied to each. The function is then considered to be improperly integrable on the whole interval if it is improperly integrable on each subinterval, and if the latter holds, then the improper integral on the whole interval is the sum of the improper integrals on the subintervals.

Example 6.4.10. In Example 5.6.6 (2) we saw that the integral $\int_{-1}^1 \frac{1}{x^2}\,dx$ cannot be evaluated directly by the Fundamental Theorem of Calculus Version II (Theorem 5.6.4). The reason that the Fundamental Theorem of Calculus Version II is not applicable to this integral is that the function is defined only on $[-1,0) \cup (0,1]$, and it has vertical asymptotes at $x = 0$, one on each side.

The correct way to evaluate this integral is to break up the domain of the function into the two intervals $[-1,0)$ and $(0,1]$, and then to evaluate each of the improper integrals $\int_{-1}^0 \frac{1}{x^2}\,dx$ and $\int_0^1 \frac{1}{x^2}\,dx$. The entire improper integral $\int_{-1}^1 \frac{1}{x^2}\,dx$ will then be convergent if and only if both of the two improper integrals $\int_{-1}^0 \frac{1}{x^2}\,dx$ and $\int_0^1 \frac{1}{x^2}\,dx$ are convergent. We saw in Example 6.4.8 (2) that $\int_0^1 \frac{1}{x^2}\,dx$ is divergent, and hence $\int_{-1}^1 \frac{1}{x^2}\,dx$ is divergent. ◊

There are some situations in which it is not possible to prove directly that an improper integral is convergent, and to compute its value, but where it is nonetheless possible to prove indirectly that the improper integral is convergent, even though an exact numerical value for the improper integral cannot be found. For the reader who is familiar with convergence tests for series, the following theorem about improper integrals should look very familiar, being the analog of the Comparison Test for series (which we will see in Section 9.3). We state the following theorem for Type 2 improper integrals, because we will need it in Section 7.4. The reader is asked to state and prove the analogous result for Type 1 improper integrals in Exercise 6.4.10.

Theorem 6.4.11 (Comparison Test for Type 2 Improper Integrals). *Let $[a,b) \subseteq \mathbb{R}$ be a non-degenerate half-open interval, and let $f,g\colon [a,b) \to \mathbb{R}$ be functions. Suppose that f and g are locally integrable, and that there is some $\delta > 0$ such that $x \in [a,b)$ and $|x-b| \leq \delta$ imply $0 \leq f(x) \leq g(x)$. If g is improperly integrable, then f is improperly integrable.*

Proof. Suppose that g is improperly integrable. There are two cases, depending upon whether $a \leq b-\delta < a$ or $a > b-\delta$. Suppose that $a \leq b-\delta$; the other case is similar, and we omit the details. Let $c = b-\delta$. Then $a \leq c < b$.

If $t \in (c,b)$, then $f|_{[c,t]}$ and $g|_{[c,t]}$ are integrable, and by Theorem 5.3.2 we know that $0 \leq \int_c^t f(x)\,dx \leq \int_c^t g(x)\,dx$. We also know that $\int_c^c f(x)\,dx = 0$ and $\int_c^c g(x)\,dx = 0$. Let $F,G\colon [c,b) \to \mathbb{R}$ be defined by

$$F(t) = \int_c^t f(x)\,dx \quad \text{and} \quad G(t) = \int_c^t g(x)\,dx$$

for all $t \in [c,b)$. Then $0 \leq F(t) \leq G(t)$ for all $t \in [c,b)$. By Exercise 5.6.4 we see that F is increasing.

Because g is improperly integrable, it follows from Exercise 6.4.6 that $g|_{[c,b)}$ is improperly integrable, and hence $\lim_{t\to b^-} G(t) = \lim_{t\to b^-} \int_c^t g(x)\,dx$ exists. We can now apply Exercise 4.5.10 (2) to deduce that $\lim_{t\to b^-} F(t)$ exists and $\lim_{t\to b^-} F(t) \leq \lim_{t\to b^-} G(t)$. Hence $\lim_{t\to b^-} \int_c^t f(x)\,dx$ exists, which means that $f|_{[c,b)}$ is improperly integrable. Using Exercise 6.4.6 again we conclude that f is improperly integrable. □

In our discussion of π in Sections 7.3 and 7.4, we will need to know that Integration by Parts and Integration by Substitution both work for Type 2 improper integrals, as we will now see. The reader who will skip those sections can also safely skip the following two theorems and proofs. The reader is asked to state and prove the analogous results for Type 1 improper integrals in Exercise 6.4.11 and Exercise 6.4.12.

In the statement of the following theorem we should properly write "$(f \circ g|_{[a,b)}) \cdot g'|_{[a,b)}$" rather than "$(f \circ g) \cdot g'$," but for the sake of readability we abuse notation and write the latter.

Theorem 6.4.12 (Integration by Substitution for Type 2 Improper Integrals). *Let $[a,b],[c,d] \subseteq \mathbb{R}$ be non-degenerate closed bounded intervals, and let $g\colon [a,b] \to [c,d]$ and $f\colon [c,d) \to \mathbb{R}$ be functions. Suppose that f is continuous, that g is bijective and differentiable and that g' is integrable. Then $(f \circ g) \cdot g'$ is improperly integrable if and only if f is improperly integrable, and if they are improperly integrable then*

$$\int_a^b f(g(x))g'(x)\,dx = \int_c^d f(x)\,dx.$$

Proof. Because g is differentiable, then it is continuous by the closed interval analog of Theorem 4.2.4. It then follows from Exercise 4.5.11 that g is strictly monotone.

Suppose that g is strictly increasing; the other case is similar, and we omit the details. It follows from Exercise 4.6.3 (1) that $[g(a),g(b)] = g([a,b]) = [c,d]$, and hence that $g(a) = c$ and $g(b) = d$. Therefore g maps $[a,b)$ bijectively onto $[c,d)$.

Let $t \in (a,b)$. Because g is strictly increasing, then so is $g|_{[a,t]}$. By Exercise 4.2.3 (5) we know that $g|_{[a,t]}$ is differentiable, and hence it is continuous. It follows from Exercise 4.6.3 (1) that $g([a,t]) = [g(a),g(t)] = [c,g(t)]$. Because g' is integrable, if follows from Theorem 5.5.6 that $g'|_{[a,t]}$ is integrable.

Because f is continuous, then $f|_{[c,g(t)]}$ is continuous by Exercise 3.3.2 (2). We can now apply Integration by Substitution for Definite Integrals (Theorem 5.7.4) to $f|_{[c,g(t)]}$ and $g|_{[a,t]}$, and we deduce that $[(f \circ g) \cdot g']|_{[a,t]}$ is integrable and

$$\int_a^t f(g(x))g'(x)\,dx = \int_c^{g(t)} f(x)\,dx.$$

It now follows that $\lim_{t\to b^-} \int_a^t f(g(x))g'(x)\,dx$ exists if and only if $\lim_{t\to b^-} \int_c^{g(t)} f(x)\,dx$ exists, and if these limits exist then

$$\lim_{t\to b^-} \int_a^t f(g(x))g'(x)\,dx = \lim_{t\to b^-} \int_c^{g(t)} f(x)\,dx.$$

By definition $\lim_{t\to b^-} \int_a^t f(g(x))g'(x)\,dx$ exists if and only if $(f \circ g) \cdot g'$ is improperly integrable, and if this limit exists then it equals $\int_a^b f(g(x))g'(x)\,dx$. We claim that $\lim_{t\to b^-} \int_c^{g(t)} f(x)\,dx$ exists if and only if $\lim_{w\to d^-} \int_c^w f(x)\,dx$ exists, and if these limits exist then they are equal. We will prove this claim shortly, but assuming that the claim is true, we then observe that by definition $\lim_{w\to d} \int_c^w f(x)\,dx$ exists if and only if f is improperly integrable, and if this limit exists then it equals $\int_c^d f(x)\,dx$. If we put all the above observations together, it follows that $(f \circ g) \cdot g'$ is improperly integrable if and only if f is improperly integrable, and if they are improperly integrable then

$$\int_a^b f(g(x))g'(x)\,dx = \int_c^d f(x)\,dx.$$

It remains to prove that $\lim_{t\to b^-} \int_c^{g(t)} f(x)\,dx$ exists if and only if $\lim_{w\to d^-} \int_c^w f(x)\,dx$ exists, and if these limits exist then they are equal. First, suppose that $\lim_{w\to d^-} \int_c^w f(x)\,dx$ exists. Let $F\colon [c,d) \to \mathbb{R}$ be defined by $F(z) = \int_c^z f(x)\,dx$ for all $z \in [c,d)$, which makes sense because f is locally integrable. Our hypothesis can then be restated by saying that $\lim_{w\to d^-} F(w)$ exists. Because g is continuous, then by the one-sided analog of Lemma 3.3.2 we know that $\lim_{t\to b^-} g(t) = g(b) = d$. We now abuse notation and think of g as a function $[a,b) \to [c,d)$, and we can then apply the one-sided analog of Theorem 3.2.12 to deduce that $\lim_{t\to b^-} (F \circ g)(t)$ exists and $\lim_{t\to b^-} (F \circ g)(t) = \lim_{w\to d^-} F(w)$, which means that $\lim_{t\to b^-} \int_c^{g(t)} f(x)\,dx = \lim_{t\to b^-} F(g(t))$ exists and $\lim_{t\to b^-} \int_c^{g(t)} f(x)\,dx = \lim_{w\to d^-} \int_c^w f(x)\,dx$.

Second, suppose that $\lim_{t\to b^-} \int_c^{g(t)} f(x)\,dx$ exists. Observe that $g|_{(a,b]}$ is strictly increasing and is continuous at b. We can therefore apply Exercise 4.6.6 (2) to deduce that $g^{-1}\colon (c,d] \to (a,b]$ is continuous at $d = g(b)$. By the one-sided analog of Lemma 3.3.2 it follows that $\lim_{w\to d^-} g^{-1}(w) = g^{-1}(d) = b$. The same type of argument used in the previous paragraph can then be used to show that $\lim_{w\to d^-} \int_c^w f(x)\,dx$ exists and $\lim_{w\to d^-} \int_c^w f(x)\,dx = \lim_{w\to d^-} \int_c^{g^{-1}(g(w))} f(x)\,dx = \lim_{t\to b^-} \int_c^{g(t)} f(x)\,dx$, and we omit the details. □

Theorem 6.4.13 (Integration by Parts for Type 2 Improper Integrals). *Let $[a,b] \subseteq \mathbb{R}$ be a non-degenerate closed bounded interval, and let $f,g\colon [a,b] \to \mathbb{R}$ be functions. Suppose that f and g are continuous on $[a,b]$, that f and g are differentiable on $[a,b)$ and that f' and g' are locally integrable on $[a,b)$. Then $f'g$ is improperly integrable if and only if fg' is improperly integrable, and if they are improperly integrable then*

$$\int_a^b f(x)g'(x)\,dx = [f(b)g(b) - f(a)g(a)] - \int_a^b f'(x)g(x)\,dx.$$

Proof. Because f and g are continuous on $[a,b]$, then by the one-sided analog of Lemma 3.3.2 we know that $\lim_{x\to b^-} f(x) = f(b)$ and $\lim_{x\to b^-} g(x) = g(b)$. By the one-sided analogs of Theorem 3.2.10 (2) (4) and Exercise 3.2.1 we deduce that $\lim_{x\to b^-} [f(x)g(x) - f(a)g(a)] = f(b)g(b) - f(a)g(a)$.

Let $t \in (a,b)$. Because f and g are differentiable on $[a,b)$, then by Exercise 4.2.3 (5) $f|_{[a,t]}$ and $g|_{[a,t]}$ are differentiable. Because f' and g' are locally integrable, then $f'|_{[a,t]}$ and $g'|_{[a,t]}$ are integrable. We can now apply Integration by Parts for Definite Integrals (Theorem 5.7.6) to $f|_{[a,t]}$ and $g|_{[a,t]}$, and we deduce that $[f'g]|_{[a,t]}$ and $[fg']|_{[a,t]}$ are integrable and

$$\int_a^t f(x)g'(x)\,dx = [f(t)g(t) - f(a)g(a)] - \int_a^t f'(x)g(x)\,dx.$$

Using the one-sided analog of Theorem 3.2.10 (2), and the fact that $\lim_{t\to b^-} [f(t)g(t) - f(a)g(a)]$ exists and $\lim_{t\to b^-} [f(t)g(t) - f(a)g(a)] = f(b)g(b) - f(a)g(a)$, we deduce that $\lim_{t\to b^-} \int_a^t f(x)g'(x)\,dx$ exists if and only if $\lim_{t\to b^-} \int_a^t f'(x)g(x)\,dx$ exists, and if these limits exist then

$$\lim_{t\to b^-} \int_a^t f(x)g'(x)\,dx = [f(b)g(b) - f(a)g(a)] - \lim_{t\to b^-} \int_a^t f'(x)g(x)\,dx.$$

We deduce immediately that $f'g$ is improperly integrable if and only if fg' is improperly integrable, and if they are improperly integrable then

$$\int_a^b f(x)g'(x)\,dx = [f(b)g(b) - f(a)g(a)] - \int_a^b f'(x)g(x)\,dx. \qquad \square$$

Reflections

It might appear as if this section is making a big deal out of very little. First, why must we define improper integrals by limits, rather than defining them directly? Let us consider the case of Type 1 improper integrals. The definition of the Riemann integral of a function on a closed bounded interval is in terms of Riemann sums, and the key observation is that the only type of sum that is guaranteed to exist is the sum of finitely many numbers, which is what a Riemann sum is. If one were to try to define a "Riemann sum" directly for a function on an unbounded interval, then either the unbounded interval would have to be subdivided into infinitely many subintervals, or at least one of the subintervals would itself have to be unbounded, and in neither case would it be certain that the sum could be evaluated. Hence, to be able to use Riemann sums, we must restrict our attention to functions on closed bounded intervals, and hence we define improper integrals as limits of regular integrals.

Second, even if we accept as reasonable the use of limits in the definition of improper integrals, why must we go to all the effort of proving the various theorems in this section? The answer is seen by analogy with the concept of series, which the reader has likely seen informally in a calculus course (and which we will discuss series in this text in Chapter 9). The tricky part in dealing with series is not defining what it means for series to be convergent in principle, but rather evaluating whether any given series is in fact convergent or divergent. We should think about improper integrals similarly, in that a number of the theorems in the present section are aimed at showing that certain improper integrals are convergent. Clearly the Comparison Test for Type 2 Improper Integrals, which is an analog of the corresponding convergence test for series, is a result of this sort, but even the last two theorems in this section can be thought of as results that tell us when certain improper integrals are convergent (with the added benefit of nice formulas for the values of these integrals when they exist).

Exercises

Exercise 6.4.1. [Used in Example 9.3.7.] Let $p \in \mathbb{R}$. Prove that the improper integral $\int_1^\infty \frac{1}{x^p}\,dx$ is convergent if and only if $p > 1$. Make use of standard properties of the graph of x^r; these properties will be proved rigorously in Exercise 7.2.16.

Exercise 6.4.2. [Used in Section 6.4.] Let $[a,\infty) \subseteq \mathbb{R}$ be a closed unbounded interval, let $f,g\colon [a,\infty) \to \mathbb{R}$ be functions and let $k \in \mathbb{R}$. Suppose that f and g are improperly integrable.

(1) Prove that $f+g$ is improperly integrable and $\int_a^\infty [f+g](x)\,dx = \int_a^\infty f(x)\,dx + \int_a^\infty g(x)\,dx$.
(2) Prove that kf is improperly integrable and $\int_a^\infty [kf](x)\,dx = k\int_a^\infty f(x)\,dx$.
(3) Prove that if $f(x) \geq g(x)$ for all $x \in [a,\infty)$, then $\int_a^\infty f(x)\,dx \geq \int_a^\infty g(x)\,dx$.

Exercise 6.4.3. [Used in Section 6.4.] Find an example of a function $f\colon [1,\infty) \to \mathbb{R}$ such that f is improperly integrable, but that f is not bounded on any interval of the form $[a,\infty)$, where $a \in [1,\infty)$. [Use Exercise 5.3.3 (3).]

Exercise 6.4.4. Let $[a,\infty) \subseteq \mathbb{R}$ be a closed unbounded interval, and let $f\colon [a,\infty) \to \mathbb{R}$ be a function. Suppose that f is improperly integrable. Prove that if $\lim_{x\to\infty} f(x)$ exists, then $\lim_{x\to\infty} f(x) = 0$. (Exercise 6.4.3 shows that $\lim_{x\to\infty} f(x)$ need not exist.)

Exercise 6.4.5. **[Used in Section 7.3 and Theorem 7.4.3.]** As for all other exercises in this section, you may use standard rules for integration, even if we have not yet proved them, but for this exercise do not use trigonometric functions and inverse trigonometric functions, because we will use this exercise in the definition of the arcsine function in Section 7.3.

(1) Prove that the improper integral $\int_0^1 \frac{1}{\sqrt{1-x}}\,dx$ is convergent.
(2) Prove that the improper integral $\int_0^1 \frac{1}{\sqrt{1-x^2}}\,dx$ is convergent.

Exercise 6.4.6. **[Used in Theorem 6.4.11.]** Let $[a,b) \subseteq \mathbb{R}$ be a non-degenerate half-open interval, let $c \in [a,b)$ and let $f\colon [a,b) \to \mathbb{R}$ be a function. Suppose that f is locally integrable. Prove that f is improperly integrable if and only if $f|_{[c,b)}$ is improperly integrable.

Exercise 6.4.7. **[Used in Section 6.4 and Theorem 7.4.3.]** Let $[a,b) \subseteq \mathbb{R}$ be a non-degenerate half-open interval, let $f,g\colon [a,b) \to \mathbb{R}$ be functions and let $k \in \mathbb{R}$. Suppose that f and g are improperly integrable.

(1) Prove that $f+g$ is improperly integrable and $\int_a^b [f+g](x)\,dx = \int_a^b f(x)\,dx + \int_a^b g(x)\,dx$.
(2) Prove that kf is improperly integrable and $\int_a^b [kf](x)\,dx = k\int_a^b f(x)\,dx$.
(3) Prove that if $f(x) \geq g(x)$ for all $x \in [a,b)$, then $\int_a^b f(x)\,dx \geq \int_a^b g(x)\,dx$.

Exercise 6.4.8. Find an example of functions $f,g\colon (0,1] \to \mathbb{R}$ such that f and g are improperly integrable, but that fg is not improperly integrable.

Exercise 6.4.9. Let $[a,b) \subseteq \mathbb{R}$ be a non-degenerate half-open interval, and let $f,g\colon [a,b) \to \mathbb{R}$ be functions. Suppose that $f(x) \geq 0$ and $g(x) > 0$ for all $x \in [a,b)$. Suppose that $\lim_{x\to b^-} \frac{f(x)}{g(x)}$ exists and $\lim_{x\to b^-} \frac{f(x)}{g(x)} > 0$. Prove that $\int_a^b f(x)\,dx$ is convergent if and only if $\int_a^b g(x)\,dx$ is convergent. (For the reader who is familiar with convergence tests for series, observe that this exercise is the analog of the Limit Comparison Test for series, which we will see in Section 9.3.)

Exercise 6.4.10. **[Used in Section 6.4 and Exercise 6.4.13.]** State and prove the analog of Theorem 6.4.11 (Comparison Test for Type 2 Improper Integrals) for Type 1 improper integrals.

Exercise 6.4.11. **[Used in Section 6.4.]** State and prove the analog of Theorem 6.4.12 (Integration by Substitution for Type 2 Improper Integrals) for Type 1 improper integrals.

Exercise 6.4.12. **[Used in Section 6.4 and Exercise 6.4.13.]** State and prove the analog of Theorem 6.4.13 (Integration by Parts for Type 2 Improper Integrals) for Type 1 improper integrals.

Exercise 6.4.13. This exercise discusses the **gamma function**, which is a generalization of the notion of factorial (discussed in Example 2.5.12) to the positive real numbers. (In fact, the gamma function is defined for most complex numbers, though we will not discuss that here.) See [Art64] for more about the gamma function.

Let $\Gamma\colon (0,\infty) \to \mathbb{R}$ be defined by

$$\Gamma(x) = \int_0^\infty e^{-t} t^{x-1}\, dt$$

for all $x \in (0,\infty)$.

Let $x \in (0,\infty)$. The integral in the definition of $\Gamma(x)$ is improper. More precisely, if we break up the integral as

$$\int_0^1 e^{-t} t^{x-1}\, dt + \int_1^\infty e^{-t} t^{x-1}\, dt,$$

then the first of these integrals is a Type 2 improper integral for each $x \in (0,1)$, and the second of these integrals is a Type 1 improper integral for each $x \in (0,\infty)$.

(1) Prove that the integral $\int_0^1 e^{-t} t^{x-1}\, dt$ is convergent.
(2) Prove that the integrals $\int_1^\infty e^{-t} t^{x-1}\, dt$ is convergent.
(3) Prove that $\Gamma(x+1) = x\Gamma(x)$.
(4) Let $n \in \mathbb{N}$. Prove that $\Gamma(n) = (n-1)!$.

[Use Exercise 6.3.1 (1), Exercise 6.4.10 and Exercise 6.4.12.]

6.5 Historical Remarks

The historical remarks for this chapter are very brief. The current chapter discusses limits to infinity of functions, which are a slight variation of regular limits of functions, and hence much of the history of the material in this chapter is subsumed in the historical remarks in Section 3.6. Moreover, limits of sequences are also a type of limit to infinity, where $x \to \infty$ (with x being a real number) is replaced with $n \to \infty$ (where n is a natural number), and hence the following historical remarks also have some overlap with the remarks in Section 8.5.

Ancient World

Parmenides of Elea (c. 515–c. 450 BCE) believed that motion is an illusion. His disciple Zeno of Elea (c. 490–c. 425 BCE) provided four famous arguments to show that there is no motion. For example, Zeno's third argument, the Arrow, says that at every instant, an arrow is in exactly one place, so that it cannot really move. A modern approach to this problem is to think of the arrow as going through infinitely many instances, and we then have to consider the indeterminate form $0 \cdot \infty$.

Medieval Period

Bhaskara II (1114–1185) believed in the infinite, and in 1150 essentially said that $\frac{a}{0} = \infty$ and $\infty + a = \infty$. These formulas are reminiscent of our rules about limits to infinity.

Seventeenth Century

Type 1 improper integrals make use of limits to infinity, and an early example of such an integral was computed by Evangelista Torricelli (1608–1647), who showed that rotating an infinite hyperbola yielded a finite volume. Torricelli claimed to be the first person to show that an infinitely large object can have finite content, though the idea may have occurred to Oresme much earlier, and perhaps to Fermat and Roberval as well.

The symbol "∞" that we now use for infinity is due to John Wallis (1616–1703) in 1659. He used this symbol with two different meanings, first as the number of lines into which a region is divided, and second as the thing to which n goes as we subdivide a region into more and more pieces. This multiplicity of meanings of the symbol ∞, and in general of meanings of the word "infinity," persists to this day, and it can only be clarified by rigorous definitions for each context where this symbol and word are used.

In 1696 Guillaume de l'Hôpital (1661–1704) published the first printed textbook on differential calculus, *Analyse des infiniment petits pour l'intelligence des lignes courbes*. He never published a textbook on integral calculus. Many of the ideas in l'Hôpital's book were due to Johann Bernoulli (1667–1748), who was paid by the nobleman l'Hôpital. This textbook includes what we now call l'Hôpital's Rule in the $\frac{0}{0}$ case, though it should presumably be called Bernoulli's Rule; the proof given for this result would not be considered rigorous today.

Eighteenth Century

Leonhard Euler (1707–1783) freely used infinitely large and infinitely small numbers (the latter being infinitesimals). He let w denote an infinitely small number, and he let $I = \frac{x}{w}$, where x is a positive real number, so that I is infinitely large. He then wrote equations such as $I - 1 = I$ and $\frac{I-1}{I} = 1$, which we would write today as $\lim_{n\to\infty} (n-1) = \infty$ and $\lim_{n\to\infty} \frac{n-1}{n} = 1$, respectively. Euler's intuition about infinity was later justified by the rigorous treatment of infinitesimals by Robinson, as mentioned at the end of Section 3.6.

Nineteenth Century

Carl Friedrich Gauss (1777–1855) did not accept infinitely large quantities, and instead used an inequality technique to prove that some limits exist, though such proofs were not rigorous by subsequent standards. He implicitly used some results that we now prove, such as the Monotone Convergence Theorem, and did not explicitly state all of the definitions he used, such as limits to infinity.

Augustin Louis Cauchy (1789–1857), who gave the first rigorous definition of definite integrals for continuous functions on closed bounded integrals, also defined Type 1 improper integrals for continuous functions on unbounded intervals, and Type 2 improper integrals for functions on closed bounded intervals that have isolated discontinuities.

7

Transcendental Functions

7.1 Introduction

In previous chapters of this book we used various standard functions, called elementary functions, to provide examples of the concepts under discussion. These functions are familiar to the reader from precalculus and calculus courses, and are found in many applications of mathematics. We are now in a position to give a rigorous treatment of the elementary functions we have been using.

The most widely used elementary functions are the linear, polynomial, rational, exponential, logarithmic and trigonometric functions. The first three of these are called algebraic functions, and the second three are called transcendental functions. The algebraic functions are simple to define; polynomial functions (including linear functions) were defined in Definition 2.5.10, and rational functions are just quotients of polynomial functions. The transcendental functions, by contrast, are much harder to define than the algebraic ones, and the definitions given in precalculus and calculus courses are often rather informal. For rigorous definitions of these functions, and proofs that they behave as expected, real analysis is needed. In fact, real analysis is even needed for a rigorous definition of x^r when r is irrational.

Contrary to the approach in precalculus courses, where power functions are defined first, and then exponential functions are defined in terms of power functions, and then logarithms are defined in terms of exponentials, here we take the standard rigorous approach, which reverses the process by starting with logarithms (defined in terms of integration), and then defining exponentials in terms of logarithms, and then defining power functions in terms of exponentials. We then define the sine and cosine functions, which are trickier to define than logarithms and exponentials. We do not discuss the other four standard trigonometric functions, because they can be expressed in terms of sine and cosine. We conclude the chapter with further discussion of the number π, which is first defined as part of our discussion of sine and cosine. Although our treatment of sine and cosine will be unrelated to logarithms and exponentials, except somewhat by analogy, in fact these trigonometric functions are related to exponentials (and hence to logarithms) via complex numbers, though it is beyond the scope of this book to discuss such matters; see any introductory complex analysis text, for example [BC09, Chapter 3], for details.

The problem with the approach to defining exponentials and logarithms in calculus courses is not the definition of exponential functions in terms of power functions, nor the definition of logarithms in terms of exponentials—both those steps are fine—the problem is in the informal definition used for power functions. Let $x \in (0,\infty)$. It is clear intuitively what x^n means for all $n \in \mathbb{N}$; see Definition 2.5.6 for a formal definition that captures this intuitive idea. It is also simple to extend this definition to all $n \in \mathbb{Z}$, as we saw in Definition 2.5.8. It is also not hard, intuitively, to define x^q for all $q \in \mathbb{Q}$. If $a,b \in \mathbb{N}$, then we could define $x^{\frac{a}{b}} = \sqrt[b]{x^a}$; to be truly rigorous here one needs to provide a rigorous definition of $\sqrt[n]{x}$ for all $x > 0$ and $n \in \mathbb{N}$, which we did in Exercise 3.5.6, though which is not usually done in a calculus class. The real problem occurs when we want to define x^r for an irrational number r. At best in a calculus class, it is stated informally that such a definition can be made using limits of sequences, where one first finds a sequence $\{c_n\}_{n=1}^{\infty}$ of rational numbers that converges to r, and one then defines x^r as the limit of the values of x^{c_n} as n goes to infinity, though of course the details of such a construction are omitted (the details would include a proof that the desired limit always exists, and that it is independent of the choice of the sequence that converges to r); at worst, it is not even stated that some sort of technically complicated definition of x^r is needed. In principle, rather than using sequences, it would be possible to formulate the definition of x^r using least upper bounds, as follows. If $x > 1$, we could define x^r to be $\text{lub}\{x^q \mid q \in \mathbb{Q} \text{ and } q < r\}$; if $0 < x < 1$, we could define x^r to be $\text{glb}\{x^q \mid q \in \mathbb{Q} \text{ and } q < r\}$. Of course, one would first have to use the Least Upper Bound Property and the Greatest Lower Bound Property to verify that this least upper bound and this greatest lower bound exist, and once that is established, it would take a bit of effort to prove that this definition of x^r behaves as one would want (for example that the function x^r is differentiable, and has the expected derivative). Such an approach can indeed be made to work, as seen in [Olm62, Sections 1102–1105]. Fortunately there is a nicer way to proceed that avoids this direct use of least upper bounds and greatest lower bounds, as we will soon see.

In our discussion of logarithms and exponentials, we will focus on the natural logarithm and e^x, and pay little attention to other logarithmic and exponential functions, because the natural logarithm and e^x are the best behaved among their peers, and there is very little reason to use the other logarithmic or exponential functions. Historically, base 10 logarithms were used for computational purposes prior to the widespread use of computers and calculators, though given the existence of such technology, there is little need for base 10 logarithms any more.

7.2 Logarithmic and Exponential Functions

We start with a discussion of logarithms. Although logarithms were invented as a tool for doing numerical calculations in the pre-computer era, a use for which logarithms are no longer needed, we now view logarithms as functions, which turn out to be very useful in a variety of applications. We will focus our attention on the "natural logarithm," which informally is the logarithm function with base e. However, because we have not yet defined the number e (we will do so later in this section), we will now give a definition of the natural logarithm function that makes no reference to bases in general or the number e in particular. Logarithms with other bases will be defined at the end of this section, just to show that it can be done. Because the natural logarithm function is the only logarithm function we will need, we will often drop the word "natural" and refer to it simply as the "logarithm function."

The idea behind the definition of the logarithm function is that whereas the function itself is somewhat tricky, its derivative is much simpler, and is a function that we can deal with using what we have previously seen. Of course, we cannot take the derivative of a function that we have not yet defined, but intuitively we know what we want the derivative of the natural logarithm function to be, namely, the function $f\colon (0,\infty) \to \mathbb{R}$ defined by $f(x) = \frac{1}{x}$ for all $x \in (0,\infty)$. Indeed, we will simply define the natural logarithm to be a function whose derivative is the function f. The key ingredients in our construction of the natural logarithm will be the Fundamental Theorem of Calculus Version I (Theorem 5.6.2), and the fact that the function f is continuous, using Example 3.3.3 (2), and hence locally integrable, as remarked after Definition 6.4.1.

Definition 7.2.1. The **natural logarithm** function is the function $\ln\colon (0,\infty) \to \mathbb{R}$ defined by

$$\ln x = \int_1^x \frac{1}{t}\,dt$$

for all $x \in (0,\infty)$. △

Observe that no analogous definition for the exponential function can be given, because the derivative of the exponential function is not anything simpler than itself. It is for that reason that we will define the exponential function in terms of the natural logarithm, and not vice versa.

We now see some of the familiar properties of the natural logarithm function.

Theorem 7.2.2.

1. *The function* $\ln$ *is differentiable, and* $\ln' x = \frac{1}{x}$ *for all* $x \in (0,\infty)$.
2. *The function* $\ln$ *is strictly increasing.*

Proof.

(1) This fact follows from the Fundamental Theorem of Calculus Version I (Theorem 5.6.2).

(2) By Part (1) of this theorem, and making use of the fact that $\frac{1}{x} > 0$ for all $x \in (0,\infty)$, we deduce that $\ln' x > 0$ for all $x \in (0,\infty)$. It then follows from Theorem 4.5.2 (2) that $\ln$ is strictly increasing. □

Because we know the derivative of ln, it is possible to sketch the graph of the function, as seen in Figure 7.2.1.

The next set of properties of the natural logarithm function makes use of Definition 2.5.6 and Definition 2.5.8.

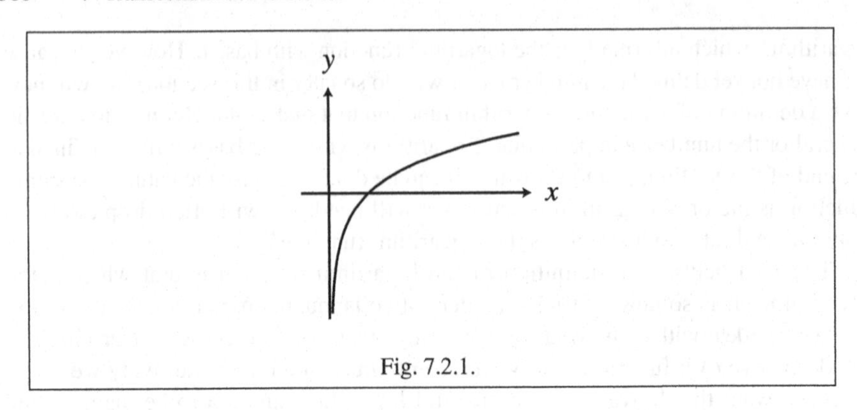

Fig. 7.2.1.

Theorem 7.2.3. *Let $x, y \in (0,\infty)$, and let $n \in \mathbb{Z}$.*

1. $\ln 1 = 0$.
2. $\ln(xy) = \ln x + \ln y$.
3. $\ln\left(\frac{x}{y}\right) = \ln x - \ln y$.
4. $\ln(x^n) = n \ln x$.

Proof.

(1) This fact follows directly from the definition of ln.

(2) Let $h\colon (0,\infty) \to \mathbb{R}$ be defined by $h(t) = \ln(xt)$ for all $t \in (0,\infty)$. Observe that h is well-defined, because $x, t \in (0,\infty)$ implies that $xt \in (0,\infty)$. We know that ln is differentiable by Theorem 7.2.2 (1), and it then follows from Example 4.2.3 (1) and the Chain Rule (Theorem 4.3.3) that h is differentiable and

$$h'(t) = \frac{1}{xt} \cdot x = \frac{1}{t} = \ln' t$$

for all $t \in (0,\infty)$. By Lemma 4.4.7 (2) there is some $C \in \mathbb{R}$ such that $h(t) = \ln t + C$ for all $t \in (0,\infty)$. Substituting $t = 1$ into this last equation yields $\ln(1) + C = h(1) = \ln(x \cdot 1) = \ln x$, and then using Part (1) of this theorem we deduce that $C = \ln x$. It follows that $h(t) = \ln t + \ln x$ for all $t \in (0,\infty)$. If we substitute $t = y$ in this last equation, we deduce that $\ln(xy) = \ln x + \ln y$.

(3) As a preliminary step, substituting $x = \frac{1}{y}$ into Part (2) of this theorem yields $\ln(1) = \ln\left(\frac{1}{y}\right) + \ln y$. Making use of Part (1) of this theorem, it follows that $\ln\left(\frac{1}{y}\right) = -\ln y$. Hence $\ln\left(\frac{x}{y}\right) = \ln\left(x \cdot \frac{1}{y}\right) = \ln x + \ln\left(\frac{1}{y}\right) = \ln x - \ln y$.

(4) First, we prove that $\ln(x^n) = n \ln x$ for all $n \in \mathbb{N}$ by induction on n. If $n = 1$ then the result is trivial. Now let $n \in \mathbb{N}$. Suppose that the result is true for n. By Part (2) of this theorem together with Definition 2.5.6 we see that $\ln(x^{n+1}) = \ln(x \cdot x^n) = \ln x + \ln(x^n) = \ln x + n \ln x = (n+1) \ln x$. Hence the result is true for $n+1$, and we deduce that the result holds for all $n \in \mathbb{N}$.

Next, we note that $x^0 = 1$ by Definition 2.5.8, and Part (1) of this theorem then implies that $\ln(x^0) = \ln 1 = 0 = 0 \cdot \ln x$. Therefore the result holds for $n = 0$.

Finally, let $m \in \mathbb{N}$. Then $x^{-m} = (x^m)^{-1} = \frac{1}{x^m}$ by Definition 2.5.8. Hence by Parts (1) and (3) of this theorem, together with what we have just proved, we see that $\ln(x^{-m}) = \ln\left(\frac{1}{x^m}\right) = \ln 1 - \ln(x^m) = 0 - m\ln x = (-m)\ln x$. Therefore the result holds for all $n \in -\mathbb{N}$. □

We will see in Theorem 7.2.14 (1) that the analog of Theorem 7.2.3 (4) holds for all $n \in \mathbb{R}$, not just for all $n \in \mathbb{Z}$, but to do so we will have to wait until we have defined x^n for all $n \in \mathbb{R}$.

We want to define the exponential function as the inverse function of ln, but in order to be sure that ln has an inverse function, we need the following lemma. Whereas it is easy to see that ln is injective, because it is strictly increasing, it is a bit trickier to show that ln is surjective, and we will need to make use of the Intermediate Value Theorem (Theorem 3.5.2).

Lemma 7.2.4. *The function* ln *is bijective.*

Proof. We know from Theorem 7.2.2 (2) that ln is strictly increasing, and it follows immediately that ln is injective.

Let $y \in \mathbb{R}$. We need to show that there is some $w \in (0,\infty)$ such that $y = \ln w$. By Exercise 7.2.1 we know that $\ln 2 > 0$, and we can therefore apply Exercise 2.6.12 to deduce that there is some $m \in \mathbb{N}$ such that $y \in [-m\ln 2, m\ln 2]$. By Theorem 7.2.3 (4) it follows that $y \in [\ln(2^{-m}), \ln(2^m)]$. If $y = \ln(2^{-m})$, then we could take $w = 2^{-m}$; observe that $2^{-m} = \left(\frac{1}{2}\right)^m > 0$, so w would be in $(0,\infty)$. If $y = \ln(2^m)$, we could take $w = 2^m$. Now suppose that $y \in (\ln(2^{-m}), \ln(2^m))$. By Theorem 7.2.2 (1) ln is differentiable, and therefore ln is continuous by Theorem 4.2.4. We can now apply the Intermediate Value Theorem (Theorem 3.5.2) to $\ln|[2^{-m}, 2^m]$ to deduce that there is some $w \in (2^{-m}, 2^m)$ such that $y = \ln w$. □

We now turn to exponential functions, which are extremely widespread in mathematics and its applications. In precalculus and calculus courses, such functions are usually defined by saying $f(x) = a^x$ for all $x \in \mathbb{R}$, where a is some positive real number. The most useful exponential function, and the one which we will refer to as "the exponential function," is the one usually defined informally as e^x. As much as this formulation of exponential functions seems reasonable intuitively, there are two problems with this approach, which are that we have not yet defined how to raise a real number to an arbitrary real number power (we have only dealt with integer powers so far), and that we have not yet defined the number e. Instead, we will define the exponential function in terms of the natural logarithm, with no mention made of the number e yet.

The crucial tool in the definition of the exponential function is the notion of an inverse function; the reader should review the discussion of inverse functions in Section 4.6.

The following definition makes sense by Lemma 7.2.4.

Definition 7.2.5. The **exponential** function is the function $\exp\colon \mathbb{R} \to (0,\infty)$ defined by $\exp = \ln^{-1}$. △

Observe that the codomain of exp is $(0,\infty)$ rather than $\mathbb{R}$, which makes no difference in practice, but was needed for the sake of exp being the properly defined inverse function of the natural logarithm.

The fact that exp is the inverse function of ln can be expressed in two equivalent ways, which are stated in the two parts of the following lemma. The second part is more commonly taught in courses such as precalculus, and is included for the sake of familiarity; the first part is more useful in proofs.

Lemma 7.2.6. *Let $x \in (0,\infty)$, and let $y \in \mathbb{R}$.*

1. $\exp(\ln x) = x$ *and* $\ln(\exp y) = y$.
2. $\ln x = y$ *if and only if* $\exp y = x$.

Proof.

(1) This part of the lemma follows immediately from the fact that $\exp = \ln^{-1}$, together with the definition of inverse functions; see [Blo10, Section 4.3] for a discussion of inverse functions.

(2) Suppose that $\ln x = y$. Then $\exp(\ln x) = \exp y$, and hence Part (1) of this lemma implies that $x = \exp y$. The proof of the other implication is similar, and we omit the details. □

As mentioned in Section 4.6, the graph of an inverse function is the reflection of the graph of the original function in the line $y = x$. We saw a sketch of the graph of ln in Figure 7.2.1, and so we immediately obtain the graph of exp, as seen in Figure 7.2.2.

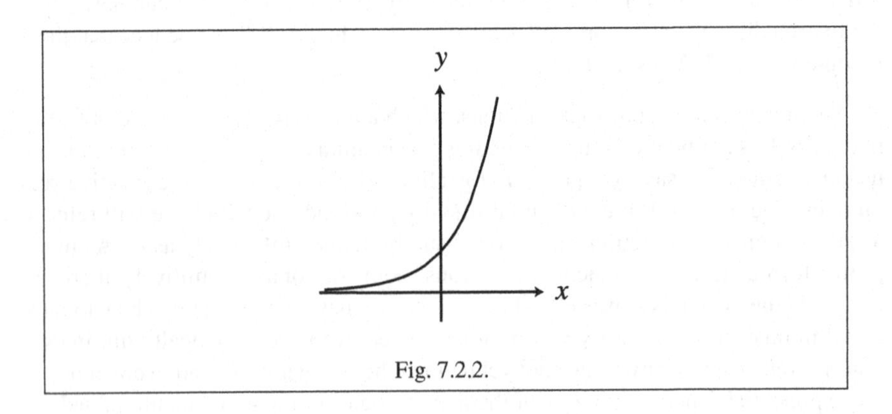

Fig. 7.2.2.

We now see some familiar properties of the exponential function.

Theorem 7.2.7.

1. *The function* exp *is bijective.*
2. *The function* exp *is differentiable, and* $\exp' x = \exp x$ *for all* $x \in \mathbb{R}$.
3. *The function* exp *is strictly increasing.*

Proof.

(1) We know that ln is bijective by Lemma 7.2.4. Given that $\exp = \ln^{-1}$, it follows from a standard fact about functions that exp is bijective; see [Blo10, Exercise 4.4.13] for details.

(2) By Theorem 7.2.2 (1) we know that ln is differentiable, and as noted in the proof of Part (2) of that theorem we know that $\ln' x > 0$ for all $x \in (0, \infty)$. Therefore Theorem 4.6.4 (3) (4) applied to ln imply that $\exp = \ln^{-1}$ is differentiable, and that

$$\exp' x = [\ln^{-1}]'(x) = \frac{1}{\ln'(\ln^{-1}(x))} = \frac{1}{\ln'(\exp x)} = \frac{1}{\frac{1}{\exp x}} = \exp x$$

for all $x \in \mathbb{R}$.

(3) Using Part (2) of this theorem, and making use of the fact that $\exp x > 0$ for all $x \in \mathbb{R}$, we see that $\exp' x > 0$ for all $x \in \mathbb{R}$. It then follows from Theorem 4.5.2 (2) that exp is strictly increasing. □

Theorem 7.2.8. *Let $x, y \in \mathbb{R}$, and let $n \in \mathbb{Z}$.*

1. $\exp 0 = 1$.
2. $\exp(x+y) = \exp x \cdot \exp y$.
3. $\exp(x-y) = \frac{\exp x}{\exp y}$.
4. $\exp(nx) = [\exp x]^n$.

Proof. We will prove Parts (1) and (2), leaving the rest to the reader in Exercise 7.2.6. In this proof we will make repeated use of Lemma 7.2.6 (1).

(1) We know by Theorem 7.2.3 (1) that $\ln 1 = 0$. Hence $\exp 0 = \exp(\ln 1) = 1$.

(2) Because ln is bijective, by Lemma 7.2.4, there are $w, z \in (0, \infty)$ such that $\ln w = x$ and $\ln z = y$. It follows from Theorem 7.2.3 (2) that $\ln(wz) = \ln w + \ln z$. Hence $\exp(x+y) = \exp(\ln w + \ln z) = \exp(\ln(wz)) = wz = \exp x \cdot \exp y$. □

We will see in Theorem 7.2.14 (2) that the analog of Theorem 7.2.8 (4) holds for all $n \in \mathbb{R}$, not just for all $n \in \mathbb{Z}$.

We are now ready to give the definition of the number e; this definition makes sense because $\ln\colon (0, \infty) \to \mathbb{R}$ is bijective.

Definition 7.2.9. The number e is the unique number in $(0, \infty)$ such that $\ln e = 1$. △

Some texts define e as

$$e = \lim_{n \to \infty} \left(1 + \frac{1}{n}\right)^n.$$

To make such a definition rigorous, it would be necessary first to define what is meant by limits of sequences, and second to prove that this particular limit exists, neither of which we have learned about yet. We will do these things later in the text, and we will indeed see that the definition of e given in Definition 7.2.9 is equivalent to the limit definition of e. More strongly, we will see in Example 8.4.3 that

$$\exp r = \lim_{n\to\infty}\left(1+\frac{r}{n}\right)^n$$

for all $r \in \mathbb{R}$. Both our definition of e and this other definition in terms of sequences are useful in certain circumstances. Of course, it is not legitimate to define the same thing in two different ways. If one wants to use both approaches to e, one has to choose one of these definitions of e, and then prove that the other definition is equivalent to the chosen one. We choose the definition of e in terms of the natural logarithm, because it requires fewer technicalities.

The following lemma shows that with respect to the integers, the exponential function is precisely what we want it to be.

Lemma 7.2.10. *Let $n \in \mathbb{Z}$. Then $\exp n = e^n$. In particular $\exp 1 = e$.*

Proof. First, we note that because $\ln e = 1$, then by Lemma 7.2.6 (1) we see that $\exp 1 = \exp(\ln e) = e$. We then use Theorem 7.2.8 (4) to see that $\exp n = \exp(1 \cdot n) = [\exp 1]^n = e^n$. □

We will see in Theorem 7.2.14 (3) that the analog of Lemma 7.2.10 holds for all $n \in \mathbb{R}$, not just for all $n \in \mathbb{Z}$.

We now turn to power functions, which have the form $f(x) = x^r$ for all $x \in (0,\infty)$, where r is a real number. Power functions are among the most widely used functions, and no discussion of functions would be complete without them. Even though power functions might appear to be a more elementary type of function than logarithms and exponentials, we have waited until now to define power functions because the intuitive way of defining power functions is difficult to make rigorous, as was mentioned in Section 7.1, but there is an alternative, and easier, way to proceed that makes use of the natural logarithm and the exponential function.

In our discussion of x^r, we restrict attention to $x > 0$ rather than all $x \in \mathbb{R}$ because we want $x^{\frac{1}{2}}$ to equal $\sqrt{x}$, and we cannot take square roots of negative numbers. We ignore $x = 0$ to avoid special cases, though we can simply define $0^r = 0$ for all $r \in \mathbb{R} - \{0\}$. We do not attempt to define "0^0," because that is an indeterminate form, similarly to $0 \cdot \infty$ and $\infty - \infty$; see Exercise 6.3.8 for evidence that 0^0 is an indeterminate form.

Definition 7.2.11. Let $r \in \mathbb{R}$. Let $p_r\colon (0,\infty) \to (0,\infty)$ be defined by $p_r(x) = \exp(r \ln x)$ for all $x \in (0,\infty)$. △

We will switch to more standard notation, and write x^r instead of $p_r(x)$. This more standard notation is not entirely proper, however, because the name of the function (in this case p_r) should not include any specific elements of the domain (in this case x). When we write p_r we write the name of the function, and when we write $p_r(x)$ we mean an element of the codomain. With the expression "x^r" there is no name of the function to write that does not include x, so it is not a proper name for a function, but we will write x^r nonetheless, because it is completely standard.

We now see some standard properties of x^r.

Theorem 7.2.12. *Let $x, y \in (0, \infty)$, and let $r, s \in \mathbb{R}$.*

1. $1^r = 1$, *and* $x^0 = 1$, *and* $x^1 = x$.
2. $x^{r+s} = x^r x^s$.
3. $x^{r-s} = \frac{x^r}{x^s}$.
4. $(x^r)^s = x^{rs}$.
5. $(xy)^r = x^r y^r$.
6. $\left(\frac{x}{y}\right)^r = \frac{x^r}{y^r}$.
7. $x^{-r} = \frac{1}{x^r}$.

Proof. We will prove Part (2), leaving the rest to the reader in Exercise 7.2.9.

(2) By Theorem 7.2.8 (2) we see that $x^r x^s = \exp(r \ln x) \cdot \exp(s \ln x) = \exp(r \ln x + s \ln x) = \exp((r+s) \ln x) = x^{r+s}$. □

In Section 7.1 an informal, albeit technically complicated, definition of $x^{\frac{a}{b}}$ was suggested, where $a, b \in \mathbb{N}$. The reader is asked in Exercise 7.2.11 (2) to show that this suggested definition is in fact a consequence of Definition 7.2.11 and Theorem 7.2.12.

Theorem 7.2.13. *Let $r \in \mathbb{R}$.*

1. *The function x^r is differentiable, and $[x^r]' = rx^{r-1}$ for all $x \in (0, \infty)$.*
2. *If $r > 0$, the function x^r is strictly increasing; if $r = 0$, the function x^r is constant; and if $r < 0$, the function x^r is strictly decreasing.*
3. *If $r \neq 0$, the function x^r is bijective.*

Proof. We will prove Part (1), leaving the rest to the reader in Exercise 7.2.10.

(1) Combining Theorem 4.3.1 (3), the Chain Rule (Theorem 4.3.3), Theorem 7.2.2 (1) and Theorem 7.2.7 (2), we deduce that the function $x^r = \exp(r \ln x)$ is differentiable, and $[x^r]' = [\exp(r \ln x)]' = \exp(r \ln x) \cdot r \cdot \frac{1}{x} = x^r \cdot r \cdot \frac{1}{x} = rx^{r-1}$ for all $x \in (0, \infty)$. □

Before proceeding any further in our discussion of x^r, we need to clarify one important matter. For each $x \in (0, \infty)$ and $n \in \mathbb{Z}$, we have actually defined the value of x^n in two different ways, once in the combination of Definition 2.5.6 and Definition 2.5.8, and once in Definition 7.2.11; the latter definition is entirely independent of the former. If our definitions are to make sense, we need to verify that they both yield the same result. Let $x \in (0, \infty)$. First, by Definition 7.2.11 together with Lemma 7.2.6 (1) and Theorem 7.2.12 (2) we see that if we use our new definition of x^r, then $x^1 = \exp(1 \cdot \ln x) = \exp(\ln x) = x$, and if $n \in \mathbb{N}$ then $x^{n+1} = x^{1+n} = x^1 \cdot x^n = x \cdot x^n$. Definition 2.5.6 was based upon Definition by Recursion (Theorem 2.5.5), and the uniqueness in that theorem implies that the definition of x^n for all $n \in \mathbb{N}$ in both Definition 2.5.6 and Definition 7.2.11 must agree. Again using Definition 7.2.11, this time with Theorem 7.2.8 (1) and Theorem 7.2.12 (7), we see that $x^0 = \exp(0 \cdot \ln x) = \exp 0 = 1$, and that if $n \in \mathbb{N}$ then $x^{-n} = \frac{1}{x^n} = (x^n)^{-1}$. Hence we see that Definition 2.5.8 agrees with Definition 7.2.11.

We are now in a position to prove some promised properties of logarithms and exponentials that involve power functions.

Theorem 7.2.14. *Let $x \in (0, \infty)$, and let $y, r \in \mathbb{R}$.*

1. $\ln(x^r) = r \ln x$.
2. $\exp(ry) = [\exp(y)]^r$.
3. $\exp r = e^r$.
4. $e^{\ln x} = x$ *and* $\ln(e^y) = y$.

Proof. We will prove Part (1), leaving the rest to the reader in Exercise 7.2.12.

(1) Using the definition of x^r and Lemma 7.2.6 (1), we see that $\ln(x^r) = \ln(\exp(r \ln x)) = r \ln x$. □

Because of Theorem 7.2.14 (3) we can now switch to the more standard "e^x" notation rather than "$\exp x$." As with the notation x^r, the notation e^x is not a proper name for a function, but we will use it nonetheless, because it is standard.

We conclude this section with a look at exponential functions and logarithm functions with bases other than e, this time starting with exponentials.

Definition 7.2.15. Let $a \in (0, \infty)$. The **exponential function with base** a is the function $\exp_a \colon \mathbb{R} \to (0, \infty)$ defined by $\exp_a x = a^x = \exp(x \ln a)$ for all $x \in \mathbb{R}$. △

By Theorem 7.2.14 (3) we note that $\exp_e = \exp$. As with the notation x^r and e^x, we will use the improper but more standard notation a^x rather than $\exp_a x$.

For any $a \in (0, \infty)$, the function a^x satisfies all the analogous properties of e^x stated in Theorem 7.2.8; some of these properties are proved in Exercise 7.2.13. The one place where there is a difference between e^x and a^x when $a \neq e$ is in the formula for the derivative. The following theorem is derived immediately from Theorem 7.2.7 (2) and the Chain Rule (Theorem 4.3.3), and we omit the details.

Theorem 7.2.16. *Let $a \in (0, \infty)$. The function a^x is differentiable, and $[a^x]' = a^x \ln a$ for all $x \in \mathbb{R}$.*

The fact that e^x has a simpler derivative than a^x when $a \neq e$ is the reason why e^x is preferred over exponential functions with other bases.

The following lemma is needed to allow us to define logarithms with arbitrary bases.

Lemma 7.2.17. *Let $a \in (0, \infty)$.*

1. *If $a > 1$, the function a^x is strictly increasing; if $a = 1$, the function a^x is constant; and if $0 < a < 1$, the function $\exp_a$ is strictly decreasing.*
2. *If $a \neq 1$, then the function $\exp_a$ is bijective.*

Proof.

(1) By Theorem 7.2.16 we know that $[a^x]' = a^x \ln a$ for all $x \in \mathbb{R}$. By the definition of ln, it is seen that $\ln a$ is positive, zero or negative if $a > 1$, or $a = 1$, or $0 < a < 1$, respectively. The desired result now follows from Theorem 4.5.2.

(2) Suppose that $a \neq 1$. Then by Part (1) of this lemma we know that $\exp_a$ is either strictly increasing or strictly decreasing. It follows that $\exp_a$ is injective.

Let $z \in (0,\infty)$. By Theorem 7.2.7 (1) we know that exp is surjective. Hence there is some $w \in \mathbb{R}$ such that $\exp w = z$. We know by Theorem 7.2.3 (1) that $\ln 1 = 0$, and by Lemma 7.2.4 that ln is injective, and $a \neq 1$ implies that $\ln a \neq 0$. It follows that $\exp_a(\frac{w}{\ln a}) = \exp(\frac{w}{\ln a} \cdot \ln a) = \exp w = z$. Therefore $\exp_a$ is surjective. □

Lemma 7.2.17 (2) allows us to make the following definition.

Definition 7.2.18. Let $a \in (0,\infty)$. Suppose that $a \neq 1$. The **logarithm function with base** a is the function $\log_a : (0,\infty) \to \mathbb{R}$ defined by $\log_a = (\exp_a)^{-1}$. △

Observe that $\log_e = \ln$. The relationship of $\log_a$ to $\exp_a$ for any a is the same as the relationship between ln and exp, as stated in the following lemma; the proof of this lemma is just like the proof of Lemma 7.2.6, and we omit the details.

Lemma 7.2.19. *Let $a, x \in (0,\infty)$, and let $y \in \mathbb{R}$. Suppose that $a \neq 1$.*

1. $a^{\log_a x} = x$ *and* $\log_a(a^y) = y$.
2. $\log_a x = y$ *if and only if* $a^y = x$.

For any $a \in (0,\infty)$ such that $a \neq 1$, the function $\log_a$ satisfies all the analogous properties of ln stated in Theorem 7.2.3; some of these properties are proved in Exercise 7.2.14 (1) (2). As seen in Part (3) of that exercise, one place where there is a difference between $\log_a$ and ln when $a \neq e$ is in the formula for the derivative. As seen in the following lemma, a logarithm function with one base is just as good as a logarithm function with any other base, and so bases other than e are not really needed. We defined logarithm functions with bases other than e simply because the reader has most likely encountered such functions previously, and we wanted to show that they can be defined rigorously.

Lemma 7.2.20. *Let $a, b, x \in (0,\infty)$. Suppose that $a \neq 1$ and $b \neq 1$. Then*

$$\log_a x = \frac{\log_b x}{\log_b a}.$$

Proof. Left to the reader in Exercise 7.2.15. □

Reflections

Students in precalculus and introductory calculus courses sometimes find the idea of logarithms confusing, though from a mathematician's perspective it is not entirely clear why that is so. Students do not find the exponential function nearly as troubling, and yet the exponential function and the natural logarithm function are not inherently very different as functions; they are inverse functions of each other, and they have rather analogous properties. From the author's experience teaching precalculus and calculus, he suspects that the difficulty students find with logarithms is because they are typically defined in such courses as the inverse functions of exponential functions, and the inverse function relationship, especially as usually stated in these courses, seems to be confusing—students often find inverse trigonometric functions similarly troubling.

Specifically, in elementary courses the exponential function is defined intuitively in terms of the number e raised to a power (which is usually explained carefully for rational powers, and usually glossed over for irrational powers). The natural logarithm function is then defined to be the function $\ln x$ that satisfies the property that $\ln x = y$ if and only if $e^y = x$, and it is this if and only if statement used as a definition that is, perhaps, the source of the confusion. This definition of $\ln x$ is somewhat abstract, and does not resemble the way most familiar functions are defined (which is in terms of explicit formulas). In the rigorous treatment of logarithms, such as found in this text, we define logarithm directly rather than as an inverse function. Of course, we then need to define the exponential function as the inverse function of logarithm, and so inverse functions cannot be avoided; it is simply that logarithm gets the initial treatment this time. Moreover, rather than phrasing the inverse function relation between logarithm and exponential as is done in elementary courses, we emphasize the equivalent conditions $\ln(e^x) = x$ and $e^{\ln x} = x$, which is much more natural when thinking of exponentials and logarithms as functions.

In addition to the fact that logarithms are discussed before exponentials, the other major difference between the treatment of these functions in a real analysis course and in elementary courses is the fact that whereas in elementary courses one thinks of the natural logarithm function as the logarithm with base e, in the real analysis approach we define $\ln x$ and e^x without any reference to a "base" (and indeed without any reference to the number e). Moreover, we use the exponential function e^x and the logarithm function $\ln x$ virtually exclusively, ignoring those with other bases; logarithms and exponentials with other bases were important historically, but are not needed in real analysis.

The reader might wonder why we did not try to define the exponential function e^x by an integral, and then define the natural logarithm as the inverse function of e^x. The reason is that we want the derivative of e^x to be itself, and therefore we cannot define it in terms of the integral of itself. By contrast, we want the derivative of $\ln x$ to be $\frac{1}{x}$, and this latter function is continuous, and hence integrable on closed bounded integrals, which is what makes the logarithm function easy to define as an integral.

Exercises

Exercise 7.2.1. **[Used in Lemma 7.2.4.]** Prove that $\ln 2 > 0$.

Exercise 7.2.2. **[Used in Example 10.4.11.]** Prove that $\ln x$ is infinitely differentiable, and that $\ln^{(n)}(x) = \frac{(-1)^{n-1}(n-1)!}{x^n}$ for all $x \in (0,\infty)$ and all $n \in \mathbb{N}$.

Exercise 7.2.3. **[Used in Exercise 8.4.13.]** Let $a, b \in (0,\infty)$. Suppose that $a < b$. Prove that $\frac{b-a}{b} < \ln\left(\frac{b}{a}\right) < \frac{b-a}{a}$. [Use Exercise 5.5.3.]

Exercise 7.2.4. Prove that $\ln(1+x) \leq x$ for all $x \in (0,\infty)$.

Exercise 7.2.5. **[Used in Example 6.3.8.]** Prove that $\lim_{x\to 0^+} \ln x = -\infty$.

Exercise 7.2.6. **[Used in Theorem 7.2.8.]** Prove Theorem 7.2.8 (3) (4).

Exercise 7.2.7. Prove that $\exp x \geq 1+x$ for all $x \in (0,\infty)$.

Exercise 7.2.8. **[Used in Example 6.4.6.]**

(1) Using only the properties of e^x stated in Section 7.2, prove that $\lim_{x\to\infty} e^{-x} = 0$.
(2) Prove that $\lim_{x\to\infty} e^{-x^2} = 0$ and $\lim_{x\to-\infty} e^{-x^2} = 0$.

Exercise 7.2.9. **[Used in Theorem 7.2.12.]** Prove Theorem 7.2.12 (1) (3) (4) (5) (6) (7).

Exercise 7.2.10. **[Used in Theorem 7.2.13.]** Prove Theorem 7.2.13 (2) (3).

Exercise 7.2.11. **[Used in Example 6.4.8, Section 7.2 and Example 10.2.8.]** Let $x \in (0,\infty)$, and let $n \in \mathbb{N}$. In Exercise 3.5.6 we defined $\sqrt[n]{x}$. The purpose of this exercise is to show that that definition is compatible with the definition of x^r given in the present section.

(1) Prove that $\sqrt[n]{x} = x^{\frac{1}{n}}$.
(2) Let $a, b \in \mathbb{N}$. Prove that $x^{\frac{a}{b}} = \sqrt[b]{x^a} = (\sqrt[b]{x})^a$.
(3) Prove that $\lim_{x\to 0^+} \sqrt[n]{x} = 0$.
(4) The method of Exercise 3.5.6 might appear to be conceptually very different from the approach to roots using fractional powers, because the former uses the Intermediate Value Theorem and the latter uses power functions, but in fact power functions ultimately rely upon the Intermediate Value Theorem as well; show where.

Exercise 7.2.12. **[Used in Theorem 7.2.14.]** Prove Theorem 7.2.14 (2) (3) (4).

Exercise 7.2.13. **[Used in Section 7.2.]** Let $a \in (0,\infty)$, and let $x, y \in \mathbb{R}$.

(1) Prove that $a^{x+y} = a^x a^y$.
(2) Prove that $a^{xy} = [a^y]^x$.

Exercise 7.2.14. **[Used in Section 7.2 and Exercise 7.2.15.]** Let $a \in (0,\infty)$, let $x, y \in (0,\infty)$ and let $r \in \mathbb{R}$. Suppose that $a \neq 1$.

(1) Prove that $\log_a(xy) = \log_a x + \log_a y$.
(2) Prove that $\log_a(x^r) = r\log_a x$.
(3) Prove that the function $\log_a$ is differentiable, and $\log_a' x = \frac{1}{\ln a}\frac{1}{x}$.

Exercise 7.2.15. **[Used in Lemma 7.2.20.]** Prove Lemma 7.2.20.

[Use Exercise 7.2.14 (2).]

Exercise 7.2.16. **[Used in Exercise 6.4.1 and Example 9.3.7.]** This exercise makes use of Exercise 6.2.15. Let $r \in \mathbb{R}$. Prove that

$$\lim_{x\to\infty} x^r = \begin{cases} \infty, & \text{if } r > 0 \\ 1, & \text{if } r = 0 \\ 0, & \text{if } r < 0. \end{cases}$$

7.3 Trigonometric Functions

In addition to the exponential and logarithmic functions, the other commonly used transcendental functions are the trigonometric functions. Although the origin of these functions is in the study of triangles, they have many other important applications both inside and outside of mathematics, for example in the study of oscillatory motion (springs, pendulum, waves and more). Our concern in this text is not with triangles at all, but rather with giving the trigonometric functions a rigorous definition, and then proving their basic properties.

There are six basic trigonometric functions, which are sine, cosine, tangent, secant, cosecant and cotangent. However, the latter four can be defined in terms of sine and cosine, and so we will restrict our attention to these two functions.

In precalculus and calculus courses the trigonometric functions are usually defined in terms of the unit circle. Conceptually this approach is the nicest way of defining the trigonometric functions, but, as usually presented, it is not completely rigorous, unless it is preceded by a rigorous discussion of arc length, which is virtually never done in that context due to the technical difficulties it would entail (as seen in Section 5.9). Instead of using the unit circle, we will give a definition of sine and cosine that is somewhat analogous to our treatment of logarithmic and exponential functions in Section 7.2, though defining the trigonometric functions is a bit more complicated.

When we discussed logarithms and exponentials, we started with the natural logarithm first because the derivative of the natural logarithm is a very simple function, namely, the function $f\colon (0,\infty) \to \mathbb{R}$ defined by $f(x) = \frac{1}{x}$ for all $x \in (0,\infty)$, which we already knew was continuous and hence integrable. By contrast, the derivative of the exponential function is not anything simpler than itself, and that is why we defined the exponential function as the inverse of the natural logarithm, and not vice versa.

A similar situation occurs with the sine function. It's derivative is not anything simpler than itself, and so we cannot define the sine function directly as an integral. However, the derivative of the arcsine function is a much simpler function than arcsine, and it does not involve trigonometric functions at all, and it is a function that we know is continuous, and hence is integrable. We will therefore start our treatment of the trigonometric functions with a definition of the arcsine function, and then define sine in terms of arcsine, and cosine in terms of sine. It is assumed that the reader is informally familiar with the arcsine function as the inverse of sine restricted to the domain $\left[-\frac{\pi}{2}, \frac{\pi}{2}\right]$; here we will give a definition of the arcsine function in terms of an integral.

Recall the definition of the natural logarithm in Definition 7.2.1. To define the arcsine function analogously we will use the integral

$$\int_0^x \frac{1}{\sqrt{1-t^2}}\,dt.$$

It might not be apparent to the reader at this point what the relation is between integrating the function $f\colon [-1,1] \to \mathbb{R}$ defined by $f(x) = \frac{1}{\sqrt{1-x^2}}$ for all $x \in (-1,1)$ and the trigonometric functions, but, as the reader will see in Section 7.4, this integral computes the arc length of part of the unit circle, which in turn is used in the informal definition of sine and cosine. We note that the above integral is a regular integral for $x \in (-1,1)$, but it is an improper integral for $x = 1$ and $x = -1$, because $\frac{1}{\sqrt{1-t^2}}$ is not defined for $t = 1$ or $t = -1$. Hence we will need to make use of improper integrals, discussed in Section 6.4, in our discussion of sine and cosine; we will also make use of the related topic of limits to infinity, see in Section 6.2. The reader who has not read these two sections from Chapter 6, but nonetheless wants to learn about the trigonometric functions, can still read the present section, though a few things will have to be accepted without proof.

Not only does the definition of sine and cosine have the complication of needing improper integrals, which did not arise in our discussion of logarithmic and exponential functions, but there is another difficulty in the definition of sine and cosine that was not encountered in the definition of logarithms and exponentials, which is that using the arcsine to define sine and cosine yields the definition of these functions only on $\left[-\frac{\pi}{2}, \frac{3\pi}{2}\right]$. To extend the definition to all $\mathbb{R}$, we need to have sine and cosine "repeat themselves" every 2π, and as our first step toward the definition of the two trigonometric functions we now consider the general idea of functions that repeat themselves. This concept will also be used in our discussion of a continuous but nowhere differentiable function in Section 10.5.

Definition 7.3.1. Let $f\colon \mathbb{R} \to \mathbb{R}$ be a function. The function f is **periodic** if there is some $P \in (0, \infty)$ such that $f(x+P) = f(x)$ for all $x \in \mathbb{R}$. The number P is called the **period** of f. △

Observe that if a function f has period P, then it has period nP for any $n \in \mathbb{N}$.

The most familiar examples of periodic functions are sine and cosine, though many other periodic functions exist. Periodic functions play an important role in some parts of mathematics and in applications of mathematics, because they are useful for describing repeating phenomena such as waves and springs.

The simplest way to construct periodic functions is to take a function with domain a non-degenerate interval, and extend it to all of $\mathbb{R}$. The following lemma states that this type of construction works.

Lemma 7.3.2. *Let $[a,b] \subseteq \mathbb{R}$ be a non-degenerate closed bounded interval, and let $h = b - a$.*

1. *Let $k\colon [a,b) \to \mathbb{R}$ be a function. Then there is a unique periodic function $f\colon \mathbb{R} \to \mathbb{R}$ with period h such that $f|_{[a,b)} = k$.*
2. *Let $g\colon [a,b] \to \mathbb{R}$ be a function. Suppose that $g(a) = g(b)$. Then there is a unique periodic function $f\colon \mathbb{R} \to \mathbb{R}$ with period h such that $f|_{[a,b]} = g$.*

Proof.

(1) We first show existence. Let $f\colon \mathbb{R} \to \mathbb{R}$ be defined as follows. Let $x \in \mathbb{R}$. By Exercise 2.6.14 (1) we know that there is a unique $n \in \mathbb{Z}$ such that $a + (n-1)h \le x < a + nh$. Hence $x - (n-1)h \in [a,b)$. We then let $f(x) = k(x - (n-1)h)$. If $x \in [a,b)$, then by the uniqueness of n it must be the case that $n = 1$, and hence $f(x) = k(x)$. Therefore $f|_{[a,b)} = k$. Next, we note that because $a + (n-1)h \le x < a + nh$, then $a + [(n+1) - 1]h \le x + h < a + (n+1)h$. It follows by the uniqueness of n that $f(x+h)$ is defined by $f(x+h) = k(x + h - [(n+1) - 1]h) = k(x - (n-1)h) = f(x)$. Hence f is periodic with period h.

We now show uniqueness. Let $p\colon \mathbb{R} \to \mathbb{R}$ be a periodic function with period h such that $p|_{[a,b)} = k$. Let $y \in \mathbb{R}$. By Exercise 2.6.14 (1) we know that there is a unique $m \in \mathbb{Z}$ such that $a+(m-1)h \leq y < a+mh$. Hence $y-(m-1)h \in [a,b)$. Using the fact that p and f are periodic with period h, and that $p|_{[a,b)} = k = f|_{[a,b)}$, we see that $p(y) = p(y-(m-1)h) = k(y-(m-1)h) = f(y-(m-1)h) = f(y)$. Hence $p = f$, and we conclude that f is unique.

(2) The proof of this part of the lemma is very similar to the proof of Part (1), and we omit the details. □

Lemma 7.3.2 is usually taken for granted, and often one just says informally "extend the function k periodically with period h" or a similar phrase. However, the proof of this lemma uses Exercise 2.6.14 (1), which relies upon Corollary 2.6.8 (1), which relies upon the Archimedean Property (Theorem 2.6.7), which in turn relies upon the Least Upper Bound Property of the real numbers. Hence, the ability to construct periodic extensions is not a trivial matter, even if it appears simple intuitively.

We can now use Lemma 7.3.2 to make the following definition.

Definition 7.3.3. Let $[a,b] \subseteq \mathbb{R}$ be a non-degenerate closed bounded interval, and let $h = b-a$.

1. Let $k\colon [a,b) \to \mathbb{R}$ be a function. The **periodic extension** of k is the unique periodic function $f\colon \mathbb{R} \to \mathbb{R}$ with period h such that $f|_{[a,b)} = k$.
2. Let $g\colon [a,b] \to \mathbb{R}$ be a function. Suppose that $g(a) = g(b)$. The **periodic extension** of g is the unique periodic function $f\colon \mathbb{R} \to \mathbb{R}$ with period h such that $f|_{[a,b]} = g$. △

The following lemma relates the behavior of a function to the behavior of its periodic extension.

Lemma 7.3.4. *Let $[a,b] \subseteq \mathbb{R}$ be a non-degenerate closed bounded interval. Let $g\colon [a,b] \to \mathbb{R}$ be a function. Suppose that $g(a) = g(b)$. Let $f\colon \mathbb{R} \to \mathbb{R}$ be the periodic extension of g.*

1. *If g is continuous, then f is continuous and bounded.*
2. *If g is differentiable, and if $g'(a) = g'(b)$, where $g'(a)$ and $g'(b)$ are one-sided derivatives, then f is differentiable, and f' is the periodic extension of g'.*

Proof. Let $h = b-a$, so that f has period h.

(1) Suppose that g is continuous. Let $x \in \mathbb{R}$. By Exercise 2.6.14 (1) we know that there is a unique $n \in \mathbb{Z}$ such that $a+(n-1)h \leq x < a+nh$. Hence $x \in (a+(n-2)h, a+nh)$. By Lemma 2.3.7 (2) there is some $\delta > 0$ such that $(x-\delta, x+\delta) \subseteq (a+(n-2)h, a+nh)$

Using the notation of Exercise 3.3.11, observe that $[a+(n-2)h, a+(n-1)h] = [a,b]+(n-2)h$ and $[a+(n-1)h, a+nh] = [a,b]+(n-1)h$. It follows from Exercise 3.3.11 that $f|_{[a+(n-2)h, a+(n-1)h]}$ and $f|_{[a+(n-1)h, a+nh]}$ are continuous. Using the Pasting Lemma (Lemma 3.3.10) we see that $f|_{[a+(n-2)h, a+nh]}$ is continuous. By Exercise 3.3.2 (2) the restriction of $f|_{[a+(n-2)h, a+nh]}$ to $(x-\delta, x+\delta)$ is continuous, which means that $f|_{(x-\delta,x+\delta)}$ is continuous. Because $\mathbb{R}\cap(x-\delta,x+\delta)=(x-\delta,x+\delta)$, it follows from Exercise 3.3.2 (1) that f is continuous at x.

We know by Corollary 3.4.6 that g is bounded. For each $x\in\mathbb{R}$, we know that $f(x)=g(y)$ for some $y\in\mathbb{R}$. It follows that f is bounded.

(2) The proof of this part of the lemma is very similar to the proof of Part (1) of the lemma, though we replace Exercise 3.3.11 with the Chain Rule (Theorem 4.3.3), and we replace Exercise 3.3.2 (2), Exercise 3.3.2 (1) and the Pasting Lemma with Exercise 4.2.3 (5), Exercise 4.2.3 (1) and Exercise 4.3.7 (1), respectively; we omit the details. □

Prior to defining arcsine, we have one more preliminary step, which is the definition of the number π. Informally, we are used to thinking of π as the ratio of the circumference of a circle to its diameter, and also as the number that appears in the well-known formula for the area of a circle in terms of its radius. We are also used to using the decimal expansion of π, which is $\pi=3.14159\ldots$. However, as much as our intuitive conception of π is correct, simply saying that π is the ratio of the circumference of a circle to its diameter is not a rigorous definition, because it would first be necessary to prove that this ratio is the same in all circles. This geometric fact about circles will be proved in Section 7.4; at present we will take the quicker route of defining π as an improper integral, which is less geometrically appealing, but allows us to proceed more directly to the definition of the sine and cosine functions.

To see that the following definition makes sense, we need some preliminary observations. Using some facts about continuity that we have encountered, the reader can verify that the function $f\colon(-1,1)\to\mathbb{R}$ defined by $f(x)=\frac{1}{\sqrt{1-x^2}}=(1-x^2)^{-\frac{1}{2}}$ for all $x\in(-1,1)$ is continuous. It then follows from Theorem 5.4.11 that f is locally integrable. Finally, Exercise 6.4.5 (2) implies that the improper integral in the following definition is convergent.

Definition 7.3.5. The number π is defined by

$$\pi=2\int_0^1\frac{1}{\sqrt{1-x^2}}\,dx.$$

△

We will not give a rigorous calculation of the numerical value of π as $3.14159\ldots$ at this point, both because we will not need it, and because we do not yet have the tools to do so. We will give a proof that π is an irrational number in Theorem 7.4.5, and a calculation of the first few digits of the decimal expansion of π in Example 10.4.17.

We are now ready to define the arcsine function, from which we will then define the sine function.

Definition 7.3.6. The **arcsine** function is the function $\arcsin\colon[-1,1]\to\mathbb{R}$ defined by

$$\arcsin x=\int_0^x\frac{1}{\sqrt{1-t^2}}\,dt$$

for all $x\in[-1,1]$. where the integral is improper when $x=1$ and $x=-1$. △

The fact that the improper integral in Definition 7.3.6 is convergent when $x = 1$ is due, as remarked above, to Exercise 6.4.5 (2); it is left to the reader to verify that the improper integral is also convergent when $x = -1$.

We now state those properties of arcsine that we will need for our construction of sine.

Lemma 7.3.7.

1. $\arcsin(-x) = -\arcsin x$ *for all* $x \in [-1, 1]$.
2. $\arcsin 0 = 0$, *and* $\arcsin 1 = \frac{\pi}{2}$, *and* $\arcsin(-1) = -\frac{\pi}{2}$.
3. *The function* arcsin *is differentiable on* $(-1, 1)$, *and* $\arcsin' x = \frac{1}{\sqrt{1-x^2}}$ *for all* $x \in (-1, 1)$.
4. $\lim_{x \to 1^-} \arcsin' x = \infty$ *and* $\lim_{x \to -1^+} \arcsin' x = \infty$.
5. *The function* arcsin *is continuous.*
6. *The function* arcsin *is strictly increasing.*
7. *The function* $\arcsin\colon [-1, 1] \to \left[-\frac{\pi}{2}, \frac{\pi}{2}\right]$ *is bijective.*

Proof.

(1) Let $x \in [-1, 1]$. Then using the substitution $u = -t$, which implies $du = -dt$, we see that

$$\arcsin(-x) = \int_0^{-x} \frac{1}{\sqrt{1-t^2}}\, dt = -\int_0^{x} \frac{1}{\sqrt{1-(-t)^2}} (-1)\, dt$$
$$= -\int_0^{x} \frac{1}{\sqrt{1-u^2}}\, du = -\arcsin x.$$

For the sake of brevity we have used the standard notation for substitution found in calculus courses. To make this substitution rigorous, it would be necessary to use Integration by Substitution for Definite Integrals (Theorem 5.7.4) when $x \in (-1, 1)$, and Integration by Substitution for Type 2 Improper Integrals (Theorem 6.4.12) when $x = 1$ and $x = -1$, where these theorems are applicable because the functions involved are continuous and differentiable as needed; we omit the details.

(2) This part of the lemma follows immediately from the definition of arcsin, the definition of π and Part (1) of this lemma.

(3) This part of the lemma follows immediately from the definition of arcsin and the Fundamental Theorem of Calculus Version I (Theorem 5.6.2).

(4) This part is left to the reader in Exercise 7.3.4.

(5) The continuity of arcsin at all points in $(-1, 1)$ follows from Part (3) of this lemma and Theorem 4.2.4. The definition of $\arcsin 1$ as an improper integral means that

$$\arcsin 1 = \int_0^1 \frac{1}{\sqrt{1-t^2}}\, dt = \lim_{y \to 1^-} \int_0^1 \frac{1}{\sqrt{1-t^2}}\, dt = \lim_{y \to 1^-} \arcsin y.$$

It follows from the one-sided analog of Lemma 3.3.2 that arcsin is continuous at $x = 1$. A similar argument works for $x = -1$, and we omit the details.

(6) We see by Part (3) of this lemma that $\arcsin' x = \frac{1}{\sqrt{1-x^2}} > 0$ for all $x \in (-1,1)$. Using Part (5) of this lemma, we can apply Theorem 4.5.2 (2) to arcsin, and it follows that arcsin is strictly increasing.

(7) This part of the lemma follows from Exercise 4.6.3 (1) together with the previous parts of this lemma. □

We are finally ready to give the definition of the sine and cosine functions. The following definition makes sense because of Lemma 7.3.7 (7), which says that the function $\arcsin\colon [-1,1] \to \left[-\frac{\pi}{2}, \frac{\pi}{2}\right]$ is bijective, and hence has an inverse function.

Definition 7.3.8. The **sine** function is the function $\sin\colon \mathbb{R} \to \mathbb{R}$ defined as follows. Let $f\colon \left[-\frac{\pi}{2}, \frac{3\pi}{2}\right] \to \mathbb{R}$ be defined by

$$f(x) = \begin{cases} \arcsin^{-1}(x), & \text{if } x \in \left[-\frac{\pi}{2}, \frac{\pi}{2}\right] \\ \arcsin^{-1}(\pi - x), & \text{if } x \in \left[\frac{\pi}{2}, \frac{3\pi}{2}\right]. \end{cases}$$

The function f is well-defined, because $\arcsin^{-1}\left(\frac{\pi}{2}\right) = \arcsin^{-1}\left(\pi - \frac{\pi}{2}\right)$. Using the fact that $f\left(\frac{3\pi}{2}\right) = \arcsin^{-1}\left(\pi - \frac{3\pi}{2}\right) = \arcsin^{-1}\left(-\frac{\pi}{2}\right) = f\left(-\frac{\pi}{2}\right)$, we can apply Definition 7.3.3 (2) to the function f, and we let $\sin\colon \mathbb{R} \to \mathbb{R}$ be the periodic extension of f. △

The following theorem states a few of the basic properties of sine.

Theorem 7.3.9.

1. $\sin 0 = 0$, *and* $\sin\left(\frac{\pi}{2}\right) = 1$, *and* $\sin\left(-\frac{\pi}{2}\right) = -1$.
2. sin *is periodic with period* 2π.
3. $\sin(\pi - x) = \sin x$ *for all* $x \in \mathbb{R}$.
4. $0 < \sin x < 1$ *for all* $x \in \left(0, \frac{\pi}{2}\right)$, *and* $-1 < \sin x < 1$ *for all* $x \in \left(-\frac{\pi}{2}, \frac{\pi}{2}\right)$, *and* $-1 \le \sin x \le 1$ *for all* $x \in \mathbb{R}$.

Proof. Left to the reader in Exercise 7.3.5. □

We now turn to the cosine function. In the approach to sine and cosine based upon the unit circle, we define each of these two functions independently, whereas here we define cosine in terms of sine. The following definition makes sense by Theorem 7.3.9 (4).

We use the common notation $\sin^2 x$ as an abbreviation for $[\sin x]^2$, and similarly for cosine.

Definition 7.3.10. The **cosine** function is the function $\cos\colon \mathbb{R} \to \mathbb{R}$ defined by

$$\cos x = \begin{cases} \sqrt{1 - \sin^2 x}, & \text{if } x \in \left[-\frac{\pi}{2} + 2\pi n, \frac{\pi}{2} + 2\pi n\right] \text{ for some } n \in \mathbb{Z} \\ -\sqrt{1 - \sin^2 x}, & \text{if } x \in \left[\frac{\pi}{2} + 2\pi n, \frac{3\pi}{2} + 2\pi n\right] \text{ for some } n \in \mathbb{Z}. \end{cases}$$

△

The following theorem states a few of the basic properties of cosine. This theorem follows immediately from Definition 7.3.10 and Theorem 7.3.9, and we omit the details.

Theorem 7.3.11.

1. $\cos 0 = 1$, *and* $\cos\left(\frac{\pi}{2}\right) = 0$, *and* $\cos\left(-\frac{\pi}{2}\right) = 0$.
2. $\cos$ *is periodic with period* 2π.
3. $0 < \cos x \leq 1$ *for all* $x \in \left(-\frac{\pi}{2}, \frac{\pi}{2}\right)$, *and* $0 \leq \cos x \leq 1$ *for all* $x \in \left[-\frac{\pi}{2}, \frac{\pi}{2}\right]$, *and* $-1 \leq \cos x \leq 1$ *for all* $x \in \mathbb{R}$.
4. $\sin^2 x + \cos^2 x = 1$ *for all* $x \in \mathbb{R}$.

Next, we look at the derivatives of sine and cosine. The idea is to use what we saw about derivatives of inverse function in Section 4.6, though we have a slight problem because the function arcsin is not differentiable at $x = 1$ and $x = -1$. Fortunately, we will be able to resolve this problem by using Theorem 6.3.9.

Theorem 7.3.12.

1. *The function* $\sin$ *is differentiable and* $\sin' = \cos$.
2. *The function* $\cos$ *is differentiable and* $\cos' = -\sin$.

Proof. We prove Part (1), leaving the remaining part to the reader in Exercise 7.3.6.

(1) First, we look at sin restricted to $\left(-\frac{\pi}{2}, \frac{\pi}{2}\right)$. By definition we know that sin restricted to $\left(-\frac{\pi}{2}, \frac{\pi}{2}\right)$ equals $\arcsin^{-1}$. From Lemma 7.3.7 (3) we know that arcsin is differentiable on $(-1,1)$, and $\arcsin' x = \frac{1}{\sqrt{1-x^2}}$ for all $x \in (-1,1)$. Because $\frac{1}{\sqrt{1-x^2}} \neq 0$ for all $x \in (-1,1)$, it follows from Theorem 4.6.4 (3) that sin restricted to $\left(-\frac{\pi}{2}, \frac{\pi}{2}\right)$ is differentiable. Moreover, if $x \in \left(-\frac{\pi}{2}, \frac{\pi}{2}\right)$, Theorem 4.6.4 (4) implies that

$$\sin' x = [\arcsin^{-1}]'(x) = \frac{1}{\arcsin'(\arcsin^{-1}(x))} = \frac{1}{\arcsin'(\sin x)}$$
$$= \frac{1}{\frac{1}{\sqrt{1-\sin^2 x}}} = \sqrt{1 - \sin^2 x} = \cos x.$$

Next, consider arcsin restricted to $(-1,1]$. It follows from various parts of Lemma 7.3.7 that $\arcsin|_{(-1,1]}$ satisfies the hypotheses of Theorem 6.3.9, and we then use that theorem to deduce that sin is differentiable on $\left(-\frac{\pi}{2}, \frac{\pi}{2}\right]$, and that $\sin'\left(\frac{\pi}{2}\right) = 0$, where $\sin'\left(\frac{\pi}{2}\right)$ is a one-sided derivative. It then follows from Theorem 7.3.11 (1) that $\sin'\left(\frac{\pi}{2}\right) = \cos\left(\frac{\pi}{2}\right)$. A similar argument shows that sin is differentiable on $\left[-\frac{\pi}{2}, \frac{\pi}{2}\right]$, and that $\sin'\left(-\frac{\pi}{2}\right) = 0 = \cos\left(-\frac{\pi}{2}\right)$; we omit the details.

Because $\sin'\left(\frac{\pi}{2}\right) = 0$, where $\sin'\left(\frac{\pi}{2}\right)$ is a one-sided derivative, it now follows from Exercise 4.3.7 (2) that sin is differentiable on $\left[-\frac{\pi}{2}, \frac{3\pi}{2}\right]$, and that $\sin' x = -\sin'(\pi - x)$ for all $x \in \left[\frac{\pi}{2}, \frac{3\pi}{2}\right]$. If $x \in \left[\frac{\pi}{2}, \frac{3\pi}{2}\right]$, then $\pi - x \in \left[-\frac{\pi}{2}, \frac{\pi}{2}\right]$, and using the previous paragraph and Theorem 7.3.9 (3), we see that $\sin' x = -\sqrt{1 - \sin^2(\pi - x)} = -\sqrt{1 - \sin^2 x} = \cos x$.

It follows from the above that $\sin'\left(\frac{3\pi}{2}\right) = -\sin'\left(\pi - \left(\frac{3\pi}{2}\right)\right) = -\sin'\left(-\frac{\pi}{2}\right) = 0 = \sin'\left(-\frac{\pi}{2}\right)$. We therefore use Lemma 7.3.4 (2) to deduce that sin is differentiable, which is Part (1) of this theorem, and that $\sin'$ is the periodic extension of the restriction of $\sin'$ to $\left[-\frac{\pi}{2}, \frac{3\pi}{2}\right]$. Because $\sin' = \cos$ on $\left[-\frac{\pi}{2}, \frac{3\pi}{2}\right]$, and because cos is the periodic extension of its restriction $\left[-\frac{\pi}{2}, \frac{3\pi}{2}\right]$, then the uniqueness in Lemma 7.3.2 (2) implies that $\sin' = \cos$ on $\mathbb{R}$. □

We conclude our discussion of the trigonometric functions with the following trigonometric identities, which can be proved geometrically without the use of calculus, but which have a very nice proof using Theorem 7.3.12.

Theorem 7.3.13. *Let $x, y \in \mathbb{R}$.*

1. $\sin(x+y) = \sin x \cos y + \cos x \sin y$.
2. $\sin(x-y) = \sin x \cos y - \cos x \sin y$.
3. $\cos(x+y) = \cos x \cos y - \sin x \sin y$.
4. $\cos(x-y) = \cos x \cos y + \sin x \sin y$.
5. $\sin(-x) = -\sin x$.
6. $\cos(-x) = \cos x$.

Proof. We prove all parts of the theorem together. Let $c \in \mathbb{R}$. Let $f\colon \mathbb{R} \to \mathbb{R}$ be defined by $f(w) = \sin w \cos(c+w) - \sin(c+w)\cos w$ for all $w \in \mathbb{R}$. Using Theorem 7.3.12, the Product Rule (Theorem 4.3.1 (4)) and the Chain Rule (Theorem 4.3.3), we see that

$$f'(w) = \cos w \cos(c+w) - \sin w \sin(c+w) - \cos(c+w)\cos w + \sin(c+w)\sin w = 0$$

for all $w \in \mathbb{R}$. It follows from Lemma 4.4.7 (1) that there is some $C \in \mathbb{R}$ such that $f(w) = C$ for all $w \in \mathbb{R}$. By Theorem 7.3.9 (1) and Theorem 7.3.11 (1) we see that $C = f(0) = \sin 0 \cos(c+0) - \sin(c+0)\cos 0 = -\sin c$. Hence $\sin w \cos(c+w) - \sin(c+w)\cos w = -\sin c$ for all $w \in \mathbb{R}$. If we let $c = -x$ and $w = x$, we deduce that $\sin x \cos 0 - \sin 0 \cos x = -\sin(-x)$, which implies that $\sin x = -\sin(-x)$, and hence $\sin(-x) = -\sin x$. If we let $c = x - y$ and $w = y$, we deduce that $\sin y \cos x - \sin x \cos y = -\sin(x-y)$, which implies that $\sin(x-y) = \sin x \cos y - \cos x \sin y$. Because this last equality holds for all $x, y \in \mathbb{R}$, we then see that $\sin(x+y) = \sin(x-(-y)) = \sin x \cos(-y) - \cos x \sin(-y) = \sin x \cos y + \cos x \sin y$.

By differentiating both sides of the equation $\sin(-x) = -\sin x$, and using the Chain Rule, we obtain $-\cos(-x) = -\cos x$, and hence $\cos(-x) = \cos x$. Finally, by differentiating both sides of the equations $\sin(x+y) = \sin x \cos y + \cos x \sin y$ and $\sin(x-y) = \sin x \cos y - \cos x \sin y$ with respect to x (thinking of y as a constant), we obtain $\cos(x+y) = \cos x \cos y - \sin x \sin y$ and $\cos(x-y) = \cos x \cos y + \sin x \sin y$. □

Reflections

The method of defining the sine and cosine functions given in this section is surprisingly complicated, and it appears to be quite different from the way one sees these functions defined in trigonometry and precalculus courses, where the trigonometric functions are based upon the unit circle. In fact, the difference between the two approaches is more a matter of style (that is, a rigorous treatment in contrast to an informal one) than of substance. If one looks at the proof of Theorem 7.4.3, it will be observed that the integral used to compute the circumference of a circle is the same integral (though with different limits of integration) as the one used to define the arcsine function in Definition 7.3.6. Essentially, the integral used to define arcsine is the rigorous replacement for using the length of an arc on the unit circle in the definition of sine and cosine. The lengthy details in the definition of sine and cosine reflect a number of technical complications that arise along the way: the fact that the integral is improper at $x = 1$, the need to define arcsine with the integral rather than sine, and the fact that we cannot obtain the entire sine function as the inverse of arcsine, and so we need to look at periodic extensions of functions.

The above comments notwithstanding, there is one substantial difference between the way we think of the sine and cosine functions in the present section and in trigonometry and precalculus courses, which is that in those elementary courses we think of sine and cosine as functions of angles, and we look at the relation of sine and cosine to triangles, whereas at present we think of sine and cosine simply as functions $\mathbb{R} \to \mathbb{R}$, with no mention of angles or triangles. If one wants to think of sine and cosine as functions of angles, then one can think of the number x in $\sin x$ and $\cos x$ as representing an angle measured in radians. It is important to stress that measuring angles in degrees has no place in calculus; degrees (and any measure of angles other than radians) are arbitrary, and do not work properly with derivatives and integrals.

In addition to the rigorous definition of the sine and cosine functions using integrals, there is another widely used rigorous definition of these functions in terms of power series; such a definition is given in Exercise 10.4.13. The definition of sine and cosine in terms of power series is quicker and easier than the definition given in the present section, though the ease is only apparent, because it relies upon various facts about power series that need to be proved rigorously; the definition used in the present section, by contrast, relies only upon what we have seen so far in this text. Moreover, the definition of sine and cosine using power series has no direct relation—within the realm of real analysis—to the informal definition of these functions using the unit circle; the power series method simply produces functions that behave the way one would expect sine and cosine to behave, for example that they have the anticipated derivatives. (There is a relation between the series definition of sine and cosine and the unit circle via the complex numbers, but that is beyond the scope of this book.) For these reasons, we have included the more cumbersome definition in the present section, especially for the reader who might not make it to the sections on power series.

Exercises

Exercise 7.3.1. Let $f: \mathbb{R} \to \mathbb{R}$ be a function. Prove that if f is continuous and periodic, then f is bounded.

Exercise 7.3.2. [Used in Exercise 7.3.5.] Let $[a,b] \subseteq \mathbb{R}$ be a non-degenerate closed bounded interval, let $h = b - a$ and let $g: [a,b] \to \mathbb{R}$ be a function. Suppose that $g(a+b-x) = g(x)$ for all $x \in \left[\frac{a+b}{2}, b\right]$. Then $g(b) = g(a+b-b) = g(a)$. By Exercise 7.3.2 (2) there is a unique periodic function $f: \mathbb{R} \to \mathbb{R}$ with period h such that $f|_{[a,b]} = g$. Prove that $f(a+b-x) = f(x)$ for all $x \in \mathbb{R}$. [Use Exercise 2.6.14.]

Exercise 7.3.3. **[Used in Section 7.4 and Exercise 7.4.3.]** Prove that $\pi > 0$.

Exercise 7.3.4. **[Used in Lemma 7.3.7.]** Prove Lemma 7.3.7 (4).

Exercise 7.3.5. **[Used in Theorem 7.3.9.]** Prove Theorem 7.3.9. Use only concepts and results stated prior to that theorem. [Use Exercise 4.6.3 (2) and Exercise 7.3.2.]

Exercise 7.3.6. **[Used in Theorem 7.3.12.]** Prove Theorem 7.3.12 (2).
[Use Exercise 4.4.7.]

Exercise 7.3.7. **[Used in Exercise 7.3.8.]** Let $x \in \mathbb{R}$.

(1) Prove that $\sin\left(\frac{\pi}{2} - x\right) = \cos x$, and that $\cos\left(\frac{\pi}{2} - x\right) = \sin x$.
(2) Prove that $\sin(2x) = 2\sin x \cos x$, and that $\cos(2x) = \cos^2 x - \sin^2 x = 2\cos^2 x - 1 = 1 - 2\sin^2 x$.

Exercise 7.3.8. **[Used in Example 10.4.17.]** Prove that $\sin\left(\frac{\pi}{6}\right) = \frac{1}{2}$. Use only what we have seen in this text; in particular, you may not use facts about triangles that we have not proved. [Use Exercise 7.3.7.]

Exercise 7.3.9. **[Used in Exercise 7.3.10 and Exercise 10.4.13.]** Using Theorem 7.3.12 we see that $\sin'' x = -\sin x$ and $\cos'' x = -\cos x$ for all $x \in \mathbb{R}$. That is, both sin and cos satisfy the differential equation $f''(x) + f(x) = 0$ for all $x \in \mathbb{R}$. (A differential equation is simply an equation that involves derivatives of functions. The reader who is not familiar with differential equations should not be concerned—we will not be using any facts from the study of differential equations in this text.) We will see in Part (3) of this exercise that this differential equation, together with particular values for $f(0)$ and $f'(0)$, completely characterize the sine and cosine functions. Our approach follows [Spi67, Chapter 15]. For this exercise, use only what has been proved in this text; in particular, do not use facts about differential equations that you have seen elsewhere but we have not proved.

Suppose that $f : \mathbb{R} \to \mathbb{R}$ is twice differentiable, and that $f''(x) + f(x) = 0$ for all $x \in \mathbb{R}$.

(1) Prove that if $f(0) = 0$ and $f'(0) = 0$, then $f(x) = 0$ for all $x \in \mathbb{R}$. The idea is to multiply both sides of the equation $f''(x) + f(x) = 0$ by something that makes the left-hand side of the equation into the derivative of something.
(2) Prove that if $f(0) = B$ and $f'(0) = C$ for some $B, C \in \mathbb{R}$, then $f(x) = C\sin x + B\cos x$ for all $x \in \mathbb{R}$.
(3) Prove that sin is the unique function $g : \mathbb{R} \to \mathbb{R}$ that satisfies $g(0) = 0$, and $g'(0) = 1$, and $g''(x) + g(x) = 0$ for all $x \in \mathbb{R}$, and that cos is the unique function $h : \mathbb{R} \to \mathbb{R}$ that satisfies $h(0) = 1$, and $h'(0) = 0$, and $h''(x) + h(x) = 0$ for all $x \in \mathbb{R}$.

Exercise 7.3.10. Use Exercise 7.3.9 (1) to give an alternative proof of Theorem 7.3.13 (1).

7.4 More about π

The definition that we gave for the number π in Definition 7.3.5 was technically convenient from the point of view of defining sine and cosine, but it is not an entirely satisfactory definition, because it does not bear a direct resemblance to the standard approach to π that we all learned when we were young, which is to think of π as the ratio between the circumference and the diameter of a circle. In the present section we will show that π as we have defined it satisfies the expected geometric properties, and so no harm was done defining π as we did. We will also see a proof that π is irrational, which is not related to the issues of circumference and area, but is a fascinating fact about π that we have the tools to prove.

Although we think of π as a number, specifically, the number $3.14159\ldots$, the symbol π also represents an idea that is more important than this numerical value. The number π is usually defined as the ratio of the circumference of a circle to its diameter. How do we know, however, that all circles have the same ratio of the circumference to the diameter? That question is often glossed over in elementary discussions of π, but if this ratio were different in different circles, then the definition of π as this ratio would make no sense. We will now use tools from real analysis to prove that this ratio is the same for all circles, and that this ratio equals π as we defined it in terms of an improper integral in Definition 7.3.5. That fact is the idea that the number π represents. The familiar formula $C = \pi D$, where C is the circumference of a circle and D is its diameter, immediately follows from this fact about π. There is, of course, another famous formula involving circles and π, namely, the formula $A = \pi r^2$, where A is the area of a circle, and r is its radius, and we will prove this formula as well. Our proofs will make use of the treatment of area and arc length in Section 5.9, and of improper integrals in Section 6.4.

To compute the area of a circle, nothing about area beyond what was stated in Section 5.9 will be needed. By contrast, we will need one additional fact about arc length that was not discussed in that section, because when we compute the circumference of a circle via integration, we will need to deal with a function whose derivative is not bounded, and we will therefore need to use Type 2 improper integrals. In particular, we will need an improper integral version of Theorem 5.9.17, which we will see after the following lemma. The reader should first review the concept of rectifiability in Section 5.9.

Lemma 7.4.1. *Let $[a,b] \subseteq \mathbb{R}$ be a non-degenerate closed bounded interval, and let $f\colon [a,b] \to \mathbb{R}$ be a function. Suppose that f is bounded, that f is continuous at b, that $f|_{[a,s]}$ is rectifiable for each $s \in (a,b)$ and that $\lim_{s\to b^-} L_a^s(f)$ exists. Then f is rectifiable and $L_a^b(f) = \lim_{s\to b^-} L_a^s(f)$.*

Proof. Suppose that f is not rectifiable. Then $\mathcal{A}_f$ is not bounded above.

Let $L = \lim_{s\to b^-} L_a^s(f)$. Then there is some $\delta > 0$ such that $s \in (a,b)$ and $b - \delta < s < b$ imply $|L_a^s(f) - L| < 1$. Because f is bounded there is some $M \in \mathbb{R}$ such that $|f(x)| \le M$ for all $x \in [a,b]$. Because $\mathcal{A}_f$ is not bounded above, there is some partition $P = \{x_0, x_1, \ldots, x_n\}$ of $[a,b]$ such that

$$C(f,P) \ge |L| + 1 + \sqrt{\delta^2 + 4M^2}.$$

Let $t \in (x_{n-1}, x_n) \cap (b-\delta, b)$.

Choose some $t \in (x_{n-1}, x_n)$ such that $|t - x_n| < \delta$. Because $x_n = b$, then $b - \delta < t < b$. Then $|L_a^t(f) - L| < 1$. By Lemma 2.3.9 (7) we deduce that $|L_a^t(f)| - |L| < 1$, and hence $|L_a^t(f)| < |L| + 1$. Let $Q = \{x_0, x_1, \ldots, x_{n-1}, t, x_n\}$. Then Q is a refinement of P, and by Lemma 5.9.14 we know that $C(f, Q) \geq C(f, P)$.

Let $R = \{x_0, x_1, \ldots, x_{n-1}, t\}$, which is a partition of $[a, t]$. Because $|f(x)| \leq M$ for all $x \in [a, b]$, it follows that $|f(x_n) - f(t)| \leq 2M$. Then

$$\sqrt{[x_n - t]^2 + [f(x_n) - f(t)]^2} < \sqrt{\delta^2 + 4M^2}.$$

Therefore

$$\begin{aligned} C(f, Q) &= \sum_{i=1}^{n-1} \sqrt{[x_i - x_{i-1}]^2 + [f(x_i) - f(x_{i-1})]^2} \\ &\quad + \sqrt{[t - x_{n-1}]^2 + [f(t) - f(x_{n-1})]^2} + \sqrt{[x_n - t]^2 + [f(x_n) - f(t)]^2} \\ &< C(f|_{[a,t]}, R) + \sqrt{\delta^2 + 4M^2}, \end{aligned}$$

and hence

$$C(f|_{[a,t]}, R) + \sqrt{\delta^2 + 4M^2} > C(f, Q) \geq C(f, P) \geq |L| + 1 + \sqrt{\delta^2 + 4M^2}.$$

It follows that $C(f|_{[a,t]}, R) > |L| + 1$. Because $f|_{[a,t]}$ is rectifiable, then $L_a^t(f) = \operatorname{lub} \mathcal{A}_{f|_{[a,t]}}$, and therefore $L_a^t(f) \geq C(f|_{[a,t]}, R) > |L| + 1$. However, we saw previously that $|L_a^t(f)| < |L| + 1$, which is a contradiction. We conclude that f is rectifiable. Therefore $L_a^b(f)$ is defined.

We now show that $L_a^b(f) = \lim_{s \to b^-} L_a^s(f)$. Let $\varepsilon > 0$. Because f is continuous at b, there is some $\delta_1 > 0$ such that $x \in [a, b]$ and $|x - b| < \delta_1$ imply $|f(x) - f(b)| < \frac{\varepsilon}{6}$.

Because $L_a^b(f) = \operatorname{lub} \mathcal{A}_f$, it follows from Lemma 2.6.5 (1) that there is a partition $Z = \{z_0, z_1, \ldots, z_n\}$ of $[a, b]$ such that $L_a^b(f) - \frac{\varepsilon}{3} < C(f, Z) \leq L_a^b(f)$. Let $\eta = \min\{\delta_1, \frac{\varepsilon}{6}, z_n - z_{n-1}\}$.

Suppose that $u \in [a, b)$ and $b - \eta < u < b$. Then $u \in (z_{n-1}, z_n)$. Let $X = \{z_0, z_1, \ldots, z_{n-1}, u, z_n\}$. Then X is a partition of $[a, b]$. Because $f|_{[a,u]}$ is rectifiable, Lemma 2.6.5 (1) implies that there is a partition Y of $[a, u]$ such that $L_a^u(f) - \frac{\varepsilon}{3} < C(f|_{[a,u]}, Y) \leq L_a^u(f)$. Let $W = X \cup Y$, and let $V = W \cap [a, u]$. Then W is a refinement of Z, and V is a refinement of Y. It follows from Lemma 5.9.14 that $C(f, W) \geq C(f, Z)$ and $C(f|_{[a,u]}, V) \geq C(f|_{[a,u]}, Y)$. Hence, using the definition of $L_a^b(f)$ and $L_a^u(f)$ as least upper bounds, we see that $L_a^b(f) - \frac{\varepsilon}{3} < C(f, W) \leq L_a^b(f)$ and $L_a^u(f) - \frac{\varepsilon}{3} < C(f|_{[a,u]}, V) \leq L_a^u(f)$. Therefore $|L_a^b(f) - C(f, W)| < \frac{\varepsilon}{3}$ and $|L_a^u(f) - C(f|_{[a,u]}, V)| < \frac{\varepsilon}{3}$.

Because $V = W \cap [a, u]$, and because the partition W has no points between u and z_n, it follows that $C(f, W) = C(f|_{[a,u]}, V) + \sqrt{[z_n - u]^2 + [f(z_n) - f(u)]^2}$. Hence $|C(f, W) - C(f|_{[a,u]}, V)| = \sqrt{[z_n - u]^2 + [f(z_n) - f(u)]^2}$. By the choice of u we know that $|u - b| < \eta$. Hence $|u - b| < \frac{\varepsilon}{6}$ and $|u - b| < \delta_1$. From the latter we deduce that $|f(u) - f(b)| < \frac{\varepsilon}{6}$. Observe that $z_n = b$. Then

$$\left|C(f,W) - C(f|_{[a,u]},V)\right| = \sqrt{[z_n - u]^2 + [f(z_n) - f(u)]^2} < \sqrt{\left(\frac{\varepsilon}{6}\right)^2 + \left(\frac{\varepsilon}{6}\right)^2} < \frac{\varepsilon}{3}.$$

Finally, we see that

$$\begin{aligned}
&\left|L_a^u(f) - L_a^b(f)\right| \\
&\quad = \left|L_a^u(f) - C(f|_{[a,u]},V) + C(f|_{[a,u]},V) - C(f,W) + C(f,W) - L_a^b(f)\right| \\
&\quad \le \left|L_a^u(f) - C(f|_{[a,u]},V)\right| \\
&\qquad\qquad + \left|C(f|_{[a,u]},V) - C(f,W)\right| + \left|C(f,W) - L_a^b(f)\right| \\
&\quad < \frac{\varepsilon}{3} + \frac{\varepsilon}{3} + \frac{\varepsilon}{3} = \varepsilon.
\end{aligned}$$

It follows that $\lim_{s \to b^-} L_a^s(f) = L_a^b(f)$. □

Theorem 7.4.2. *Let $[a,b] \subseteq \mathbb{R}$ be a non-degenerate closed bounded interval, and let $f\colon [a,b] \to \mathbb{R}$ be a function. Suppose that f is continuous on $[a,b]$ and continuously differentiable on (a,b), and that f' is bounded on (a,s) for any $s \in (a,b)$. If $\sqrt{1+[f']^2}$ is improperly integrable, then f is rectifiable and*

$$L_a^b(f) = \int_a^b \sqrt{1 + [f'(x)]^2}\, dx,$$

where the integral is improper.

Proof. Suppose that $\sqrt{1+[f']^2}$ is improperly integrable.

If $s \in (a,b)$, then $f|_{[a,s]}$ satisfies the hypotheses of Theorem 5.9.17, and it follows from that theorem that $f|_{[a,s]}$ is rectifiable and

$$L_a^s(f) = \int_a^s \sqrt{1 + [f'(x)]^2}\, dx.$$

Because $\sqrt{1+[f']^2}$ is improperly integrable, then

$$\lim_{s \to b^-} L_a^s(f) = \lim_{s \to b^-} \int_a^s \sqrt{1 + [f']^2}\, dx = \int_a^b \sqrt{1 + [f'(x)]^2}\, dx,$$

and so in particular $\lim_{s \to b^-} L_a^s(f)$ exists.

By Theorem 3.4.6 the function f is bounded. By hypothesis f is continuous at b. We can therefore use Lemma 7.4.1 to deduce that f is rectifiable and

$$L_a^b(f) = \lim_{s \to b^-} L_a^s(f) = \int_a^b \sqrt{1 + [f'(x)]^2}\, dx. \qquad \square$$

We are now ready to look at circumferences of circles. Let $(a,b) \in \mathbb{R}^2$, and let $r \in (0,\infty)$. The circle of radius r centered at (a,b) is given by the equation $(x-a)^2 + (y-b)^2 = r^2$; this equation is just a restatement of the Pythagorean Theorem. The following theorem shows that the ratio of the circumference to the diameter is the same for all circles, and that this ratio equals π, as we defined it in Definition 7.3.5.

Theorem 7.4.3. *Let $(a,b) \in \mathbb{R}^2$, and let $r \in (0,\infty)$. If C is the circumference of the circle of radius r centered at (a,b), and if D is the diameter of this circle, then $\frac{C}{D} = \pi$.*

Proof. Let C be the circumference of the circle of radius r centered at (a,b), and let D be the diameter of this circle. Then $D = 2r$.

To compute C, we first solve the equation $(x-a)^2 + (y-b)^2 = r^2$ for y, and we obtain $y = b \pm \sqrt{r^2 - (x-a)^2}$, which yields two functions, namely, the one with the positive square root, which describes the upper semicircle, and the one with the negative square root, which describes the lower semicircle. Because of the top-to-bottom symmetry of the circle, in order to find the circumference of the circle, it will suffice to find the arc length of the upper semicircle and multiply it by 2. (A proof of this fact makes use of Exercise 5.9.11 and Exercise 5.9.12; the details are left to the reader.) Because of the left-to-right symmetry of the upper semicircle, in order to find the circumference of the circle, it will suffice to find the arc length of the upper right quarter of the circle and multiply it by 4. (A proof of this fact makes use of Exercise 5.9.14 and Exercise 5.9.13; again, the details are left to the reader.)

Let $h\colon [a,a+r] \to \mathbb{R}$ be defined by $h(x) = b + \sqrt{r^2 - (x-a)^2}$ for all $x \in [a,a+r]$. We will show that h is rectifiable, and it will then follow that $C = 4L_a^{a+r}(h)$.

It is left to the reader to verify that h is continuous on $[a,a+r]$ and continuously differentiable on $(a,a+r)$. If $s \in (a,a+r)$, then by Exercise 3.3.2 (2) we know that h' is continuous on $[a,s]$, and Corollary 3.4.6 then implies that h' is bounded on $[a,s]$, and hence on (a,s). It can also be verified that $\sqrt{1+[h'(x)]^2} = \frac{r}{\sqrt{r^2-(x-a)^2}}$ for all $x \in [a,a+r)$; again the details are left to the reader. We would like to apply Theorem 7.4.2 to the function h, and so we need to show that $\sqrt{1+[h']^2}$ is improperly integrable.

Let $g\colon [a,a+r] \to [0,1]$ be defined by $g(x) = \frac{x-a}{r}$ for all $x \in [a,a+r]$, and let $f\colon [0,1) \to \mathbb{R}$ be defined by $f(x) = \frac{r}{\sqrt{1-x^2}}$ for all $x \in [0,1)$. It is left to the reader to verify that f is continuous, that g is strictly increasing and differentiable, that g' is integrable, that $g(a) = 0$ and $g(a+r) = 1$. Moreover, the reader can verify that $f(g(x))g'(x) = \frac{r}{\sqrt{r^2-(x-a)^2}} = \sqrt{1+[h'(x)]^2}$ for all $x \in [a,a+r)$. We now use Integration by Substitution for Improper Integrals (Theorem 6.4.12) to deduce that $\sqrt{1+[h']^2} = (f \circ g) \cdot g'$ is improperly integrable if and only if f is improperly integrable, and if they are improperly integrable then

$$\int_a^{a+r} \sqrt{1+[h'(x)]^2}\,dx = \int_a^{a+r} f(g(x))g'(x)\,dx = \int_0^1 f(x)\,dx = \int_0^1 \frac{r}{\sqrt{1-x^2}}\,dx.$$

By combining Exercise 6.4.5 (2) and Exercise 6.4.7 (2) we see that f is improperly integrable, which implies that $\sqrt{1+[h']^2}$ is improperly integrable and

$$\int_a^{a+r} \sqrt{1+[h'(x)]^2}\,dx = \int_0^1 \frac{r}{\sqrt{1-x^2}}\,dx = r\int_0^1 \frac{1}{\sqrt{1-x^2}}\,dx.$$

We can now apply Theorem 7.4.2 to the function h, and it follows that h is rectifiable and

$$L_a^{a+r}(h) = \int_a^{a+r} \sqrt{1+[h'(x)]^2}\,dx = r\int_0^1 \frac{1}{\sqrt{1-x^2}}\,dx.$$

Therefore

$$C = 4r\int_0^1 \frac{1}{\sqrt{1-x^2}}\,dx,$$

and we conclude that

$$\frac{C}{D} = 2\int_0^1 \frac{1}{\sqrt{1-x^2}}\,dx = \pi,$$

where the last equality holds by Definition 7.3.5. □

We now turn to the area of a circle. In the proof of the following theorem we will use Integration by Substitution and Integration by Parts. For the sake of brevity we will use the standard notation found in calculus courses for these techniques of integration, rather than the proper formulations stated in Theorem 5.7.4, Theorem 5.7.6 and Theorem 6.4.13.

Theorem 7.4.4. *Let $(a,b) \in \mathbb{R}^2$, and let $r \in (0,\infty)$. If A is the area of the circle of radius r centered at (a,b), then $A = \pi r^2$.*

Proof. Let A be the area of the circle of radius r centered at (a,b). As in the proof of Theorem 7.4.3, the circle of radius r centered at (a,b) is given by the equation $(x-a)^2+(y-b)^2 = r^2$, which yields $y = b \pm \sqrt{r^2-(x-a)^2}$. As was the case for the circumference, the symmetry of the circle allows us to find the area of the circle by finding the area of the upper right quarter of the circle and multiplying it by 4. (A proof of this fact makes use of Exercise 5.9.3, Exercise 5.9.4, Exercise 5.9.5 and Exercise 5.9.6; the details are left to the reader.)

Let $h\colon [a,a+r] \to \mathbb{R}$ be defined by $h(x) = b+\sqrt{r^2-(x-a)^2}$ for all $x \in [a,a+r]$. As in the proof of Theorem 7.4.3, the function h is continuous. Hence h is integrable by Theorem 5.4.11. By Theorem 5.9.11 we know that the area of the upper right quarter of the circle is $\int_a^{a+r} \sqrt{r^2-(x-a)^2}\,dx$. Hence

$$A = 4\int_a^{a+r} \sqrt{r^2-(x-a)^2}\,dx.$$

It is left to the reader to verify that the substitution $w = \frac{x-a}{r}$ yields

$$A = 4r^2\int_0^1 \sqrt{1-w^2}\,dw.$$

To make this substitution rigorous, it would be necessary to use Theorem 5.7.4, though we omit the details, which are straightforward given that all functions involved are continuous and differentiable where needed.

We now use Integration by Parts, though we need to make use of the improper integration version of it, which is given in Theorem 6.4.13, because not all functions involved are defined for $w=1$. Using $u=\sqrt{1-w^2}$ and $dv=dw$, we see that

$$\begin{aligned}\int_0^1 \sqrt{1-w^2}\,dw &= \left[w\sqrt{1-w^2}\right]_0^1 + \int_0^1 \frac{w^2}{\sqrt{1-w^2}}\,dw\\ &= 0+\int_0^1 \frac{w^2}{\sqrt{1-w^2}}\,dw = \int_0^1 \frac{1}{\sqrt{1-w^2}}\,dw - \int_0^1 \frac{1-w^2}{\sqrt{1-w^2}}\,dw\\ &= \int_0^1 \frac{1}{\sqrt{1-w^2}}\,dw - \int_0^1 \sqrt{1-w^2}\,dw.\end{aligned}$$

Solving for the original integral we obtain

$$\int_0^1 \sqrt{1-w^2}\,dw = \frac{1}{2}\int_0^1 \frac{1}{\sqrt{1-w^2}}\,dw.$$

Hence the area of a circle of radius r centered at (a,b) is given by

$$\begin{aligned}A = 4r^2\int_0^1 \sqrt{1-w^2}\,dw &= 4r^2\cdot\frac{1}{2}\int_0^1 \frac{1}{\sqrt{1-w^2}}\,dw\\ &= r^2\cdot 2\int_0^1 \frac{1}{\sqrt{1-u^2}}\,du = \pi r^2,\end{aligned}$$

where the last equality holds by Definition 7.3.5. □

Observe that in the proof of Theorem 7.4.4 we never actually evaluated any integrals. All we did was relate the area of a circle to its circumference, when both of these numbers are expressed as integrals.

We conclude this section with a discussion of a very different aspect of π, which is the fact that it is an irrational number. We follow, with added details, the proof in [Jef73, Appendix III], which says about this proof: “The following was set as an example in the Mathematics Preliminary Examination at Cambridge in 1945 by Dame Mary Cartwright, but she has not traced its origin.” Nor, it appears, was the proof published by its creator. (Another proof of the irrationality of π, which is better known, is in [Niv47], though the exposition is quite terse; a variant of that proof, from [Spi67, Chapter 16], is seen in Exercise 7.4.3.) These proofs that π is irrational are all the type of proof where it is hard to get any good intuition about what is going on, and one simply follows the proof step by step to see how it goes.

Our definition of π was given in Definition 7.3.5, but rather than using that definition directly in the proof that π is irrational, we use the following facts about π, sine and cosine:

(1) $\pi > 0$;
(2) the function sin is differentiable and $\sin' = \cos$, and the function cos is differentiable and $\cos' = -\sin$;
(3) $\sin\left(\frac{\pi}{2}\right) = 1$, and $\sin\left(-\frac{\pi}{2}\right) = -1$, and $\cos 0 = 1$, and $\cos\left(\frac{\pi}{2}\right) = 0$, and $\cos\left(-\frac{\pi}{2}\right) = 0$;

(4) $0 \leq \cos x \leq 1$ for all $x \in \left[-\frac{\pi}{2}, \frac{\pi}{2}\right]$.

Property (1) is given in Exercise 7.3.3; Property (2) is given in Theorem 7.3.12; and Properties (3) and (4) are given in Theorem 7.3.9 and Theorem 7.3.11. The reader who has not read Section 7.3 should nonetheless be familiar, at least informally, with these properties of π, sine and cosine, and can therefore read the proof of the following theorem without having to go back and read any of Section 7.3. This proof also uses the notion of factorials, which we defined in Example 2.5.12.

Theorem 7.4.5. *The number π is irrational.*

Proof. Let $g\colon [-1,1] \to \mathbb{R}$ be defined by $g(x) = \cos\left(\frac{\pi x}{2}\right)$ for all $x \in [-1,1]$, and for each $n \in \mathbb{N} \cup \{0\}$ let $f_n\colon [-1,1] \to \mathbb{R}$ be defined by $f_n(x) = \frac{(1-x^2)^n}{n!}$ for all $x \in [-1,1]$. It is left to the reader to verify that these functions satisfy the following properties; this verification makes use of Properties (2), (3) and (4) listed prior to this proof.

(a) f is twice differentiable.
(b) $f_n(1) = 0$, and $f_n(-1) = 0$, and $f_n(0) \neq 0$, for all $n \in \mathbb{N} \cup \{0\}$ such that $n \geq 1$.
(c) $f_n'(1) = 0$, and $f_n'(-1) = 0$, for all $n \in \mathbb{N} \cup \{0\}$ such that $n \geq 2$.
(d) $f_0(x) = k$ for all $x \in [-1,1]$, for some $k \in \mathbb{Z}$.
(e) $f_1''(x) = r$ for all $x \in [-1,1]$, for some $r \in \mathbb{Z}$.
(f) $f_n'' = p_n f_{n-1} + q_n f_{n-2}$ for some $p_n, q_n \in \mathbb{Z}$, for all $n \in \mathbb{N}$ such that $n \geq 2$.
(g) $0 \leq f_n(x) \leq \frac{1}{n!}$ for all $x \in [-1,1]$, for all $n \in \mathbb{N} \cup \{0\}$.
(h) g is twice differentiable.
(i) $g(1) = 0$, and $g(-1) = 0$, and $g(0) \neq 0$, and $g'(1) - g'(-1) = \frac{\pi}{c}$ for some $c \in \mathbb{Z} - \{0\}$.
(j) $g'' = \frac{\pi^2}{cd} g$ for some $d \in \mathbb{Z} - \{0\}$.
(k) $0 \leq g(x) \leq 1$ for $x \in [-1,1]$.

For the rest of this proof we will use only the above properties of g and f_n, and not the definitions of these functions.

By Property (j) we see that $\left[\frac{cd}{\pi^2} g'\right]' = \frac{cd}{\pi^2} g'' = g$, and hence $\frac{cd}{\pi^2} g'$ is an antiderivative of g.

Suppose that π is rational. Then by Property (1) listed prior to this proof and Lemma 2.4.12 (2) there are $a, b \in \mathbb{N}$ such that $\pi = \frac{a}{b}$.

Let $n \in \mathbb{N}$. By Properties (a) and (h) we see that each of f_n and g is differentiable, and hence by Theorem 4.2.4 we know that each of these functions is continuous. By Corollary 3.3.6 it follows that $f_n g$ is continuous, and hence by Theorem 5.4.11 we know that $f_n g$ is integrable. Let

$$J_n = a^{2n+1} \int_{-1}^{1} f_n(x) g(x)\, dx.$$

We now show that $J_n \in \mathbb{Z}$ for all $n \in \mathbb{N} \cup \{0\}$. The proof is by induction on n, using the variant of induction stated in Theorem 2.5.4; we start the induction at $n = 0$ rather than $n = 1$, but that is not a problem. In this proof by induction we will repeatedly use the fact that $\pi = \frac{a}{b}$, and what we saw above about an antiderivative of g. We will also repeatedly use Integration by Parts for Definite Integrals (Theorem 5.7.6), though for the sake of brevity we will use the standard notation for Integration by Parts found in calculus courses. It is left to the reader to use Properties (a), (e), (f), (h) and (j) to verify that the hypotheses of Integration by Parts for Definite Integrals hold as needed.

First, using Properties (d) and (i), we see that

$$\begin{aligned} J_0 &= a^{2\cdot 0+1}\int_{-1}^{1} f_0(x)g(x)\,dx = ak\int_{-1}^{1} g(x)\,dx = ak\left[\frac{cd}{\pi^2}g'\right]_{-1}^{1} \\ &= ak\frac{cd}{\pi^2}[g'(1)-g'(-1)] = ak\frac{cd}{\pi^2}\frac{\pi}{c} = kdb. \end{aligned}$$

Hence $J_0 \in \mathbb{Z}$.

Second, this time using Properties (b), (i) and (e), we see that

$$\begin{aligned} J_1 &= a^{2\cdot 1+1}\int_{-1}^{1} f_1(x)g(x)\,dx = a^3\left\{\left[f_1(x)\frac{cd}{\pi^2}g'(x)\right]_{-1}^{1} - \int_{-1}^{1} f_1'(x)\frac{cd}{\pi^2}g'(x)\,dx\right\} \\ &= a^3\left\{0 - \frac{cd}{\pi^2}\int_{-1}^{1} f_1'(x)g'(x)\,dx\right\} \\ &= -a^3\frac{cd}{\pi^2}\left\{[f_1'(x)g(x)]_{-1}^{1} - \int_{-1}^{1} f_1''(x)g(x)\,dx\right\} \\ &= -a^3\frac{cd}{\pi^2}\left\{0 - r\int_{-1}^{1} g(x)\,dx\right\} = a^3\frac{cd}{\pi^2}r\left[\frac{cd}{\pi^2}g'(x)\right]_{-1}^{1} \\ &= a^3\frac{c^2d^2}{\pi^4}r\,[g'(1)-g'(-1)] = a^3\frac{c^2d^2}{\pi^4}r\frac{\pi}{c} = cd^2rb^3. \end{aligned}$$

Hence $J_1 \in \mathbb{Z}$.

Now let $n \in \mathbb{N}\cup\{0\}$, and suppose that $n \geq 2$. Suppose further that $J_k \in \mathbb{Z}$ for all $k \in \{0,\ldots,n-1\}$. Using Properties (b), (c) and (f) we see that

$$\begin{aligned} J_n &= a^{2n+1}\int_{-1}^{1} f_n(x)g(x)\,dx = a^{2n+1}\left\{\left[f_n(x)\frac{cd}{\pi^2}g'(x)\right]_{-1}^{1} - \int_{-1}^{1} f_n'(x)\frac{cd}{\pi^2}g'(x)\,dx\right\} \\ &= a^{2n+1}\left\{0 - \frac{cd}{\pi^2}\int_{-1}^{1} f_n'(x)g'(x)\,dx\right\} \\ &= -a^{2n+1}\frac{cd}{\pi^2}\left\{[f_n'(x)g(x)]_{-1}^{1} - \int_{-1}^{1} f_n''(x)g(x)\,dx\right\} \\ &= -a^{2n+1}\frac{cd}{\pi^2}\left\{0 - \int_{-1}^{1}[p_nf_{n-1}(x)+q_nf_{n-2}(x)]g(x)\,dx\right\} \\ &= a^2\frac{cd}{\pi^2}\left\{p_na^{2n-1}\int_{-1}^{1} f_{n-1}(x)g(x)\,dx + q_na^2a^{2n-3}\int_{-1}^{1} f_{n-2}(x)g(x)\,dx\right\} \\ &= b^2cd\left\{p_nJ_{n-1}+q_na^2J_{n-2}\right\}. \end{aligned}$$

By the inductive hypothesis we know that $J_{n-1} \in \mathbb{Z}$ and $J_{n-2} \in \mathbb{Z}$, and it follows that $J_n \in \mathbb{Z}$, which completes the inductive step. Hence $J_n \in \mathbb{Z}$ for all $n \in \mathbb{N} \cup \{0\}$.

Let $n \in \mathbb{N} \cup \{0\}$. By Properties (g) and (k) we see that $0 \leq a^{2n+1} f_n(x)g(x) \leq \frac{a^{2n+1}}{n!}$ for all $x \in [-1,1]$, and by Properties (b) and (i) we see that $a^{2n+1} f_n(0)g(0) > 0$. Using Theorem 5.3.2 (3) and Exercise 5.5.7 we deduce that $0 < J_n \leq \frac{2a^{2n+1}}{n!}$.

Next, we claim that there is some $k \in \mathbb{N}$ such that $\frac{a^{2k}}{k!} < \frac{1}{2a}$. This claim can be proved most easily using sequences and series, which will be discussed in detail in Chapters 8 and 9. More specifically, let $c = a^2$. The final remark in Example 9.5.2 (2) implies that $\lim_{n\to\infty} \frac{a^{2n}}{n!} = \lim_{n\to\infty} \frac{c^n}{n!} = 0$, which means intuitively that $\frac{a^{2n}}{n!}$ can be made as small as desired if n is made sufficiently large, and that in turn can be seen to imply the claim by using the definition of the convergence of sequences given in Section 8.2. Although Example 9.5.2 (2) comes later in the text, nothing in the present section is used in that example, and there is no fear of circular reasoning if we use that example in the present proof. For the reader who will not read Section 9.5, or who cannot wait until then to see all of the details of the present proof, an alternative (and ad hoc) proof of the claim that does not use sequences is found in Exercise 7.4.2, where again we think of c as a^2.

It follows from the above claim that $\frac{2a^{2k+1}}{k!} < 1$. We saw above that $0 < J_k \leq \frac{2a^{2k+1}}{k!}$, and therefore $0 < J_n < 1$. On the other hand, we saw that J_k is an integer, which is a contradiction to Theorem 2.4.10 (2). Hence π is irrational. □

It should be mentioned that π is not only an irrational number, but it is in fact a "transcendental number," which means that it is not the root of any polynomial with rational coefficients. By comparison, the number $\sqrt{2}$ is also irrational, but it is not transcendental, because it is a root of the equation $x^2 - 2 = 0$. A real number that is the root of a polynomial equation with rational coefficients is called an "algebraic number." See [Ste04, Section 24.3] for a proof that π is transcendental.

Finally, we note that there is a glaring omission in our treatment of π in this section, which is that we have not provided a method for computing the first few digits of the decimal expansion of π. One could in principle try to approximate π by drawing a circle very carefully, measuring the circumference and diameter, and dividing the former by the latter; of course, such a method is not very satisfying. There are many methods for computing the first few (or many) digits of the decimal expansion of π; one such method, which is based upon series, will be seen in Example 10.4.17. For more information about computing the digits of the decimal expansion of π, and about π in general, see [AH01] or [EL04]; see [BBB04] for historical sources about π.

Reflections

The material in this section is the least central to the study of real analysis of any section in this book. Most introductory courses in real analysis do not give a thorough discussion of the trigonometric functions, and in particular do not discuss the number π in detail. However, given that we defined the trigonometric functions in Section 7.3, and therefore had to define π in the process, it would be a pity not to discuss the interesting and familiar facts about π discussed in the present section, especially because they are proved using some results from real analysis that were seen previously in this text.

That the proof of the irrationality of π is tricky is not surprising, but it is somewhat surprising that the proofs of the familiar formulas for the circumference and the area of a circle take as much effort as they do. In fact, the only reason it is surprising is that these familiar formulas are all too familiar—we learn them at a very early age, so early that we accept them as true simply because we were told so by our teachers. In particular, the reason that the proofs of the formulas for the circumference and the area of a circle are trickier than might at first be expected is because of the appearance of improper integrals in the definition of π and in the formula for the arc length of a circle.

Exercises

Exercise 7.4.1. In Section 7.3 we defined the number π, and the arcsine function, using improper integrals. The purpose of this exercise is to show that it is possible to define π and arcsine using integrals but avoiding improper integrals. Hence the sine and cosine functions can be defined without using improper integrals. This alternative approach is slightly longer than the approach we used in Section 7.3, and hence we used the latter approach in the text for the sake of brevity.

Let $A\colon (-1,1) \to \mathbb{R}$ be defined by

$$A(x) = \int_0^x \frac{1}{\sqrt{1-t^2}}\,dt$$

for all $x \in (-1,1)$. Observe that the function A is the restriction of the arcsine function as defined in Definition 7.3.6 to $(-1,1)$. Because the domain of A does not include 1 and -1, then A is defined without improper integration. Using the same ideas as in the proof of Lemma 7.3.7, it can be verified that $A(-x) = -A(x)$ for all $x \in (-1,1)$, and that A is differentiable and strictly increasing; we omit the details.

(1) Prove that the function A is bounded. Do not use the arcsine function in your proof, because we are going to provide an alternative definition of that function subsequently in this exercise.

(2) By Part (1) of this exercise, the Least Upper Bound Property and the Greatest Lower Bound Property we see that $A((-1,1))$ has a least upper bound and a greatest lower bound. Prove that $\operatorname{glb} A((-1,1)) = -\operatorname{lub} A((-1,1))$. [Use Exercise 2.6.5.]

(3) We now let π be defined by $\pi = 2\operatorname{lub} A((-1,1))$, where this least upper bound exists as observed in Part (2). Prove that this definition of π is equivalent to Definition 7.3.5. [Use Exercise 4.5.9 and Exercise 4.5.10 (1).]

(4) Using the previous parts of this exercise we now let the **arcsine** function be the function $\arcsin\colon [-1,1] \to \mathbb{R}$ defined by

$$\arcsin x = \begin{cases} -\frac{\pi}{2}, & \text{if } x = -1 \\ A(x), & \text{if } x \in (-1, 1) \\ \frac{\pi}{2}, & \text{if } x = 1. \end{cases}$$

Prove that the function arcsin as defined above is continuous. (The other standard properties of arcsine can also be proved using this definition.) [Use Exercise 4.6.5 (1).]

Exercise 7.4.2. [Used in Theorem 7.4.5 and Exercise 7.4.3.] Let $c, p \in (0, \infty)$. We prove that there is some $n \in \mathbb{N}$ such that $\frac{c^n}{n!} < p$.

(1) By using Corollary 2.6.8 (1) there is some $k \in \mathbb{Z}$ such that

$$\frac{1}{\ln 2} \cdot \ln\left(\frac{c^{2c}}{(2c)!p}\right) < k.$$

By choosing a larger value of k if necessary, we may suppose that $k \in \mathbb{N}$. Prove that

$$\frac{c^{2c}}{(2c)!} < 2^k p.$$

(2) Let $n = 2c + k$. Prove that $\frac{c^n}{n!} < p$.

Exercise 7.4.3. [Used in Section 7.4.] In this exercise we give an alternative proof, from [Spi67, Chapter 16], that π is irrational. As in the proof of Theorem 7.4.5, here too we need the definition of $n!$, given in Example 2.5.12, and we need only a few facts about π, sin and cos, which this time are:

(1) $\pi > 0$;
(2) the function sin is differentiable and $\sin' = \cos$, and the function cos is differentiable and $\cos' = -\sin$;
(3) $\sin 0 = 0$, and $\sin \pi = 0$, and $\cos 0 = 1$, and $\cos \pi = -1$;
(4) $0 < \sin x \leq 1$ for $x \in (0, \pi)$.

Property (1) is given in Exercise 7.3.3; Property (2) is given in Theorem 7.3.12; and Properties (3) and (4) follow from Theorem 7.3.9 and Theorem 7.3.11.

Parts (1), (2) and (3) of this exercise are preliminaries; the actual proof that π is irrational starts after that. Let $n \in \mathbb{N}$.

(1) Let $p \colon \mathbb{R} \to \mathbb{R}$ be a polynomial function of the form

$$p(x) = c_n x^n + c_{n+1} x^{n+1} + c_{n+2} x^{n+2} + \cdots + c_{2n} x^{2n}$$

for all $x \in \mathbb{R}$, where $c_n, c_{n+1}, \ldots, c_{2n} \in \mathbb{Z}$. Prove the following properties.
(i) $p^{(i)}(0) = 0$ for all $i \in \{0, \ldots, n-1\}$.
(ii) $p^{(i)}(0)$ is an integer that is divisible by $n!$ for all $i \in \{n, n+1, \ldots, 2n+1\}$.
(iii) If $p(1-x) = p(x)$ for all $x \in \mathbb{R}$, then $p^{(i)}(1) = 0$ for all $i \in \{0, \ldots, n-1\}$ and $p^{(i)}(1)$ is an integer that is divisible by $n!$ for all $i \in \{n, n+1, \ldots, 2n+1\}$.

(2) Let $f\colon \mathbb{R} \to \mathbb{R}$ be defined by

$$f(x) = \frac{x^n(1-x)^n}{n!}$$

for all $x \in \mathbb{R}$. Prove that f satisfies the following properties.
  (a) $f(0) = 0$, and $f(\frac{1}{2}) > 0$, and $f(1) = 0$.
  (b) f is infinitely differentiable.
  (c) $f^{(2n+2)}$ is the zero function.
  (d) $f^{(i)}(0)$ and $f^{(i)}(1)$ are integers for all $i \in \mathbb{N} \cup \{0\}$.
  (e) $0 < f(x) < \frac{1}{n!}$ for all $x \in (0,1)$.

(3) Prove that there is a function $g\colon \mathbb{R} \to \mathbb{R}$ that satisfies the following properties.
  (f) g is twice differentiable.
  (g) $g'' = -\pi^2 g$.
  (h) $g(0) = 0$, and $g(1) = 0$, and $g'(0) = \pi$, and $g'(1) = -\pi$.
  (i) $0 \le g(x) \le 1$ for $x \in [0,1]$.

(4) Suppose that π is rational. Then by Corollary 2.4.14 we know that π^2 is rational. Because $\pi > 0$, then $\pi^2 > 0$. By Lemma 2.4.12 (2) we know that $\pi^2 = \frac{a}{b}$ for some $a, b \in \mathbb{N}$. Let $h\colon \mathbb{R} \to \mathbb{R}$ be defined by

$$h = \sum_{i=0}^{n} (-1)^i a^{n-i} b^i \left[f^{(2i+1)} g - f^{(2i)} g' \right].$$

Prove that $h' = \pi^2 a^n f g$. It follows that h is an antiderivative of $\pi^2 a^n f g$.

(5) Prove that $\pi a^n f g$ satisfies the hypotheses of the Fundamental Theorem of Calculus Version II (Theorem 5.6.4), and then use that theorem to prove that

$$\int_0^1 \pi a^n f(x) g(x)\, dx = \sum_{i=0}^{n} (-1)^i a^{n-i} b^i \left[f^{(2i)}(1) + f^{(2i)}(0) \right].$$

By Property (d) it follows that $\int_0^1 \pi a^n f(x) g(x)\, dx$ is an integer.

(6) It follows from Properties (e) and (i) that $0 \le \pi a^n f(x) g(x) < \frac{\pi a^n}{n!}$ for all $x \in [0,1]$. By an argument similar to one used in the proof of Theorem 7.4.5, and also found in Exercise 7.4.2, there is some $k \in \mathbb{N}$ such that $\frac{a^k}{k!} < \frac{1}{\pi}$. Prove that $0 < \int_0^1 \pi a^k f(x) g(x)\, dx < 1$. By Part (5) of this exercise, where the choice of n was arbitrary, we know that $\int_0^1 \pi a^k f(x) g(x)\, dx$ is an integer, which is a contradiction to Theorem 2.4.10 (2). Hence π is irrational.
[Use Exercise 5.5.7.]

7.5 Historical Remarks

The material in this chapter starts with logarithmic and exponential functions, then trigonometric functions, and ends with an additional look at the number π. This order allows us to proceed from the less complicated material to the more complicated, though historically this order is backwards. The number π, or at least approximations to it, are found in many ancient cultures, for the understandable reason that it was important to find the circumference and area of a circle; trigonometric concepts (originally viewed geometrically, prior to the development of the function concept) also have their roots in the ancient world; logarithms are a much later invention. We will discuss the history of each of these three concepts separately, in the proper historical order.

The Number π

We are familiar with the number π from two famous formulas, namely, the formulas $C = \pi D$ and $A = \pi r^2$. It was recognized very early historically that the ratio $\frac{C}{D}$ is the same in all circles; it was also known early that the ratio $\frac{A}{r^2}$ is the same in all circles. However, it was apparently less widely known that these two ratios were the same; Euclid (c. 325–c. 265 BCE), for example, did not state this equality in the *Elements*. The ancient Babylonians and Chinese, on the other hand, knew that the area of a circle is half the circumference times half the diameter, a fact that in essence says that the two ratios are equal. Archimedes (287–212 BCE), in *The Measurement of the Circle* of around 250 BCE, proved the equivalent fact that the area of a circle is equal to the area of a right triangle with base equal to the circumference and height equal to the radius.

Mathematics originated in practical considerations, and before there was theoretical discussion of the meaning and various uses of the number π, there were approximations of its value via the ratio $\frac{C}{D}$. A number of ancient civilizations used the value of 3 as an approximation to π. For example, in I Kings 7:23 it states "And he made the molten sea of ten cubits from brim to brim, round in compass, and the height thereof was five cubits; and a line of thirty cubits did compass it round about" (translation from [MM]), which would imply that π is 3. However, much better approximations were obtained quite early in the ancient world.

The ancient Babylonians, in the period 1900–1600 BCE, had the approximation $\frac{25}{8}$ for π, and the ancient Egyptians, in the Rhind Papyrus of around 1850 BCE or earlier, had the approximation $(\frac{16}{9})^2$. Both of these approximations for π are within 0.02 of the correct value. In ancient India, from 600 BCE or earlier, they had the approximation $(\frac{9785}{5568})^2$ for π. The approximation $\sqrt{10}$ was apparently known in ancient India no later than 150 BCE. The approximation $\sqrt{2}+\sqrt{3}$ is attributed to Plato (427–347 BCE).

Archimedes, in *The Measurement of the Circle*, gave the first systematic method for finding approximations fof π, as opposed to previous approximations, which were arrived at by experimental methods. Archimedes' method was to find the perimeters of regular polygons that are circumscribed about a circle and regular polygons that are inscribed in it, providing upper bounds and lower bounds for the circumference of the circle. He used regular polygons with up to 96 sides, leading to π being between $3\frac{10}{71}$ and $3\frac{1}{7}$, the latter yielding the approximation $\frac{22}{7}$ (which is so commonly used that it is sometimes mistakenly thought to equal π). Archimedes' method, with some variation, was the main method (at least in the West) for approximating π for almost two millennia, and after Archimedes it was used with polygons with ever more sides to obtain ever greater accuracy.

In 150, Claudius Ptolemy (c. 85–c. 165) had the approximation $\frac{377}{120}$, which is better than $\frac{22}{7}$. In ancient China various approximations to π were known. For example, the approximation $\frac{7854}{2500}$, which is better than $\frac{22}{7}$, was given by Liu Hui (c. 220–c. 280) in around 263, using inscribed polygons, though similarly to Archimedes providing both upper and lower bounds. Liu Hui was aided by the fact that he had the decimal system for writing numbers, which was not available in ancient Greece. An even better approximation to π was found by Zu Chongzhi (429–500), improving the accuracy of Liu Hui by two decimal places; this approximation appears to have been the best approximation for the next 800 years. In India, Aryabhata (476–550), also known as Aryabhata I, had the approximation $\frac{62832}{20000}$ in *Aryabhatiya* of 499.

Leonardo of Pisa (1170–1250), also known as Fibonacci, without making use of Archimedes' work, used a 96-sided polygon to obtain the approximation $\frac{864}{275}$. Not much progress was made in approximating π in medieval Europe, though better results were obtained elsewhere. Ghiyath al-Din Jamshid Mas'ud al-Khashi (1390–1450), around 1430, used a polygon with $3 \cdot 2^{28}$ sides to arrive at an approximation that is good to 16 decimal places. Additionally, a number of series for π were developed in India by the 15th century, at least 100 years earlier than such series were found in Europe. Among these series were $\frac{\pi}{4} = 1 - \frac{1}{3} + \frac{1}{5} - \frac{1}{7} + \cdots$, with the correction factor $\frac{\pi}{4} = 1 - \frac{1}{3} + \frac{1}{5} - \frac{1}{7} + \cdots \mp \frac{1}{p-1} \pm \frac{p}{2(p^2+1)}$ that improves the accuracy, and $\pi = \sqrt{12}(1 - \frac{1}{3 \cdot 3^1} + \frac{1}{5 \cdot 3^2} - \frac{1}{7 \cdot 3^3} + \cdots)$.

A new method for computing π was introduced in 1579 by François Viète (1540–1603), who combined Archimedes' method with trigonometry, and used a polygon with $3 \cdot 2^{17}$ sides, to arrive at an approximation that is good to 9 decimal places. Better results were subsequently obtained with similar methods using polygons with more sides, eventually obtaining 39 decimal place accuracy in 1630. In the mid-17th century other variants of the method of inscribed and circumscribed polygons were made, with more complicated procedures for going from one polygon to the next; such approaches were taken, for example, by James Gregory (1638–1675) and René Descartes (1596–1650).

A revolution in computing π took place in mid-17th-century Europe due to the introduction of methods based upon series and calculus, rather than approximating circles with polygons. First, there were infinite products for π by Viète in 1593 and John Wallis (1616–1703) in 1655, and a continued fraction formula for π by William Brouncker (1620–1684) in 1658. Series for π were found simultaneously with the advent of calculus. In 1665–1666 Isaac Newton (1643–1727) found the series $\pi = \frac{3\sqrt{3}}{4} + 24(\frac{1}{12} - \frac{1}{5 \cdot 2^5} - \frac{1}{28 \cdot 2^7} - \frac{1}{72 \cdot 2^9} - \cdots)$. Gottfried von Leibniz (1646–1716), in late 1673 or early 1674, found the power series for the arctangent function and used it to obtain the series $\pi = 4 \arctan 1 = 4(1 - \frac{1}{3} + \frac{1}{5} - \frac{1}{7} + \cdots)$, which was known earlier in India.

Leibniz' series for π converges very slowly, but a much faster converging series was found by John Machin (1680–1752) in 1706 using the formula $\pi = 16 \arctan \frac{1}{5} - 4 \arctan \frac{1}{239}$ together with the power series for arctangent. Using this method, Machin obtained the first 100 decimal places of π. Variants of this arctangent method were used until around 1970, first by hand, and eventually by computer.

Although the symbol "π" is a Greek letter, the use of this symbol to denote the ratio of the circumference to the diameter of a circle is due not to ancient Greek mathematicians, but rather to William Jones (1675–1749) in 1706. However, given that the mathematical ideas about π in Jones' book are due to Machin, as Jones himself said, it is possible that the symbol π is due to Machin as well. The use of π as we now use it did not catch on from Jones' work, and other symbols were used at the time, but Leonhard Euler (1707–1783) used π in a paper in 1736, and again in the influential textbook *Introductio in analysin infinitorum* of 1748, and the symbol became widespread thereafter.

Euler provided a number of nice results involving π. For example, he proved in 1736 that $\sum_{n=1}^{\infty} \frac{1}{n^2} = \frac{\pi^2}{6}$, a result that had eluded great mathematicians such as Leibniz, Jakob Bernoulli (1654–1705) and Johann Bernoulli (1667–1748). In *Introductio in analysin infinitorum* Euler also proved the formula $e^{ix} = \cos x + i \sin x$ in, which is very important in complex analysis, and from which the famous formula $e^{i\pi} + 1 = 0$ is deduced, showing a relation between the two important numbers π and e.

The irrationality of π was suspected in the 15th century, but it was first proved only in 1766 by Johann Lambert (1728–1777). The transcendence of π was proved by Ferdinand von Lindemann (1852–1939) in 1882, a fact that implied that the ancient Greek quest to square the circle with straightedge and compass was doomed to fail.

A number of algorithms for computing π, by series approximations and other methods, were developed prior to the computer era. Around 1800 Carl Friedrich Gauss (1777–1855) invented a particularly fast algorithm for computing π based upon the arithmetic-geometric mean. The best approximation of π prior to the computer era appears to have been 1120 decimal places. With the aid of computers, the world record today is billions of decimal places. At first computers used older algorithms, such as Machin's arctangent method, but eventually newer methods were developed, both algorithms specific to computing the decimal expansion of π, including a rediscovery of Gauss' algorithm, as well as a faster method of multiplying numbers, known as Fast Fourier Transform multiplication, which speeded up any algorithm that involved multiplication, including those for computing π.

Trigonometry

Whereas today we think of trigonometry in terms of the six trigonometric functions, the subject started out rather differently. For example, the ancient Greeks, who originally did not have angle measure, did not use these functions. Nonetheless, Propositions 12–13 in Book II of Euclid's *Elements* are equivalent to what we call the Law of Cosines, and as such represent trigonometric ideas.

The need for something analogous to the trigonometric functions, and for the associated tables of values needed for calculations in an era without computing technology, arose in the study of astronomy because of the use of angles as coordinates for heavenly bodies, an idea due to the ancient Babylonians. Hipparchus of Nicaea (190–120 BCE), also known as Hipparchus of Rhodes, was among the first ancient

Greeks to use the Babylonian system of measuring angles by dividing the circle into 360°, and he was probably the first person to compile a trigonometric table. This table, rather than giving values of a function as we think of it today, gave the lengths of chords subtended by arcs in a given circle (in a unit circle the length of the arc represents the angle via radian measure, and the length of the chord is twice the sine of half the angle subtended by the arc).

The most important work on trigonometry in the ancient world was in *Mathematiki Syntaxis* of around 150 by Ptolemy. Because of its importance this work was referred to as *Megisti Syntaxis*, which means "Greatest Compilation," and that title became *al-magisti* in Arabic, which was then Latinized into *Almagest*, the common name for the work. The *Almagest* is a treatise on astronomy, stating what we now call the Ptolemaic system, but it includes trigonometry for use in astronomical calculations. Ptolemy had a trigonometric table (also consisting of lengths of chords in a circle), and had a theorem that included as a special case the equivalent of our formula for $\sin(x-y)$. He also had the equivalent of our half-angle formula for sine, and the equivalent of the Law of Sines. Ptolemy constructed his trigonometric table using his geometric equivalents of such formulas.

The sine function was invented in India, where, as in ancient Greece, trigonometric ideas were used for astronomical calculations. Indian trigonometry appears to have been influenced by ancient Greek work on this subject; later European trigonometry would, in turn, be influenced by the development of Indian trigonometry beyond what was imported from Greece.

In the Siddhantas, including the Paitamaha Siddhanta of the 5th century and the later Surya Siddhanta, rather than using tables of the lengths of chords subtended by arcs (corresponding to angles), there were more convenient tables of lengths of half the chords of double the angles; for a unit circle such lengths would be the same as the values of the sine function we use today. Other trigonometric functions, such as cosine, arcsine, tangent and secant, were also considered. Our word "sine" derives from the Sanskrit word "jiva," which was rendered "jiba" in Arabic; Robert of Chester (who translated mathematics from Arabic into Latin in the 12th century) thought the word was "jaib," which means bay or inlet in Arabic, and so he translated it as "sinus" in Latin, which means inlet, bosom or any welcoming fold (the fact that Arabic is sometimes written without vowels might have been the source of this error).

Aryabhata, in the *Aryabhatiya* of 499, had a sine table. Varahamihira (505–587), in the *Pancasiddhantika* of 575, had a more accurate sine table than Aryabhata, as well as a cosine table, and some identities relating these two functions. Brahmagupta (598–670), in the *Brahmasphuta Siddhanta* of 628, had ideas that implied the Law of Sines. Bhaskara II (1114–1185), also known as Bhaskaracharya, showed the equivalent of the fact that the derivative of sine is cosine, as well as the equivalent of our formulas for $\sin(x+y)$ and $\sin(x-y)$.

The earliest equivalent of a tangent table appeared in China, in the *Ta yen li* of Yi Xing (683–727), who was influenced by Indian astronomy and trigonometry.

Arab mathematics was influenced by both the ancient Greeks and the Indians. In trigonometry the Arab mathematicians at first used both chord tables as in the *Almagest* and sine tables as in India, though they eventually settled on sine tables. All six trigonometric functions, viewed geometrically, were used in the Arab world by the 9th century.

Similarly to the spread of the Indian place-value system, Indian trigonometry came to Europe via the Arab world, for example through the work on astronomy and trigonometry of Abu Abdallah Mohammad ibn Jabir al-Battani (c. 850–929), also known as Albategnius, whose *Kitab al-Zij* was translated into Latin. Mohammad Abu'l-Wafa Al-Buzjani (940–998) had all six trigonometric functions and relations between them, double-angle and half-angle formulas, and a sine table with angles at $0.25°$ intervals and with accuracy equivalent to eight decimal places. The trigonometric functions in India were defined in circles of arbitrary size, whereas in the Arab world they were generally defined using the unit circle, as we now do. The *Treatise on the Quadrilateral* by Nasir al-Din al-Tusi (1201–1274) was the first treatise on trigonometry (planar and spherical) in its own right, independent of astronomy.

The first treatise on trigonometry in its own right in Europe was *De triangulis omnimodis* of 1464 by Regiomontanus (1436–1476), also known as Johann Müller of Königsberg, who may have been influenced by the work of al-Tusi. Regiomontanus had many results about right triangles, the Law of Sines (with proof), many examples of solving problems with triangles, and some results on spherical trigonometry. This work did not have the tangent function, though another work of Regiomontanus, *Tabulae directionum* of 1467, had it, as well as a sine table using sexagesimal numbers; in 1468 Regiomontanus had a sine table using decimals.

The famous text of Nicolaus Copernicus (1473–1543), *De revolutionibus orbium coelestium* of 1543, contained not only his important work on astronomy, but also some sections on trigonometry, this material having been previously published separately in 1542. Copernicus' work on trigonometry was possibly influenced by Ptolemy and Regiomontanus. Copernicus' student Georg Joachim Rheticus (1514–1574) took ideas of Regiomontanus and Copernicus and his own ideas and wrote the extensive treatise *Opus palatinum de triangulis* (completed and published in 1596 after Rheticus' death), which included tables for all six trigonometric functions; he defined sine and cosine in terms of right triangles, rather than in terms of a circle.

Viète, in works of 1571 and 1593, had tables of all six trigonometric functions, solved problems with triangles and had trigonometric identities. One of the identities Viète had was $\sin x + \sin y = 2\sin\frac{x+y}{2}\cos\frac{x-y}{2}$, which can be rewritten as $\sin(A+B) + \sin(A-B) = 2\sin A\cos B$, which in turn was used to convert products of numbers into sums prior to the invention of logarithms.

Prior to Euler the trigonometric functions were thought of not as functions in the modern sense but as lines (or the lengths of lines) related to the unit circle; Euler was the first person to view the trigonometric functions as functions per se. In 1739 Euler discussed harmonic oscillators, and the sine function was a solution of a differential equation; in 1743 he used the sine and cosine functions in his method for solving linear differential equations in general. All of this work led to the trigonometric functions being viewed as yet another type of transcendental function. Euler's textbook *Introductio in analysin infinitorum* of 1748 gave the first systematic treatment of the trigonometric functions as we know them today. Euler introduced our modern definitions of sine and cosine in terms of the unit circle. He had various trigonometric identities, including $\sin^2 x + \cos^2 x = 1$ and the formula for $\sin(x+y)$. He obtained the Maclaurin series for sine and cosine, and also the identity $e^{ix} = \cos x + i\sin x$.

Logarithms and Exponentials

Logarithms and exponentials arose much later than the trigonometric functions, presumably because they did not appear naturally in a topic of interest to the ancient world in the way that trigonometry arose as a tool for astronomy.

Thomas Bradwardine, from Merton College at Oxford, in 1328 used something that is equivalent to what we now call an exponential function in his attempt to resolve a matter relating to Aristotle's views on force and resistance. Nicole Oresme (1323–1382) subsequently explored such functions. He did not have a satisfactory way of writing exponentials, but he seemed to understand how to manipulate exponents with both integer and fractional powers, and raised the issue of the meaning of irrational powers.

In the late 16th century, developments in fields such as astronomy and navigation led to the need for more accurate computing. For example, 15-place tables of trigonometric functions were published in 1596 and 1613. John Napier (1550–1617) invented logarithms strictly for the purpose of making multiplication and division computationally easier. He had worked on developing his logarithms for many years, and he was, at least in part, inspired to complete and publish his approach after hearing about the use of trigonometric functions to convert products into sums and differences, which was reported to him by someone who had apparently heard about it while visiting the observatory of Tycho Brahe (1546–1601). Napier published his approach in 1614, after which logarithms gained rapid acceptance as a computational tool. Napier's work was particularly noteworthy because fractional powers and the exponential notation had not yet been developed, and the decimal system of notation for fractions, though developed, had not yet been widely accepted; Napier's use of decimal notation was very influential in the widespread adoption of this system in the 17th century. The idea that a table of exponents and powers of a number (for example 2) allowed multiplication to be done via addition was known prior to Napier, but such a table had increasingly large gaps, and could not be used in practice. Napier had the idea of producing a table without such gaps using arithmetic and geometric sequences; his approach involved thinking about the motion of points on a line. Similar ideas were developed by Jost Bürgi (1552–1632) at the same time as Napier, though the latter published his ideas first. Napier's definition of the logarithm of a number was not exactly the same as ours, and it was Henry Briggs (1561–1630) who reformulated Napier's approach (initially in consultation with Napier), and developed the common logarithms (base 10) we know today. Briggs published a preliminary table of common logarithms in 1617, and an expanded version in 1624. Briggs' table was completed and published by others in 1628, and these tables gained widespread use.

Grégoire de Saint-Vincent (1584–1667), in 1647, showed some properties of the area under the curve $xy = 1$, and in 1649 Alfonso Antonio de Sarasa (1618–1667) observed that Grégoire de Saint-Vincent's properties implied that the area function for the curve $y = \frac{1}{x}$ has the same addition-to-multiplication property as logarithm, which was close to understanding what we now express by saying that $\ln x$ is the antiderivative of $\frac{1}{x}$.

Wallis, in his *Arithmetica infinitorum* of 1656, was the first person to use, and understand, fractional exponents as we do today.

Newton computed the power series for $\ln(1+x)$ in the mid-1660s, but he did not publish it at the time. The series was published by Nicolaus Mercator (1620–1687), not to be confused with the inventor of the Mercator projection Gerardus Mercator, in *Logarithmotechnia* of 1668, where he used the observation of Sarasa about the curve $y = \frac{1}{x}$, as well as ideas of Wallis. Mercator used the term "natural logarithm," and gave the correct ratio between natural logarithms and common logarithms.

Leibniz worked out, and Johann Bernoulli explicitly stated, the derivatives of logarithms.

Euler's *Introductio in analysin infinitorum* of 1748 gave the first systematic treatment of logarithms as we now know them. Euler was the first person to think of $\log_a x$ as the number y such that $a^y = x$. He obtained the Maclaurin series for a^x using the binomial series, and he then defined $\log_a x$ in terms of a^x. He defined e as $e = 1 + \frac{1}{1!} + \frac{1}{2!} + \cdots$. He had the idea that $e^x = (1 + \frac{x}{I})^I$, where I is an infinitely large number, and this was Euler's way of saying what we now phrase as $e^x = \lim_{n\to\infty} (1 + \frac{x}{n})^n$.

8

Sequences

8.1 Introduction

In a typical calculus course, sequences are usually treated very briefly, and their role is primarily as a prelude to the study of series. In real analysis, by contrast, sequences assume a much more important role. We will certainly use sequences in our study of series in Chapter 9, but, as will be seen in the present chapter, we will prove some substantial and important theorems about sequences in their own right, such as the Monotone Convergence Theorem (Corollary 8.3.4) and the Bolzano–Weierstrass Theorem (Theorem 8.3.9). As was the case for the important theorems concerning continuity, derivatives and integrals that we saw in previous chapters, the important theorems concerning sequences rely upon the Least Upper Bound Property of the real numbers.

In many real analysis texts the study of sequences precedes the study of limits of functions. In such an approach, the proofs of some of the important theorems involving continuity, derivatives and integrals make use of sequences, rather than directly using the Least Upper Bound Property. In this text we have placed the study of sequences after the study of continuity, derivatives and integrals, and have offered proofs for those topics that more directly make use of the Least Upper Bound Property, both to highlight the role of the Least Upper Bound Property, and to keep to a minimum the technical tools needed to study continuity, derivatives and integrals. Be that as it may, sequences are a central topic in real analysis, no matter where they are placed in a textbook. We will see a few applications of sequences in Section 8.4.

8.2 Sequences

The reader is familiar, at least informally, with the notion of a sequence of real numbers, for example the sequence

$$\frac{1}{2}, \frac{1}{4}, \frac{1}{8}, \frac{1}{16}, \ldots.$$

Intuitively, a sequence of real numbers is a collection of real numbers of which there is a first, a second, a third and so on, with one real number for each element of $\mathbb{N}$. It is important to distinguish between the term "sequence" and the related but not identical term "series," which is the sum of a sequence, for example

$$\frac{1}{2}+\frac{1}{4}+\frac{1}{8}+\frac{1}{16}+\cdots.$$

In colloquial usage, the words "sequence" and "series" are often used interchangeably, but in mathematical terminology the two concepts are distinct, and these two words should be used in their precise meanings. (Using mathematical terminology, the final playoffs in American baseball should be called the "world sequence" rather than the "world series.")

Although we think of a sequence informally as a collection of numbers of which there is a first, a second, a third and so on, that is not a rigorous definition—any statement with "etc." is not entirely rigorous. The basis of the following definition is that we use the natural numbers as the model set of which there is intuitively a first, second, third and so on. Then, we can select a first, second, third and so on, elements of an arbitrary set by using a function from the natural numbers to that set. In other words, although we informally write a sequence of real numbers as $a_1, a_2, \ldots$, the formal definition of such a sequence is a function $f\colon \mathbb{N} \to \mathbb{R}$, where we think of $f(1)$ as the first element of the sequence, of $f(2)$ as the second element of the sequence and so on; that is, we can think of the sequence as given by $a_n = f(n)$ for all $n \in \mathbb{N}$. As seen in the following definition, there can be sequences in any non-empty set.

Definition 8.2.1. Let A be a non-empty set. A **sequence** in A is a function $f\colon \mathbb{N} \to A$. If $f\colon \mathbb{N} \to A$ is a sequence, and if $a_i = f(i)$ for all $i \in \mathbb{N}$, then we will write either $a_1, a_2, a_3, \ldots$ or $\{a_n\}_{n=1}^{\infty}$ to denote the sequence. Each number a_n, where $n \in \mathbb{N}$, is called a **term** of the sequence $\{a_n\}_{n=1}^{\infty}$. A sequence in the set $\mathbb{R}$ is also called a sequence of real numbers. △

It is important to recognize that a sequence is not just a set of elements, but is a countably infinite collection of elements in a given order. A set, by contrast, even if it is countably infinite, does not have an order to its elements. In other words, we distinguish between a sequence $\{a_n\}_{n=1}^{\infty}$, and the set of terms of the sequence $\{a_n \mid n \in \mathbb{N}\}$. For example, if we let $\{a_n\}_{n=1}^{\infty}$ be the sequence $\{(-1)^n\}_{n=1}^{\infty}$, then $\{a_n\}_{n=1}^{\infty}$ has infinitely many terms, but the set $\{a_n \mid n \in \mathbb{N}\} = \{-1, 1\}$ has only two elements. Even if no two terms of a sequence are equal to each other, the sequence is still not the same as its set of elements. For example, the two sequences $\frac{1}{2}, \frac{1}{4}, \frac{1}{8}, \frac{1}{16}, \ldots$ and $\frac{1}{4}, \frac{1}{2}, \frac{1}{8}, \frac{1}{6}, \ldots$ are different as sequences, but they have the same sets of elements.

In Section 3.2 we discussed the notion of the limit of a function as $x \to c$, using the ε–δ definition. We now give the analogous definition of the limit of a sequence. Actually, this definition more closely resembles the definition of Type 1 limits to infinity discussed in Section 6.2 than the definition of ordinary limits of functions seen in Section 3.2. However, it is not assumed in the present section that the reader is familiar with Section 6.2, except as a useful analogy for the final definition in this section.

Informally, as is often stated in calculus courses, the intuitive idea of a limit of a sequence $\{a_n\}_{n=1}^{\infty}$ as n goes to infinity is that the value of a_n gets closer and closer to a number L as the value of n gets larger and larger. Of course, not every sequence has a limit. It is important to stress that when we say "n goes to infinity" we mean only that n gets larger and larger; there is no real number "∞" to which n is getting closer and closer.

As we did for limits of functions, we measure "arbitrary closeness" with an arbitrarily chosen positive number, often denoted with a symbol such as ε. We rephrase the first part of the expression "the value of a_n gets closer and closer to a number L as the value of n gets larger and larger" by using ε to denote our measure of closeness. The crucial idea of a limit of a sequence existing is: if, for every possible choice of $\varepsilon > 0$, no matter how small, we can show that for all n sufficiently large, the value of a_n will be within distance ε of L, then we will say that the limit of $\{a_n\}_{n=1}^{\infty}$ is L. We will use $N \in \mathbb{N}$ to denote the measure of largeness of n. Then if for each possible choice of $\varepsilon > 0$, no matter how small, we can show that there is some $N \in \mathbb{N}$ such that for all n at least as large as N, then a_n will be within ε distance of L, we will say that the limit of $\{a_n\}_{n=1}^{\infty}$ as n goes to infinity is L. To say that a_n is within distance ε of L is to say that $|a_n - L| < \varepsilon$. We then see that the rigorous way to say "the value of a_n gets closer and closer to a number L as the value of n gets larger and larger" is to say that for each $\varepsilon > 0$, there is some $N \in \mathbb{N}$ such that for all $n \in \mathbb{N}$ such that $n \geq N$, it is the case that $|a_n - L| < \varepsilon$. As seen in Figure 8.2.1, where the values of $a_1, a_2, \ldots$ are represented by dots, the expression "for all $n \in \mathbb{N}$ such that $a \geq N$, it is the case that $|a_n - L| < \varepsilon$" can be viewed graphically by saying that a_n is within a band of width 2ε centered at L whenever $n \in \mathbb{N}$ and $n \geq N$.

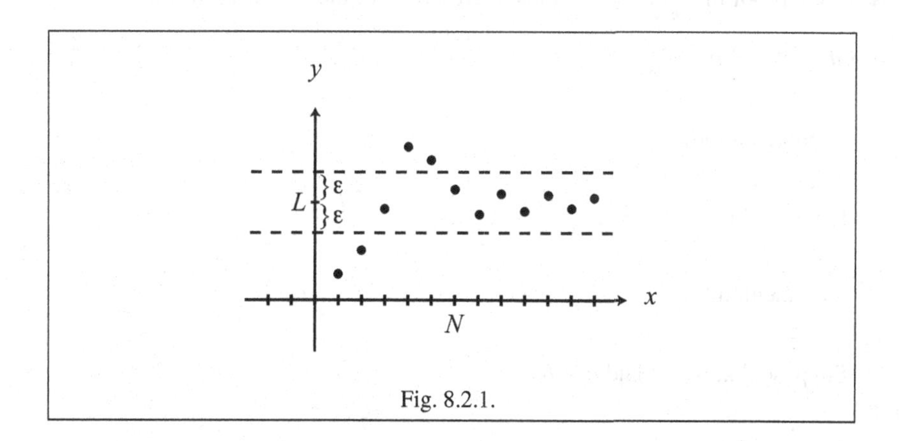

Fig. 8.2.1.

The reader will notice that our use of "ε" in the above discussion is the same as our use of "ε" in the discussion of limits of functions in Section 3.2, though we replaced the "δ" (which will often be small when ε is small) with an integer "N" (which will often be large when ε is small). Other than this replacement of the usually small "δ" with the usually large "N," the following definition is virtually the same as Definition 3.2.1. In other words, limits of functions and limits of sequences are different in that they deal with functions that have different domains (open intervals in $\mathbb{R}$ versus the natural numbers), but they work the same way in the codomain (which is $\mathbb{R}$ in both cases).

Definition 8.2.2. Let $\{a_n\}_{n=1}^{\infty}$ be a sequence in $\mathbb{R}$, and let $L \in \mathbb{R}$. The number L is the **limit** of $\{a_n\}_{n=1}^{\infty}$, written

$$\lim_{n\to\infty} a_n = L,$$

if for each $\varepsilon > 0$, there is some $N \in \mathbb{N}$ such that $n \in \mathbb{N}$ and $n \geq N$ imply $|a_n - L| < \varepsilon$. If $\lim_{n\to\infty} a_n = L$, we also say that $\{a_n\}_{n=1}^{\infty}$ **converges** to L. If $\{a_n\}_{n=1}^{\infty}$ converges to some real number, we say that $\{a_n\}_{n=1}^{\infty}$ is **convergent**; otherwise we say that $\{a_n\}_{n=1}^{\infty}$ is **divergent**. △

Similarly to the definition of the limit of a function, here too the order of the quantifiers is crucial. The definition of the limit of a sequence could be written in logical symbols as

$$(\forall\varepsilon > 0)(\exists N \in \mathbb{N})[(n \in \mathbb{N} \land n \geq N) \to |a_n - L| < \varepsilon].$$

The order of the quantifiers cannot be changed. If we want to prove that $\lim_{n\to\infty} a_n = L$, the proof must start by choosing an arbitrary $\varepsilon > 0$. Next, after possible argumentation, a value of $N \in \mathbb{N}$ must be given, where N may depend upon ε. We then choose an arbitrary $n \in \mathbb{N}$ such that $n \geq N$. Finally, again after possible argumentation, we must deduce that $|a_n - L| < \varepsilon$. It is important that the arbitrary choices are indeed arbitrary. A typical proof that $\lim_{n\to\infty} a_n = L$ must therefore have the following form:

Proof. Let $\varepsilon > 0$.

⋮

(argumentation)

⋮

Let $N \in \mathbb{N}$ be such that ....

⋮

(argumentation)

⋮

Suppose that $n \in \mathbb{N}$ and $n \geq N$.

⋮

(argumentation)

⋮

Therefore $|a_n - L| < \varepsilon$. □

Such proofs are often called "ε–N proofs." If you feel comfortable with ε–δ proofs, then you should have no trouble with ε–N proofs.

Just as the definition of limits of sequences is very similar to the definition of limits of functions, so too many of the lemmas, theorems and proofs in this section are very similar to the corresponding results about limits of functions found in Section 3.2.

For our first lemma, observe that it is not stated in Definition 8.2.2 that the number "L" in the definition is unique. However, it turns out that if $\lim_{n\to\infty} a_n = L$ for some $L \in \mathbb{R}$, then there is only one such number L. In other words, if a sequence has a limit, that means there is a *single* number L that a_n is getting closer and closer to as n gets larger; if there is no such number, then there is no limit.

Lemma 8.2.3. *Let $\{a_n\}_{n=1}^{\infty}$ be a sequence in $\mathbb{R}$. If $\lim_{n\to\infty} a_n = L$ for some $L \in \mathbb{R}$, then L is unique.*

Proof. Suppose that $\lim_{n\to\infty} a_n = L_1$ and $\lim_{n\to\infty} a_n = L_2$ for some $L_1, L_2 \in \mathbb{R}$ such that $L_1 \neq L_2$. Let $\varepsilon = \frac{|L_1 - L_2|}{2}$. Then $\varepsilon > 0$. Hence there is some $N_1 \in \mathbb{N}$ such that $n \in \mathbb{N}$ and $n \geq N_1$ imply $|a_n - L_1| < \varepsilon$, and there is some $N_2 \in \mathbb{N}$ that $n \in \mathbb{N}$ and $n \geq N_2$ imply $|a_n - L_2| < \varepsilon$. Let $N = \max\{N_1, N_2\}$. Then $N \geq N_1$ and $N \geq N_2$, and hence

$$\begin{aligned}|L_1 - L_2| &= |L_1 - a_N + a_N - L_2| \leq |L_1 - a_N| + |a_N - L_2| \\ &= |a_N - L_1| + |a_N - L_2| < \varepsilon + \varepsilon = 2\varepsilon = 2\frac{|L_1 - L_2|}{2} = |L_1 - L_2|,\end{aligned}$$

which is a contradiction. We deduce that if $\lim_{n\to\infty} a_n = L$ for some $L \in \mathbb{R}$, then L is unique. □

Because of Lemma 8.2.3 we can refer to "the" limit of a sequence, if the limit exists.

Example 8.2.4. In some parts of this example, we will first do some scratch work prior to the actual proof. As always, it is important to avoid confusing the scratch work with the proof.

(1) Let $c \in \mathbb{R}$, and let $\{a_n\}_{n=1}^{\infty}$ be the constant sequence defined by $a_n = c$ for all $n \in \mathbb{N}$. We will prove that $\lim_{n\to\infty} a_n = c$; we could write this limit as $\lim_{n\to\infty} c = c$. Let $\varepsilon > 0$. Let $N = 1$. Suppose that $n \in \mathbb{N}$ and $n \geq N$. Then

$$|a_n - c| = |c - c| = 0 < \varepsilon.$$

(2) We will prove that $\lim_{n\to\infty} \frac{1}{n} = 0$.

Scratch Work We will work backwards for our scratch work. We want to find $N \in \mathbb{N}$ such that $n \in \mathbb{N}$ and $n \geq N$ imply $\left|\frac{1}{n} - 0\right| < \varepsilon$, which is the same as $\frac{1}{n} < \varepsilon$. So, it will be a good choice to pick some $N \in \mathbb{N}$ such that $\frac{1}{N} < \varepsilon$.

Actual Proof Let $\varepsilon > 0$. By Corollary 2.6.8 (2) there is some $N \in \mathbb{N}$ such that $\frac{1}{N} < \varepsilon$. Suppose that $n \in \mathbb{N}$ and $n \geq N$. Then

$$\left|\frac{1}{n} - 0\right| = \left|\frac{1}{n}\right| = \frac{1}{n} \leq \frac{1}{N} < \varepsilon.$$

(3) We will prove that $\lim_{n\to\infty} \frac{n^2}{2n^2+1} = \frac{1}{2}$.

Scratch Work We want to find $N \in \mathbb{N}$ such that $n \in \mathbb{N}$ and $n \geq N$ imply $\left|\frac{n^2}{2n^2+1} - \frac{1}{2}\right| < \varepsilon$, which is the same as $\left|\frac{-1}{4n^2+2}\right| < \varepsilon$, which is equivalent to $\frac{1}{4n^2+2} < \varepsilon$, which in turn is the same as $\frac{1}{\varepsilon} < 4n^2+2$. Solving for n we obtain $n > \frac{1}{2}\sqrt{\frac{1}{\varepsilon} - 2}$. If $\frac{1}{\varepsilon} \geq 2$, which is the same as $\varepsilon \leq \frac{1}{2}$, then we will want to use $N > \frac{1}{2}\sqrt{\frac{1}{\varepsilon} - 2}$. If $\frac{1}{\varepsilon} < 2$, which is the same as $\varepsilon > \frac{1}{2}$, then we cannot use $\sqrt{\frac{1}{\varepsilon} - 2}$, but instead we observe that $\frac{1}{\varepsilon} < 2 < 4n^2+2$ for all $n \in \mathbb{N}$, so we can choose an arbitrary value for N.

Actual Proof Let $\varepsilon > 0$. There are two cases. First, suppose that $\varepsilon > \frac{1}{2}$. Let $N = 1$. Suppose that $n \in \mathbb{N}$ and $n \geq N$. Then

$$\left|\frac{n^2}{2n^2+1} - \frac{1}{2}\right| = \left|\frac{-1}{4n^2+2}\right| = \frac{1}{4n^2+2} < \frac{1}{2} < \varepsilon.$$

Second, suppose that $\varepsilon \leq \frac{1}{2}$. By Corollary 2.6.8 (1) there is some $N \in \mathbb{N}$ such that

$$N > \frac{1}{2}\sqrt{\frac{1}{\varepsilon} - 2}.$$

Suppose that $n \in \mathbb{N}$ and $n \geq N$. Then

$$n > \frac{1}{2}\sqrt{\frac{1}{\varepsilon} - 2},$$

which implies, with some rearranging, that

$$\frac{1}{4n^2+2} < \varepsilon.$$

Then

$$\left|\frac{n^2}{2n^2+1} - \frac{1}{2}\right| = \frac{1}{4n^2+2} < \varepsilon.$$

(4) We will prove that $\{(-1)^n\}_{n=1}^{\infty}$ is divergent. Suppose that $\lim_{n\to\infty} (-1)^n = L$ for some $L \in \mathbb{R}$. Let $\varepsilon = \frac{1}{2}$. Then there is some $N \in \mathbb{N}$ such that $n \in \mathbb{N}$ and $n \geq N$ imply $|(-1)^n - L| < \frac{1}{2}$. Choose $n_1, n_2 \in \mathbb{N}$ such that $n_1 \geq N$ and n_1 is odd, and that $n_2 \geq N$ and n_2 is even. Then

$$\begin{aligned} 2 &= |(-1) - 1| = |(-1)^{n_1} - (-1)^{n_2}| = |(-1)^{n_1} - L + L - (-1)^{n_2}| \\ &\leq |(-1)^{n_1} - L| + |L - (-1)^{n_2}| < \frac{1}{2} + \frac{1}{2} = 1, \end{aligned}$$

which is a contradiction. We conclude that the sequence is divergent. ◊

A very simple, but very useful, observation about limits of sequences is that if finitely many terms of a sequence are changed, it does not change whether or not the sequence is convergent, and if the sequence is convergent, it does not change the limit of the sequence. A proof of this fact is given in Exercise 8.2.3.

For our next definition, recall from Section 1.7 or 2.2 the definition of a subset of $\mathbb{R}$ being bounded.

Definition 8.2.5. Let $\{a_n\}_{n=1}^\infty$ be a sequence in $\mathbb{R}$. The sequence $\{a_n\}_{n=1}^\infty$ is **bounded above, bounded below** or **bounded** if the set $\{a_n \mid n \in \mathbb{N}\}$ is bounded above, bounded below or bounded, respectively. △

By Exercise 2.3.11 we know that a sequence $\{a_n\}_{n=1}^\infty$ is bounded if and only if there is some $M \in \mathbb{R}$ such that $|a_n| \leq M$ for all $n \in \mathbb{N}$; it is always possible to choose M so that $M > 0$.

In contrast to a number of other theorems and lemmas in this section, which have exact analogs for limits of functions, the following lemma has only a partial analog. Whereas it is possible for a limit of the form $\lim_{x\to c} f(x)$ to exist even though the function f is not bounded, the discrete nature of the natural numbers leads to the fact that any convergent sequence is bounded, as we now prove. The closest analog of the following lemma for limits of functions is Lemma 3.2.7.

Lemma 8.2.6. *Let $\{a_n\}_{n=1}^\infty$ be a sequence in $\mathbb{R}$. If $\{a_n\}_{n=1}^\infty$ is convergent, then $\{a_n\}_{n=1}^\infty$ is bounded.*

Proof. Suppose that $\{a_n\}_{n=1}^\infty$ is convergent. Let $L = \lim_{n\to\infty} a_n$. Then there is some $N \in \mathbb{N}$ such that $n \in \mathbb{N}$ and $n \geq N$ imply $|a_n - L| < 1$, which by Lemma 2.3.9 (7) implies $|a_n| - |L| < 1$, and hence $|a_n| < |L| + 1$. Let

$$M = \max\{|a_1|, |a_2|, \ldots, |a_{N-1}|, |L| + 1\}.$$

It then follows that $|a_k| \leq M$ for all $k \in \mathbb{N}$. Therefore $\{a_n\}_{n=1}^\infty$ is bounded. □

Although Lemma 8.2.6 shows that a convergent sequence is bounded, the converse is not true. For example, the sequence $\{(-1)^n\}_{n=1}^\infty$ is clearly bounded, but we saw in Example 8.2.4 (4) that it is divergent.

The following lemma is the analog of Lemma 3.2.8. The reader is asked in Exercise 8.2.6 to show that this hypothesis of boundedness in the following lemma cannot be dropped.

Lemma 8.2.7. *Let $\{a_n\}_{n=1}^\infty$ and $\{b_n\}_{n=1}^\infty$ be sequences in $\mathbb{R}$. Suppose that $\lim_{n\to\infty} a_n = 0$, and that $\{b_n\}_{n=1}^\infty$ is bounded. Then $\lim_{n\to\infty} a_n b_n = 0$.*

Proof. Let $\varepsilon > 0$. Because $\{b_n\}_{n=1}^\infty$ is bounded, there is some $M \in \mathbb{R}$ such that $|b_n| \leq M$ for all $n \in \mathbb{N}$; we may assume that $M > 0$. Because $\lim_{n\to\infty} a_n = 0$, there is some $N \in \mathbb{N}$ such that $n \in \mathbb{N}$ and $n \geq N$ imply $|a_n - 0| < \frac{\varepsilon}{M}$. Suppose that $n \in \mathbb{N}$ and $n \geq N$. Then

$$|a_n b_n - 0| = |a_n b_n| = |a_n| \cdot |b_n| < \frac{\varepsilon}{M} \cdot M = \varepsilon.$$ □

Example 8.2.8. By Example 8.2.4 (2) we know that $\lim_{n\to\infty} \frac{1}{n} = 0$. The sequence $\{\sin n\}_{n=1}^{\infty}$ is bounded, because $|\sin x| \leq 1$ for all $x \in \mathbb{R}$ by Theorem 7.3.9 (4). Lemma 8.2.7 then implies that $\lim_{n\to\infty} \frac{\sin n}{n} = 0$. ◊

We now see that limits of sequences behave nicely with respect to the addition, subtraction, multiplication and division of the terms of sequences.

Theorem 8.2.9. *Let $\{a_n\}_{n=1}^{\infty}$ and $\{b_n\}_{n=1}^{\infty}$ be sequences in $\mathbb{R}$, and let $k \in \mathbb{R}$. Suppose that $\{a_n\}_{n=1}^{\infty}$ and $\{b_n\}_{n=1}^{\infty}$ are convergent.*

1. *$\{a_n + b_n\}_{n=1}^{\infty}$ is convergent and $\lim_{n\to\infty} (a_n + b_n) = \lim_{n\to\infty} a_n + \lim_{n\to\infty} b_n$.*
2. *$\{a_n - b_n\}_{n=1}^{\infty}$ is convergent and $\lim_{n\to\infty} (a_n - b_n) = \lim_{n\to\infty} a_n - \lim_{n\to\infty} b_n$.*
3. *$\{ka_n\}_{n=1}^{\infty}$ is convergent and $\lim_{n\to\infty} ka_n = k \lim_{n\to\infty} a_n$.*
4. *$\{a_n b_n\}_{n=1}^{\infty}$ is convergent and $\lim_{n\to\infty} a_n b_n = \left[\lim_{n\to\infty} a_n\right] \cdot \left[\lim_{n\to\infty} b_n\right]$.*
5. *If $\lim_{n\to\infty} b_n \neq 0$, then $\left\{\frac{a_n}{b_n}\right\}_{n=1}^{\infty}$ is convergent and $\lim_{n\to\infty} \frac{a_n}{b_n} = \frac{\lim_{n\to\infty} a_n}{\lim_{n\to\infty} b_n}$.*

Proof. We will prove Part (4), leaving the rest to the reader in Exercise 8.2.11.

The proofs of the various parts of this theorem are analogous to the proofs of the corresponding parts of Theorem 3.2.10.

(4) Let $L = \lim_{n\to\infty} a_n$ and $M = \lim_{n\to\infty} b_n$. Let $\varepsilon > 0$. By Lemma 8.2.6 we know that $\{b_n\}_{n=1}^{\infty}$ is bounded. Hence there is some $B \in \mathbb{R}$ such that $|b_n| \leq B$ for all $n \in \mathbb{N}$. We may assume that $B > 0$. Then $B + |L| > 0$. There is some $N_1 \in \mathbb{N}$ such that $n \in \mathbb{N}$ and $n \geq N_1$ imply $|a_n - L| < \frac{\varepsilon}{B+|L|}$, and there is some $N_2 \in \mathbb{N}$ such that $n \in \mathbb{N}$ and $n \geq N_2$ imply $|b_n - M| < \frac{\varepsilon}{B+|L|}$. Let $N = \max\{N_1, N_2\}$. Suppose that $n \in \mathbb{N}$ and $n \geq N$. Then

$$\begin{aligned} |a_n b_n - LM| &= |a_n b_n - b_n L + b_n L - LM| \leq |b_n| \cdot |a_n - L| + |L| \cdot |b_n - M| \\ &< B \cdot \frac{\varepsilon}{B + |L|} + |L| \cdot \frac{\varepsilon}{B + |L|} = \varepsilon. \end{aligned}$$ □

Theorem 8.2.9 has both theoretical and practical uses, a simple example of the latter being the following example.

Example 8.2.10. We will prove that $\lim_{n\to\infty} \frac{n}{n+1} = 1$. Although it would be possible to provide an ε–N proof for this limit, we will use Theorem 8.2.9 to give an easier proof. Observe that if $n \in \mathbb{N}$, then

$$\frac{n}{n+1} = \frac{1}{1 + \frac{1}{n}}.$$

We saw in Example 8.2.4 (1) (2) that $\lim_{n\to\infty} 1 = 1$ and $\lim_{n\to\infty} \frac{1}{n} = 0$. We can then apply Theorem 8.2.9 (1) (5) to deduce that

$$\lim_{n\to\infty} \frac{n}{n+1} = \lim_{n\to\infty} \frac{1}{1 + \frac{1}{n}} = \frac{1}{1+0} = 1.$$ ◊

Our next result is the analog of Theorem 3.2.13, with the one modification that it is sufficient if the term-by-term comparison of the two sequences starts after finitely many terms.

Theorem 8.2.11. *Let $\{a_n\}_{n=1}^{\infty}$ and $\{b_n\}_{n=1}^{\infty}$ be sequences in $\mathbb{R}$. Suppose that there is some $N \in \mathbb{N}$ such that $n \in \mathbb{N}$ and $n \geq N$ imply $a_n \leq b_n$. If $\{a_n\}_{n=1}^{\infty}$ and $\{b_n\}_{n=1}^{\infty}$ are convergent, then $\lim_{n\to\infty} a_n \leq \lim_{n\to\infty} b_n$.*

Proof. Suppose that $\{a_n\}_{n=1}^{\infty}$ and $\{b_n\}_{n=1}^{\infty}$ are convergent. Let $L = \lim_{n\to\infty} a_n$ and $M = \lim_{n\to\infty} b_n$. Suppose that $M < L$. Let $\varepsilon = \frac{L-M}{2}$. Then $\varepsilon > 0$. Hence there is some $N_1 \in \mathbb{N}$ such that $n \in \mathbb{N}$ and $n \geq N_1$ imply $|a_n - L| < \varepsilon$, and there is some $N_2 \in \mathbb{N}$ such that $n \in \mathbb{N}$ and $n \geq N_2$ imply $|b_n - M| < \varepsilon$. Let $P = \max\{N, N_1, N_2\}$. Then $|a_P - L| < \varepsilon$ and $|b_P - M| < \varepsilon$. It follows that $L - \varepsilon < a_P < L + \varepsilon$ and $M - \varepsilon < b_P < M + \varepsilon$, and hence

$$b_P < M + \varepsilon = M + \frac{L-M}{2} = \frac{L+M}{2} = L - \frac{L-M}{2} = L - \varepsilon < a_P,$$

which is a contradiction to the fact that $a_n \leq b_n$ for all $n \in \mathbb{N}$. Therefore $L \leq M$. □

The following theorem provides a convenient way to find the limit of a sequence by "trapping it" between two sequences that have limits that can be dealt with more easily.

Theorem 8.2.12 (Squeeze Theorem for Sequences). *Let $\{a_n\}_{n=1}^{\infty}$, $\{b_n\}_{n=1}^{\infty}$ and $\{c_n\}_{n=1}^{\infty}$ be sequences in $\mathbb{R}$. Suppose that there is some $N \in \mathbb{N}$ such that $n \in \mathbb{N}$ and $n \geq N$ imply $a_n \leq b_n \leq c_n$. If $\{a_n\}_{n=1}^{\infty}$ and $\{c_n\}_{n=1}^{\infty}$ are convergent and $\lim_{n\to\infty} a_n = \lim_{n\to\infty} c_n$, then $\{b_n\}_{n=1}^{\infty}$ is convergent and $\lim_{n\to\infty} b_n = \lim_{n\to\infty} a_n = \lim_{n\to\infty} c_n$.*

Proof. Suppose that $\{a_n\}_{n=1}^{\infty}$ and $\{c_n\}_{n=1}^{\infty}$ are convergent and $\lim_{n\to\infty} a_n = \lim_{n\to\infty} c_n$. Let $L = \lim_{n\to\infty} a_n = \lim_{n\to\infty} c_n$. Let $\varepsilon > 0$. Then there is some $N_1 \in \mathbb{N}$ such that $n \in \mathbb{N}$ and $n \geq N_1$ imply $|a_n - L| < \varepsilon$, and there is some $N_2 \in \mathbb{N}$ such that $n \in \mathbb{N}$ and $n \geq N_2$ imply $|c_n - L| < \varepsilon$. Let $P = \max\{N, N_1, N_2\}$. Suppose that $n \in \mathbb{N}$ and $n \geq P$. Then $a_n \leq b_n \leq c_n$, and $|a_n - L| < \varepsilon$ and $|c_n - L| < \varepsilon$. It follows that $L - \varepsilon < a_n < L + \varepsilon$ and $L - \varepsilon < c_n < L + \varepsilon$, and hence

$$L - \varepsilon < a_n \leq b_n \leq c_n < L + \varepsilon.$$

Therefore $|b_n - L| < \varepsilon$. □

The following useful example is, in part, an application of the Squeeze Theorem for Sequences (Theorem 8.2.12).

Example 8.2.13. Let $r \in \mathbb{R}$. We want to examine the convergence or divergence of the sequence $\{r^n\}_{n=1}^{\infty}$. There are six cases.

First, suppose that $0 < r < 1$. We will show that $\{r^n\}_{n=1}^{\infty}$ is convergent, and that $\lim_{n\to\infty} r^n = 0$. It follows from the hypothesis on r that $\frac{1}{r} > 1$. Let $q = \frac{1}{r} - 1$. Then

$q > 0$ and $\frac{1}{r} = 1 + q$. It now follows from Exercise 2.5.13 (1) that if $n \in \mathbb{N}$ then $\frac{1}{r^n} = \left(\frac{1}{r}\right)^n = (1+q)^n \geq 1 + nq > nq$. Hence $0 < r^n < \frac{1}{q}\frac{1}{n}$ for all $n \in \mathbb{N}$. It follows from Example 8.2.4 (1) that $\lim_{n\to\infty} 0 = 0$, and from Example 8.2.4 (2) and Theorem 8.2.9 (3) that $\lim_{n\to\infty} \frac{1}{q}\frac{1}{n} = 0$. Because $0 < r^n < \frac{1}{q}\frac{1}{n}$ for all $n \in \mathbb{N}$, then by the Squeeze Theorem for Sequences (Theorem 8.2.12) we deduce that $\lim_{n\to\infty} r^n = 0$.

Second, suppose that $r > 1$. We will show that $\{r^n\}_{n=1}^{\infty}$ is divergent. By using Lemma 8.2.6, it will suffice to show that $\{r^n\}_{n=1}^{\infty}$ is not bounded. Let $M \in \mathbb{R}$. If $M \leq 0$, then $r^n > M$ for all $n \in \mathbb{N}$. Now suppose that $M > 0$. It follows from the hypothesis on r that $0 < \frac{1}{r} < 1$, and hence by the previous paragraph we know that $\lim_{n\to\infty} \left(\frac{1}{r}\right)^n = 0$. Therefore there is some $N \in \mathbb{N}$ such that $n \in \mathbb{N}$ and $n \geq N$ imply $\left|\left(\frac{1}{r}\right)^n - 0\right| < \frac{1}{M}$. It follows that if $n \in \mathbb{N}$ and $n \geq N$, then $r^n = |r^n| > M$. Hence $\{r^n\}_{n=1}^{\infty}$ is not bounded above, and therefore it is not bounded.

Third, suppose that $r = 1$ or $r = 0$. Then $\lim_{n\to\infty} r^n = \lim_{n\to\infty} 1 = 1$ or $\lim_{n\to\infty} r^n = \lim_{n\to\infty} 0 = 0$ by Example 8.2.4 (1).

Fourth, suppose that $-1 < r < 0$. Then $0 < |r| < 1$, and by the first case, we see that $\lim_{n\to\infty} |r^n| = \lim_{n\to\infty} |r|^n = 0$. It then follows from Exercise 8.2.13 (2) that $\lim_{n\to\infty} r^n = 0$.

Fifth, suppose that $r = -1$. Then $\{r^n\}_{n=1}^{\infty}$ is divergent by Example 8.2.4 (4).

Sixth, suppose that $r < -1$. We will show that $\{r^n\}_{n=1}^{\infty}$ is divergent, once again by showing that $\{r^n\}_{n=1}^{\infty}$ is not bounded. Let $M \in \mathbb{R}$. Suppose that $M > 0$; the case where $M \leq 0$ is similar, and we omit the details. It follows from the hypothesis on r that $|r| > 1$. Using the argument given in the second case, we know that there is some $N \in \mathbb{N}$ such that $n \in \mathbb{N}$ and $n \geq N$ imply $|r|^n > M$. It follows that if $n \in \mathbb{N}$ and $n \geq N$ and n is even, then $r^n = |r|^n > M$. Hence $\{r^n\}_{n=1}^{\infty}$ is not bounded.

Putting all of the above cases together, we see that $\{r^n\}_{n=1}^{\infty}$ is convergent if and only if $-1 < r \leq 1$. If $-1 < r < 1$ then $\lim_{n\to\infty} r^n = 0$, and if $r = 1$ then $\lim_{n\to\infty} r^n = 1$. ◊

Similarly to the Type 2 limits to infinity for functions that were discussed in Section 6.2, it is also possible to define what it means for a sequence to diverge to infinity or negative infinity.

Definition 8.2.14. Let $\{a_n\}_{n=1}^{\infty}$ be a sequence in $\mathbb{R}$. The sequence $\{a_n\}_{n=1}^{\infty}$ **diverges to infinity**, written

$$\lim_{n\to\infty} a_n = \infty,$$

if for each $P \in \mathbb{R}$, there is some $N \in \mathbb{N}$ such that $n \in \mathbb{N}$ and $n \geq N$ imply $a_n > P$. The sequence $\{a_n\}_{n=1}^{\infty}$ **diverges to negative infinity**, written

$$\lim_{n\to\infty} a_n = -\infty,$$

if for each $Q \in \mathbb{R}$, there is some $N \in \mathbb{N}$ such that $n \in \mathbb{N}$ and $n \geq N$ imply $a_n < Q$. △

Observe that we say that a sequence "diverges to infinity," and not that it "converges to infinity," because convergence always means convergence to a real number, and there is no real number that is infinite.

Reflections

Nowhere in an introductory real analysis course is the difference in emphasis between such a course and a calculus course more immediately apparent than in the role of sequences. In a calculus course, sequences per se receive very brief treatment, usually only as minimally needed for use as partial sums of series. In a real analysis course, by contrast, sequences have a very important role to play—even in the present text, where sequences are located in the same place as in a calculus course, as opposed to some real analysis texts, which locate sequences earlier. The reason for the difference in the role of sequences in a real analysis course and a calculus course is the importance of sequences as a tool for rigorous proofs, as seen in the applications of sequences found in Section 8.4, though this collection of applications is but a sampling of the wide use of sequences. Ultimately, the value of sequences is that by using theorems such as the Monotone Convergence Theorem and the Bolzano–Weierstrass Theorem, both found in Section 8.3 and both equivalent to the Least Upper Bound Property of the real numbers, we have additional—and sometimes simpler—ways of using that property.

Having stressed the importance of sequences, the most immediate impression one gets upon first encounter with the introductory material about sequences in the present section is its similarity to the material concerning limits of functions in Section 3.2. In fact, both limits of sequences and limits of functions are special cases of a more general type of limit based upon the idea of directed sets; see [Bea97] for details. Hence, it is not a coincidence that some of the definitions, theorems and proofs concerning limits of sequences are so similar to their analogs for limits of functions. If the reader finds the material about limits of sequences easier than limits of functions, that is in no small measure because the reader has already gained experience with limits of functions, and hence, in contrast to the reader's initial encounter with that material, which was likely her first encounter with ε–δ arguments, by now the variant of such arguments seen in the present section are nothing new. Moreover, after some of the tricky proofs we saw in Chapter 5, the proofs in the present section are indeed easy by comparison. It is only when we get to the more substantial theorems about sequences in Section 8.3 that we start to see tricky proofs.

Exercises

Exercise 8.2.1. **[Used in Example 9.2.4.]** Use only the definition of limits of sequences for each of the following proofs.

(1) Prove that $\left\{\frac{1}{\sqrt{n}}\right\}_{n=1}^{\infty}$ is convergent.

(2) Prove that $\left\{\frac{3n+4}{5n-1}\right\}_{n=1}^{\infty}$ is convergent.

(3) Prove that $\{n\}_{n=1}^{\infty}$ is divergent.

Exercise 8.2.2. **[Used in Example 9.4.5, Example 9.5.7 and Example 10.5.2.]** Let $b, c \in \mathbb{R}$. Suppose that $b \neq c$. Let $\{a_n\}_{n=1}^{\infty}$ be defined by

$$a_n = \begin{cases} b, & \text{if } n \text{ is even} \\ c, & \text{if } n \text{ is odd.} \end{cases}$$

Prove that $\{a_n\}_{n=1}^{\infty}$ is divergent.

Exercise 8.2.3. [Used in Section 8.2.] Let $\{a_n\}_{n=1}^{\infty}$ and $\{b_n\}_{n=1}^{\infty}$ be sequences in $\mathbb{R}$. Suppose that there is some $N \in \mathbb{N}$ such that $n \in \mathbb{N}$ and $n \geq N$ imply $a_n = b_n$. Prove that $\{a_n\}_{n=1}^{\infty}$ is convergent if and only if $\{b_n\}_{n=1}^{\infty}$ is convergent, and if they are convergent then $\lim_{n\to\infty} a_n = \lim_{n\to\infty} b_n$.

Exercise 8.2.4. [Used in Example 9.2.4, Theorem 9.2.5, Theorem 9.4.15 and Corollary 10.4.15.] Let $\{a_n\}_{n=1}^{\infty}$ be a sequence in $\mathbb{R}$. Let $r \in \mathbb{N}$, and let $\{b_n\}_{n=1}^{\infty}$ be the sequence defined by $b_n = a_{n+r}$ for all $n \in \mathbb{R}$. Prove that $\{a_n\}_{n=1}^{\infty}$ is convergent if and only if $\{b_n\}_{n=1}^{\infty}$ is convergent, and if they are convergent then $\lim_{n\to\infty} a_n = \lim_{n\to\infty} b_n$.

Exercise 8.2.5. [Used in Exercise 8.2.11, Theorem 9.4.7, Theorem 9.4.15 and Lemma 10.4.13.] Let $\{a_n\}_{n=1}^{\infty}$ be a sequence in $\mathbb{R}$, and let $L \in \mathbb{R}$. Using only the definition of limits of sequences, prove that $\{a_n\}_{n=1}^{\infty}$ is convergent and $\lim_{n\to\infty} a_n = L$ if and only if $\{a_n - L\}_{n=1}^{\infty}$ is convergent and $\lim_{n\to\infty} (a_n - L) = 0$.

Exercise 8.2.6. [Used in Section 8.2.] Find an example of sequences $\{a_n\}_{n=1}^{\infty}$ and $\{b_n\}_{n=1}^{\infty}$ in $\mathbb{R}$ such that $\{a_n\}_{n=1}^{\infty}$ is convergent and $\lim_{n\to\infty} a_n = 0$, and that $\{a_n b_n\}_{n=1}^{\infty}$ is divergent.

Exercise 8.2.7. Let $\{a_n\}_{n=1}^{\infty}$ be a sequence in $\mathbb{R}$. Suppose that $\{a_n\}_{n=1}^{\infty}$ is convergent. Prove that if $\lim_{n\to\infty} a_n > 0$, then there is some $M > 0$ and some $N \in \mathbb{N}$ such that $n \in \mathbb{N}$ and $n \geq N$ imply $a_n > M$. (This exercise is the analog for sequences of the Sign-Preserving Property for Limits (Theorem 3.2.4).)

Exercise 8.2.8. [Used in Theorem 9.4.15.] Let $\{a_n\}_{n=1}^{\infty}$ and $\{b_n\}_{n=1}^{\infty}$ be sequences in $\mathbb{R}$. Suppose that $\{a_n\}_{n=1}^{\infty}$ and $\{b_n\}_{n=1}^{\infty}$ are convergent. Prove that if $\lim_{n\to\infty} a_n = \lim_{n\to\infty} b_n$, then $\{\min\{a_n, b_n\}\}_{n=1}^{\infty}$ is convergent and $\lim_{n\to\infty} \min\{a_n, b_n\} = \lim_{n\to\infty} a_n = \lim_{n\to\infty} b_n$.

Exercise 8.2.9. [Used in Theorem 8.4.1, Exercise 8.4.9, Exercise 8.4.14 and Exercise 8.4.15.] Let $\{a_n\}_{n=1}^{\infty}$ and $\{b_n\}_{n=1}^{\infty}$ be sequences in $\mathbb{R}$, and let $L \in \mathbb{R}$. Suppose that $\{b_n\}_{n=1}^{\infty}$ is convergent and $\lim_{n\to\infty} b_n = 0$, and that there is some $N \in \mathbb{N}$ such that $n \in \mathbb{N}$ and $n \geq N$ imply $|a_n - L| \leq b_n$. Prove that $\{a_n\}_{n=1}^{\infty}$ is convergent and $\lim_{n\to\infty} a_n = L$.

Exercise 8.2.10. [Used in Example 8.4.3 and Example 9.3.7.] Let $\{a_n\}_{n=1}^{\infty}$ be a sequence in $\mathbb{R}$, and let $f\colon [1, \infty) \to \mathbb{R}$ be a function. Suppose that $f(n) = a_n$ for all $n \in \mathbb{N}$.

(1) Prove that if $\lim_{x\to\infty} f(x)$ exists, then $\{a_n\}_{n=1}^{\infty}$ is convergent and $\lim_{n\to\infty} a_n = \lim_{x\to\infty} f(x)$.

(2) This part of the exercise makes use of Exercise 6.2.15. Prove that if $\lim_{x\to\infty} f(x) = \infty$, then $\lim_{n\to\infty} a_n = \infty$.

(3) Find an example of a sequence $\{a_n\}_{n=1}^{\infty}$ and a function f that satisfy the hypotheses of this exercise, and such that $\{a_n\}_{n=1}^{\infty}$ is convergnet and $\lim_{x\to\infty} f(x)$ does not exist.

Exercise 8.2.11. **[Used in Theorem 8.2.9.]** Prove Theorem 8.2.9 (1) (2) (3) (5).

[Use Exercise 8.2.5.]

Exercise 8.2.12. **[Used in Example 9.2.4.]** Let $\{a_n\}_{n=1}^{\infty}$ and $\{b_n\}_{n=1}^{\infty}$ be sequences in $\mathbb{R}$, and let $k \in \mathbb{R}$. Suppose that $\{a_n\}_{n=1}^{\infty}$ is divergent and $\{b_n\}_{n=1}^{\infty}$ is convergent.

(1) Prove that $\{a_n + b_n\}_{n=1}^{\infty}$ is divergent.
(2) Prove that $\{a_n - b_n\}_{n=1}^{\infty}$ is divergent.
(3) Prove that if $k \neq 0$, then $\{ka_n\}_{n=1}^{\infty}$ is divergent.
(4) Find an example of sequences $\{c_n\}_{n=1}^{\infty}$ and $\{d_n\}_{n=1}^{\infty}$ in $\mathbb{R}$ such that $\{c_n\}_{n=1}^{\infty}$ and $\{d_n\}_{n=1}^{\infty}$ are divergent and $\{c_n + d_n\}_{n=1}^{\infty}$ is convergent.

Exercise 8.2.13. **[Used in Example 8.2.13, Theorem 9.4.15, Exercise 10.3.5, Corollary 10.4.15 and Theorem 10.5.2.]** Let $\{a_n\}_{n=1}^{\infty}$ be a sequence in $\mathbb{R}$.

(1) Prove that if $\{a_n\}_{n=1}^{\infty}$ is convergent, then $\{|a_n|\}_{n=1}^{\infty}$ is convergent and $\lim_{n\to\infty} |a_n| = |\lim_{n\to\infty} a_n|$.
(2) Prove that $\{a_n\}_{n=1}^{\infty}$ is convergent and $\lim_{n\to\infty} a_n = 0$ if and only if $\{|a_n|\}_{n=1}^{\infty}$ is convergent and $\lim_{n\to\infty} |a_n| = 0$.
(3) Find an example of a sequence $\{b_n\}_{n=1}^{\infty}$ in $\mathbb{R}$ such that $\{|b_n|\}_{n=1}^{\infty}$ is convergent and $\{b_n\}_{n=1}^{\infty}$ is divergent.

Exercise 8.2.14. **[Used in Section 8.3.]** Let $r \in \mathbb{R}$. Prove that there is a sequence $\{q_n\}_{n=1}^{\infty}$ in $\mathbb{Q}$ such that $\lim_{n\to\infty} q_n = r$.

Exercise 8.2.15. **[Used in Example 9.5.2 and Example 9.2.4.]** Using only Definition 8.2.14, prove that $\lim_{n\to\infty} n = \infty$.

Exercise 8.2.16. **[Used in Theorem 10.5.2.]** Let $r \in (1, \infty)$. Prove that $\lim_{n\to\infty} r^n = \infty$.

Exercise 8.2.17. **[Used in Example 9.2.4, Example 9.5.2 and Theorem 10.5.2.]** Let $\{a_n\}_{n=1}^{\infty}$ and $\{b_n\}_{n=1}^{\infty}$ be sequences in $\mathbb{R}$, and let $k \in \mathbb{R}$. Suppose that $\lim_{n\to\infty} a_n = \infty$ and $\{b_n\}_{n=1}^{\infty}$ is convergent.

(1) Prove that $\lim_{n\to\infty} (a_n + b_n) = \infty$.
(2) Prove that $\lim_{n\to\infty} (a_n + k) = \infty$.
(3) Prove that if $k > 0$ then $\lim_{n\to\infty} ka_n = \infty$.

Exercise 8.2.18. **[Used in Example 9.2.4 and Theorem 10.5.2.]** Let $\{a_n\}_{n=1}^{\infty}$ and $\{b_n\}_{n=1}^{\infty}$ be sequences in $\mathbb{R}$. Suppose that $\lim_{n\to\infty} a_n = \infty$.

(1) Suppose that there is some $N \in \mathbb{N}$ such that $n \in \mathbb{N}$ and $n \geq N$ imply $a_n \leq b_n$. Prove that $\lim_{n\to\infty} b_n = \infty$.
(2) Suppose that there is some $N \in \mathbb{N}$ such that $n \in \mathbb{N}$ and $n \geq N$ imply $a_n \leq |b_n|$. Prove that $\{b_n\}_{n=1}^{\infty}$ is divergent.

8.3 Three Important Theorems

The basic properties of sequences that we saw in Section 8.2 rely only upon the algebraic properties of the real numbers; that is, these properties would also hold for sequences in the set of rational numbers. We now turn to three important theorems about sequences, all of which rely upon the Least Upper Bound Property of the real numbers.

As motivation for what we will see in this section, recall Theorem 5.4.7, which gave a characterization of integrability that does not require a guess as to the value of the integral. That theorem stated, essentially, that if all the Riemann sums of a function on a non-degenerate closed bounded interval became closer and closer to each other for partitions with smaller and smaller norms, then the function is integrable. Similarly, it would be nice to be able to prove that a sequence is convergent without having to guess the value of the limit of the sequence, because in some situations it is not feasible to make such a guess. In particular, we will now see the (much simpler) analog of Theorem 5.4.7 for sequences, which states intuitively that a sequence is convergent if and only if the terms of the sequence get closer and closer to each other as n gets larger and larger. This characterization of the convergence of sequences, called the Cauchy Completeness Theorem, will be given in Corollary 8.3.16 below. (There is an analogous result for limits of functions given in Exercise 3.2.18, but the version for sequences is more widely known, and much more widely used.) Before we will be ready to prove that result, however, we will need some other important concepts and theorems about sequences; these other theorems do not provide complete characterizations of the convergence of sequences, but they state some conditions that guarantee convergence, and are useful in their own right.

The first of the three important theorems of this section, called the Monotone Convergence Theorem, needs the following definition, which is analogous to Definition 4.5.1.

Definition 8.3.1. Let $\{a_n\}_{n=1}^{\infty}$ be a sequence in $\mathbb{R}$.

1. The sequence $\{a_n\}_{n=1}^{\infty}$ is **increasing** if $n < m$ implies $a_n \leq a_m$ for all $n, m \in \mathbb{N}$.
2. The sequence $\{a_n\}_{n=1}^{\infty}$ is **strictly increasing** if $n < m$ implies $a_n < a_m$ for all $n, m \in \mathbb{N}$.
3. The sequence $\{a_n\}_{n=1}^{\infty}$ is **decreasing** if $n < m$ implies $a_n \geq a_m$ for all $n, m \in \mathbb{N}$.
4. The sequence $\{a_n\}_{n=1}^{\infty}$ is **strictly decreasing** if $n < m$ implies $a_n > a_m$ for all $n, m \in \mathbb{N}$.
5. The sequence $\{a_n\}_{n=1}^{\infty}$ is **monotone** if it is either increasing or decreasing.
6. The sequence $\{a_n\}_{n=1}^{\infty}$ is **strictly monotone** if it is either strictly increasing or strictly decreasing. △

Some books use the terms "non-decreasing" and "increasing" to mean what we call "increasing" and "strictly increasing," respectively, and similarly for decreasing. As was the case with functions, there is no definitive terminology here.

Example 8.3.2.

(1) We will show that the sequence $\left\{\frac{n+1}{n+2}\right\}_{n=1}^{\infty}$ is strictly increasing, and hence it is

strictly monotone. Let us denote this sequence by $\{a_n\}_{n=1}^{\infty}$. One approach is as follows. Let $n, m \in \mathbb{N}$, and suppose that $n < m$. Then $a_m - a_n = \frac{m+1}{m+2} - \frac{n+1}{n+2} = \frac{m-n}{(n+1)(n+2)} > 0$, and therefore $a_n < a_m$. Hence $\{a_n\}_{n=1}^{\infty}$ is strictly increasing. Another approach is to consider the function $f\colon [1, \infty) \to \mathbb{R}$ defined by $f(x) = \frac{x+1}{x+2}$ for all $x \in [1, \infty)$. Then $f'(x) = \frac{1}{(x+2)^2}$ for all $x \in (1, \infty)$. Hence $f'(x) > 0$ for all $x \in (1, \infty)$, and it follows from Theorem 4.5.2 (2) that f is strictly increasing. By restricting f to the natural numbers, we see that $\{a_n\}_{n=1}^{\infty}$ is strictly increasing.

(2) The sequence $\{(-1)^n\}_{n=1}^{\infty}$ is not monotone; we omit the details. ◊

The following theorem, though not very difficult to prove, will immediately imply the Monotone Convergence Theorem (Corollary 8.3.4). For the proof of Theorem 8.3.3, it is important to keep in mind the difference between a sequence $\{a_n\}_{n=1}^{\infty}$, and the set of all of its terms $\{a_n \mid n \in \mathbb{N}\}$. The set $\{a_n \mid n \in \mathbb{N}\}$ is not a sequence, and cannot have a limit, though it can have a least upper bound or a greatest lower bound. The idea of this theorem is that if a sequence is increasing, then it can do one of two things, namely, either increase without bound, in which case the sequence converges to infinity, or not increase without bound, in which case the sequence is bounded above, and it would then seem plausible that there is some number to which the sequence converges.

Theorem 8.3.3. *Let $\{a_n\}_{n=1}^{\infty}$ be a sequence in $\mathbb{R}$.*

1. *If $\{a_n\}_{n=1}^{\infty}$ is increasing and bounded above, then $\{a_n\}_{n=1}^{\infty}$ is convergent and*
$$\lim_{n\to\infty} a_n = \operatorname{lub}\{a_n \mid n \in \mathbb{N}\}.$$
2. *If $\{a_n\}_{n=1}^{\infty}$ is increasing and not bounded above, then $\lim_{n\to\infty} a_n = \infty$.*
3. *If $\{a_n\}_{n=1}^{\infty}$ is decreasing and bounded below, then $\{a_n\}_{n=1}^{\infty}$ is convergent and*
$$\lim_{n\to\infty} a_n = \operatorname{glb}\{a_n \mid n \in \mathbb{N}\}.$$
4. *If $\{a_n\}_{n=1}^{\infty}$ is decreasing and not bounded below, then $\lim_{n\to\infty} a_n = -\infty$.*

Proof. We will prove Part (1), leaving Part (2) to the reader in Exercise 8.3.4; the other two parts are similar, and we omit the details.

(1) Suppose that $\{a_n\}_{n=1}^{\infty}$ is increasing and bounded above. Let $A = \{a_n \mid n \in \mathbb{N}\}$. Then A is bounded above. Clearly $A \neq \emptyset$, and hence the Least Upper Bound Property implies that A has a least upper bound. Let $\varepsilon > 0$. By Lemma 2.6.5 (1) there is some $a_N \in A$ such that $\operatorname{lub} A - \varepsilon < a_N \leq \operatorname{lub} A$. Suppose that $n \in \mathbb{N}$ and $n \geq N$. Because $\{a_n\}_{n=1}^{\infty}$ is increasing, it follows that $\operatorname{lub} A - \varepsilon < a_N \leq a_n \leq \operatorname{lub} A$, and hence $|a_n - \operatorname{lub} A| < \varepsilon$. Therefore $\lim_{n\to\infty} a_n = \operatorname{lub} A$. □

The following corollary is an immediate consequence of Theorem 8.3.3 and Lemma 8.2.6.

Corollary 8.3.4 (Monotone Convergence Theorem). *Let $\{a_n\}_{n=1}^{\infty}$ be a sequence in $\mathbb{R}$. Suppose that $\{a_n\}_{n=1}^{\infty}$ is monotone. Then $\{a_n\}_{n=1}^{\infty}$ is convergent if and only if $\{a_n\}_{n=1}^{\infty}$ is bounded.*

The Monotone Convergence Theorem (Corollary 8.3.4) is useful for proving other theorems, for example some results about the convergence of series in Section 9.3. Moreover, this theorem is nice in that it is essentially the sequence analog of the Least Upper Bound Property. We will see in Theorem 8.3.17 below that the Monotone Convergence Theorem is equivalent to the Least Upper Bound Property; in fact, some people use the former in the axioms for the real numbers instead of the latter.

As nice as the Monotone Convergence Theorem is in principle, however, it is not very useful from the point of view of proving that specific sequences are convergent, because most sequences are not monotone. We now turn to another approach to finding convergent sequences. Consider the sequence $\{(-1)^n\}_{n=1}^{\infty}$, which is certainly not monotone. We notice, however, that even though this sequence is not convergent, it "contains" two convergent sequences inside it, namely, the collection of all terms that have value 1, and the collection of all terms that have value -1. In order to state our next important theorem, the Bolzano–Weierstrass Theorem, given as Theorem 8.3.9 below, we first need to discuss the general notion of a sequence contained in another sequence.

Consider the sequence $\left\{\frac{1}{n}\right\}_{n=1}^{\infty}$, which we can write out as

$$\frac{1}{1}, \frac{1}{2}, \frac{1}{3}, \frac{1}{4}, \ldots. \tag{8.3.1}$$

We can find many sequences that are "contained" in this sequence, for example the sequences

$$\frac{1}{1}, \frac{1}{3}, \frac{1}{5}, \frac{1}{7}, \ldots$$

and

$$\frac{1}{2}, \frac{1}{4}, \frac{1}{8}, \frac{1}{16}, \ldots. \tag{8.3.2}$$

A sequence is not just a countably infinite set, but it is a countably infinite set listed in a specific order, and it is important to note that for one sequence to be "contained" in another the order of the terms must be preserved. For example, we do not consider the sequence

$$\frac{1}{4}, \frac{1}{2}, \frac{1}{8}, \frac{1}{6}, \ldots$$

as being "contained" as a sequence in the sequence given in Equation 8.3.1, even though it is contained as a subset. In order to preserve the order of the original sequence, we need to look at the subscripts of the terms of the sequences under consideration. For example, if the sequence in Equation 8.3.1 is denoted $\{a_n\}_{n=1}^{\infty}$, then we could denote the sequence given in Equation 8.3.2 by $\{a_{2^n}\}_{n=1}^{\infty}$; the reason that the order of the terms in the original sequence is preserved is that the sequence of subscripts $\{2^n\}_{n=1}^{\infty}$ is strictly increasing. This idea is formalized in the following definition.

Definition 8.3.5. Let $\{a_n\}_{n=1}^{\infty}$ be a sequence in $\mathbb{R}$. Suppose that $\{a_n\}_{n=1}^{\infty}$ is defined by a function $f\colon \mathbb{N} \to \mathbb{R}$ such that $f(n) = a_n$ for all $n \in \mathbb{N}$. Let $g\colon \mathbb{N} \to \mathbb{N}$ be a function. Suppose that g is strictly increasing. Then the sequence defined by the function $f \circ g\colon \mathbb{N} \to \mathbb{R}$ is a **subsequence** of $\{a_n\}_{n=1}^{\infty}$. The sequence defined by $f \circ g$ is written as $\left\{a_{n_k}\right\}_{k=1}^{\infty}$, where $g(k) = n_k$ for all $k \in \mathbb{N}$. △

Although Definition 8.3.5 is phrased in terms of the function g, in practice we will virtually never explicitly mention this function, and we will define subsequences by writing expressions of the form $\left\{a_{n_k}\right\}_{k=1}^{\infty}$. Observe that the function g in Definition 8.3.5 is strictly increasing if and only if the sequence $\{n_k\}_{k=1}^{\infty}$ is a strictly increasing sequence in $\mathbb{N}$.

Example 8.3.6.

(1) Let $\{a_n\}_{n=1}^{\infty}$ be a sequence in $\mathbb{R}$. Taking every other term of this sequence starting with the first yields the subsequence $\{a_{2n-1}\}_{n=1}^{\infty}$. Formally, using Definition 8.3.5, we use the function $g\colon \mathbb{N} \to \mathbb{N}$ defined by $g(n) = 2n - 1$ for all $n \in \mathbb{N}$ to define this subsequence, though we do so only to show that it can be done, and from now on we will not write the function g explicitly.

(2) Let $\{b_n\}_{n=1}^{\infty}$ be defined by $b_n = (-1)^n$ for all $n \in \mathbb{N}$. We saw in Example 8.2.4 (4) that $\{b_n\}_{n=1}^{\infty}$ is divergent. The subsequence $\{b_{2n}\}_{n=1}^{\infty}$ is the sequence that is constantly 1, and this subsequence is convergent by Example 8.2.4 (1). Hence, we see that a sequence can be divergent, but have a convergent subsequence.

(3) The sequence $\{n\}_{n=1}^{\infty}$ is divergent, and every subsequence of it is also divergent. Hence, not every sequence has a convergent subsequence.

(4) Let $\{c_n\}_{n=1}^{\infty}$ be defined by

$$c_n = \begin{cases} 1, & \text{if } n \text{ is even} \\ n, & \text{if } n \text{ is odd.} \end{cases}$$

The sequence $\{c_n\}_{n=1}^{\infty}$ is divergent, and it is not bounded, but in contrast to the sequence in Part (3) of this example, the sequence $\{c_n\}_{n=1}^{\infty}$ has a convergent subsequence, which is $\{c_{2n}\}_{n=1}^{\infty}$. ◊

The following lemma says that if a sequence is convergent, then so is every subsequence.

Lemma 8.3.7. *Let $\{a_n\}_{n=1}^{\infty}$ be a sequence in $\mathbb{R}$, and let $\left\{a_{n_k}\right\}_{k=1}^{\infty}$ be a subsequence of $\{a_n\}_{n=1}^{\infty}$. If $\{a_n\}_{n=1}^{\infty}$ is convergent, then $\left\{a_{n_k}\right\}_{k=1}^{\infty}$ is convergent and $\lim_{k\to\infty} a_{n_k} = \lim_{n\to\infty} a_n$.*

Proof. Suppose that $\{a_n\}_{n=1}^{\infty}$ is convergent. Let $\varepsilon > 0$. Let $L = \lim_{n\to\infty} a_n$. Then there is some $N \in \mathbb{N}$ such that $n \in \mathbb{N}$ and $n \geq N$ imply $|a_n - L| < \varepsilon$. Suppose that $k \in \mathbb{N}$ and $k \geq N$. Then by Exercise 8.3.5 we see that $n_k \geq k \geq N$. Hence $|a_{n_k} - L| < \varepsilon$. It follows that $\lim_{k\to\infty} a_{n_k} = L$. □

In Example 8.3.6 (2) we saw a divergent sequence that has a convergent subsequence, and hence Lemma 8.3.7 cannot be made into an if and only if statement.

In Example 8.3.6 (4) we saw a divergent sequence that is not bounded and that has a convergent subsequence, and in Example 8.3.6 (3) we saw a divergent sequence that is not bounded and that does not have a convergent subsequence. In Example 8.3.6 (2) we saw a divergent sequence that is bounded and that has a convergent subsequence. The reader will have noticed that missing from Example 8.3.6 is a divergent sequence that is bounded and that does not have a convergent subsequence. It turns out, as we will see by the Bolzano–Weierstrass Theorem, which is the second of our important theorems and which is stated as Theorem 8.3.9 below, that no such example exists. We start with the following somewhat surprising lemma. Refer to Example 2.5.13 for the definition of a greatest element of a set.

Lemma 8.3.8. *Let $\{a_n\}_{n=1}^{\infty}$ be a sequence in $\mathbb{R}$. Then $\{a_n\}_{n=1}^{\infty}$ has a monotone subsequence.*

Proof. For each $k \in \mathbb{N}$, let $T_k = \{a_n \mid n \in \mathbb{N} \text{ and } n \geq k\}$. There are two cases.

First, suppose that T_k has a greatest element for each $k \in \mathbb{N}$. We define a sequence $\{n_k\}_{k=1}^{\infty}$ in $\mathbb{N}$ using Definition by Recursion as follows. Let $n_1 \in \mathbb{N}$ be such that a_{n_1} is the greatest element of T_1. (Although the set T_1 has a unique greatest element by Exercise 2.5.17, there may be more than one $i \in \mathbb{N}$ such that a_i is the greatest element of T_1, in which case we let n_1 be one such number; to avoid choosing, it would be possible to select the smallest possible n_1 by the Well-Ordering Principle (Theorem 1.2.10, Axiom 1.4.4 or Theorem 2.4.6).) Let $n_2 \in \mathbb{N}$ be such that $n_2 \geq n_1 + 1$ and a_{n_2} is the greatest element of T_{n_1+1}. Then $n_2 > n_1$. Similarly, there is some $n_3 \in \mathbb{N}$ such that $n_3 > n_2$ and a_{n_3} is the greatest element of T_{n_2+1}. Continuing in this way, we define a sequence $\{n_k\}_{k=1}^{\infty}$ in $\mathbb{N}$ that is strictly increasing, and such that a_{n_k} is the greatest element of $T_{n_{k-1}+1}$ for all $k \in \mathbb{N}$ such that $k \geq 2$.

We have therefore defined a subsequence $\{a_{n_k}\}_{k=1}^{\infty}$ of $\{a_n\}_{n=1}^{\infty}$. Let $i, j \in \mathbb{N}$. Suppose that $i < j$. Then $j \geq 2$, and hence a_{n_j} is the greatest element of $T_{n_{j-1}+1}$. There are now two subcases. First, suppose that $i = 1$. Then a_{n_i} is the greatest element of T_1. Because $n_{j-1} + 1 > 1$, then $T_1 \supseteq T_{n_{j-1}+1}$. It follows from Exercise 2.5.18 (1) that $a_{n_i} \geq a_{n_j}$. Second, suppose that $i > 1$. Then $i \geq 2$, and hence a_{n_i} is the greatest element of $T_{n_{i-1}+1}$. Because $\{n_k\}_{k=1}^{\infty}$ is a strictly increasing subsequence of $\mathbb{N}$, then $i < j$ implies that $n_{i-1} + 1 < n_{j-1} + 1$, and hence $T_{n_{i-1}+1} \supseteq T_{n_{j-1}+1}$. It follows from Exercise 2.5.18 (1) again that $a_{n_i} \geq a_{n_j}$. By putting together the two subcases, we see that $\{a_{n_k}\}_{k=1}^{\infty}$ is decreasing, and is therefore monotone.

Second, suppose that there is some $r \in \mathbb{N}$ such that T_r does not have a greatest element. We define a sequence $\{m_k\}_{k=1}^{\infty}$ in $\mathbb{N}$ using Definition by Recursion as follows. Let $m_1 = r$. Then $a_{m_1} \in T_{m_1} = T_r$, and by hypothesis a_{m_1} is not a greatest element of T_{m_1}. Therefore there is some $m_2 \in \mathbb{N}$ such that $m_2 \geq m_1 + 1$ and $a_{m_2} > a_{m_1}$. (As before, the choice of m_2 is not necessarily unique; to avoid choosing, it would be possible to select the smallest possible m_2 by the Well-Ordering Principle.) Hence $m_2 > m_1$. Then $T_{m_1} \supseteq T_{m_2}$ and $T_{m_1} - T_{m_2} = \{a_{m_1}, a_{m_1+1}, \ldots, a_{m_2-1}\}$, which is a finite set. Using Exercise 2.5.18 (2) we see that T_{m_2} has no greatest element. Similarly, there is some $m_3 \in \mathbb{N}$ such that $m_3 > m_2$ and $a_{m_3} > a_{m_2}$. Continuing in this way, we define a sequence $\{m_k\}_{k=1}^{\infty}$ in $\mathbb{N}$ that is strictly increasing, and such that the subsequence $\{a_{m_k}\}_{k=1}^{\infty}$ of $\{a_n\}_{n=1}^{\infty}$ is strictly increasing. Hence $\{a_{m_k}\}_{k=1}^{\infty}$ is monotone. □

We are now ready to prove the second of our important theorems, which is both interesting in its own right, and is a very useful tool in real analysis.

Theorem 8.3.9 (Bolzano–Weierstrass Theorem). *Let $\{a_n\}_{n=1}^{\infty}$ be a sequence in $\mathbb{R}$. If $\{a_n\}_{n=1}^{\infty}$ is bounded, then $\{a_n\}_{n=1}^{\infty}$ has a convergent subsequence.*

Proof. Suppose that $\{a_n\}_{n=1}^{\infty}$ is bounded. By Lemma 8.3.8 we know that $\{a_n\}_{n=1}^{\infty}$ has a monotone subsequence. Because $\{a_n\}_{n=1}^{\infty}$ is bounded then so is the monotone subsequence, and hence this subsequence is convergent by the Monotone Convergence Theorem (Corollary 8.3.4). □

Another proof of the Bolzano–Weierstrass Theorem (Theorem 8.3.9), which uses the Nested Interval Theorem (Theorem 8.4.7) instead of the Monotone Convergence Theorem (Corollary 8.3.4), is given in Exercise 8.4.9; the reader in encouraged to try that exercise after reading Section 8.4, because it is based upon a very nice idea.

Whereas the Bolzano–Weierstrass Theorem is stated in terms of sequences, the essential idea of the theorem, which is that an infinite collection of points that are confined to an appropriate region must get arbitrarily close to some single point, holds in more general situations. This more general idea is related to the topological notion of "compactness"; see [Mun00, Section 28] for details.

We are now ready for the characterization of the convergence of sequences promised at the beginning of this section. We start with a precise definition of the notion of the terms of a sequence getting closer and closer to each other.

Definition 8.3.10. Let $\{a_n\}_{n=1}^{\infty}$ be a sequence in $\mathbb{R}$. The sequence $\{a_n\}_{n=1}^{\infty}$ is a **Cauchy sequence** if for each $\varepsilon > 0$, there is some $N \in \mathbb{N}$ such that $n, m \in \mathbb{N}$ and $n, m \geq N$ imply $|a_n - a_m| < \varepsilon$. △

Example 8.3.11.

(1) We will prove that $\left\{\frac{1}{n}\right\}_{n=1}^{\infty}$ is a Cauchy sequence. Let $\varepsilon > 0$. By Corollary 2.6.8 (2) there is some $N \in \mathbb{N}$ such that $\frac{1}{N} < \varepsilon$. Suppose that $n, m \in \mathbb{N}$ and $n, m \geq N$. Without loss of generality, assume that $m \geq n$. Then $|m - n| < m$, and therefore

$$\left|\frac{1}{n} - \frac{1}{m}\right| = \frac{|m-n|}{nm} < \frac{m}{nm} = \frac{1}{n} \leq \frac{1}{N} < \varepsilon.$$

(2) We will prove that $\{(-1)^n\}_{n=1}^{\infty}$ is not a Cauchy sequence. Let $\varepsilon = 1$. Let $N \in \mathbb{N}$ be an integer. Choose $n, m \in \mathbb{N}$ such that $n, m \geq N$, and that n is even and m is odd. Then

$$|(-1)^n - (-1)^m| = |1 - (-1)| = 2 > \varepsilon.$$

We therefore see that for the given ε, there is no "N" that works, and hence $\{(-1)^n\}_{n=1}^{\infty}$ is not a Cauchy sequence. ◊

What is the relation between a sequence being a Cauchy sequence and being convergent? One aspect of this relationship is quite simple. If a sequence is convergent, then its terms get closer and closer to a single number, which implies that the terms must also get closer and closer to each other.

Theorem 8.3.12. *Let $\{a_n\}_{n=1}^{\infty}$ be a sequence in $\mathbb{R}$. If $\{a_n\}_{n=1}^{\infty}$ is convergent, then $\{a_n\}_{n=1}^{\infty}$ is a Cauchy sequence.*

Proof. Suppose that $\{a_n\}_{n=1}^{\infty}$ is convergent. Let $L = \lim_{n\to\infty} a_n$. Let $\varepsilon > 0$. There is some $N \in \mathbb{N}$ such that $n \in \mathbb{N}$ and $n \geq N$ imply $|a_n - L| < \frac{\varepsilon}{2}$. Suppose that $n,m \in \mathbb{N}$ and $n,m \geq N$. Then

$$|a_n - a_m| = |a_n - L + L - a_m| \leq |a_n - L| + |L - a_m| < \frac{\varepsilon}{2} + \frac{\varepsilon}{2} = \varepsilon. \qquad \square$$

The proof of Theorem 8.3.12 does not use anything other than the definitions of convergence and Cauchy sequences, and basic algebraic properties of the real numbers. In particular, the Least Upper Bound Property was not used. Hence, this theorem would still be true if the real numbers were replaced with the rational numbers. On the other hand, the converse to Theorem 8.3.12 is definitely not true for the rational numbers. For example, let $\{t_n\}_{n=1}^{\infty}$ be a sequence of rational numbers that converges to $\sqrt{2}$; such a sequence exists by Exercise 8.2.14. Because a sequence of rational numbers is also a sequence of real numbers, and because $\{t_n\}_{n=1}^{\infty}$ converges in $\mathbb{R}$ to $\sqrt{2}$, then it follows from Theorem 8.3.12 that this sequence is a Cauchy sequence when viewed as a sequence in $\mathbb{R}$. It is then also a Cauchy sequence when viewed as a sequence in $\mathbb{Q}$. However, because $\sqrt{2}$ is not rational, as we know by Theorem 2.6.11, it follows that $\{t_n\}_{n=1}^{\infty}$ is not convergent as a sequence in $\mathbb{Q}$, even though it is a Cauchy sequence in $\mathbb{Q}$.

In contrast to the situation in $\mathbb{Q}$, we will see in Theorem 8.3.15 below that any Cauchy sequence in $\mathbb{R}$ is convergent in $\mathbb{R}$, a fact that, as expected, ultimately relies upon the Least Upper Bound Property. We break up the proof of Theorem 8.3.15 into two lemmas, which we now state and prove. The proofs of these lemmas do not make use of the Least Upper Bound Property, but the proof of Theorem 8.3.15 will use the Bolzano–Weierstrass Theorem, which relies upon the Monotone Convergence Theorem, which in turn uses the Least Upper Bound Property in its proof.

Lemma 8.3.13. *Let $\{a_n\}_{n=1}^{\infty}$ be a sequence in $\mathbb{R}$. Suppose that $\{a_n\}_{n=1}^{\infty}$ is a Cauchy sequence. If $\{a_n\}_{n=1}^{\infty}$ has a convergent subsequence $\{a_{n_k}\}_{k=1}^{\infty}$, then $\{a_n\}_{n=1}^{\infty}$ is convergent and $\lim_{n\to\infty} a_n = \lim_{k\to\infty} a_{n_k}$.*

Proof. Suppose that $\{a_{n_k}\}_{k=1}^{\infty}$ is a convergent subsequence of $\{a_n\}_{n=1}^{\infty}$. Let $L = \lim_{k\to\infty} a_{n_k}$. Let $\varepsilon > 0$. Because $\{a_n\}_{n=1}^{\infty}$ is a Cauchy sequence, there is some $N \in \mathbb{N}$ such that $n,m \in \mathbb{N}$ and $n,m \geq N$ imply $|a_n - a_m| < \frac{\varepsilon}{2}$. Because $\lim_{k\to\infty} a_{n_k} = L$, there is some $M \in \mathbb{N}$ such that $k \in \mathbb{N}$ and $k \geq M$ imply $|a_k - L| < \frac{\varepsilon}{2}$. Let $P = \max\{M,N\}$. Suppose that $n \in \mathbb{N}$ and $n \geq P$. Then $n \geq N$. By Exercise 8.3.5 we know that $n_P \geq P$, and hence $n_P \geq M$ and $n_P \geq N$. Then

$$|a_n - L| = |a_n - a_{n_P} + a_{n_P} - L| \leq |a_n - a_{n_P}| + |a_{n_P} - L| < \frac{\varepsilon}{2} + \frac{\varepsilon}{2} = \varepsilon. \qquad \square$$

Lemma 8.3.14. *Let $\{a_n\}_{n=1}^{\infty}$ be a sequence in $\mathbb{R}$. If $\{a_n\}_{n=1}^{\infty}$ is a Cauchy sequence, then $\{a_n\}_{n=1}^{\infty}$ is bounded.*

Proof. Left to the reader in Exercise 8.3.15. □

We are now ready for our main result about Cauchy sequences.

Theorem 8.3.15. *Let $\{a_n\}_{n=1}^{\infty}$ be a sequence in $\mathbb{R}$. If $\{a_n\}_{n=1}^{\infty}$ is a Cauchy sequence, then $\{a_n\}_{n=1}^{\infty}$ is convergent.*

Proof. Suppose that $\{a_n\}_{n=1}^{\infty}$ is a Cauchy sequence. By Lemma 8.3.14 we know that $\{a_n\}_{n=1}^{\infty}$ is bounded. By the Bolzano–Weierstrass Theorem (Theorem 8.3.9) we know that $\{a_n\}_{n=1}^{\infty}$ has a convergent subsequence. By Lemma 8.3.13 we deduce that $\{a_n\}_{n=1}^{\infty}$ is convergent. □

The following corollary, the third of our important theorems, is simply a combination of Theorem 8.3.12 and Theorem 8.3.15.

Corollary 8.3.16 (Cauchy Completeness Theorem). *Let $\{a_n\}_{n=1}^{\infty}$ be a sequence in $\mathbb{R}$. Then $\{a_n\}_{n=1}^{\infty}$ is convergent if and only if $\{a_n\}_{n=1}^{\infty}$ is a Cauchy sequence.*

The question of whether or not Cauchy sequences are convergent is important not only for the real numbers, but also in the more general context of metric spaces, where Cauchy sequences are not always convergent, and where such a space is called "complete" if all Cauchy sequences are convergent; see [Mun00, Sections 20 and 43] for details.

As mentioned at the beginning of this section, our three main theorems rely upon the Least Upper Bound Property. We will now see that two of these theorems, namely, the Monotone Convergence Theorem (Corollary 8.3.4) and the Bolzano–Weierstrass Theorem (Theorem 8.3.9), are each equivalent to the Least Upper Bound Property. For a discussion of what we mean by "equivalent" in this context, see the discussion in Section 3.5 prior to Lemma 3.5.3. Somewhat surprisingly, the Cauchy Completeness Theorem (Corollary 8.3.16) does not imply the Least Upper Bound Property. In the proof that the Monotone Convergence Theorem and the Bolzano–Weierstrass Theorem imply the Least Upper Bound Property, we will make use of the fact that $\lim_{n\to\infty} \left(\frac{1}{2}\right)^{n-1} = 0$, and the proof of that fact, given in Example 8.2.13, made use of the fact that $\lim_{n\to\infty} \frac{1}{n} = 0$. This last fact was proved in Example 8.2.4 (2), and that proof made use of Corollary 2.6.8 (2), which in turn was a consequence of the Archimedean Property (Theorem 2.6.7). If an ordered field does not satisfy the Archimedean Property, then by Exercise 8.3.16 we see that $\lim_{n\to\infty} \frac{1}{n}$ does not converge to 0, which may seem strange, but it demonstrates the importance of the Archimedean Property, and that this property should not be taken for granted. The proof of the Archimedean Property that we saw in Section 2.6 made use of the Least Upper Bound Property, though, as seen in Exercise 8.3.18, the Archimedean Property can also be proved using the Monotone Convergence Theorem. Hence, when proving that the Monotone Convergence Theorem implies the Least Upper Bound Property, we can make use of the Archimedean Property and any of its consequences.

However, and this is the surprising part, it is not the case that the Cauchy Completeness Theorem implies the Archimedean Property, because there are examples of ordered fields in which every Cauchy sequence is convergent, but for which the Archimedean Property does not hold; see [GO03, Chapter 1, Example 7] for such an example. Because the Least Upper Bound Property implies the Archimedean Property, we deduce that the Cauchy Completeness Theorem is not equivalent to the Least Upper Bound Property. (It is the case that the Cauchy Completeness Theorem together with the Archimedean Property are equivalent to the Least Upper Bound Property; see [Olm62, Appendix Section 5] for a proof.) However, we note that our reliance on the Least Upper Bound Property in the proof of the Cauchy Completeness Theorem was necessary, and was not simply a matter of convenience (the reader is urged to trace the proof back to the Least Upper Bound Property). There exist ordered fields that do not satisfy the Cauchy Completeness Theorem (for example the rational numbers), and hence any proof of the Cauchy Completeness Theorem for the real numbers must ultimately rely upon some aspect of the real numbers beyond the axiom for an ordered field, and the only axiom for the real numbers other than that of an ordered field is the Least Upper Bound Property.

Theorem 8.3.17. *The following are equivalent.*

a. The Least Upper Bound Property.
b. The Monotone Convergence Theorem.
c. The Bolzano–Weierstrass Theorem.

Proof. We have already seen that the axioms of the real numbers, that is, the axiom for an ordered field together with the Least Upper Bound Property, imply the Monotone Convergence Theorem and the Bolzano–Weierstrass Theorem. We also know that the Monotone Convergence Theorem implies the Bolzano–Weierstrass Theorem, as can be seen by examining the proof of the latter. It can also be seen that the Bolzano–Weierstrass Theorem implies the Monotone Convergence Theorem, by combining the former with Exercise 8.3.8. Hence the Bolzano–Weierstrass Theorem is equivalent to the Monotone Convergence Theorem, and to complete the proof, it will suffice to show that the Monotone Convergence Theorem together with the axiom for an ordered field imply the Least Upper Bound Property.

The proof is by contradiction. Suppose that F is an ordered field that satisfies the Monotone Convergence Theorem, but does not satisfy the Least Upper Bound Property. We note by Exercise 8.3.18 that F satisfies the Archimedean Property.

Let a, b, A, P and Q be as in Lemma 3.5.3. By Parts (1) and (2) of that lemma we know that $P \cup Q = [a,b]$, and $P \cap Q = \emptyset$, and $a < b$, and $A \cap [a,b] \subseteq P$, and $a \in P$, and $b \in Q$.

By Exercise 8.3.17 there is a family $\{[a_n,b_n]\}_{n=1}^{\infty}$ of closed bounded intervals in $[a,b]$ such that for each $i \in \mathbb{N}$ the following three conditions hold: (1) $a_i \in P$ and $b_i \in Q$; (2) $[a_{i+1},b_{i+1}] \subseteq [a_i,b_i]$; and (3) $b_i - a_i = \frac{b-a}{2^{i-1}}$. By Condition (2) we know that $a_i \le a_{i+1}$ for all $i \in \mathbb{N}$, and therefore by Exercise 8.3.1 we see that $\{a_n\}_{n=1}^{\infty}$ is increasing. Hence $\{a_n\}_{n=1}^{\infty}$ is bounded below by a_1. Because $[a_n,b_n] \subseteq [a,b]$ for all $n \in \mathbb{N}$, then $\{a_n\}_{n=1}^{\infty}$ is bounded above by b. Therefore $\{a_n\}_{n=1}^{\infty}$ is bounded. We will show that $\{a_n\}_{n=1}^{\infty}$ is divergent, which will contradict the fact that the Monotone Convergence Theorem holds for F.

Suppose that $\{a_n\}_{n=1}^{\infty}$ is convergent. Let $L = \lim_{n\to\infty} a_n$. Let $n \in \mathbb{N}$. For each $k \in \mathbb{N}$ such that $k \geq n$, we observe that $a_n \leq a_k$ because $\{a_n\}_{n=1}^{\infty}$ is increasing, and $a_k < b_n$ by Lemma 3.5.3 (3), making use of the fact that $a_k \in P$ and $b_n \in Q$. Therefore $a_k \in [a_n, b_n]$ for all $k \in \mathbb{N}$ such that $k \geq n$. Because the convergence of a sequence is not changed by dropping finitely many terms of the sequence, we use Theorem 8.2.11 to deduce that $L \in [a_n, b_n]$. It follows that $L \in \bigcap_{n=1}^{\infty} [a_n, b_n]$.

Because $[a_n, b_n] \subseteq [a, b]$ for all $n \in \mathbb{N}$, it follows that $L \in [a, b]$. Hence $L \in P$ or $L \in Q$. First, suppose that $L \in Q$. Then L is an upper bound of A by the definition of Q. Let $\varepsilon > 0$. Using Example 8.2.13, which holds for F because F satisfies the Archimedean Property, together with Theorem 8.2.9 (3), we deduce that $\lim_{n\to\infty} (b_n - a_n) = \lim_{n\to\infty} (b-a)\left(\frac{1}{2}\right)^{n-1} = 0$. It follows that there is some $N \in \mathbb{N}$ such that $n \in \mathbb{N}$ and $n \geq N$ imply $|(b_n - a_n) - 0| < \varepsilon$. In particular $b_N - a_N < \varepsilon$. Because $L \in [a_N, b_N]$, it follows that $L - a_N < \varepsilon$. Because $a_N \in P$, then by Lemma 3.5.3 (4) there is some $x \in A$ such that $a_N < x$. Because L is an upper bound of A, then $a_N < x \leq L$. It follows that $L - x < \varepsilon$. We now use Exercise 2.6.6 to deduce that $L = \text{lub} A$, which is a contradiction, because A has no least upper bound. Second, suppose that $L \in P$. By Lemma 3.5.3 (4) there is some $y \in P$ such that $L < y$. Let $\eta = y - L$. Then $\eta > 0$. Using the same argument as before, there is some $K \in \mathbb{N}$ such that $b_K - a_K < \eta$. Because $L \in [a_K, b_K]$, it follows that $b_K - L < \eta$. Therefore $b_K - L < y - L$, and hence $b_K < y$. However, we know $y \in P$ and $b_K \in Q$, which is a contradiction to Lemma 3.5.3 (3). We conclude that $\{a_n\}_{n=1}^{\infty}$ is divergent, which is the desired contradiction to the Monotone Convergence Theorem. □

Reflections

The three important theorems in this section are somewhat analogous, in both their content and uses, to the two important theorems in Section 3.5. Similarly to those two theorems, the Extreme Value Theorem and the Intermediate Value Theorem, the three theorems of the present section, the Monotone Convergence Theorem, the Bolzano–Weierstrass Theorem and the Cauchy Completeness Theorem, are also existence theorems, because they each state that a certain sequence (which could be a subsequence of the original sequence) is convergent, and that is the same as saying that there exists a real number to which the sequence converges. Just as the two theorems of Section 3.5 rely upon the Least Upper Bound Property of the real numbers (and in fact are equivalent to it), so too do the three theorems of the present sections rely upon the Least Upper Bound Property (two of the theorems are equivalent to the Least Upper Bound Property, and one necessarily relies upon it).

Moreover, just as the two theorems of Section 3.5 are used to prove other useful theorems in real analysis (for example the Extreme Value Theorem is used to prove Rolle's Theorem, which in turn is used to prove the Mean Value Theorem), the three theorems of the present section can also be used to prove substantial theorems in real analysis. Indeed, some texts in real analysis discuss sequences prior to the treatment of derivatives and integrals, and use these three theorems (and other theorems about sequences) in proofs that we do without sequences. For example, in our proof of Theorem 5.4.7 (which is the main technical tool we use for proving various properties of integrals), we make use of the Least Upper Bound Property via the No Gap Lemma (Lemma 2.6.6), whereas in some texts this theorem is proved using one of the theorems about sequences that rely upon the Least Upper Bound Property.

Exercises

Exercise 8.3.1. **[Used in Theorem 8.3.17, Theorem 8.4.7 and Theorem 9.4.15.]** Let $\{a_n\}_{n=1}^{\infty}$ be a sequence in $\mathbb{R}$. Suppose that $a_n \leq a_{n+1}$ for all $n \in \mathbb{N}$. Prove that $\{a_n\}_{n=1}^{\infty}$ is increasing. This result might seem obvious, but a proof is needed.

Exercise 8.3.2. Using only the definition of strictly increasing, prove that $\left\{\frac{3n-1}{n+3}\right\}_{n=1}^{\infty}$ is strictly increasing.

Exercise 8.3.3. **[Used in Theorem 9.3.6.]** Let $f\colon [1,\infty) \to \mathbb{R}$ be a function. Suppose that f is increasing. Prove that $\lim_{x\to\infty} f(x)$ exists if and only if the sequence $\{f(n)\}_{n=1}^{\infty}$ is convergent.

Exercise 8.3.4. **[Used in Theorem 8.3.3.]** Prove Theorem 8.3.3 (2).

Exercise 8.3.5. **[Used in Lemma 8.3.7 and Lemma 8.3.13.]** Let $\{a_n\}_{n=1}^{\infty}$ be a sequence in $\mathbb{R}$, and let $\{a_{n_k}\}_{k=1}^{\infty}$ be a subsequence of $\{a_n\}_{n=1}^{\infty}$. Prove that $n_k \geq k$ for all $k \in \mathbb{N}$. Equivalently, let $g\colon \mathbb{N} \to \mathbb{N}$ be a strictly increasing function. Prove that $g(k) \geq k$ for all $k \in \mathbb{N}$.

Exercise 8.3.6. **[Used in Exercise 9.2.7 and Theorem 9.3.8.]** Let $\{a_n\}_{n=1}^{\infty}$ be a sequence in $\mathbb{R}$. Prove that $\{a_n\}_{n=1}^{\infty}$ is convergent if and only if $\{a_{2n}\}_{n=1}^{\infty}$ and $\{a_{2n-1}\}_{n=1}^{\infty}$ are both convergent and $\lim_{n\to\infty} a_{2n} = \lim_{n\to\infty} a_{2n-1}$, and that if these conditions hold then $\lim_{n\to\infty} a_n = \lim_{n\to\infty} a_{2n} = \lim_{n\to\infty} a_{2n-1}$.

Exercise 8.3.7. **[Used in Exercise 9.4.6.]** Let $\{a_n\}_{n=1}^{\infty}$ be a sequence in $\mathbb{R}$. Suppose that $\{a_{n_k}\}_{k=1}^{\infty}$ is a subsequence of $\{a_n\}_{n=1}^{\infty}$ such that if $k \in \mathbb{N}$, then $a_{n_k} = a_{n_k+1} = a_{n_k+2} = \cdots = a_{n_{k+1}-1}$. Prove that $\{a_n\}_{n=1}^{\infty}$ is convergent, divergent or diverges to infinity if and only if $\{a_{n_k}\}_{k=1}^{\infty}$ is convergent, divergent or diverges to infinity, respectively.

Exercise 8.3.8. **[Used in Theorem 8.3.17 and Exercise 9.3.1.]** Let $\{a_n\}_{n=1}^{\infty}$ be a sequence in $\mathbb{R}$, and let $\{a_{n_k}\}_{k=1}^{\infty}$ be a subsequence of $\{a_n\}_{n=1}^{\infty}$. Suppose that $\{a_n\}_{n=1}^{\infty}$ is monotone. Prove that if $\{a_{n_k}\}_{k=1}^{\infty}$ is convergent, then $\{a_n\}_{n=1}^{\infty}$ is convergent.

Exercise 8.3.9. Let $\{a_n\}_{n=1}^{\infty}$ be a sequence in $\mathbb{R}$.

(1) Suppose that $\{a_n\}_{n=1}^{\infty}$ is bounded and divergent. Prove that $\{a_n\}_{n=1}^{\infty}$ has at least two convergent subsequences that converge to different numbers.
(2) Suppose that $\{a_n\}_{n=1}^{\infty}$ is bounded. Suppose that all convergent subsequences of $\{a_n\}_{n=1}^{\infty}$ have the same limit. Prove that $\{a_n\}_{n=1}^{\infty}$ is convergent.
(3) Find an example of a sequence $\{b_n\}_{n=1}^{\infty}$ in $\mathbb{R}$, and a number $M \in \mathbb{R}$, such that if $\{b_{n_k}\}_{k=1}^{\infty}$ is a convergent subsequence of $\{b_n\}_{n=1}^{\infty}$ then $\lim_{k\to\infty} b_{n_k} = M$, but that $\{b_n\}_{n=1}^{\infty}$ is divergent.

Exercise 8.3.10. Let $\{a_n\}_{n=1}^{\infty}$ be a sequence in $\mathbb{R}$, and let $\{a_{n_k}\}_{k=1}^{\infty}$ be a subsequence of $\{a_n\}_{n=1}^{\infty}$. Prove that if $\lim_{n\to\infty} a_n = \infty$, then $\lim_{k\to\infty} a_{n_k} = \infty$.

Exercise 8.3.11. Using only the definition of Cauchy sequences, prove that $\left\{\frac{1}{\sqrt{n}}\right\}_{n=1}^{\infty}$ is a Cauchy sequence.

Exercise 8.3.12. Using only the definition of Cauchy sequences, prove that $\left\{\frac{n}{2n+1}\right\}_{n=1}^{\infty}$ is a Cauchy sequence.

Exercise 8.3.13. Let $\{a_n\}_{n=1}^{\infty}$ be a Cauchy sequence in $\mathbb{R}$. Using only the definition of Cauchy sequences, prove that $\{|a_n|\}_{n=1}^{\infty}$ is a Cauchy sequence.

Exercise 8.3.14. Let $\{a_n\}_{n=1}^{\infty}$ and $\{b_n\}_{n=1}^{\infty}$ be Cauchy sequences in $\mathbb{R}$. Let $k \in \mathbb{R}$.

(1) Using only the definition of Cauchy sequences, prove that $\{a_n + b_n\}_{n=1}^{\infty}$ is a Cauchy sequence.
(2) Using only the definition of Cauchy sequences, prove that $\{ka_n\}_{n=1}^{\infty}$ is a Cauchy sequence.
(3) Using only the definition of Cauchy sequences and Lemma 8.3.14, prove that $\{a_n b_n\}_{n=1}^{\infty}$ is a Cauchy sequence.

Exercise 8.3.15. [Used in Lemma 8.3.14.] Prove Lemma 8.3.14.

Exercise 8.3.16. [Used in Section 8.3.] Let F be an ordered field. Suppose that F does not satisfy the Archimedean Property. Prove that $\lim_{n\to\infty} \frac{1}{n}$ does not converge to 0.

Exercise 8.3.17. [Used in Theorem 8.3.17.] Let F be an ordered field, and let $[a,b] \subseteq F$ be a non-degenerate closed bounded interval. Suppose that there are subsets $P, Q \subseteq [a,b]$ such that $a \in P$ and $b \in Q$, and that $P \cup Q = [a,b]$ and $P \cap Q = \emptyset$. Prove that there is a family $\{[a_n, b_n]\}_{n=1}^{\infty}$ of closed bounded intervals in $[a,b]$ such that for each $i \in \mathbb{N}$ the following three conditions hold: (1) $a_i \in P$ and $b_i \in Q$; (2) $[a_{i+1}, b_{i+1}] \subseteq [a_i, b_i]$; and (3) $b_i - a_i = \frac{b-a}{2^{i-1}}$.

Exercise 8.3.18. [Used in Section 8.3 and Theorem 8.3.17.] The proof of the Archimedean Property (Theorem 2.6.7) made use of the Least Upper Bound Property. Modify the proof of the Archimedean Property so that it relies upon the Monotone Convergence Theorem (Corollary 8.3.4) instead of the Least Upper Bound Property.

8.4 Applications of Sequences

Sequences have many uses in real analysis, and throughout mathematics. In particular, there are a number of connections between sequences and some of the topics we have seen in previous chapters, and in this section we will see a few such applications of sequences. The material in this section might seem haphazard, and indeed it is; the purpose of this section is to show the wide range of uses of sequences, not to build toward any one specific result.

We start with the relation of the limits of sequences to the limits of functions. Most of the theorems and proofs about limits of sequences in Section 8.2 are very similar to the analogous theorems and proofs about limits of functions in Section 3.2. In fact, limits of sequences and limits of functions are not just analogous, but they are concretely related to each other, as seen in the following theorem.

Theorem 8.4.1 (Sequential Characterization of Limits). *Let $I \subseteq \mathbb{R}$ be an open interval, let $c \in I$, let $f: I - \{c\} \to \mathbb{R}$ be a function and let $L \in \mathbb{R}$. Then $\lim_{x\to c} f(x) = L$ if and only if $\lim_{n\to\infty} f(c_n) = L$ for every sequence $\{c_n\}_{n=1}^{\infty}$ in $I - \{c\}$ such that $\lim_{n\to\infty} c_n = c$.*

Proof. Suppose that $\lim_{x\to c} f(x) = L$. Let $\{c_n\}_{n=1}^{\infty}$ be a sequence in $I - \{c\}$ such that $\lim_{n\to\infty} c_n = c$. Let $\varepsilon > 0$. Because $\lim_{x\to c} f(x) = L$, there is some $\delta > 0$ such that $x \in I - \{c\}$ and $|x - c| < \delta$ imply $|f(x) - L| < \varepsilon$. Because $\lim_{n\to\infty} c_n = c$, there is some $N \in \mathbb{N}$ such that $n \in \mathbb{N}$ and $n \geq N$ imply $|c_n - c| < \delta$. Suppose that $n \in \mathbb{N}$ and $n \geq N$. Then $|c_n - c| < \delta$, and we know by definition that $c_n \in I - \{c\}$. Hence $|f(c_n) - L| < \varepsilon$. It follows that $\lim_{n\to\infty} f(c_n) = L$.

Next, suppose that $\lim_{n\to\infty} f(c_n) = L$ for every sequence $\{c_n\}_{n=1}^{\infty}$ in $I - \{c\}$ such that $\lim_{n\to\infty} c_n = c$. Suppose also that $\lim_{x\to c} f(x) \neq L$. Then there is some $\varepsilon > 0$ such that for all $\delta > 0$, it is not the case that $x \in I - \{c\}$ and $|x - c| < \delta$ imply $|f(x) - L| < \varepsilon$. Let $n \in \mathbb{N}$. Then it is not the case that $x \in I - \{c\}$ and $|x - c| < \frac{1}{n}$ imply $|f(x) - L| < \varepsilon$. Hence, there is some $x_n \in I - \{c\}$ such that $|x_n - c| < \frac{1}{n}$ and $|f(x_n) - L| \geq \varepsilon$. We have therefore defined a sequence $\{x_n\}_{n=1}^{\infty}$ in $I - \{c\}$. It follows from Exercise 8.2.9 that $\lim_{n\to\infty} x_n = c$, where we use the sequence $\{b_n\}_{n=1}^{\infty}$ defined by $b_n = \frac{1}{n}$ for all $n \in \mathbb{N}$, and $N = 1$. However, the fact that $|f(x_n) - L| \geq \varepsilon$ for all $n \in \mathbb{N}$ means that $\lim_{n\to\infty} f(x_n) \neq L$, which is a contradiction. Hence $\lim_{x\to c} f(x) = L$. □

It is important to observe in the statement of Theorem 8.4.1 that in order to know that the limit of a function exists, it is necessary to know something about the limits of all appropriate sequences; just knowing that the condition holds for one such sequence is not sufficient to imply that the limit of the function exists, as is seen in Exercise 8.4.1.

Just as sequences can be used to detect the existence of limits of functions, sequences can similarly be used to detect continuity.

Corollary 8.4.2 (Sequential Characterization of Continuity). *Let $I \subseteq \mathbb{R}$ be an open interval, and let $f: I \to \mathbb{R}$ be a function. Then f is continuous if and only if*

$$\lim_{n\to\infty} f(c_n) = f(\lim_{n\to\infty} c_n)$$

for every sequence $\{c_n\}_{n=1}^{\infty}$ in I such that $\{c_n\}_{n=1}^{\infty}$ is convergent and $\lim_{n\to\infty} c_n \in I$.

Proof. Left to the reader in Exercise 8.4.2. □

We now use the Sequential Characterization of Continuity (Corollary 8.4.2), together with l'Hôpital's Rule for $\frac{0}{0}$ (Theorem 6.3.5), to provide justification for a well-known formula for the value of the number e.

Example 8.4.3. Recall our definition of the number e in Definition 7.2.9. That is certainly a correct definition, but it looks rather different from another common definition of e that is often used in high school algebra, which is

$$e = \lim_{n\to\infty}\left(1+\frac{1}{n}\right)^n.$$

We now show that this limit exists, and that it equals the number e as we defined it in Definition 7.2.9.

Let $r \in \mathbb{R}$. Then, by the one-sided analog of l'Hôpital's Rule for $\frac{0}{0}$ (Theorem 6.3.5) we see that

$$\lim_{x\to 0+}\frac{\ln(1+rx)}{x} = \lim_{x\to 0+}\frac{\frac{r}{1+rx}}{1} = r.$$

Next, by Exercise 6.3.9 (1), we deduce that

$$\lim_{t\to\infty}\ln\left(1+\frac{r}{t}\right)^t = \lim_{t\to\infty} t\ln\left(1+\frac{r}{t}\right) = \lim_{x\to 0+}\frac{1}{x}\ln\left(1+\frac{r}{\frac{1}{x}}\right) = \lim_{x\to 0+}\frac{\ln(1+rx)}{x} = r.$$

It now follows from Exercise 8.2.10 (1) that

$$\lim_{n\to\infty}\ln\left(1+\frac{r}{n}\right)^n = r.$$

By Theorem 7.2.7 (2) and Theorem 4.2.4 we know that e^x is a continuous function. We can now use Theorem 7.2.14 (4) together with the Sequential Characterization of Continuity (Corollary 8.4.2) to deduce that

$$\lim_{n\to\infty}\left(1+\frac{r}{n}\right)^n = \lim_{n\to\infty} e^{\ln\left(1+\frac{r}{n}\right)^n} = e^{\lim_{n\to\infty}\ln\left(1+\frac{r}{n}\right)^n} = e^r.$$

The above equation gives a nice characterization of the exponential function. In particular, if we choose $r = 1$, and use Theorem 7.2.12 (1), we see that

$$e = e^1 = \lim_{n\to\infty}\left(1+\frac{1}{n}\right)^n. \qquad \lozenge$$

Our next application of sequences is to integration. Recall that in the definition of integrals, given in Definition 5.2.4, it was necessary to consider all possible partitions and all possible representative sets of these partitions to verify that a function is integrable. In practice, doing so can be quite cumbersome. Is it possible to consider only some of the partitions and representative sets? The following theorem shows that if we already know in principle that a function is integrable, for example if it is continuous using Theorem 5.4.11, then the value of the integral can be computed as the limit of a single sequence of Riemann sums, using an appropriately chosen sequence of partitions and representative sets.

Theorem 8.4.4. *Let $[a,b] \subseteq \mathbb{R}$ be a non-degenerate closed bounded interval, and let $f\colon [a,b] \to \mathbb{R}$ be a function. Suppose that f is integrable. Let $\{P_n\}_{n=1}^{\infty}$ be a sequence* *of partitions of* $[a,b]$ *such that* $\lim_{n\to\infty} \|P_n\| = 0$. *For each* $n \in \mathbb{N}$, *let* T_n *be a representative set of* P_n. *Then*

$$\int_a^b f(x)\,dx = \lim_{n\to\infty} S(f,P_n,T_n).$$

Proof. Let $\varepsilon > 0$. Because f is integrable, there is some $\delta > 0$ such that if Q is a partition of $[a,b]$ with $\|Q\| < \delta$, and if S is a representative set of Q, then

$$\left| S(f,Q,S) - \int_a^b f(x)\,dx \right| < \varepsilon.$$

Because $\lim_{n\to\infty} \|P_n\| = 0$, there is some $N \in \mathbb{N}$ such that $n \in \mathbb{N}$ and $n \geq N$ imply $|\|P_n\| - 0| < \delta$, which is equivalent to $\|P_n\| < \delta$. Suppose that $m \in \mathbb{N}$ and $m \geq N$. Then $\|P_m\| < \delta$, and hence

$$\left| S(f,P_m,T_m) - \int_a^b f(x)\,dx \right| < \varepsilon.$$

It follows that

$$\lim_{n\to\infty} S(f,P_n,T_n) = \int_a^b f(x)\,dx. \qquad \square$$

It is important to note that if we do not already know that a given function is integrable, then looking at only a single sequence of partitions and representative sets as in Theorem 8.4.4 does not suffice to prove integrability. For example, in Example 5.2.6 (3), a sequence of Riemann sums with representative sets that have rational numbers will have one limit, but a sequence of Riemann sums with representative sets that have irrational numbers will have a different limit.

The following example illustrates the use of Theorem 8.4.4 in computing the value of an integral.

Example 8.4.5. We saw in Theorem 5.4.11 that all continuous functions on non-degenerate closed bounded intervals are integrable. Hence we know that $\int_0^2 x^2\,dx$ exists. We can then use Theorem 8.4.4 to compute the value of this integral as follows.

Let $f\colon [0,2] \to \mathbb{R}$ be defined by $f(x) = x^2$ for all $x \in [0,2]$. Let $n \in \mathbb{N}$. Let the partition P_n of $[0,2]$, and the representative set T_n of P_n, be defined as in Example 5.2.3 (1). It was seen in that example that $\|P_n\| = \frac{2}{n}$ and $S(f,P_n,T_n) = \frac{4(n+1)(2n+1)}{3n^2}$. Hence by Example 8.2.4 (2) and Theorem 8.2.9 (3) we see that $\lim_{n\to\infty} \|P_n\| = 0$. Theorem 8.4.4 and Theorem 8.2.9 then imply that

$$\begin{aligned}\int_0^2 x^2\,dx &= \lim_{n\to\infty} S(f,P_n,T_n)\\ &= \lim_{n\to\infty} \frac{4(n+1)(2n+1)}{3n^2} = \lim_{n\to\infty} \frac{4(1+\frac{1}{n})(2+\frac{1}{n})}{3} = \frac{8}{3}.\end{aligned}$$

As seen by Example 5.6.6 (1), which uses the Fundamental Theorem of Calculus Version II (Theorem 5.6.4), our use of Theorem 8.4.4 yields the correct value of the integral. Our point here, however, is not simply to obtain the correct value of the integral, but rather to show that in some cases it is possible to use limits of Riemann sums to compute an integral directly, without making use of the Fundamental Theorem of Calculus, which is nice to see because the definition of integrals is not about antiderivatives per se. Of course, the method we have used here to compute an integral is very cumbersome, and would not be feasible if we did not happen to have a very convenient summation formula to use. Ultimately, we could not compute the value of many integrals without the Fundamental Theorem of Calculus. ◊

We now turn to a different use of Theorem 8.4.4, which is to give an intuitive explanation for the formula for the average value of a function.

Example 8.4.6. We learn at an early age how to compute the average value of a finite collection of numbers, which is by adding the numbers and then dividing by how many numbers there are. It is not as obvious how to find the average value of an infinite collection of numbers. In particular, let $[a,b] \subseteq \mathbb{R}$ be a non-degenerate closed bounded interval, and let $f\colon [a,b] \to \mathbb{R}$ be a function. We want to find the average value of the function f, which means that we want to find something that would be called the average value of the numbers in the set $f([a,b])$, which is an infinite set.

One way to proceed would be to take finite samples of the elements of $f([a,b])$, compute the average value of each sample, and then take the limit of these average values as the sizes of the samples go to infinity, if such a limit exists. More specifically, let $n \in \mathbb{N}$. We then select a collection of n elements of $[a,b]$, which we denote $T_n = \{t_1^n, t_2^n, \ldots, t_n^n\}$, where the superscript n indicates that the numbers are in T_n. The average value of the set $\{f(t_1^n), f(t_2^n), \ldots, f(t_n^n)\}$ is

$$\frac{f(t_1^n) + f(t_2^n) + \cdots + f(t_n^n)}{n}.$$

We then wish to look at

$$\lim_{m\to\infty} \frac{f(t_1^m) + f(t_2^m) + \cdots + f(t_m^m)}{m},$$

in the hope that this limit will exist. Even if the limit exists, for it to be meaningful we would need to know that all such limits are equal no matter what numbers were selected for each T_n. Unfortunately, if we do not place some restrictions on the choice of the numbers $t_1^n, t_2^n, \ldots, t_n^n$, then they might all be clustered in one part of the interval $[a,b]$, so that $f(t_1^n), f(t_2^n), \ldots, f(t_n^n)$ might not accurately represent the function. To remedy this problem, we need to ensure that the numbers $t_1^n, t_2^n, \ldots, t_n^n$ are reasonably evenly spaced, though not necessarily exactly evenly spaced. One way to ensure that the numbers $t_1^n, t_2^n, \ldots, t_n^n$ are roughly spread out is to let $P_n = \{x_0^n, x_1^n, x_2^n, \ldots, x_n^n\}$ be the partition of $[a,b]$ with n intervals of length $\frac{b-a}{n}$, and to insist that $t_i^n \in [x_{i-1}^n, x_i^n]$ for all $i \in \{1, \ldots, n\}$. In other words, we suppose that T_n is a representative set of P_n. Then

$$\frac{f(t_1^n) + f(t_2^n) + \cdots + f(t_n^n)}{n} = \frac{1}{b-a} \sum_{i=1}^{n} f(t_i^n) \frac{b-a}{n} = \frac{1}{b-a} \sum_{i=1}^{n} f(t_i^n)(x_i^n - x_{i-1}^n)$$

$$= \frac{1}{b-a} S(f, P_n, T_n).$$

Now suppose that f is integrable. It is evident that $\lim_{m\to\infty} \|P_m\| = 0$, and it then follows from Theorem 8.2.9 (3) and Theorem 8.4.4 that

$$\lim_{m\to\infty} \frac{f(t_1^m) + f(t_2^m) + \cdots + f(t_m^m)}{m} = \lim_{m\to\infty} \frac{1}{b-a} S(f, P_m, T_m)$$
$$= \frac{1}{b-a} \lim_{m\to\infty} S(f, P_m, T_m) = \frac{1}{b-a} \int_a^b f(x)\,dx.$$

Hence, when f is integrable, we deduce that

$$\lim_{m\to\infty} \frac{f(t_1^m) + f(t_2^m) + \cdots + f(t_m^m)}{m}$$

exists and equals $\frac{1}{b-a}\int_a^b f(x)\,dx$, for any choice of sets T_n subject to the restriction discussed above. It then makes sense to define the **average value** of an integrable function to be $\frac{1}{b-a}\int_a^b f(x)\,dx$. ◊

For our next two applications of sequences, we will need the following theorem, which is about certain "sequences" of closed bounded intervals in $\mathbb{R}$, by which we mean families of closed bounded intervals indexed by the natural numbers. This theorem, known as the Nested Interval Theorem, can be thought of as the sequence version of the No Gap Lemma (Lemma 2.6.6), and indeed our proof of the Nested Interval Theorem will make use of the No Gap Lemma. More accurately, one should think of the No Gap Lemma as the non-sequence version of the Nested Interval Theorem, given that the latter is well-known and widely used and the former is neither. A more common proof of the Nested Interval Theorem uses the Monotone Convergence Theorem (Corollary 8.3.4), or, more precisely, Theorem 8.3.3, which is the substance of the Monotone Convergence Theorem; that proof is left to the reader in Exercise 8.4.8.

Theorem 8.4.7 (Nested Interval Theorem). *Let $\{[a_n, b_n]\}_{n=1}^\infty$ be a family of closed bounded intervals in $\mathbb{R}$. Suppose that $[a_{i+1}, b_{i+1}] \subseteq [a_i, b_i]$ for all $i \in \mathbb{N}$.*

1. *$\bigcap_{n=1}^\infty [a_n, b_n] \neq \emptyset$.*
2. *If $\lim_{n\to\infty} (b_n - a_n) = 0$, then $\bigcap_{n=1}^\infty [a_n, b_n] = \{c\}$, where $c = \lim_{n\to\infty} a_n = \lim_{n\to\infty} b_n$.*

Proof. We prove both parts of the theorem together. By hypothesis we know that $a_i \leq a_{i+1}$ and $b_{i+1} \leq b_i$ for all $i \in \mathbb{N}$. It then follows from Exercise 8.3.1, and the analog of that exercise for decreasing sequences, that $\{a_n\}_{n=1}^\infty$ is increasing and $\{b_n\}_{n=1}^\infty$ is decreasing.

Let $A = \{a_n \mid n \in \mathbb{N}\}$, and let $B = \{b_n \mid n \in \mathbb{N}\}$. Let $i, j \in \mathbb{N}$. If $i = j$, then clearly $a_i \leq b_j$. If $i < j$, then $a_i \leq a_j \leq b_j \leq b_i$, and hence $a_i \leq b_j$. If $i > j$, a similar argument shows that $a_i \leq b_j$. Therefore A and B satisfy the hypothesis of the No Gap Lemma (Lemma 2.6.6). From Part (1) of that lemma, we know that A has a least upper bound and B has a greatest lower bound, and $\text{lub}A \leq \text{glb}B$. Let $a = \text{lub}A$, and let

$b = \text{glb}B$. If $n \in \mathbb{N}$, then $a_n \le a \le b \le b_n$, and hence $[a,b] \subseteq [a_n,b_n]$. It follows that $[a,b] \subseteq \bigcap_{n=1}^{\infty}[a_n,b_n]$. Because $[a,b] \ne \emptyset$, then $\bigcap_{n=1}^{\infty}[a_n,b_n] \ne \emptyset$, which proves Part (1) of the theorem.

Next, let $x \in \mathbb{R} - [a,b]$. First, suppose that $x < a$. Let $\varepsilon = a - x$. Then $\varepsilon > 0$. By Lemma 2.6.5 (1) there is some $a_k \in A$ such that $a - \varepsilon < a_k \le a$, which implies that $a - (a - x) < a_k \le a$, and hence $x < a_k \le a$. It follows that $x \notin [a_k, b_k]$, and hence $x \notin \bigcap_{n=1}^{\infty}[a_n,b_n]$. Second, suppose that $x > b$. A similar argument shows that $x \notin \bigcap_{n=1}^{\infty}[a_n,b_n]$, and we omit the details. By contrapositive, we deduce that $\bigcap_{n=1}^{\infty}[a_n,b_n] \subseteq [a,b]$. It follows that $\bigcap_{n=1}^{\infty}[a_n,b_n] = [a,b]$.

Now suppose that $\lim_{n\to\infty}(b_n - a_n) = 0$. Let $\varepsilon > 0$. Then there is some $N \in \mathbb{N}$ such that $n \in \mathbb{N}$ and $n \ge N$ imply $|(b_n - a_n) - 0| < \varepsilon$. In particular $b_N - a_N < \varepsilon$. It follows from Part (2) of the No Gap Lemma that $a = \text{lub}A = \text{glb}B = b$. Hence $\bigcap_{n=1}^{\infty}[a_n,b_n] = [a,b] = \{a\}$. Moreover, suppose $n \in \mathbb{N}$ and $n \ge N$. Because $a \in [a_n,b_n]$, then $|a_n - a| \le |b_n - a_n| < \varepsilon$ and $|b_n - a| \le |b_n - a_n| < \varepsilon$. It follows that $a = \lim_{n\to\infty} a_n$ and $a = \lim_{n\to\infty} b_n$. We have therefore proved Part (2) of the theorem. □

We note that the Nested Interval Theorem does not hold for the set of rational numbers, for example because of the existence of a family of intervals $\{[a_n,b_n]\}_{n=1}^{\infty}$ where $\{a_n\}_{n=1}^{\infty}$ and $\{b_n\}_{n=1}^{\infty}$ are sequences of rational numbers that converge to $\sqrt{2}$, the former from below and the latter from above.

Our next application of sequences is a nice proof that the set of real numbers is uncountable. The reader is likely to be familiar with the famous proof of the uncountability of $\mathbb{R}$ known as "Cantor's diagonal argument"; see [Blo10, Theorem 6.7.3] for this proof. The proof is very clever, and is important historically, though it does have a problem, which is that it makes use of the fact that every real number can be represented in decimal notation, and that such representation is unique if decimal representations that eventually become the number 9 repeating are not allowed. The existence of such decimal representation of real numbers can be proved, as we did in Section 2.8, but the proof is not trivial, and therefore a complete treatment of Cantor's diagonal argument is much more involved than it might at first appear. Moreover, whereas the decimal representation of real numbers is extremely useful from a computational perspective, it is not conceptually at the heart of the real numbers, and it would be nice to have a proof of the uncountability of $\mathbb{R}$ that is more directly related to the fundamental properties of the real numbers. The following proof of the uncountability of $\mathbb{R}$, which is also due to Georg Cantor (1845–1918) and in fact precedes his more famous diagonal argument, makes use of the Nested Interval Theorem (Theorem 8.4.7), but not decimals. This proof is a special case of a more general theorem in topology, as in [Mun00, pp. 176–177]. For this proof, we assume that the reader is familiar with basic properties of countable sets. In particular, we use the fact that a set A is countable if and only if there is a surjective function $f: \mathbb{N} \to A$; see [Blo10, Sections 6.5–6.6] for a proof of this fact, and for general information about countable and uncountable sets.

Theorem 8.4.8. *The set* $\mathbb{R}$ *is uncountable.*

Proof. Suppose that $\mathbb{R}$ is countable. Then there is a surjective function $f: \mathbb{N} \to \mathbb{R}$.

By Exercise 2.3.10 there is a non-degenerate open bounded interval $(a_1, b_1) \subseteq \mathbb{R}$ such that $f(1) \notin [a_1, b_1]$. Similarly, there is a non-degenerate open bounded interval $(a_2, b_2) \subseteq (a_1, b_1)$ such that $f(2) \notin [a_2, b_2]$. Continuing in this way, we use Definition by Recursion to define a family $\{[a_n, b_n]\}_{n=1}^{\infty}$ of non-degenerate closed bounded intervals in $\mathbb{R}$ such that $f(n) \notin [a_n, b_n]$ and $(a_{n+1}, b_{n+1}) \subseteq (a_n, b_n)$ for all $n \in \mathbb{N}$. Hence $[a_{n+1}, b_{n+1}] \subseteq [a_n, b_n]$ for all $n \in \mathbb{N}$. It then follows from the Nested Interval Theorem (Theorem 8.4.7) that $\bigcap_{n=1}^{\infty} [a_n, b_n] \neq \emptyset$. Let $y \in \bigcap_{n=1}^{\infty} [a_n, b_n]$. Because $f(n) \notin [a_n, b_n]$ for all $n \in \mathbb{N}$, it follows that $y \neq f(n)$ for all $n \in \mathbb{N}$. Hence f is not surjective, which is a contradiction. We deduce that $\mathbb{R}$ is uncountable. □

As another application of sequences, we now provide the details that were missing from Example 5.8.2 (5), which was part of our discussion of sets of measure zero.

Example 8.4.9. As part of our discussion of sets of measure zero in Section 5.8, it was stated without proof in Example 5.8.2 (5) that there are uncountable subsets of $\mathbb{R}$ that have measure zero. We now show a famous example of such a set, which is called the Cantor set. To define the Cantor set, we first use Definition by Recursion to define a family $\{C_n\}_{n=1}^{\infty}$ of subsets of the interval $[0,1]$. Let C_1 be defined by removing the "open middle third" from $[0,1]$; that is, let

$$C_1 = \left[0, \tfrac{1}{3}\right] \cup \left[\tfrac{2}{3}, 1\right].$$

Let C_2 be defined by removing the open middle third from each of the two disjoint intervals in C_1; that is, let

$$C_2 = \left[0, \tfrac{1}{9}\right] \cup \left[\tfrac{2}{9}, \tfrac{1}{3}\right] \cup \left[\tfrac{2}{3}, \tfrac{7}{9}\right] \cup \left[\tfrac{8}{9}, 1\right].$$

Let C_3 be defined by removing the open middle third from each of the four disjoint intervals in C_2; that is, let

$$C_3 = \left[0, \tfrac{1}{27}\right] \cup \left[\tfrac{2}{27}, \tfrac{1}{9}\right] \cup \left[\tfrac{2}{9}, \tfrac{7}{27}\right] \cup \left[\tfrac{8}{27}, \tfrac{1}{3}\right] \cup \left[\tfrac{2}{3}, \tfrac{19}{27}\right] \cup \left[\tfrac{20}{27}, \tfrac{7}{9}\right] \cup \left[\tfrac{8}{9}, \tfrac{25}{27}\right] \cup \left[\tfrac{26}{27}, 1\right].$$

Continuing this way, we define a family of sets $\{C_n\}_{n=1}^{\infty}$ such that C_n that is the union of 2^n closed bounded intervals of length $\frac{1}{3^n}$, and $C_{n+1} \subseteq C_n$, for all $n \in \mathbb{N}$. See Figure 8.4.1 for an illustration of C_1, C_2 and C_3.

The Cantor set is the set C defined by $C = \bigcap_{n=1}^{\infty} C_n$. The set C is not empty, because for each $n \in \mathbb{N}$, the 2^n intervals in C_n have 2^{n+1} endpoints, and each of these endpoints is also in C_k for all $k \in \mathbb{N}$ such that $k \geq n$, and so these endpoints are all in C.

We now show that C has measure zero, as defined in Definition 5.8.1. Let $\varepsilon > 0$. By Example 8.2.13 we know that $\lim_{n \to \infty} \left(\frac{2}{3}\right)^n = 0$. Then there is some $N \in \mathbb{N}$ such that $n \in \mathbb{N}$ and $n \geq N$ imply $|\left(\frac{2}{3}\right)^n - 0| < \frac{\varepsilon}{3}$. Hence $2^N \cdot \frac{3}{3^N} < \varepsilon$. The set C_N is the union of 2^N closed intervals of the form $\left[\frac{a-1}{3^N}, \frac{a}{3^N}\right]$, where the values of a are 2^N of the elements of $\{1, 2, \ldots, 3^N\}$; we do not need the precise list of possible values of a, but only the number of such values. Each closed interval of the form $\left[\frac{a-1}{3^N}, \frac{a}{3^N}\right]$ is contained in the

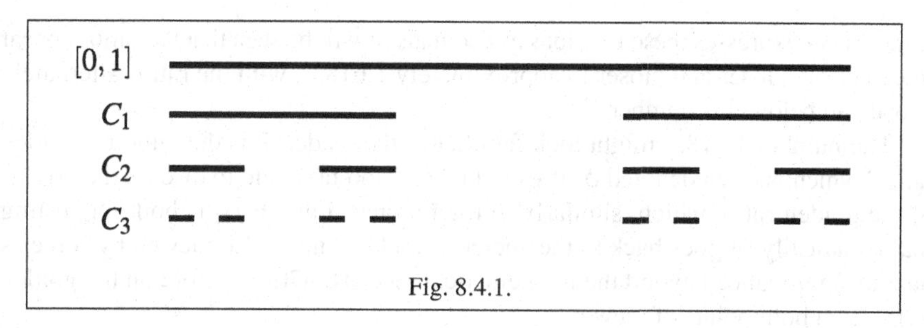

Fig. 8.4.1.

open interval $\left(\frac{a-2}{3^N}, \frac{a+1}{3^N}\right)$, which has length $\frac{3}{3^N}$. Hence C_N is contained in the union of 2^N open intervals of length $\frac{3}{3^N}$. Because C is a subset of C_N, it follows that C is contained in the union of 2^N open intervals of length $\frac{3}{3^N}$. The sum of the lengths of these 2^N open intervals is $2^N \cdot \frac{3}{3^N} < \varepsilon$. It follows that C has measure zero.

There are two standard approaches to proving that C is uncountable. One approach, which is left to the reader in Exercise 8.4.10, uses the Nested Interval Theorem (Theorem 8.4.7), and is similar to the proof of Theorem 8.4.8. Another approach, which we will not discuss, makes use of the base 3 representation of real numbers, often called the "ternary expansion" of real numbers; see [TBB01, Section 6.5.2] or [Sto01, Section 3.3] for details. ◊

We conclude this section with an application of the Cauchy Completeness Theorem (Corollary 8.3.16) to the Fibonacci numbers.

Example 8.4.10. The Fibonacci numbers are the terms of the widely studied sequence that starts

$$1, 1, 2, 3, 5, 8, 13, 21, 34, 55, 89, 144\ldots.$$

Formally, the Fibonacci numbers are the terms of the unique sequence $\{F_n\}_{n=1}^{\infty}$ that results from using Definition by Recursion with the conditions $F_1 = 1$ and $F_2 = 1$, and $F_{n+2} = F_n + F_{n+1}$ for all $n \in \mathbb{N}$. That such a formula defines a unique sequence follows from Exercise 2.5.20.

The Fibonacci numbers have a long history, and are well studied by mathematicians. These numbers have many interesting mathematical properties, as seen in [Knu73, Section 1.2.8 and exercises], [GKP94, Section 6.6] and [HHP97, Chapter 3]. The Fibonacci numbers are also viewed by some authors as having something of a spiritual significance, the validity of which the author of this text is not qualified to judge; the interested reader might wish to consult [Gar87] or [Hun70].

Our interest in the Fibonacci numbers concerns convergence. Clearly the Fibonacci numbers do not converge as a sequence, because the sequence of Fibonacci numbers is not bounded, and Lemma 8.2.6 states that any convergent sequence is bounded. However, it is possible to construct an interesting convergent sequence out of the Fibonacci numbers by considering the sequence of ratios of the Fibonacci numbers with their predecessors, which is

$$\frac{1}{1}, \frac{2}{1}, \frac{3}{2}, \frac{5}{3}, \frac{8}{5}, \frac{13}{8}, \frac{21}{13}, \ldots.$$

If the reader expresses these fractions as decimals, it will be seen that the ratios appear to be getting closer and closer to approximately 1.618..., with the ratios alternately above and below this number.

The number 1.618... might look familiar to the reader; it is the famous "golden ratio," which is often denoted ϕ. It would take us too far afield to discuss the origins of the golden ratio, which, similarly to the Fibonacci numbers, is both interesting mathematically (it goes back to the ancient Greeks), and is also viewed by some as having significance beyond the mathematical. See [Hun70] for more on the golden ratio, from both points of view.

The simplest way to define ϕ in modern terms is to view it as the unique positive solution of the quadratic equation

$$x^2 - x - 1 = 0. \tag{8.4.1}$$

This equation does not give any indication as to the origin of the golden ratio, but it is the aspect of the golden ratio that we need at present. By applying the quadratic formula to Equation 8.4.1, it is seen that $\phi = \frac{1+\sqrt{5}}{2}$, which is approximately 1.618.... By Theorem 2.6.11 we know that $\sqrt{5}$ is irrational, and it follows that ϕ is irrational.

Our goal is to show that

$$\lim_{n\to\infty} \frac{F_{n+1}}{F_n} = \phi.$$

The standard "proof" of this fact, which one often sees in elementary treatments of the subject is, as follows. "Suppose that $\lim_{n\to\infty} \frac{F_{n+1}}{F_n} = L$. Then

$$\begin{aligned}
1+L = 1 + \lim_{n\to\infty} \frac{F_{n+1}}{F_n} &= \lim_{n\to\infty} \frac{F_n}{F_n} + \lim_{n\to\infty} \frac{F_{n+1}}{F_n} = \lim_{n\to\infty} \frac{F_n + F_{n+1}}{F_n} \\
&= \lim_{n\to\infty} \frac{F_{n+2}}{F_n} = \lim_{n\to\infty} \left(\frac{F_{n+2}}{F_{n+1}} \cdot \frac{F_{n+1}}{F_n} \right) \\
&= \left(\lim_{n\to\infty} \frac{F_{n+2}}{F_{n+1}} \right) \left(\lim_{n\to\infty} \frac{F_{n+1}}{F_n} \right) = L^2.
\end{aligned}$$

Hence $1 + L = L^2$, which is equivalent to $L^2 - L - 1 = 0$. Therefore L satisfies Equation 8.4.1. Because all of the F_n are positive then L cannot be negative, and it follows that L is the positive root of Equation 8.4.1, which means that $L = \phi$."

Before reading further, the reader is encouraged to find the flaw in the above proof.

As with many incomplete proofs, the above "proof" is indeed the proof of something—it is simply not the complete proof of what we want to show. What the above proof shows is that if $\lim_{n\to\infty} \frac{F_{n+1}}{F_n}$ exists, then it must equal ϕ. The flaw in the above proof is that it does not give any reason for us to believe that $\lim_{n\to\infty} \frac{F_{n+1}}{F_n}$ exists. We will now fill in this gap in the proof by showing that $\left\{ \frac{F_{n+1}}{F_n} \right\}_{n=1}^{\infty}$ is a Cauchy sequence, and it will then follow from the Cauchy Completeness Theorem (Corollary 8.3.16) that $\lim_{n\to\infty} \frac{F_{n+1}}{F_n}$ exists.

Let $\varepsilon > 0$. By Corollary 2.6.8 (2) there is some $N \in \mathbb{N}$ such that $\frac{1}{N} < \varepsilon$. By taking a larger value of N if necessary, we may assume that $N \geq 5$. Suppose that $n, m \in \mathbb{N}$ and $n, m \geq N$. If $n = m$ then clearly

$$\left| \frac{F_{n+1}}{F_n} - \frac{F_{m+1}}{F_m} \right| = 0 < \varepsilon.$$

Now suppose that $n \neq m$. Without loss of generality, assume that $m > n$. Then, using Exercise 8.4.12 (5), Exercise 2.5.8 and Exercise 8.4.12 (2) in that order, we see that

$$\begin{aligned}
&\left| \frac{F_{n+1}}{F_n} - \frac{F_{m+1}}{F_m} \right| \\
&\quad = \left| \left(\frac{F_{n+1}}{F_n} - \frac{F_{n+2}}{F_{n+1}} \right) + \left(\frac{F_{n+2}}{F_{n+1}} - \frac{F_{n+3}}{F_{n+2}} \right) + \cdots + \left(\frac{F_m}{F_{m-1}} - \frac{F_{m+1}}{F_m} \right) \right| \\
&\quad = \left| \frac{(F_{n+1})^2 - F_{n+2}F_n}{F_nF_{n+1}} + \frac{(F_{n+2})^2 - F_{n+3}F_{n+1}}{F_{n+1}F_{n+2}} + \cdots + \frac{(F_m)^2 - F_{m+1}F_{m-1}}{F_{m-1}F_m} \right| \\
&\quad = \left| \frac{(-1)^{n+2}}{F_nF_{n+1}} + \frac{(-1)^{n+3}}{F_{n+1}F_{n+2}} + \cdots + \frac{(-1)^{m+1}}{F_{m-1}F_m} \right| \\
&\quad = \left| \frac{1}{F_nF_{n+1}} - \frac{1}{F_{n+1}F_{n+2}} + \cdots + (-1)^{m-n-1} \frac{1}{F_{m-1}F_m} \right| \\
&\quad \leq \left| \frac{1}{F_nF_{n+1}} \right| \leq \frac{1}{F_N} \leq \frac{1}{N} < \varepsilon.
\end{aligned}$$

◊

Reflections

This section has a number of "fun" topics, for example the Cantor set and the Fibonacci numbers, which help show the diverse uses of sequences. The reader might wonder, however, about the absence of such a concentration of "fun" topics in any other section of this text. Are other aspects of real analysis inherently less enjoyable than sequences? The answer, fortunately, is no. There are many fun and useful applications of the material (such as derivatives and integrals) that was discussed in the previous chapters of this text, but the reader has undoubtedly seen such applications in a calculus course, and so it is not necessary to include such material here. The treatment of sequences in most introductory calculus courses, by contrast, is quite cursory, and so the reader might not have seen how valuable sequences are prior to studying real analysis, and hence the inclusion of the topics discussed in the present section.

Moreover, in many real analysis texts sequences are treated before derivatives and integrals, with proofs about the latter two topics making use of sequences. In such texts, the importance of sequences as a tool in real analysis is immediately apparent. In this text, by contrast, where sequences were placed after the discussion of derivatives and integrals in order to emphasize the role of the Least Upper Bound Property in proofs about those topics, it was necessary to include some additional brief topics to demonstrate the value of sequences, and that presented a good opportunity for some topics that are particularly fun.

Exercises

Exercise 8.4.1. [Used in Section 8.4.] Find an example of a function $f\colon \mathbb{R} \to \mathbb{R}$ for which $\lim_{x\to 0} f(x)$ does not exist, and yet there is a sequence $\{c_n\}_{n=1}^{\infty}$ in $\mathbb{R} - \{0\}$ such that $\lim_{n\to\infty} c_n = 0$ and $\lim_{n\to\infty} f(c_n)$ exists.

Exercise 8.4.2. [Used in Corollary 8.4.2.] Prove Corollary 8.4.2.

Exercise 8.4.3. [Used in Theorem 10.5.2.] Let $I \subseteq \mathbb{R}$ be an open interval, let $c \in I$ and let $f\colon I \to \mathbb{R}$ be a function. Suppose that there is a sequence $\{a_n\}_{n=1}^{\infty}$ in $\mathbb{R} - \{0\}$ such that $c + a_n \in I$ for all $n \in \mathbb{N}$, that $\lim_{n\to\infty} a_n = 0$ and that the sequence

$$\left\{ \frac{f(c+a_n) - f(c)}{a_n} \right\}_{n=1}^{\infty}$$

is divergent. Prove that f is not differentiable at c.

Exercise 8.4.4. Let $A \subseteq \mathbb{R}$ be a non-empty set. Prove that if A has a least upper bound, then there is a sequence $\{a_n\}_{n=1}^{\infty}$ in A such that $\lim_{n\to\infty} a_n = \operatorname{lub} A$.

Exercise 8.4.5. Let $[a,b] \subseteq \mathbb{R}$ be a non-degenerate closed bounded interval, and let $f\colon [a,b] \to \mathbb{R}$ be a function. Suppose that f is continuous. Prove that there is some $c \in [a,b]$ such that $f(c)$ equals the average value of f.

Exercise 8.4.6. [Used in Exercise 8.4.7.] Let $[a,b] \subseteq \mathbb{R}$ be a non-degenerate closed bounded interval, let $f\colon [a,b] \to \mathbb{R}$ be a function, let $\{[a_n,b_n]\}_{n=1}^{\infty}$ be a family of closed bounded intervals in $[a,b]$ and let $\{e_n\}_{n=1}^{\infty}$ be a sequence in $\mathbb{R}$. Suppose that $\lim_{n\to\infty} e_n = 0$, and that $n \in \mathbb{N}$ and $x, y \in [a_n, b_n]$ imply $|f(x) - f(y)| < e_n$.

(1) Suppose that $[a_{i+1}, b_{i+1}] \subseteq (a_i, b_i)$ for all $i \in \mathbb{N}$. Let $c \in \bigcap_{n=1}^{\infty} [a_n, b_n]$. Prove that f is continuous at c.

(2) Suppose that the hypothesis of Part (1) of this exercise is changed to the weaker hypothesis $[a_{i+1}, b_{i+1}] \subseteq [a_i, b_i]$ for all $i \in \mathbb{N}$. Does the conclusion still hold? Give a proof or a counterexample.

Exercise 8.4.7. [Used in Section 3.3.] The purpose of this exercise is to prove that there is no function $g\colon [0,1] \to \mathbb{R}$ that is continuous at every rational number in $[0,1]$, and discontinuous at every irrational number in $[0,1]$. This proof is due to Vito Volterra (1860–1940).

Suppose that there is a function $g\colon [0,1] \to \mathbb{R}$ that is continuous at every rational number in $[0,1]$, and discontinuous at every irrational number in $[0,1]$. Let $s\colon [0,1] \to \mathbb{R}$ be the function given in Example 3.3.3 (7). We saw that s is discontinuous at every rational number in $[0,1]$, and continuous at every irrational number in $[0,1]$.

(1) Let $p_1 \in \mathbb{Q} \cap (0,1)$. Prove that there is a non-degenerate closed bounded interval $[c_1, d_1] \subseteq (0,1)$ such that $p_1 \in (c_1, d_1)$, and that $x, y \in [c_1, d_1]$ implies $|s(x) - s(y)| < 1$.

(2) By Theorem 2.6.13 (2) there is some $q_1 \in (\mathbb{R} - \mathbb{Q}) \cap [c_1, d_1]$. Prove that there is a non-degenerate closed bounded interval $[a_1, b_1] \subseteq (c_1, d_1)$ such that $q_1 \in (a_1, b_1)$, and that $x, y \in [a_1, b_1]$ implies $|g(x) - g(y)| < 1$. Observe that $x, y \in [a_1, b_1]$ implies $|s(x) - s(y)| < 1$.

(3) Prove that there is a family of closed bounded intervals $\{[a_n, b_n]\}_{n=1}^{\infty}$ in $[a, b]$ such that $[a_{i+1}, b_{i+1}] \subseteq (a_i, b_i)$ for all $i \in \mathbb{N}$, and that $n \in \mathbb{N}$ and $x, y \in [a_n, b_n]$ imply $|s(x) - s(y)| < \frac{1}{n}$ and $|g(x) - g(y)| < \frac{1}{n}$.

(4) Obtain a contradiction. [Use Exercise 8.4.6 (1).]

Exercise 8.4.8. [Used in Section 8.4.] Give an alternative proof of the Nested Interval Theorem (Theorem 8.4.7) that relies upon Theorem 8.3.3 instead of the No Gap Lemma.

Exercise 8.4.9. [Used in Section 8.3.] The purpose of this exercise is to give a proof of the Bolzano–Weierstrass Theorem (Theorem 8.3.9) that uses the Nested Interval Theorem (Theorem 8.4.7) instead of the Monotone Convergence Theorem (Corollary 8.3.4).

Let $\{a_n\}_{n=1}^{\infty}$ be a sequence in $\mathbb{R}$. Suppose that $\{a_n\}_{n=1}^{\infty}$ is bounded. Hence there are $s_0, t_0 \in \mathbb{R}$ such that $s_0 \le a_k \le t_0$ for all $k \in \mathbb{N}$. We divide $[s_0, t_0]$ into two subintervals $[s_0, \frac{s_0+t_0}{2}]$ and $[\frac{s_0+t_0}{2}, t_0]$. It must be the case that at least one of these two subintervals contains a_k for infinitely many $k \in \mathbb{N}$. Choose a subinterval for which this condition holds; it does not matter which one is chosen if the condition holds for both subintervals. Rename the chosen subinterval $[s_1, t_1]$. Then $[s_1, t_1] \subseteq [s_0, t_0]$, and that $t_1 - s_1 = \frac{t_0 - s_0}{2}$. Continuing in this way, we use Definition by Recursion to define a family $\{[s_n, t_n]\}_{n=1}^{\infty}$ of non-degenerate closed bounded intervals in $\mathbb{R}$ such that for each $n \in \mathbb{N}$, the following three conditions hold: (1) the interval $[s_n, t_n]$ contains a_k for infinitely many $k \in \mathbb{N}$; (2) $[s_{n+1}, t_{n+1}] \subseteq [s_n, t_n]$; and (3) $t_{n+1} - s_{n+1} = \frac{t_n - s_n}{2}$.

Use the above idea to prove the Bolzano–Weierstrass Theorem.

[Use Exercise 8.2.9.]

Exercise 8.4.10. [Used in Example 8.4.9 and Exercise 8.4.11.] Use the Nested Interval Theorem (Theorem 8.4.7) to prove that the Cantor set, as defined in Example 8.4.9, is uncountable.

Exercise 8.4.11. [Used in Section 5.8.] Find an example of a function $f\colon [0,1] \to \mathbb{R}$ such that f is integrable, but that the set of numbers at which the function is discontinuous is uncountable. [Use Exercise 2.6.13, Exercise 3.3.2 and Exercise 8.4.10.]

Exercise 8.4.12. [Used in Example 8.4.10 and Exercise 9.5.7.] This exercise has some properties of the sequence of Fibonacci numbers $\{F_n\}_{n=1}^{\infty}$.

(1) Prove that $F_n \le 2^{n-1}$ for all $n \in \mathbb{N}$.

(2) Prove that $F_n \ge n$ for all $n \in \mathbb{N}$ such that $n \ge 5$.

(3) Prove that $F_1 + F_2 + \cdots + F_n = F_{n+2} - 1$ for all $n \in \mathbb{N}$.

(4) Prove that ${F_1}^2 + {F_2}^2 + \cdots + {F_n}^2 = F_n F_{n+1}$ for all $n \in \mathbb{N}$.

(5) Prove that $(F_{n+1})^2 - F_{n+2} F_n = (-1)^{n+2}$ for all $n \in \mathbb{N}$.

Exercise 8.4.13. **[Used in Exercise 9.3.10.]** For each $n \in \mathbb{N}$, let

$$\gamma_n = \frac{1}{1} + \frac{1}{2} + \cdots + \frac{1}{n} - \ln n.$$

Using what the reader is asked to prove below, it follows from Theorem 8.3.3 (3) that the sequence $\{\gamma_n\}_{n=1}^{\infty}$ is convergent. The limit of this sequence, which is often denoted γ, is know as Euler's constant (or the Euler–Mascheroni constant), and has been widely studied; see [Hav03] for more about this number. We will make use of the sequence $\{\gamma_n\}_{n=1}^{\infty}$ in Exercise 9.3.10, where we find the exact value of the sum of the alternating harmonic series.

(1) Prove that $\{\gamma_n\}_{n=1}^{\infty}$ is strictly decreasing.
(2) Prove that $\{\gamma_n\}_{n=1}^{\infty}$ is bounded below by 0.

[Use Exercise 7.2.3.]

Exercise 8.4.14. Suppose we are given an equation of the form $f(x) = 0$, for some function $f : C \to \mathbb{R}$, where $C \subseteq \mathbb{R}$ is a non-degenerate interval. Suppose further that we know, for example by looking at the graph of f, that the equation $f(x) = 0$ has a solution, that is, there exists a number $r \in C$ such that $f(r) = 0$. If we cannot find the exact value of r, then the next best thing would be to approximate it. One general method of approximating the solution of an equation is iteration, which means starting with a guess for the solution, then modifying the guess in order to obtain what is hopefully a better approximation of the solution, then modifying that and so on. More precisely, we use Definition by Recursion to define a sequence $\{x_n\}_{n=1}^{\infty}$ in C, with the hope that the sequence will converge to r. In practice, we never compute the whole sequence, but only finitely many of its terms, and, if all goes well, the value of x_n for sufficiently large n will be a good approximation of r.

In this exercise we discuss the Bisection Method, which gives one way of defining the sequence $\{x_n\}_{n=1}^{\infty}$. The advantages of the Bisection Method are that it is simple to understand and implement, and that it requires only that the function be continuous, not necessarily differentiable; the disadvantage is that it is slower than some other methods, for example Newton's Method (discussed in Exercise 8.4.15), where slower in this context means that for the same amount of accuracy, more terms of the sequence are needed. See [Est02, Chapter 13] or [CK07, Section 3.1] for details about the Bisection Method.

The Bisection Method works as follows. Suppose that f is continuous. Suppose further that $C = [a,b]$, and that $f(a)$ and $f(b)$ are non-zero and have different signs; that is, suppose that $f(a) > 0$ and $f(b) < 0$, or $f(a) < 0$ and $f(b) > 0$. Let $s_1 = a$, let $t_1 = b$ and let $x_1 = \frac{s_1+t_1}{2}$. If $f(x_1) = 0$, we have found what we are looking for, and we stop. Now suppose that $f(x_1) \neq 0$. Then $f(x_1)$ has a different sign from precisely one of $f(s_1)$ or $f(t_1)$. If $f(x_1)$ has a different sign from $f(s_1)$, then we let $s_2 = s_1$ and $t_2 = x_1$; if $f(x_1)$ has a different sign from $f(t_1)$, then we let $s_2 = x_1$ and $t_2 = t_1$. In either case, we let $x_2 = \frac{s_2+t_2}{2}$. Continuing in this way, we use Definition by Recursion to define a family $\{[s_n, t_n]\}_{n=1}^{\infty}$ of non-degenerate closed bounded intervals in $\mathbb{R}$, and a sequence $\{x_n\}_{n=1}^{\infty}$ in $\mathbb{R}$, such that for each $n \in \mathbb{N}$, the following four conditions hold:

(1) $f(s_n)$ and $f(t_n)$ are non-zero and have different signs; (2) $[s_{n+1}, t_{n+1}] \subseteq [s_n, t_n]$; (3) $t_{n+1} - s_{n+1} = \frac{t_n - s_n}{2}$; and (4) $x_n = \frac{s_n + t_n}{2}$ and $f(x_n) \neq 0$.

Prove the following theorem.

Let $[a,b] \subseteq \mathbb{R}$ be a non-degenerate closed bounded interval, and let $f\colon [a,b] \to \mathbb{R}$ be a function. Suppose that f is continuous, and that either $f(a) > 0$ and $f(b) < 0$, or $f(a) < 0$ and $f(b) > 0$. Then $\{x_n\}_{n=1}^{\infty}$ is convergent, and if $r = \lim_{n\to\infty} x_n$ then $f(r) = 0$.

[Use Exercise 8.2.9.]

Exercise 8.4.15. Newton's Method is a very useful method for finding approximate solutions of equations. (Newton did not actually propose the method as we now know it; the modern form of the method is due to Joseph Raphson (1648–1715) and Thomas Simpson (1710–1761).)Newton's Method is based upon iteration. Before continuing with this exercise, read the first paragraph of Exercise 8.4.14 for the general idea of iteration in the context of solving an equation; we will use the notation of that exercise. Newton's Method is one way of defining the sequence $\{x_n\}_{n=1}^{\infty}$ discussed in the first paragraph of Exercise 8.4.14.

The advantage of Newton's Method is that it is fast, in the sense of needing relatively few terms in the sequence to obtain the desired amount of accuracy; the disadvantages are that it requires that the function is twice differentiable and that it satisfies some additional hypotheses, and that a poor choice of x_1 can cause problems. See [Est02, Chapter 31] or [CK07, Section 3.2] for details about Newton's Method, see [HH99, Section 2.8] for a discussion of the rapidity of convergence of Newton's Method and see [Fal03, Section 14.5] for the relation between Newton's Method and fractals.

Newton's Method works as follows. Suppose that f is differentiable, and that $f'(x) \neq 0$ for all $x \in C$. We choose $x_1 \in C$ arbitrarily, though hopefully close to r. We then define the sequence $\{x_n\}_{n=1}^{\infty}$ by the formula

$$x_{n+1} = x_n - \frac{f(x_n)}{f'(x_n)} \tag{8.4.2}$$

for all $n \in \mathbb{N}$.

The above definition of $\{x_n\}_{n=1}^{\infty}$ has two potential problems. First, even if $x_n \in C$ for some $n \in \mathbb{N}$, it is not immediately evident that $x_{n+1} \in C$, and without that it would not be possible to define x_{n+2}. Second, even if x_n is defined for all $n \in \mathbb{N}$, it is not evident that the sequence $\{x_n\}_{n=1}^{\infty}$ is convergent, not to mention that it converges to a solution of the equation. Fortunately, we will see that if f is sufficiently well behaved, and if x_1 is chosen sufficiently close to a solution, then everything works out well. Our theorem is as follows.

Let $[a,b] \subseteq \mathbb{R}$ be a non-degenerate closed bounded interval, and let $f\colon [a,b] \to \mathbb{R}$ be a function. Suppose that f is twice differentiable, that f'' is bounded, that f' is bounded away from zero (see Definition 5.5.3), and that either $f(a) > 0$ and $f(b) < 0$, or $f(a) < 0$ and $f(b) > 0$. Then the following two facts hold.

(a) There is a unique $r \in (a,b)$ such that $f(r) = 0$.

(b) There is a non-degenerate open bounded interval $I \subseteq [a,b]$ such that $r \in I$, and such that for any $x_1 \in I$, the sequence $\{x_n\}_{n=1}^{\infty}$ defined by Equation 8.4.2 exists in I, and $\{x_n\}_{n=1}^{\infty}$ is convergent, and $\lim_{n\to\infty} x_n = r$.

The proof of the theorem will be done in steps, starting with the geometric motivation for the formula in Equation 8.4.2 before the actual proof.

(1) Let $n \in \mathbb{N}$. Suppose that $x_n \in [a,b]$ has been defined. Because f' is bounded away from zero, then the tangent line to the graph of f at the point $(x_n, f(x_n))$ will intersect the x-axis. Prove that this point of intersection is x_{n+1}. See Figure 8.4.2.

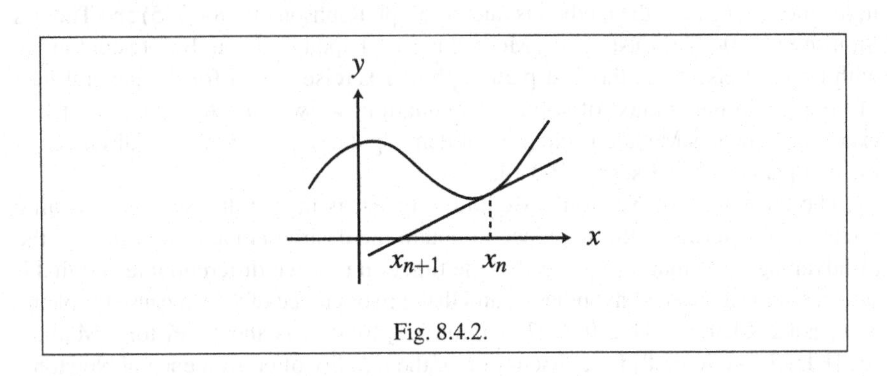

Fig. 8.4.2.

(2) Prove Part (a) of this exercise.

(3) By hypothesis on f, there is some $M \in \mathbb{R}$ such that $|f''(x)| \le M$ for all $x \in [a,b]$, and there is some $P > 0$ such that $|f(x)| \ge P$ for all $x \in [a,b]$. We may assume that $M > 0$. Because $r \in (a,b)$, then by Lemma 2.3.7 (2) there is some $\delta > 0$ such that $(r-\delta, r+\delta) \subseteq (a,b)$. By taking a smaller value of δ if necessary, we may assume that $\delta < \frac{2P}{M}$. Let $I = (r-\delta, r+\delta)$.

Let $x_1 \in I$. Clearly $|x_1 - r| \le \left(\frac{M\delta}{2P}\right)^1 |x_1 - r|$. Let $k \in \mathbb{N}$. Suppose that $x_k \in I$ has been defined, and that $|x_k - r| \le \left(\frac{M\delta}{2P}\right)^k |x_1 - r|$. Let x_{k+1} be defined by Equation 8.4.2. By Taylor's Theorem (Theorem 4.4.6), using $n = 1$, and $x = r$ and $c = x_k$, there is some p strictly between r and x_k (except that $p = x_k$ when $r = x_k$) such that

$$f(r) = f(x_k) + f'(x_k)(r - x_k) + \frac{f''(p)}{2}(r - x_k)^2.$$

Prove that $|x_{k+1} - r| \le \frac{M}{2P}|x_k - r|^2$.

(4) Using Part (3) of this exercise, prove that $|x_{k+1} - r| < \delta$ and that $|x_{k+1} - r| \le \left(\frac{M\delta}{2P}\right)^{k+1} |x_1 - r|$. It follows that $x_{k+1} \in I$.

(5) It follows from Definition by Recursion, using Parts (3) and (4) of this exercise, that $\{x_n\}_{n=1}^{\infty}$ is defined in I, and that $|x_n - r| \le \left(\frac{M\delta}{2P}\right)^n |x_1 - r|$ for all $n \in \mathbb{N}$.

Prove that $\{x_n\}_{n=1}^{\infty}$ is convergent, and that $\lim_{n\to\infty} x_n = r$, which proves Part (b) of this exercise. [Use Exercise 8.2.9.]

8.5 Historical Remarks

Similarly to the historical remarks concerning limits to infinity in Section 6.5, the historical remarks for the present chapter are very brief. This chapter discusses sequences, and limits of sequences are a type of limit to infinity, and are also a variation of ordinary limits of functions, and hence much of the history of the material in this chapter overlaps with the history discussed in Sections 3.6 and 6.5.

Ancient World

Zeno of Elea (c. 490–c. 425 BCE) gave four arguments, which have been preserved in Book VI of the *Physics* of Aristotle (384–322 BCE), to show that there is no motion. For example, Zeno's first argument, the Dichotomy, states that for an object to move from one place to another, it must first pass through the midpoint, and then through the midpoint of what remains, ad infinitum, and so the destination will never be reached, presumably because it is not thought possible to traverse an infinite number of points in a finite amount of time; hence motion is not possible. Though this argument is not about sequences per se, in fact it is essentially about the sequence $\frac{1}{2}, \frac{1}{4}, \frac{1}{8}, \frac{1}{16}, \ldots$, those lengths being the distance to each successive midpoint, and proportional to the time to traverse such lengths while traveling at uniform speed. From a modern perspective the resolution of Zeno's argument is that the series $\frac{1}{2} + \frac{1}{4} + \frac{1}{8} + \frac{1}{16} + \cdots$ is convergent, and in fact the sum of the series is 1. The second of Zeno's arguments, the Achilles, is similar. Zeno's arguments had an influence on subsequent discussion of infinitesimals and the infinite, and in particular it appears to have been the reason (or one of the reasons) that the ancient Greeks did not use limits of sequences.

Limits of sequences arose naturally in some geometry problems such as area and volume calculations, where the region is approximated by a sequence of polygons or polyhedra with an increasing number of sides, but because the ancient Greeks did not take limits, they developed the method of exhaustion as an alternative way to handle such situations. This method, which is attributed to Eudoxus of Cnidus (408–355 BCE), was used by Euclid (c. 325–c. 265 BCE) in the *Elements*, and was brought to its peak by Archimedes (287–212 BCE). In modern terms, the method of exhaustion avoids "letting n go to infinity" in the limit of a sequence by using a double proof by contradiction (referred to as reductio ad absurdum), which is complicated in practice, and has the drawback that it can only prove that a sequence converges to a number, but it cannot produce the number itself. Although today we can find the limit of a sequence directly without using the complicated arguments of reductio ad absurdum, it is important to observe that when we use the ε–N definition of $\lim_{n\to\infty} a_n$, we also avoid dealing explicitly with something called infinity (even though for the sake of brevity we use the symbolic notation "$n \to \infty$"), and so in a sense the modern approach to limits of sequences recaptures something of what the ancient Greeks wanted. The modern approach, however, is much more convenient to use.

The method of compression of Archimedes, used in area and volume problems, is a reductio ad absurdum method that is similar to the following modern formulation concerning sequences. Let $s, c \in \mathbb{R}$, and let $\{u_n\}_{n=1}^{\infty}$ and $\{l_n\}_{n=1}^{\infty}$ be sequences in $\mathbb{R}$ with positive terms. Suppose that $l_n \le s \le u_n$ and $l_n \le c \le u_n$ for all $n \in \mathbb{N}$. The difference form of the method of compression is equivalent to proving that $s = c$ by proving that for each $\varepsilon > 0$, the inequality $u_n - l_n < \varepsilon$ holds for all sufficiently large $n \in \mathbb{N}$; the quotient form is equivalent to proving that $s = c$ by proving that for each $\alpha > 1$, the inequality $\frac{u_n}{l_n} < \alpha$ holds for all sufficiently large $n \in \mathbb{N}$.

Medieval Period

In opposition to Aristotle, who did not believe in the existence of the infinite (but only in a potential infinite), and whose work was very influential in medieval Europe, Gregory of Rimini (1300-1358) maintained that God could create an infinite stone by creating equal-sized stones at each of the times $t = 0, \frac{1}{2}, \frac{3}{4}, \frac{7}{8}, \ldots$. That is, God could create sequences.

Eighteenth Century

Leonhard Euler (1707–1783) seems to have viewed sequences as he did other functions; that is, he viewed sequences as given by appropriate formulas, rather than as arbitrary functions with domain $\mathbb{N}$, which is how we understand sequences today. From Euler's perspective, if a sequence $\{a_n\}_{n=1}^{\infty}$ is given, it would define a unique function $f\colon \mathbb{R} \to \mathbb{R}$, simply by using the same formula. Carl Friedrich Gauss (1777–1855) viewed sequences as we do today, and in particular understood that a sequence does not specify a unique function $f\colon \mathbb{R} \to \mathbb{R}$. However, Gauss' treatment of sequences was not entirely rigorous, and he implicitly used some results that we now prove, such as the Monotone Convergence Theorem.

Nineteenth Century

In their attempt at providing rigorous proofs of some basic facts about continuity, Bernard Bolzano (1781–1848) and Augustin Louis Cauchy (1789–1857) made use of what we now call the Cauchy Completeness Theorem, though they could not prove it because they lacked the axiomatic properties of the real numbers. Bolzano did provide a proof that the Cauchy Completeness Theorem implied the Least Upper Bound Property, using the idea of bisection (similar to the method of Exercise 8.4.9). Cauchy's proof of the Intermediate Value Theorem relied implicitly upon the Monotone Convergence Theorem, and explicitly on the fact that a continuous function works nicely with respect to convergent sequences (as stated in Corollary 8.4.2). In the 1860s Karl Weierstrass (1815–1897) used a bisection argument similar to Bolzano's to prove a version of what we now call the Bolzano–Weierstrass Theorem for bounded infinite sets (rather than for bounded sequences as in the version of this theorem in the present text).

Richard Dedekind (1831–1916), using his construction of the real numbers from the rational numbers in *Stetigkeit und irrationale Zahlen* of 1872 (originally formulated in lectures in 1858), provided what was probably the first rigorous proof of the Monotone Convergence Theorem. Such a proof was not possible without a rigorous treatment of the real numbers.

9

Series

9.1 Introduction

Now that we have seen sequences of real numbers in Chapter 8, we turn to the related notion of series of real numbers. We have already encountered series informally, for example in Section 5.8, where we discussed sets of measure zero. We now give a rigorous treatment of series.

A series is a sum of countably many numbers, added up in the given order. Although series might appear innocuous at first glance, because they seem to resemble finite sums, in fact infinite sums are much trickier than finite ones, as we will see, for example, in Section 9.4.

One of the most important type of series is power series, which are an extremely useful tool in a wide range of mathematics and its applications, including numerical calculations. We commence our treatment of power series in this chapter, where we think of power series as a certain type of series with a "variable." However, to make full use of power series we will need to approach them from a different perspective, and we will therefore put off completing our discussion of power series until Section 10.4.

9.2 Series

Intuitively, a series is the sum of a sequence. For example, if we are given the sequence

$$\frac{1}{2}, \frac{1}{4}, \frac{1}{8}, \frac{1}{16}, \ldots,$$

we can form the series

$$\frac{1}{2} + \frac{1}{4} + \frac{1}{8} + \frac{1}{16} + \cdots,$$

which can also be written as

$$\sum_{n=1}^{\infty} \frac{1}{2^n}.$$

More generally, we have the following definition.

Definition 9.2.1. A **series** in $\mathbb{R}$ (also called a series of real numbers) is a formal sum

$$\sum_{n=1}^{\infty} a_n = a_1 + a_2 + a_3 + \cdots,$$

where $\{a_n\}_{n=1}^{\infty}$ is a sequence in $\mathbb{R}$. Each number a_n, where $n \in \mathbb{N}$, is called a **term** of the series $\sum_{n=1}^{\infty} a_n$. $\triangle$

The expression "formal sum" means simply a sequence of numbers with addition symbols written between the terms of the sequence. Of course, while we can certainly write down a formal sum of infinitely many numbers, we have to ask whether such an infinite sum actually means anything, beyond being just symbols written on the page; not everything that can be written down actually means something. In particular, addition for real numbers has been defined for only two real numbers at a time. Using Definition by Recursion it is possible to extend the definition of addition to the sum of any finite collection of real numbers; see Exercise 2.5.19 for details. However, there is no direct way to extend the definition of addition of real numbers to infinitely many real numbers. Indeed, there ought not to be such a definition, because it is evident that not every infinite collection of real numbers adds up to a real number. For example, consider the sequence $1^2, 2^2, 3^2, \cdots$. We can formally write down the infinite sum $1^2 + 2^2 + 3^2 + \cdots$, but there is no hope that this sum will equal a real number. On the other hand, it turns out that some infinite sums of real numbers do equal real numbers.

To obtain an intuitive feel for how an infinite collection of numbers can add up to a finite amount, let us view the situation backwards. Consider a unit square, as shown in the first row of Figure 9.2.1. The area of the square is 1. Cut the square into two equal pieces, each of which has height $\frac{1}{2}$, and rearrange the pieces as shown in the second row of the figure. Each piece has area $\frac{1}{2}$, and the combined area of the two pieces is still 1, which we can express by writing $1 = \frac{1}{2} + \frac{1}{2}$. Next, we cut one of the rectangles with height $\frac{1}{2}$ into two pieces and rearrange, as shown in the third row of the figure; in this case the area is $1 = \frac{1}{2} + \frac{1}{4} + \frac{1}{4}$. In the fourth and fifth row of the figure we see the result of continuing this process two more times, the last one with area $1 = \frac{1}{2} + \frac{1}{4} + \frac{1}{8} + \frac{1}{16} + \frac{1}{16}$. Continuing in this fashion, it seems plausible that

$$1 = \frac{1}{2} + \frac{1}{4} + \frac{1}{8} + \frac{1}{16} + \cdots.$$

This equation turns out to be correct; we will see a proof in Example 9.2.4 (1).

Intuitively, an infinite sum adds up to a finite amount if the terms of the series go to zero sufficiently fast, though it is hard to make the intuitive notion of "going to zero sufficiently fast" into a rigorous definition. For our formal definition of the convergence of series we use a different idea, which is that to attempt to find the sum of a series, we can add the first two terms, then the first three terms, then the first four terms and so on, and see what happens as a result of this process. In this way, we reduce the question of the convergence of a series to a question of the convergence of a certain sequence. It is important to note that in this process, we always add the terms of the series in the given order, rather than all at once; as we will see in our discussion of rearrangements of series in Section 9.4, sticking to the given order of the terms of a series is important.

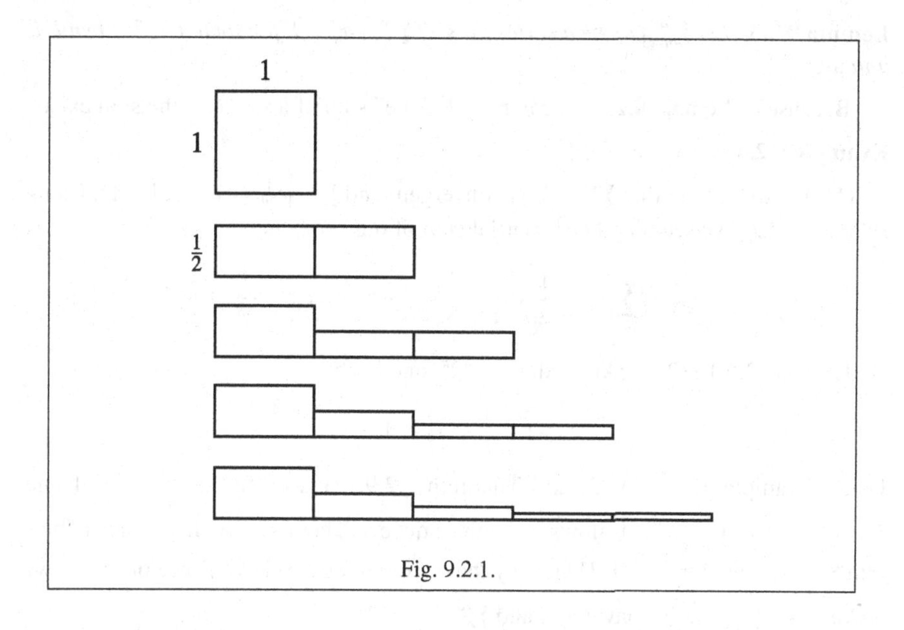

Fig. 9.2.1.

Definition 9.2.2. Let $\sum_{n=1}^{\infty} a_n$ be a series in $\mathbb{R}$.

1. For each $k \in \mathbb{N}$, the kth **partial sum** of $\sum_{n=1}^{\infty} a_n$, denoted s_k, is defined by $s_k = \sum_{i=1}^{k} a_i$. The **sequence of partial sums** of $\sum_{n=1}^{\infty} a_n$ is the sequence $\{s_n\}_{n=1}^{\infty}$.
2. Let $L \in \mathbb{R}$. The number L is the **sum** of $\sum_{n=1}^{\infty} a_n$, written
$$\sum_{n=1}^{\infty} a_n = L,$$
if the sequence of partial sums $\{s_n\}_{n=1}^{\infty}$ is convergent and $\lim_{n\to\infty} s_n = L$. If $\sum_{n=1}^{\infty} a_n = L$, we also say that $\sum_{n=1}^{\infty} a_n$ **converges** to L. If $\sum_{n=1}^{\infty} a_n$ converges to some real number, we say that $\sum_{n=1}^{\infty} a_n$ is **convergent**; otherwise we say that $\sum_{n=1}^{\infty} a_n$ is **divergent**.
3. The series $\sum_{n=1}^{\infty} a_n$ **diverges to infinity**, written
$$\sum_{n=1}^{\infty} a_n = \infty,$$
if $\lim_{n\to\infty} s_n = \infty$. The series $\sum_{n=1}^{\infty} a_n$ **diverges to negative infinity**, written
$$\sum_{n=1}^{\infty} a_n = -\infty,$$

if $\lim_{n\to\infty} s_n = -\infty$. △

The following lemma, which states that if a series is convergent then its sum is unique, follows immediately from Lemma 8.2.3, and we omit the proof.

Lemma 9.2.3. *Let $\sum_{n=1}^{\infty} a_n$ be a series in $\mathbb{R}$. If $\sum_{n=1}^{\infty} a_n = L$ for some $L \in \mathbb{R}$, then L is unique.*

Because of Lemma 9.2.3 we can refer to "the" sum of a series, if the sum exists.

Example 9.2.4.

(1) We will prove that $\sum_{n=1}^{\infty} \frac{1}{2^n}$ is convergent and $\sum_{n=1}^{\infty} \frac{1}{2^n} = 1$. Let $k \in \mathbb{N}$. Using Exercise 2.5.6, we see that the kth partial sum of our series is

$$s_k = \sum_{i=1}^{k} a_i = \frac{1}{2} + \frac{1}{4} + \frac{1}{8} + \cdots + \frac{1}{2^k} = 1 - \frac{1}{2^k}.$$

By Exercise 2.5.13 (2) we know that $k < 2^k$, and hence

$$1 - \frac{1}{k} < s_k < 1.$$

Using Example 8.2.4 (1) (2) and Theorem 8.2.9 (2), we see that $\lim_{n\to\infty} 1 = 1$ and $\lim_{n\to\infty} \left(1 - \frac{1}{n}\right) = 1$. It now follows from the Squeeze Theorem for Sequences (Theorem 8.2.12) that $\lim_{n\to\infty} s_n = 1$. Hence, by the definition of the convergence of series, we deduce that $\sum_{n=1}^{\infty} \frac{1}{2^n}$ is convergent and $\sum_{n=1}^{\infty} \frac{1}{2^n} = 1$.

(2) We will prove that the series $\sum_{n=1}^{\infty} (-1)^{n-1}$ is divergent. Let $k \in \mathbb{N}$. The kth partial sum of our series is

$$s_k = 1 + (-1) + 1 + \cdots + (-1)^{k-1} = \begin{cases} 1, & \text{if } k \text{ is even} \\ 0, & \text{if } k \text{ is odd.} \end{cases}$$

It can be proved by an ε–N proof similar to the proof in Example 8.2.4 (4) that $\{s_n\}_{n=1}^{\infty}$ is divergent. However, we can avoid such a proof by observing that $s_n = \frac{1}{2}\left[1 + (-1)^{n-1}\right]$ for all $n \in \mathbb{N}$. If the sequence $\{s_n\}_{n=1}^{\infty}$ is convergent, then it would follow from Theorem 8.2.9 that $\{(-1)^n\}_{n=1}^{\infty} = \{1 - 2s_n\}_{n=1}^{\infty}$ is convergent, which is a contradiction to Example 8.2.4 (4). Hence $\{s_n\}_{n=1}^{\infty}$ is divergent, which means that $\sum_{n=1}^{\infty} (-1)^{n-1}$ is divergent.

(3) We will prove that $\sum_{n=1}^{\infty} \frac{2}{n(n+1)} = 2$. Observe that if $n \in \mathbb{N}$, then $\frac{2}{n(n+1)} = \frac{2}{n} - \frac{2}{n+1}$. Let $k \in \mathbb{N}$. The kth partial sum of our series is

$$s_k = \left(\frac{2}{1} - \frac{2}{2}\right) + \left(\frac{2}{2} - \frac{2}{3}\right) + \left(\frac{2}{3} - \frac{2}{4}\right) + \cdots + \left(\frac{2}{k} - \frac{2}{k+1}\right) = 2 - \frac{2}{n+1}.$$

Example 8.2.4 (2), Theorem 8.2.9 (3) and Exercise 8.2.4 imply that $\lim_{n\to\infty} \frac{2}{n+1} = 0$, and it then follows from Example 8.2.4 (1) and Theorem 8.2.9 (2) that $\lim_{n\to\infty} s_n = 2$. Hence $\sum_{n=1}^{\infty} \frac{2}{n(n+1)}$ is convergent and $\sum_{n=1}^{\infty} \frac{2}{n(n+1)} = 2$. This series is an example of what is called a "telescoping series."

(4) Let $a, r \in \mathbb{R}$. The series $\sum_{n=1}^{\infty} ar^{n-1}$ is called a **geometric series**. This type of series is widely used in mathematics and its applications. If $a = 0$ then the series is constantly zero, which is convergent for all values of r; from now on we will assume that $a \neq 0$. Writing out the terms of the series, we see that

$$\sum_{n=1}^{\infty} ar^{n-1} = a + ar + ar^2 + ar^3 + \cdots.$$

Observe that the ratio of each term in the series to the previous term is r; such a constant ratio characterizes geometric series.

Let $k \in \mathbb{N}$. The kth partial sum of our series is

$$s_k = a + ar + ar^2 + \cdots + ar^{k-1}.$$

If $r = 1$, then clearly $s_k = ka$. If $r \neq 1$, then Exercise 2.5.12 (3) states that

$$s_k = \frac{a(1-r^k)}{1-r}.$$

Is the sequence $\{s_n\}_{n=1}^{\infty}$ convergent? It depends upon the value of r. First, suppose that $r = 1$. Then $\{s_n\}_{n=1}^{\infty}$ is given by $\{na\}_{n=1}^{\infty}$, and this sequence is divergent by Exercise 8.2.1 (3) and Exercise 8.2.12 (3).

Second, suppose that $r \neq 1$. Then $\{s_n\}_{n=1}^{\infty} = \left\{\frac{a(1-r^n)}{1-r}\right\}_{n=1}^{\infty}$. We saw in Example 8.2.13 that the sequence $\{r^n\}_{n=1}^{\infty}$ is convergent if and only if $-1 < r \leq 1$, and that if $-1 < r < 1$ then $\lim_{n\to\infty} r^n = 0$. Using Example 8.2.4 (1), Theorem 8.2.9 and Exercise 8.2.12, we deduce that $\left\{\frac{a(1-r^n)}{1-r}\right\}_{n=1}^{\infty}$ is convergent if and only if $-1 < r < 1$ (we are assuming here that $r \neq 1$), and that if $-1 < r < 1$ then $\lim_{n\to\infty} s_n = \frac{a(1-0)}{1-r} = \frac{a}{1-r}$.

Therefore $\sum_{n=1}^{\infty} ar^{n-1}$ is convergent if and only if $-1 < r < 1$, and if $-1 < r < 1$ then $\sum_{n=1}^{\infty} ar^{n-1} = \frac{a}{1-r}$.

(5) We will prove that the series $\sum_{n=1}^{\infty} \frac{1}{n}$ is divergent. This series is known as the **harmonic series**, and it is an interesting series even though it is not convergent; see the comment at the end of Example 9.3.7. This proof of divergence is due to Nicole Oresme (1323–1382).

Let $k \in \mathbb{N}$. The kth partial sum of our series is

$$s_k = \frac{1}{1} + \frac{1}{2} + \frac{1}{3} + \cdots + \frac{1}{k}.$$

There is no simple formula for s_k, but it turns out that we can prove that $\{s_n\}_{n=1}^{\infty}$ is divergent even without such a formula.

The hard work for our proof was done in Exercise 2.5.7, which states that if $n \in \mathbb{N}$, then $s_{2^n} \geq \frac{n+2}{2}$. It follows from Exercise 8.2.15, Example 8.2.4 (1) and Exercise 8.2.17 that $\lim_{n\to\infty} \frac{n+2}{2} = \infty$, and it then follows from Exercise 8.2.18 (1) that $\lim_{n\to\infty} s_{2^n} = \infty$. In particular, we deduce that $\{s_{2^n}\}_{n=1}^{\infty}$ is divergent. By Lemma 8.3.7 we see that $\{s_n\}_{n=1}^{\infty}$ is divergent, and hence the series $\sum_{n=1}^{\infty} \frac{1}{n}$ is divergent. $\lozenge$

We need the following two very simple observations about series. First, a series does not have to start with $n = 1$. If $k \in \mathbb{Z}$, we can just as well consider series of the form $\sum_{n=k}^{\infty} a_n$. For simplicity, we will state all results about series starting with $n = 1$, except in some situations where starting with $n = 0$ is more convenient. Second, the convergence of a series is unaffected by changing, or dropping, finitely many terms of the series; see Exercise 9.2.4 and Exercise 9.2.5 for details.

The series seen in Example 9.2.4 are actually atypical, in that we were able to find nice formulas for the partial sums for most of these series. In general, however, it is not possible in practice to find useful formulas for the partial sums of most series, and hence other, more indirect, tests for convergence are used. There are a variety of such tests, some of which will be seen in Sections 9.3 and 9.4. The simplest such test, which we now state, is one that yields only divergence, not convergence. The idea of this test is that if a series is convergent, then the sequence of partial sums is convergent, and hence the partial sums get closer and closer to each other, and hence the terms of the series must go to zero.

Theorem 9.2.5 (Divergence Test). *Let $\sum_{n=1}^{\infty} a_n$ be a series in $\mathbb{R}$. If $\{a_n\}_{n=1}^{\infty}$ does not converge to 0, then $\sum_{n=1}^{\infty} a_n$ is divergent.*

Proof. Suppose that $\sum_{n=1}^{\infty} a_n$ is convergent. Let $L = \sum_{n=1}^{\infty} a_n$. Let $\{s_n\}_{n=1}^{\infty}$ be the sequence of partial sums of $\sum_{n=1}^{\infty} a_n$. Then $\lim_{n\to\infty} s_n = L$. Let $\varepsilon > 0$. Then there is some $N \in \mathbb{N}$ such that $n \in \mathbb{N}$ and $n \geq N$ imply $|s_n - L| < \frac{\varepsilon}{2}$. Suppose that $n \in \mathbb{N}$ and $n \geq N + 1$. Then $n - 1 \geq N$, and therefore

$$|a_n - 0| = |s_n - s_{n-1}| = |s_n - L + L - s_{n-1}| \leq |s_n - L| + |L - s_{n-1}| < \frac{\varepsilon}{2} + \frac{\varepsilon}{2} = \varepsilon.$$

It follows that $\lim_{n\to\infty} a_n = 0$. □

The Divergence Test (Theorem 9.2.5), though quite simple, is often the source of student error in calculus courses, because of the failure to distinguish between the contrapositive and the converse of a statement. The contrapositive of the Divergence Test says that if $\sum_{n=1}^{\infty} a_n$ is convergent, then $\{a_n\}_{n=1}^{\infty}$ converges to 0, and that is certainly true. However, it is definitely not the case that if $\{a_n\}_{n=1}^{\infty}$ converges to 0, then $\sum_{n=1}^{\infty} a_n$ must be convergent, as seen in Example 9.2.4 (5).

We conclude this section with the following basic result about the convergence of series.

Theorem 9.2.6. *Let $\sum_{n=1}^{\infty} a_n$ and $\sum_{n=1}^{\infty} b_n$ be series in $\mathbb{R}$, and let $k \in \mathbb{R}$. Suppose that $\sum_{n=1}^{\infty} a_n$ and $\sum_{n=1}^{\infty} b_n$ are convergent.*

1. *$\sum_{n=1}^{\infty} (a_n + b_n)$ is convergent and $\sum_{n=1}^{\infty} (a_n + b_n) = \sum_{n=1}^{\infty} a_n + \sum_{n=1}^{\infty} b_n$.*
2. *$\sum_{n=1}^{\infty} (a_n - b_n)$ is convergent and $\sum_{n=1}^{\infty} (a_n - b_n) = \sum_{n=1}^{\infty} a_n - \sum_{n=1}^{\infty} b_n$.*
3. *$\sum_{n=1}^{\infty} ka_n$ is convergent and $\sum_{n=1}^{\infty} ka_n = k\sum_{n=1}^{\infty} a_n$.*

Proof. We will prove Part (1); the other parts are similar, and we omit the details.

(1) Let $\{s_n\}_{n=1}^{\infty}$, and $\{t_n\}_{n=1}^{\infty}$ and $\{u_n\}_{n=1}^{\infty}$ be the sequences of partial sums of $\sum_{n=1}^{\infty} a_n$, and $\sum_{n=1}^{\infty} b_n$ and $\sum_{n=1}^{\infty} (a_n + b_n)$, respectively. Because $\sum_{n=1}^{\infty} a_n$ and $\sum_{n=1}^{\infty} b_n$

are convergent, then $\{s_n\}_{n=1}^\infty$ and $\{t_n\}_{n=1}^\infty$ are convergent, and $\lim_{n\to\infty} s_n = \sum_{n=1}^\infty a_n$ and $\lim_{n\to\infty} t_n = \sum_{n=1}^\infty b_n$. Using the Associative and Commutative Laws for Addition, which work for all *finite* sums, we see that

$$u_k = \sum_{i=1}^{k}(a_i + b_i) = \sum_{i=1}^{k} a_i + \sum_{i=1}^{k} b_i = s_k + t_k$$

for all $k \in \mathbb{N}$. It now follows from Theorem 8.2.9 (1) that

$$\lim_{n\to\infty} u_n = \lim_{n\to\infty} (s_n + t_n) = \lim_{n\to\infty} s_n + \lim_{n\to\infty} t_n = \sum_{n=1}^{\infty} a_n + \sum_{n=1}^{\infty} b_n.$$

Hence $\sum_{n=1}^\infty (a_n + b_n)$ is convergent and $\sum_{n=1}^\infty (a_n + b_n) = \sum_{n=1}^\infty a_n + \sum_{n=1}^\infty b_n$. □

The reader might wonder why there is no mention of products and quotients of series in Theorem 9.2.6. The answer is that they do not work out as nicely as sums and differences. We will not discuss quotients of series in this text, because there is no nice formula for such quotients even for nicely behaved series, but we will discuss products of series very briefly here, and in more detail in Section 9.4 when we have an additional concept at our disposal. Consider the equation $\sum_{n=1}^\infty (a_n + b_n) = \sum_{n=1}^\infty a_n + \sum_{n=1}^\infty b_n$ from Theorem 9.2.6 (1). We want to look at the analog for products of series of each side of this equation. Suppose that $\sum_{n=1}^\infty a_n$ and $\sum_{n=1}^\infty b_n$ are convergent series. Is $\sum_{n=1}^\infty a_n b_n$ necessarily convergent? The answer, as seen in Example 9.2.7 below, is in general no. In Exercise 9.3.5 (1), it is seen that if $a_n \ge 0$ and $b_n \ge 0$ for all $n \in \mathbb{N}$, then $\sum_{n=1}^\infty a_n b_n$ is convergent. However, as seen in Part (3) of that exercise, even if $\sum_{n=1}^\infty a_n b_n$ is convergent, it is not necessarily equal to $[\sum_{n=1}^\infty a_n] \cdot [\sum_{n=1}^\infty b_n]$. We then ask, separately from the question of the convergence of $\sum_{n=1}^\infty a_n b_n$, whether there is a nice formula for $[\sum_{n=1}^\infty a_n] \cdot [\sum_{n=1}^\infty b_n]$? As seen in Section 9.4, it turns out that in some cases the answer is yes, and in other cases the answer is no.

Example 9.2.7. In Exercise 9.4.1 it will be seen that the series $\sum_{n=1}^\infty (-1)^{n-1} \frac{1}{\sqrt{n}}$ is convergent. If we take both $\sum_{n=1}^\infty a_n$ and $\sum_{n=1}^\infty b_n$ to be this series, then $\sum_{n=1}^\infty a_n b_n = \sum_{n=1}^\infty \frac{1}{n}$, and we saw in Example 9.2.4 (5) that $\sum_{n=1}^\infty \frac{1}{n}$ is divergent. Also, the series $\sum_{n=1}^\infty \frac{a_n}{b_n} = \sum_{n=1}^\infty 1$ is divergent. ◊

Reflections

In a typical second-semester calculus course, part of the semester is devoted to "sequences and series." In practice, sequences are discussed only insofar as they are needed to define the convergence of series via partial sums, and to some extent series are discussed only insofar as they are needed to determine the intervals of convergence of power series. That approach to series is appropriate for such courses, because from the applied point of view of introductory calculus courses it is power series, as opposed to series in general, that is the most important tool for applications. For example, power series are useful in finding solutions of differential equations.

From the perspective of real analysis, on the other hand, while we still use series in the service of power series, series are studied in their own right as well, there being some very interesting and surprising results about series, for example Theorem 9.4.15 about rearrangements of conditionally convergent series.

Exercises

Exercise 9.2.1. Prove that $\sum_{n=1}^{\infty} \frac{1}{(3n-1)(3n+2)}$ is convergent, and find its sum.

Exercise 9.2.2. [Used in Exercise 9.3.2, Lemma 9.5.3 and Theorem 10.4.4.] Let $\sum_{n=1}^{\infty} a_n$ be a series in $\mathbb{R}$. Prove that if $\sum_{n=1}^{\infty} a_n$ is convergent, then $\{a_n\}_{n=1}^{\infty}$ is bounded.

Exercise 9.2.3. [Used in Example 10.4.6.] Let $\sum_{n=1}^{\infty} a_n$ be a series in $\mathbb{R}$, and let $k \in \mathbb{R}$. Suppose that $k \neq 0$. Prove that if $\sum_{n=1}^{\infty} a_n$ is divergent, then $\sum_{n=1}^{\infty} ka_n$ is divergent.

Exercise 9.2.4. [Used in Section 9.2 and Exercise 9.5.5.] Let $\sum_{n=1}^{\infty} a_n$ be a series in $\mathbb{R}$, and let $k \in \mathbb{N}$. Prove that $\sum_{n=k}^{\infty} a_n$ is convergent if and only if $\sum_{n=1}^{\infty} a_n$ is convergent.

Exercise 9.2.5. [Used in Section 9.2 and Theorem 9.3.2.] Let $\sum_{n=1}^{\infty} a_n$ and $\sum_{n=1}^{\infty} b_n$ be series in $\mathbb{R}$. Suppose that there is some $P \in \mathbb{N}$ such that $n \in \mathbb{N}$ and $n \geq P$ imply $a_n = b_n$. Prove that $\sum_{n=1}^{\infty} a_n$ is convergent if and only if $\sum_{n=1}^{\infty} b_n$ is convergent.

Exercise 9.2.6. [Used in Theorem 10.4.18.] Let $\sum_{n=1}^{\infty} a_n$ and $\sum_{n=1}^{\infty} b_n$ be series in $\mathbb{R}$. Suppose that $a_n \leq b_n$ for all $n \in \mathbb{N}$, and that $\sum_{n=1}^{\infty} a_n$ and $\sum_{n=1}^{\infty} b_n$ are convergent.

(1) Prove that $\sum_{n=1}^{\infty} a_n \leq \sum_{n=1}^{\infty} b_n$.
(2) Suppose that $a_p < b_p$ for some $p \in \mathbb{N}$. Prove that $\sum_{n=1}^{\infty} a_n < \sum_{n=1}^{\infty} b_n$.

Exercise 9.2.7. [Used in Example 9.4.11.] Let $\sum_{n=1}^{\infty} a_n$ be a series. Suppose that $\sum_{n=1}^{\infty} a_n$ is convergent. Prove that the series

$$0 + a_1 + 0 + a_2 + 0 + a_3 + 0 + a_4 + 0 + \cdots$$

is convergent, and that it has the same sum as $\sum_{n=1}^{\infty} a_n$. This result might seem obvious, but a proof is needed. [Use Exercise 8.3.6.]

Exercise 9.2.8. Let $\sum_{n=1}^{\infty} a_n$ be a series in $\mathbb{R}$.

(1) Prove that if $\sum_{n=1}^{\infty} a_n$ is convergent, then $\sum_{n=1}^{\infty} (a_n - a_{n+1})$ is convergent.
(2) If $\sum_{n=1}^{\infty} (a_n - a_{n+1})$ is convergent, does that necessarily imply that $\sum_{n=1}^{\infty} a_n$ is convergent? Give a proof or a counterexample.

Exercise 9.2.9. [Used in Section 9.6.] It was proved in Example 9.2.4 (5) that the harmonic series is divergent. The purpose of this exercise is to give another proof of that fact. This proof is due to Pietro Mengoli (1626–1686). Let $\{s_n\}_{n=1}^{\infty}$ be the sequence of partial sums of the harmonic series.

(1) Prove that $\frac{1}{n-1} + \frac{1}{n} + \frac{1}{n+1} > \frac{3}{n}$ for all $n \in \mathbb{N}$ such that $n \geq 2$.
(2) Prove that $s_{3n+1} > 1 + s_n$ for all $n \in \mathbb{N}$.
(3) Suppose that $\{s_n\}_{n=1}^{\infty}$ is convergent, and derive a contradiction.

Exercise 9.2.10. Let $r \in \mathbb{R}$. Suppose that $|r| < 1$. Does the series $\sum_{n=1}^{\infty} \frac{1}{1+r^n}$ converge or diverge? Justify your answer.

Exercise 9.2.11. [Used in Exercise 9.3.11, Theorem 9.4.7 and Theorem 9.4.12.] Let $\sum_{n=1}^{\infty} a_n$ be a series in $\mathbb{R}$. Prove that $\sum_{n=1}^{\infty} a_n$ is convergent if and only if for each $\varepsilon > 0$, there is some $N \in \mathbb{N}$ such that $n, m \in \mathbb{N}$ and $n > m \geq N$ imply $\left|\sum_{k=m+1}^{n} a_k\right| < \varepsilon$.

Exercise 9.2.12. [Used in Theorem 9.4.15.] Let $\sum_{n=1}^{\infty} a_n$ be a series in $\mathbb{R}$, and let $k \in \mathbb{N}$. Suppose that $\sum_{n=1}^{\infty} a_n = \infty$. Prove that $\sum_{n=k}^{\infty} a_n = \infty$.

9.3 Convergence Tests

As we saw in Section 9.2, a series is convergent, by definition, if the sequence of partial sums of the series is convergent. In practice, however, it is often very difficult to find an explicit formula for the partial sums of a given series, and therefore it is not always possible to verify whether a series is convergent or not by appealing directly to the definition of convergence. Fortunately, there are a number of convergence tests that can be used to determine whether or not various series converge. As with techniques of integration, no one convergence test treats all series, and in practice, for a particular series, one has to examine the various convergence tests to decide which one seems most likely to be helpful in the given situation.

The disadvantage of the various convergence tests is that they tell us only whether or not a series is convergent, but not what the sum of the series is if it is convergent. However, knowing that a series is convergent in principle, even without knowing the sum, is better than not knowing anything about the convergence of the series.

In this section we will see some well-known convergence tests. Most of these tests concern series with non-negative terms. At the heart of our treatment of series with non-negative terms is the following lemma, which is very easy to prove given our previous work.

Lemma 9.3.1. *Let $\sum_{n=1}^{\infty} a_n$ be a series in $\mathbb{R}$. Suppose that $a_n \geq 0$ for all $n \in \mathbb{N}$, or that $a_n \leq 0$ for all $n \in \mathbb{N}$. Let $\{s_n\}_{n=1}^{\infty}$ be the sequence of partial sums of $\sum_{n=1}^{\infty} a_n$. Then $\sum_{n=1}^{\infty} a_n$ is convergent if and only if $\{s_n\}_{n=1}^{\infty}$ is bounded.*

Proof. Suppose that $a_n \geq 0$ for all $n \in \mathbb{N}$. Then the sequence $\{s_n\}_{n=1}^{\infty}$ is increasing, and the lemma then follows immediately from the Monotone Convergence Theorem (Corollary 8.3.4). The other case is similar, and we omit the details. □

We start our discussion of convergence tests with the Comparison Test, which is perhaps the easiest convergence test to use. The idea of this test is that if a series with non-negative terms is known to be convergent, then any other series with non-negative terms, and which is term-by-term no greater than the convergent series, will also be convergent.

Theorem 9.3.2 (Comparison Test). *Let $\sum_{n=1}^{\infty} a_n$ and $\sum_{n=1}^{\infty} b_n$ be series in $\mathbb{R}$. Suppose that $a_n \geq 0$ and $b_n \geq 0$ for all $n \in \mathbb{N}$, and that there is some $N \in \mathbb{N}$ such that $n \in \mathbb{N}$ and $n \geq N$ imply $a_n \leq b_n$.*

1. *If $\sum_{n=1}^{\infty} b_n$ is convergent, then $\sum_{n=1}^{\infty} a_n$ is convergent.*
2. *If $\sum_{n=1}^{\infty} a_n$ is divergent, then $\sum_{n=1}^{\infty} b_n$ is divergent.*

Proof. We will prove Part (1); Part (2), which is the contrapositive of Part (1), then follows immediately.

(1) Suppose that $\sum_{n=1}^{\infty} b_n$ is convergent.

First, suppose that $N = 1$. Then $a_n \le b_n$ for all $n \in \mathbb{N}$. Let $\{s_n\}_{n=1}^{\infty}$ and $\{t_n\}_{n=1}^{\infty}$ be the sequences of partial sums of $\sum_{n=1}^{\infty} a_n$ and $\sum_{n=1}^{\infty} b_n$, respectively. Then $\{t_n\}_{n=1}^{\infty}$ is convergent. It follows from Lemma 8.2.6 that $\{t_n\}_{n=1}^{\infty}$ is bounded. Hence, there is some $M \in \mathbb{R}$ such that $|t_n| \le M$ for all $n \in \mathbb{N}$. Because $0 \le a_n \le b_n$ for all $n \in \mathbb{N}$, it follows that $0 \le s_n \le t_n$ for all $n \in \mathbb{N}$. Hence $|s_n| \le M$ for all $n \in \mathbb{N}$, which means that $\{s_n\}_{n=1}^{\infty}$ is bounded. Lemma 9.3.1 then implies that $\{s_n\}_{n=1}^{\infty}$ is convergent, and therefore $\sum_{n=1}^{\infty} a_n$ is convergent.

Second, suppose that $N > 1$. Let $\{c_n\}_{n=1}^{\infty}$ be defined by

$$c_n = \begin{cases} 0, & \text{if } n < N \\ a_n, & \text{if } n \ge N. \end{cases}$$

Then $0 \le c_n \le b_n$ for all $n \in \mathbb{N}$. Using the previous paragraph, we see that $\sum_{n=1}^{\infty} c_n$ is convergent. Exercise 9.2.5 then implies that $\sum_{n=1}^{\infty} a_n$ is convergent. □

Example 9.3.3. In Example 9.2.4 (5) we saw that $\sum_{n=1}^{\infty} \frac{1}{n}$ is divergent. Clearly $\frac{1}{n} < \frac{1}{n-0.3}$ for all $n \in \mathbb{N}$, and therefore by the Comparison Test (Theorem 9.3.2) it follows that $\sum_{n=1}^{\infty} \frac{1}{n-0.3}$ is divergent. On the other hand, we cannot use the Comparison Test to evaluate the convergence or divergence of $\sum_{n=1}^{\infty} \frac{1}{n+0.3}$, because $\frac{1}{n+0.3} < \frac{1}{n}$ for all $n \in \mathbb{N}$, and that inequality is "the wrong way" to use the Comparison Test. That is, the Comparison Test says that a series that is term-by-term smaller than a convergent series is convergent, and that a series that is term-by-term greater than a divergent series is divergent, but it does not say anything about a series that is term-by-term smaller than a divergent series, or term-by-term greater than a convergent series. On the other hand, it would seem intuitively reasonable that the series $\sum_{n=1}^{\infty} \frac{1}{n+0.3}$ ought to behave similarly to $\sum_{n=1}^{\infty} \frac{1}{n}$, because as n becomes very large the 0.3 becomes negligible. In fact, these two series do behave similarly; it is simply that the Comparison Test does not help us in this situation. ◊

The following convergence test allows us to evaluate the convergence of the series in Example 9.3.3 for which the Comparison Test (Theorem 9.3.2) does not apply. This new convergence test also involves a comparison between two series, but it does not require a hypothesis as strict as $a_n \le b_n$ for all $n \in \mathbb{N}$ such that n is greater than or equal to a given integer. The idea of this convergence test, called the Limit Comparison Test, is that if the ratios of the terms of two series behave nicely as n goes to infinity, then the two series have comparable convergence properties.

The statement of the following theorem makes implicit use of the fact that if $\{c_n\}_{n=1}^{\infty}$ is a sequence in $\mathbb{R}$, and $c_n \ge 0$ for all $n \in \mathbb{N}$, and $\{c_n\}_{n=1}^{\infty}$ is convergent, then $\lim_{n\to\infty} c_n \ge 0$, which follows from Example 8.2.4 (1) and Theorem 8.2.11.

Theorem 9.3.4 (Limit Comparison Test). *Let $\sum_{n=1}^{\infty} a_n$ and $\sum_{n=1}^{\infty} b_n$ be series in $\mathbb{R}$. Suppose that $a_n \geq 0$ and $b_n > 0$ for all $n \in \mathbb{N}$, and that $\left\{\frac{a_n}{b_n}\right\}_{n=1}^{\infty}$ is convergent or it diverges to infinity. Let $L = \lim_{n\to\infty} \frac{a_n}{b_n}$.*

1. *Suppose that $L \in (0,\infty)$. Then $\sum_{n=1}^{\infty} a_n$ is convergent if and only if $\sum_{n=1}^{\infty} b_n$ is convergent.*
2. *Suppose that $L = 0$. If $\sum_{n=1}^{\infty} b_n$ is convergent then $\sum_{n=1}^{\infty} a_n$ is convergent.*
3. *Suppose that $L = \infty$. If $\sum_{n=1}^{\infty} b_n$ is divergent then $\sum_{n=1}^{\infty} a_n$ is divergent.*

Proof. We will prove Part (1), leaving the rest to the reader in Exercise 9.3.2.

(1) Because $L \in (0,\infty)$, then $\frac{L}{2} > 0$. Hence there is some $N \in \mathbb{N}$ such that $n \in \mathbb{N}$ and $n \geq N$ imply $|\frac{a_n}{b_n} - L| < \frac{L}{2}$. Hence $n \in \mathbb{N}$ and $n \geq N$ imply $\frac{b_n L}{2} < a_n < \frac{3b_n L}{2}$. By the Comparison Test (Theorem 9.3.2) it follows that if $\sum_{n=1}^{\infty} a_n$ is convergent then $\sum_{n=1}^{\infty} \frac{b_n L}{2}$ is convergent, and that if $\sum_{n=1}^{\infty} \frac{3b_n L}{2}$ is convergent then $\sum_{n=1}^{\infty} a_n$ is convergent. However, we can use Theorem 9.2.6 (3) to see that $\sum_{n=1}^{\infty} \frac{b_n L}{2}$ and $\sum_{n=1}^{\infty} \frac{3b_n L}{2}$ are each convergent if and only if $\sum_{n=1}^{\infty} b_n$ is convergent. It now follows that $\sum_{n=1}^{\infty} a_n$ is convergent if and only if $\sum_{n=1}^{\infty} b_n$ is convergent. □

It is seen in Exercise 9.3.3 that Theorem 9.3.4 (2) (3) cannot be made into if and only if statements.

Example 9.3.5. In Example 9.3.3 we conjectured that $\sum_{n=1}^{\infty} \frac{1}{n+0.3}$ is divergent, though we did not yet have the tools to prove it. We can now prove this result by applying the Limit Comparison Test (Theorem 9.3.4) to $\sum_{n=1}^{\infty} \frac{1}{n}$ and $\sum_{n=1}^{\infty} \frac{1}{n+0.3}$. Using Example 8.2.4 (1) (2) and Theorem 8.2.9 (1) (3) we see that

$$L = \lim_{n\to\infty} \frac{\frac{1}{n}}{\frac{1}{n+0.3}} = \lim_{n\to\infty} \frac{n+0.3}{n} = \lim_{n\to\infty} \left(1 + \frac{0.3}{n}\right) = 1.$$

It follows from the Limit Comparison Test that $\sum_{n=1}^{\infty} \frac{1}{n+0.3}$ is convergent if and only if $\sum_{n=1}^{\infty} \frac{1}{n}$ is convergent. We saw in Example 9.2.4 (5) that $\sum_{n=1}^{\infty} \frac{1}{n}$ is divergent, and hence $\sum_{n=1}^{\infty} \frac{1}{n+0.3}$ is divergent. ◊

The Comparison Test (Theorem 9.3.2) and the Limit Comparison Test (Theorem 9.3.4) are very easy to use, but they have one major drawback, which is that to show that one series is convergent, we need to find another series for comparison whose convergence is known, and finding that other series for comparison is not always easy in practice. We now turn to another convergence test, called the Integral Test, which has limited use, but is very effective in some situations, and it does not require a second series for comparison. The Integral Test relies upon Type 1 improper Integrals, as discussed in Section 6.4. The idea of this convergence test is seen in Figure 9.3.1. If the terms of a series $\sum_{n=1}^{\infty} a_n$ are non-negative and decreasing, and if a well-behaved function $f\colon [1,\infty) \to \mathbb{R}$ can be found such that $f(n) = a_n$ for all $n \in \mathbb{N}$, then the value of the improper integral $\int_1^{\infty} f(x)\,dx$ is closely related to the sum of the areas of the rectangles shown in the figure, and that sum is precisely $\sum_{n=1}^{\infty} a_n$, because all of the rectangles have width 1.

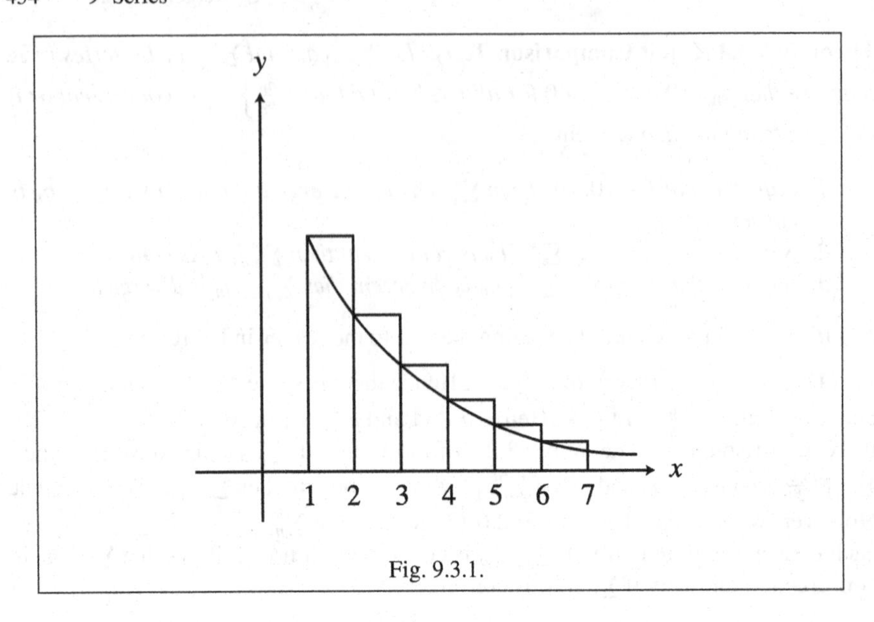

Fig. 9.3.1.

Theorem 9.3.6 (Integral Test). *Let $\sum_{n=1}^{\infty} a_n$ be a series in $\mathbb{R}$, and let $f\colon [1,\infty) \to \mathbb{R}$ be a function. Suppose that f is continuous and decreasing, that $f(x) \geq 0$ for all $x \in [1,\infty)$, and that $f(n) = a_n$ for all $n \in \mathbb{N}$. Then $\sum_{n=1}^{\infty} a_n$ is convergent if and only if $\int_1^{\infty} f(x)\,dx$ is convergent.*

Proof. Because $f(x) \geq 0$ for all $x \in [1,\infty)$, then $a_n = f(n) \geq 0$ for all $n \in \mathbb{N}$. Let $\{s_n\}_{n=1}^{\infty}$ be the sequence of partial sums of $\sum_{n=1}^{\infty} a_n$. It follows from Lemma 9.3.1 that $\sum_{n=1}^{\infty} a_n$ is convergent if and only if $\{s_n\}_{n=1}^{\infty}$ is bounded.

Because f is monotone, it follows from Exercise 5.4.12 that f is locally integrable. Let $F\colon [1,\infty) \to \mathbb{R}$ be defined by $F(x) = \int_1^x f(t)\,dt$ for all $x \in [1,\infty)$. By Exercise 5.6.4 we see that F is increasing. Hence $\{F(n)\}_{n=1}^{\infty}$ is increasing. It follows from Exercise 8.3.3 that $\lim_{x\to\infty} F(x)$ exists if and only if the sequence $\{F(n)\}_{n=1}^{\infty}$ is convergent. In other words, the improper integral $\int_1^{\infty} f(x)\,dx$ is convergent if and only if $\{F(n)\}_{n=1}^{\infty}$ is convergent.

By Theorem 5.3.2 (1) we know that $F(n) \geq 0$ for all $n \in \mathbb{N}$. Because $\{F(n)\}_{n=1}^{\infty}$ is increasing, the Monotone Convergence Theorem (Corollary 8.3.4) implies that $\{F(n)\}_{n=1}^{\infty}$ is convergent if and only if it is bounded. Hence $\int_1^{\infty} f(x)\,dx$ is convergent if and only if $\{F(n)\}_{n=1}^{\infty}$ is bounded.

We now show that $\{s_n\}_{n=1}^{\infty}$ is bounded if and only if $\{F(n)\}_{n=1}^{\infty}$ is bounded, which, together with what we saw above, will prove the theorem.

Let $k \in \mathbb{N}$. Because f is decreasing, then $f(k) \geq f(x) \geq f(k+1)$ for all $x \in [k,k+1]$, and hence by Theorem 5.3.2 (3) we see that $f(k+1)\cdot 1 \leq \int_k^{k+1} f(x)\,dx \leq f(k)\cdot 1$.

Let $n \in \mathbb{N}$. Then

$$\sum_{i=1}^{n-1} f(i+1) \leq \sum_{i=1}^{n-1} \int_i^{i+1} f(x)\,dx \leq \sum_{i=1}^{n-1} f(i).$$

By Theorem 5.5.7, extended to n subintervals by induction, it follows that

$$\sum_{i=1}^{n} f(i) - f(1) = \sum_{i=1}^{n-1} f(i+1) \leq \int_1^n f(x)\,dx \leq \sum_{i=1}^{n-1} f(i).$$

Hence

$$s_n - a_1 \leq F(n) \leq s_{n-1},$$

and it follows that

$$s_n \leq F(n) + a_1 \quad \text{and} \quad F(n) \leq s_n.$$

We deduce that $\{s_n\}_{n=1}^{\infty}$ is bounded if and only if $\{F(n)\}_{n=1}^{\infty}$ is bounded. □

Example 9.3.7. Let $p \in \mathbb{R}$. The series $\sum_{n=1}^{\infty} \frac{1}{n^p}$ is called a ***p*-series**. Whether or not the series $\sum_{n=1}^{\infty} \frac{1}{n^p}$ is convergent depends upon the value of p.

First, suppose that $p = 1$. Then $\sum_{n=1}^{\infty} \frac{1}{n^p}$ is the harmonic series $\sum_{n=1}^{\infty} \frac{1}{n}$, which was shown to be divergent in Example 9.2.4 (5).

Second, suppose that $p = 0$. Then $\sum_{n=1}^{\infty} \frac{1}{n^p}$ is the constant series $\sum_{n=1}^{\infty} 1$, which is divergent by the Divergence Test (Theorem 9.2.5).

Third, suppose that $p > 0$ and $p \neq 1$. We will use the Integral Test (Theorem 9.3.6) to test the convergence of $\sum_{n=1}^{\infty} \frac{1}{n^p}$. Let $f\colon [1,\infty) \to \mathbb{R}$ be defined by $f(x) = x^{-p}$ for all $x \in [1,\infty)$. Then $f(n) = \frac{1}{n^p}$ for all $n \in \mathbb{N}$. We see by Definition 7.2.11 that $f(x) > 0$ for all $x \in [1,\infty)$. By Theorem 7.2.13 (1) we know that f is differentiable, and $f'(x) = -px^{-p-1} \leq 0$ for all $x \in (1,\infty)$. It follows from Theorem 4.2.4 that f is continuous, and it follows from Theorem 4.5.2 (3) that f is strictly decreasing. We have therefore verified that the Integral Test is applicable, and it follows from that convergence test and Exercise 6.4.1 that $\sum_{n=1}^{\infty} \frac{1}{n^p}$ is convergent if $p > 1$, and divergent if $0 < p < 1$.

Fourth, suppose that $p < 0$. Then $-p > 0$, and therefore by Exercise 7.2.16 we see that $\lim_{x\to\infty} x^{-p} = \infty$. It follows from Exercise 8.2.10 (2) that $\lim_{n\to\infty} \frac{1}{n^p} = \lim_{n\to\infty} n^{-p} = \infty$. The Divergence Test (Theorem 9.2.5) then implies that $\sum_{n=1}^{\infty} \frac{1}{n^p}$ is divergent.

Putting the various cases together, we see that $\sum_{n=1}^{\infty} \frac{1}{n^p}$ is convergent if and only if $p > 1$. For example, the series $\sum_{n=1}^{\infty} \frac{1}{n^2}$ is convergent, and the series $\sum_{n=1}^{\infty} \frac{1}{\sqrt{n}}$ is divergent. Moreover, we see that among the p-series, the harmonic series is right at the "boundary" of divergence and convergence, and it is therefore an interesting series even though it is divergent, and it is a useful series for the Comparison Test (Theorem 9.3.2) and the Limit Comparison Test (Theorem 9.3.4). ◊

For the final convergence test of this section, we look at series with alternating positive and negative terms. The idea of this convergence test is as follows. Suppose that we have a series of the form $\sum_{n=1}^{\infty} (-1)^{n-1} a_n$, where we assume that $a_n > 0$ for all $n \in \mathbb{N}$, that $\{a_n\}_{n=1}^{\infty}$ is decreasing and that $\lim_{n\to\infty} a_n = 0$. Let $\{s_n\}_{n=1}^{\infty}$ be the sequence of partial sums of $\sum_{n=1}^{\infty} s_n$. Then $s_1 = a_1$, and $s_2 = a_1 - a_2$ is less than or equal to s_1, and $s_3 = a_1 - a_2 + a_3$ is greater than or equal to s_2 but less than or equal to s_1 and so on, as seen in Figure 9.3.2. It then appears, intuitively, as if the terms of $\{s_n\}_{n=1}^{\infty}$ are getting closer and closer to something, which we will indeed prove using the Nested

Interval Theorem (Theorem 8.4.7). Also, we note that whereas the following theorem is stated for series of the form $\sum_{n=1}^{\infty}(-1)^{n-1}a_n$, the analogous result also holds for series of the form $\sum_{n=1}^{\infty}(-1)^{n}a_n$.

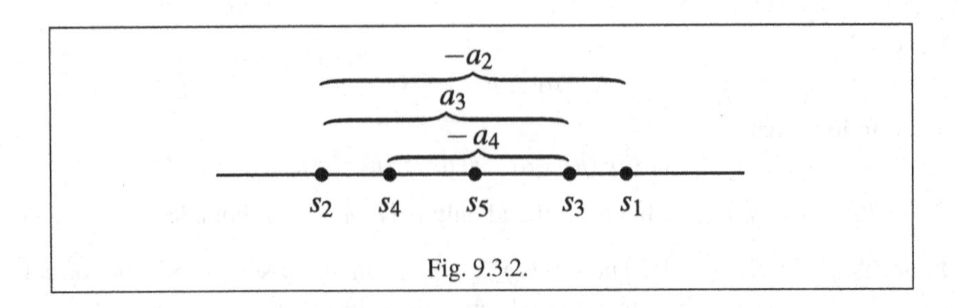

Fig. 9.3.2.

Not only is the following convergence test different from the previous convergence tests in that it is not about series with non-negative terms, but it has another distinctive feature. In general, when we prove that a series is convergent by using a convergence test, for example the Comparison Test (Theorem 9.3.2) or the Limit Comparison Test (Theorem 9.3.4), we do not learn from the convergence test what the sum of the series equals. In the case of alternating series, however, although the following convergence test does not tell us the exact value of the sum of the series, Part (2) of the theorem gives us a way to estimate the sum of the series very easily. For this reason alternating series are particularly convenient to work with.

Theorem 9.3.8 (Alternating Series Test). *Let $\{a_n\}_{n=1}^{\infty}$ be a sequence in $\mathbb{R}$. Suppose that $a_n > 0$ for all $n \in \mathbb{N}$, that $\{a_n\}_{n=1}^{\infty}$ is decreasing, and that $\lim_{n\to\infty} a_n = 0$.*

1. *The series $\sum_{n=1}^{\infty}(-1)^{n-1}a_n$ is convergent.*
2. *Let $L = \sum_{n=1}^{\infty}(-1)^{n-1}a_n$, and let $\{s_n\}_{n=1}^{\infty}$ be the sequence of partial sums of $\sum_{n=1}^{\infty}(-1)^{n-1}a_n$. Then $|L - s_n| \le a_{n+1}$ for all $n \in \mathbb{N}$.*

Proof. We prove both parts of the theorem together. Let $m \in \mathbb{N}$. By hypothesis we know that $a_{2m} \ge a_{2m+1} \ge a_{2m+2} > 0$. Because $s_{2m+1} = s_{2m-1} - a_{2m} + a_{2m+1}$ and $s_{2m+2} = s_{2m+1} - a_{2m+2} = s_{2m} + a_{2m+1} - a_{2m+2}$, it follows that $s_{2m} \le s_{2m+2} \le s_{2m+1} \le s_{2m-1}$. We therefore see that $\{[s_{2n}, s_{2n-1}]\}_{n=1}^{\infty}$ is a family of closed bounded intervals in $\mathbb{R}$ such that $\left[s_{2(i+1)}, s_{2(i+1)-1}\right] \subseteq [s_{2i}, s_{2i-1}]$ for all $i \in \mathbb{N}$. We know by hypothesis that $\lim_{n\to\infty} a_n = 0$, and it then follows from Lemma 8.3.7 that $\{a_{2n}\}_{n=1}^{\infty}$ is convergent and that $\lim_{n\to\infty} a_{2n} = 0$. Hence $\lim_{n\to\infty}(s_{2n-1} - s_{2n}) = \lim_{n\to\infty} a_{2n} = 0$.

We can now apply both parts of the Nested Interval Theorem (Theorem 8.4.7) to $\{[s_{2n}, s_{2n-1}]\}_{n=1}^{\infty}$ to deduce that $\bigcap_{n=1}^{\infty}[s_{2n}, s_{2n-1}] = \{c\}$, where $c = \lim_{n\to\infty} s_{2n} = \lim_{n\to\infty} s_{2n-1}$.

By Exercise 8.3.6 it follows that $\{s_n\}_{n=1}^{\infty}$ is convergent and $\lim_{n\to\infty} s_n = c$. Hence $\sum_{n=1}^{\infty}(-1)^{n-1}a_n$ is convergent, and $\sum_{n=1}^{\infty}(-1)^{n-1}a_n = c$. To conform to the notation of Part (2) of the theorem, we rename c as L.

For each $k \in \mathbb{N}$, we know that $L \in [s_{2k+2}, s_{2k+1}] \subseteq [s_{2k}, s_{2k-1}]$, and hence $s_{2k} \leq s_{2k+2} \leq L \leq s_{2k+1} \leq s_{2k-1}$. Let $n \in \mathbb{N}$. If $n = 2j$ for some $j \in \mathbb{N}$, then $s_{2j} \leq L \leq s_{2j+1}$, and hence $|L - s_n| = |L - s_{2j}| \leq |s_{2j} - s_{2j+1}| = a_{2j+1} = a_{n+1}$. If $n = 2j - 1$ for some $j \in \mathbb{N}$, then $s_{2j} \leq L \leq s_{2j-1}$, and hence $|L - s_n| = |L - s_{2j-1}| \leq |s_{2j-1} - s_{2j}| = a_{2j} = a_{n+1}$. In either case, we deduce that $|L - s_n| \leq a_{n+1}$. □

Example 9.3.9. We will prove that the series $\sum_{n=1}^{\infty} (-1)^{n-1} \frac{1}{n}$ is convergent. This series is known as the **alternating harmonic series**. It is evident that $\frac{1}{n} > 0$ for all $n \in \mathbb{N}$, and that $\left\{\frac{1}{n}\right\}_{n=1}^{\infty}$ is decreasing. We saw in Example 8.2.4 (2) that $\lim_{n \to \infty} \frac{1}{n} = 0$. We can therefore use Part (1) of the Alternating Series Test (Theorem 9.3.8) to deduce that $\sum_{n=1}^{\infty} (-1)^{n-1} \frac{1}{n}$ is convergent.

Let $L = \sum_{n=1}^{\infty} (-1)^{n-1} \frac{1}{n}$. Suppose that we want to estimate the value of L to within two decimal places. That is, we want to find a real number which is no farther from L than 0.005. Let $\{s_n\}_{n=1}^{\infty}$ be the sequence of partial sums of $\sum_{n=1}^{\infty} (-1)^{n-1} \frac{1}{n}$. By Part (2) of the Alternating Series Test we see that $|L - s_{201}| \leq \frac{1}{201} < \frac{1}{200} = 0.005$. Hence s_{201} is the desired approximation of L, and a simple numerical calculation yields

$$s_{201} = \frac{1}{1} - \frac{1}{2} + \frac{1}{3} - \cdots + \frac{1}{201} \approx 0.69.$$

The precise value of L is given in Exercise 9.3.10, using a slightly devious method. ◊

Reflections

The various convergence tests in this section should be familiar to the reader from a calculus course, and they have been included here so that the reader can see that this material is susceptible to a rigorous treatment. From the point of view of the overall development of real analysis, however, these tests, while very useful for concrete computations with series, are not as important as the material in the sections preceding and following the present one. On the other hand, the convergence tests in the present section are useful for producing examples involving series, and even the most detailed treatment of real analysis, or any part of mathematics, that is lacking nice examples is hard to understand, or worse.

Exercises

Exercise 9.3.1. [Used in Exercise 9.3.6, Exercise 9.3.7 and Lemma 9.4.14.] The purpose of this exercise is to refine the statement of Lemma 9.3.1. Let $\sum_{n=1}^{\infty} a_n$ be a series in $\mathbb{R}$. Suppose that $a_n \geq 0$ for all $n \in \mathbb{N}$. Let $\{s_n\}_{n=1}^{\infty}$ be the sequence of partial sums of $\sum_{n=1}^{\infty} a_n$.

(1) Prove that either $\sum_{n=1}^{\infty} a_n$ is convergent or $\sum_{n=1}^{\infty} a_n = \infty$.
(2) Prove that if there is some $M \in \mathbb{R}$ such that $s_n \leq M$ for all $n \in \mathbb{N}$, then $\sum_{n=1}^{\infty} a_n$ is convergent and $\sum_{n=1}^{\infty} a_n \leq M$.
(3) Prove that $\sum_{n=1}^{\infty} a_n$ is convergent if and only if $\{s_n\}_{n=1}^{\infty}$ has a bounded subsequence. [Use Exercise 8.3.8.]

Exercise 9.3.2. [Used in Theorem 9.3.4.] Prove Theorem 9.3.4 (2) (3).

[Use Exercise 9.2.2.]

Exercise 9.3.3. [Used in Section 9.3.] Give examples to show that Theorem 9.3.4 (2) (3) cannot be made into if and only if statements.

Exercise 9.3.4. Let $\sum_{n=1}^{\infty} a_n$ be a series in $\mathbb{R}$. Suppose that $a_n \geq 0$ for all $n \in \mathbb{N}$. Prove that if $\sum_{n=1}^{\infty} a_n$ is convergent then $\sum_{n=1}^{\infty} (a_{2n-1} + a_{2n})$ is convergent.

Exercise 9.3.5. [Used in Section 9.2.] Let $\sum_{n=1}^{\infty} a_n$ and $\sum_{n=1}^{\infty} b_n$ be series in $\mathbb{R}$. Suppose that $a_n \geq 0$ and $b_n \geq 0$ for all $n \in \mathbb{N}$.

(1) Prove that if $\sum_{n=1}^{\infty} a_n$ and $\sum_{n=1}^{\infty} b_n$ are convergent, then $\sum_{n=1}^{\infty} a_n b_n$ is convergent.

(2) In order to guarantee that $\sum_{n=1}^{\infty} a_n b_n$ is convergent, is it necessary that both $\sum_{n=1}^{\infty} a_n$ and $\sum_{n=1}^{\infty} b_n$ are convergent? If yes, explain why. If not, what weaker hypotheses on $\sum_{n=1}^{\infty} a_n$ and $\sum_{n=1}^{\infty} b_n$ would suffice?

(3) Find an example of series $\sum_{n=1}^{\infty} c_n$ and $\sum_{n=1}^{\infty} d_n$ in $\mathbb{R}$ such that $c_n \geq 0$ and $d_n \geq 0$ for all $n \in \mathbb{N}$, that $\sum_{n=1}^{\infty} c_n$ and $\sum_{n=1}^{\infty} d_n$ are convergent and that $\sum_{n=1}^{\infty} c_n d_n \neq [\sum_{n=1}^{\infty} c_n] \cdot [\sum_{n=1}^{\infty} d_n]$.

Exercise 9.3.6. [Used in Lemma 5.8.3.] Let $\left\{\sum_{n=1}^{\infty} a_n^k\right\}_{k=1}^{\infty}$ be a sequence of series in $\mathbb{R}$, and let $\sum_{n=1}^{\infty} b_n$ be a series in $\mathbb{R}$. Suppose that $a_n^k \geq 0$ for all $n, k \in \mathbb{N}$, that $\sum_{n=1}^{\infty} b_n$ is convergent and that for each $k \in \mathbb{N}$ the series $\sum_{n=1}^{\infty} a_n^k$ is convergent and $\sum_{n=1}^{\infty} a_n^k \leq b_k$. Let $f\colon \mathbb{N} \to \mathbb{N} \times \mathbb{N}$ be a bijective function. Such a function exists because $\mathbb{N} \times \mathbb{N}$ is countably infinite, which is a standard fact about the cardinality of the number systems; see [Blo10, Sections 6.5–6.7] for details. Let $\sum_{n=1}^{\infty} c_n$ be defined as follows. For each $i \in \mathbb{N}$, we have $f(i) = (n_i, k_i)$ for some $n_i, k_i \in \mathbb{N}$, and then let $c_i = a_{n_i}^{k_i}$. Prove that $\sum_{n=1}^{\infty} c_n$ is convergent and $\sum_{n=1}^{\infty} c_n \leq \sum_{n=1}^{\infty} b_n$. [Use Exercise 9.3.1 (2).]

Exercise 9.3.7. The purpose of this exercise is to prove and apply a convergence test known as the Cauchy Condensation Test. This convergence test generalizes the idea used in Example 9.2.4 (5), where we saw that the harmonic series is divergent by looking at partial sums of the form s_{2^n}.

(1) Let $\sum_{n=1}^{\infty} a_n$ be series in $\mathbb{R}$. Suppose that $a_n \geq 0$ for all $n \in \mathbb{N}$, and that $\{a_n\}_{n=1}^{\infty}$ is decreasing. Prove that $\sum_{n=1}^{\infty} a_n$ is convergent if and only if $\sum_{n=1}^{\infty} 2^n a_{2^n}$ is convergent. [Use Exercise 9.3.1 (3).]

(2) Let $p \in \mathbb{R}$. In Example 9.3.7 we saw that $\sum_{n=1}^{\infty} \frac{1}{n^p}$ is convergent if and only if $p > 1$. Give an alternative (and simpler) proof of this fact using Part (1) of this exercise.

Exercise 9.3.8. [Used in Section 2.8 and Exercise 9.3.9.] In Section 2.8 it was proved, using the Least Upper Bound Property but not series, that every real number has a unique base p representation, where p is a natural number greater than 1. Now that we have the use of series at our disposal, the material in Section 2.8 can be simplified in a number of ways. You can do the present exercise even if you have not read Section 2.8.

Let $p \in \mathbb{N}$. Suppose that $p > 1$. Let $\{a_n\}_{n=1}^{\infty}$ be a sequence in $\{0, \dots, p-1\}$, where we are using the notation of Definition 2.5.3. We can replace Definition 2.8.4, which uses least upper bounds, with the definition of $\sum_{i=1}^{\infty} a_i p^{-i}$ as a series.

The purpose of this exercise is to provide simplified proofs of Lemma 2.8.3 and Lemma 2.8.5 using what we have learned about series.

(1) Prove that $\sum_{i=1}^{\infty} a_i p^{-i}$ is convergent.
(2) Prove that $\sum_{i=1}^{\infty} a_i p^{-i} = \text{lub}\{\sum_{i=1}^{n} a_i p^{-i} \mid n \in \mathbb{N}\}$.
(3) Prove that $0 \le \sum_{i=1}^{\infty} a_i p^{-i} \le 1$.
(4) Prove that $\sum_{i=1}^{\infty} a_i p^{-i} = 0$ if and only if $a_i = 0$ for all $i \in \mathbb{N}$.
(5) Prove that $\sum_{i=1}^{\infty} a_i p^{-i} = 1$ if and only if $a_i = p-1$ for all $i \in \mathbb{N}$.
(6) Let $m \in \mathbb{N}$. Suppose that $m > 1$. Prove that $\sum_{i=1}^{\infty} a_i p^{-i} \ge \sum_{i=1}^{m-1} a_i p^{-i}$, where equality holds if and only if $a_i = 0$ for all $i \in \mathbb{N}$ such that $i \ge m$.
(7) Let $m \in \mathbb{N}$. Suppose that $m > 1$, and that $a_{m-1} \ne p-1$. Prove that

$$\sum_{i=1}^{\infty} a_i p^{-i} \le \sum_{i=1}^{m-2} a_i p^{-i} + \frac{a_{m-1}+1}{p^{m-1}},$$

where equality holds if and only if $a_i = p-1$ for all $i \in \mathbb{N}$ such that $i \ge m$.

Exercise 9.3.9. [Used in Section 2.8.] This exercise makes use of Exercise 9.3.8. Read the statements of all the definitions, lemmas and theorems in Section 2.8, though not the proofs. Then read the proof of Theorem 2.8.10. Using the definition of $\sum_{i=1}^{\infty} a_i p^{-i}$ as a series as given in Exercise 9.3.8, simplify the proof of Theorem 2.8.10 as much as possible.

Exercise 9.3.10. [Used in Example 9.3.9 and Example 9.4.11.] It was seen in Example 9.3.9 that the alternating harmonic series $\sum_{n=1}^{\infty} (-1)^{n-1} \frac{1}{n}$ is convergent, and an estimate was given for the sum of the series. In this exercise we make use of Exercise 8.4.13 to find the exact value of the sum of this series.

Let $\{s_n\}_{n=1}^{\infty}$ be the sequence of partial sums of $\sum_{n=1}^{\infty} (-1)^{n-1} \frac{1}{n}$. Let $\{\gamma_n\}_{n=1}^{\infty}$ be the sequence defined in Exercise 8.4.13. It was proved in that exercise that $\{\gamma_n\}_{n=1}^{\infty}$ is convergent.

(1) Prove that $s_{2n} = \gamma_{2n} - \gamma_n + \ln 2$ for all $n \in \mathbb{N}$.
(2) Prove that $\sum_{n=1}^{\infty} (-1)^{n-1} \frac{1}{n} = \ln 2$.

Exercise 9.3.11. Let $\{a_n\}_{n=1}^{\infty}$ and $\{b_n\}_{n=1}^{\infty}$ be sequences in $\mathbb{R}$. Suppose that $a_n > 0$ for all $n \in \mathbb{N}$, that $\{a_n\}_{n=1}^{\infty}$ is decreasing, that $\lim_{n\to\infty} a_n = 0$ and that the sequence of partial sums of the series $\sum_{n=1}^{\infty} b_n$ is bounded. Prove that the series $\sum_{n=1}^{\infty} a_n b_n$ is convergent. This result, known as Dirichlet's Test, is a generalization of the Alternating Series Test (Theorem 9.3.8). [Use Exercise 5.7.6 (1) and Exercise 9.2.11.]

9.4 Absolute Convergence and Conditional Convergence

The first question to be asked about a series is whether or not it is convergent, but it turns out that even among convergent series there are differences in the nature of the convergence. We know from the Divergence Test (Theorem 9.2.5) that if a series is convergent, then the terms of the series go to zero. Intuitively, the distinction that we need to make among convergent series is that for some series the terms go to zero so rapidly that even the series of absolute values of the terms is convergent, whereas for other series the terms go to zero more slowly, and the series is convergent only by virtue of cancellation between positive and negative terms. We do not have a formal way of defining how rapidly the terms of the series go to zero, but we can still distinguish between these two types of convergent series as follows.

Definition 9.4.1. Let $\sum_{n=1}^{\infty} a_n$ be a series in $\mathbb{R}$. The series $\sum_{n=1}^{\infty} a_n$ is **absolutely convergent** if $\sum_{n=1}^{\infty} |a_n|$ is convergent. The series $\sum_{n=1}^{\infty} a_n$ is **conditionally convergent** if $\sum_{n=1}^{\infty} a_n$ is convergent but not absolutely convergent. △

Example 9.4.2.

(1) For a series with non-negative terms there is no difference between convergence and absolute convergence, and hence any convergent series with non-negative terms, for example the series in Example 9.2.4 (1), is absolutely convergent.

(2) We saw in Example 9.3.9 that the alternating harmonic series $\sum_{n=1}^{\infty} (-1)^{n-1}\frac{1}{n}$ is convergent. On the other hand, the series $\sum_{n=1}^{\infty} |(-1)^{n-1}\frac{1}{n}| = \sum_{n=1}^{\infty} \frac{1}{n}$ is the harmonic series, which was shown to be divergent in Example 9.2.4 (5). Therefore $\sum_{n=1}^{\infty} (-1)^{n-1}\frac{1}{n}$ is not absolutely convergent, but it is conditionally convergent. ◇

We saw in Example 9.4.2 (2) that a series can be convergent but not absolutely convergent, and hence such a series is conditionally convergent. On the other hand, we see in the following theorem that if a series is absolutely convergent then it is convergent.

Theorem 9.4.3. *Let $\sum_{n=1}^{\infty} a_n$ be a series in $\mathbb{R}$. If $\sum_{n=1}^{\infty} a_n$ is absolutely convergent, then $\sum_{n=1}^{\infty} a_n$ is convergent.*

Proof. Suppose that $\sum_{n=1}^{\infty} a_n$ is absolutely convergent. Then the series $\sum_{n=1}^{\infty} |a_n|$ is convergent. It follows from Theorem 9.2.6 (3) that $\sum_{n=1}^{\infty} 2|a_n|$ is convergent.

Let $\sum_{n=1}^{\infty} b_n$ be defined by $b_n = |a_n| - a_n$ for all $n \in \mathbb{N}$. It is straightforward to verify that $0 \le b_n \le 2|a_n|$ for all $n \in \mathbb{N}$. Because $\sum_{n=1}^{\infty} 2|a_n|$ is convergent, it follows from the Comparison Test (Theorem 9.3.2) that $\sum_{n=1}^{\infty} b_n$ is convergent. Because $a_n = |a_n| - b_n$ for all $n \in \mathbb{N}$, Theorem 9.2.6 (2) implies that $\sum_{n=1}^{\infty} a_n$ is convergent. □

Rather than simply categorizing a given series as convergent or divergent, we can now categorize it as either absolutely convergent, conditionally convergent or divergent.

We have one more convergence test for series, which we put off until the present section because it shows not only convergence, but absolute convergence. This convergence test is convenient in that it does not require a second series for comparison, in contrast to the Comparison Test (Theorem 9.3.2) and the Limit Comparison Test (Theorem 9.3.4). On the other hand, we will see there are some useful series, such as the alternating harmonic series, that cannot be evaluated with this new convergence test. The intuitive idea of this test is as follows. We already know about the convergence of geometric series, as discussed in Example 9.2.4 (4). A series $\sum_{n=1}^{\infty} a_n$ is a geometric series if $\frac{a_{n+1}}{a_n}$ is constant for all $n \in \mathbb{N}$. Suppose that a series $\sum_{n=1}^{\infty} a_n$ is not necessarily a geometric series, so that $\frac{a_{n+1}}{a_n}$ is not necessarily constant, but suppose that $\frac{a_{n+1}}{a_n}$ gets closer and closer to some number as n gets larger. That would mean that as n gets larger, the series becomes more and more similar to a geometric series, and so intuitively this series should behave similarly to a geometric series; hence the convergence or divergence of the series should depend upon whether $\left|\lim_{n\to\infty} \frac{a_{n+1}}{a_n}\right|$ is less than 1 or greater than 1. The following theorem confirms this intuitive idea, though for convenience we are able to put the absolute value inside the limit.

Theorem 9.4.4 (Ratio Test). *Let $\sum_{n=1}^{\infty} a_n$ be a series in $\mathbb{R}$. Suppose that $a_n \neq 0$ for all $n \in \mathbb{N}$, and that $\left\{\left|\frac{a_{n+1}}{a_n}\right|\right\}_{n=1}^{\infty}$ is convergent or it diverges to infinity. Let $L = \lim_{n\to\infty} \left|\frac{a_{n+1}}{a_n}\right|$.*

1. *If $L \in [0,1)$, then $\sum_{n=1}^{\infty} a_n$ is absolutely convergent.*
2. *If $L \in (1,\infty)$ or $L = \infty$, then $\sum_{n=1}^{\infty} a_n$ is divergent.*

Proof. We will prove Part (1), leaving the remaining part to the reader in Exercise 9.4.8.

(1) Suppose that $L \in [0,1)$. Let $\varepsilon = \frac{1-L}{2}$. Then $\varepsilon > 0$. Hence there is some $N \in \mathbb{N}$ such that $n \in \mathbb{N}$ and $n \geq N$ imply $\left|\frac{|a_{n+1}|}{|a_n|} - L\right| < \varepsilon$. Suppose that $n \in \mathbb{N}$ and $n \geq N$. Then $\left|\frac{|a_{n+1}|}{|a_n|} - L\right| < \varepsilon$, which implies that $L - \varepsilon < \frac{|a_{n+1}|}{|a_n|} < L + \varepsilon$, and hence $\frac{|a_{n+1}|}{|a_n|} < L + \frac{1-L}{2} = \frac{1+L}{2}$. Let $r = \frac{1+L}{2}$. Then $|a_{n+1}| < r|a_n|$.

In particular, we deduce that $|a_{N+1}| < r|a_N|$. Because it is also the case that $|a_{N+2}| < r|a_{N+1}|$, we deduce that $|a_{N+2}| < r^2|a_N|$. It can then be proved by induction that $|a_{N+k}| < r^k|a_N|$ for all $k \in \mathbb{N}$; the details are left to the reader. It follows that $|a_p| < \frac{|a_N|}{r^N} r^p$ for all $p \in \mathbb{N}$ such that $p \geq N$.

Because $L \in [0,1)$, then $r \in \left[\frac{1}{2}, 1\right)$. The series $\sum_{n=1}^{\infty} \frac{|a_N|}{r^N} r^n$ is a geometric series and it is convergent, as discussed in Example 9.2.4 (4). Because $0 < |a_p| < \frac{|a_N|}{r^N} r^p$ for all $p \in \mathbb{N}$ such that $p \geq N$, it follows from the Comparison Test (Theorem 9.3.2) that $\sum_{n=1}^{\infty} |a_n|$ is convergent, which means that $\sum_{n=1}^{\infty} a_n$ is absolutely convergent. □

The reader will have noticed that the Ratio Test (Theorem 9.4.4) does not discuss the case where $L = 1$. We will see in Example 9.4.5 (2) below that it is not possible to predict convergence or divergence in that case.

Example 9.4.5.

(1) We will show that the series $\sum_{n=1}^{\infty} \frac{2^n}{n!}$ is absolutely convergent. Clearly all the terms of the series are non-zero, so we can use the Ratio Test (Theorem 9.4.4). Using Example 8.2.4 (2) and Theorem 8.2.9 (3) we see that

$$L = \lim_{n\to\infty}\left|\frac{a_{n+1}}{a_n}\right| = \lim_{n\to\infty}\left|\frac{\frac{2^{n+1}}{(n+1)!}}{\frac{2^n}{n!}}\right| = \lim_{n\to\infty}\left|\frac{2}{n}\right| = \lim_{n\to\infty}\frac{2}{n} = 0.$$

It then follows from Part (1) of the Ratio Test that $\sum_{n=1}^{\infty}\frac{2^n}{n!}$ is absolutely convergent.

(2) We examine the convergence or divergence of the two series $\sum_{n=1}^{\infty}\frac{1}{n}$ and $\sum_{n=1}^{\infty}\frac{1}{n^2}$. For the former, we use Example 8.2.10 to see that

$$L = \lim_{n\to\infty}\left|\frac{\frac{1}{n+1}}{\frac{1}{n}}\right| = \lim_{n\to\infty}\left|\frac{n}{n+1}\right| = 1.$$

A similar computation shows that the corresponding limit for the second series is also 1; we omit the details. We now see why the Ratio Test does not treat the case $L = 1$, because the series $\sum_{n=1}^{\infty}\frac{1}{n}$ is divergent by Example 9.2.4 (5), and the series $\sum_{n=1}^{\infty}\frac{1}{n^2}$ is convergent by Example 9.3.7. That is, both convergence and divergence can occur when $L = 1$.

(3) In Part (2) of this example we saw that the Ratio Test failed to determine convergence or divergence when $\lim_{n\to\infty}\left|\frac{a_{n+1}}{a_n}\right| = 1$. We now see a different way in which the Ratio Test can fail, which is when the limit $\lim_{n\to\infty}\left|\frac{a_{n+1}}{a_n}\right|$ does not exist. Let $\sum_{n=1}^{\infty} c_n$ be the series

$$\frac{1}{2^0 3^1} + \frac{1}{2^2 3^2} + \frac{1}{2^2 3^3} + \frac{1}{2^4 3^4} + \frac{1}{2^4 3^5} + \cdots,$$

which is given by

$$c_n = \begin{cases} \frac{1}{2^n 3^n}, & \text{if } n \text{ is even} \\ \frac{1}{2^{n-1} 3^n}, & \text{if } n \text{ is odd.} \end{cases}$$

To apply the Ratio Test, we observe that

$$\left|\frac{c_{n+1}}{c_n}\right| = \begin{cases} \frac{1}{3}, & \text{if } n \text{ is even} \\ \frac{1}{12}, & \text{if } n \text{ is odd.} \end{cases}$$

By Exercise 8.2.2 we know that $\left\{\left|\frac{c_{n+1}}{c_n}\right|\right\}_{n=1}^{\infty}$ is divergent, and hence we cannot use the Ratio Test to evaluate the convergence of $\sum_{n=1}^{\infty} c_n$. On the other hand, it is simple to see that the series $\sum_{n=1}^{\infty} c_n$ is convergent, by using the Comparison Test (Theorem 9.3.2) with the geometric series $\sum_{n=1}^{\infty}\frac{1}{2^{n-1}3^n}$. ◊

Although it might appear at first to the reader that the difference between absolutely convergent series and conditionally convergent series is merely a technicality, there is in fact a big difference between the behavior of these two types of series. We now see two topics that highlight just how very different these two types of series are.

Our first topic concerns the multiplication of series. As was proved in Theorem 9.2.6 (1), the convergence of the sum of two convergent series works out nicely;

more precisely, we saw that $\sum_{n=1}^{\infty}(a_n+b_n)=\sum_{n=1}^{\infty}a_n+\sum_{n=1}^{\infty}b_n$ whenever $\sum_{n=1}^{\infty}a_n$ and $\sum_{n=1}^{\infty}b_n$ are convergent. By contrast, the situation for the product of convergent series is not as simple.

What interests us about products of series is not the analog of $\sum_{n=1}^{\infty}(a_n+b_n)$, which would be $\sum_{n=1}^{\infty}a_nb_n$ and which is not particularly useful to us, but rather the analog of $\sum_{n=1}^{\infty}a_n+\sum_{n=1}^{\infty}b_n$, which is $[\sum_{n=1}^{\infty}a_n]\cdot[\sum_{n=1}^{\infty}b_n]$. In particular, we would like to know whether there is a convenient way to compute the product of two series, because multiplying each term in one series by each term in the other series, as is done when we multiply two finite sums, would be tricky with series because of their infinite nature.

As an example of multiplying finite sums, let us look at the product $[a_0+a_1+a_2]\cdot[b_0+b_1+b_2]$. This product is expanded by multiplying every term in $a_0+a_1+a_2$ with every term in $b_0+b_1+b_2$, resulting in

$$\begin{aligned}&[a_0+a_1+a_2]\cdot[b_0+b_1+b_2]\\&\qquad=a_0b_0+a_0b_1+a_0b_2+a_1b_0+a_1b_1+a_1b_2+a_2b_0+a_2b_1+a_2b_2.\end{aligned}$$

For convenience, we can rearrange the terms of this product by grouping elements by the sums of their subscripts, which yields

$$\begin{aligned}&[a_0+a_1+a_2]\cdot[b_0+b_1+b_2]\\&\qquad=a_0b_0+(a_0b_1+a_1b_0)+(a_0b_2+a_1b_1+a_2b_0)+(a_1b_2+a_2b_1)+a_2b_2.\end{aligned}$$

We will use this idea of grouping elements by the sums of their subscripts when we treat products of series; the first three terms in the above equation are the typical ones, whereas the last two terms do not quite fit the pattern, because, in contrast to series, the sums $a_0+a_1+a_2$ and $b_0+b_1+b_2$ stopped at the subscript $n=2$.

For notational convenience, it will be simpler to consider the product of two series that start with $n=0$ rather than $n=1$, that is, series of the form $\sum_{n=0}^{\infty}a_n$. There is no loss of generality in having our series start with $n=0$, because any series of the form $\sum_{n=1}^{\infty}a_n$ can be rewritten as $\sum_{n=0}^{\infty}a_{n+1}$.

The following theorem shows that products of series work out exactly as expected when at least one of the series is absolutely convergent. We start with a definition.

Definition 9.4.6. Let $\sum_{n=0}^{\infty}a_n$ and $\sum_{n=0}^{\infty}b_n$ be series in $\mathbb{R}$. The **Cauchy product** of $\sum_{n=0}^{\infty}a_n$ and $\sum_{n=0}^{\infty}b_n$ is the series $\sum_{n=0}^{\infty}e_n$ defined by $e_n=\sum_{k=0}^{n}a_kb_{n-k}$ for all $n\in\mathbb{N}\cup\{0\}$. △

Observe that the Cauchy product of any two series in $\mathbb{R}$ is always defined, even if the original series are divergent.

Theorem 9.4.7. *Let $\sum_{n=0}^{\infty}a_n$ and $\sum_{n=0}^{\infty}b_n$ be series in $\mathbb{R}$. Let $\sum_{n=0}^{\infty}e_n$ be the Cauchy product of $\sum_{n=0}^{\infty}a_n$ and $\sum_{n=0}^{\infty}b_n$. Suppose that $\sum_{n=0}^{\infty}a_n$ and $\sum_{n=0}^{\infty}b_n$ are convergent, and that $\sum_{n=0}^{\infty}a_n$ or $\sum_{n=0}^{\infty}b_n$ is absolutely convergent. Then $\sum_{n=0}^{\infty}e_n$ is convergent and $[\sum_{n=0}^{\infty}a_n]\cdot[\sum_{n=0}^{\infty}b_n]=\sum_{n=0}^{\infty}e_n$.*

Proof. Without loss of generality, assume that $\sum_{n=0}^{\infty} a_n$ is absolutely convergent.

Let $\{s_n\}_{n=0}^{\infty}$, and $\{t_n\}_{n=0}^{\infty}$ and $\{u_n\}_{n=1}^{\infty}$ be the sequences of partial sums of $\sum_{n=0}^{\infty} a_n$, and $\sum_{n=0}^{\infty} b_n$ and $\sum_{n=0}^{\infty} e_n$, respectively. Let $A = \sum_{n=0}^{\infty} a_n$ and $B = \sum_{n=0}^{\infty} b_n$. Hence $\lim_{n\to\infty} s_n = A$ and $\lim_{n\to\infty} t_n = B$.

As a preliminary step, let $\{h_n\}_{n=0}^{\infty}$ be defined by $h_n = \sum_{j=1}^{n} a_j[t_{n-j} - B]$ for all $n \in \mathbb{N} \cup \{0\}$. We will show that $\{h_n\}_{n=0}^{\infty}$ is convergent and $\lim_{n\to\infty} h_n = 0$. Let $\varepsilon > 0$. Because $\lim_{k\to\infty} t_k = B$, then by Exercise 8.2.5 we know that $\{t_k - B\}_{n=0}^{\infty}$ is convergent and $\lim_{k\to\infty} (t_k - B) = 0$. It follows from Lemma 8.2.6 that $\{t_k - B\}_{k=0}^{\infty}$ is bounded. Therefore there is some $Q \in \mathbb{R}$ such that $|t_k - B| \le Q$ for all $k \in \mathbb{N} \cup \{0\}$. We may assume that $Q > 0$. Because $\sum_{n=1}^{\infty} a_n$ is absolutely convergent, then $\sum_{n=0}^{\infty} |a_n|$ is convergent. Let $P = \sum_{n=0}^{\infty} |a_n| + 1$. Then $P > 0$.

Because $\lim_{k\to\infty} t_k = B$, there is some $N \in \mathbb{N}$ such that $k \in \mathbb{N} \cup \{0\}$ and $k \ge N$ imply $|t_k - B| < \frac{\varepsilon}{2P}$. Because $\sum_{n=0}^{\infty} |a_n|$ is convergent, then by Exercise 9.2.11 there is some $M \in \mathbb{N}$ such that $n, m \in \mathbb{N} \cup \{0\}$ and $n > m \ge M$ imply $\left|\sum_{k=m+1}^{n} |a_k|\right| < \frac{\varepsilon}{2Q}$. Let $J = \max\{N, M\}$, and let $K = 2J$.

Suppose that $n \in \mathbb{N} \cup \{0\}$ and $n \ge K$. Using Exercise 2.5.3, we see that

$$\begin{aligned}
|h_n - 0| &= \left|\sum_{j=0}^{n} a_j[t_{n-j} - B]\right| \le \sum_{j=0}^{n} |a_j| \cdot |t_{n-j} - B| \\
&= \sum_{j=0}^{J} |a_j| \cdot |t_{n-j} - B| + \sum_{j=J+1}^{n} |a_j| \cdot |t_{n-j} - B| \\
&< \frac{\varepsilon}{2P} \sum_{j=0}^{J} |a_j| + Q \sum_{j=J+1}^{n} |a_j| < \frac{\varepsilon}{2P} \sum_{j=0}^{\infty} |a_j| + Q \cdot \frac{\varepsilon}{2Q} \\
&< \frac{\varepsilon}{2P} \cdot P + Q \cdot \frac{\varepsilon}{2Q} = \varepsilon.
\end{aligned}$$

We deduce that $\{h_n\}_{n=1}^{\infty}$ is convergent and $\lim_{n\to\infty} h_n = 0$.

Let $n \in \mathbb{N} \cup \{0\}$. Then

$$\begin{aligned}
u_n &= \sum_{j=0}^{n} e_j = \sum_{j=0}^{n} \sum_{k=0}^{j} a_k b_{j-k} = \sum_{k=0}^{n} a_k \sum_{j=k}^{n} b_{j-k} = \sum_{k=0}^{n} a_k t_{n-k} \\
&= \sum_{k=0}^{n} a_k [t_{n-k} - B] + \sum_{k=0}^{n} a_k B = h_n + s_n B.
\end{aligned}$$

We know that $\lim_{n\to\infty} s_n = A$ and $\lim_{n\to\infty} h_n = 0$, and therefore by Theorem 8.2.9 we deduce that $\{u_n\}_{n=0}^{\infty}$ is convergent and $\lim_{n\to\infty} u_n = \lim_{n\to\infty} h_n + \lim_{n\to\infty} s_n B = AB$. We conclude that $\sum_{n=0}^{\infty} e_n$ is convergent and $\sum_{n=0}^{\infty} e_n = [\sum_{n=0}^{\infty} a_n] \cdot [\sum_{n=0}^{\infty} b_n]$. □

Example 9.4.8. Let $\sum_{n=0}^{\infty} a_n = \sum_{n=0}^{\infty} b_n = \sum_{n=0}^{\infty} \frac{1}{2^{n+1}}$. This series can also be written as $\sum_{n=1}^{\infty} \frac{1}{2^n}$, and it is a geometric series, as discussed in Example 9.2.4 (4). By that example we know that $\sum_{n=0}^{\infty} a_n$ and $\sum_{n=0}^{\infty} b_n$ are convergent and

$$\sum_{n=0}^{\infty} a_n = \sum_{n=0}^{\infty} b_n = \frac{\frac{1}{2}}{1-\frac{1}{2}} = 1.$$

Because all of the terms of this series are positive, then it is absolutely convergent.

Let $\sum_{n=0}^{\infty} e_n$ be the Cauchy product of $\sum_{n=0}^{\infty} a_n$ and $\sum_{n=0}^{\infty} b_n$. If $n \in \mathbb{N} \cup \{0\}$, then

$$e_n = \sum_{k=0}^{n} a_k b_{n-k} = \sum_{k=0}^{n} \frac{1}{2^{k+1}} \frac{1}{2^{(n-k)+1}} = \sum_{k=0}^{n} \frac{1}{2^{n+2}} = \frac{n+1}{2^{n+2}}.$$

It now follows from Theorem 9.4.7 that $\sum_{n=0}^{\infty} e_n = \sum_{n=0}^{\infty} \frac{n+1}{2^{n+2}} = \sum_{n=1}^{\infty} \frac{n}{2^{n+1}}$ is convergent and

$$\sum_{n=1}^{\infty} \frac{n}{2^{n+1}} = \sum_{n=0}^{\infty} e_n = \left[\sum_{n=0}^{\infty} a_n\right] \cdot \left[\sum_{n=0}^{\infty} b_n\right] = 1 \cdot 1 = 1.$$

It would have been straightforward to use the Ratio Test (Theorem 9.4.4) to show that $\sum_{n=1}^{\infty} \frac{n}{2^{n+1}}$ is absolutely convergent, but that would not have told us what the sum of the series is. ◊

It is not possible to strengthen the conclusion of Theorem 9.4.7 and say that the Cauchy product of $\sum_{n=0}^{\infty} a_n$ and $\sum_{n=0}^{\infty} b_n$ is absolutely convergent, because, as seen in Exercise 9.4.10, there are series $\sum_{n=0}^{\infty} a_n$ and $\sum_{n=0}^{\infty} b_n$ such that $\sum_{n=0}^{\infty} a_n$ is absolutely convergent and $\sum_{n=0}^{\infty} b_n$ is conditionally convergent, and the Cauchy product of $\sum_{n=0}^{\infty} a_n$ and $\sum_{n=0}^{\infty} b_n$ is conditionally convergent. However, as seen in Exercise 9.4.11, if $\sum_{n=0}^{\infty} a_n$ and $\sum_{n=0}^{\infty} b_n$ are both absolutely convergent, then the Cauchy product of $\sum_{n=0}^{\infty} a_n$ and $\sum_{n=0}^{\infty} b_n$ is absolutely convergent.

It is not possible to weaken the hypotheses of Theorem 9.4.7 and say only that $\sum_{n=0}^{\infty} a_n$ and $\sum_{n=0}^{\infty} b_n$ are convergent, because, as seen in the first part of the following example, due to Augustin Louis Cauchy (1789–1857), there are series $\sum_{n=0}^{\infty} a_n$ and $\sum_{n=0}^{\infty} b_n$ such that $\sum_{n=0}^{\infty} a_n$ and $\sum_{n=0}^{\infty} b_n$ are conditionally convergent, and the Cauchy product of $\sum_{n=0}^{\infty} a_n$ and $\sum_{n=0}^{\infty} b_n$ is divergent. On the other hand, as seen in the second part of the example, the Cauchy product can be convergent (and in fact absolutely convergent) even if the series $\sum_{n=0}^{\infty} a_n$ and $\sum_{n=0}^{\infty} b_n$ are poorly behaved.

Example 9.4.9.

(1) Let $\sum_{n=0}^{\infty} a_n = \sum_{n=0}^{\infty} b_n = \sum_{n=0}^{\infty} (-1)^n \frac{1}{\sqrt{n+1}}$. This series can also be written as $\sum_{n=1}^{\infty} (-1)^{n-1} \frac{1}{\sqrt{n}}$, and it is proved in Exercise 9.4.1 that this series is conditionally convergent. Let $\sum_{n=0}^{\infty} e_n$ be the Cauchy product of $\sum_{n=0}^{\infty} a_n$ and $\sum_{n=0}^{\infty} b_n$. Let $n \in \mathbb{N} \cup \{0\}$. Then

$$e_n = \sum_{k=0}^{n} a_k b_{n-k} = \sum_{k=0}^{n} \frac{(-1)^k}{\sqrt{k+1}} \frac{(-1)^{n-k}}{\sqrt{(n-k)+1}} = (-1)^n \sum_{k=0}^{n} \frac{1}{\sqrt{(k+1)[(n-k)+1]}}.$$

It is left to the reader to verify that $(k+1)[(n-k)+1] \leq (n+1)^2$ for all $k \in \{0,\ldots,n\}$. Hence $\frac{1}{\sqrt{(k+1)[(n-k)+1]}} \geq \frac{1}{n+1}$ for all $k \in \{0,\ldots,n\}$. Therefore $|e_n| \geq \sum_{k=0}^{n} \frac{1}{n+1} = 1$. It follows that $\{e_n\}_{n=0}^{\infty}$ does not converge to 0, and by the Divergence Test (Theorem 9.2.5) we conclude that $\sum_{n=0}^{\infty} e_n$ is divergent; the fact that the series starts at $n = 0$

rather than $n = 1$ makes no difference for the Divergence Test. Hence, the conclusion of Theorem 9.4.7 does not hold for $\sum_{n=0}^{\infty} a_n$ and $\sum_{n=0}^{\infty} b_n$.

(2) Let

$$\sum_{n=0}^{\infty} c_n = 1 + 1 + 2 + 2^2 + 2^3 + 2^4 + \cdots,$$

and

$$\sum_{n=0}^{\infty} d_n = -1 + 1 + 1 + 1 + 1 + \cdots.$$

The series $\sum_{n=0}^{\infty} c_n$ and $\sum_{n=0}^{\infty} d_n$ are divergent. Let $\sum_{n=0}^{\infty} f_n$ be the Cauchy product of $\sum_{n=0}^{\infty} c_n$ and $\sum_{n=0}^{\infty} d_n$ (which is defined for any two series, convergent or not). Then $f_0 = -1$, and if $n \in \mathbb{N} \cup \{0\}$ then

$$\begin{aligned} f_n &= \sum_{k=0}^{n} c_k d_{n-k} = 1 \cdot 1 + 1 \cdot 1 + 2 \cdot 1 + 2^2 \cdot 1 + \cdots + 2^{n-2} \cdot 1 + 2^{n-1} \cdot (-1) \\ &= 1 + \frac{1 \cdot (1 - 2^{n-1})}{1 - 2} - 2^{n-1} = 0, \end{aligned}$$

where the equality before last follows from Exercise 2.5.12 (3). It is evident that $\sum_{n=0}^{\infty} f_n$ is absolutely convergent. ◊

We saw in Example 9.4.9 (1) that if $\sum_{n=0}^{\infty} a_n$ and $\sum_{n=0}^{\infty} b_n$ are conditionally convergent, then it is not necessarily the case that the Cauchy product of $\sum_{n=0}^{\infty} a_n$ and $\sum_{n=0}^{\infty} b_n$ is convergent. However, if $\sum_{n=0}^{\infty} a_n$ and $\sum_{n=0}^{\infty} b_n$ are conditionally convergent and the Cauchy product of $\sum_{n=0}^{\infty} a_n$ and $\sum_{n=0}^{\infty} b_n$ is convergent, then, as the reader is asked to show in Exercise 10.4.11, the sum of the Cauchy product must equal $\left[\sum_{n=0}^{\infty} a_n\right] \cdot \left[\sum_{n=0}^{\infty} b_n\right]$.

Example 9.4.9 (1) shows that conditionally convergent series are not as well behaved as absolutely convergent series, at least with respect to the Cauchy product of series. We now turn to an even more surprising difference between absolutely convergent series and conditionally convergent series, which is the issue of rearranging the terms of a series. Actually, the problem that we saw in regard to the Cauchy product of two conditionally convergent series is a special case of the more general issue of the rearrangement of series, because the way the Cauchy product of $\sum_{n=0}^{\infty} a_n$ and $\sum_{n=0}^{\infty} b_n$ works is by regrouping terms according to the sums of their subscripts, and that regrouping involves a rearrangement of the order of terms of the form $a_i b_j$.

We start our discussion of rearrangements of series with the analogous question for finite sums. First, we need to remind ourselves what the sum of finitely many numbers, for example $3 + 7 + 2$, means. In principle, addition is defined for only two numbers at a time. However, as we saw in Exercise 2.5.19, it is possible to use Definition by Recursion to define $a_1 + a_2 + \cdots + a_n$ for any $n \in \mathbb{N}$ and any $a_1, \ldots, a_n \in \mathbb{R}$. Once we have that definition, it is possible to make use of the Commutative and Associative Laws for Addition to rearrange the order of any finite sum without changing the value of the sum. For example, we see that $3+7+2 = (3+7)+2 = (7+3)+2 = 7+(3+2) = 7+(2+3) = 7+2+3$.

Does rearranging the terms of a series, which is an infinite sum, change the value of the sum of the series? Of course, that question is only relevant to convergent series. The answer to this question is not at all obvious, because the definition of the sum of a series is very much dependent upon the order of the terms; if the terms of a series are rearranged, then the sequence of partial sums is changed. In fact, as seen in Example 9.4.11 below, the sum of a series can change if the terms of the series are rearranged. To state that example, we need the following definition.

Definition 9.4.10. Let $\sum_{n=1}^{\infty} a_n$ and $\sum_{n=1}^{\infty} b_n$ be series in $\mathbb{R}$. The series $\sum_{n=1}^{\infty} b_n$ is a **rearrangement** of $\sum_{n=1}^{\infty} a_n$ if there is a bijective function $f\colon \mathbb{N} \to \mathbb{N}$ such that $b_n = a_{f(n)}$ for all $n \in \mathbb{N}$. △

A bijective function from a set to itself is often called a "permutation" of the set, though we will not use that terminology.

Example 9.4.11. It was seen in Example 9.4.2 (2) that the alternating harmonic series $\sum_{n=1}^{\infty} (-1)^{n-1}\frac{1}{n}$ is conditionally convergent. Let $S = \sum_{n=1}^{\infty} (-1)^{n-1}\frac{1}{n}$. Writing out the terms of the series, we see that

$$S = 1 - \frac{1}{2} + \frac{1}{3} - \frac{1}{4} + \frac{1}{5} - \frac{1}{6} + \frac{1}{7} - \frac{1}{8} + \cdots.$$

It follows from Theorem 9.2.6 (3) that

$$\frac{S}{2} = \frac{1}{2} - \frac{1}{4} + \frac{1}{6} - \frac{1}{8} + \frac{1}{10} - \frac{1}{12} + \frac{1}{14} - \frac{1}{16} + \cdots.$$

Using Exercise 9.2.7 we see that

$$\frac{S}{2} = 0 + \frac{1}{2} + 0 - \frac{1}{4} + 0 + \frac{1}{6} + 0 - \frac{1}{8} + 0 + \frac{1}{10} + 0 - \frac{1}{12} + 0 + \frac{1}{14} + 0 - \frac{1}{16} + \cdots.$$

By adding the above series to the alternating harmonic series, and then using Theorem 9.2.6 (1), we deduce that

$$\frac{3S}{2} = 1 + 0 + \frac{1}{3} - \frac{1}{2} + \frac{1}{5} + 0 + \frac{1}{7} - \frac{1}{4} + \frac{1}{9} + 0 \cdots.$$

By the obvious variant of Exercise 9.2.7 we conclude that

$$\frac{3S}{2} = 1 + \frac{1}{3} - \frac{1}{2} + \frac{1}{5} + \frac{1}{7} - \frac{1}{4} + \frac{1}{9} + \cdots.$$

This last series is a rearrangement of the alternating harmonic series, where the pattern for the rearrangement is that after every two terms with odd denominators we have one term with an even denominator. Though in practice we rarely write out a rearrangement of a series using the formal definition given in Definition 9.4.10, if we wished to do so in the present case, we could denote the alternating harmonic series by $\sum_{n=1}^{\infty} a_n$, and then the rearrangement given above is the series $\sum_{n=1}^{\infty} a_{f(n)}$, where $f\colon \mathbb{N} \to \mathbb{N}$ is defined by

$$f(n) = \begin{cases} 4k-3, & \text{if } n = 3k-2 \text{ for some } k \in \mathbb{N} \\ 4k-1, & \text{if } n = 3k-1 \text{ for some } k \in \mathbb{N} \\ 2k, & \text{if } n = 3k \text{ for some } k \in \mathbb{N}. \end{cases}$$

Finally, we see by Exercise 9.3.10 (2) that $S = \ln 2$, and therefore $S \neq 0$ by Theorem 7.2.3 (1) and Lemma 7.2.4. Hence $\frac{3S}{2} \neq S$, which means that a rearrangement of a convergent series can be convergent and have a different sum from the original series. ◊

The strange behavior seen in Example 9.4.11 occurred with a conditionally convergent series. The following theorem shows that no such problem can occur with an absolutely convergent series.

Theorem 9.4.12. *Let $\sum_{n=1}^{\infty} a_n$ be a series in $\mathbb{R}$. Suppose that $\sum_{n=1}^{\infty} a_n$ is absolutely convergent. If $\sum_{n=1}^{\infty} b_n$ is a rearrangement of $\sum_{n=1}^{\infty} a_n$, then $\sum_{n=1}^{\infty} b_n$ is absolutely convergent and $\sum_{n=1}^{\infty} b_n = \sum_{n=1}^{\infty} a_n$.*

Proof. Suppose that $\sum_{n=1}^{\infty} b_n$ is a rearrangement of $\sum_{n=1}^{\infty} a_n$. First, we show that $\sum_{n=1}^{\infty} b_n$ is absolutely convergent. Let $\{u_n\}_{n=1}^{\infty}$ and $\{v_n\}_{n=1}^{\infty}$ be the sequences of partial sums of $\sum_{n=1}^{\infty} |a_n|$ and $\sum_{n=1}^{\infty} |b_n|$, respectively. By hypothesis we know that $\{u_n\}_{n=1}^{\infty}$ is convergent, and by Lemma 9.3.1 we therefore know that $\{u_n\}_{n=1}^{\infty}$ is bounded. Hence, there is some $M \in \mathbb{R}$ such that $|u_n| \leq M$ for all $n \in \mathbb{N}$.

Let $k \in \mathbb{N}$. Because $\sum_{n=1}^{\infty} b_n$ is a rearrangement of $\sum_{n=1}^{\infty} a_n$, there is a bijective function $f\colon \mathbb{N} \to \mathbb{N}$ such that $b_n = a_{f(n)}$ for all $n \in \mathbb{N}$. Let $J = \max\{f(1), \ldots, f(k)\}$. It follows that $0 \leq v_k = |b_1| + \cdots + |b_k| = |a_{f(1)}| + \cdots + |a_{f(k)}| \leq |a_1| + \cdots + |a_J| = u_J \leq M$. It follows that $|v_n| \leq M$ for all $n \in \mathbb{N}$, and therefore $\{v_n\}_{n=1}^{\infty}$ is bounded. By Lemma 9.3.1 we deduce that $\sum_{n=1}^{\infty} |b_n|$ is convergent, and therefore $\sum_{n=1}^{\infty} b_n$ is absolutely convergent.

Next, we show that $\sum_{n=1}^{\infty} b_n = \sum_{n=1}^{\infty} a_n$. Let $\{s_n\}_{n=1}^{\infty}$ and $\{t_n\}_{n=1}^{\infty}$ be the sequences of partial sums of $\sum_{n=1}^{\infty} a_n$ and $\sum_{n=1}^{\infty} b_n$, respectively. Because $\sum_{n=1}^{\infty} a_n$ and $\sum_{n=1}^{\infty} b_n$ are absolutely convergent, then $\sum_{n=1}^{\infty} a_n$ and $\sum_{n=1}^{\infty} b_n$ are convergent by Theorem 9.4.3. Hence $\{s_n\}_{n=1}^{\infty}$ and $\{t_n\}_{n=1}^{\infty}$ are convergent. We will show that $\{t_n - s_n\}_{n=1}^{\infty}$ is convergent and that $\lim_{n\to\infty} (t_n - s_n) = 0$. It will then follow from Theorem 8.2.9 (1) that $\sum_{n=1}^{\infty} b_n = \lim_{n\to\infty} t_n = \lim_{n\to\infty} [s_n + (t_n - s_n)] = \sum_{n=1}^{\infty} a_n + 0 = \sum_{n=1}^{\infty} a_n$.

Let $\varepsilon > 0$. Because $\sum_{n=1}^{\infty} |a_n|$ is convergent, then by Exercise 9.2.11 there is some $N \in \mathbb{N}$ such that $n, m \in \mathbb{N}$ and $n > m \geq N$ imply $|\sum_{k=m+1}^{n} |a_k|| < \varepsilon$, which means that $\sum_{k=m+1}^{n} |a_i| < \varepsilon$.

Let $P = \max\{f^{-1}(1), \ldots, f^{-1}(N)\}$ and $Q = \max\{f(1), \ldots, f(P)\}$. Then we see that $\{1, \ldots, N\} \subseteq \{f(1), \ldots, f(P)\} \subseteq \{1, \ldots, Q\}$. Hence $N \leq P \leq Q$.

Suppose that $m \in \mathbb{N}$ and $m \geq Q + 1$. Then

$$|(t_m - s_m) - 0| = \left| \sum_{j=1}^{m} b_j - \sum_{i=1}^{m} a_i \right| = \left| \sum_{j=1}^{m} a_{f(j)} - \sum_{i=1}^{m} a_i \right|$$

$$= \left| \left[\sum_{j=1}^{P} a_{f(j)} + \sum_{j=P+1}^{m} a_{f(j)} \right] - \left[\sum_{i=1}^{N} a_i + \sum_{i=N+1}^{m} a_i \right] \right|.$$

Because $\{1,\ldots,N\} \subseteq \{f(1),\ldots,f(P)\}$, then in this last expression, each term of the form a_i in the summation $\sum_{i=1}^{N} a_i$ will cancel out with a term of the form $a_{f(j)}$ in the summation $\sum_{j=1}^{P} a_{f(j)}$. Some other terms might also cancel out, and what remains inside the absolute value after all possible canceling is a summation of the form $\sum_{i\in V} a_i - \sum_{i\in W} a_i$, for some sets $V, W \subseteq \{N+1,\ldots,Q+1\}$ such that $V \cap W = \emptyset$. Using Exercise 2.5.3, we see that

$$|(t_m - s_m) - 0| = \left| \sum_{i\in V} a_i - \sum_{i\in W} a_i \right| \le \sum_{i\in V} |a_i| + \sum_{i\in W} |a_i| \le \sum_{i=N+1}^{Q+1} |a_i| < \varepsilon,$$

where the last inequality holds because $Q+1 > N$. We conclude that $\{t_n - s_n\}_{n=1}^{\infty}$ is convergent and $\lim_{n\to\infty} (t_n - s_n) = 0$. □

Comparing Theorem 9.4.12 and Example 9.4.11 shows that rearrangements of conditionally convergent series are not as well behaved as rearrangements of absolutely convergent series. In fact, as will be seen in Theorem 9.4.15 below, rearrangements of conditionally convergent series behave even worse than might be imagined from just the evidence of Example 9.4.11. We start with the following definition and lemma, which help clarify the difference between absolutely convergent and conditionally convergent series.

Definition 9.4.13. Let $a \in \mathbb{R}$. The **positive part** of a, denoted a^+, and the **negative part** of a, denoted a^-, are defined by

$$a^+ = \begin{cases} a, & \text{if } a \ge 0 \\ 0, & \text{if } a < 0 \end{cases} \quad \text{and} \quad a^- = \begin{cases} 0, & \text{if } a \ge 0 \\ -a, & \text{if } a < 0. \end{cases}$$ △

Lemma 9.4.14. *Let $\sum_{n=1}^{\infty} a_n$ be a series in $\mathbb{R}$.*

1. *The series $\sum_{n=1}^{\infty} a_n$ is absolutely convergent if and only if $\sum_{n=1}^{\infty} a_n^+$ and $\sum_{n=1}^{\infty} a_n^-$ are convergent, and if they are convergent then $\sum_{n=1}^{\infty} a_n = \sum_{n=1}^{\infty} a_n^+ - \sum_{n=1}^{\infty} a_n^-$.*
2. *If $\sum_{n=1}^{\infty} a_n$ is conditionally convergent, then $\sum_{n=1}^{\infty} a_n^+ = \infty$ and $\sum_{n=1}^{\infty} a_n^- = \infty$.*

Proof. Observe that for each $n \in \mathbb{N}$, we can evaluate a_n^+ and a_n^- by the equations $a_n^+ = \frac{1}{2}|a_n| + \frac{1}{2}a_n$ and $a_n^- = \frac{1}{2}|a_n| - \frac{1}{2}a_n$. Hence $a_n = a_n^+ - a_n^-$ and $|a_n| = a_n^+ + a_n^-$.

(1) Suppose that $\sum_{n=1}^{\infty} a_n$ is absolutely convergent. Then $\sum_{n=1}^{\infty} |a_n|$ is convergent, and by Theorem 9.4.3 we know that $\sum_{n=1}^{\infty} a_n$ is convergent. Using the preliminary observation about a_n^+ and a_n^-, it follows from Theorem 9.2.6 that $\sum_{n=1}^{\infty} a_n^+$ is convergent and $\sum_{n=1}^{\infty} a_n^+ = \frac{1}{2}\sum_{n=1}^{\infty} |a_n| + \frac{1}{2}\sum_{n=1}^{\infty} a_n$, and that $\sum_{n=1}^{\infty} a_n^-$ is convergent and $\sum_{n=1}^{\infty} a_n^- = \frac{1}{2}\sum_{n=1}^{\infty} |a_n| - \frac{1}{2}\sum_{n=1}^{\infty} a_n$. Solving these two equations for $\sum_{n=1}^{\infty} a_n$ yields $\sum_{n=1}^{\infty} a_n = \sum_{n=1}^{\infty} a_n^+ - \sum_{n=1}^{\infty} a_n^-$.

Next, suppose that $\sum_{n=1}^{\infty} a_n^+$ and $\sum_{n=1}^{\infty} a_n^+$ are convergent. Again using the preliminary observation about a_n^+ and a_n^- and Theorem 9.2.6, we deduce that $\sum_{n=1}^{\infty} |a_n|$

is convergent and $\sum_{n=1}^{\infty} |a_n| = \sum_{n=1}^{\infty} a_n^+ + \sum_{n=1}^{\infty} a_n^-$. Because $\sum_{n=1}^{\infty} |a_n|$ is convergent, then $\sum_{n=1}^{\infty} a_n$ is absolutely convergent.

(2) Suppose that $\sum_{n=1}^{\infty} a_n$ is conditionally convergent. Suppose also that at least one of $\sum_{n=1}^{\infty} a_n^+ = \infty$ and $\sum_{n=1}^{\infty} a_n^- = \infty$ is false. We consider the case where $\sum_{n=1}^{\infty} a_n^+ = \infty$ is false; the other case is similar, and we omit the details. Because $a_n^+ \geq 0$ for all $n \in \mathbb{N}$, then Exercise 9.3.1 (1) implies that $\sum_{n=1}^{\infty} a_n^+$ is convergent. Using the preliminary observation about a_n^+ and a_n^-, we see that $a_n^- = a_n^+ - a_n$ for all $n \in \mathbb{N}$, and it follows from Theorem 9.2.6 that $\sum_{n=1}^{\infty} a_n^-$ is convergent. Part (1) of this lemma then implies that $\sum_{n=1}^{\infty} a_n$ is absolutely convergent, which is a contradiction. □

The following somewhat surprising theorem, due to Georg Friedrich Bernhard Riemann (1826–1866), shows that rearrangements of conditionally convergent series are as poorly behaved as possible.

Theorem 9.4.15. *Let $\sum_{n=1}^{\infty} a_n$ be a series in $\mathbb{R}$. Suppose that $\sum_{n=1}^{\infty} a_n$ is conditionally convergent.*

1. *Let $x \in \mathbb{R}$. Then there is a rearrangement $\sum_{n=1}^{\infty} d_n$ of $\sum_{n=1}^{\infty} a_n$ such that $\sum_{n=1}^{\infty} d_n$ is convergent and $\sum_{n=1}^{\infty} d_n = x$.*
2. *There is a rearrangement $\sum_{n=1}^{\infty} c_n$ of $\sum_{n=1}^{\infty} a_n$ such that $\sum_{n=1}^{\infty} c_n$ is divergent.*

Proof. We will prove Part (1), leaving the remaining part to the reader in Exercise 9.4.13.

(1) If $\sum_{n=1}^{\infty} a_n$ has only finitely many non-zero terms, then it would be absolutely convergent, as the reader can verify. Hence $\sum_{n=1}^{\infty} a_n$ has infinitely many non-zero terms. By Exercise 9.4.6 we may remove all terms that are zero from $\sum_{n=1}^{\infty} a_n$ without changing the convergence or divergence of $\sum_{n=1}^{\infty} a_n$ or any rearrangement of it. Hence, we may assume without loss of generality that the series $\sum_{n=1}^{\infty} a_n$ has no zero terms. By Lemma 9.4.14 (2), we know that $\sum_{n=1}^{\infty} a_n^+ = \infty$ and $\sum_{n=1}^{\infty} a_n^- = \infty$. Hence, it must be the case that each of these two series has infinitely many non-zero terms. Let $\sum_{n=1}^{\infty} b_n$ and $\sum_{n=1}^{\infty} c_n$ be obtained from $\sum_{n=1}^{\infty} a_n^+$ and $\sum_{n=1}^{\infty} a_n^-$, respectively by removing all terms that are zero. By Exercise 9.4.6 again it follows that $\sum_{n=1}^{\infty} b_n = \infty$ and $\sum_{n=1}^{\infty} c_n = \infty$. Each term of $\sum_{n=1}^{\infty} a_n$ is found in precisely one of $\sum_{n=1}^{\infty} b_n$ or $\sum_{n=1}^{\infty} c_n$. We will construct the desired rearrangement of $\sum_{n=1}^{\infty} a_n$ by arranging the combined terms of $\sum_{n=1}^{\infty} b_n$ and $\sum_{n=1}^{\infty} c_n$, as described below.

As a preliminary step, we define two sequences $\{p_n\}_{n=1}^{\infty}$ and $\{q_n\}_{n=1}^{\infty}$ in $\mathbb{N}$, and two sequences $\{y_n\}_{n=1}^{\infty}$ and $\{z_n\}_{n=1}^{\infty}$ in $\mathbb{R}$, using joint Definition by Recursion, by which we mean that we will first define p_1, q_1, y_1 and z_1, and we will then define all four of p_{k+1}, q_{k+1}, y_{k+1} and z_{k+1} in terms of p_k, q_k, y_k and z_k, for all $k \in \mathbb{N}$. (This type of joint Definition by Recursion works because it is really a single Definition by Recursion in the set $\mathbb{N} \times \mathbb{N} \times \mathbb{R} \times \mathbb{R}$.) For convenience, we will use the convention that any summation of the form $\sum_c^d$ with $c > d$ is taken to be zero.

First, because $\sum_{n=1}^{\infty} b_n = \infty$, there is some $p \in \mathbb{N}$ such that $\sum_{i=1}^{p} b_i > x$. By the Well-Ordering Principle (Theorem 1.2.10, Axiom 1.4.4 or Theorem 2.4.6), we can find the smallest such natural number p, which we will call p_1. Hence $\sum_{i=1}^{p_1-1} b_i \leq x < \sum_{i=1}^{p_1} b_i$.

Let $y_1 = \sum_{i=1}^{p_1} b_i$. It follows that $y_1 - b_{p_1} \le x < y_1$, and therefore $0 < y_1 - x \le b_{p_1}$. Similarly, because $\sum_{n=1}^{\infty} c_n = \infty$, there is a smallest $q_1 \in \mathbb{N}$ such that $\sum_{i=1}^{q_1} c_i > y_1 - x$. Hence $\sum_{i=1}^{q_1} c_i > y_1 - x \ge \sum_{i=1}^{q_1-1} c_i$. Let $z_1 = y_1 - \sum_{i=1}^{q_1} c_i$. It follows that $0 < x - z_1 \le c_{q_1}$.

Second, suppose that we have defined p_k, q_k, y_k and z_k for some $k \in \mathbb{N}$, and that $0 < y_k - x \le b_{p_k}$ and $0 < x - z_k \le c_{q_k}$. We define $p_{k+1}, q_{k+1} \in \mathbb{N}$ and $y_{k+1}, z_{k+1} \in \mathbb{R}$ as follows. By Exercise 9.2.12 we see that $\sum_{n=p_k+1}^{\infty} b_n = \infty$ and $\sum_{n=q_k+1}^{\infty} c_n = \infty$. Then there is a smallest $p_{k+1} \in \mathbb{N}$ such that $p_{k+1} > p_k$ and $\sum_{i=p_k+1}^{p_{k+1}} b_i > x - z_k$. Hence $\sum_{i=p_k+1}^{p_{k+1}} b_i > x - z_k \ge \sum_{i=p_k+1}^{p_{k+1}-1} b_i$. Let $y_{k+1} = z_k + \sum_{i=p_k+1}^{p_{k+1}} b_i$. It follows that $0 < y_{k+1} - x \le b_{p_{k+1}}$. There is also a smallest $q_{k+1} \in \mathbb{N}$ such that $q_{k+1} > q_k$ and $\sum_{i=q_k+1}^{q_{k+1}} c_i > y_{k+1} - x$. Hence $\sum_{i=q_k+1}^{q_{k+1}} c_i > y_{k+1} - x \ge \sum_{i=q_k+1}^{q_{k+1}-1} c_i$. Let $z_{k+1} = y_{k+1} - \sum_{i=q_k+1}^{q_{k+1}} c_i$. It follows that $0 < x - z_{k+1} \le c_{q_{k+1}}$.

We have now defined sequences $\{p_n\}_{n=1}^{\infty}$, $\{q_n\}_{n=1}^{\infty}$, $\{y_n\}_{n=1}^{\infty}$ and $\{z_n\}_{n=1}^{\infty}$, such that $0 < y_n - x \le b_{p_n}$ and $0 < x - z_n \le c_{q_n}$ for all $n \in \mathbb{N}$. Moreover, it is seen by the above construction that $p_n < p_{n+1}$, and $q_n < q_{n+1}$ for all $n \in \mathbb{N}$, and it follows from the analog of Exercise 8.3.1 for strictly increasing sequences that $\{p_n\}_{n=1}^{\infty}$ and $\{q_n\}_{n=1}^{\infty}$ are strictly increasing.

Because $\sum_{n=1}^{\infty} a_n$ is convergent, we know by the Divergence Test (Theorem 9.2.5) that $\lim_{n\to\infty} a_n = 0$. It follows from Exercise 8.2.13 (2) that $\lim_{n\to\infty} |a_n| = 0$. By the definition of $\{b_n\}_{n=1}^{\infty}$ and $\{c_n\}_{n=1}^{\infty}$ we see that both of these sequences are subsequences of $\{|a_n|\}_{n=1}^{\infty}$, and we deduce from Lemma 8.3.7 that $\lim_{n\to\infty} b_n = 0$ and $\lim_{n\to\infty} c_n = 0$. Because $\{b_{p_n}\}_{n=1}^{\infty}$ and $\{c_{q_n}\}_{n=1}^{\infty}$ are subsequences of $\{b_n\}_{n=1}^{\infty}$ and $\{c_n\}_{n=1}^{\infty}$, respectively, using Lemma 8.3.7 again we conclude that $\lim_{n\to\infty} b_{p_n} = 0$ and $\lim_{n\to\infty} c_{q_n} = 0$.

The inequalities $0 < y_n - x \le b_{p_n}$ and $0 < x - z_n \le c_{q_n}$ for all $n \in \mathbb{N}$, together with Example 8.2.4 (1) and the Squeeze Theorem for Sequences (Theorem 8.2.12), now imply that $\lim_{n\to\infty} (y_n - x) = 0$ and $\lim_{n\to\infty} (x - z_n) = 0$. By Exercise 8.2.5 we deduce that $\lim_{n\to\infty} y_n = x$ and $\lim_{n\to\infty} z_n = x$.

We now define a rearrangement $\sum_{n=1}^{\infty} d_n$ of $\sum_{n=1}^{\infty} a_n$ by

$$\sum_{n=1}^{\infty} d_n = b_1 + \cdots + b_{p_1} - c_1 - \cdots - c_{q_1} + b_{p_1+1} + \cdots + b_{p_2} - c_{q_1+1} - \cdots - c_{q_2} + \cdots.$$

We will show that $\sum_{n=1}^{\infty} d_n$ is convergent and $\sum_{n=1}^{\infty} d_n = x$. Let $\{s_n\}_{n=1}^{\infty}$ be the sequence of partial sums of $\sum_{n=1}^{\infty} d_n$. Because the terms in the sequences $\{b_n\}_{n=1}^{\infty}$ and $\{c_n\}_{n=1}^{\infty}$ are all positive, a look at the definition of the sequences $\{y_n\}_{n=1}^{\infty}$ and $\{z_n\}_{n=1}^{\infty}$ shows that $s_1 = b_1$, and that the values of s_n are increasing until $s_{p_1} = y_1$; the values of s_n are then decreasing until $s_{p_1+q_1} = z_1$; the values of s_n are then increasing until $s_{p_1+q_1+p_2} = y_2$; the values of s_n are then decreasing until $s_{p_1+q_1+p_2+q_2} = z_2$, and so on. It follows that if $n = p_1 + q_1 + \cdots + q_k + i$ for some $k \in \mathbb{N} \cup \{0\}$ and $i \in \{1, \ldots, p_k\}$, then $z_{k-1} \le s_n < y_k$ (where for convenience we let $p_1 + q_1 + p_0 + q_0 = 0$ and $z_0 = 0$), and if $n = p_1 + q_1 + \cdots + p_k + i$ for some $k \in \mathbb{N} \cup \{0\}$ and $i \in \{1, \ldots, q_k\}$, then $z_k \le s_n < y_k$ (where for convenience we let $p_1 + q_1 + p_0 = 0$). Every s_n falls precisely into one of these two cases, and we can put the two cases together to deduce that $\min\{z_{k-1}, z_k\} \leq s_n \leq y_k$ for all $n \in \mathbb{N}$. We saw above that $\lim_{n\to\infty} y_n = x$ and $\lim_{n\to\infty} z_n = x$. It follows from Exercise 8.2.4 that $\lim_{n\to\infty} z_{n-1} = x$, and it follows from Exercise 8.2.8 that $\{\min\{z_{n-1}, z_n\}\}_{n=1}^{\infty}$ is convergent and $\lim_{n\to\infty} \min\{z_{n-1}, z_n\} = x$. Finally, the inequality $\min\{z_{k-1}, z_k\} \leq s_n \leq y_k$ for all $n \in \mathbb{N}$, together with the Squeeze Theorem for Sequences (Theorem 8.2.12), imply that $\lim_{n\to\infty} s_n = x$, from which we conclude that $\sum_{n=1}^{\infty} d_n = x$. □

Reflections

Other than the discussion of the Ratio Test, which is indispensable for working with power series, much of this section (for example, the discussion of products and rearrangements of series) is skipped in some introductory real analysis courses. Doing so is unfortunate, however, because this material contains the one truly surprising result, as well as the most substantial proof, in this chapter, namely, Theorem 9.4.15 and its proof. It is often these surprises that make mathematics as interesting as it is—it would be boring to study any subject where everything was already known on an intuitive basis beforehand. Moreover, if everything that seemed intuitively true turned out to be indeed true, we could simply do everything on an intuitive basis and be sure we were correct. It is the existence of counterintuitive facts that requires us to proceed with utmost rigor.

Exercises

Exercise 9.4.1. **[Used in Example 9.2.7 and Example 9.4.9.]** Prove that the series $\sum_{n=1}^{\infty} (-1)^{n-1} \frac{1}{\sqrt{n}}$ is conditionally convergent.

Exercise 9.4.2. **[Used in Exercise 9.4.9 and Exercise 10.4.10.]** Let $\sum_{n=1}^{\infty} a_n$ be a series in $\mathbb{R}$. Prove that if $\sum_{n=1}^{\infty} a_n$ is absolutely convergent, then $\left|\sum_{n=1}^{\infty} a_n\right| \leq \sum_{n=1}^{\infty} |a_n|$.

Exercise 9.4.3. Let $\sum_{n=1}^{\infty} a_n$ be a series in $\mathbb{R}$. Prove that if $\sum_{n=1}^{\infty} a_n$ is absolutely convergent, then $\sum_{n=1}^{\infty} (a_n)^2$ is convergent.

Exercise 9.4.4. Find an example of a series $\sum_{n=1}^{\infty} a_n$ such that $\sum_{n=1}^{\infty} a_n$ is convergent, but such that $\sum_{n=1}^{\infty} (a_n)^2$ is divergent.

Exercise 9.4.5. **[Used in Theorem 10.4.4.]** Let $a, r \in \mathbb{R}$. Suppose that $a \neq 0$. Prove that the series $\sum_{n=1}^{\infty} nar^{n-1}$ is convergent if and only if $|r| < 1$.

Exercise 9.4.6. **[Used in Theorem 9.4.15.]** Let $\sum_{n=1}^{\infty} b_n$ be a series in $\mathbb{R}$. Suppose that $\sum_{n=1}^{\infty} b_n$ has infinitely many non-zero terms. Let $\sum_{n=1}^{\infty} c_n$ be the series obtained by removing all terms that are zero from $\sum_{n=1}^{\infty} b_n$. Prove that $\sum_{n=1}^{\infty} b_n$ is absolutely convergent, conditionally convergent, divergent or diverges to infinity if and only if $\sum_{n=1}^{\infty} c_n$ is absolutely convergent, conditionally convergent, divergent or diverges to infinity, respectively. [Use Exercise 8.3.7.]

Exercise 9.4.7. Let $\sum_{n=1}^{\infty} a_n$ be a series in $\mathbb{R}$, and let $\{b_n\}_{n=1}^{\infty}$ be a sequence in $\mathbb{R}$.

Prove that if $\sum_{n=1}^{\infty} a_n$ is absolutely convergent and $\{b_n\}_{n=1}^{\infty}$ is bounded, then $\sum_{n=1}^{\infty} a_n b_n$ is absolutely convergent.

Exercise 9.4.8. **[Used in Theorem 9.4.4.]** Prove Theorem 9.4.4 (2).

Exercise 9.4.9. **[Used in Exercise 9.5.9.]** Let $\sum_{n=1}^{\infty} a_n$ be a series in $\mathbb{R}$. Suppose that $\sum_{n=1}^{\infty} a_n$ is absolutely convergent, and that for each $Q \in \mathbb{N}$, the series $\sum_{n=1}^{\infty} \frac{a_n}{Q^n}$ is convergent and $\sum_{n=1}^{\infty} \frac{a_n}{Q^n} = 0$. Prove that $a_n = 0$ for all $n \in \mathbb{N}$.

To prove the result, suppose to the contrary that $a_n \neq 0$ for some $n \in \mathbb{N}$. By the Well-Ordering Principle (Theorem 1.2.10, Axiom 1.4.4 or Theorem 2.4.6), there is a smallest $p \in \mathbb{N}$ such that $a_p \neq 0$. Let $\varepsilon > 0$. Prove that $|a_p| < \varepsilon$, and derive a contradiction. [Use Exercise 9.4.2.]

Exercise 9.4.10. **[Used in Section 9.4.]** Find an example of series $\sum_{n=0}^{\infty} a_n$ and $\sum_{n=0}^{\infty} b_n$ in $\mathbb{R}$ such that $\sum_{n=0}^{\infty} a_n$ is absolutely convergent and $\sum_{n=0}^{\infty} b_n$ is conditionally convergent, and that the Cauchy product of $\sum_{n=0}^{\infty} a_n$ and $\sum_{n=0}^{\infty} b_n$ is conditionally convergent.

Exercise 9.4.11. **[Used in Section 9.4.]** Let $\sum_{n=0}^{\infty} a_n$ and $\sum_{n=0}^{\infty} b_n$ be series in $\mathbb{R}$. Let $\sum_{n=0}^{\infty} e_n$ be the Cauchy product of $\sum_{n=0}^{\infty} a_n$ and $\sum_{n=0}^{\infty} b_n$. Suppose that $\sum_{n=0}^{\infty} a_n$ and $\sum_{n=0}^{\infty} b_n$ are absolutely convergent. Prove that $\sum_{n=0}^{\infty} e_n$ is absolutely convergent.

There is no need to construct a proof analogous to that of Theorem 9.4.7; in fact, make use of that theorem applied to $\sum_{n=0}^{\infty} |a_n|$ and $\sum_{n=0}^{\infty} |b_n|$. [Use Exercise 2.5.3.]

Exercise 9.4.12. It was seen in Example 9.4.2 (2) that the alternating harmonic series $\sum_{n=1}^{\infty} (-1)^{n-1} \frac{1}{n}$ is conditionally convergent. It follows from Theorem 9.4.15 (1) that there is a rearrangement of this series that has sum equal to 0. Using the method of the proof of that theorem, write out the first 15 terms of such a rearrangement. Is there a pattern to the rearrangement? It is not necessary to prove that the pattern holds.

Exercise 9.4.13. **[Used in Theorem 9.4.15.]** Prove Theorem 9.4.15 (2).

9.5 Power Series as Functions

Up till now in this chapter we have looked at series of numbers, for example

$$\frac{1}{1} + \frac{1}{2} + \frac{1}{3} + \cdots.$$

We now turn to series with a "variable" in them. There are many different types of such series, for example

$$\frac{1}{x} + \frac{2}{x^2} + \frac{3}{x^3} + \cdots \quad \text{and} \quad \sin \pi x + \sin 2\pi x + \sin 3\pi x + \cdots.$$

Although both of the above types of series are useful (the former in complex analysis and the latter in physics, among other uses), we will restrict our attention to the simplest and most widely used form of series with a "variable," which are power series, for example, the series

$$1+\frac{x}{2}+\frac{x^2}{3}+\frac{x^3}{4}+\frac{x^4}{5}+\frac{x^5}{6}+\cdots.$$

Power series are a generalization of polynomials.

As we stated when we defined polynomials in Section 2.5, there is actually no such thing as a "variable." The symbol "x" that we used in the above series simply represents an element of $\mathbb{R}$. We do not happen to know the numerical value of the element "x," but that does not make it any more variable than any other letter such as "c" that represents an element in $\mathbb{R}$. As such, the above series with "x" are also series with terms that are numbers. To avoid the issue of "variables," we will view power series as functions, analogously to the definition of polynomials in Definition 2.5.10.

Definition 9.5.1. Let $A \subseteq \mathbb{R}$ be a set, and let $f\colon A \to \mathbb{R}$ be a function. The function f is a **power series** if there is some $a \in \mathbb{R}$ and a sequence $\{c_n\}_{n=0}^{\infty}$ in $\mathbb{R}$ such that $\sum_{n=0}^{\infty} c_n(x-a)^n$ is convergent for all $x \in A$ and $f(x) = \sum_{n=0}^{\infty} c_n(x-a)^n$ for all $x \in A$. △

We will often consider power series where $a = 0$, that is, power series of the form $\sum_{n=0}^{\infty} c_n x^n$.

Although we defined power series as functions in Definition 9.5.1, in practice it is customary to define a power series simply by saying "let $\sum_{n=0}^{\infty} c_n(x-a)^n$ be a power series" without referring to the power series as a function, and in particular without stating a domain and a codomain. Fortunately, no real problem arises from this informal way of defining power series, and so for convenience we will use it, though with the following caveat. If we say "let $\sum_{n=0}^{\infty} c_n(x-a)^n$ be a power series," then we are to think of such an expression as defining a function, where the domain is assumed to be the set $A = \{x \in \mathbb{R} \mid \sum_{n=0}^{\infty} c_n(x-a)^n \text{ is convergent}\}$, which makes sense when we realize that "x" is not a "variable," and where the codomain is assumed to be $\mathbb{R}$. We could also take any subset of A to be the domain, but we will not do so unless otherwise stated.

We now examine a few examples of power series, each presented in the customary way stated above; for each of these power series we will find the domain. A very useful tool for finding the domain of a power series is the Ratio Test (Theorem 9.4.4), which was stated for series of numbers, but which holds for power series as well, because for each value of "x" in the domain, the power series is a series with terms that are numbers.

Example 9.5.2.

(1) We want to find the domain of the power series $\sum_{n=0}^{\infty} x^n$. This series has the form $\sum_{n=0}^{\infty} c_n(x-a)^n$, where $a = 0$, and where $c_n = 1$ for all $n \in \mathbb{N} \cup \{0\}$. Let $x \in \mathbb{R}$. The series $\sum_{n=0}^{\infty} x^n$ is a geometric series, as discussed in Example 9.2.4 (4), because the series has the form $\sum_{n=1}^{\infty} ar^{n-1}$, where $a = 1$ and $r = x$. We saw in Example 9.2.4 (4) that a geometric series $\sum_{n=1}^{\infty} ar^{n-1}$ is convergent if and only if $-1 < r < 1$, and that if $-1 < r < 1$ then $\sum_{n=1}^{\infty} ar^{n-1} = \frac{a}{1-r}$. Hence the power series $\sum_{n=0}^{\infty} x^n$ is convergent if and only if $-1 < x < 1$, and if $-1 < x < 1$ then $\sum_{n=0}^{\infty} x^n = \frac{1}{1-x}$. The domain of $\sum_{n=0}^{\infty} x^n$ is therefore the open interval $(-1,1)$.

(2) Recall the definition of factorials given in Example 2.5.12. We want to find the domain of the power series $\sum_{n=0}^{\infty} \frac{x^n}{n!}$. Let $x \in \mathbb{R}$. We will use the Ratio Test (Theorem 9.4.4) to evaluate whether or not the series $\sum_{n=1}^{\infty} \frac{x^n}{n!}$ is convergent. However, the hypotheses of the Ratio Test include the requirement that the terms of the series are never zero, so we need to consider two cases. First, suppose that $x = 0$. Then the power series is $\sum_{n=0}^{\infty} \frac{0^n}{n!}$, which is absolutely convergent. Second, suppose that $x \neq 0$. Using Example 8.2.4 (2), Exercise 8.2.4 and Theorem 8.2.9 (3), we see that

$$L = \lim_{n\to\infty} \left| \frac{\frac{x^{n+1}}{(n+1)!}}{\frac{x^n}{n!}} \right| = \lim_{n\to\infty} \frac{|x|}{n+1} = 0.$$

It follows from the Ratio Test that $\sum_{n=0}^{\infty} \frac{x^n}{n!}$ is absolutely convergent. Putting the two cases together, we deduce that the domain of $\sum_{n=0}^{\infty} \frac{x^n}{n!}$ is $\mathbb{R}$, and that the power series is absolutely convergent for all $x \in \mathbb{R}$.

As a consequence of the above calculation, we use the Divergence Test (Theorem 9.2.5) to deduce that $\lim_{n\to\infty} \frac{x^n}{n!} = 0$ for all $x \in \mathbb{R}$. This fact is used in the proofs of Theorem 7.4.5 and Corollary 10.4.15. Normally we use sequences to prove things about series, but in this instance it worked out nicely the other way around.

(3) We want to find the domain of the power series $\sum_{n=1}^{\infty} \frac{(x-3)^n}{n5^n}$. Let $x \in \mathbb{R}$. First, suppose that $x = 3$. Then the power series is $\sum_{n=1}^{\infty} \frac{0^n}{n5^n}$, which is absolutely convergent. Second, suppose that $x \neq 3$. Using Example 8.2.10 and Theorem 8.2.9 (3), we see that

$$L = \lim_{n\to\infty} \left| \frac{\frac{(x-3)^{n+1}}{(n+1)5^{n+1}}}{\frac{(x-3)^n}{n5^n}} \right| = \lim_{n\to\infty} \frac{n}{n+1} \frac{|x-3|}{5} = \frac{|x-3|}{5}.$$

It follows from the Ratio Test that $\sum_{n=1}^{\infty} \frac{(x-3)^n}{n5^n}$ is absolutely convergent when $\frac{|x-3|}{5} < 1$, and is divergent when $\frac{|x-3|}{5} > 1$. That is, the power series is absolutely convergent when $-2 < x < 8$, and is divergent when $x < -2$ and when $x > 8$. Hence the domain of $\sum_{n=1}^{\infty} \frac{(x-3)^n}{n5^n}$ contains the open interval $(-2, 8)$, and does not intersect the set $(-\infty, -2) \cup (8, \infty)$.

It remains to be verified whether $\sum_{n=1}^{\infty} \frac{(x-3)^n}{n5^n}$ is convergent or divergent for each of $x = -2$ and $x = 8$. The Ratio Test will not work for these two values of x, because they are the values of x that yield $L = 1$, which is precisely when the Ratio Test is inconclusive. We therefore examine the two cases $x = 8$ and $x = -2$ individually. If $x = 8$, the power series $\sum_{n=1}^{\infty} \frac{(x-3)^n}{n5^n}$ is $\sum_{n=1}^{\infty} \frac{1}{n}$, which is the harmonic series, and which is divergent by Example 9.2.4 (5). If $x = 2$, the power series $\sum_{n=1}^{\infty} \frac{(x-3)^n}{n5^n}$ is $\sum_{n=1}^{\infty} \frac{(-1)^n}{n}$, which is -1 times the alternating harmonic series, and which is convergent by Example 9.4.2 (2) and Theorem 8.2.9 (3).

Putting the various cases together, we see that the domain of $\sum_{n=1}^{\infty} \frac{(x-3)^n}{n5^n}$ is $[-2, 8)$.

(4) We want to find the domain of the power series $\sum_{n=1}^{\infty} n!x^n$. Let $x \in \mathbb{R}$. First, suppose that $x = 0$. Then the power series is $\sum_{n=1}^{\infty} n!0^n$, which is absolutely convergent.

Second, suppose that $x \neq 0$. Using Exercise 8.2.15 and Exercise 8.2.17 (3), we see that

$$L = \lim_{n\to\infty} \left| \frac{(n+1)!x^{n+1}}{n!x^n} \right| = \lim_{n\to\infty} n|x| = \infty.$$

If follows from the Ratio Test that $\sum_{n=1}^{\infty} n!x^n$ is divergent for all $x \in \mathbb{R} - \{0\}$. Therefore the domain of $\sum_{n=1}^{\infty} n!x^n$ is $\{0\}$. ◊

We observe that in all of the parts of Example 9.5.2, the domain of the series was an interval in $\mathbb{R}$, where we think of $\mathbb{R}$ as the interval $(-\infty,\infty)$, and we think of $\{0\}$ as the interval $[0,0]$. This observation turns out not to be coincidence. Rather remarkably, no matter how strange a power series might seem, its domain must be an interval, as we will prove in Theorem 9.5.4 below. We start with a lemma.

Lemma 9.5.3. *Let $\sum_{n=0}^{\infty} c_n x^n$ be a power series in $\mathbb{R}$, and let $p,q \in \mathbb{R}$. If $\sum_{n=0}^{\infty} c_n q^n$ is convergent, and if $|p| < |q|$, then $\sum_{n=0}^{\infty} c_n p^n$ is absolutely convergent.*

Proof. Suppose that $\sum_{n=0}^{\infty} c_n q^n$ is convergent, and that $|p| < |q|$. Observe that $q \neq 0$. Because $\sum_{n=0}^{\infty} c_n q^n$ is convergent, then by Exercise 9.2.2 we know that $\{c_n q^n\}_{n=0}^{\infty}$ is bounded. Hence there is some $M \in \mathbb{R}$ such that $|c_n q^n| \leq M$ for all $n \in \mathbb{N} \cup \{0\}$. Then

$$|c_n p^n| = |c_n q^n| \cdot \left|\frac{p}{q}\right|^n \leq M \cdot \left|\frac{p}{q}\right|^n$$

for all $n \in \mathbb{N} \cup \{0\}$. The series $\sum_{n=0}^{\infty} M \cdot \left|\frac{p}{q}\right|^n$ is a geometric series, as discussed in Example 9.2.4 (4), and by that exercise the series is convergent because $\left|\frac{p}{q}\right| < 1$. The Comparison Test (Theorem 9.3.2) now implies that $\sum_{n=0}^{\infty} |c_n p^n|$ is convergent, which means that $\sum_{n=0}^{\infty} c_n p^n$ is absolutely convergent. □

Theorem 9.5.4. *Let $\sum_{n=0}^{\infty} c_n (x-a)^n$ be a power series in $\mathbb{R}$. Then precisely one of the following holds:*

(1) the power series is absolutely convergent for all $x \in \mathbb{R}$;
(2) the power series is convergent only for $x = a$, where it is absolutely convergent;
(3) there is some $R \in (0,\infty)$ such that the power series is absolutely convergent for all $x \in (a-R, a+R)$, and the power series is divergent for all $x \in \mathbb{R} - [a-R, a+R]$.

Proof. It is evident that no two of the three options hold simultaneously.

Suppose that Options (1) and (2) do not hold. We will show that Option (3) holds. First, suppose that $a = 0$. Hence the power series is $\sum_{n=0}^{\infty} c_n x^n$. Because Options (1) and (2) do not hold, there are $r, p \in \mathbb{R}$ such that $\sum_{n=0}^{\infty} c_n r^n$ is not absolutely convergent, and such that $p \neq 0$ and $\sum_{n=0}^{\infty} c_n p^n$ is convergent. Let $q = |r| + 1$. Then $|q| > |r|$, and by Lemma 9.5.3 it must be the case that $\sum_{n=0}^{\infty} c_n q^n$ is divergent.

Let

$$S = \left\{ x \in \mathbb{R} \mid \sum_{n=0}^{\infty} c_n x^n \text{ is convergent} \right\}.$$

Then $p \in S$ and $q \notin S$. Hence $S \neq \emptyset$. Let $y \in \mathbb{R}$. Suppose that $|y| > |q|$. It follows from Lemma 9.5.3 that $\sum_{n=0}^{\infty} c_n y^n$ is divergent. Hence $S \subseteq [-|q|, |q|]$, and therefore S is bounded. The Least Upper Bound Property implies that S has a least upper bound. Let $R = \operatorname{lub} S$. Because $|q|$ is an upper bound of S, then $R \leq |q|$. We know that $\sum_{n=0}^{\infty} c_n p^n$ is convergent, and hence it follows from Lemma 9.5.3 that $\sum_{n=0}^{\infty} c_n \left(\frac{|p|}{2}\right)^n$ is convergent. Therefore $\frac{|p|}{2} \in S$. Because $p \neq 0$ we deduce that $R > 0$.

Let $x \in (-R, R)$. Then $|x| < R$, and hence $|x|$ is not an upper bound of S. Therefore there is some $z \in S$ such that $|x| < z$. Clearly $z > 0$, and hence $|x| < |z|$. By the definition of S we know $\sum_{n=0}^{\infty} c_n z^n$ is convergent. By Lemma 9.5.3 it follows that $\sum_{n=0}^{\infty} c_n x^n$ is absolutely convergent.

Now let $y \in \mathbb{R} - [-R, R]$. Then $|y| > R$. Suppose that $\sum_{n=0}^{\infty} c_n y^n$ is convergent. Let $t = \frac{R+|y|}{2}$. Then $0 < R < t < |y|$. It follows that $|t| < |y|$, and therefore by Lemma 9.5.3 we see that $\sum_{n=0}^{\infty} c_n t^n$ is convergent. Hence $t \in S$, which is a contradiction to the fact that $R = \operatorname{lub} S$. We conclude that $\sum_{n=0}^{\infty} c_n y^n$ is divergent.

We now turn to the general case, which involves the power series $\sum_{n=0}^{\infty} c_n (x-a)^n$, where a is not necessarily zero. By the previous case, we see that there is some $R \in (0, \infty)$ such that this series is absolutely convergent when $x - a \in (-R, R)$, and the series is divergent when $x - a \in \mathbb{R} - [-R, R]$. It follows immediately that the series is absolutely convergent for all $x \in (a - R, a + R)$, and the series is divergent for all $x \in \mathbb{R} - [a - R, a + R]$. □

It is important to note that in Case (3) of Theorem 9.5.4, nothing is said about convergence at $x = a - R$ and $x = a + R$. In fact, anything can happen at these points. That is, there are power series that converge precisely on $(a - R, a + R)$, there are power series that converge precisely on $(a - R, a + R]$, there are power series that converge precisely on $[a - R, a + R)$ and there are power series that converge precisely on $[a - R, a + R]$. What is remarkable in Theorem 9.5.4 is that in all possible cases, the set of real numbers for which the power series converges is precisely an interval of some form, as opposed to some more complicated type of subset of $\mathbb{R}$, which leads to the following definition.

Definition 9.5.5. Let $\sum_{n=0}^{\infty} c_n (x-a)^n$ be a power series in $\mathbb{R}$. The **interval of convergence** of the power series is the set $\{x \in \mathbb{R} \mid \sum_{n=0}^{\infty} c_n (x-a)^n \text{ is convergent}\}$. The **radius of convergence** of the power series is defined as follows:

(1) if the power series is absolutely convergent for all $x \in \mathbb{R}$, the radius of convergence is $R = \infty$;
(2) if the power series is convergent only for $x = a$, the radius of convergence is $R = 0$;
(3) if there is some $R \in (0, \infty)$ such that the power series is absolutely convergent for all $x \in (a - R, a + R)$, and the power series is divergent for all $x \in \mathbb{R} - [a - R, a + R]$, the radius of convergence is R. △

For convenience, we will consider the symbol ∞ to be greater than any real number, though ∞ should not be thought of as a real number, but simply as a useful symbol.

Suppose that a power series has radius of convergence R. If we say that "$R > 0$," we mean that R is a positive real number or $R = \infty$; if we say that "$R < \infty$," we mean that R is a positive real number or $R = 0$. Observe that $R > 0$ if and only if the interval of convergence is non-degenerate. Also, if $R = \infty$, then we will write $(a - R, a + R)$ to mean $(-\infty, \infty) = \mathbb{R}$, which allows us to avoid special cases.

Example 9.5.6. We restate the results of Example 9.5.2 in terms of radius of convergence and interval of convergence. For Part (1) of that example, the radius of convergence is 1 and the interval of convergence is $(-1, 1)$; for Part (2), the radius of convergence is ∞ and the interval of convergence is $(-\infty, \infty)$; for Part (3), the radius of convergence is 5 and the interval of convergence is $[-2, 8)$; for Part (4), the radius of convergence is 0 and the interval of convergence is $[0, 0]$. ◊

Theorem 9.5.4 tells us that in principle the domain of any power series is an interval, but it does not tell us how to find the interval of convergence. In practice, the most common way to find the interval of convergence of a power series is what we did in Example 9.5.2 (3), which is to use the Ratio Test (Theorem 9.4.4) to find the radius of convergence, and then to check the endpoints of the potential interval of convergence individually. However, as seen in the following example, there are some power series for which the Ratio Test does not work, though such power series are not encountered very often in practice.

Example 9.5.7. We want to find the radius of convergence and the interval of convergence of the power series

$$1 + x + \frac{x^2}{2^2} + \frac{x^3}{2^2} + \frac{x^4}{2^4} + \frac{x^5}{2^4} + \cdots.$$

We can write this power series as $\sum_{n=0}^{\infty} c_n x^n$, where

$$c_n = \begin{cases} \frac{1}{2^n}, & \text{if } n \text{ is even} \\ \frac{1}{2^{n-1}}, & \text{if } n \text{ is odd.} \end{cases}$$

Let $x \in \mathbb{R}$. If $x = 0$, then it is evident that $\sum_{n=0}^{\infty} c_n x^n$ is convergent. Now suppose that $x \neq 0$. If we attempt to use the Ratio Test (Theorem 9.4.4) to evaluate whether or not $\sum_{n=0}^{\infty} c_n x^n$ is convergent, we would see that

$$\left| \frac{c_{n+1} x^{n+1}}{c_n x^n} \right| = \begin{cases} |x|, & \text{if } n \text{ is even} \\ \frac{|x|}{4}, & \text{if } n \text{ is odd.} \end{cases}$$

By Exercise 8.2.2 we know that $\left\{ \left| \frac{c_{n+1} x^{n+1}}{c_n x^n} \right| \right\}_{n=1}^{\infty}$ is divergent, because we are assuming that $x \neq 0$, and hence we cannot use the Ratio Test to evaluate the convergence of $\sum_{n=0}^{\infty} c_n x^n$.

Fortunately, it is still possible to determine where $\sum_{n=0}^{\infty} c_n x^n$ is convergent, as follows. Let $x \in \mathbb{R}$. We first look at the series $\sum_{n=0}^{\infty} |c_n x^n| = \sum_{n=0}^{\infty} c_n |x|^n$, to test for absolute convergence. Observe that if $n \in \mathbb{N} \cup \{0\}$, then

$$\frac{|x|^n}{2^n} \le c_n|x|^n \le \frac{|x|^n}{2^{n-1}}.$$

The series $\sum_{n=0}^{\infty} \frac{|x|^n}{2^n}$ and $\sum_{n=0}^{\infty} \frac{|x|^n}{2^{n-1}}$ are both geometric series with $r = \frac{|x|}{2}$, as discussed in Example 9.2.4 (4), and therefore both of these series are convergent if and only if $\frac{|x|}{2} < 1$. Hence, these two series are convergent if and only if $x \in (-2,2)$. It then follows from the Comparison Test (Theorem 9.3.2) that the series $\sum_{n=0}^{\infty} |c_n x^n|$ is convergent if and only if $x \in (-2,2)$, which means that $\sum_{n=0}^{\infty} c_n x^n$ is absolutely convergent if and only if $x \in (-2,2)$. Theorem 9.5.4 then implies that the radius of convergence of $\sum_{n=0}^{\infty} c_n x^n$ is 2.

It remains to be verified whether $\sum_{n=0}^{\infty} c_n x^n$ is convergent or divergent for each of $x = -2$ and $x = 2$. If $x = 2$, the power series is

$$1+2+\frac{2^2}{2^2}+\frac{2^3}{2^2}+\frac{2^4}{2^4}+\frac{2^5}{2^4}+\cdots = 1+2+1+2+1+2+\cdots,$$

and this series is divergent by the Divergence Test (Theorem 9.2.5). If $x = -2$, the power series is

$$1+(-2)+\frac{2^2}{2^2}+\frac{-2^3}{2^2}+\frac{2^4}{2^4}+\frac{-2^5}{2^4}+\cdots = 1-2+1-2+1-2+\cdots,$$

and again this series is divergent.

We therefore see that the interval of convergence of $\sum_{n=0}^{\infty} c_n x^n$ is $(-2,2)$. ◊

The calculation of the interval of convergence in Example 9.5.7 was based upon the Comparison Test (Theorem 9.3.2), which we could use because of the lucky (and rare) resemblance between the power series and certain geometric series. It would be nice to have a more systematic way of computing the interval of convergence of power series in those cases where the Ratio Test (Theorem 9.4.4) fails. A commonly used method for such calculations involves the Root Test, which in turn uses the notion of the limit superior of a sequence. We will not discuss either of these topics, because we will not otherwise need them. See [Sto01, Section 2.5] for limit superior, [Sto01, Section 7.1] for the Root Test in general and [Sto01, Section 8.7] for the use of the Root Test for finding the radius of convergence of power series.

When working with polynomials, the reader has probably made use of the fact that the coefficients of a polynomial are unique; that is, if two polynomials are equal, then their coefficients are equal. We now state the analogous result for power series, as long as the power series have positive radius of convergence.

Theorem 9.5.8. *Let $\sum_{n=0}^{\infty} c_n(x-a)^n$ and $\sum_{n=0}^{\infty} d_n(x-a)^n$ be power series in $\mathbb{R}$. Suppose that each of $\sum_{n=0}^{\infty} c_n(x-a)^n$ and $\sum_{n=0}^{\infty} d_n(x-a)^n$ has a positive radius of convergence. Let I_c and I_d be the intervals of convergence of $\sum_{n=0}^{\infty} c_n(x-a)^n$ and $\sum_{n=0}^{\infty} d_n(x-a)^n$, respectively. If $\sum_{n=0}^{\infty} c_n(x-a)^n = \sum_{n=0}^{\infty} d_n(x-a)^n$ for all $x \in I_c \cap I_d$, then $c_n = d_n$ for all $n \in \mathbb{N} \cup \{0\}$.*

Proof. Left to the reader in Exercise 9.5.9. □

We conclude this section with two theorems about the convergence of sums, differences and products of power series. The first theorem follows immediately from Theorem 9.2.6, and we omit the proof.

Theorem 9.5.9. *Let $\sum_{n=0}^{\infty} c_n(x-a)^n$ and $\sum_{n=0}^{\infty} d_n(x-a)^n$ be power series in $\mathbb{R}$, let $k \in \mathbb{R}$ and let $r \in \mathbb{N}$. Let I_c and I_d be the intervals of convergence of $\sum_{n=0}^{\infty} c_n(x-a)^n$ and $\sum_{n=0}^{\infty} d_n(x-a)^n$, respectively.*

1. *$\sum_{n=0}^{\infty} (c_n+d_n)(x-a)^n$ is convergent for all $x \in I_c \cap I_d$, and*

$$\sum_{n=0}^{\infty} (c_n+d_n)(x-a)^n = \sum_{n=0}^{\infty} c_n(x-a)^n + \sum_{n=0}^{\infty} d_n(x-a)^n$$

   for all $x \in I_c \cap I_d$.

2. *$\sum_{n=0}^{\infty} (c_n-d_n)(x-a)^n$ is convergent for all $x \in I_c \cap I_d$, and*

$$\sum_{n=0}^{\infty} (c_n-d_n)(x-a)^n = \sum_{n=0}^{\infty} c_n(x-a)^n - \sum_{n=0}^{\infty} d_n(x-a)^n$$

   for all $x \in I_c \cap I_d$.

3. *$\sum_{n=0}^{\infty} kc_n(x-a)^{n+r}$ is convergent for all $x \in I_c$, and*

$$\sum_{n=0}^{\infty} kc_n(x-a)^{n+r} = k(x-a)^r \sum_{n=0}^{\infty} c_n(x-a)^n$$

   for all $x \in I_c$.

Our next theorem, concerning products of power series, is less straightforward than Theorem 9.5.9. Intuitively, it would be reasonable to expect that multiplying power series is analogous to multiplying polynomials. For example, we see that

$$\begin{aligned}&[a_0+a_1x+a_2x^2]\cdot[b_0+b_1x+b_2x^2]\\&= a_0b_0+(a_0b_1+a_1b_0)x+(a_0b_2+a_1b_1+a_2b_0)x^2+(a_1b_2+a_2b_1)x^3+a_2b_2x^4.\end{aligned}$$

The first three terms of this product are the typical ones, whereas the x^3 and x^4 terms have coefficients that do not quite fit the pattern, because the original polynomials stopped at x^2, in contrast to power series. The product of two power series works analogously, with one caveat. The issue is the difference between absolute convergence and conditional convergence. As we saw in Section 9.4, absolutely convergent series of numbers are better behaved than conditionally convergent series with such terms, and in particular the multiplication of absolutely convergent series is better behaved than the multiplication of conditionally convergent series. Hence, as we now see, the product of two power series behaves nicely if we restrict attention to the interval of convergence with the endpoints removed, to ensure absolute convergence.

The proof of the following theorem appears simple, but that is because we did the hard work for it in the proof of Theorem 9.4.7.

Theorem 9.5.10. *Let $\sum_{n=0}^{\infty} c_n(x-a)^n$ and $\sum_{n=0}^{\infty} d_n(x-a)^n$ be power series in $\mathbb{R}$. Let R_c and R_d be the radii of convergence of $\sum_{n=0}^{\infty} c_n(x-a)^n$ and $\sum_{n=0}^{\infty} d_n(x-a)^n$, respectively. Let $R = \min\{R_c, R_d\}$. For each $n \in \mathbb{N}$, let $e_n = \sum_{k=0}^{n} c_k d_{n-k}$. The power series $\sum_{n=0}^{\infty} e_n(x-a)^n$ is absolutely convergent for all $x \in (a-R, a+R)$, and $[\sum_{n=0}^{\infty} c_n(x-a)^n] \cdot [\sum_{n=0}^{\infty} d_n(x-a)^n] = \sum_{n=0}^{\infty} e_n(x-a)^n$ for all $x \in (a-R, a+R)$.*

Proof. If $R = 0$ then we are concerned only with convergence at $x = a$, and the result is trivial in that case. Now assume that $R > 0$. Let $y \in (a-R, a+R)$. Then $\sum_{n=0}^{\infty} c_n(y-a)^n$ and $\sum_{n=0}^{\infty} d_n(y-a)^n$ are absolutely convergent. Because $\sum_{k=0}^{n} c_k(y-a)^k \cdot d_{n-k}(y-a)^{n-k} = e_n(y-a)^n$ for all $n \in \mathbb{N} \cup \{0\}$, it follows from Theorem 9.4.7 that $\sum_{n=0}^{\infty} e_n(y-a)^n$ is convergent and that $[\sum_{n=0}^{\infty} c_n(y-a)^n] \cdot [\sum_{n=0}^{\infty} d_n(y-s)^n] = \sum_{n=0}^{\infty} e_n(y-a)^n$.

Because $\sum_{n=0}^{\infty} e_n(x-a)^n$ is convergent for all $x \in (a-R, a+R)$, then the radius of convergence of $\sum_{n=0}^{\infty} e_n(x-a)^n$ must be at least R. It follows that $\sum_{n=0}^{\infty} e_n(x-a)^n$ is absolutely convergent for all $x \in (a-R, a+R)$. □

We will see further results about power series, for example how to differentiate and integrate them, in Section 10.4; we have to wait until then because such facts about power series rely upon the notion of uniform convergence of series of functions, which is discussed in Section 10.3.

Reflections

This section on power series will probably leave the reader feeling unsatisfied—none of the really useful things one learns about power series, such as representing a function by its Taylor series, are touched upon in this section, and they are instead found in Section 10.4. That delay is due to the fact that in order to prove those useful results about power series, it will first be necessary to develop the notion of uniformly convergent sequences and series of functions; that topic will be discussed in Sections 10.2 and 10.3. However, in order to see which aspects of power series make use of uniform convergence and which do not, and in order to keep Section 10.4 to a manageable size, we have included in the present section as much about power series as could be said without making use of uniform convergence.

Exercises

Exercise 9.5.1. Let $\sum_{n=0}^{\infty} c_n x^n$ be a power series in $\mathbb{R}$. Suppose that this power series is convergent for $x = 9$ and is divergent for $x = -12$. For each of the following statements, say whether it is true, false, or not determined by the given information.

(1) The series is absolutely convergent for $x = 7$.
(2) The series is convergent for $x = -9$.
(3) The series is convergent for $x = 10$.
(4) The series is convergent for $x = 15$.

Exercise 9.5.2. Find the radius of convergence and interval of convergence of the following power series.

(1) $\sum_{n=0}^{\infty} \frac{(-1)^n x^n}{\sqrt{n}}$.
(2) $\sum_{n=0}^{\infty} \frac{x^n}{n^2}$.
(3) $\sum_{n=0}^{\infty} \frac{(x-7)^n}{3^n}$.

Exercise 9.5.3. **[Used in Example 10.4.11.]** Prove that the interval of convergence of the power series $\sum_{n=1}^{\infty} \frac{(-1)^{n-1}}{n}(x-1)^n$ is $(0,2]$.

Exercise 9.5.4. For each of the following intervals, either find an example of a power series that has the interval as its interval of convergence, or explain why the interval is not the interval of convergence of any power series.

(1) $(-3,3]$.
(2) $[3,11]$.
(3) $[4,\infty)$.

Exercise 9.5.5. Let $\sum_{n=0}^{\infty} c_n(x-a)^n$ be a power series in $\mathbb{R}$, and let $r \in \mathbb{N}$. Prove that $\sum_{n=0}^{\infty} c_n(x-a)^n$ and $\sum_{n=0}^{\infty} c_{n+r}(x-a)^n$ have the same interval of convergence. [Use Exercise 9.2.4.]

Exercise 9.5.6. Let $\sum_{n=0}^{\infty} c_n(x-a)^n$ and $\sum_{n=0}^{\infty} d_n(x-a)^n$ be power series in $\mathbb{R}$. Suppose that $|c_n| \leq |d_n|$ for all $n \in \mathbb{N} \cup \{0\}$.

(1) Prove that the radius of convergence of $\sum_{n=0}^{\infty} d_n(x-a)^n$ is less than or equal to the radius of convergence of $\sum_{n=0}^{\infty} c_n(x-a)^n$.
(2) Is it necessarily the case that the interval of convergence of $\sum_{n=0}^{\infty} d_n(x-a)^n$ is a subset of the interval of convergence of $\sum_{n=0}^{\infty} c_n(x-a)^n$? Give a proof or a counterexample.
(3) Find an example of power series $\sum_{n=0}^{\infty} b_n x^n$ and $\sum_{n=0}^{\infty} e_n x^n$ such that $|b_n| < |e_n|$ for all $n \in \mathbb{N} \cup \{0\}$, and that the two series have the same intervals of convergence.

Exercise 9.5.7. **[Used in Exercise 10.4.9.]** In this exercise we use the sequence of Fibonacci numbers, denoted $\{F_n\}_{n=1}^{\infty}$, which was defined in Example 8.4.10.

(1) Prove that the radius of convergence of the power series $\sum_{n=0}^{\infty} F_{n+1}x^n$ is at least $\frac{1}{2}$. [Use Exercise 8.4.12 (1).]
(2) Let I be the interval of convergence of $\sum_{n=0}^{\infty} F_{n+1}x^n$, and let $f\colon I \to \mathbb{R}$ be defined by $f(x) = \sum_{n=0}^{\infty} F_{n+1}x^n$ for all $x \in I$. Prove that $f(x) - xf(x) - x^2 f(x) = 1$ for all $x \in I$.

Exercise 9.5.8. Let $\sum_{n=0}^{\infty} c_n(x-a)^n$ be a power series in $\mathbb{R}$. Suppose that the sequence $\{c_n\}_{n=1}^{\infty}$ is bounded. Prove that $\sum_{n=0}^{\infty} c_n(x-a)^n$ has radius of convergence at least 1.

Exercise 9.5.9. **[Used in Theorem 9.5.8.]** Prove Theorem 9.5.8. [Use Exercise 9.4.9.]

9.6 Historical Remarks

The reader might view the study of series as separate from the core ideas of calculus such as derivatives and integrals, and it might therefore appear as if the history of series might be separate from the historical remarks seen in previous chapters. In fact, derivatives and integrals involve infinite processes (via infinitesimals or limits), as do series (which are infinite sums), and hence, as the reader will see, the history of series involves many of the same mathematicians encountered in the history of derivatives and integrals.

Ancient World

Arithmetic and geometric series were known in India in the 4th century BCE, and they were further explored in the period 300–1350 AD.

Although Euclid (c. 325–c. 265 BCE) did not consider infinite series, there is a formula (expressed in words) for the partial sums of geometric series in Proposition 35 of Book IX of the *Elements*. In modern notation, if $a_1 + a_2 + a_3 + \cdots$ is a geometric series, Euclid's formula is $\frac{a_2 - a_1}{a_1} = \frac{a_{n+1} - a_1}{s_n}$, where s_n is the nth partial sum; this formula is equivalent to the formula for the partial sum given in Example 9.2.4 (4). Also, Proposition 1 of Book X of the *Elements*, which is the basis for the method of exhaustion, is essentially a fact about certain partial sums of series becoming as close as desired to a given number, though Euclid did not think of it in that way.

Archimedes (287–212 BCE), in the *Quadrature of the Parabola*, showed a geometric equivalent of the sum $\sum_{n=0}^{\infty} \frac{1}{4^n} = \frac{4}{3}$; this sum of a geometric series might be considered the first sum of a series in Europe, though Archimedes did not think in terms of infinite sums.

Medieval Period

The scholars at Merton College at Oxford in the 14th century were led from their philosophical interests to various infinite series. One of them, Richard Swineshead, also known as Suiseth and Calculator, argued in *Liber calculationum* of around 1350 that $\sum_{n=1}^{\infty} \frac{n}{2^n} = 2$ (which is proved in Example 9.4.8, though with both sides of the equation divided by 2). Swineshead, whose approach was verbal and not rigorous, viewed this problem in terms of a body with uniform motion for half a period of time, then twice the velocity for the next quarter period, then three times the velocity for an eighth period, ad infinitum, and he argued that the total distance traveled would be four times the distance covered in the first half period. This sum of a series appears to be the first sum of a series in Europe since the geometric series summed by Archimedes mentioned above.

Nicole Oresme (1323–1382), around 1350, discussed geometric series, and also proved that the harmonic series is divergent by the proof commonly used today (which is given in Example 9.2.4 (5)). Oresme gave a clever picture proof of the sum of twice Swineshead's series, making use of geometric series. The methods of Swineshead and Oresme were further developed in the 15th and 16th centuries, though the significance of this study of series was not the particular results about series that were obtained, but the progress in the gradual acceptance of infinite processes by mathematicians. Series became important in mathematics only from the 17th century, as a way to avoid the method of exhaustion in area and volume calculations.

Renaissance

François Viète (1540–1603) showed an understanding of geometric series, as part of his work on the area of the circle in 1593. For example, he computed $\sum_{n=0}^{\infty} \frac{1}{4^n} = \frac{4}{3}$, and showed that this sum was related to Archimedes' geometric computation of the area of the parabola. Viète had the intuitive idea of the convergence of series being based upon the partial sums.

Seventeenth Century

A good early account of the summation of geometric series is due to Grégoire de Saint-Vincent (1584–1667) in *Opus geometricum quadraturae circuli et sectionum coni* of 1647. He had the intuitive idea of the convergence of series being based upon the partial sums, he used geometric series to provide a method for finding the areas of conics, and he used such series to resolve Zeno's Achilles paradox. Leibniz studied this work at the recommendation of Huygens.

Evangelista Torricelli (1608–1647) made use of geometric series in area problems. He proved, geometrically, that if $S = a + ar + ar^2 + \cdots$, then $\frac{S}{a} = \frac{a}{a-ar}$, which is equivalent to our familiar formula $S = \frac{a}{1-r}$, assuming that $a \neq 0$.

Pietro Mengoli (1626–1686), a student of Cavalieri who was influenced by Grégoire de Saint-Vincent and Torricelli, contributed to the development of series in 1650. He had two axioms about series, which stated in modern terms are: (1) if $\sum_{n=1}^{\infty} a_n = \infty$, then for any $x > 0$, there is some $N \in \mathbb{N}$ such that $\sum_{n=1}^{N} a_n > x$; and (2) if a series with positive terms is convergent, then any rearrangement is convergent, and has the same sum as the original series (which is proved in Theorem 9.4.12). Mengoli deduced, among other things, that if the sequence of partial sums of a series with positive terms is bounded then the series is convergent (which is proved in Lemma 9.3.1). Mengoli obtained sums such as $\sum_{n=1}^{\infty} \frac{1}{n(n+1)} = 1$ and $\sum_{n=1}^{\infty} \frac{1}{n(n+1)(n+2)} = \frac{1}{4}$. He also gave a different proof than Oresme that the harmonic series is divergent (this proof is given in Exercise 9.2.9).

John Wallis (1616–1703), in *Arithmetica infinitorum* of 1656, used clever (though far from rigorous, even to his contemporaries) analogical thinking to derive an infinite product formula for $\frac{\pi}{2}$. His method used Pascal's triangle, and hence the binomial coefficients, and this work helped inspire Newton to discover the binomial series, which he then used in his version of calculus. In *Mathesis universalis* of 1657, Wallis attempted to arithmetize Euclid's *Elements*. For example, he wrote the first n terms of a geometric series in the way we write them today as $A, AR, \ldots, AR^{n-1}$, and he proved that the sum of these terms is $\frac{VR-A}{R-1}$, where $V = AR^{n-1}$. Frans van Schooten (1615–1660) published another algebraic treatment of geometric series in 1657.

Newton and Leibniz

Prior to their work on calculus, both Isaac Newton (1643–1727) and Gottfried von Leibniz (1646–1716), whose approaches were quite different, first looked at some questions about series; Newton looked at the binomial series, which is an example of power series, whereas Leibniz initially looked at series of numbers rather than power series.

Leibniz's first accomplishment in mathematics was, in answer to a question suggested by Huygen's, showing that $\frac{1}{1}+\frac{1}{3}+\frac{1}{6}+\frac{1}{10}+\cdots+\frac{1}{\frac{n(n+1)}{2}}+\cdots=2$ (which is proved in Example 9.2.4 (3)). Leibniz' approach was to look at the series of differences of consecutive terms. It was pointed out to Leibniz that this result was essentially known already, though he had not been aware of that fact. We might speculate that if Leibniz had been aware that this series was known, then he would presumably not have looked at it, and then he might not have started thinking about the relation of sums and differences, and it was this relation that inspired his work on derivatives and integrals, which he viewed as a kind of difference and sum, respectively. In general, Leibniz handled series in a formal way, and was willing to work with divergent series, though he managed to obtain some correct results when using such series. Leibniz formulated and proved the Alternating Series Test, including the part about the difference between the partial sums and the limit, in an unpublished work of around 1676, and more completely in a letter to Johann Bernoulii in 1713.

Eighteenth Century

During the 17th and 18th centuries it was assumed that whatever held for finite sums also held for infinite sums. For example, because $\sum_{n=1}^{k}(a_n+b_n)=\sum_{n=1}^{k}a_n+\sum_{n=1}^{k}b_n$ for any finite sums, it was assumed that $\sum_{n=1}^{\infty}(a_n+b_n)=\sum_{n=1}^{\infty}a_n+\sum_{n=1}^{\infty}b_n$ automatically held for series. Similarly, because the terms of a finite sum can be rearranged without changing the value of the sum, it was assumed that the same held for series, though we now know that that is not true for conditionally convergent series (as seen in a special case in Example 9.4.11, and more generally in Theorem 9.4.15).

In the 18th century, the approach to series, and more generally to real analysis, changed in the direction of separating analysis from geometry. The approach to series became more formal, and questions of convergence were considered for the purpose of specific applications, but not in the derivation of general facts about series. As such, divergent series were allowed, though sometimes with strange results. On the one hand, Jakob Bernoulli (1654–1705) noted in 1696 that $1-1+1-1+1-\cdots$ had no sum. On the other hand, Daniel Bernoulli (1700–1782) said in 1724 that substituting $x=1$ into the expression $\frac{1}{1+x}=1-x+x^2-x^3+\cdots$ yielded $1-1+1-1+1-\cdots=\frac{1}{2}$, ignoring the interval of convergence. He also said that $1+2+4+8+16+\cdots=-1$ by arguing that if $S=1+2+4+8+16+\cdots$, then $S-1=2+4+8+16+\cdots$, so $\frac{S-1}{2}=1+2+4+8+16+\cdots$, which meant that $\frac{S-1}{2}=S$, which implied that $S=-1$; the flaw, of course, is the assumption that there is such a number S. Such issues were debated during this period.

However, even if some strange special cases were deduced using the formal approach to series, such special cases were not considered sufficient to invalidate the general results, which did not seem to be viewed as applicable to all cases. Indeed, the formal approach allowed for the discovery of some very useful results, for example the theory of generating functions of Pierre-Simon Laplace (1749–1827). Power series, and also trigonometric series, were employed in this period in the solution of differential equations. Some of the mathematicians who helped develop series in the mid-18th century were Abraham de Moivre (1667–1754), Daniel Bernoulli, James Stirling (1692–1770) and Colin Maclaurin (1698–1746).

The work of Leonhard Euler (1707–1783) on series was very influential. Among many other results, Euler proved that the sum of the alternating harmonic series is $\ln 2$ (which is proved in Exercise 9.3.10), and he stated and used, in informal terms, the idea that if a series $\sum_{n=1}^{\infty} a_n$ with positive terms is convergent, then $\sum_{n=p}^{k} a_n$ can be made as small as desired for large enough p and all $k \geq p$ (which is proved in Exercise 9.2.11). Similarly to his predecessors, Euler thought that series could be properly manipulated even if divergent.

Nineteenth Century

The formal approach to series of the 18th century was replaced early in the 19th century with an approach to series that avoided formal manipulation, and instead recognized the importance of convergence (especially for power series and other series of functions).

An early work that involved convergence was by Anastácio da Cunha (1744–1787) in 1782, who gave a somewhat imprecise statement that seems to be similar to the idea a series is convergent if its sequence of partial sums is a Cauchy sequence. However, da Cunha worked in isolation in Portugal, and his ideas were not widely noticed.

Joseph Fourier (1768–1830) helped bring attention to the importance of the convergence of series via his work on trigonometric series in *Théorie analytique de la chaleur*, submitted in 1807 but published only in 1822. Fourier used what we now call Fourier series to solve the partial differential equation that describes the propagation of heat, and he was very careful to specify for which values of x his series of functions represented the original function. The first person to publish a paper with this new approach to series was Carl Friedrich Gauss (1777–1855) in 1812, who said that one should restrict attention to where a power series converged. Gauss' idea of when $\sum_{n=0}^{\infty} c_n (x-a)^n$ converged was when $\lim_{n \to \infty} c_n = 0$, which we now know is not the correct approach, but nonetheless he correctly asserted that the issue of convergence was important. Bernard Bolzano (1781–1848) took this new approach in 1816, but similarly to his other ideas, his work on series was not widely known and did not influence subsequent developments.

The first systematic study of the convergence of series, and the first systematic exposition of this new approach, were due to Augustin Louis Cauchy (1789–1857) in the textbooks *Cours d'analyse a l'École Royal Polytechnique* of 1821 and *Résumé des leçons a l'École Royal Polytechnique* of 1823. Cauchy, who stressed the need for rigor in real analysis, explicitly stated that divergent series have no sums, and that power series should be used only on their intervals of convergence. Cauchy proved that power series converge precisely on intervals, and that the power series expansion of a function is unique (by a proof similar to Euler). Cauchy separated the question of convergence of series from finding their sums if convergent. Cauchy's definition of the convergence of a series (of numbers) was exactly the one we use today, though his definition of limit for the sequence of partial sums, though intuitively correct, was still informal. Cauchy used his definition to show, for example, that $\sum_{n=0}^{\infty} x^n$ is convergent if and only if $|x| < 1$. Cauchy asserted that a series is convergent if and only if, in modern terms, its sequence of partial sums is a Cauchy sequence; he did not seem to realize that he needed the Cauchy Completeness Theorem to prove this equivalence. Cauchy proved a number of results that had, in the 18th century, been thought true simply because they were the analogs for series of facts about finite sums, for example the fact that if $\sum_{n=1}^{\infty} a_n$ and $\sum_{n=1}^{\infty} b_n$ are convergent, then $\sum_{n=1}^{\infty} (a_n + b_n)$ is convergent and $\sum_{n=1}^{\infty} (a_n + b_n) = \sum_{n=1}^{\infty} a_n + \sum_{n=1}^{\infty} b_n$. Similarly to Gauss, Cauchy stressed the importance of convergence tests, because he felt that one needed to know if a series was convergent before using it, and he proved, among others, the Ratio Test (which had been used by Gauss), the Root Test, the Alternating Series Test and the Integral Test.

10

Sequences and Series of Functions

10.1 Introduction

In this chapter, our final one, we bring together a number of ideas that we saw in previous chapters. We learned about sequences of numbers in Chapter 8, and series of numbers in Chapter 9. In the present chapter we study sequences and series of functions, which we will then use to further our study of power series (which was commenced in Section 9.5), and finally to construct a continuous but nowhere differentiable function, a fitting point at which to conclude our study of introductory real analysis.

Whereas our primary interest here is in series of functions, of which power series are an example, our main technicalities, which concern the notion of uniform convergence, are to be found in Section 10.2, which treats sequences of functions.

10.2 Sequences of Functions

In Section 8.2 we saw the notion of a sequence of numbers, for example $\{3^n\}_{n=1}^{\infty}$. We now turn to the analogous notion of a sequence of functions, for example the sequence $\{x^n\}_{n=1}^{\infty}$, where for each $n \in \mathbb{N}$ we think of x^n as a shorthand notation for the function $f_n \colon \mathbb{R} \to \mathbb{R}$ defined by $f_n(x) = x^n$ for $x \in \mathbb{R}$.

Although we saw some of the basic properties of power series in Section 9.5, there are some substantial results about such series that we did not prove in that section, due to the lack of some important tools. In order to prove these results about power series, we need to change the way we view them. Consider, for example, the power series

$$1 + x + x^2 + x^3 + x^4 + x^5 + \cdots.$$

In our original discussion of power series in Section 9.5, we thought of such a power series as a collection of series of numbers, one series for each value of x. Now, by contrast, we want to view this power series as a single series, the terms of which form the sequence of functions $\{x^n\}_{n=1}^{\infty}$. This shift in point of view might seem minor at first, but it will turn out to be very important.

As was the case for series of numbers, in order to understand series of functions, we first need to learn about sequences of functions.

Definition 10.2.1. Let $A \subseteq \mathbb{R}$ be a non-empty set. Let $\mathcal{F}(A,\mathbb{R})$ denote the set of all functions $A \to \mathbb{R}$. A **sequence** of functions $A \to \mathbb{R}$ is a function $F: \mathbb{N} \to \mathcal{F}(A,\mathbb{R})$. If $F: \mathbb{N} \to \mathcal{F}(A,\mathbb{R})$ is a sequence of functions, and if $f_i = F(i)$ for all $i \in \mathbb{N}$, then we write $\{f_n: A \to \mathbb{R}\}_{n=1}^{\infty}$, or $\{f_n\}_{n=1}^{\infty}$ when there is no ambiguity, to denote the sequence of functions. Each function f_n, where $n \in \mathbb{N}$, is called a **term** of the sequence $\{f_n\}_{n=1}^{\infty}$. △

Suppose that $\{f_n\}_{n=1}^{\infty}$ is a sequence of functions $A \to \mathbb{R}$. What would it mean to say that $\{f_n\}_{n=1}^{\infty}$ converges to a function $f: A \to \mathbb{R}$? In contrast to the situation for sequences of numbers, where there is only one plausible notion of convergence, it turns out that for sequences of functions there is more than one possible way to define convergence, and the different definitions are not equivalent. We will discuss two approaches, called pointwise convergence and uniform convergence. The former type of convergence is the most simple definition of convergence, but the latter type of convergence is much more nicely behaved.

We start our discussion with pointwise convergence, where we simply examine the convergence of the sequence $\{f_n(x)\}_{n=1}^{\infty}$ for each $x \in A$ separately.

Definition 10.2.2. Let $A \subseteq \mathbb{R}$ be a non-empty set, let $\{f_n\}_{n=1}^{\infty}$ be a sequence of functions $A \to \mathbb{R}$ and let $f: A \to \mathbb{R}$ be a function. The sequence of functions $\{f_n\}_{n=1}^{\infty}$ **converges pointwise** to f if $\lim_{n\to\infty} f_n(x) = f(x)$ for all $x \in A$. If $\{f_n\}_{n=1}^{\infty}$ converges pointwise to some function $A \to \mathbb{R}$, we say that $\{f_n\}_{n=1}^{\infty}$ is **pointwise convergent**. △

Before seeing some examples of pointwise convergence, we need the following lemma, which shows, as expected, that if a sequence of functions converges pointwise to a function, then that function is unique.

Lemma 10.2.3. *Let $A \subseteq \mathbb{R}$ be a non-empty set, and let $\{f_n\}_{n=1}^{\infty}$ be a sequence of functions $A \to \mathbb{R}$. If $\{f_n\}_{n=1}^{\infty}$ converges pointwise to f for some function $f: A \to \mathbb{R}$, then f is unique.*

Proof. Suppose that $\{f_n\}_{n=1}^{\infty}$ converges pointwise to f for some function $f: A \to \mathbb{R}$. Let $x \in A$. Then $\lim_{n\to\infty} f_n(x) = f(x)$. By Lemma 8.2.3 we know that $\lim_{n\to\infty} f_n(x)$ is unique. Hence f is unique. □

Example 10.2.4.

(1) For each $n \in \mathbb{N}$, let $f_n: [-1,1] \to \mathbb{R}$ be defined by $f_n(x) = 1 - \frac{x^2}{n}$ for all $x \in [-1,1]$. Let $f: [-1,1] \to \mathbb{R}$ be defined by $f(x) = 1$ for all $x \in [-1,1]$. Let $y \in [-1,1]$. Using Example 8.2.4 (1) (2), and Theorem 8.2.9, we see that $\lim_{n\to\infty} f_n(y) = \lim_{n\to\infty} (1 - \frac{y^2}{n}) = 1 = f(y)$. It follows that $\{f_n\}_{n=1}^{\infty}$ converges pointwise to f.

(2) For each $n \in \mathbb{N}$, let $g_n\colon [0,1] \to \mathbb{R}$ be defined by $g_n(x) = x^n$ for all $x \in [0,1]$. Let $g\colon [0,1] \to \mathbb{R}$ be defined by

$$g(x) = \begin{cases} 0, & \text{if } x \in [0,1) \\ 1, & \text{if } x = 1. \end{cases}$$

Let $z \in [0,1]$. Then $\lim_{n\to\infty} g_n(z) = \lim_{n\to\infty} z^n$. It follows from Example 8.2.13 that if $z \in [0,1)$ then $\lim_{n\to\infty} g_n(z) = 0$, and that if $z = 1$ then $\lim_{n\to\infty} g_n(z) = 1$. Hence $\lim_{n\to\infty} g_n(z) = g(z)$. It follows that $\{g_n\}_{n=1}^{\infty}$ converges pointwise to g.

For each $n \in \mathbb{N}$, the function g_n is differentiable by Exercise 4.3.5 and Exercise 4.2.3 (5), and therefore g_n is continuous by Theorem 4.2.4. However, the function g is neither differentiable nor continuous. Hence, a sequence of continuous functions can converge pointwise to a discontinuous function, and a sequence of differentiable functions can converge pointwise to a non-differentiable function. Another way of expressing this problem with respect to continuity is by noting that $\lim_{x\to 1^-} \lim_{n\to\infty} g_n(x) \neq \lim_{n\to\infty} \lim_{x\to 1^-} g_n(x)$. In general in real analysis, it cannot always be assumed that the order of limits can be interchanged, unless there is a specific theorem that justifies doing so in the given situation. (Another situation involving sequences of functions in which limits cannot be interchanged is seen in Exercise 10.2.10.)

(3) For each $n \in \mathbb{N}$, let $h_n\colon [0,1] \to \mathbb{R}$ be defined by

$$h_n(x) = \begin{cases} 1, & \text{if } x \in \mathbb{Q} \cap [0,1] \text{ and } x = \frac{a}{b} \\ & \qquad \text{for some } a \in \mathbb{N} \cup \{0\} \text{ and } b \in \{1,\ldots,n\} \\ 0, & \text{otherwise.} \end{cases}$$

Let $h\colon [0,1] \to \mathbb{R}$ be defined by

$$h(x) = \begin{cases} 1, & \text{if } x \in \mathbb{Q} \cap [0,1] \\ 0, & \text{otherwise.} \end{cases}$$

The reader is asked in Exercise 10.2.1 (1) to prove that $\{h_n\}_{n=1}^{\infty}$ converges pointwise to h.

Let $n \in \mathbb{N}$. There are only finitely many numbers $x \in [0,1] \cap \mathbb{Q}$ such that $x = \frac{a}{b}$ for some $a \in \mathbb{N} \cup \{0\}$ and $b \in \{1,\ldots,n\}$. It follows that h_n is zero except at finitely many points, and hence h_n is integrable by Exercise 5.3.3 (2). On the other hand, the function h is not integrable, as was seen in Example 5.2.6 (3). Hence, a sequence of integrable functions can converge pointwise to a non-integrable function.

(4) For each $n \in \mathbb{N}$ such that $n \geq 2$, let $p_n\colon [0,1] \to \mathbb{R}$ be defined by

$$p_n(x) = \begin{cases} n^2 x, & \text{if } x \in [0, \frac{1}{n}) \\ 2n - n^2 x, & \text{if } x \in [\frac{1}{n}, \frac{2}{n}) \\ 0, & \text{if } x \in [\frac{2}{n}, 1]. \end{cases}$$

See Figure 10.2.1 for the graph of p_n. Let $p\colon [0,1] \to \mathbb{R}$ be defined by $p(x) = 0$ for all $x \in [0,1]$. The reader is asked in Exercise 10.2.1 (2) to prove that $\{p_n\}_{n=2}^{\infty}$ converges pointwise to p.

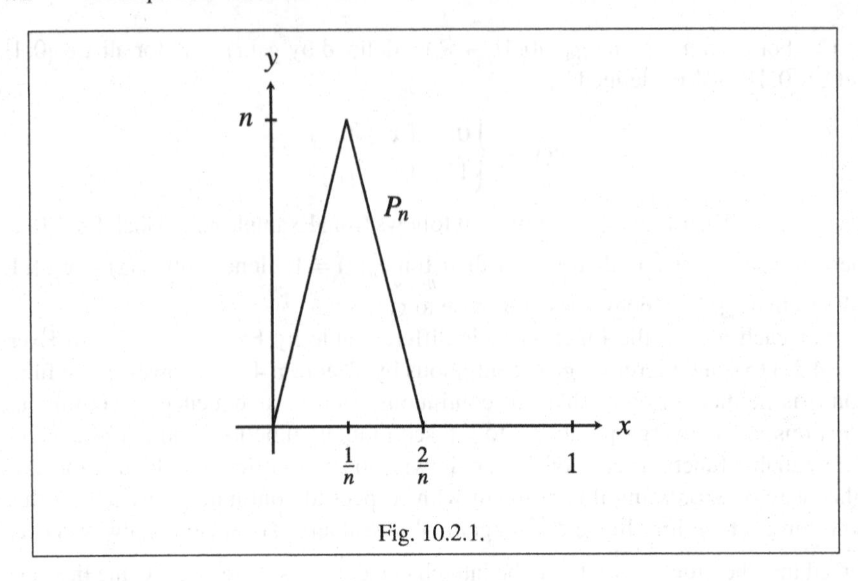

Fig. 10.2.1.

For each $n \in \mathbb{N}$, it can be verified that the function p_n is continuous, and hence it is integrable by Theorem 5.4.11, and that $\int_0^1 p_n(x)\,dx = 1$; the details are left to the reader. We know by Example 5.2.6 (1) that p is integrable and $\int_0^1 p(x)\,dx = 0$. Hence, even though $\{p_n\}_{n=1}^{\infty}$ converges pointwise to p, the sequence of numbers $\left\{\int_0^1 p_n(x)\,dx\right\}_{n=1}^{\infty}$ does not converge to $\int_0^1 p(x)\,dx$, which can be written as

$$\lim_{n\to\infty} \int_0^1 p_n(x)\,dx \neq \int_0^1 \lim_{n\to\infty} p_n(x)\,dx. \qquad \Diamond$$

As we saw in Example 10.2.4 (2)–(4), pointwise convergence of sequences of functions does not always behave nicely with respect to fundamental properties such as continuity, differentiability and integrability. The problem, it turns out, is not that continuity, differentiability and integrability are too tricky for convergence of sequences of functions to handle, but rather that pointwise convergence is not necessarily the best way to define convergence of sequences of functions.

To obtain a closer look at pointwise convergence, we can express the definition of pointwise convergence using logical symbols as

$$(\forall x \in A)(\forall \varepsilon > 0)(\exists N \in \mathbb{N})[(n \in \mathbb{N} \wedge n \geq N) \to |f_n(x) - f(x)| < \varepsilon].$$

As always, the order of the quantifiers is crucial. Because we are first given x and ε, and we then show that there exists an appropriate N, the choice of N can depend upon both x and ε. Intuitively, the reason the choice of N might depend upon x as well as ε is because the sequence $\{f_n(x)\}_{n=1}^{\infty}$ might converge to $f(x)$ "faster" for some values of x and "slower" for some other values of x. It turns out that it is precisely this difference in the rate of convergence (speaking informally) for different x that leads to the problems with the relation between pointwise convergence and continuity, differentiability and integrability. As we will see below, we can avoid this sort of problem if we consider a different type of convergence of sequences of functions, in which for a given ε, the same N works for all x. In logical symbols, we want this new type of convergence to be defined by

$$(\forall \varepsilon > 0)(\exists N \in \mathbb{N})(\forall x \in A)[(n \in \mathbb{N} \wedge n \geq N) \rightarrow |f_n(x) - f(x)| < \varepsilon].$$

We now state this definition properly.

Definition 10.2.5. Let $A \subseteq \mathbb{R}$ be a non-empty set, let $\{f_n\}_{n=1}^{\infty}$ be a sequence of functions $A \rightarrow \mathbb{R}$ and let $f: A \rightarrow \mathbb{R}$ be a function. The sequence of functions $\{f_n\}_{n=1}^{\infty}$ **converges uniformly** to f for each $\varepsilon > 0$, there is some $N \in \mathbb{N}$ such that $n \in \mathbb{N}$ and $n \geq N$ imply $|f_n(x) - f(x)| < \varepsilon$ for all $x \in A$. If $\{f_n\}_{n=1}^{\infty}$ converges uniformly to some function $A \rightarrow \mathbb{R}$, we say that $\{f_n\}_{n=1}^{\infty}$ is **uniformly convergent**. △

The intuitive idea of a sequence of functions $\{f_n\}_{n=1}^{\infty}$ converging uniformly to a function f is that for any $\varepsilon > 0$, if n is sufficiently large then the graph of f_n will be within a band that is distance ε above and below the graph of f; see Figure 10.2.2, where the solid line is the graph of f, the dashed lines indicate the edges of the band that is distance ε above and below the graph of f, and the dotted line is the graph of f_n.

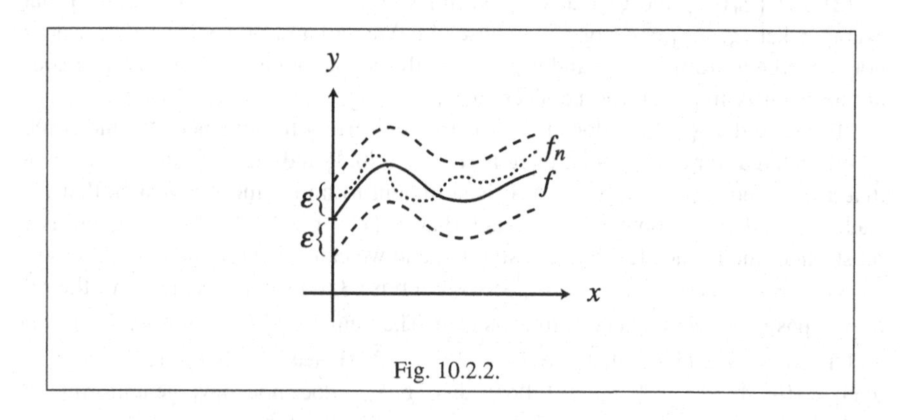

Fig. 10.2.2.

The first of the following two lemmas is derived immediately from the definition of pointwise convergence and uniform convergence, and the second is derived from the first together with Lemma 10.2.3; we omit the details.

Lemma 10.2.6. *Let $A \subseteq \mathbb{R}$ be a non-empty set, let $\{f_n\}_{n=1}^{\infty}$ be a sequence of functions $A \rightarrow \mathbb{R}$ and let $f: A \rightarrow \mathbb{R}$ be a function. If $\{f_n\}_{n=1}^{\infty}$ converges uniformly to f, then $\{f_n\}_{n=1}^{\infty}$ converges pointwise to f.*

Lemma 10.2.7. *Let $A \subseteq \mathbb{R}$ be a non-empty set, and let $\{f_n\}_{n=1}^{\infty}$ be a sequence of functions $A \rightarrow \mathbb{R}$. If $\{f_n\}_{n=1}^{\infty}$ converges uniformly to f for some function $f: A \rightarrow \mathbb{R}$, then f is unique.*

We now see in the second part of the following example that pointwise convergence does not necessarily imply uniform convergence.

Example 10.2.8.

(1) For each $n \in \mathbb{N}$, let $k_n\colon [-1,1] \to \mathbb{R}$ be defined by $k_n(x) = \frac{x^n}{n}$ for all $x \in [-1,1]$. Let $k\colon [-1,1] \to \mathbb{R}$ be defined by $k(x) = 0$ for all $x \in [0,1]$. We will prove that $\{k_n\}_{n=1}^{\infty}$ converges uniformly to k. Let $\varepsilon > 0$. By Corollary 2.6.8 (2) there is some $N \in \mathbb{N}$ such that $\frac{1}{N} < \varepsilon$. Suppose that $n \in \mathbb{N}$ and $n \geq N$. Let $x \in [-1,1]$. Then $|k_n(x) - k(x)| = |\frac{x^n}{n} - 0| \leq \frac{|x^n|}{n} \leq \frac{1}{n} \leq \frac{1}{N} < \varepsilon$.

Although we said that uniform convergence is supposed to work better than pointwise convergence with respect to continuity, differentiability and integrability, and indeed we will see that that is true, even uniform continuity is not perfect when it comes to differentiability.

Observe that k_n is differentiable for each $n \in \mathbb{N}$, and that k is differentiable. However, if $n \in \mathbb{N}$, then $k_n'(x) = x^{n-1}$ for all $x \in [0,1]$, where we use one-sided derivatives at the endpoints of the closed interval. Hence $\lim_{n\to\infty} k_n'(1) = \lim_{n\to\infty} 1 = 1 \neq 0 = k'(1)$. Moreover, using Example 8.2.4 (4) we see that $\lim_{n\to\infty} k_n'(-1) = \lim_{n\to\infty} (-1)^{n-1}$ does not exist. Hence, even though $\{k_n\}_{n=1}^{\infty}$ converges uniformly to k, the sequence $\{k_n'\}_{n=1}^{\infty}$ does not converge pointwise to k'.

(2) Let $\{g_n\}_{n=1}^{\infty}$ and g be as defined in Example 10.2.4 (2). It was seen in that example that $\{g_n\}_{n=1}^{\infty}$ converges pointwise to g. We will now prove that $\{g_n\}_{n=1}^{\infty}$ does not converge uniformly to g, and hence we will see that pointwise convergence does not necessarily imply uniform convergence.

To prove that $\{g_n\}_{n=1}^{\infty}$ does not converge uniformly to g we need to find some $\varepsilon > 0$ such that for each $N \in \mathbb{N}$, there is some $x \in [0,1]$ and there is some $n \in \mathbb{N}$ such that $n \geq N$ and $|g_n(x) - g(x)| \geq \varepsilon$. It is sufficient to find some $\varepsilon > 0$ such that for each $n \in \mathbb{N}$, there is some $x \in [0,1]$ such that $|g_n(x) - g(x)| \geq \varepsilon$ (this statement is a bit stronger than is needed, but it is simpler, and we can show it in the present case).

Let $\varepsilon = \frac{1}{2}$. Let $n \in \mathbb{N}$. Let $x = \frac{1}{\sqrt[n]{2}}$; see Exercise 3.5.6 for the existence of the nth root of positive real numbers. It follows from Theorem 7.2.12 (1), Exercise 7.2.11 (1) and Theorem 7.2.13 (2) that $1 = 1^{\frac{1}{n}} < 2^{\frac{1}{n}} = \sqrt[n]{2}$. Hence $x \in (0,1)$. Then $|g_n(x) - g(x)| = |(\frac{1}{\sqrt[n]{2}})^n - 0| = \frac{1}{2} \geq \varepsilon$. It follows that $\{g_n\}_{n=1}^{\infty}$ does not converge uniformly to g.

Graphically, the fact that $\{g_n\}_{n=1}^{\infty}$ does not converge uniformly to g is seen in Figure 10.2.3, where the graphs of g_2, g_4, g_6 and g_8 are shown. Observe that for any $\varepsilon \in (0,1)$, the band that is distance ε above and below the graph of g on the half-open interval $[0,1)$ will not contain the graph of any of the functions g_n, no matter how large n is. ◊

The following theorem is the analog of the Cauchy Completeness Theorem (Corollary 8.3.16) for uniform convergence of sequences of functions.

Theorem 10.2.9 (Cauchy Criterion for Uniform Convergence). *Let $A \subseteq \mathbb{R}$ be a non-empty set, and let $\{f_n\}_{n=1}^{\infty}$ be a sequence of functions $A \to \mathbb{R}$. Then $\{f_n\}_{n=1}^{\infty}$ is* *uniformly convergent if and only if for each $\varepsilon > 0$, there is some $N \in \mathbb{N}$ such that $n, m \in \mathbb{N}$ and $n, m \geq N$ imply $|f_n(x) - f_m(x)| < \varepsilon$ for all $x \in A$.*

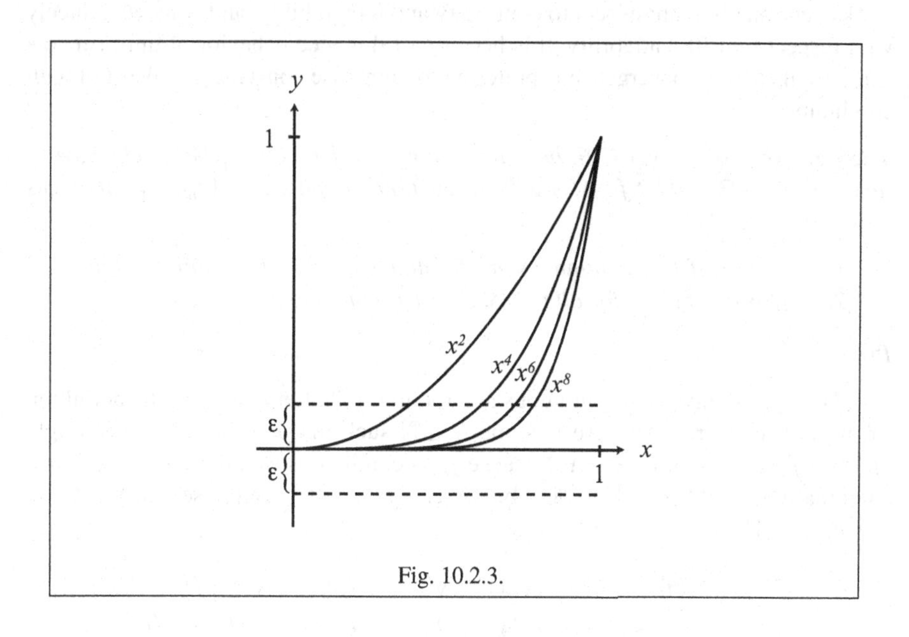

Fig. 10.2.3.

Proof. First, suppose that $\{f_n\}_{n=1}^{\infty}$ is uniformly convergent. Then there is a function $f: A \to \mathbb{R}$ such that $\{f_n\}_{n=1}^{\infty}$ converges uniformly to f. Let $\varepsilon > 0$. By the definition of uniform convergence, there is some $N \in \mathbb{N}$ such that $r \in \mathbb{N}$ and $r \geq N$ imply $|f_r(x) - f(x)| < \frac{\varepsilon}{2}$ for all $x \in A$. Suppose that $n, m \in \mathbb{N}$ and $n, m \geq N$. Let $x \in A$. Then

$$\begin{aligned}|f_n(x) - f_m(x)| &= |f_n(x) - f(x) + f(x) - f_m(x)| \\ &\leq |f_n(x) - f(x)| + |f(x) - f_m(x)| < \frac{\varepsilon}{2} + \frac{\varepsilon}{2} = \varepsilon.\end{aligned}$$

Second, suppose that for each $\varepsilon > 0$, there is some $M \in \mathbb{N}$ such that $n, m \in \mathbb{N}$ and $n, m \geq M$ imply $|f_n(x) - f_m(x)| < \varepsilon$ for all $x \in A$. Let $g: A \to \mathbb{R}$ be defined as follows. Let $y \in A$. The hypothesis on $\{f_n\}_{n=1}^{\infty}$ implies that the sequence $\{f_n(y)\}_{n=1}^{\infty}$ is a Cauchy sequence. The Cauchy Completeness Theorem (Corollary 8.3.16) implies that $\{f_n(y)\}_{n=1}^{\infty}$ is convergent. Let $g(y) = \lim_{n\to\infty} f_n(y)$.

Let $\varepsilon > 0$. There is some $M \in \mathbb{N}$ such that $n, m \in \mathbb{N}$ and $n, m \geq M$ imply $|f_n(x) - f_m(x)| < \frac{\varepsilon}{2}$ for all $x \in A$. Suppose that $k \in \mathbb{N}$ and $k \geq M$. Let $z \in A$. We know that $g(z) = \lim_{n\to\infty} f_n(z)$, and hence there is some $P \in \mathbb{N}$ such that $m \in \mathbb{N}$ and $m \geq P$ imply $|f_m(z) - g(z)| < \frac{\varepsilon}{2}$. Let $Q = \max\{M, P\}$. Then

$$\begin{aligned}|f_k(z) - g(z)| &= |f_k(z) - f_Q(z) + f_Q(z) - g(z)| \leq |f_k(z) - f_Q(z)| + |f_Q(z) - g(z)| \\ &< \frac{\varepsilon}{2} + \frac{\varepsilon}{2} = \varepsilon.\end{aligned}$$

We conclude that $\{f_n\}_{n=1}^{\infty}$ converges uniformly to g. □

We now show that in contrast to pointwise convergence, uniform convergence works very nicely with respect to continuity and integrability, and somewhat nicely with respect to differentiability. It is because of this nice behavior of uniform convergence that such convergence is preferred to pointwise convergence. We start with continuity.

Theorem 10.2.10. *Let $A \subseteq \mathbb{R}$ be a non-empty set, let $\{f_n\}_{n=1}^{\infty}$ be a sequence of functions $A \to \mathbb{R}$ and let $f\colon A \to \mathbb{R}$ be a function. Suppose that $\{f_n\}_{n=1}^{\infty}$ converges uniformly to f.*

1. *Let $c \in A$. If f_n is continuous at c for all $n \in \mathbb{N}$, then f is continuous at c.*
2. *If f_n is continuous for all $n \in \mathbb{N}$, then f is continuous.*

Proof.

(1) Suppose that f_n is continuous at c for all $n \in \mathbb{N}$. Let $\varepsilon > 0$. By the definition of uniform convergence there is some $N \in \mathbb{N}$ such that $n \in \mathbb{N}$ and $n \geq N$ imply $|f_n(x) - f(x)| < \frac{\varepsilon}{3}$ for all $x \in A$. Because f_N is continuous at c, there is some $\delta > 0$ such that $x \in A$ and $|x - c| < \delta$ imply $|f_N(x) - f_N(c)| < \frac{\varepsilon}{3}$. Suppose that $x \in A$ and $|x - c| < \delta$. Then

$$\begin{aligned}|f(x) - f(c)| &= |f(x) - f_N(x) + f_N(x) - f_N(c) + f_N(c) - f(c)| \\ &\leq |f(x) - f_N(x)| + |f_N(x) - f_N(c)| + |f_N(c) - f(c)| \\ &< \frac{\varepsilon}{3} + \frac{\varepsilon}{3} + \frac{\varepsilon}{3} = \varepsilon.\end{aligned}$$

(2) This part of the theorem follows immediately from Part (1) of this theorem. □

Although uniform convergence is used in Theorem 10.2.10 to prove that f is continuous, it is more than is minimally needed to prove continuity at a single number c in the domain of f; a condition equivalent to the continuity of f at a single number is given in Exercise 10.2.11.

We now turn to integrability, which works as nicely with respect to uniform convergence as does continuity.

Theorem 10.2.11. *Let $[a,b] \subseteq \mathbb{R}$ be a non-degenerate closed bounded interval, let $\{f_n\}_{n=1}^{\infty}$ be a sequence of functions $[a,b] \to \mathbb{R}$ and let $f\colon [a,b] \to \mathbb{R}$ be a function. Suppose that $\{f_n\}_{n=1}^{\infty}$ converges uniformly to f. If f_n is integrable for all $n \in \mathbb{N}$, then f is integrable and*

$$\lim_{n\to\infty} \int_a^b f_n(x)\,dx = \int_a^b f(x)\,dx.$$

Proof. Suppose that f_n is integrable for all $n \in \mathbb{N}$. We first show that $\left\{\int_a^b f_n(x)\,dx\right\}_{n=1}^{\infty}$ is a Cauchy sequence. Let $\varepsilon > 0$. Because $\{f_n\}_{n=1}^{\infty}$ converges uniformly to f, it follows from the Cauchy Criterion for Uniform Convergence (Theorem 10.2.9) that there is some $N \in \mathbb{N}$ such that $n, m \in \mathbb{N}$ and $n, m \geq N$ imply $|f_n(x) - f_m(x)| < \frac{\varepsilon}{2(b-a)}$ for all $x \in [a,b]$. Suppose that $n, m \in \mathbb{N}$ and $n, m \geq N$. Then by Theorem 5.3.1 (2), Theorem 5.5.5 and Theorem 5.3.2 (3) we see that

$$\left| \int_a^b f_n(x)\,dx - \int_a^b f_m(x)\,dx \right| = \left| \int_a^b [f_n(x) - f_m(x)]\,dx \right| \leq \int_a^b |f_n(x) - f_m(x)|\,dx$$
$$\leq \frac{\varepsilon}{2(b-a)} \cdot (b-a) = \frac{\varepsilon}{2} < \varepsilon.$$

It follows that $\left\{ \int_a^b f_n(x)\,dx \right\}_{n=1}^{\infty}$ is a Cauchy sequence.

The Cauchy Completeness Theorem (Corollary 8.3.16) implies that the sequence $\left\{ \int_a^b f_n(x)\,dx \right\}_{n=1}^{\infty}$ is convergent. Let $L = \lim_{n\to\infty} \int_a^b f_n(x)\,dx$. We now show that f is integrable and $\int_a^b f(x)\,dx = L$, which will complete the proof.

Let $\eta > 0$. Because $\{f_n\}_{n=1}^{\infty}$ converges uniformly to f, there is some $M \in \mathbb{N}$ such that $n \in \mathbb{N}$ and $n \geq M$ imply $|f_n(x) - f(x)| < \frac{\eta}{3(b-a)}$ for all $x \in [a,b]$. Because $\lim_{n\to\infty} \int_a^b f_n(x)\,dx = L$, there is some $K \in \mathbb{N}$ such that $n \in \mathbb{N}$ and $n \geq K$ imply $|\int_a^b f_n(x)\,dx - L| < \frac{\eta}{3}$. Let $J = \max\{M, K\}$. Because f_J is integrable, there is some $\delta > 0$ such that if P is a partition of $[a,b]$ with $\|P\| < \delta$, and if T is a representative set of P, then $|S(f_J, P, T) - \int_a^b f_J(x)\,dx| < \frac{\eta}{3}$.

Let R be a partition of $[a,b]$ with $\|R\| < \delta$, and let V be a representative set of P. By Exercise 5.2.2, we see that

$$|S(f_J, R, V) - S(f, R, V)| \leq \frac{\eta}{3(b-a)} \cdot (b-a) = \frac{\eta}{3}.$$

Then

$$\begin{aligned}
&|S(f,R,V) - L| \\
&= \left| S(f,R,V) - S(f_J,R,V) + S(f_J,R,V) - \int_a^b f_J(x)\,dx + \int_a^b f_J(x)\,dx - L \right| \\
&\leq |S(f,R,V) - S(f_J,R,V)| + \left| S(f_J,R,V) - \int_a^b f_J(x)\,dx \right| + \left| \int_a^b f_J(x)\,dx - L \right| \\
&< \frac{\eta}{3} + \frac{\eta}{3} + \frac{\eta}{3} = \varepsilon.
\end{aligned}$$

Therefore f is integrable and $\int_a^b f(x)\,dx = L$. □

The displayed equation in Theorem 10.2.11 can be written as

$$\lim_{n\to\infty} \int_a^b f_n(x)\,dx = \int_a^b \lim_{n\to\infty} f_n(x)\,dx,$$

which is often stated informally by saying that "the limit passes through the integral sign." We need to proceed with caution in such situations, however, because we saw in Example 10.2.4 (4) that the limit does not pass through the integral sign in general, and does so only under suitable hypotheses (namely, uniform convergence).

As seen in Example 10.2.8 (1), differentiability does not work as nicely with respect to uniform convergence as do continuity and integrability. For that reason, the following theorem has stronger hypotheses than Theorem 10.2.10 and Theorem 10.2.11.

Theorem 10.2.12. *Let $I \subseteq \mathbb{R}$ be a non-degenerate open bounded interval, and let $\{f_n\}_{n=1}^{\infty}$ be a sequence of functions $I \to \mathbb{R}$. Suppose that f_n is differentiable for all $n \in \mathbb{N}$, that $\{f_n'\}_{n=1}^{\infty}$ is uniformly convergent and that $\{f_n(c)\}_{n=1}^{\infty}$ is convergent for some $c \in I$. Then there is a function $f\colon I \to \mathbb{R}$ such that f is differentiable, that $\{f_n\}_{n=1}^{\infty}$ converges uniformly to f and that $\{f_n'\}_{n=1}^{\infty}$ converges uniformly to f'.*

Proof. We start with a preliminary step. Let $n, m \in \mathbb{N}$, and let $x, y \in I$. Suppose that $x \neq y$. Because f_n and f_m are differentiable on I, then Theorem 4.3.1 (2) implies that $f_n - f_m$ is differentiable on I, and Theorem 4.2.4 then implies that $f_n - f_m$ is continuous on I. We can therefore apply the Mean Value Theorem (Theorem 4.4.4) to $f_n - f_m$ restricted to the closed bounded interval from x to y (we do not know which of x or y is larger, but it does not matter). Hence there is some d strictly between x and y such that

$$f_n'(d) - f_m'(d) = \frac{[f_n(x) - f_m(x)] - [f_n(y) - f_m(y)]}{x - y}. \tag{10.2.1}$$

We now show that $\{f_n\}_{n=1}^{\infty}$ is uniformly convergent. Let $I = (a, b)$. Let $\varepsilon > 0$. Because $\{f_n(c)\}_{n=1}^{\infty}$ is convergent, then by the Cauchy Completeness Theorem (Corollary 8.3.16) we know that $\{f_n(c)\}_{n=1}^{\infty}$ is a Cauchy sequence. Hence there is some $N \in \mathbb{N}$ such that $n, m \in \mathbb{N}$ and $n, m \geq N$ imply $|f_n(c) - f_m(c)| < \frac{\varepsilon}{2}$. Because $\{f_n'\}_{n=1}^{\infty}$ is uniformly convergent, then by the Cauchy Criterion for Uniform Convergence (Theorem 10.2.9) there is some $M \in \mathbb{N}$ such that $n, m \in \mathbb{N}$ and $n, m \geq M$ imply $|f_n'(x) - f_m'(x)| < \frac{\varepsilon}{2(b-a)}$ for all $x \in I$. Let $P = \max\{N, M\}$.

Suppose that $n, m \in \mathbb{N}$ and $n, m \geq P$. Let $z \in I$. There are two cases. First, suppose that $z = c$. Then $|f_n(z) - f_m(z)| = |f_n(c) - f_m(c)| < \frac{\varepsilon}{2} < \varepsilon$. Second, suppose that $z \neq c$. By the preliminary step, using $x = z$ and $y = c$, there is some q strictly between z and c such that Equation 10.2.1 holds, using $d = q$. Then

$$\begin{aligned} |f_n(z) - f_m(z)| &= |[f_n(z) - f_m(z)] - [f_n(c) - f_m(c)] + [f_n(c) - f_m(c)]| \\ &\leq |[f_n(z) - f_m(z)] - [f_n(c) - f_m(c)]| + |f_n(c) - f_m(c)| \\ &= |f_n'(q) - f_m'(q)| \cdot |z - c| + |f_n(c) - f_m(c)| \\ &< \frac{\varepsilon}{2(b-a)} \cdot (b-a) + \frac{\varepsilon}{2} = \varepsilon. \end{aligned}$$

Putting the two cases together, we can apply the Cauchy Criterion for Uniform Convergence to $\{f_n\}_{n=1}^{\infty}$, and we deduce that $\{f_n\}_{n=1}^{\infty}$ is uniformly convergent. Hence there is a function $f\colon I \to \mathbb{R}$ such that $\{f_n\}_{n=1}^{\infty}$ converges uniformly to f.

Let $p \in I$. We will show that f is differentiable at p, and that $\lim_{n\to\infty} f_n'(p) = f'(p)$. First, however, we need another preliminary step. For each $n \in \mathbb{N}$, let $g_n\colon I \to \mathbb{R}$ be defined by

$$g_n(x) = \begin{cases} \frac{f_n(x)-f_n(p)}{x-p}, & \text{if } x \in I - \{p\} \\ f_n'(p), & \text{if } x = p. \end{cases}$$

We have therefore defined a sequence of functions $\{g_n\}_{n=1}^{\infty}$. We will show that $\{g_n\}_{n=1}^{\infty}$ is uniformly convergent.

Let $\eta > 0$. Because $\{f_n'\}_{n=1}^{\infty}$ is uniformly convergent, then by the Cauchy Criterion for Uniform Convergence there is some $Q \in \mathbb{N}$ such that $n, m \in \mathbb{N}$ and $n, m \geq Q$ imply $|f_n'(x) - f_m'(x)| < \eta$ for all $x \in I$. Suppose that $n, m \in \mathbb{N}$ and $n, m \geq Q$. Let $w \in I$. There are two cases. First, suppose that $w = p$. Then $|g_n(w) - g_m(w)| = |g_n(p) - g_m(p)| = |f_n'(p) - f_m'(p)| < \eta$. Second, suppose that $w \neq p$. By the first preliminary step, using $x = w$ and $y = p$, there is some r strictly between w and p such that Equation 10.2.1 holds, using $d = r$. Then

$$\begin{aligned} |g_n(w) - g_m(w)| &= \left| \frac{f_n(w) - f_n(p)}{w - p} - \frac{f_m(w) - f_m(p)}{w - p} \right| \\ &= \left| \frac{[f_n(w) - f_m(w)] - [f_n(p) - f_m(p)]}{w - p} \right| = |f_n'(r) - f_m'(r)| < \eta. \end{aligned}$$

Putting the two cases together, we can apply the Cauchy Criterion for Uniform Convergence to $\{g_n\}_{n=1}^{\infty}$, and we deduce that $\{g_n\}_{n=1}^{\infty}$ is uniformly convergent. Hence there is a function $g\colon I \to \mathbb{R}$ such that $\{g_n\}_{n=1}^{\infty}$ converges uniformly to g.

Let $k \in \mathbb{N}$. Because f_k is differentiable at p, we know that

$$\lim_{x\to p} \frac{f_k(x) - f_k(p)}{x - p}$$

exists and equals $f_k'(p)$. It therefore follows from Lemma 3.3.2 that g_k is continuous at p. We deduce from Theorem 10.2.10 (1) that g is continuous at p. It follows from Lemma 3.3.2 that $\lim_{x\to p} g(x) = g(p)$.

Because $\{f_n'\}_{n=1}^{\infty}$ is uniformly convergent, then by Lemma 10.2.6 we know that $\{f_n'\}_{n=1}^{\infty}$ is pointwise convergent, and hence $\{f_n'(p)\}_{n=1}^{\infty}$ is convergent. It then follows from Theorem 8.2.9 that

$$\begin{aligned} \lim_{x\to p} \frac{f(x) - f(p)}{x - p} &= \lim_{x\to p} \lim_{n\to\infty} \frac{f_n(x) - f_n(p)}{x - p} = \lim_{x\to p} \lim_{n\to\infty} g_n(x) = \lim_{x\to p} g(x) \\ &= g(p) = \lim_{n\to\infty} g_n(p) = \lim_{n\to\infty} f_n'(p). \end{aligned}$$

We deduce that f is differentiable at p and $f'(p) = \lim_{n\to\infty} f_n'(p)$.

It follows that f is differentiable, and that $\{f_n'\}_{n=1}^{\infty}$ converges pointwise to f'. Because $\{f_n'\}_{n=1}^{\infty}$ is uniformly convergent by hypothesis, we then use Lemma 10.2.3 and Lemma 10.2.6 to conclude that $\{f_n'\}_{n=1}^{\infty}$ converges uniformly to f'. □

The conclusion of Theorem 10.2.12 can be summarized by writing

$$\lim_{n\to\infty} f_n'(x) = \left[\lim_{n\to\infty} f_n(x)\right]'$$

for all $x \in I$, though, as before, we need to proceed with caution when writing such expressions, because they hold only under certain hypotheses.

The reader might be puzzled about the hypothesis that $\{f_n(c)\}_{n=1}^{\infty}$ is convergent for some $c \in I$ in the statement of Theorem 10.2.12; it seems strange to require such a hypothesis for only one number in I. This hypothesis cannot be dropped, however, as the reader is asked to show in Exercise 10.2.8.

Reflections

The material in this section might appear upon first encounter to be somewhat dry and technical. However, this section contains some examples of counterintuitive behavior, namely, the existence of a sequence of continuous functions that converges (pointwise) to a discontinuous function (and similarly for integrable or differentiable functions), and seeing such counterintuitive behavior is, in addition to our use of rigorous proofs, one of the features that distinguishes real analysis from introductory calculus. Moreover, the distinction between pointwise convergence and uniform convergence of sequences of functions, and the better behavior of the latter, is crucial for the remaining sections of this chapter, which include discussion of power series and a continuous but nowhere differentiable function.

Exercises

Exercise 10.2.1. [Used in Example 10.2.4.]

(1) Let $\{h_n\}_{n=1}^{\infty}$ and h be the functions given in Example 10.2.4 (3). Prove that $\{h_n\}_{n=1}^{\infty}$ converges pointwise to h.
(2) Let $\{p_n\}_{n=2}^{\infty}$ and p be the functions given in Example 10.2.4 (4). Prove that $\{p_n\}_{n=2}^{\infty}$ converges pointwise to p.

Exercise 10.2.2.

(1) Let $\{h_n\}_{n=1}^{\infty}$ and h be the functions given in Example 10.2.4 (3). Using only the definition of uniform convergence, prove that $\{h_n\}_{n=1}^{\infty}$ is not uniformly convergent.
(2) Let $\{p_n\}_{n=1}^{\infty}$ and p be the functions given in Example 10.2.4 (4). Using only the definition of uniform convergence, prove that $\{p_n\}_{n=1}^{\infty}$ is not uniformly convergent.

Exercise 10.2.3. For each $n \in \mathbb{N}$, let $f_n\colon [0,1] \to \mathbb{R}$ be defined by $f_n(x) = \frac{1}{1+x^n}$ for all $x \in [0,1]$. Is $\{f_n\}_{n=1}^{\infty}$ pointwise convergent, uniformly convergent or neither? Prove your answer.

Exercise 10.2.4. Let $A \subseteq \mathbb{R}$ be a non-empty set, let $\{f_n\}_{n=1}^{\infty}$ and $\{g_n\}_{n=1}^{\infty}$ be sequences of functions $A \to \mathbb{R}$, let $f, g\colon A \to \mathbb{R}$ be functions and let $k \in \mathbb{R}$. Suppose that $\{f_n\}_{n=1}^{\infty}$ and $\{g_n\}_{n=1}^{\infty}$ converge pointwise to f and g, respectively.

(1) Prove that $\{f_n + g_n\}_{n=1}^{\infty}$ converges pointwise to $f + g$.
(2) Prove that $\{kf_n\}_{n=1}^{\infty}$ converges pointwise to kf.
(3) Prove that $\{f_n g_n\}_{n=1}^{\infty}$ converges pointwise to fg.

Exercise 10.2.5. Let $A \subseteq \mathbb{R}$ be a non-empty set, let $\{f_n\}_{n=1}^{\infty}$ be a sequence of functions $A \to \mathbb{R}$ and let $f \colon A \to \mathbb{R}$ be a function. Suppose that $\{f_n\}_{n=1}^{\infty}$ converges pointwise to f.

(1) Suppose that f_n is increasing for each $n \in \mathbb{N}$. Is f necessarily increasing? Give a proof or a counterexample.
(2) Suppose that f_n is bounded for each $n \in \mathbb{N}$. Is f necessarily bounded? Give a proof or a counterexample.

Exercise 10.2.6. [Used in Exercise 10.3.7.] Let $A \subseteq \mathbb{R}$ be a non-empty set, let $\{f_n\}_{n=1}^{\infty}$ and $\{g_n\}_{n=1}^{\infty}$ be sequences of functions $A \to \mathbb{R}$, let $f, g \colon A \to \mathbb{R}$ be functions and let $k \in \mathbb{R}$. Suppose that $\{f_n\}_{n=1}^{\infty}$ and $\{g_n\}_{n=1}^{\infty}$ converge uniformly to f and g, respectively.

(1) Prove that $\{f_n + g_n\}_{n=1}^{\infty}$ converges uniformly to $f + g$.
(2) Prove that $\{kf_n\}_{n=1}^{\infty}$ converges uniformly to kf.
(3) Prove that if f_n is bounded for each $n \in \mathbb{N}$, then f is bounded.
(4) Prove that if f is bounded, then there are $N \in \mathbb{N}$ and $M \in \mathbb{R}$ such that $|f_n(x)| \leq M$ for all $x \in A$ and all $n \in \mathbb{N}$ such that $n \geq N$.
(5) Prove that if f and g are bounded, then $\{f_n g_n\}_{n=1}^{\infty}$ converges uniformly to fg.
(6) Find an example of $\{f_n\}_{n=1}^{\infty}$ and $\{g_n\}_{n=1}^{\infty}$ such that $\{f_n g_n\}_{n=1}^{\infty}$ does not converge uniformly to fg.

Exercise 10.2.7. Let $A \subseteq \mathbb{R}$ be a non-empty set, let $\{f_n\}_{n=1}^{\infty}$ be a sequence of functions $A \to \mathbb{R}$, let $f \colon A \to \mathbb{R}$ be a function and let $g \colon \mathbb{R} \to \mathbb{R}$ be a function. Prove that if $\{f_n\}_{n=1}^{\infty}$ converges uniformly to f, and if g is uniformly continuous, then $\{g \circ f_n\}_{n=1}^{\infty}$ converges uniformly to $g \circ f$.

Exercise 10.2.8. [Used in Section 10.2.] Find an example of a sequence $\{f_n\}_{n=1}^{\infty}$ of functions $(0,1) \to \mathbb{R}$ such that f_n is differentiable for each $n \in \mathbb{N}$, that $\{f_n'\}_{n=1}^{\infty}$ is uniformly convergent and that $\{f_n\}_{n=1}^{\infty}$ is not pointwise convergent.

Exercise 10.2.9. Let $C \subseteq \mathbb{R}$ be a non-degenerate closed bounded interval, and let $g, h \colon C \to \mathbb{R}$ be functions. Suppose that g and h are continuous. The **distance** from g to h, denoted $\|g - h\|$, is defined by $\|g - h\| = \text{lub}\{|g(x) - h(x)| \mid x \in C\}$; this least upper bound exists by the Extreme Value Theorem (Theorem 3.5.1).

Let $\{f_n\}_{n=1}^{\infty}$ be a sequence of functions $C \to \mathbb{R}$, and let $f \colon C \to \mathbb{R}$ be a function. Suppose that f_n is continuous for all $n \in \mathbb{N}$, and that f is continuous. Prove that $\{f_n\}_{n=1}^{\infty}$ converges uniformly to f if and only if $\lim_{n \to \infty} \|f_n - f\| = 0$.

Exercise 10.2.10. [Used in Example 10.2.4.] A sequence of functions is a collection of functions indexed by the natural numbers. It is also possible to have a collection of functions that is doubly indexed by the natural numbers, which is equivalent to being indexed by $\mathbb{N} \times \mathbb{N}$. Such a collection is denoted $\{f_{n,m}\}_{n,m=1}^{\infty}$. Find an example of such a doubly indexed collection of functions $\mathbb{R} \to \mathbb{R}$ such that for each $n \in \mathbb{N}$ the sequence $\{f_{n,m}\}_{m=1}^{\infty}$ is pointwise convergent and for each $m \in \mathbb{N}$ the sequence $\{f_{n,m}\}_{n=1}^{\infty}$ is pointwise convergent, and that for some $x \in \mathbb{R}$ the sequence $\left\{\lim_{m\to\infty} f_{n,m}(x)\right\}_{n=1}^{\infty}$ is convergent and the sequence $\left\{\lim_{n\to\infty} f_{n,m}(x)\right\}_{m=1}^{\infty}$ is convergent, but $\lim_{n\to\infty}\lim_{m\to\infty} f_{n,m}(x) \neq \lim_{m\to\infty}\lim_{n\to\infty} f_{n,m}(x)$.

Exercise 10.2.11. **[Used in Section 10.2.]** Let $A \subseteq \mathbb{R}$ be a non-empty set, let $c \in A$, let $\{f_n\}_{n=1}^{\infty}$ be a sequence of functions $A \to \mathbb{R}$ and let $f\colon A \to \mathbb{R}$ be a function. Suppose that $\{f_n\}_{n=1}^{\infty}$ converges pointwise to f, and that f_n is continuous at c for all $n \in \mathbb{N}$. Prove that f is continuous at c if and only if for each $\varepsilon > 0$ and each $M \in \mathbb{N}$, there is some $\delta > 0$ and some $p \in \mathbb{N}$ such that $p \geq M$ and that $x \in A$ and $|x - c| < \delta$ imply $|f_p(x) - f(x)| < \varepsilon$.

Exercise 10.2.12. The purpose of this exercise is to show that if the hypotheses of Theorem 10.2.12 are slightly strengthened, then a simpler proof of the conclusion of that theorem can be given. This simpler proof does not resemble the proof of Theorem 10.2.12, but rather uses the Fundamental Theorem of Calculus, both versions.

Let $I \subseteq \mathbb{R}$ be a non-degenerate open bounded interval, and let $\{f_n\}_{n=1}^{\infty}$ be a sequence of functions $I \to \mathbb{R}$. Suppose that f_n is continuously differentiable for all $n \in \mathbb{N}$, that $\{f_n'\}_{n=1}^{\infty}$ is uniformly convergent and that $\{f_n(c)\}_{n=1}^{\infty}$ is convergent for some $c \in I$. Prove that there is a function $f\colon I \to \mathbb{R}$ such that f is differentiable, and $\{f_n\}_{n=1}^{\infty}$ converges uniformly to f, and $\{f_n'\}_{n=1}^{\infty}$ converges uniformly to f'. (Observe that Theorem 10.2.12 has the weaker hypothesis that f_n is required to be only differentiable, rather than continuously differentiable, for all $n \in \mathbb{N}$.)

To prove the result, let $I = (a, b)$, let $L = \lim_{n\to\infty} f_n(c)$ and let $h\colon I \to \mathbb{R}$ be the function such that $\{f_n'\}_{n=1}^{\infty}$ converges uniformly to h. Because f_n' is continuous for all $n \in \mathbb{N}$, then Theorem 10.2.10 (2) implies that h is continuous, and hence h is integrable by Theorem 5.4.11. Let $f\colon I \to \mathbb{R}$ be defined by $f(x) = L + \int_c^x h(t)\,dt$ for all $x \in I$. Prove that f has the desired properties.

10.3 Series of Functions

Having looked at sequences of functions in Section 10.2, we now turn to series of functions. Analogously to series of numbers, a series of functions is a formal infinite sum of functions, which might or might not actually add up to a function.

Definition 10.3.1. Let $A \subseteq \mathbb{R}$ be a non-empty set. A **series** of functions $A \to \mathbb{R}$ is a formal sum

$$\sum_{n=1}^{\infty} f_n = f_1 + f_2 + f_3 + \cdots,$$

where $\{f_n\}_{n=1}^{\infty}$ is a sequence of functions $A \to \mathbb{R}$. Each function f_n, where $n \in \mathbb{N}$, is called a **term** of the series $\sum_{n=1}^{\infty} f_n$. △

Analogously to series of numbers, the convergence of series of functions is defined in terms of the sequence of partial sums. Hence, just as we have both pointwise convergence and uniform convergence of sequences of functions, we will also have pointwise convergence and uniform convergence of series of functions.

Definition 10.3.2. Let $A \subseteq \mathbb{R}$ be a non-empty set, and let $\sum_{n=1}^{\infty} f_n$ be a series of functions $A \to \mathbb{R}$.

1. For each $k \in \mathbb{N}$, the kth **partial sum** of $\sum_{n=1}^{\infty} f_n$, denoted s_k, is defined by $s_k = \sum_{i=1}^{k} f_i$. The **sequence of partial sums** of $\sum_{n=1}^{\infty} f_n$ is the sequence $\{s_n\}_{n=1}^{\infty}$.
2. The series of functions $\sum_{n=1}^{\infty} f_n$ is **pointwise convergent** if the sequence of partial sums $\{s_n\}_{n=1}^{\infty}$ is pointwise convergent. If $\{s_n\}_{n=1}^{\infty}$ converges pointwise to a function $f\colon A \to \mathbb{R}$, we say that $\sum_{n=1}^{\infty} f_n$ **converges pointwise** to f.
3. The series of functions $\sum_{n=1}^{\infty} f_n$ is **uniformly convergent** if the sequence of partial sums $\{s_n\}_{n=1}^{\infty}$ is uniformly convergent. If $\{s_n\}_{n=1}^{\infty}$ uniformly converges to a function $f\colon A \to \mathbb{R}$, we say that $\sum_{n=1}^{\infty} f_n$ **converges uniformly** to f. $\triangle$

As with series of numbers, we note that a series of functions does not have to start with $n = 1$, and that the convergence of a series of functions is unaffected by changing, or dropping, finitely many terms of the series.

As with the convergence of sequences of functions, it is similarly the case that uniform convergence of series of functions behaves nicer than pointwise convergence of series of functions.

The following two lemmas are derived immediately from Lemma 10.2.6 and Lemma 10.2.3, respectively, and we omit the proofs.

Theorem 10.3.3. *Let $A \subseteq \mathbb{R}$ be a non-empty set, let $\sum_{n=1}^{\infty} f_n$ be a series of functions $A \to \mathbb{R}$ and let $f\colon A \to \mathbb{R}$ be a function. If $\sum_{n=1}^{\infty} f_n$ converges uniformly to f, then $\sum_{n=1}^{\infty} f_n$ converges pointwise to f.*

Lemma 10.3.4. *Let $A \subseteq \mathbb{R}$ be a non-empty set, and let $\sum_{n=1}^{\infty} f_n$ be a series of functions $A \to \mathbb{R}$. If $\sum_{n=1}^{\infty} f_n$ converges pointwise or converges uniformly to f for some function $f\colon A \to \mathbb{R}$, then f is unique.*

We will see in Example 10.3.7 that pointwise convergence of a series of functions does not necessarily imply uniform convergence. However, just as it is not always easy to prove that a series of numbers is convergent directly by the definition, which is why we need the various convergence tests given in Sections 9.3 and 9.4, it is also not always easy to prove uniform convergence of a series of functions directly by the definition. Hence, before giving an example of uniform convergence of a series of functions, we prove the following two convergence tests. The first of our tests is analogous to the Divergence Test for series of numbers (Theorem 9.2.5). We use the term "zero function" to refer to any function that is constantly zero (so that the codomain of the function must be a subset of $\mathbb{R}$).

Theorem 10.3.5 (Divergence Test for Series of Functions). *Let $A \subseteq \mathbb{R}$ be a non-empty set, and let $\sum_{n=1}^{\infty} f_n$ be a series of functions $A \to \mathbb{R}$.*

1. *If $\{f_n\}_{n=1}^{\infty}$ does not converge pointwise to the zero function, then $\sum_{n=1}^{\infty} f_n$ is not pointwise convergent.*
2. *If $\{f_n\}_{n=1}^{\infty}$ does not converge uniformly to the zero function, then $\sum_{n=1}^{\infty} f_n$ is not uniformly convergent.*

Proof. Left to the reader in Exercise 10.3.3. $\square$

The following convergence test for series of functions is somewhat analogous to the Comparison Test for series of numbers (Theorem 9.3.2), though instead of comparing one series of functions with another series of functions, we will compare a series of functions with a series of numbers.

Theorem 10.3.6 (Weierstrass M-Test). *Let $A \subseteq \mathbb{R}$ be a non-empty set, and let $\sum_{n=1}^{\infty} f_n$ be a series of functions $A \to \mathbb{R}$. Suppose that for each $k \in \mathbb{N}$, there is some $M_k \in \mathbb{R}$ such that $|f_k(x)| \leq M_k$ for all $x \in A$. If $\sum_{n=1}^{\infty} M_n$ is convergent, then $\sum_{n=1}^{\infty} f_n$ is uniformly convergent.*

Proof. Suppose that $\sum_{n=1}^{\infty} M_n$ is convergent. Observe that $M_k \geq 0$ for all $k \in \mathbb{N}$. Let $\{s_n\}_{n=1}^{\infty}$ be the sequence of partial sums of $\sum_{n=1}^{\infty} f_n$, and let $\{t_n\}_{n=1}^{\infty}$ be the sequence of partial sums of $\sum_{n=1}^{\infty} M_n$.

Let $\varepsilon > 0$. By hypothesis the sequence $\{t_n\}_{n=1}^{\infty}$ is convergent, and it follows from the Cauchy Completeness Theorem (Corollary 8.3.16) that $\{t_n\}_{n=1}^{\infty}$ is a Cauchy sequence. Hence there is some $N \in \mathbb{N}$ such that $n, m \in \mathbb{N}$ and $n, m \geq N$ imply $|t_n - t_m| < \varepsilon$.

Suppose that $n, m \in \mathbb{N}$ and $n, m \geq N$. Let $x \in A$. If $n = m$, then $|s_n(x) - s_m(x)| = 0 < \varepsilon$. Now suppose that $n \neq m$. Without loss of generality, assume that $n > m$. Using Exercise 2.5.3, and the fact that $M_k \geq 0$ for all $k \in \mathbb{N}$, we see that

$$\begin{aligned} |s_n(x) - s_m(x)| &= \left| \sum_{i=1}^{n} f_i(x) - \sum_{i=1}^{m} f_i(x) \right| = \left| \sum_{i=m+1}^{n} f_i(x) \right| \leq \sum_{i=m+1}^{n} |f_i(x)| \leq \sum_{i=m+1}^{n} M_i \\ &= \left| \sum_{i=m+1}^{n} M_i \right| = \left| \sum_{i=1}^{n} M_i - \sum_{i=1}^{m} M_i \right| = |t_n - t_m| < \varepsilon. \end{aligned}$$

It now follows from the Cauchy Criterion for Uniform Convergence (Theorem 10.2.9) that $\{s_n\}_{n=1}^{\infty}$ is uniformly convergent, and hence $\sum_{n=1}^{\infty} f_n$ is uniformly convergent. □

Example 10.3.7. For each $n \in \mathbb{N} \cup \{0\}$, let $f_n \colon (-1,1) \to \mathbb{R}$ be defined by $f_n(x) = x^n$ for all $x \in (-1,1)$. We want to examine the convergence of the series $\sum_{n=0}^{\infty} f_n$, which can be written as $\sum_{n=0}^{\infty} x^n$.

First, we show that the series $\sum_{n=0}^{\infty} f_n$ is pointwise convergent. Let $y \in (-1,1)$. By Example 9.2.4 (4) we see that the series of numbers $\sum_{n=0}^{\infty} y^n$ is a geometric series, that it is convergent and that $\sum_{n=0}^{\infty} y^n = \frac{1}{1-y}$. It follows that $\sum_{n=0}^{\infty} f_n$ converges pointwise to the function $f \colon (-1,1) \to \mathbb{R}$ defined by $f(x) = \frac{1}{1-x}$ for all $x \in (-1,1)$.

Next, we show that $\sum_{n=0}^{\infty} f_n$ is not uniformly convergent. Let $g \colon (-1,1) \to \mathbb{R}$ be defined by $g(x) = 0$ for all $x \in (-1,1)$. The same argument used in Example 10.2.8 (2) can be used to show that $\{f_n\}_{n=0}^{\infty}$ does not converge uniformly to g. It now follows from the Divergence Test for Series of Functions (Theorem 10.3.5) that $\sum_{n=0}^{\infty} f_n$ is not uniformly convergent. Hence, pointwise convergence of a series of functions does not necessarily imply uniform convergence.

Finally, let $b \in (0,1)$. For each $n \in \mathbb{N}$, let $h_n = f_n|_{[-b,b]}$. We will show that $\sum_{n=0}^{\infty} h_n$ is uniformly convergent. For each $n \in \mathbb{N}$, let $M_n = b^n$. Using Lemma 2.3.9 (5), which can be extended by recursion to any finite product of numbers, and Exercise 2.5.12

(2), we see that if $n \in \mathbb{N}$ and $x \in [-b,b]$, then $0 \leq |x| \leq b$, which implies that $|h_n(x)| = |x^n| = |x|^n \leq b^n = M_n$.

Also, observe that $\sum_{n=0}^{\infty} M_n = \sum_{n=0}^{\infty} b^n$ is convergent, again because it is a geometric series, and $|b| < 1$. It now follows from the Weierstrass M-Test (Theorem 10.3.6) that $\sum_{n=0}^{\infty} h_n$ is uniformly convergent. We could also state this result by saying that $\sum_{n=0}^{\infty} f_n$ "is uniformly convergent on $[-b,b]$." $\lozenge$

The series $\sum_{n=0}^{\infty} f_n$ in Example 10.3.7 is a power series, as defined in Section 9.5. We now see in the following theorem that the behavior of this example is typical of all power series. This theorem, which is a nice application of the Weierstrass M-Test (Theorem 10.3.6), will be important in our further study of power series in Section 10.4.

Theorem 10.3.8. *Let $\sum_{n=0}^{\infty} c_n(x-a)^n$ be a power series in $\mathbb{R}$. Let R be the radius of convergence of $\sum_{n=0}^{\infty} c_n(x-a)^n$. Suppose that $R > 0$. If $P \in (0,R)$, then $\sum_{n=0}^{\infty} c_n(x-a)^n$ is uniformly convergent on $[a-P, a+P]$.*

Proof. Let $P \in (0,R)$. Then $a+P \in (a-R, a+R)$, and hence $\sum_{n=0}^{\infty} c_n(x-a)^n$ is absolutely convergent at $x = a+P$. Therefore $\sum_{n=0}^{\infty} c_n P^n = \sum_{n=0}^{\infty} c_n((a+P)-a)^n$ is absolutely convergent, which means that $\sum_{n=0}^{\infty} |c_n| P^n$ is convergent.

Let $k \in \mathbb{N} \cup \{0\}$ and $x \in [a-P, a+P]$. Then $|x-a| \leq P$, and it follows from Lemma 2.3.9 (5) (extended by recursion to any finite product of numbers) and Exercise 2.5.12 (2) that $|c_k(x-a)^k| \leq |c_k| P^k$. We can now use the Weierstrass M-Test (Theorem 10.3.6) with $M_n = |c_n| P^n$ for all $n \in \mathbb{N} \cup \{0\}$, and we deduce that $\sum_{n=0}^{\infty} c_n(x-a)^n$ is uniformly convergent on $[a-P, a+P]$. (The Weierstrass M-Test was stated for series that start at $n = 1$ rather than $n = 0$, but that makes no difference.) $\square$

We conclude this section by showing that the expected behavior of uniformly convergent series of functions with respect to continuity, differentiability and integrability holds.

Theorem 10.3.9. *Let $A \subseteq \mathbb{R}$ be a non-empty set, let $\sum_{n=1}^{\infty} f_n$ be a series of functions $A \to \mathbb{R}$ and let $f \colon A \to \mathbb{R}$ be a function. Suppose that $\sum_{n=1}^{\infty} f_n$ converges uniformly to f.*

1. *Let $c \in A$. If f_n is continuous at c for all $n \in \mathbb{N}$, then f is continuous at c.*
2. *If f_n is continuous for all $n \in \mathbb{N}$, then f is continuous.*

Proof. Let $\{s_n\}_{n=1}^{\infty}$ be the sequence of partial sums of $\sum_{n=1}^{\infty} f_n$, so that $\{s_n\}_{n=1}^{\infty}$ converges uniformly to f. For each $k \in \mathbb{N}$, we know that s_k is continuous, because it is the sum of finitely many continuous functions. (In Theorem 3.3.5 (1) we saw that the sum of two continuous functions is continuous, and that result can be extended by recursion to any finite sum.) Both parts of this theorem now follow immediately from Theorem 10.2.10 applied to $\{s_n\}_{n=1}^{\infty}$. $\square$

The next two theorems show that uniformly convergent series of functions, subject to the appropriate hypotheses, can be integrated and differentiated term by term.

The proofs of these two theorems are completely analogous to the proof of Theorem 10.3.9, this time relying upon Theorem 10.2.11 and Theorem 10.2.12, and we omit the details. The unpleasant hypotheses in Theorem 10.3.11 correspond to the analogous hypotheses in Theorem 10.2.12.

Theorem 10.3.10. *Let $[a,b] \subseteq \mathbb{R}$ be a non-degenerate closed bounded interval, let $\sum_{n=1}^{\infty} f_n$ be a series of functions $[a,b] \to \mathbb{R}$ and let $f\colon [a,b] \to \mathbb{R}$ be a function. Suppose that $\sum_{n=1}^{\infty} f_n$ converges uniformly to f. If f_n is integrable for all $n \in \mathbb{N}$, then f is integrable and*

$$\int_a^b f(x)\,dx = \sum_{n=1}^{\infty} \int_a^b f_n(x)\,dx.$$

Theorem 10.3.11. *Let $I \subseteq \mathbb{R}$ be a non-degenerate open bounded interval, and let $\sum_{n=1}^{\infty} f_n$ be a series of functions $I \to \mathbb{R}$. Suppose that f_n is differentiable for all $n \in \mathbb{N}$, that $\sum_{n=1}^{\infty} f_n'$ is uniformly convergent and that $\sum_{n=1}^{\infty} f_n(c)$ is convergent for some $c \in I$. Then there is a function $f\colon I \to \mathbb{R}$ such that f is differentiable, and $\sum_{n=1}^{\infty} f_n$ converges uniformly to f, and $\sum_{n=1}^{\infty} f_n'$ converges pointwise to f'.*

The displayed equation in Theorem 10.3.10 can be written as

$$\int_a^b \left[\sum_{n=1}^{\infty} f_n(x)\right] dx = \sum_{n=1}^{\infty} \int_a^b f_n(x)\,dx,$$

and the conclusion of Theorem 10.3.11 can be summarized by writing

$$\left[\sum_{n=1}^{\infty} f_n(x)\right]' = \sum_{n=1}^{\infty} f_n'(x)$$

for all $x \in I$, though caution is needed when writing both of these equations, because they hold only under suitable hypotheses.

Reflections

This section, in contrast to the previous one, contains no surprises. That is not due to the fact that series of functions are in some way better behaved than sequences of functions, but rather it is due to the fact that the same fundamental issue, namely, the difference between pointwise convergence and uniform convergence, is the basis of the behavior of both sequences and series of functions. Because the distinction between these two types of convergence was first discussed in the context of sequences of functions, the surprising behavior associated with this distinction was also discussed there, and there is no need to repeat it in the present section. As we will see in the next two sections, it is series of functions rather than sequences of functions that are of greater immediate use to us. Of course, as with series of numbers, so too the convergence of series of functions cannot be treated if the convergence of sequences of functions has not been previously discussed.

Exercises

Exercise 10.3.1. For each $n \in \mathbb{N}$, let $f_n \colon [1,\infty) \to \mathbb{R}$ be defined by $f_n(x) = 3^{-nx}$ for all $x \in [1,\infty)$. Prove that $\sum_{n=1}^{\infty} f_n = \sum_{n=1}^{\infty} 3^{-nx}$ is uniformly convergent.

Exercise 10.3.2. Let $\sum_{n=1}^{\infty} a_n$ be a series in $\mathbb{R}$. For each $n \in \mathbb{N}$, let $g_n \colon [-1,1] \to \mathbb{R}$ be defined by $g_n(x) = a_n x^n$ for all $x \in [-1,1]$. Prove that if $\sum_{n=1}^{\infty} a_n$ is absolutely convergent, then $\sum_{n=1}^{\infty} g_n = \sum_{n=1}^{\infty} a_n x^n$ is uniformly convergent.

Exercise 10.3.3. **[Used in Theorem 10.3.5.]** Prove Theorem 10.3.5.

Exercise 10.3.4. Let $A \subseteq \mathbb{R}$ be a non-empty set, and let $\sum_{n=1}^{\infty} f_n$ and $\sum_{n=1}^{\infty} g_n$ be series of functions $A \to \mathbb{R}$. Suppose that $f_n(x) \geq 0$ and $g_n(x) \geq 0$ for all $x \in A$ and all $n \in \mathbb{N}$, and that there is some $N \in \mathbb{N}$ such that $n \in \mathbb{N}$ and $n \geq N$ imply $f_n(x) \leq g_n(x)$ for all $x \in A$. Prove that if $\sum_{n=1}^{\infty} g_n$ is uniformly convergent, then $\sum_{n=1}^{\infty} f_n$ is uniformly convergent. (This exercise is the analog for uniform convergence of the Comparison Test (Theorem 9.3.2).)

Exercise 10.3.5. Let $[a,b] \subseteq \mathbb{R}$ be a non-degenerate closed bounded interval, and let $\sum_{n=1}^{\infty} f_n$ be a series of functions $[a,b] \to \mathbb{R}$. Suppose f_n is increasing for all $n \in \mathbb{N}$, that $f_n(x) > 0$ for all $x \in A$ and all $n \in \mathbb{N}$, that the sequence $\{f_n(x)\}_{n=1}^{\infty}$ is decreasing for all $x \in A$. Prove that if $\sum_{n=1}^{\infty} (-1)^{n-1} f_n$ is pointwise convergent, then $\sum_{n=1}^{\infty} (-1)^{n-1} f_n$ is uniformly convergent. [Use Exercise 8.2.13 (2).]

Exercise 10.3.6.

(1) Let $A \subseteq \mathbb{R}$ be a non-empty set, and let $\sum_{n=1}^{\infty} f_n$ be a series of functions $A \to \mathbb{R}$. Suppose that $i, j \in \mathbb{N}$ and $i \neq j$ imply that $\{x \in A \mid f_i(x) \neq 0\} \cap \{x \in A \mid f_j(x) \neq 0\} = \emptyset$, and that for each $k \in \mathbb{N}$, there is some $M_k \in \mathbb{R}$ such that $|f_k(x)| \leq M_k$ for all $x \in A$. Prove that if $\{M_n\}_{n=1}^{\infty}$ is convergent and $\lim_{n\to\infty} M_n = 0$, then $\sum_{n=1}^{\infty} f_n$ is uniformly convergent.

(2) Find an example of a series $\sum_{n=1}^{\infty} g_n$ of functions $[0,1] \to \mathbb{R}$ such that $\sum_{n=1}^{\infty} g_n$ is uniformly convergent, and that $\sum_{n=1}^{\infty} g_n$ does not satisfy the hypotheses of the M-Test (Theorem 10.3.6).

Exercise 10.3.7. The reader is undoubtedly familiar with the importance of differential equations in many applications of mathematics. In addition to their practical use, however, differential equations can also be studied from a theoretical point of view. The most fundamental theoretical question about differential equations is to find criteria that guarantee that certain of them have solutions, and even better unique solutions; such an existence question is quite distinct from the practical question of how to find formulas for the solutions of specific differential equations.

We are interested here in ordinary differential equations, which are differential equations with a single variable. The standard formulation of an ordinary differential equation with initial condition is $y' = f(x,y)$ and $y(a) = b$, where f is some appropriate function of two variables. For example, if $f(x,y) = 5y$ for all (x,y) in some appropriate subset of $\mathbb{R}^2$, and if $a = 0$ and $b = 3$, then the differential equation with initial condition becomes $y' = 5y$ and $y(0) = 3$, which, as the reader might recognize, has solution $y = 3e^{5x}$.

Because we do not discuss functions of more than one variable in this text, we cannot treat such a general formulation of differential equations; see [BD09, Section 2.8] or [Str00, Section 11.1] for a discussion of the general case. However, we can handle at present differential equations of the form $y' = f(y)$. The example $y' = 5y$ fits into this restricted formulation. The initial condition for our type of differential equation remains of the form $y(a) = b$.

The purpose of this exercise is to prove that differential equations of the above type with initial conditions always have solutions if f is sufficiently well behaved, and if we are willing to restrict our attention to a sufficiently small interval containing a. This result is a special case of a more general existence and uniqueness theorem due to Charles Emile Picard (1856–1941); our proof uses the method known as Picard iteration, and includes all the ingredients of the proof of the more general existence result. Our theorem is as follows:

Let $I \subseteq \mathbb{R}$ be an open interval, let $a \in \mathbb{R}$, let $b \in I$ and let $f\colon I \to \mathbb{R}$ be a function. Suppose that there is some $K \in \mathbb{R}$ such that $|f(x) - f(y)| \le K|x-y|$ for all $x, y \in I$. Then there is some $\delta > 0$ and a function $g\colon (a-\delta, a+\delta) \to I$ such that $g'(x) = f(g(x))$ for all $x \in (a-\delta, a+\delta)$ and $g(a) = b$.

As discussed in Exercise 3.4.5, a function that satisfies the hypothesis involving the number K in this theorem is said to satisfy a **Lipschitz condition**.

The proof of the theorem will be done in steps. We start with a few preliminary observations. First, it must be the case that $K \ge 0$, and we may assume that $K > 0$. Second, it follows from Exercise 3.4.5 (1) and Lemma 3.4.2 that f is continuous. Third, by Lemma 2.3.7 (2) there is some $\delta > 0$ such that $[b-\eta, b+\eta] \subseteq I$. Then Corollary 3.4.6 implies that there is some $M \in \mathbb{R}$ such that $|f(x)| \le M$ for all $x \in [b-\eta, b+\eta]$. We may assume that $M > 0$.

Let $\delta = \min\{\frac{\eta}{M}, \frac{1}{2K}\}$. Then $(b - M\delta, b + M\delta) \subseteq [b-\eta, b+\eta] \subseteq I$.

(1) We define a sequence $\{g_n\}_{n=1}^{\infty}$ of functions $(a-\delta, a+\delta) \to I$ using Definition by Recursion as follows. Let $g_1(x) = b$ for all $x \in (a-\delta, a+\delta)$. Because $f \circ g_1$ is constant, then it is continuous by Example 3.3.3 (1). Hence the restriction of $f \circ g_1$ to each non-degenerate closed bounded interval in $(a-\delta, a+\delta)$ is continuous by Exercise 3.3.2 (2), and it follows from Theorem 5.4.11 that $f \circ g_1$ is integrable on every non-degenerate closed bounded interval in $(a-\delta, a+\delta)$. Let

$$g_2(x) = b + \int_a^x f(g_1(t))\,dt$$

for all $x \in (a-\delta, a+\delta)$. By Theorem 5.3.1 (4) and Definition 5.5.8 we see that $g_2(x) = b + f(b)(x-a)$ for all $x \in (a-\delta, a+\delta)$. Then g_2 is continuous by Example 3.3.3 (1), and

$$g_2(x) \in (b - |f(b)|\delta, b + |f(b)|\delta) \subseteq (b - M\delta, b + M\delta) \subseteq I$$

for all $x \in (a-\delta, a+\delta)$.

Now suppose that we have defined g_n for some $n \in \mathbb{N}$, and that g_n is continuous, and $g_n(x) \in I$ for all $x \in (a-\delta, a+\delta)$. Then $f \circ g_n$ is continuous by Theorem 3.3.8 (3), and, as above, it follows that $f \circ g_n$ is integrable on every non-degenerate closed bounded interval in $(a-\delta, a+\delta)$. Let

$$g_{n+1}(x) = b + \int_a^x f(g_n(t))\,dt$$

for all $x \in (a-\delta, a+\delta)$. Prove that g_{n+1} is continuous, and that $g_{n+1}(x) \in I$ for all $x \in (a-\delta, a+\delta)$. We have then defined $\{g_n\}_{n=1}^{\infty}$.

(2) Prove that $|g_{n+1}(x) - g_n(x)| \le \frac{M}{K}(K\delta)^n$ for all $x \in (a-\delta, a+\delta)$ and all $n \in \mathbb{N}$.

(3) Prove that the series $\sum_{n=1}^{\infty}(g_{n+1} - g_n)$ is uniformly convergent.

(4) Prove that the sequence $\{g_n\}_{n=1}^{\infty}$ is uniformly convergent.

[Use Exercise 10.2.6 (1).]

(5) By Part (4) of this exercise $\{g_n\}_{n=1}^{\infty}$ converges uniformly to some function $g\colon (a-\delta, a+\delta) \to \mathbb{R}$. Prove that g is continuous, that $g(x) \in I$ for all $x \in (a-\delta, a+\delta)$, that $f \circ g$ is integrable on every non-degenerate closed bounded interval in $(a-\delta, a+\delta)$ and that $g(a) = b$.

(6) Prove that

$$g(x) = b + \int_a^x f(g(t))\,dt$$

for all $x \in (a-\delta, a+\delta)$. [Use Exercise 10.2.9.]

(7) Prove that g is differentiable and $g'(x) = f(g(x))$ for all $x \in (a-\delta, a+\delta)$.

10.4 Functions as Power Series

We first encountered power series in Section 9.5, where we viewed power series in the context of series in general. We now revisit power series, which we can study in more depth than before by making use of the tools we developed in the previous sections of this chapter; in particular, we make use of the notion of uniform convergence of series of functions.

Not only do we have more tools at our disposal in the present section in comparison with Section 9.5, but, no less important, we adopt two changes in our point of view toward power series. First, as we mentioned at the beginning of Section 10.2, rather than viewing a power series such as

$$1 + x + x^2 + x^3 + x^4 + x^5 + \cdots$$

as a collection of series of numbers, one series for each value of x, we now view power series as a single series of functions. Second, whereas in Section 9.5 we started with power series, which we then viewed as functions, we now want to start with functions, such as e^x and $\sin x$, that might not initially be given in terms of power series, and we want to see if such functions can be represented as power series. In the process of discussing this matter, we will also answer some questions left unfinished in Section 9.5, such as whether power series, when viewed as functions, are continuous, differentiable and integrable.

In keeping with our new point of view, we commence our current study of power series with the following definition.

Definition 10.4.1. Let $A \subseteq \mathbb{R}$ be a set, let $a \in A$ and let $f\colon A \to \mathbb{R}$ be a function. The function f is **represented by a power series** centered at a if there is a power series $\sum_{n=0}^{\infty} c_n(x-a)^n$ with non-degenerate interval of convergence I such that $I \subseteq A$ and $f(x) = \sum_{n=0}^{\infty} c_n(x-a)^n$ for all $x \in I$. △

Observe in Definition 10.4.1 that the power series $\sum_{n=0}^{\infty} c_n(x-a)^n$ is required to have a non-degenerate interval of convergence, which, as noted in Section 9.5, is equivalent to the requirement that the power series has positive radius of convergence. A power series with zero radius of convergence would represent a function at only one point, and as such would not be of any use. Moreover, note that in Definition 10.4.1 it is not required that the interval of convergence of the power series equals the domain of the function, but only that the interval of convergence is a subset of the domain. Although it is nice when the interval of convergence of the power series equals the domain of the function, as happens in some cases, we will see in Example 10.4.3 (1) that that is not always possible.

As the reader might expect, if a function is represented by a power series, then that power series representation is unique. The following theorem is just a reformulation of Theorem 9.5.8 in the terminology of the present section, and we omit the proof.

Theorem 10.4.2. *Let $A \subseteq \mathbb{R}$ be a set, let $a \in A$ and let $f\colon A \to \mathbb{R}$ be a function. If f is represented by a power series centered at a, then the power series is unique.*

We are now ready for our first examples of functions represented by power series.

Example 10.4.3.

(1) Let $f\colon \mathbb{R} - \{1\} \to \mathbb{R}$ be defined by $f(x) = \frac{1}{1-x}$ for all $x \in \mathbb{R} - \{1\}$. We saw in Example 9.5.2 (1) that the power series $\sum_{n=0}^{\infty} x^n$ has interval of convergence $(-1,1)$, and that $\sum_{n=0}^{\infty} x^n = \frac{1}{1-x}$ for all $x \in (-1,1)$. Hence f is represented by the power series $\sum_{n=0}^{\infty} x^n$, which is centered at 0.

This power series representation of f might appear to be less than satisfying, because the interval of convergence of the power series is only a small part of the domain of the function f, and the reader might wonder if there is some other power series representation of f centered at 0 that has a larger interval of convergence than $(-1,1)$. Unfortunately, there is no such power series representation, because Theorem 10.4.2 says that if a function has a power series representation centered at a number, that representation is unique, so we can do no better for our function f than we have already done. Actually, it should not be too surprising that the function f cannot be represented by a power series centered at 0 that has a larger interval of convergence. The function f has a vertical asymptote at $x = 1$, and there is no possibility of extending f to a continuous function defined on all of $\mathbb{R}$. As we will see in Corollary 10.4.5, if a function is represented by a power series, then the function is continuous on the interior of the interval of convergence of the power series. Because our function f cannot be extended to a continuous function at $x = 1$, and because intervals of convergence of power series are symmetric about the number a by Theorem 9.5.4 (except possibly at the endpoints), then we see that there is no hope of finding a power series representation for f centered at 0 with a radius of convergence larger than 1.

(2) In Part (1) of this example we saw that $\frac{1}{1-x} = \sum_{n=0}^{\infty} x^n$ for all $x \in (-1,1)$. We can use this power series representation to obtain power series representations of other functions. For example, we see that

$$\frac{1}{1+3x^2} = \frac{1}{1-(-3x^2)} = \sum_{n=0}^{\infty} (-3x^2)^n = \sum_{n=0}^{\infty} (-1)^n 3^n x^{2n}$$

for all $x \in \mathbb{R}$ such that $-1 < -3x^2 < 1$, which means for all $x \in \left(-\frac{1}{\sqrt{3}}, \frac{1}{\sqrt{3}}\right)$.

We can find many other power series representations by starting with the power series representation for $\frac{1}{1-x}$ and making other substitutions for x. However, doing so is not always entirely trivial. Suppose that we want to find the power series representation for the function $g\colon \mathbb{R} - \{-3\} \to \mathbb{R}$ defined by $g(x) = \frac{1}{x+3}$ for all $x \in \mathbb{R} - \{-3\}$. Which substitution should we use with $\frac{1}{1-x}$ to obtain $\frac{1}{x+3}$? One straightforward approach would be

$$\frac{1}{x+3} = \frac{1}{1-[-(x+2)]} = \sum_{n=0}^{\infty} [-(x+2)]^n = \sum_{n=0}^{\infty} (-1)^n (x+2)^n$$

for all $x \in \mathbb{R}$ such that $-1 < -(x+2) < 1$, which means for all $x \in (-3,-1)$. This power series representation is centered at -2. Another approach would be

$$\frac{1}{x+3} = \frac{1}{3} \cdot \frac{1}{1-\left(-\frac{x}{3}\right)} = \frac{1}{3}\sum_{n=0}^{\infty} \left(-\frac{x}{3}\right)^n = \sum_{n=0}^{\infty} (-1)^n \frac{x^n}{3^{n+1}}$$

for all $x \in \mathbb{R}$ such that $-1 < -\frac{x}{3} < 1$, which means for all $x \in (-3,3)$. This latter power series representation, which is centered at 0, has a larger interval of convergence than our previous attempt, and it is therefore preferable. The reason the latter power series has a larger interval of convergence than the former is that in the latter the number about which the power series is centered is farther from $x = -3$, where the function g has a vertical asymptote. ◊

The method for finding power series representations in Example 10.4.3 has limited use, and will not help us find power series representations of some familiar functions such as e^x and $\sin x$. We will find the power series representation for these functions later in this section, after we have proved some theorems about power series. The main technical tool used to prove these theorems is Theorem 10.3.8, which will allow us to use the nice properties of uniformly convergent series that we saw in Section 10.3. We start by proving that functions represented by power series can be differentiated and integrated term by term, except possibly at the endpoints of the interval of convergence. Recall from Section 9.5 that if the radius of convergence of a power series is $R = \infty$, then we will write $(a-R, a+R)$ to mean $(-\infty, \infty) = \mathbb{R}$, which allows us to avoid special cases.

Theorem 10.4.4. *Let $A \subseteq \mathbb{R}$ be a set, let $a \in A$ and let $f\colon A \to \mathbb{R}$ be a function. Suppose that f is represented by a power series $\sum_{n=0}^{\infty} c_n(x-a)^n$. Let R be the radius of convergence of $\sum_{n=0}^{\infty} c_n(x-a)^n$.*

1. *The power series* $\sum_{n=1}^{\infty} nc_n(x-a)^{n-1}$ *has radius of convergence* R. *The function* f *is differentiable on* $(a-R, a+R)$, *and* $f'(x) = \sum_{n=1}^{\infty} nc_n(x-a)^{n-1}$ *for all* $x \in (a-R, a+R)$.
2. *The power series* $\sum_{n=0}^{\infty} c_n \frac{(x-a)^{n+1}}{n+1}$ *has radius of convergence* R. *The function* f *is integrable on any closed subinterval of* $(a-R, a+R)$, *and* $\int_a^x f(t)\,dt = \sum_{n=0}^{\infty} c_n \frac{(x-a)^{n+1}}{n+1}$ *for all* $x \in (a-R, a+R)$.

Proof. We will prove Part (1), leaving the remaining part to the reader in Exercise 10.4.3. We follow [Pow94].

(1) Let P be the radius of convergence of the power series $\sum_{n=1}^{\infty} nc_n(x-a)^{n-1}$. We first show that $P = R$.

If $P = 0$, then $P \leq R$. Now suppose that $P > 0$. Let $y \in (a-P, a+P)$. Then $\sum_{n=1}^{\infty} nc_n(y-a)^{n-1}$ is absolutely convergent, which means that $\sum_{n=1}^{\infty} |nc_n(y-a)^{n-1}|$ is convergent. By Theorem 9.2.6 (3) we see that $\sum_{n=1}^{\infty} |y-a| \cdot |nc_n(y-a)^{n-1}|$ is convergent, and hence that $\sum_{n=1}^{\infty} |nc_n(y-a)^n|$ is convergent. If $n \in \mathbb{N}$, then $|c_n(y-a)^n| \leq |nc_n(y-a)^n|$. It follows from the Comparison Test (Theorem 9.3.2) that $\sum_{n=1}^{\infty} |c_n(y-a)^n|$ is convergent, which means that $\sum_{n=1}^{\infty} c_n(y-a)^n$ is absolutely convergent. Hence $\sum_{n=0}^{\infty} c_n(y-a)^n$ is absolutely convergent. Therefore $y \in [a-R, a+R]$. It follows that $(a-P, a+P) \subseteq [a-R, a+R]$, which implies that $P \leq R$.

We know that $R > 0$. Let $z \in (a-R, a+R)$. By Exercise 2.3.9 there is some $T \in (0, R)$ such that $z \in (a-T, a+T)$. Let $w = a+T$. Then $|z-a| < |w-a|$ and $w \in (a-R, a+R)$. Hence $|z-a| < |w-a|$, and that $\sum_{n=0}^{\infty} c_n(w-a)^n$ is absolutely convergent. Therefore $\sum_{n=1}^{\infty} |c_n(w-a)^n|$ is convergent. It follows from Exercise 9.2.2 that $\{|c_n(w-a)^n|\}_{n=1}^{\infty}$ is bounded. Hence there is some $M \in \mathbb{R}$ such that $|c_n(w-a)^n| \leq M$ for all $n \in \mathbb{N}$.

If $n \in \mathbb{N}$, then

$$|nc_n(z-a)^{n-1}| = n \cdot \frac{|c_n(w-a)^n|}{|w-a|} \cdot \left|\frac{z-a}{w-a}\right|^{n-1} \leq n \cdot \frac{M}{|w-a|} \cdot \left|\frac{z-a}{w-a}\right|^{n-1}.$$

Because $\left|\frac{z-a}{w-a}\right| < 1$, it follows from Exercise 9.4.5 that $\sum_{n=1}^{\infty} n \cdot \frac{M}{|w-a|} \cdot \left|\frac{z-a}{w-a}\right|^{n-1}$ is convergent. We then use the Comparison Test (Theorem 9.3.2) to see that $\sum_{n=1}^{\infty} |nc_n(z-a)^{n-1}|$ is convergent, and hence $\sum_{n=1}^{\infty} nc_n(z-a)^{n-1}$ is absolutely convergent. Therefore $z \in [a-P, a+P]$. It follows that $(a-R, a+R) \subseteq [a-P, a+P]$, which implies that $R \leq P$. We conclude that $P = R$.

Let $u \in (a-R, a+R)$. By Exercise 2.3.9 there is some $Q \in (0, R)$ such that $u \in (a-Q, a+Q)$. By Theorem 10.3.8 we know that $\sum_{n=0}^{\infty} c_n(x-a)^n$ is uniformly convergent on $[a-Q, a+Q]$. Because $\sum_{n=1}^{\infty} nc_n(x-a)^{n-1}$ has radius of convergence R, it also follows that $\sum_{n=1}^{\infty} nc_n(x-a)^{n-1}$ is uniformly convergent on $[a-Q, a+Q]$. We can now apply Theorem 10.3.11 to the restriction of $\sum_{n=0}^{\infty} c_n(x-a)^n$ to $(a-Q, a+Q)$, where the value of c in the statement of the theorem can be anything in $(a-Q, a+Q)$. We deduce that there is some function $g\colon (a-Q, a+Q) \to \mathbb{R}$ such that g is differentiable, and the restriction of $\sum_{n=0}^{\infty} c_n(x-a)^n$ to $(a-Q, a+Q)$ converges uniformly to g, and the restriction of $\sum_{n=0}^{\infty} [c_n(x-a)^n]' = \sum_{n=1}^{\infty} nc_n(x-a)^{n-1}$ to

$(a-Q,a+Q)$ converges pointwise to g'. However, we know by hypothesis that $\sum_{n=0}^{\infty} c_n(x-a)^n$ converges uniformly to f, and it follows immediately that the restriction of $\sum_{n=0}^{\infty} c_n(x-a)^n$ to $(a-Q,a+Q)$ converges uniformly to $f|_{(a-Q,a+Q)}$. By Lemma 10.3.4 we deduce that $f|_{(a-Q,a+Q)} = g$. Hence $\sum_{n=1}^{\infty} nc_n(x-a)^{n-1} = f'(x)$ for all $x \in (a-Q,a+Q)$. In particular, $f'(u) = \sum_{n=1}^{\infty} nc_n(u-a)^{n-1}$. It follows that $f'(x) = \sum_{n=1}^{\infty} nc_n(x-a)^{n-1}$ for all $x \in (a-R,a+R)$. □

The following corollary is an immediate consequence of Theorem 10.4.4 (1) and Theorem 4.2.4, and we omit the proof.

Corollary 10.4.5. *Let $A \subseteq \mathbb{R}$ be a set, let $a \in A$ and let $f\colon A \to \mathbb{R}$ be a function. Suppose that f is represented by a power series $\sum_{n=0}^{\infty} c_n(x-a)^n$. Let R be the radius of convergence of $\sum_{n=0}^{\infty} c_n(x-a)^n$. Then f is continuous on $(a-R,a+R)$.*

Theorem 10.4.4 does not make any claims about the endpoints of the interval of convergence of the power series, and that is because, as we see in the first part of the following example, the convergence or divergence of the original power series at the endpoints does not necessarily imply the convergence or divergence at the endpoints of the derivative or integral.

Example 10.4.6.

(1) We saw in Example 9.5.2 (1) that the power series $\sum_{n=0}^{\infty} x^n$ has interval of convergence $(-1,1)$, and hence radius of convergence 1, and that $\sum_{n=0}^{\infty} x^n = \frac{1}{1-x}$ for all $x \in (-1,1)$. It then follows from Theorem 10.4.4 (1) that $\sum_{n=1}^{\infty} nx^{n-1}$ has radius of convergence 1, and that

$$\sum_{n=1}^{\infty} nx^{n-1} = \left[\frac{1}{1-x}\right]' = \frac{1}{(1-x)^2}$$

for all $x \in (-1,1)$. It is left to the reader to verify that the interval of convergence of the power series $\sum_{n=1}^{\infty} nx^{n-1}$ is $(-1,1)$, and so in this case differentiation does not change the interval of convergence.

It follows from Theorem 10.4.4 (2) that $\sum_{n=0}^{\infty} \frac{x^{n+1}}{n+1}$ has radius of convergence 1, and that

$$\sum_{n=0}^{\infty} \frac{x^{n+1}}{n+1} = \int_0^x \frac{1}{1-t}\, dt$$

for all $x \in (-1,1)$. Using Integration by Substitution for Definite Integrals (Theorem 5.7.4) with the substitution $u = 1-t$, and Definition 7.2.1, we see that

$$\int_0^x \frac{1}{1-t}\, dt = -\int_1^{1-x} \frac{1}{u}\, du = -\ln(1-x)$$

for all $x \in (-1,1)$. Hence

$$\ln(1-x) = \sum_{n=0}^{\infty} \frac{-x^{n+1}}{n+1} \tag{10.4.1}$$

for all $x \in (-1,1)$. Let us examine the convergence of this power series at the endpoints of the interval $(-1,1)$. If $x = 1$, then the power series $\sum_{n=0}^{\infty} \frac{-x^{n+1}}{n+1}$ is $\sum_{n=1}^{\infty} \frac{-1}{n}$, which is -1 times the harmonic series, and which is divergent by Example 9.2.4 (5) and Exercise 9.2.3. If $x = -1$, then the power series $\sum_{n=0}^{\infty} \frac{-x^{n+1}}{n+1}$ is $\sum_{n=1}^{\infty} \frac{(-1)^{n-1}}{n}$, which is the alternating harmonic series, and which is convergent by Example 9.3.9. Hence the interval of convergence of $\sum_{n=0}^{\infty} \frac{x^{n+1}}{n+1}$ is $[-1,1)$, which is not the same as the interval of convergence of $\sum_{n=0}^{\infty} x^n$, even though both power series have the same radius of convergence.

We will see in Exercise 10.4.10 (5) that the power series formula for $\ln(1-x)$ in Equation 10.4.1 also holds for $x = -1$.

(2) The statement of Theorem 10.4.4 seems very straightforward, though in fact it should not be taken for granted. By the Comparison Test (Theorem 9.3.2) it can be verified that the series $\sum_{n=1}^{\infty} \frac{\sin nx}{n^2}$ is absolutely convergent for all $x \in \mathbb{R}$. However, if we use term-by-term differentiation on this series, we obtain the series $\sum_{n=1}^{\infty} \frac{\cos nx}{n}$. This new series is divergent when $x = 2\pi k$, for all $k \in \mathbb{Z}$, because for such values of x the series is $\sum_{n=1}^{\infty} \frac{1}{n}$, which is the harmonic series, and which was proved to be divergent in Example 9.2.4 (5). Hence, even though the radius of convergence of a power series is preserved under term-by-term differentiation and integration, that is not necessarily true for all series with "variables." ◊

The following corollary, which will be very important to us in finding the power series representation of functions, is an immediate consequence of Theorem 10.4.4 (1), and we omit the proof.

Corollary 10.4.7. *Let $A \subseteq \mathbb{R}$ be a set, let $a \in A$ and let $f : A \to \mathbb{R}$ be a function. Suppose that f is represented by a power series $\sum_{n=0}^{\infty} c_n(x-a)^n$. Let R be the radius of convergence of $\sum_{n=0}^{\infty} c_n(x-a)^n$. Then f is infinitely differentiable on $(a-R, a+R)$.*

We saw in Example 4.2.5 (2) a function $g : \mathbb{R} \to \mathbb{R}$ that is differentiable, but such that g' is not differentiable. It follows from Corollary 10.4.7 that the function g cannot be represented by a power series.

We saw in Example 10.4.3 and Example 10.4.6 (1) a few instances of functions that are represented by power series, though these examples were essentially found by luck, because the power series $\sum_{n=0}^{\infty} x^n$ happens to be a geometric series. It would be nice to have a more systematic way of finding power series representations for those functions that can be so represented, and we will see such a method shortly. We start with the following theorem, the statement of which makes implicit use of Corollary 10.4.7. Recall the definition of factorials given in Example 2.5.12.

Theorem 10.4.8. *Let $A \subseteq \mathbb{R}$ be a set, let $a \in A$ and let $f : A \to \mathbb{R}$ be a function. Suppose that f is represented by a power series $\sum_{n=0}^{\infty} c_n(x-a)^n$. Then*

$$c_n = \frac{f^{(n)}(a)}{n!}$$

for all $n \in \mathbb{N} \cup \{0\}$.

Proof. Let R be the radius of convergence of $\sum_{n=0}^{\infty} c_n(x-a)^n$. We know that $R > 0$, and that $f(x) = \sum_{n=0}^{\infty} c_n(x-a)^n$ for all $x \in (a-R, a+R)$. By Theorem 10.4.4 (1) we know that $f'(x) = \sum_{n=1}^{\infty} nc_n(x-a)^{n-1}$ for all $x \in (a-R, a+R)$. It can be verified by induction that if $k \in \mathbb{N}$, then

$$f^{(k)}(x) = \sum_{n=k}^{\infty} n(n-1)(n-2)\cdots(n-k+1)c_n(x-a)^{n-k}$$

for all $x \in (a-R, a+R)$; the details are left to the reader. Hence $f^{(k)}(a) = k(k-1)(k-2)\cdots(k-k+1)c_k = k!c_k$ for each $k \in \mathbb{N}$, and the desired formula follows immediately. □

We note that Theorem 10.4.8 gives an alternative proof of Theorem 10.4.2.

The following definition gives us a convenient notation for the power series described in Theorem 10.4.8, and for the partial sums of this power series.

Definition 10.4.9. Let $I \subseteq \mathbb{R}$ be an open interval, let $a \in I$ and let $f: I \to \mathbb{R}$ be a function. Suppose that f is infinitely differentiable at a. For each $k \in \mathbb{N} \cup \{0\}$, the kth **Taylor polynomial** of f centered at a is the polynomial function $T_k^{f,a}: I \to \mathbb{R}$ defined by

$$T_k^{f,a}(x) = \sum_{n=0}^{k} \frac{f^{(n)}(a)}{n!}(x-a)^n$$

for all $x \in I$. The **Taylor series** of f centered at a is the series

$$T^{f,a}(x) = \sum_{n=0}^{\infty} \frac{f^{(n)}(a)}{n!}(x-a)^n.$$

When $a = 0$, the kth Taylor polynomial of f centered at a is called the kth **Maclaurin polynomial** of f, and the Taylor series of f centered at a is called the **Maclaurin series** of f. △

Observe that the Taylor polynomials of a function are the partial sums of its Taylor series. Moreover, these polynomials are quite interesting and useful in their own right. Suppose that $f: I \to \mathbb{R}$ is a function, where $I \subseteq \mathbb{R}$ is an open interval, and let $a \in I$. Suppose that f is infinitely differentiable at a. Let $n \in \mathbb{N}$. It would be nice to find a polynomial, denoted p_n, such that p_n is the best possible approximation of f by a polynomial of degree n. Of course, there could be a number of possible choices of criteria for what we might mean by "best possible approximation," but a very commonly used criterion is that we would want p_n to agree with f at a, and we would want the kth derivative of p_n to agree with the kth derivative of f at a for all $k \in \{1, 2, \ldots, n\}$. It turns out that if p_n satisfies this criterion, then it must in fact equal the nth Taylor polynomial of f centered at a; the proof is virtually the same as the proof of Theorem 10.4.8, except that we start with a polynomial rather than a power series, and we omit the details. Hence, the Taylor polynomials are, in many cases, very useful for approximating the value of the original function, at least near the point a. In general, the higher the value of n, the better the approximation of f. For example, if

$f\colon \mathbb{R} \to \mathbb{R}$ is defined by $f(x) = \sin x$ for all $x \in \mathbb{R}$, then the graph of f, and the graphs of its Maclaurin polynomials for $n \in \{1,3,5,7\}$, are seen in Figure 10.4.1. It should be noted, however, that whereas Taylor polynomials are useful for approximating functions in some situations, for modern computing they are in general not as useful as other polynomial approximations of functions, or some non-polynomial algorithms (such as the CORDIC algorithm for approximating trigonometric and logarithmic functions) that are particularly suited to computer and calculator architecture; see [Mul06] for details.

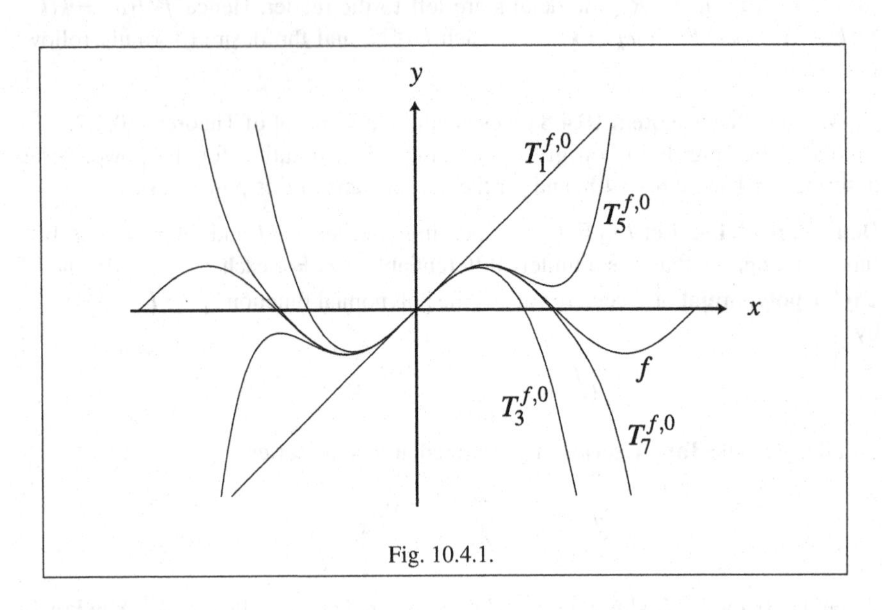

Fig. 10.4.1.

The following corollary is simply a restatement of Theorem 10.4.8, using the terminology of Definition 10.4.9.

Corollary 10.4.10. *Let $A \subseteq \mathbb{R}$ be a set, let $a \in A$ and let $f\colon A \to \mathbb{R}$ be a function. Suppose that f is represented by a power series $\sum_{n=0}^{\infty} c_n (x-a)^n$. Then $\sum_{n=0}^{\infty} c_n (x-a)^n$ is the Taylor series of f centered at a.*

It is important to stress that Corollary 10.4.10 does not say that every function is represented by its Taylor series, but only that if a function has a power series representation, then that representation must be the Taylor series. Indeed, as we will see in the fourth part of the following example, while an infinitely differentiable function always has a Taylor series centered at any number a in its domain, such a function is not always represented by its Taylor series centered at a.

Example 10.4.11.

(1) Let $f\colon \mathbb{R} \to \mathbb{R}$ be defined by $f(x) = e^x$ for all $x \in \mathbb{R}$. We compute the Maclaurin series for f. It follows from Theorem 7.2.7 (2) that f is infinitely differentiable, and that $f^{(n)}(x) = e^x$ for all $x \in \mathbb{R}$ and all $n \in \mathbb{N} \cup \{0\}$. By Theorem 7.2.8 (1) we see that

$f^{(n)}(0) = 1$ for all $n \in \mathbb{N} \cup \{0\}$. Hence the Maclaurin series of f is

$$T^{f,0}(x) = \sum_{n=0}^{\infty} \frac{f^{(n)}(0)}{n!} x^n = \sum_{n=0}^{\infty} \frac{x^n}{n!}.$$

We saw in Example 9.5.2 (2) that the interval of convergence of this power series is $\mathbb{R}$.

Is the function f represented by its Maclaurin series? More specifically, does e^x equal its Maclaurin series for all $x \in \mathbb{R}$? The answer will turn out to be yes, though we will have to wait until Example 10.4.16 to see a proof of this fact.

(2) Let $g\colon (0,\infty) \to \mathbb{R}$ be defined by $g(x) = \ln x$ for all $x \in (0,\infty)$. The function g does not have a Maclaurin series, because g is not defined at 0, but we will compute the Taylor series for g centered at $a = 1$. We know by Exercise 7.2.2 that g is infinitely differentiable, and that $g^{(n)}(x) = \frac{(-1)^{n-1}(n-1)!}{x^n}$ for all $n \in \mathbb{N}$ and all $x \in (0,\infty)$. It follows that $g^{(n)}(1) = (-1)^{n-1}(n-1)!$ for all $n \in \mathbb{N}$. By Theorem 7.2.3 (1) we see that $g^{(0)}(1) = 0$. Hence the Taylor series of g centered at 1 is

$$T^{g,1}(x) = \sum_{n=0}^{\infty} \frac{g^{(n)}(1)}{n!}(x-1)^n = \sum_{n=1}^{\infty} \frac{(-1)^{n-1}(n-1)!}{n!}(x-1)^n = \sum_{n=1}^{\infty} \frac{(-1)^{n-1}}{n}(x-1)^n.$$

We saw in Exercise 9.5.3 that the interval of convergence of this power series is $(0,2]$.

Is the function g represented by its Taylor series centered at $a = 1$? The answer is yes, and in contrast to Part (1) of this example, we can prove most of this fact now, because of what we saw in Example 10.4.6 (1). It is left to the reader to verify that if $x = 1 - u$ is substituted in Equation 10.4.1, it follows that $\ln u = \sum_{n=1}^{\infty} \frac{(-1)^{n-1}}{n}(u-1)^n$ for all $u \in (0,2)$, which means that $g(x)$ equals its Taylor series for all $x \in (0,2)$. It takes additional effort to show that $g(x)$ equals its Taylor series at $x = 2$; see Exercise 10.4.10 (5) for details. We deduce that the function g is represented by its Taylor series. However, it is important to observe that such representation does not hold for all x in the domain of g, which is $(0,\infty)$, because the Taylor series is not convergent on all of the domain of g. There is no way to avoid this problem, because by Corollary 10.4.10 the function g cannot be represented by any other power series.

(3) Let $r \in \mathbb{R}$, and let $p\colon (-1,1) \to \mathbb{R}$ be defined by $p(x) = (1+x)^r$ for all $x \in (-1,1)$.

We compute the Maclaurin series for p. It follows from Theorem 7.2.13 (1) together with the standard rules for differentiation (discussed in Section 4.3) that p is infinitely differentiable, and that if $x \in (-1,1)$, then

$$p^{(k)}(x) = \begin{cases} r(r-1)\cdots(r-k+1)(1+x)^{r-k}, & \text{if } k \in \mathbb{N} \\ (1+x)^r, & \text{if } k = 0. \end{cases}$$

By Theorem 7.2.12 (1) we see that

$$p^{(k)}(0) = \begin{cases} r(r-1)\cdots(r-k+1), & \text{if } k \in \mathbb{N} \\ 1, & \text{if } k = 0. \end{cases}$$

For convenience, we use the following standard notation. Let $a \in \mathbb{R}$, and let $k \in \mathbb{N} \cup \{0\}$. The **binomial coefficient** $\binom{a}{k}$ is defined by

$$\binom{a}{k} = \begin{cases} \frac{a(a-1)\cdots(a-k+1)}{k!}, & \text{if } k \in \mathbb{N} \\ 1, & \text{if } k = 0. \end{cases}$$

We note that if $a \in \mathbb{N} \cup \{0\}$ and $k \in \{0, \ldots, a\}$, then this definition of $\binom{a}{k}$ agrees with the standard definition of $\binom{a}{k}$ found in the Binomial Theorem, counting problems, Pascal's triangle and elsewhere; see [Blo10, Section 7.7] for a very brief discussion of these topics.

Using the above notation, we see that the Maclaurin series of p is

$$T^{p,0}(x) = \sum_{n=0}^{\infty} \frac{p^{(n)}(0)}{n!} x^n = \sum_{n=0}^{\infty} \binom{r}{n} x^n.$$

This power series is called the **binomial series** for r. It is seen in Exercise 10.4.5 (3) that if $r \in \mathbb{N} \cup \{0\}$ then the radius of convergence of the binomial series is ∞, and if $r \notin \mathbb{N} \cup \{0\}$ then the radius of convergence of the binomial series is 1; in either case the interval of convergence contains $(-1, 1)$. (In those cases where the radius of convergence is 1, the convergence or divergence of the binomial series at $x = 1$ and $x = -1$ depends upon the value of r; see [How66] for details.)

Is the function p represented by its Maclaurin series? The answer is yes, and we can prove this fact now, though not by the straightforward method of Part (2) of this example, but rather by a clever trick that is available for this particular power series. Let $q\colon (-1, 1) \to \mathbb{R}$ be defined by $q(x) = (1+x)^{-r} \sum_{n=0}^{\infty} \binom{r}{n} x^n$ for all $x \in (-1, 1)$. It follows from Theorem 10.4.4 (1) together with the Product Rule (Theorem 4.3.1 (4)) that q is differentiable. It is left to the reader in Exercise 10.4.5 (4) to verify that $q'(x) = 0$ for all $x \in (-1, 1)$. It is straightforward to verify that $q(0) = 1$, and we then use Lemma 4.4.7 (1) to deduce that $q(x) = 1$ for all $x \in (-1, 1)$. It follows that $p(x)$ equals its Maclaurin series for all $x \in (-1, 1)$, and hence p is represented by its Maclaurin series.

(4) The following example is due to Augustin Louis Cauchy (1789–1857). Let $h\colon \mathbb{R} \to \mathbb{R}$ be defined by

$$h(x) = \begin{cases} e^{-\frac{1}{x^2}}, & \text{if } x \neq 0 \\ 0, & \text{if } x = 0. \end{cases}$$

The hard work for this example was done in Exercise 6.3.10 (3), where it was proved that h is infinitely differentiable, and that $h^{(n)}(0) = 0$ for all $n \in \mathbb{N}$. It follows that the Maclaurin series of h is the series that is constantly zero, which clearly has radius of convergence $R = \infty$. However, we know that $h(x) \neq 0$ for all $x \in \mathbb{R} - \{0\}$, which follows from the fact that the exponential function has codomain $(0, \infty)$. Hence, even though the function h is infinitely differentiable, it is not represented by its Maclaurin series. ◊

We saw in Example 10.4.11 (2) (3) two functions that were represented by their Taylor series. However, the proofs in these two cases were very particular to the functions under consideration. We now turn to a more broadly applicable method for proving that functions are represented by their Taylor series, stated in Corollary 10.4.15 below; this method also does not work in all cases, but it is nonetheless quite useful, for example for the function in Example 10.4.11 (1), as we will see in Example 10.4.16. We start with the following definition and lemma, which reformulate this issue in a convenient way.

Definition 10.4.12. Let $I \subseteq \mathbb{R}$ be an open interval, let $a \in I$ and let $f\colon I \to \mathbb{R}$ be a function. Suppose that f is infinitely differentiable at a. For each $n \in \mathbb{N} \cup \{0\}$, the nth **Taylor polynomial remainder** of f centered at a is the function $R_n^{f,a}\colon I \to \mathbb{R}$ defined by $R_n^{f,a} = f - T_n^{f,a}$. △

The following lemma is deduced immediately from Definition 10.4.12, the fact that the Taylor polynomials are the partial sums of the Taylor series, and Exercise 8.2.5; we omit the proof.

Lemma 10.4.13. *Let $I \subseteq \mathbb{R}$ be an open interval, let $f\colon I \to \mathbb{R}$ be a function and let $a \in I$. Suppose that f is infinitely differentiable at a. Let $x \in I$. Then $f(x) = T^{f,a}(x)$ if and only if $\lim_{n\to\infty} R_n^{f,a}(x) = 0$.*

At first glance, Lemma 10.4.13 may not appear to be very helpful, because it looks as if we are just renaming things; that is, figuring out whether or not the nth Taylor polynomial remainder of a function converges to zero does not appear to be any easier than figuring out whether or not the Taylor series converges to the function. However, as we see in the following theorem, corollary and example, there is a convenient expression for the nth Taylor polynomial remainder of a function, and this expression turns out to be easy to work with for some specific functions. This theorem, named after Lagrange when stated in its present form, is actually just a restatement of what we earlier called Taylor's Theorem (Theorem 4.4.6), though this time with domain an open interval, and hence we state it without proof.

Theorem 10.4.14 (Lagrange Form of the Remainder Theorem). *Let $I \subseteq \mathbb{R}$ be an open interval, let $a \in I$, let $f\colon I \to \mathbb{R}$ be a function and let $n \in \mathbb{N} \cup \{0\}$. Suppose that f is $n+1$ times differentiable. Let $x \in I$. Then there is some p strictly between x and a (except that $p = a$ when $x = a$) such that*

$$R_n^{f,a}(x) = \frac{f^{(n+1)}(p)}{(n+1)!}(x-a)^{n+1}.$$

Corollary 10.4.15. *Let $I \subseteq \mathbb{R}$ be an open interval, let $a \in I$ and let $f\colon I \to \mathbb{R}$ be a function. Suppose that f is infinitely differentiable, and that there is some $M \in \mathbb{R}$ such that $|f^{(n)}(x)| \le M^n$ for all $x \in I$ and all $n \in \mathbb{N} \cup \{0\}$. Then*

$$f(x) = \sum_{n=0}^{\infty} \frac{f^{(n)}(a)}{n!}(x-a)^n$$

for all $x \in I$.

Proof. Let $x \in I$. Let $n \in \mathbb{N} \cup \{0\}$. By the Lagrange Form of the Remainder Theorem (Theorem 10.4.14) there is some p between x and a such that

$$R_n^{f,a}(x) = \frac{f^{(n+1)}(p)}{(n+1)!}(x-a)^{n+1}.$$

Hence

$$0 \le \left|R_n^{f,a}(x)\right| = \left|\frac{f^{(n+1)}(p)}{(n+1)!}(x-a)^{n+1}\right| \le \frac{[M(x-a)]^{n+1}}{(n+1)!}.$$

By the final remark in Example 9.5.2 (2) and Exercise 8.2.4 we know that $\lim_{n\to\infty} \frac{[M(x-a)]^{n+1}}{(n+1)!} = 0$. It then follows from the Squeeze Theorem for Sequences (Theorem 8.2.12) that $\left\{\left|R_n^{f,a}(x)\right|\right\}_{n=1}^{\infty}$ is convergent and $\lim_{n\to\infty}\left|R_n^{f,a}(x)\right| = 0$. By Exercise 8.2.13 (2) we deduce that $\lim_{n\to\infty} R_n^{f,a}(x) = 0$. It now follows from Lemma 10.4.13 that $f(x) = T^{f,a}(x)$. □

Example 10.4.16. We saw in Example 10.4.11 (1) that the Maclaurin series for e^x is $\sum_{n=0}^{\infty} \frac{x^n}{n!}$, and we saw in Example 9.5.2 (2) that the interval of convergence of this power series is $\mathbb{R}$. We want to show that e^x equals its Maclaurin series for all $x \in \mathbb{R}$. We cannot apply Corollary 10.4.15 directly to e^x, because there does not exist an "M" as in the statement of that corollary that works for all $x \in \mathbb{R}$, but we can apply the corollary to the restriction of e^x to any bounded open interval in $\mathbb{R}$.

Let $w \in \mathbb{R}$. Then there is some $c \in (0,\infty)$ such that $w \in (-c,c)$. By Theorem 7.2.7 (3) and Theorem 7.2.8 (1) we see that $e^c > e^0 = 1$. Moreover, by Theorem 7.2.7 (3) again and Exercise 2.5.1, it follows that if $y \in (-c,c)$ and $n \in \mathbb{N}$, then $|f^{(n)}(y)| = |e^y| = e^y \le e^c \le [e^c]^n$. Corollary 10.4.15 now implies that $e^y = \sum_{n=0}^{\infty} \frac{y^n}{n!}$ for all $y \in (-c,c)$. In particular, we deduce that $e^w = \sum_{n=0}^{\infty} \frac{w^n}{n!}$. Therefore $e^x = \sum_{n=0}^{\infty} \frac{x^n}{n!}$ for all $x \in \mathbb{R}$.

As a special case of the above, we see that

$$e = e^1 = 1 + \frac{1}{1!} + \frac{1}{2!} + \frac{1}{3!} + \cdots. \tag{10.4.2}$$

Additionally, we can obtain the Maclaurin series for other useful functions by substituting into the Maclaurin series for e^x. For example, we see that

$$e^{-x^2} = \sum_{n=0}^{\infty} \frac{(-x^2)^n}{n!} = \sum_{n=0}^{\infty} \frac{(-1)^n x^{2n}}{n!}$$

for all $x \in \mathbb{R}$. Although we did not obtain this power series representation of e^{-x^2} directly by Definition 10.4.9, this power series must be the Maclaurin series nonetheless because of Corollary 10.4.10. It is very nice that we can compute the Maclaurin series of e^{-x^2} by this indirect method, because attempting to do so by Definition 10.4.9 would be quite tedious, because the derivatives of e^{-x^2} are messy. ◊

We conclude this section with the following two applications of power series, the first to the number π, and the second to the number e.

Example 10.4.17. In Section 7.4 we saw various properties of the number π, but one thing we were not able to do in that section is something very important from a computational point of view, which is to provide the first few digits of the decimal expansion of π. We know by Theorem 7.4.5 that π is an irrational number, and so whereas it is possible to compute finitely many digits in the decimal expansion of π, we cannot write down the entire decimal expansion. Of course, everyone learns at a young age that $\pi = 3.14159\ldots$, but we want to prove that fact rigorously based upon the definition of π given in Definition 7.3.5. There are a number of ways to approach this problem, the simplest being with series. The most familiar way to write π as a series is

$$\pi = 4\left(1 - \frac{1}{3} + \frac{1}{5} - \frac{1}{7} + \cdots\right). \tag{10.4.3}$$

We will present a different series for π, both because the above series is proved using the arctangent function, which we have not discussed, and because this series converges extremely slowly, and therefore many terms are needed to obtain accuracy for even the first few digits in the decimal expansion of π. Instead, we provide a different way to write π as a series, which converges much faster than the above series, and which uses the arcsine function, which was discussed in detail in Section 7.3, rather than arctangent.

We find the Maclaurin series for arcsin as follows. Using the Binomial series given in Example 10.4.11 (3), we see that

$$\begin{aligned}\frac{1}{\sqrt{1+x}} &= (1+x)^{-\frac{1}{2}} = \sum_{n=0}^{\infty}\binom{-\frac{1}{2}}{n}x^n \\ &= 1 + \frac{-\frac{1}{2}}{1!}x + \frac{(-\frac{1}{2})(-\frac{1}{2}-1)}{2!}x^2 + \frac{(-\frac{1}{2})(-\frac{1}{2}-1)(-\frac{1}{2}-2)}{3!}x^3 + \cdots \\ &= 1 - \frac{1}{2\cdot 1!}x + \frac{1\cdot 3}{2^2\cdot 2!}x^2 - \frac{1\cdot 3\cdot 5}{2^3\cdot 3!}x^3 + \cdots\end{aligned}$$

for all $x \in (-1,1)$. Using Lemma 7.3.7 (3) we see that

$$\arcsin' x = \frac{1}{\sqrt{1-x^2}} = \frac{1}{\sqrt{1+(-x^2)}} = 1 + \frac{1}{2\cdot 1!}x^2 + \frac{1\cdot 3}{2^2\cdot 2!}x^4 + \frac{1\cdot 3\cdot 5}{2^3\cdot 3!}x^6 + \cdots$$

for all $x \in (-1,1)$. It now follows from Lemma 7.3.7, the Fundamental Theorem of Calculus Version II (Theorem 5.6.4) and Theorem 10.4.4 (2) that

$$\arcsin x = \int_0^x \arcsin' t\, dt = x + \frac{1}{2\cdot 1!\cdot 3}x^3 + \frac{1\cdot 3}{2^2\cdot 2!\cdot 5}x^5 + \frac{1\cdot 3\cdot 5}{2^3\cdot 3!\cdot 7}x^7 + \cdots$$

for all $x \in (-1,1)$.

We know by Exercise 7.3.8 that $\sin\left(\frac{\pi}{6}\right) = \frac{1}{2}$, and hence $\frac{\pi}{6} = \arcsin\left(\frac{1}{2}\right)$. Therefore

$$\pi = 6\arcsin\left(\frac{1}{2}\right) = 6\left(\frac{1}{2} + \frac{1}{2\cdot 1!\cdot 3\cdot 2^3} + \frac{1\cdot 3}{2^2\cdot 2!\cdot 5\cdot 2^5} + \frac{1\cdot 3\cdot 5}{2^3\cdot 3!\cdot 7\cdot 2^7} + \cdots\right).$$

Summing the first 10 terms of this series yields $\pi \approx 3.141592623\ldots$, which is correct up to the seventh decimal place. By contrast, summing the first 10 terms of the series in Equation 10.4.3 yields $3.041839619\ldots$, which is a much worse approximation of π.

The problem of computing the digits of the decimal expansion of π has a long history, and continues to be the subject of research; see [AH01] for more information. $\Diamond$

We saw in Theorem 7.4.5 that the number π is irrational. We now have the tools to prove that the number e is irrational as well.

Theorem 10.4.18. *The number e is irrational.*

Proof. Suppose that e is rational. By Definition 7.2.9 we know that $e > 0$, and hence by Lemma 2.4.12 (2) there are $a, b \in \mathbb{N}$ such that $e = \frac{a}{b}$. Let

$$q = e - (1 + \frac{1}{1!} + \frac{1}{2!} + \frac{1}{3!} + \cdots \frac{1}{b!}).$$

Then $b!q$ is an integer, because each term in q is a fraction that has a denominator with factors that are all factors in $b!$.

By Equation 10.4.2, we see that

$$q = \frac{1}{(b+1)!} + \frac{1}{(b+2)!} + \frac{1}{(b+3)!} + \cdots.$$

Using Exercise 9.2.6 (2) and Example 9.2.4 (4), and noting that $b \geq 1$ and so $\left|\frac{1}{b+1}\right| < 1$, we see that

$$\begin{aligned} b!q &= b!\left(\frac{1}{(b+1)!} + \frac{1}{(b+2)!} + \frac{1}{(b+3)!} + \cdots\right) \\ &= \frac{1}{b+1} + \frac{1}{(b+1)(b+2)} + \frac{1}{(b+1)(b+2)(b+3)} + \cdots \\ &< \frac{1}{(b+1)^1} + \frac{1}{(b+1)^2} + \frac{1}{(b+1)^3} + \cdots \\ &= \frac{\frac{1}{(b+1)}}{1 - \frac{1}{(b+1)}} = \frac{1}{b} \leq 1, \end{aligned}$$

It is evident that $q > 0$, and because $b! \geq 1$, then $b!q > 0$. Hence $0 < b!q < 1$, which is a contradiction to Theorem 2.4.10 (2). We conclude that e is irrational. $\square$

Similarly to what was stated about the number π in Section 7.4, the number e is not only an irrational number, but it is in fact a transcendental number, which means that it is not the root of any polynomial with rational coefficients; see [Spi67, Chapter 20] for a proof.

Reflections

The reader who has taken a standard introductory calculus sequence should be familiar with many of the results and examples of power series discussed in this section, and hence this section might appear to consist of a lot of effort aimed at proving some elementary and familiar results. In fact, the material in this section should not be taken for granted. It is hard to overestimate the importance of power series in both the history and applications of calculus. The relative brevity of the proofs in this section should not fool the reader into thinking that the material in this section is simple; the proofs involve some substantial ideas, and they appear brief only because the hard work was already done in the previous sections of this chapter.

Exercises

Exercise 10.4.1. [Used in Exercise 10.4.9.] Let $c \in \mathbb{R} - \{0\}$. Let $f\colon \mathbb{R} - \{c\} \to \mathbb{R}$ be defined by $f(x) = \frac{1}{x-c}$ for all $x \in \mathbb{R} - \{0\}$.

(1) Find the Maclaurin series for f.
(2) What is the interval of convergence of this Maclaurin series?
(3) Prove that f equals its Maclaurin series for all x in the interval of convergence.

Exercise 10.4.2. Let $p\colon \mathbb{R} \to \mathbb{R}$ be a polynomial function. What is the Maclaurin series of p?

Exercise 10.4.3. [Used in Theorem 10.4.4.] Prove Theorem 10.4.4 (2).

Exercise 10.4.4. Give a direct proof of Corollary 10.4.5 using Theorem 10.3.8, but not using Theorem 10.4.4. [Use Exercise 2.3.9.]

Exercise 10.4.5. [Used in Example 10.4.11.] This exercise uses Example 10.4.11 (3).

(1) Let $a \in \mathbb{R}$, and let $k \in \mathbb{N} \cup \{0\}$. Prove that $\binom{a}{k+1} = \binom{a}{k}\frac{a-k}{k+1}$.
(2) Let $a \in \mathbb{R}$, and let $k \in \mathbb{N} \cup \{0\}$. Prove that if $a \notin \mathbb{N} \cup \{0\}$ then $\binom{a}{k} \neq 0$, and if $a \in \mathbb{N} \cup \{0\}$ and $k > a$ then $\binom{a}{k} = 0$.
(3) Prove that if $r \in \mathbb{N} \cup \{0\}$ then the radius of convergence of the binomial series is ∞, and if $r \notin \mathbb{N} \cup \{0\}$ then the radius of convergence of the binomial series is 1.
(4) Prove that $q'(x) = 0$ for all $x \in (-1,1)$.

Exercise 10.4.6. [Used in Exercise 10.4.13.]

(1) Prove that the Maclaurin series for $\sin x$ is $\sum_{n=0}^{\infty} \frac{(-1)^n x^{2n+1}}{(2n+1)!}$.
(2) Prove that $\sin x$ equals its Maclaurin series for all $x \in \mathbb{R}$.

Exercise 10.4.7. Let $f\colon \mathbb{R} \to \mathbb{R}$ be a function. Suppose that f is represented by a Maclaurin series $\sum_{n=0}^{\infty} c_n x^n$. Find a condition on the sequence $\{c_n\}_{n=1}^{\infty}$ that is equivalent to the condition that $f(-x) = f(x)$ for all $x \in \mathbb{R}$, and prove the equivalence. A function that satisfies this condition is called an **even** function.

Exercise 10.4.8. Let $A \subseteq \mathbb{R}$ be a set, let $a \in A$ and let $f\colon A \to \mathbb{R}$ be a function. Suppose that f is represented by a power series $\sum_{n=0}^{\infty} c_n(x-a)^n$. Let R be the radius of convergence of $\sum_{n=0}^{\infty} c_n(x-a)^n$. Find a condition on the sequence $\{c_n\}_{n=1}^{\infty}$ that is equivalent to f having a local maximum at a, and prove the equivalence.

Exercise 10.4.9. In this exercise we use the sequence of Fibonacci numbers, denoted $\{F_n\}_{n=1}^{\infty}$, which was defined in Example 8.4.10. The purpose of this exercise is to use power series to prove that

$$F_n = \frac{1}{\sqrt{5}}\left[\left(\frac{1+\sqrt{5}}{2}\right)^n - \left(\frac{1-\sqrt{5}}{2}\right)^n\right]$$

for all $n \in \mathbb{N}$. (There are other proofs of this formula that do not involve power series, for example the proof in [Blo10, Exercise 6.4.12].) This formula is known as Binet's formula, though it is also attributed to the earlier mathematicians Daniel Bernoulli and Leonhard Euler. Using the notation ϕ for the "golden ratio," as discussed in Example 8.4.10, we can write Binet's formula as $F_n = \frac{1}{\sqrt{5}}[\phi^n - (-1/\phi)^n]$ for all $n \in \mathbb{N}$.

(1) Find the two roots r_1 and r_2 of the polynomial x^2+x-1, and find numbers $A, B \in \mathbb{R}$ such that $\frac{1}{1-x-x^2} = \frac{A}{x-r_1} + \frac{B}{x-r_2}$. This last expression is the partial fraction decomposition of $\frac{1}{1-x-x^2}$.
(2) Use Part (1) of this exercise together with Exercise 10.4.1 to find the Maclaurin series of $\frac{1}{1-x-x^2}$.
(3) It follows from Exercise 9.5.7 that $\frac{1}{1-x-x^2}$ is represented by the power series $\sum_{n=0}^{\infty} F_{n+1}x^n$. Use Theorem 10.4.2 together with Part (2) of this exercise to derive Binet's formula.

Exercise 10.4.10. **[Used in Example 10.4.6, Example 10.4.11 and Exercise 10.4.11.]** Let $A \subseteq \mathbb{R}$ be a set, let $a \in A$ and let $f\colon A \to \mathbb{R}$ be a function. Suppose that f is represented by a power series $\sum_{n=0}^{\infty} c_n(x-a)^n$. Let R be the radius of convergence of $\sum_{n=0}^{\infty} c_n(x-a)^n$. Suppose that $R \neq \infty$. We know by hypothesis that $f(x) = \sum_{n=0}^{\infty} c_n(x-a)^n$ for all $x \in (a-R, a+R)$, because the interval of convergence of $\sum_{n=0}^{\infty} c_n(x-a)^n$ contains $(a-R, a+R)$. Additionally, we know by Corollary 10.4.5 that f is continuous on $(a-R, a+R)$. It might be the case that the interval of convergence of $\sum_{n=0}^{\infty} c_n(x-a)^n$ contains the endpoints $x = a-R$ or $x = a+R$. If it does, is f necessarily continuous at these endpoints? The purpose of this exercise is to prove Abel's Theorem, which says that the answer is yes. More specifically, given the above hypotheses, Abel's Theorem states that if $\sum_{n=0}^{\infty} c_n(x-a)^n$ is convergent at $x = a+R$, then $\lim_{x\to(a+R)^-} f(x) = \sum_{n=0}^{\infty} c_n R^n$. A similar result holds for the left endpoint $x = a-R$, and we omit the details.

The proof of the theorem will be done in steps, starting with a special case.

(1) Suppose that $a = 0$ and $R = 1$. Suppose that $\sum_{n=0}^{\infty} c_n x^n$ is convergent at $x = a+R = 1$. Then $\sum_{n=0}^{\infty} c_n$ is convergent. Let $L = \sum_{n=0}^{\infty} c_n$. Let $\{s_n\}_{n=0}^{\infty}$ be the sequence of partial sums of $\sum_{n=0}^{\infty} c_n$. Let $y \in (-1,1)$, and let $p \in \mathbb{N}$. Prove that $\sum_{n=0}^{p} c_n y^n = (1-y)\sum_{n=0}^{p-1} s_n y^n + s_p y^p$. [Use Exercise 5.7.6.]

(2) Assume the hypotheses of Part (1) of this exercise. Prove that $f(y) = (1-y)\sum_{n=0}^{\infty} s_n y^n$.

(3) Assume the hypotheses of Part (1) of this exercise. By Example 9.2.4 (4) we know that $\sum_{n=0}^{\infty} y^n = \frac{1}{1-y}$, and hence $L = (1-y)\sum_{n=0}^{\infty} Ly^n$. Let $\varepsilon > 0$. Then there is some $N \in \mathbb{N}$ such that $n \in \mathbb{N}$ and $n \geq N$ imply $|s_n - L| < \frac{\varepsilon}{2}$. Let $Q = \sum_{n=0}^{N} |s_n - L| + 1$. Then $Q > 0$. Let $\delta = \min\{\frac{\varepsilon}{2Q}, 1\}$. Prove that if $x \in (-1,1)$ and $1-\delta < x < 1$, then $|f(x) - L| < \varepsilon$. It will follow that $\lim_{x\to 1^-} f(x) = L = \sum_{n=0}^{\infty} c_n$. [Use Exercise 9.4.2.]

(4) In Part (3) of this exercise we proved the theorem in the particular case where $a = 0$ and $R = 1$. Deduce the general case from this special case.

(5) Here is an application of Abel's Theorem. In Example 10.4.11 (2) we saw that

$$\ln x = \sum_{n=1}^{\infty} \frac{(-1)^{n-1}}{n}(x-1)^n$$

for all $x \in (0,2)$, and we saw that the interval of convergence of this power series is $(0,2]$. Prove that this power series formula for $\ln x$ holds for $x = 2$. Deduce that $\sum_{n=1}^{\infty} (-1)^{n-1}\frac{1}{n} = \ln 2$. (Another proof of this last fact was given in Exercise 9.3.10 (2).)

Exercise 10.4.11. **[Used in Section 9.4.]** Let $\sum_{n=0}^{\infty} c_n$ and $\sum_{n=0}^{\infty} d_n$ be series in $\mathbb{R}$. Let $\sum_{n=0}^{\infty} e_n$ be the Cauchy product of $\sum_{n=0}^{\infty} c_n$ and $\sum_{n=0}^{\infty} d_n$. Suppose that $\sum_{n=0}^{\infty} c_n$ and $\sum_{n=0}^{\infty} d_n$ are convergent. Prove that if $\sum_{n=0}^{\infty} e_n$ is convergent, then $[\sum_{n=0}^{\infty} c_n] \cdot [\sum_{n=0}^{\infty} d_n] = \sum_{n=0}^{\infty} e_n$. [Use Exercise 10.4.10.]

Exercise 10.4.12. In Example 10.4.16 it was seen that $e^x = \sum_{n=0}^{\infty} \frac{x^n}{n!}$ for all $x \in \mathbb{R}$. Although it is a straightforward calculation to find the Maclaurin series for e^x, a rigorous treatment of the subject requires a rigorous definition of the exponential function and proofs of the elementary properties of this function, which in turn requires a rigorous treatment of the natural logarithm function, as we saw in Section 7.2.

The purpose of this exercise is to provide an alternative definition of the exponential function by using the above power series as the definition of the function, and then using this definition to prove some of the standard properties of the exponential function. This approach can be used as the formal definitions of the exponential function for someone who has not read Section 7.2. Both approaches to the definition of the exponential function ultimately involve the same power series (in one case as the definition, and in the other case as a consequence of the definition), and hence they both yield the same function. The approach used in this exercise might appear to be less laborious than the definitions used in Section 7.2, though the brevity is somewhat illusory, because the use of power series requires our first having proved various results about such series; moreover, the use of power series as a definition of a function, while convenient in the present case, is not a definition that promotes an intuitive understanding of the function.

For this exercise, do not use anything from Section 7.2.

We know by Example 9.5.2 (2) that the interval of convergence of $\sum_{n=0}^{\infty} \frac{x^n}{n!}$ is $\mathbb{R}$. Let $E\colon \mathbb{R} \to \mathbb{R}$ be defined by $E(x) = \sum_{n=0}^{\infty} \frac{x^n}{n!}$ for all $x \in \mathbb{R}$.

(1) Prove that $E(0) = 1$, and that $E' = E$.
(2) It is evident that $E(x) > 0$ for all $x \in [0,\infty)$. Suppose that $E(p) \leq 0$ for some $p \in (-\infty,0)$. Obtain a contradiction by using Exercise 3.5.8 with $[a,b] = [p,0]$ and $r = \frac{1}{2}$, and the Mean Value Theorem (Theorem 4.4.4) on $[p, \text{glb} S]$, where S is the set defined in Exercise 3.5.8. Deduce that $E(x) > 0$ for all $x \in \mathbb{R}$.
(3) Let $z \in \mathbb{R}$, and let $g\colon \mathbb{R} \to \mathbb{R}$ be defined by $g(x) = \frac{E(x+z)}{E(x)E(z)}$ for all $x \in \mathbb{R}$. Prove that $g(x) = 1$ for all $x \in \mathbb{R}$. (Use Parts (1) and (2) of this exercise rather than the definition of E directly.) Deduce that $E(x+y) = E(x)E(y)$ for all $x,y \in \mathbb{R}$.

Exercise 10.4.13. This exercise is the analog for sine and cosine of what we saw for the exponential function in Exercise 10.4.12.

In Exercise 10.4.6 it was seen that $\sin x = \sum_{n=0}^{\infty} \frac{(-1)^n x^{2n+1}}{(2n+1)!}$ for all $x \in \mathbb{R}$. A similar argument can be used to show that $\cos x = \sum_{n=0}^{\infty} \frac{(-1)^n x^{2n}}{(2n)!}$ for all $x \in \mathbb{R}$; we omit the details. Although it is a straightforward calculation to find the Maclaurin series for $\sin x$ and $\cos x$, a rigorous treatment of the subject requires a rigorous definition of the sine and cosine functions and proofs of the elementary properties of these two functions, which we saw in Section 7.3, but which were a bit more complicated than might be expected.

The purpose of this exercise is to provide an alternative definition of sine and cosine by using the power series as the definitions of the functions, and then using these definitions to prove some of the standard properties of sine and cosine. Similarly to Exercise 10.4.12, this approach can be used as the formal definitions of sine and cosine for someone who has not read Section 7.3.

For this exercise, do not use anything from Section 7.3, except where otherwise noted.

(1) Prove that the interval of convergence for each of the power series $\sum_{n=0}^{\infty} \frac{(-1)^n x^{2n}}{(2n)!}$ and $\sum_{n=0}^{\infty} \frac{(-1)^n x^{2n+1}}{(2n+1)!}$ is $\mathbb{R}$. Let $S, C\colon \mathbb{R} \to \mathbb{R}$ be defined by $S(x) = \sum_{n=0}^{\infty} \frac{(-1)^n x^{2n+1}}{(2n+1)!}$ and $C(x) = \sum_{n=0}^{\infty} \frac{(-1)^n x^{2n}}{(2n)!}$ for all $x \in \mathbb{R}$.
(2) Prove that $S(0) = 0$ and $C(0) = 1$, and that $S(-x) = -S(x)$ and $C(-x) = C(x)$ for all $x \in \mathbb{R}$.
(3) Prove that $S' = C$ and that $C' = -S$. Deduce that $S''(x) + S(x) = 0$ and $C''(x) + C(x) = 0$ for all $x \in \mathbb{R}$.
(4) Prove that $S^2(x) + C^2(x) = 1$ for all $x \in \mathbb{R}$. (Use Parts (2) and (3) of this exercise rather than the definitions of S and C directly.)
(5) Prove that $S(x) > 0$ for all $x \in (0,2)$. Deduce that C is strictly decreasing on $(0,2)$.
(6) Prove that there is a unique $r \in (1,2)$ such that $C(r) = 0$. Define the number π by $\pi = 2r$.
(7) Let $y \in \mathbb{R}$. Then let $f\colon \mathbb{R} \to \mathbb{R}$ be defined by $f(x) = S(x+y) - S(x)C(y) - C(x)S(y)$ for all $x \in \mathbb{R}$. Prove that $f''(x) + f(x) = 0$ for all $x \in \mathbb{R}$, and that $f(0) = 0$ and $f'(0) = 0$. Use Exercise 7.3.9 (1) (which does not refer to sine and cosine at all) to deduce that $f(x) = 0$ for all $x \in \mathbb{R}$. Deduce that $S(x+y) =$

$S(x)C(y)+C(x)S(y)$ for the given value of y, and for all $x \in \mathbb{R}$. Because y was arbitrarily chosen, we therefore see that $S(x+y)=S(x)C(y)+C(x)S(y)$ for all $x,y \in \mathbb{R}$. Deduce that $C(x+y)=C(x)C(y)-S(x)S(y)$ for all $x,y \in \mathbb{R}$. (This part of the exercise is identical to the proof of Exercise 7.3.10.)

(8) Prove that $S(\frac{\pi}{2})=1$.

(9) Prove that $S(x+\frac{\pi}{2})=C(x)$ and $C(x+\frac{\pi}{2})=-S(x)$ for all $x \in \mathbb{R}$. Deduce that $S(x+\pi)=-S(x)$ and $C(x+\pi)=-C(x)$ for all $x \in \mathbb{R}$. Deduce that $S(x+2\pi)=S(x)$ and $C(x+2\pi)=C(x)$ for all $x \in \mathbb{R}$.

10.5 A Continuous but Nowhere Differentiable Function

Our final section of the book takes us back to an issue that we first considered much earlier, which is the relation between differentiability and continuity. We know from Theorem 4.2.4 that if a function is differentiable everywhere, it must be continuous everywhere. On the other hand, we know from Example 4.2.3 (3) that a function $f\colon \mathbb{R} \to \mathbb{R}$ can be continuous everywhere but not differentiable everywhere. If a function is continuous everywhere, how large can the set of numbers at which the function is not differentiable be? The function in Example 4.2.3 (3), which is the absolute value function, is differentiable everywhere except at a single number. A function such as a "sawtooth function," as seen in Figure 10.5.1, is everywhere except at a discrete countable set of numbers. Are there continuous functions with even worse non-differentiability?

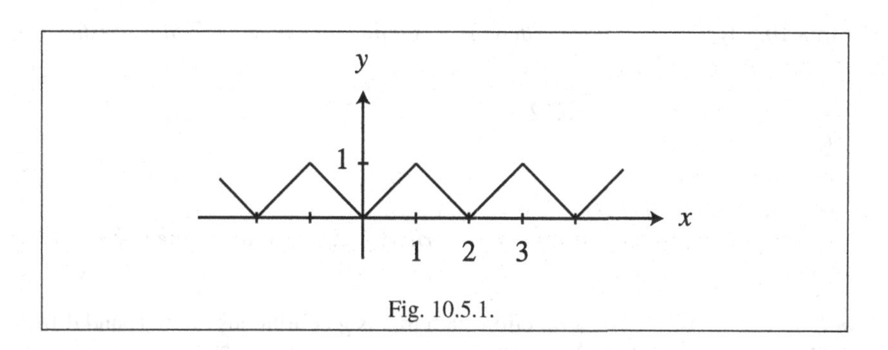

Fig. 10.5.1.

Intuitively, it might appear that for any continuous function, between any two numbers at which the function is not differentiable there must be numbers (not to mention whole intervals) at which the function is differentiable. Rather astonishingly, as we will see in Theorem 10.5.2 below, it turns out there exist functions that are continuous everywhere, but differentiable nowhere. The statement of this theorem would have been understandable when we first discussed derivatives in Section 4.2, but the proof involves series of functions, and hence we had to wait until now to see it. The first example of a continuous but nowhere differentiable function is due to Bernard Bolzano (1781–1848) in the 1830s, but, as was the case for his other mathematical work, it did not receive attention from the mathematical community. The first publicized example of such a function was due to Karl Weierstrass (1815–1897) in the 1870s, and it astonished the mathematical community at the time. We present a more recent variant of this example in the proof of Theorem 10.5.2.

For our construction, we will need the concept of a periodic function, as discussed in Section 7.3. For the reader who has skipped that section, it is not necessary to read the entire section now, but only from Definition 7.3.1 to Lemma 7.3.4.

The basic idea of the construction of a continuous but nowhere differentiable function is to start with an appropriately chosen periodic function $f\colon \mathbb{R} \to \mathbb{R}$, and then to define a new function $h\colon \mathbb{R} \to \mathbb{R}$ by

$$h(x) = \sum_{n=0}^{\infty} \left(\frac{3}{4}\right)^n f(4^n x)$$

for all $x \in \mathbb{R}$. For each $n \in \mathbb{N}$, the factor of 4^n makes the function $f(4^n x)$ have a period that is 4^n times the period of f, and it is these increasing periods that, intuitively, make the function h nowhere differentiable; the factor of $\left(\frac{3}{4}\right)^n$ is designed to make the series of functions be uniformly convergent, which is what makes the function h continuous. Of course, for this construction to work we will need to start with an appropriately chosen function f, the existence of which is shown by the following lemma. This lemma is more general than we need, because we need only one example of the type of function given by the lemma, but if one wants to put in all of the details for the existence of one such function, it is no more effort to prove the lemma more generally, and doing so helps clarify what the real issues are.

Lemma 10.5.1. *Let $K \in (0, \infty)$. Then there is a function $f\colon \mathbb{R} \to \mathbb{R}$ that satisfies the following properties.*

a. f is periodic with period 2;
b. f is continuous;
c. f is bounded;
d. if $x, y \in \mathbb{R}$, then $|f(x) - f(y)| \leq K|x - y|$;
e. if $x, y \in \mathbb{R}$ and there is no integer strictly between x and y, then $\frac{2K}{3}|x - y| \leq |f(x) - f(y)|$.

Proof. Let $g\colon [0,1] \to \mathbb{R}$ be a function such that is g continuous on $[0,1]$ and differentiable on $(0,1)$, and that $\frac{2K}{3} \leq |g'(x)| \leq K$ for all $x \in (0,1)$. There are many such functions, for example the function defined by $g(x) = Kx$ for all $x \in [0,1]$; the reader is asked to find another example of such a function g in Exercise 10.5.1. Any choice of such a function g will work.

Let $h\colon [-1,1] \to \mathbb{R}$ be defined by $h(x) = g(|x|)$ for all $x \in [-1,1]$. It follows from Exercise 3.3.1 (2) and Theorem 3.3.8 (3) that h is continuous. If $x \in [-1,1]$, then $h(-x) = g(|-x|) = g(|x|) = h(x)$; in particular we see that $h(-1) = h(1)$.

By Lemma 7.3.2 (2) there is a unique periodic function $f\colon \mathbb{R} \to \mathbb{R}$ with period 2 such that $f|_{[-1,1]} = h$. Hence Part (a) holds. By Lemma 7.3.4 (1) the function f is continuous and bounded, and hence Parts (b) and (c) hold.

We now prove that Part (d) holds. We start by observing that Exercise 4.4.6 implies that the analog of Part (d) holds for g.

Next, we show that the analog of Part (d) holds for h. Let $w,z \in [-1,1]$. Then $|h(w) - h(z)| = |g(|w|) - g(|z|)| \le K||w| - |z||$. By Lemma 2.3.9 (7) we see that $|w| - |z| \le |w - z|$ and $|z| - |w| \le |z - w| = |w - z|$, and hence $||w| - |z|| \le |w - z|$. Therefore $|h(w) - h(z)| \le K|w - z|$.

We now show that Part (d) holds for f. Let $x, y \in \mathbb{R}$. If $x = y$, then $|f(x) - f(y)| = 0 \le K|x - y|$. Now suppose that $x \ne y$. Without loss of generality, assume that $x < y$. We claim that there are $\hat{x}, \hat{y} \in [-1,1]$ such that $f(\hat{x}) = f(x)$ and $f(\hat{y}) = f(y)$ and $|\hat{x} - \hat{y}| \le |x - y|$. It will then follow that $|f(x) - f(y)| = |f(\hat{x}) - f(\hat{y})| = |h(\hat{x}) - h(\hat{y})| \le K|\hat{x} - \hat{y}| \le K|x - y|$, and the proof of Part (d) will be complete.

By Exercise 2.6.14 (1) there are unique $n, m \in \mathbb{Z}$ such that $(-1) + 2(m-1) \le x < -1 + 2m$ and $(-1) + 2(n-1)2 \le y < (-1) + 2n$. It follows $x - 2(m-1) \in [-1,1)$ and $y - 2(n-1) \in [-1,1)$. Let $\hat{y} = y - 2(n-1)$, let $x' = x - 2(n-1)$ and let $\tilde{x} = x - 2(m-1)$. Then $\tilde{x}, \hat{y} \in [-1,1)$, and $\hat{y} - x' = y - x$. Because f is periodic with period 2, then $f(\tilde{x}) = f(x') = f(x)$ and $f(\hat{y}) = f(y)$.

There are three cases. First, suppose that $|x - y| \ge 2$. Let $\hat{x} = \tilde{x}$. Then $f(\hat{x}) = f(x)$. Because $\hat{x}, \hat{y} \in [-1,1)$, then $|\hat{x} - \hat{y}| < 2 \le |x - y|$.

Second, suppose that $x' \in [-1,1)$. Let $\hat{x} = x'$. Then $f(\hat{x}) = f(x)$, and $|\hat{x} - \hat{y}| = |x' - \hat{y}| = |x - y|$.

Third, suppose that $|x - y| < 2$ and $x' \notin [-1,1)$. Let $\hat{x} = -x' - 2$. Because $x < y$ then $x' < \hat{y}$. Because $\hat{y} \in [-1,1)$ and $|x' - \hat{y}| = |x - y| < 2$, then $-3 < x' < -1$. It follows that $-1 < \hat{x} < 1$, which means that $\hat{x} \in [-1,1]$. Using what we saw above about h, and the fact that f is periodic with period 2, we see that $f(\hat{x}) = h(\hat{x}) = h(-\hat{x}) = h(x'+2) = f(x'+2) = f(x') = f(x)$. Also, we observe that $-2 < x'+1 < 0$ and $0 \le \hat{y}+1 < 2$. Then $|\hat{x} - \hat{y}| = |(-x'-2) - \hat{y}| = |-(x'+1) - (\hat{y}+1)| \le |-(x'+1)| + |\hat{y}+1| = -(x'+1) + (\hat{y}+1) = \hat{y} - x' = y - x = |x - y|$.

Finally, we prove that Part (e) holds. Let $x, y \in \mathbb{R}$. Suppose that there is no integer strictly between x and y. If $x = y$ then Part (e) is trivially true, so suppose that $x \ne y$. Without loss of generality, assume that $x < y$. Let x' and $\hat{y}$ be as in the proof of Part (d) of this lemma. Then $\hat{y} \in [-1,1)$ and $x' < \hat{y}$.

We claim that there is no integer strictly between x' and $\hat{y}$. Suppose to the contrary that $x' < p < \hat{y}$ for some $p \in \mathbb{Z}$. Then $x = x' + 2(n-1) < p + 2(n-1) < \hat{y} + 2(n-1) = y$, which is a contradiction to the fact that there is no integer strictly between x and y.

There are now two cases. First, suppose that $\hat{y} \in (-1,1)$. If it were the case that $x' < -1$, then we would have $x' < -1 < \hat{y}$, which is a contradiction. Hence $-1 \le x'$, and therefore $x' \in [-1,1)$. It cannot be the case $x' < 0 < \hat{y}$, and therefore either $x', \hat{y} \in [-1,0]$ or $x', \hat{y} \in [0,1)$.

Suppose that $x', \hat{y} \in [0,1)$. Because g is continuous on $[0,1]$ and differentiable on $(0,1)$, we can apply the Mean Value Theorem (Theorem 4.4.4) to $g|_{[x',\hat{y}]}$, and we deduce that there is some $c \in (x', \hat{y})$ such that

$$g'(c) = \frac{g(\hat{y}) - g(x')}{\hat{y} - x'}.$$

Because $\frac{2K}{3} \le |g'(c)|$, it follows that

$$\frac{2K}{3} \leq \left| \frac{g(\hat{y}) - g(x')}{\hat{y} - x'} \right|.$$

Therefore $\frac{2K}{3}|\hat{y} - x'| \leq |g(\hat{y}) - g(x')|$. We know that $\hat{y} - x' = y - x$, and that $f(x) = f(x') = h(x') = g(x')$ and $f(y) = f(\hat{y}) = h(\hat{y}) = g(\hat{y})$. It follows that $\frac{2K}{3}|y - x| \leq |f(y) - f(x)|$, which is Part (e).

Next, suppose that $x', \hat{y} \in [-1, 0]$. Then $-x', -\hat{y} \in [0, 1]$. By the same argument as in the previous paragraph, we see that $\frac{2K}{3}|(-\hat{y}) - (-x')| \leq |g(-\hat{y}) - g(-x')|$. We then observe $|(-\hat{y}) - (-x')| = |y - x|$, and that $f(x) = f(x') = h(x') = g(-x')$ and $f(y) = f(\hat{y}) = h(\hat{y}) = g(-\hat{y})$. Hence $\frac{2K}{3}|y - x| \leq |f(y) - f(x)|$, which again is Part (e).

Second, suppose that $\hat{y} = -1$. Let $y' = \hat{y} + 2$, and let $x'' = x' + 2$. Then $y' = 1$ and $x'' < y'$. The desired result now follows from an argument similar to the previous case, but with x'' and y' replacing x' and $\hat{y}$, and we omit the details. □

We now come to the grand finale of this text.

Theorem 10.5.2. *There is a function $h\colon \mathbb{R} \to \mathbb{R}$ that is continuous but nowhere differentiable.*

Proof. We follow [Rud76, p. 154], which is a variant of [McC53], which in turn is related to [vdW30]. Let $K \in (0, \infty)$, and let $f\colon \mathbb{R} \to \mathbb{R}$ be a function the existence of which is guaranteed by Lemma 10.5.1. By Part (c) of that lemma there is some $B \in \mathbb{R}$ such that $|f(x)| \leq B$ for all $x \in \mathbb{R}$.

For each $n \in \mathbb{N} \cup \{0\}$, let $f_n\colon \mathbb{R} \to \mathbb{R}$ be defined by $f_n(x) = \left(\frac{3}{4}\right)^n f(4^n x)$ for all $x \in \mathbb{R}$. Then $|f_n(x)| \leq B\left(\frac{3}{4}\right)^n$ for all $x \in \mathbb{R}$ and all $n \in \mathbb{N} \cup \{0\}$. By Example 9.2.4 (4) we know that $\sum_{n=0}^{\infty} B\left(\frac{3}{4}\right)^n$ is convergent; the fact that we start at $n = 0$ rather than $n = 1$ makes no difference. It now follows from the Weierstrass M-Test (Theorem 10.3.6) that $\sum_{n=1}^{\infty} f_n$ is uniformly convergent.

Let $h\colon \mathbb{R} \to \mathbb{R}$ be defined by $h = \sum_{n=1}^{\infty} f_n$, which means that

$$h(x) = \sum_{n=0}^{\infty} \left(\frac{3}{4}\right)^n f(4^n x)$$

for all $x \in \mathbb{R}$.

By Part (b) of Lemma 10.5.1 the function f is continuous. It then follows from Example 3.3.3 (1), Theorem 3.3.5 (3) and Theorem 3.3.8 (3) that f_n is continuous for each $n \in \mathbb{N} \cup \{0\}$. By Theorem 10.3.9 we deduce that h is continuous.

Let $x \in \mathbb{R}$. We will show that there is a sequence $\{d_n\}_{n=1}^{\infty}$ in $\mathbb{R}$ such that $\lim_{n \to \infty} d_n = 0$, and that the sequence

$$\left\{ \frac{h(x + d_n) - h(x)}{d_n} \right\}_{n=1}^{\infty}$$

is divergent. It will then follow from Exercise 8.4.3 that h is not differentiable at x, and we will conclude that h is nowhere differentiable.

Let $n \in \mathbb{N}$. We define d_n as follows. By Exercise 2.4.2 we know that at most one of $(4^n x - \frac{1}{2}, 4^n x)$ and $(4^n x, 4^n x + \frac{1}{2})$ contains an integer. If $(4^n x, 4^n x + \frac{1}{2})$ does not contain an integer, then let $d_n = \frac{1}{2 \cdot 4^n}$. If $(4^n x, 4^n x + \frac{1}{2})$ contains an integer, then $(4^n x - \frac{1}{2}, 4^n x)$ does not contain an integer, and let $d_n = -\frac{1}{2 \cdot 4^n}$. It follows that there is no integer strictly between $4^n x$ and $4^n(x + d_n) = 4^n x \pm \frac{1}{2}$, where the choice of plus or minus depends upon the definition of d_n. Using Example 8.2.13 and Theorem 8.2.9 (3) we see that $\lim_{n\to\infty} |d_n| = \lim_{n\to\infty} \frac{1}{2 \cdot 4^n} = 0$, and it then follows from Exercise 8.2.13 (2) that $\lim_{n\to\infty} d_n = 0$.

Let $k \in \mathbb{N}$. We examine the value of $|f(4^k(x + d_n)) - f(4^k x)|$. There are three cases.

First, suppose that $k < n$. By Part (d) of Lemma 10.5.1 we see that $|f(4^k(x + d_n)) - f(4^k x)| \leq K|4^k(x + d_n) - 4^k x| = K|4^k d_n| = \frac{K}{2 \cdot 4^{n-k}}$.

Second, suppose that $k = n$. Because there is no integer strictly between $4^n x$ and $4^n(x + d_n)$, it follows from Part (e) of Lemma 10.5.1 that $|f(4^n(x + d_n)) - f(4^n x)| \geq \frac{2K}{3}|4^n(x + d_n) - 4^n x| = \frac{2K}{3}\frac{1}{2} = \frac{K}{3}$.

Third, suppose that $k > n$. Then $|4^k(x + d_n) - 4^k x| = |4^k d_n| = \frac{4^{k-n}}{2}$, which is an integer multiple of 2. By Part (a) of Lemma 10.5.1 the function f is periodic with period 2, and hence $|f(4^k(x + d_n)) - f(4^k x)| = 0$.

Using the above values for $|f(4^k(x+d_n)) - f(4^k x)|$, and using Theorem 9.2.6 (2), Lemma 2.3.9 (7), Exercise 2.5.3 and Exercise 2.5.12 (3), we see that

$$\begin{aligned}
\left|\frac{h(x+d_n)-h(x)}{d_n}\right| &= \frac{1}{|d_n|}\left|\sum_{k=0}^{\infty}\left(\frac{3}{4}\right)^k f(4^k(x+d_n)) - \sum_{k=0}^{\infty}\left(\frac{3}{4}\right)^k f(4^k x)\right| \\
&= \frac{1}{|d_n|}\left|\sum_{k=0}^{\infty}\left(\frac{3}{4}\right)^k \left[f(4^k(x+d_n)) - f(4^k x)\right]\right| \\
&= \frac{1}{|d_n|}\left|\left(\frac{3}{4}\right)^n [f(4^n(x+d_n)) - f(4^n x)] + \sum_{k=0}^{n-1}\left(\frac{3}{4}\right)^k \left[f(4^k(x+d_n)) - f(4^k x)\right]\right| \\
&\geq \frac{1}{|d_n|}\left(\frac{3}{4}\right)^n |f(4^n(x+d_n)) \\
&\qquad - f(4^n x)| - \frac{1}{|d_n|}\sum_{k=0}^{n-1}\left(\frac{3}{4}\right)^k |f(4^k(x+d_n)) - f(4^k x)| \\
&\geq 2 \cdot 4^n \left(\frac{3}{4}\right)^n \frac{K}{3} - 2 \cdot 4^n \sum_{k=0}^{n-1}\left(\frac{3}{4}\right)^k \frac{K}{2 \cdot 4^{n-k}} \\
&= \frac{2K}{3}3^n - K\sum_{k=0}^{n-1} 3^k = \frac{2K}{3}3^n - K\frac{1-3^n}{1-3} = \frac{K}{6}3^n + \frac{K}{2}.
\end{aligned}$$

It follows from Exercise 8.2.16 and Exercise 8.2.17 that $\lim_{n\to\infty}\left(\frac{K}{6}3^n + \frac{K}{2}\right) = \infty$, and we then use Exercise 8.2.18 (2) to deduce that

$$\left\{\frac{h(x+d_n)-h(x)}{d_n}\right\}_{n=1}^{\infty}$$

is divergent. □

It is not possible to draw the graph of the function h defined in the proof of Theorem 10.5.2, but it is possible to draw the graphs of the partial sums of h. In Figure 10.5.2 we see the first four partial sums of h restricted to $[0,2]$, where the function $f: \mathbb{R} \to \mathbb{R}$ used in the definition of h is chosen to be the "sawtooth function" seen in Figure 10.5.1; that is, the function f is the periodic function with period 2 that equals the absolute value function on $[-1,1]$.

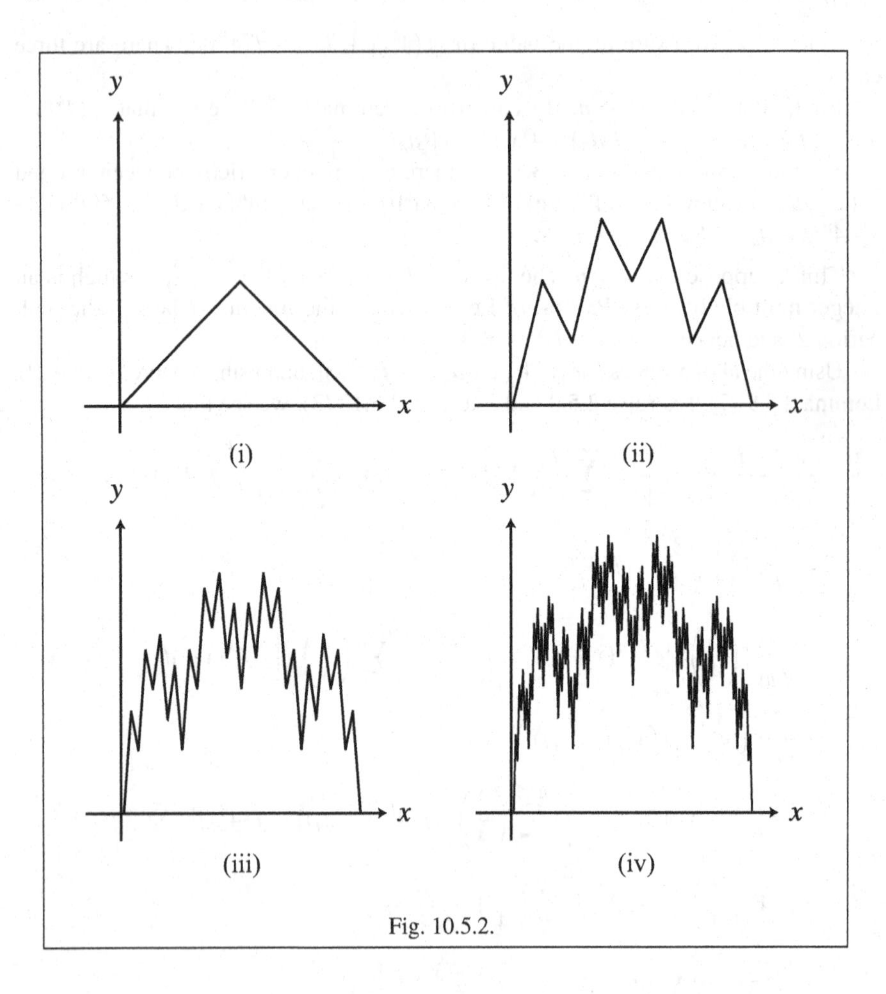

Fig. 10.5.2.

Reflections

We have now come to the end of this book, though of course not to the end of the story we are telling. This book is only an introduction to real analysis, and while we have covered most of those aspects of real analysis treated in a single-variable calculus course, the bulk of real analysis awaits the reader in subsequent texts, including both the study of more advanced aspects of single-variable real analysis, for example

Lebesgue measure and integration, and the study of real analysis in more general settings, for example in $\mathbb{R}^n$ and metric spaces. Moreover, some of the topics covered in this text lead naturally to issues that arise in other branches of mathematics, including complex analysis, point set topology, probability and differential equations. The reader is encouraged to pursue the study of these fields.

Whatever branch of mathematics one chooses to pursue, the existence of counter-intuitive functions such as the one given in Theorem 10.5.2 shows why we need to use all the rigor available to us, because things do not always work out as our intuition might tell us. Of course, that is what makes mathematics so interesting—if everything worked as our intuition told us, then there would be no surprises. As the reader might imagine, a further study of mathematics will yield many additional surprises as well.

Exercises

Exercise 10.5.1. **[Used in Lemma 10.5.1.]** Let $K \in (0, \infty)$. Find an example of a function $r\colon [0,1] \to \mathbb{R}$ such that r is continuous on $[0,1]$ and differentiable on $(0,1)$, that $\frac{2K}{3} \le |r'(x)| \le K$ for all $x \in (0,1)$ and that r does not have a constant derivative. Give an explicit formula for r.

Exercise 10.5.2. The definition of the function h in the proof of Theorem 10.5.2 is rather tricky, and the reader might wonder whether there is a simpler way to define a continuous but nowhere differentiable function. It seems unlikely that it would be possible to define such a function without some sort of limit, but would it at least be possible to give a simpler construction? Of course, if a substantially simpler construction were known, it would be used instead of the very standard construction given above. Nonetheless, it might occur to the reader to try the following idea.

Let $f\colon \mathbb{R} \to \mathbb{R}$ be the "sawtooth function" seen in Figure 10.5.1, and let $\{f_n\}_{n=1}^{\infty}$ be the sequence of functions $\mathbb{R} \to \mathbb{R}$ defined by $f_n(x) = f(nx)$ for all $x \in \mathbb{R}$ and for all $n \in \mathbb{N}$. Intuitively, the function f_n is similar to f, except that it oscillates n times faster; in other words, whereas each "tooth" in the graph of f has width 2 and height 1, each "tooth" in the graph of f_n has width $\frac{2}{n}$ and height 1. The function f_n is continuous for each $n \in \mathbb{N}$, but as n gets larger, the function f_n has more and more numbers at which it is not differentiable. Hence, one might wonder if the sequence $\{f_n\}_{n=1}^{\infty}$ converges pointwise to a function $f\colon [0,1] \to \mathbb{R}$ that is continuous but nowhere differentiable. Prove that that does not happen; that is, prove that either $\{f_n\}_{n=1}^{\infty}$ converges pointwise to a function that is not continuous, or that $\{f_n\}_{n=1}^{\infty}$ converges pointwise to a function that is continuous everywhere and differentiable somewhere, or that $\{f_n\}_{n=1}^{\infty}$ is not pointwise convergent.

Exercise 10.5.3. **[Used in Example 5.9.16.]** The purpose of this exercise is to show that there exists a continuous function that is not rectifiable; the latter concept was discussed in Section 5.9. Let $h\colon \mathbb{R} \to \mathbb{R}$ be the function defined in the proof of Theorem 10.5.2, where the function $f\colon \mathbb{R} \to \mathbb{R}$ used in the definition of h is chosen to be the "sawtooth function" seen in Figure 10.5.1; that is, the function f is the periodic function with period 2 that equals the absolute value function on $[-1,1]$. Then $h|_{[0,1]}$ is continuous by Theorem 10.5.2 and Exercise 3.3.2 (2).

(1) For each $n \in \mathbb{N} \cup \{0\}$, let $s_n \colon [0,1] \to \mathbb{R}$ be defined by $s_n(x) = \sum_{i=0}^{n} \left(\frac{3}{4}\right)^n f(4^n x)$ for all $x \in \mathbb{R}$. Let $n \in \mathbb{N} \cup \{0\}$. The graph of s_n is a polygon. Prove that the ratio of the length of the graph of s_{n+1} to the length of the graph of s_n is greater than 2.

(2) Prove that the length of the graph of s_n is the polygonal sum of $h|_{[0,1]}$ with respect to some partition of $[0,1]$.

(3) Prove that $h|_{[0,1]}$ is not rectifiable.

10.6 Historical Remarks

Whereas calculus courses today typically treat series, and in particular power series, after first covering the basics of differentiation and integration (though some real analysis texts put series, but not power series, first), in fact some discoveries about power series predate the invention of calculus, and indeed might have contributed to that invention. Moreover, whereas the modern approach to the material in the present chapter starts with a discussion of sequences and series of functions in general, and only then turns to the study of power series as an application of the properties of series of functions, from a historical perspective power series were studied, and used as a computational tool, long before the rigorous study of sequences and series of functions.

Renaissance

Power series for $\sin x$, $\cos x$ and $\arctan x$ appeared in *Tantrasamgraha-vyakhya* of around 1530, which was a commentary on *Tantrasamgraha* of around 1501 by Nilakantha Somayaji (1444–1544). It was recognized in *Tantrasamgraha-vyakhya* that the power series for $\arctan x$ is convergent only when $|x| \leq 1$. A derivation of these power series is found in *Yuktibhasa* of 1550 by Jyesthadeva (c. 1500–c. 1575), which is based mainly on *Tantrasamgraha-vyakhya*, and which attributes the series for $\arctan x$ to Madhava of Sangamagramma (1340–1425).

Seventeenth Century

Nicolaus Mercator (1620–1687), not to be confused with the inventor of the Mercator projection Gerardus Mercator, had the power series $\ln(1+x) = x - \frac{x^2}{2} + \frac{x^3}{3} - \cdots$ in *Logarithmotechnia* of 1668. Isaac Newton (1643–1727) had found this series earlier, but Mercator was the first to publish it. James Gregory (1638–1675), in *Exercitationes Geometricae* of 1668, and Edmond Halley (1656–1742), in *A most compendius and facile method for constructing logarithms, exemplified and demonstrated from the nature of numbers, without any regard to the hyperbola, with a speedy method for finding the number from the logarithm given* of 1697, gave the series $\ln\left(\frac{1+x}{1-x}\right) = 2x + \frac{2x^3}{3} + \frac{2x^5}{5} + \cdots$, which converges faster than Mercator's series. In the early 1670s Gregory stated the first few terms of various power series, including those for $\tan x$, $\arcsin x$ and $\ln \sec x$, though he did not give proofs.

Newton and Leibniz

Power series were an important part of Newton's approach to calculus. In fact, his first significant mathematical discovery, made in the mid-1660s before he invented calculus, was the binomial series, which built upon the ideas of John Wallis (1616–1703) in *Arithmetica infinitorum* of 1656. Besides the value of the series itself, which Newton used for some calculations in his calculus, the binomial series was important in that it helped make infinite processes acceptable. Moreover, although Wallis referred to fractional and negative exponents, he did not really use them, and Newton was the first to do so.

In *De analysi per aequationes numero terminorum infinitas* of 1669 (published only in 1711), Newton, in part spurred by Mercator's publication of *Logarithmotechnia* of 1668, discussed power series, essentially stating without proof that power series can be manipulated as we manipulate finite sums, for example that they can be integrated term-by-term. Newton used his methods to find the power series for e^x, $\sin x$ and $\cos x$, the latter two by using geometric ideas and term-by-term integration to find the power series for $y = \arcsin x$ and $y = \arccos x$, and then solving for x in terms of y. An important idea for Newton was the analogy between infinite decimal fractions, the widespread use of which was fairly recent, and infinite series; Newton used ideas from arithmetic (for example long division) and applied them to find the power series of some functions. In the 1691–1692 draft of *De Quadratura Curvarum*, though omitted from the final version that was printed as an appendix to *The Opticks* of 1704, Newton gave the first explicit statement of the general formula for Taylor series, though without proof; as with other aspects of his work, this formula was first published by someone else.

Although Newton did not have our notion of convergence, and in particular he did not have the concept of the interval of convergence for power series, he seemed to be aware that convergence was an issue. For example, in using power series to solve a differential equation, he noted that the method worked for small values of x, which perhaps implicitly recognized that the power series was not convergent for all values of x. On the other hand, Newton's manipulation of series, similarly to Leibniz, was mostly formal, without regard to convergence. Moreover, both Newton and Leibniz thought geometrically in terms of curves rather than functions (the idea of which came later), and so if a power series gave a geometrically meaningful answer, the convergence was implicitly assumed. That is, Newton considered convergence only when applying the power series, but not in the preliminary manipulations.

Gottfried von Leibniz (1646–1716) was also concerned with power series. In 1691 Leibniz gave power series for $\ln(1+x)$, $\arctan x$, $\sin x$, $\cos x - 1$ and $e^x - 1$. In 1693 he used power series to solve differential equations by substituting the power series $\sum_{n=0}^{\infty} c_n (x-a)^n$ into the differential equation and then finding a recurrence relation for the coefficients. The basis for this method, as used by Leibniz, is that a power series $\sum_{n=0}^{\infty} c_n (x-a)^n$ is zero on an open interval if and only if $a_n = 0$ for all $n \in \mathbb{N} \cup \{0\}$ (which follows from Theorem 9.5.8).

Eighteenth Century

A number of mathematicians, including Gregory, Newton, Leibniz and others, essentially knew the formula for Taylor series before Brook Taylor (1685–1731) was the first to publish it in 1715. Taylor used an interpolation formula of Gregory and Newton to justify the formula. In 1717 James Stirling (1692–1770) gave the proof of Taylor series (actually just Maclaurin series) that we use today via successive differentiation. Colin Maclaurin (1698–1746) also derived Taylor series by successive differentiation, and he applied such series to the study of maxima and minima, in *Treatise of Fluxions* of 1742, which was the first systematic account of Newton's approach to calculus, and which was written to defend that approach from Berkeley's criticism of the use of infinitesimals.

Power series were an important tool for representing functions in the 18th century. Such series were thought of as infinite polynomials, and it was assumed that they behaved essentially the same way as polynomials, which allowed, for example, term-by-term differentiation and integration of power series. Convergence was not the primary focus when dealing with power series, and mathematicians felt free to manipulate power series even outside the interval of convergence.

Joseph-Louis Lagrange (1736–1813), in his attempt to sidestep Berkeley's criticism of infinitesimals, took an approach to calculus based upon the representation of functions by power series; it was assumed that every function (as yet a loosely defined concept) could be so represented, and it was only later that Cauchy showed that not every function could be represented by a power series on an open interval. Lagrange proved that if a function is represented as a power series, then the power series is the Taylor series of the function, and he gave what we now call the Lagrange form of the remainder for Taylor polynomials.

Leonhard Euler (1707–1783), in the influential textbook *Introductio in analysin infinitorum* of 1748, introduced the modern definition of logarithms as the inverse of the exponential functions, and the modern definition of sine and cosine in terms of the unit circle, and he then computed the power series of e^x, $\ln(1+x)$, $\sin x$ and $\cos x$, though not via the formula for Taylor series, but by clever use of the binomial series.

Nineteenth Century

Bernard Bolzano (1781–1848) gave the first example of a continuous but nowhere differentiable function in the 1830s; as with other aspects of Bolzano's work, this example did not attract the attention of contemporary mathematicians.

Although Taylor series were known and used in the period when calculus was first invented, it was Augustin Louis Cauchy (1789–1857), in *Résumé des leçons a l'École Royal Polytechnique* of 1823, who first proved that some examples of Taylor series actually converged to the original functions. Cauchy gave a criterion for such convergence, using an integral form of the remainder, and he applied his method to prove that the Taylor series for $\sin x$ and $\cos x$ converge to these functions for all x. Cauchy gave an example of a non-zero function h with a zero Taylor series (which is in Example 10.4.11 (4)); he then pointed out that two different functions can have the same Taylor series by considering any function f with a non-zero Taylor series, and then comparing f and $f+h$. Hence, in contrast to Lagrange, Cauchy stated that it is not possible to use functions and Taylor series interchangeably. Cauchy attempted to prove that series of functions can be integrated term-by-term, something that had been previously assumed because of the general assumption that what works for finite sums works for series. However, in this proof, as in some other proofs, Cauchy implicitly used uniform convergence, which his definition of convergence did not explicitly state.

Cauchy incorrectly asserted that a convergent series of continuous functions is itself continuous in *Cours d'analyse a l'École Royal Polytechnique* of 1821. Niels Henrik Abel (1802–1829) read Cauchy's definition of convergence of series to mean pointwise convergence of series of functions, rather than uniform convergence as Cauchy understood it, and he then found an example of a series of continuous functions that converges pointwise to a discontinuous function. Abel also gave the first rigorous proof of the binomial series in 1826.

Karl Weierstrass (1815–1897) clarified the difference between pointwise convergence and uniform convergence of series of functions, and showed that termwise differentiation and integration works nicely with uniform convergence, but not with pointwise convergence. In his lectures in 1872, Weierstrass gave an example of a continuous but nowhere differentiable function, and, in contrast to Bolzano's earlier example, this one was widely seen. Weierstrass' example is similar, though not identical, to the one we use in this text (which is the standard such example used today).

Bibliography

[A'C] Norbert A'Campo, *A natural construction for the real numbers*. arXiv:math.GN/0301015.

[Apo67] Tom M. Apostol, *Calculus. Vol. I: One-Variable Calculus, with an Introduction to Linear Algebra*, 2nd ed., Blaisdell, Waltham, MA, 1967.

[AH01] Jörg Arndt and Christoph Haenel, *Pi—Unleashed*, 2nd ed., Springer-Verlag, Berlin, 2001. Translated from the 1998 German original by Catriona Lischka and David Lischka.

[Art64] Emil Artin, *The Gamma Function*, Translated by Michael Butler. Athena Series: Selected Topics in Mathematics, Holt, Rinehart and Winston, New York, 1964.

[Bar69] Margaret E. Baron, *The Origins of the Infinitesimal Calculus*, Pergamon Press, Oxford, 1969.

[Bar96] Robert G. Bartle, *Return to the Riemann integral*, Amer. Math. Monthly **103** (1996), no. 8, 625–632.

[Bea97] Alan F. Beardon, *Limits: A New Approach to Real Analysis*, Springer-Verlag, New York, 1997.

[BBB04] Lennart Berggren, Jonathan Borwein, and Peter Borwein, *Pi: A Source Book*, 3rd ed., Springer-Verlag, New York, 2004.

[BML] Garrett Birkhoff and Saunders Mac Lane, *A Survey of Modern Algebra*, 3rd ed., Macmillan, New York.

[Blo00] Ethan D. Bloch, *Proofs and Fundamentals: A First Course in Abstract Mathematics*, Birkhäuser, Boston, 2000.

[Blo10] ———, *Proofs and Fundamentals: A First Course in Abstract Mathematics*, 2nd ed., Springer-Verlag, New York, 2010.

[Bol78] Vladimir G. Boltianskiĭ, *Hilbert's Third Problem*, V. H. Winston & Sons, Washington, DC, 1978. Translated from the Russian by Richard A. Silverman; With a foreword by Albert B. J. Novikoff; Scripta Series in Mathematics.

[BD09] William Boyce and Richard DiPrima, *Elementary Differential Equations and Boundary Value Problems*, 9th ed., John Wiley & Sons, New York, 2009.

[Boy91] Carl Boyer, *A History of Mathematics*, 2nd ed., John Wiley & Sons, New York, 1991.

[Boy49] Carl B. Boyer, *The History of the Calculus and Its Conceptual Development*, Dover, New York, 1949.

[BC09] James W. Brown and Ruel V. Churchill, *Complex Variables and Applications*, 8th ed., McGraw-Hill, New York, 2009.

[Bur67] Claude W. Burrill, *Foundations of Real Numbers*, McGraw-Hill, New York, 1967.

[CK07] E. Ward Cheney and David R. Kincaid, *Numerical Mathematics and Computing*, 6th ed., Brooks/Cole, Belmont, CA, 2007.

[Coo05] Roger Cooke, *The History of Mathematics: A Brief Course*, 2nd ed., Wiley-Interscience [John Wiley & Sons], Hoboken, NJ, 2005.

[DSW94] Martin Davis, Ron Sigal, and Elaine Weyuker, *Computability, Complexity, and Languages*, 2nd ed., Academic Press, San Diego, 1994.

[Dea66] Richard Dean, *Elements of Abstract Algebra*, John Wiley & Sons, New York, 1966.

[Ded63] Richard Dedekind, *Essays on the Theory of Numbers. I: Continuity and Irrational Numbers. II: The Nature and Meaning of Numbers*, Authorized translation by Wooster Woodruff Beman, Dover, New York, 1963.

[DKO+] James Douglas, Rony Kirollos, Ben Odgers, Ross Street, and Nguyen Hanh Vo, *The Efficient Real Numbers*, `http://www.math.mq.edu.au/~street/EffR.pdf`.

[Edw79] C. H. Edwards Jr., *The Historical Development of the Calculus*, Springer-Verlag, New York, 1979.

[End72] Herbert Enderton, *A Mathematical Introduction to Logic*, Academic Press, New York, 1972.

[Est02] Donald Estep, *Practical Analysis in One Variable*, Springer-Verlag, New York, 2002.

[EL04] Pierre Eymard and Jean-Pierre Lafon, *The Number π*, American Mathematical Society, Providence, RI, 2004. Translated from the 1999 French original by Stephen S. Wilson.

[Fal03] Kenneth Falconer, *Fractal Geometry: Mathematical Foundations and Applications*, 2nd ed., John Wiley & Sons, Hoboken, NJ, 2003.

[Fer08] Giovanni Ferraro, *The Rise and Development of the Theory of Series up to the Early 1820s*, Sources and Studies in the History of Mathematics and Physical Sciences, Springer, New York, 2008.

[Gar87] Trudi Garland, *Fascinating Fibonaccis*, Dale Seymour, Palo Alto, 1987.

[GO03] Bernard R. Gelbaum and John M. H. Olmsted, *Counterexamples in Analysis*, Dover, New York, 2003. Corrected reprint of the second (1965) edition.

[Gil87] Leonard Gillman, *Writing Mathematics Well*, Mathematical Association of America, Washington, DC, 1987.

[Gol98] Robert Goldblatt, *Lectures on the Hyperreals: An Introduction to Nonstandard Analysis*, Springer-Verlag, New York, 1998.

[Gor02] Russell Gordon, *Real Analysis: A First Course*, 2nd ed., Addison-Wesley, Boston, 2002.

[GKP94] Ronald Graham, Donald Knuth, and Oren Patashnik, *Concrete Mathematics*, 2nd ed., Addison-Wesley, Reading, MA, 1994.

[Hav03] Julian Havil, *Gamma: Exploring Euler's Constant*, Princeton University Press, Princeton, NJ, 2003.

[HK03] James M. Henle and Eugene M. Kleinberg, *Infinitesimal Calculus*, Dover, New York, 2003. Reprint of the 1979 original [MIT Press, Cambridge, MA].

[Hig98] Nicholas J. Higham, *Handbook of Writing for the Mathematical Sciences*, 2nd ed., Society for Industrial and Applied Mathematics (SIAM), Philadelphia, 1998.

[HHP97] Peter Hilton, Derek Holton, and Jean Pedersen, *Mathematical Reflections*, Springer-Verlag, New York, 1997.

[How66] Aughtum Howard, *Classroom Notes: On the convergence of the binomial series*, Amer. Math. Monthly **73** (1966), no. 7, 760–761.

[HH99] John Hamal Hubbard and Barbara Burke Hubbard, *Vector Calculus, Linear Algebra, and Differential Forms: A Unified Approach*, Prentice Hall, Upper Saddle River, NJ, 1999.

[Hun70] H. E. Huntley, *The Divine Proportion*, Dover, New York, 1970.

[Jac85] Nathan Jacobson, *Basic Algebra. I*, 2nd ed., W. H. Freeman and Company, New York, 1985.

[Jah03] Hans Niels Jahnke (ed.), *A History of Analysis*, History of Mathematics, vol. 24, American Mathematical Society, Providence, RI, 2003. Translated from the German.

[Jef73] Harold Jeffreys, *Scientific Inference*, 3rd ed., Cambridge University Press, Cambridge, 1973.

[Kas80] Toni Kasper, *Integration in finite terms: the Liouville theory*, Math. Mag. **53** (1980), no. 4, 195–201.

[Kat98] Victor J. Katz, *A History of Mathematics: An Introduction*, 2nd ed., Addison-Wesley, Reading, MA, 1998.

[Kei] Jerome Keisler, *Elementary Calculus: An Approach Using Infinitesimals*, `http://www.math.wisc.edu/~keisler/calc.html`.

[Knu73] Donald E. Knuth, *The Art of Computer Programming, Volume 1: Fundamental Algorithms*, 2nd ed., Addison-Wesley, Reading, MA, 1973.

[KLR89] Donald Knuth, Tracy Larrabee, and Paul Roberts, *Mathematical Writing*, Mathematical Association of America, Washington, DC, 1989.

[Lay01] Steven Lay, *Analysis with an Introduction to Proof*, Prentice Hall, Upper Saddle River, NJ, 2001.

[MZ94] Elena Anne Marchisotto and Gholam-Ali Zakeri, *An invitation to integration in finite terms*, College Math. J. **25** (1994), no. 4, 295–308.

[Mar61] Angelo Margaris, *Successor axioms for the integers*, Amer. Math. Monthly **68** (1961), 441–444.

[McC53] John McCarthy, *An everywhere continuous nowhere differentiable function*, Amer. Math. Monthly **60** (1953), 709.

[MM] Mechon Mamre, *The Hebrew Bible According to the Jewish Publication Society 1917 Edition*, http://www.mechon-mamre.org/p/pt/pt09a07.htm.

[Men73] Elliott Mendelson, *Number Systems and the Foundations of Analysis*, Academic Press, New York, 1973.

[Mor87] Ronald P. Morash, *Bridge to Abstract Mathematics*, Random House, New York, 1987.

[Mul06] Jean-Michel Muller, *Elementary Functions: Algorithms and Implementation*, 2nd ed., Birkhäuser, Boston, 2006.

[Mun00] J. R. Munkres, *Topology*, 2nd ed., Prentice Hall, Upper Saddle River, NJ, 2000.

[New] Isaac Newton, *Extracts from the Works of Isaac Newton*, http://www.maths.tcd.ie/pub/HistMath/People/Newton/.

[Niv47] Ivan Niven, *A simple proof that π is irrational*, Bull. Amer. Math. Soc. **53** (1947), 509.

[OR] John O'Connor and Edmund Robertson, *The MacTutor History of Mathematics archive*, http://www-history.mcs.st-andrews.ac.uk/.

[Olm62] John M. H. Olmsted, *The Real Number System*, Appleton-Century-Crofts, New York, 1962.

[Pak] Igor Pak, *Lectures on Discrete and Polyhedral Geometry*, http://www.math.ucla.edu/~pak/book.htm.

[Pow94] Malcolm Pownall, *Real Analysis*, Wm. C. Brown, Dubuque, 1994.

[Rob86] Eric Roberts, *Thinking Recursively*, John Wiley & Sons, New York, 1986.

[Rob84] Fred Roberts, *Applied Combinatorics*, Prentice Hall, Englewood Cliffs, NJ, 1984.

[Ros05] Kenneth H. Rosen, *Elementary Number Theory*, 5th ed., Addison-Wesley, Reading, MA, 2005.

[Ros68] Maxwell Rosenlicht, *Introduction to Analysis*, Dover, New York, 1968.

[Ros72] ———, *Integration in finite terms*, Amer. Math. Monthly **79** (1972), 963–972.

[Ros80] Kenneth A. Ross, *Another approach to Riemann-Stieltjes integrals*, Amer. Math. Monthly **87** (1980), no. 8, 660–662.

[Ros10] Sheldon Ross, *A First Course in Probability*, 8th ed., Prentice Hall, Upper Saddle River, NJ, 2010.

[Rud53] Walter Rudin, *Principles of Mathematical Analysis*, McGraw-Hill, New York, 1953.

[Rud76] ———, *Principles of Mathematical Analysis*, 3rd ed., McGraw-Hill, New York, 1976.

[Spi65] Michael Spivak, *Calculus on Manifolds*, Benjamin, New York, 1965.

[Spi67] ———, *Calculus*, Benjamin, New York, 1967.

[SHSD73] N. E. Steenrod, P. R. Halmos, M. M. Schiffer, and J. A. Dieudonné, *How to Write Mathematics*, American Mathematical Society, Providence, 1973.

[Ste04] Ian Stewart, *Galois Theory*, 3rd ed., Chapman & Hall/CRC, Boca Raton, FL, 2004.

[Sto01] Manfred Stoll, *Introduction to Real Analysis*, 2nd ed., Addison-Wesley, Boston, 2001.

[Sto79] Robert R. Stoll, *Set Theory and Logic*, Dover, New York, 1979. Corrected reprint of the 1963 edition.

[Str] Ross Street, *Update on the Efficient Reals*, `http://www.math.mq.edu.au/~street/reals.pdf`.

[Str00] Robert S. Strichartz, *The Way of Analysis*, revised ed., Jones and Bartlett, Boston, 2000.

[TBB01] Brian S. Thomson, Judith B. Bruckner, and Andrew M. Bruckner, *Elementary Real Analysis*, Prentice Hall, Upper Saddle River, NJ, 2001.

[Tre03] William F. Trench, *Introduction to Real Analysis*, Prentice Hall, Upper Saddle River, NJ, 2003.

[Tri95] Claude Tricot, *Curves and Fractal Dimension*, Springer-Verlag, New York, 1995. With a foreword by Michel Mendès France; Translated from the 1993 French original.

[vdW30] B. L. van der Waerden, *Ein einfaches Beispiel einer nicht-differenzierbaren stetigen Funktion*, Math. Z. **32** (1930), no. 1, 474–475 (German).

[Vau95] Robert Vaught, *Set Theory*, 2nd ed., Birkhäuser, Boston, 1995.

[Wad00] William R. Wade, *An Introduction to Analysis*, 2nd ed., Prentice Hall, Upper Saddle River, NJ, 2000.

Index

△
△ △
Ethan
Bloch was
born in 1956,
and spent part of
his childhood in Con-
necticut and part in Is-
rael. He received a B.A. in
mathematics in 1978 from Reed
College, where he developed a firm
belief in the value of a liberal arts education, and a Ph.D. in mathematics in 1983
from Cornell University, under the supervision
of Prof. David Henderson. He was an Instructor
at the University of Utah for three years, and arrived
at Bard College in 1986, where he has, very fortunately,
been ever since. He is married and has two children; his
family, his work and travel to Israel more than fill his time.

This text was written using TEXShop on a Mac. The style file is
svmono from Springer Verlag, and the fonts are mathptmx (a
free version of Times Roman with mathematical symbols)
and pzc (Zapf Chancery). Commutative diagrams were
made using the DCpic package. Most figures were
drawn with Adobe Illustrator, exported as encapsu-
lated postscript files, and converted to portable
document format by Preview; a few figures were drawn using Mathematica,
and then modified with Adobe Illustrator. The labels for the figures were typeset in LATEXiT,
and exported as encapsu-
lated postscript files.
This colophon was
made with the
shapepar
package.
▽ ▽
▽